George Grätzer

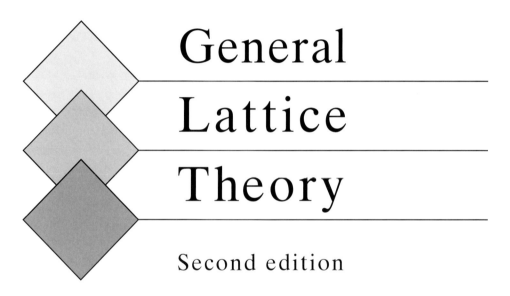

General

Lattice

Theory

Second edition

New appendices with B.A. Davey, R. Freese, B. Ganter,
M. Greferath, P. Jipsen, H.A. Priestley, H. Rose, E.T. Schmidt,
S.E. Schmidt, F. Wehrung, R. Wille

Birkhäuser Verlag
Basel · Boston · Berlin

Author:

George Grätzer
Department of Mathematics
University of Manitoba
Winnipeg, Manitoba R3T 2N2
Canada

2000 Mathematics Subject Classification 06-01, 06-02

A CIP catalogue record for this book is available from the Library of Congress, Washington D.C., USA

Bibliographic information published by Die Deutsche Bibliothek
Die Deutsche Bibliothek lists this publication in the Deutsche Nationalbibliografie;
detailed bibliographic data is available in the Internet at <http://dnb.ddb.de>.

ISBN 3-7643-6996-5 Birkhäuser Verlag, Basel – Boston – Berlin

© 2003 Birkhäuser Verlag, P.O. Box 133, CH-4010 Basel, Switzerland
Member of the BertelsmannSpringer Publishing Group
Printed on acid-free paper produced of chlorine-free pulp. TCF ∞
Cover design: Merry Obrecht Sawdey, Minneapolis, USA
Printed in Germany
ISBN 3-7643-6996-5

9 8 7 6 5 4 3 2 1   www.birkhauser.ch

To my family, Cathy, Tom, and David,
and to the memory of my father, József

Contents

Preface and Acknowledgment

A book that is more than twelve years in the making has a long history and its final form is shaped by many. It all started in my formative years as a mathematician, 1955–1961, when I worked with E. T. Schmidt. We often commented upon the need for a two—or three—volume work on lattice theory that would treat the subject in depth. We felt, however, that the time was not ripe for such a project. For instance, no such work would be complete without presenting at least one example of a nondistributive uniquely complemented lattice. We did not know how to do it without reproducing the almost thirty pages of the famous proof of R. P. Dilworth. We also thought that much more had to be learned about free lattices and varieties of lattices before the project could be attempted.

In 1962, I wrote a proposal for a volume on lattice theory that would survey the whole field in depth. Apart from doing the research necessary for the proposal, no writing was done on this book. M. H. Stone offered to publish a book on universal algebra in the D. Van Nostrand University Series in Higher Mathematics and I concentrated on that book until the end of 1967.

Maybe because mathematicians in general (or I, in particular) are like hobbits (according to J. R. R. Tolkien [1954]: "Hobbits delighted in such things if they were accurate: they liked to have books filled with things they already knew, set out fair and square with no contradictions.") or maybe because I felt that the need for an in depth book on lattice theory had not yet been satisfied, I started in 1968 on this book. In the academic year 1968–1969, I gave a course on lattice theory and I wrote a set of lecture notes. The present first two chapters are based on those notes.

This material was augmented by a chapter on pseudocomplemented distributive lattices and published under the title *Lattice Theory: First Concepts and Distributive Lattices* in 1971. The introduction of this book promised a companion volume on general lattices.

A number of research breakthroughs in the sixties now supplied me with the material (including almost all of Chapters V and VI) I needed to complete the project. But then it became apparent that a complete revision of my plans was in order.

While back in the late fifties it seemed reasonable to try to give a complete picture of lattice theory, this became patently infeasible in the seventies. For instance, in 1958 there was one paper on Stone algebras; by 1974 there were more than fifty papers on Stone algebras and related problems. A number of books have appeared dealing with specialized aspects of lattice theory and with various applications.

Besides, my experience with the writing of *Universal Algebra* taught me not to stray too far away from my research interests. Thus it was decided that while I try to include all the basic material and research methods, the illustrations will be chosen, as far as possible, from fields in which I have some personal interest.

Another change took place in the publishing field. For the second volume it became desirable to choose a publisher with a greater interest in monographs. The new arrangement made it necessary to produce a volume that does not depend on the previous publication. That is why most of the first two chapters of the former book are reproduced here (augmented by a new section, several new exercises, with updated Further Topics and References, and with a new set of Problems), thus giving the reader a self-contained book.

The work on this new book started in 1972 and then continued with an advanced course on lattice theory at the University of Manitoba in 1973–1974. The lecture notes of this course form the basis of most of Chapters III–VI.

I am grateful to my students who took the courses in 1968–1969 or in 1973–1974 and to my colleagues who attended for their helpful criticisms and for many simplified proofs. A corrected version of the first set of notes was read by R. Balbes, P. Burmeister, M. I. Gould, J. H. Hoffman, K. M. Koh, H. Lakser, S. M. Lee, R. Padmanabhan, P. Penner, and C. R. Platt. B. Jónsson, reading the manuscript of *First Concepts* for the publisher, offered many useful suggestions. The first part was proofread by R. Antonius, J. A. Gerhard, K. M. Koh, W. A. Lampe, R. W. Quackenbush, and I. Rival.

Many readers, in particular D. D. Miller, sent me corrections to *First Concepts*; this made it possible for me to improve some parts of the former book that are being reproduced here.

The second set of notes was distributed widely and I am grateful to all who offered corrections, in particular, to K. A. Baker, C. C. Chen, M. I. Gould, D. Haley, K. M. Koh, V. B. Lender, G. H. Wenzel, and B. Wolk.

In the proofreading of the present volume I was assisted by M. E. Adams, K. A. Baker, R. Beazer, J. Berman, B. A. Davey, J. A. Gerhard, M. I. Gould, D. Haley, D. Kelly, C. R. Platt, and G. H. Wenzel.

A great deal of organizational work was necessary in the distribution of manuscripts and the collation of corrections; this was faithfully carried out by R. Padmanabhan.

M. E. Adams undertook the arduous task of getting the manuscript ready for the publisher.

I received help from various individuals in specific areas, including M. Doob (matroids), I. Rival (exercises on combinatorial topics), R. Venkataraman (partially ordered vector spaces), and B. Wolk (projective geometry).

Thanks are due to the National Research Council of Canada for sponsoring my research and to Professor N. S. Mendelsohn for creating a very good environment for work. Mrs. M. McTavish did an excellent job of typing and retyping the manuscript.

Finally, I would like to thank the members and the many visitors of my seminar who, over a period of eight years, have been lecturing an average of four hours a week, 52 weeks a year, in an attempt to teach me lattice theory. Without their help I could not even have tried.

Despite the improvements so generously offered by so many, I am sure my original work can still be recognized: all the remaining mistakes are my own.

Preface to the Second Edition

In 20 years, tremendous progress has been made in Lattice Theory. Nevertheless, the change is in the superstructure not in the foundation. Accordingly, I decided to leave the book unchanged and add appendices to record the change.

In the first appendix: Retrospective, I briefly review developments from the point of view of this book, specifically, the major results of the last 20 years and solutions of the problems proposed in this book. It is remarkable how many difficult problems have been solved!

I was lucky in getting an exceptional group of people to write the other appendices: Brian A. Davey and Hilary A. Priestley on distributive lattices and duality, Friedrich Wehrung on continuous geometries, Marcus Greferath and Stefan E. Schmidt on projective lattice geometries, Peter Jipsen and Henry Rose on varieties, Ralph Freese on free lattices, Bernhard Ganter and Rudolf Wille on formal concept analysis; Thomas Schmidt collaborated with me on congruence lattices. Many of these same people are responsible for the definitive books on the same subjects.

I changed very little in the book proper. The diagrams have been redrawn and the book was typeset in LaTeX. To bring the notation up-to-date, I substituted $\operatorname{Con} L$ for $C(L)$, $\operatorname{Id} L$ for $I(L)$, and so on.

Almost 200 mathematicians helped me with this project, from correcting typos to writing long essays on the topics that should go into Retrospective. The last section of Retrospective lists the major contributors. My deeply felt thanks to all of them.

Winnipeg, Manitoba
February 1998

George Grätzer
gratzer@cc.umanitoba.ca
http://www.maths.umanitoba.ca/homepages/gratzer.html/

Introduction

In the first half of the nineteenth century, George Boole's attempt to formalize propositional logic led to the concept of Boolean algebras. While investigating the axiomatics of Boolean algebras at the end of the nineteenth century, Charles S. Pierce and Ernst Schröder found it useful to introduce the lattice concept. Independently, Richard Dedekind's research on ideals of algebraic numbers led to the same discovery. In fact, Dedekind also introduced modularity, a weakened form of distributivity. Although some of the early results of these mathematicians and of Edward V. Huntington are very elegant and far from trivial, they did not attract the attention of the mathematical community.

It was Garrett Birkhoff's work in the mid-thirties that started the general development of lattice theory. In a brilliant series of papers, he demonstrated the importance of lattice theory and showed that it provides a unifying framework for hitherto unrelated developments in many mathematical disciplines. Birkhoff himself, Valère Glivenko, Karl Menger, John von Neumann, Oystein Ore, and others had developed enough of this new field for Birkhoff to attempt to "sell" it to the general mathematical community, which he did with astonishing success in the first edition of his *Lattice Theory*. The further development of the subject matter can best be followed by comparing the first, second, and third editions of his book (G. Birkhoff [1940], [1948], and [1967]).

The goal of the present volume can be stated very simply: to discuss in depth the basics of general lattice theory. In other words, I tried to include what I consider the most important results and research methods of all of lattice theory. To treat the rudimentary results in depth and still keep the size of the volume from getting out of hand, I had to omit a great deal. I excluded many important chapters of lattice theory that have grown into research fields on their own. The reader will find appropriate references to these throughout this book. It is hoped that even those whose main interest lies in areas not treated here in detail will find this volume useful by obtaining from this book the background in lattice theory so necessary in allied fields.

In my view, distributive lattices have played a many faceted role in the development of lattice theory. Historically, lattice theory started with (Boolean) distributive lattices; as a result, the theory of distributive lattices is one of the most extensive and most satisfying chapters of lattice theory. Distributive lattices have provided the motivation for many results in general lattice theory. Many conditions on lattices and on elements and ideals of lattices are weakened forms of distributivity. Therefore, a thorough knowledge of distributive lattices is indispensable for work in lattice theory. Finally, in many applications the condition of distributivity is imposed on lattices arising in various areas of mathematics, especially algebras.

This viewpoint moved me to break with the traditional approach to lattice theory, which proceeds from partially ordered sets to general lattices, semimodular lattices, modular lattices, and, finally, to distributive lattices. That is why distributive lattices are treated as a first priority in this book. This approach has the added advantage that the reader (or the student in the classroom) reaches interesting and deep results early in the book.

Chapter I gives a concise development of the basic concepts of lattice theory. Diagrams are emphasized because I believe that an important part of learning lattice theory is the acquisition of skill in drawing diagrams. This point of view is stressed throughout the book by about 130 diagrams (heeding Alice's advice: "and what is the use of a book without pictures or conversations", L. Carroll [1865]); the reader would be well advised to draw many times more while reading the book. A special feature of this chapter is a detailed development of free lattices generated by a partial lattice over an arbitrary variety; this is one of the most important research tools of lattice theory.

Chapter II develops distributive lattices including representation theorems, congruences, congruence lattices of general lattices, Boolean algebras, and topological representations. The last section is a brief introduction to the theory of distributive lattices with pseudocomplementation. While the theory of distributive lattices is developed in detail, the reader should keep in mind that the purpose of this chapter is, basically, to serve as a model for the rest of lattice theory.

In Chapter III, we discuss congruences and ideals of general lattices. The various types of ideals discussed all imitate to some extent the behavior of ideals in distributive lattices.

After giving the basic facts concerning modular and semimodular lattices, Chapter IV investigates in detail the connection between lattice theory and geometry. We develop the theory of geometric lattices, in particular direct decompositions and geometric lattices arising out of geometries and graphs. As an important example, we investigate partition lattices. The last section deals with complemented modular lattices and projective geometries.

Chapters V and VI deal with two new areas of investigation. Varieties of lattices is one of the most promising new fields. In Chapter V most of the basic facts are presented along with some more specialized methods. Chapter VI grew out of an investigation of free lattices. It intends to prove that almost all the results on free lattices can be obtained within the framework of free products of lattices. In addition, free products can be used to construct interesting examples of lattices.

The exercises, which number almost 900, form an integral part of the book. The Bibliography contains over 750 entries; it is not, however, a comprehensive bibliography of this field. With a few exceptions, it contains only items referred to in the text. The 193 research problems, the "Further Topics and References" at the end of each chapter, and the Concluding Remarks should be of help to those who are interested in further reading and research in lattice theory.

The abandonment of the traditional structure of a lattice theory book means that concepts and notations are more evenly introduced throughout. A very detailed index and the Table of Notation should help the reader in finding where a concept or notation is first introduced.

The abbreviation "iff" stands for "if and only if". More difficult exercises are marked by *. "Theorem 10" refers to Theorem 10 of the same section, "Theorem 5.10" refers to Theorem 10 of Section 5 of the same chapter, whereas "Theorem I.5.10" refers to Theorem 10 of Section 5 in Chapter I. Similarly, "Exercise III.2.6" means Exercise 6 of Section III.2. References to the Bibliography are given in the form "J. Jakubik [1957]"; this refers to a paper (or book) by J. Jakubik published in 1957. Such references as "[1957a]" and "[1957b]" indicate that the Bibliography contains more than one work by the author published in that year. Problem I.29 is Problem 29 of Chapter I. References to the New Bibliography are given in standard form, B. Jónsson [337].

Winnipeg, Manitoba
September 1975

George Grätzer

First Concepts

1. Two Definitions of Lattices

1.1 Posets

Whereas the arithmetical properties of the set of reals $\mathbb{R}$ can be expressed in terms of addition and multiplication, the order theoretic, and thus the topological, properties are expressed in terms of the ordering relation $\leq$. The basic properties of this relation are as follows.

For all a, b, $c \in \mathbb{R}$, we have:

(P1) Reflexivity: $a \leq a$.

(P2) Antisymmetry: $a \leq b$ and $b \leq a$ imply that $a = b$.

(P3) Transitivity: $a \leq b$ and $b \leq c$ imply that $a \leq c$.

(P4) Linearity: $a \leq b$ or $b \leq a$.

There are many examples of binary relations sharing properties (P1)–(P4) with the ordering relation of reals, and there are even more enjoying properties (P1)–(P3). This fact, by itself, would not justify the introduction of a new concept. However, it has been shown that many basic concepts and results about the reals depend only on (P1)–(P3), and these can be profitably used whenever we have a relation satisfying (P1)–(P3). Relations satisfying (P1)–(P3) are called *partial ordering relations*, and sets equipped with such relations are called *partially ordered sets* or *poset*.

To make the definitions formal, let us start with two sets A, B and form the set $A \times B$ of all ordered pairs $\langle a, b \rangle$, with $a \in A$, $b \in B$. If $A = B$, we write A^2 for $A \times A$. Then a *binary relation* ϱ on A can simply be defined as a subset of A^2. The elements a, $b \in A$ are *in relation* with respect to ϱ iff $\langle a, b \rangle \in \varrho$. For $\langle a, b \rangle \in \varrho$, we shall also write $a \varrho b$, or $a \equiv b$ (ϱ). Binary relations will be denoted by small Greek letters or by special symbols.

This formal definition can be compared with the intuitive one: A binary relation ϱ on A is a "rule" that decides whether or not $a \varrho b$, for any given a, $b \in A$. Of course, any such rule will determine the set $\{ \langle a, b \rangle \mid a \varrho b, \ a, \ b \in A \}$, and this set determines ϱ, so we might as well regard ϱ as being identical with this set.

A *partially ordered set* $\langle A; \varrho \rangle$ consists of a nonempty set A and a binary relation ϱ on A such that ϱ satisfies properties (P1)–(P3). Note that these can be stated as follows: For all a, b, $c \in A$, $\langle a, a \rangle \in \varrho$; $\langle a, b \rangle$, $\langle b, a \rangle \in \varrho$ imply that $a = b$; $\langle a, b \rangle$, $\langle b, c \rangle \in \varrho$ imply that $\langle a, c \rangle \in \varrho$.

If ϱ satisfies (P1)–(P3), ϱ is a *partial ordering relation*, and will usually be denoted by $\leq$. Also, $a \geq b$ will mean $b \leq a$. Sometimes we shall say that A (rather than $\langle A; \leq \rangle$) is a poset, meaning that the partial ordering is understood. This is an ambiguous, although widely accepted, practice.

A poset $\langle A; \leq \rangle$ that also satisfies (P4) is called a *chain* (also called *fully ordered set, linearly ordered set,* and so on).

If $\langle A; \leq \rangle$ is a poset, a, $b \in A$, then a and b are *comparable* if $a \leq b$ or $b \leq a$. Otherwise, a and b are *incomparable*, in notation, $a \parallel b$. A chain is, therefore, a poset in which there are no incomparable elements.

Let $\langle A; \leq \rangle$ be a poset and let B be a nonempty subset of A. Then there is a natural partial order $\leq_B$ on B induced by $\leq$: for a, $b \in B$, $a \leq_B b$ iff $a \leq b$; we call $\langle B; \leq_B \rangle$ (or simply, $\langle B; \leq \rangle$) a *subposet* of $\langle A; \leq \rangle$.

An *unordered poset* is one in which $a \parallel b$, for all $a \neq b$.

A *chain C in a poset P* is a nonempty subset, which, as a subposet, is a chain. An *antichain C in a poset P* is a nonempty subset which, as a subposet, is unordered.

The *length, $l(C)$,* of a finite chain C is $|C| - 1$. A poset P is said to be *of length n* (in formula, $l(P) = n$), where n is a natural number, iff there is a chain in P of length n and all chains in P are of length $\leq n$. A *poset P is of finite length* iff it is of length n, for some natural number n.

Let $\mathrm{P}(X)$ denote the set of all subsets of a set X, partially ordered by $\subseteq$. If X has n elements, then $l(\mathrm{P}(X)) = n$.

The *width* of a poset P is n, where n is a natural number, iff there is an antichain in P of n elements and all antichains in P have $\leq n$ elements.

Next we define inf and sup in an arbitrary poset P (that is, $\langle P; \leq \rangle$), similarly as it is done for reals.

Let $H \subseteq P$ and $a \in P$. Then a is an *upper bound* of H iff $h \leq a$, for all $h \in H$. An upper bound a of H is the *least upper bound* of H or *supremum of*

H iff, for any upper bound b of H, we have $a \leq b$. We shall write $a = \sup H$ or $a = \bigvee H$. (The notations $a = \mathrm{l.u.b.}\, H$, $a = \sum H$ are also common in the literature).

To show the uniqueness of the supremum, let a_0 and a_1 be both suprema of H; then $a_0 \leq a_1$, since a_1 is an upper bound and a_0 is a supremum. Similarly, $a_1 \leq a_0$; thus $a_0 = a_1$ by (P2).

Let $\varnothing$ be the empty set. Let a be the least upper bound of $\varnothing$, $a = \sup \varnothing$. For every $b \in P$, we have $h \leq b$, for all $h \in \varnothing$ (since there is no such h) and so every $b \in P$ is an upper bound of $\varnothing$. Thus $a \leq b$, for all $b \in P$. We conclude that $\sup \varnothing$ exists iff P has a smallest element. We call $\sup \varnothing$ the *zero* of P and denote it by 0.

The concepts of *lower bound* and *greatest lower bound* or *infimum* are similarly defined; the latter is denoted by $\inf H$ or $\bigwedge H$. (The notations $\mathrm{g.l.b.}\, H$, $\prod H$ are also used in the literature.) The uniqueness is proved as in the last but one paragraph. Observe that $\inf \varnothing$ exists iff P has a largest element. We call $\inf \varnothing$ the *unit* (or *identity*) and denote it by 1.

The adverb "similarly" in the paragraph introducing infima can be given a very concrete meaning. Let $\langle P; \leq \rangle$ be a poset. The notation $a \geq b$ (meaning $b \leq a$) can also be regarded as a definition of a binary relation on P. This binary relation $\geq$ satisfies (P1)–(P3); as an example, let us check (P2). If $a \geq b$ and $b \geq a$, then, by the definition of $\geq$, we have $b \leq a$ and $a \leq b$; using (P2) for $\leq$, we conclude that $a = b$. (P1) and (P3) are equally trivial. Thus $\langle P; \geq \rangle$ is also a poset, called the *dual* of $\langle P; \leq \rangle$. Now, if Φ is a "statement" about posets, and if in Φ we replace all occurrences of $\leq$ by $\geq$, we get the *dual* of Φ.

Duality Principle. *If a statement Φ is true in all posets, then its dual is also true in all posets.*

This is true simply because Φ holds for $\langle P; \leq \rangle$ iff the dual of Φ holds for $\langle P; \geq \rangle$, which also ranges over all posets.

As an example, take for Φ the statement: "If $\sup H$ exists, then it is unique." We get as its dual: "If $\inf H$ exists, then it is unique." The dual of "$\langle P; \leq \rangle$ has a zero" is "$\langle P; \geq \rangle$ has a unit".

It is hard to imagine that anything as trivial as the Duality Principle could yield anything profound, and it does not; but it can save a lot of work.

1.2   Lattices

A poset $\langle L; \leq \rangle$ is a *lattice* if $\inf\{a, b\}$ and $\sup\{a, b\}$ exist, for all $a, b \in L$.

In other words, lattice theory singles out a special type of poset for detailed investigation. To make such a definition worthwhile, it must be shown that this class of posets is a very useful class, that there are many such posets in various branches of mathematics (analysis, topology, logic, algebra, geometry, and so on), and that a general study of these posets will lead to a better understanding of the behavior of the examples. This was done in the first edition of G. Birkhoff's

Lattice Theory [1940]. As we go along, we shall see many examples, most of them in the exercises. For a general survey of lattices in mathematics, see G. Birkhoff [1967] and H. H. Crapo and G.-C. Rota [1970].

We shall take the usefulness of lattice theory for granted (this is a less touchy subject now than it formerly was) and hope that the reader will like it for its intrinsic beauty.

To make the definition of a lattice less arbitrary, we note that an equivalent definition is the following:

> *A poset $\langle L; \leq \rangle$ is a lattice iff* inf *H and* sup *H exist, for any finite nonempty subset H of L.*

Proof. It is enough to prove that the first definition implies the second. So let $\langle L; \leq \rangle$ satisfy the first definition and let $H \subseteq L$ be nonempty and finite. If $H = \{a\}$, then inf $H =$ sup $H = a$ follows from the reflexivity of $\leq$ and the definitions of inf and sup. Let $H = \{a, b, c\}$. To show that inf H exists, set $d = \inf\{a, b\}$, $e = \inf\{c, d\}$; we claim that $e = $ inf H. First of all, $a \geq d$, $b \geq d$, and $c \geq e$, $d \geq e$; therefore (by transitivity) $x \geq e$, for all $x \in H$. Secondly, if f is a lower bound of H, then $a \geq f$, $b \geq f$, and thus $d \geq f$; also $c \geq f$, so that c, $d \geq f$, therefore $e \geq f$, since $e = \inf\{c, d\}$. Thus e is the infimum of H.

If $H = \{a_0, \dots, a_{n-1}\}$, $n \geq 1$, then

$$\inf\{\dots \inf\{a_0, a_1\}, \dots, a_{n-1}\}$$

is the infimum of H, by an inductive argument, whose steps are analogous to those in the preceding paragraph.

By duality (in other words, by applying the Duality Principle), we conclude that sup H exists.   ◻

A lattice need not have unit or zero, so inf ∅ and sup ∅ may not exist.

This simple proof can be varied to yield a large number of equally trivial statements about lattices and partially ordered sets, in general. Some of these will be stated as exercises and used later. To make the use of the Duality Principle legitimate for lattices, note:

> *If $\langle P; \leq \rangle$ is a lattice, so is its dual $\langle P; \geq \rangle$.*

Thus the Duality Principle applies to lattices.

We shall use the notations

$$a \wedge b = \inf\{a, b\},$$
$$a \vee b = \sup\{a, b\},$$

and call $\wedge$ the *meet* and $\vee$ the *join*. In lattices, they are both *binary operations*, which means that they can be applied to a pair of elements a, b of L to yield again an element of L. Thus $\wedge$ is a map of L^2 into L and so is $\vee$, a remark that might fail to be very illuminating at this point.

The previous proof yields that

$$(\ldots((a_0 \wedge a_1) \wedge a_2) \ldots) \wedge a_{n-1} = \inf\{a_0, \ldots, a_{n-1}\},$$

and there is a similar formula for sup. Now observe that the right-hand side does not depend on the way the elements a_i are listed. Thus $\wedge$ and $\vee$ are idempotent, commutative, and associative—that is, they satisfy the following:

(L1) Idempotency:

$$a \wedge a = a,$$
$$a \vee a = a.$$

(L2) Commutativity:

$$a \wedge b = b \wedge a,$$
$$a \vee b = b \vee a.$$

(L3) Associativity:

$$(a \wedge b) \wedge c = a \wedge (b \wedge c),$$
$$(a \vee b) \vee c = a \vee (b \vee c).$$

These properties of the operations are also called the *idempotent identities*, *commutative identities*, and *associative identities*, respectively.

As always in algebra, associativity makes it possible to write

$$a_0 \wedge a_1 \wedge \cdots \wedge a_{n-1}$$

without using parentheses (and the same for $\vee$).

There is another pair of rules that connect $\wedge$ and $\vee$. To derive them, note that if $a \leq b$, then $\inf\{a, b\} = a$; that is, $a \wedge b = a$, and conversely. Thus

$$a \leq b \quad \text{iff} \quad a \wedge b = a.$$

By duality (and by interchanging a and b), we have

$$a \leq b \quad \text{iff} \quad a \vee b = b.$$

Applying the "only if" part of the first rule to a and $a \vee b$, and that of the second rule to $a \wedge b$ and a, we get

(L4) *Absorption identities*:

$$a \wedge (a \vee b) = a,$$
$$a \vee (a \wedge b) = a.$$

Now we are faced with the crucial question: Do we know enough about $\wedge$ and $\vee$ so that lattices can be characterized purely in terms of the properties of $\wedge$ and $\vee$?

Two comments are in order. It is obvious that $\leq$ can be characterized by $\wedge$ and $\vee$ (in fact, by either of them); therefore, obtaining such a characterization is only a matter of persistence. More importantly, why should we try to get such a characterization? To rephrase the question, why should we want to characterize $\langle L; \leq \rangle$ as $\langle L; \wedge, \vee \rangle$, which is an algebra—that is, a set equipped with operations (in this case, two binary operations)? Note that $\leq$ is a subset of L^2, whereas $\wedge$ and $\vee$ are maps from L^2 into L.

The answer is simple: We want such a characterization because if we can treat lattices as algebras, then all the concepts and methods of universal algebra will become applicable. The usefulness of treating lattices as algebras will soon become clear.

An algebra $\langle L; \wedge, \vee \rangle$ is called a *lattice* iff L is a nonempty set, $\wedge$ and $\vee$ are binary operations on L, both $\wedge$ and $\vee$ are idempotent, commutative, and associative, and they satisfy the two absorption identities. The following theorem states that a lattice as an algebra and a lattice as a poset are "equivalent" concepts. (The word "equivalent" will not be defined.)

Theorem 1.

(i) Let the poset $\mathfrak{L} = \langle L; \leq \rangle$ be a lattice. Set

$$a \wedge b = \inf\{a, b\},$$
$$a \vee b = \sup\{a, b\}.$$

Then the algebra $\mathfrak{L}^a = \langle L; \wedge, \vee \rangle$ is a lattice.

(ii) Let the algebra $\mathfrak{L} = \langle L; \wedge, \vee \rangle$ be a lattice. Set

$$a \leq b \quad \text{iff} \quad a \wedge b = a.$$

Then $\mathfrak{L}^p = \langle L; \leq \rangle$ is a poset, and the poset $\mathfrak{L}^p$ is a lattice.

(iii) Let the poset $\mathfrak{L} = \langle L; \leq \rangle$ be a lattice. Then $(\mathfrak{L}^a)^p = \mathfrak{L}$.

(iv) Let the algebra $\mathfrak{L} = \langle L; \wedge, \vee \rangle$ be a lattice. Then $(\mathfrak{L}^p)^a = \mathfrak{L}$.   ⎯

Remark. (i) and (ii) describe how we pass from a poset to an algebra and back, whereas (iii) and (iv) state that going there and back takes us back to where we started.

Proof.
(i) This has already been proved.

(ii) We set $a \leq b$ to mean that $a \wedge b = a$. Now $\leq$ is reflexive since $\wedge$ is idempotent; $\leq$ is antisymmetric since $a \leq b$ and $b \leq a$ mean that $a \wedge b = a$ and $b \wedge a = b$, which, by the commutativity of $\wedge$, imply that $a = a \wedge b = b \wedge a = b$; $\leq$ is transitive, since if $a \leq b$ and $b \leq c$, then $a = a \wedge b$ and $b = b \wedge c$, and so

$$a = a \wedge b = a \wedge (b \wedge c)$$

($\wedge$ is associative)

$$= (a \wedge b) \wedge c = a \wedge c,$$

that is, $a \leq c$. Thus $\langle L; \leq \rangle$ is a poset. To prove that $\langle L; \leq \rangle$ is a lattice, we shall verify that $a \wedge b = \inf\{a, b\}$ and $a \vee b = \sup\{a, b\}$ (these are not definitions). Indeed, $a \wedge b \leq a$, since

$$(a \wedge b) \wedge a = a \wedge (b \wedge a) = a \wedge (a \wedge b) = (a \wedge a) \wedge b = a \wedge b,$$

using the associativity, commutativity, and idempotency of $\wedge$; similarly, $a \wedge b \leq b$. Now if $c \leq a$ and $c \leq b$, that is, $c \wedge a = c$ and $c \wedge b = c$, then

$$c \wedge (a \wedge b) = (c \wedge a) \wedge b = c \wedge b = c;$$

thus $a \wedge b = \inf\{a, b\}$. Finally, $a \leq a \vee b$ and $b \leq a \vee b$, because $a = a \wedge (a \vee b)$ and $b = b \wedge (a \vee b)$ by the first absorption identity; if $a \leq c$ and $b \leq c$, that is, $a = a \wedge c$ and $b = b \wedge c$, then $a \vee c = (a \wedge c) \vee c = c$ and $b \vee c = c$ (by the second absorption identity). Thus

$$(a \vee b) \wedge c = (a \vee b) \wedge (a \vee c) = (a \vee b) \wedge (a \vee (b \vee c))$$
$$= (a \vee b) \wedge ((a \vee b) \vee c) = a \vee b,$$

that is, $a \vee b \leq c$, completing the proof of $a \vee b = \sup\{a, b\}$.

(iii) It is enough to observe that the partial orderings of $\mathcal{L}$ and $(\mathcal{L}^a)^p$ are identical to get (iii).

(iv) The proof of (iv) is similar to the proof of (iii).   □

The proof of Theorem 1, and even the statement of Theorem 1, are subject to criticism. To begin with, in the definition of a lattice as an algebra, idempotency is redundant. The last step of the proof of (ii) can be made neater by first proving that

$$a = a \wedge b \quad \text{iff} \quad b = a \vee b.$$

Theorem 1 should be preceded by a similar theorem for "semilattices". All these questions will be dealt with in the exercises that follow this section.

Finally, note that for lattices as algebras, the Duality Principle takes on the following very simple form.

> Let Φ be a statement about lattices expressed in terms of $\wedge$ and $\vee$. The dual of Φ is the statement we get from Φ by interchanging $\wedge$ and $\vee$. If Φ is true for all lattices, then the dual of Φ is also true for all lattices.

To prove this we have only to observe that if $\mathfrak{L} = \langle L; \wedge, \vee \rangle$, then the dual of $\mathfrak{L}^p$ is $(\langle L; \vee, \wedge \rangle)^p$.

Most of the time, Φ involves $\leq$, and maybe 0 and 1, in addition to $\wedge$ and $\vee$. When dualizing such a Φ we interchange $\wedge$ and $\vee$, replace $\leq$ by $\geq$, and interchange 0 and 1.

Exercises

Posets

1. Define $x < y$ to mean $x \leq y$ and $x \neq y$. Prove that, in a partially ordered set, $x < x$ for no x, and that $x < y$, $y < z$ imply that $x < z$.

2. Let the binary relation $<$ satisfy the conditions of Exercise 1. Define $x \leq y$ to mean $x < y$ or $x = y$. Then show that $\leq$ is a partial ordering.

3. Prove the following extension of antisymmetry:

   If $x_0 \leq x_1 \leq \cdots \leq x_{n-1} \leq x_0$, then $x_0 = x_1 = \cdots = x_{n-1}$.

4. Enumerate all partial orderings on a five-element set.

5. Let $\leq$ be a partial ordering on A and let B be a subset of A. For $a, b \in B$, set $a \leq_B b$ iff $a \leq b$. Prove that $\leq_B$ is a partial ordering on B.

6. Let A be a set and let P be the set of all partial orderings on A. For ϱ, $\sigma \in P$, set $\varrho \leq \sigma$ iff $a\,\varrho\,b$ implies that $a\,\sigma\,b$ (for all $a, b \in A$). Prove that $\langle P; \leq \rangle$ is a poset.

7. Find an example of a poset in which inf $\varnothing$ does not exist.

8. Prove that if inf H exists for all nonempty subsets H of a poset P, then sup $\varnothing$ also exists in P.

9. Prove that the following are examples of posets:

   (i) Let $A = P(X)$, the set of all subsets of a set X; let $X_0 \leq X_1$ mean that $X_0 \subseteq X_1$ (set inclusion).

   (ii) Let A be the set of all real valued functions defined on X; for f, $g \in A$, set $f \leq g$ iff $f(x) \leq g(x)$ for all $x \in X$.

   (iii) Let A be the set of all continuous, concave, real-valued functions defined on the real interval; define $f \leq g$ as in (ii).

   (iv) Let A be the set of all open sets of a topological space; define $\leq$ as in (i).

   (v) Let A be the set of all human beings; let $a < b$ mean that a is a descendant of b.

Lattices as Posets

10. Which of the examples in Exercise 9 are lattices? For those that are lattices, compute $a \wedge b$ and $a \vee b$.

11. Show that every chain is a lattice.

12. Let A be the set of all subgroups (normal subgroups) of a group G; for $X, Y \in A$, set $X \leq Y$ to mean $X \subseteq Y$. Prove that $\langle A; \leq \rangle$ is a lattice; compute $X \wedge Y$ and $X \vee Y$.

13. Let $\langle P; \leq \rangle$ be a poset in which inf H exists *for all* $H \subseteq P$. Show that $\langle P; \leq \rangle$ is a lattice. (Hint: For $a, b \in P$, let H be the set of all upper bounds of $\{a, b\}$. Prove that $\sup\{a, b\} = \inf H$.) Relate this to Exercise 12.

Semilattices as Posets

14. A poset is a *meet-semilattice* (dually, *join-semilattice*) iff $\inf\{a, b\}$ (dually, $\sup\{a, b\}$) exists for any two elements a and b. Prove that the dual of a meet-semilattice is a join-semilattice, and conversely.

15. Let A be the set of finitely generated subgroups of a group G, partially ordered under set inclusion (as in Exercise 12). Prove that $\langle A; \subseteq \rangle$ is a join-semilattice, but not necessarily a lattice.

16. Let C be the set of all continuous, strictly-convex, real-valued functions defined on the real interval $[0, 1]$. For $f, g \in C$, set $f \leq g$ iff $f(x) \leq g(x)$, for all $x \in [0, 1]$. Prove that $\langle C; \leq \rangle$ is a meet-semilattice, but not a join-semilattice.

17. Show that a poset $\langle P; \leq \rangle$ is a lattice iff it is both a meet- and join-semilattice.

Semilattices as Algebras

18. Let $\langle A; \circ \rangle$ be an algebra with one binary operation $\circ$. The algebra $\langle A; \circ \rangle$ is a *semilattice* iff $\circ$ is idempotent, commutative, and associative. Let $\langle P; \leq \rangle$ be a join-semilattice. Show that the algebra $\langle P; \vee \rangle$ is a semilattice, where $a \vee b = \sup\{a, b\}$. State the analogous result for meet-semilattices.

19. Let the algebra $\langle A; \circ \rangle$ be a semilattice. Define the binary relations $\leq_\wedge$ and $\leq_\vee$ on A as follows: $a \leq_\wedge b$ iff $a = a \circ b$; $a \leq_\vee b$ iff $b = a \circ b$. Prove that $\langle A; \leq_\wedge \rangle$ is a poset, as a poset it is a meet-semilattice, and $a \wedge b = a \circ b$; that $\langle A; \leq_\vee \rangle$ is a poset, as a poset it is a join-semilattice, and $a \vee b = a \circ b$; and that the dual of $\langle A; \leq_\wedge \rangle$ is $\langle A; \leq_\vee \rangle$.

Semilattices

20. Prove the following statements:

  (i)   Let the poset $\mathfrak{A} = \langle A; \leq \rangle$ be a join-semilattice. Set $a \vee b = \sup\{a, b\}$. Then the algebra $\mathfrak{A}^a = \langle A; \vee \rangle$ is a semilattice.

  (ii) Let the algebra $\mathfrak{A} = \langle A; \circ \rangle$ be a semilattice. Set $a \leq b$ iff $a \circ b = b$. Then $\mathfrak{A}^p = \langle A; \leq \rangle$ is a poset, and the poset $\mathfrak{A}^p$ is a join-semilattice.

  (iii) Let the poset $\mathfrak{A} = \langle A; \leq \rangle$ be a join-semilattice. Then $(\mathfrak{A}^a)^p = \mathfrak{A}$.

  (iv) Let the algebra $\mathfrak{A} = \langle A; \circ \rangle$ be a semilattice. Then $(\mathfrak{A}^p)^a = \mathfrak{A}$.

21. Formulate and prove Theorem 1 for meet-semilattices.

Lattices as Algebras

22. Prove that the absorption identities imply the idempotency of $\wedge$ and $\vee$. (Hint: simplify $a \wedge (a \vee (a \wedge a))$ in two ways to yield $a = a \wedge a$.)

23. Let the algebra $\langle A; \wedge, \vee \rangle$ be a lattice. Define $a \leq_\wedge b$ iff $a \wedge b = a$; and define $a \leq_\vee b$ iff $a \vee b = b$. Prove that $a \leq_\wedge b$ iff $a \leq_\vee b$.

24. Prove that the algebra $\langle A; \wedge, \vee \rangle$ is a lattice iff $\langle A; \wedge \rangle$ and $\langle A; \vee \rangle$ are semilattices and $a = a \wedge b$ is equivalent to $b = a \vee b$. Verify that if $\langle A; \wedge, \vee_1 \rangle$ and $\langle A; \wedge, \vee_2 \rangle$ are both lattices, then $\vee_1$ is the same as $\vee_2$.

25. Are the three identities defining semilattices independent (that is, none follows from the others)?

*26. Prove that an algebra $\langle A; \wedge, \vee \rangle$ is a lattice iff it satisfies the two identities:

$$a = (b \wedge a) \vee a,$$

$$(((a \wedge b) \wedge c) \vee d) \vee e = (((b \wedge c) \wedge a) \vee e) \vee ((f \vee d) \wedge d).$$

(See J. A. Kalman [1968]. The first definition of lattices by two identities was found by R. Padmanabhan in 1967, Notices Amer. Math. Soc. **14**, No. 67T-468, and published in full in R. Padmanabhan [1969]. J. A. Kalman's two identities are slightly improved versions of those of R. Padmanabhan. R. N. McKenzie [1970] proved the existence of a single identity characterizing lattices.)

Miscellany

27. Let ϱ be a binary relation on the set A. The *transitive extension of* ϱ is a binary relation $\hat{\varrho}$ defined by the following rule: $a \, \hat{\varrho} \, b$ iff there exists a sequence $a = x_0, x_1, \ldots, x_n = b$ with $x_i \, \varrho \, x_{i+1}$, for $i = 0, \ldots, n-1$. Show that for a reflexive relation ϱ, the transitive extension $\hat{\varrho}$ is a partial ordering relation iff $\hat{\varrho}$ is antisymmetric. Express this condition in terms of ϱ.

28. Let ϱ be a reflexive and transitive binary relation (*quasi-ordering relation*) on the nonempty set A. Call $B \subseteq A$ a *block* iff $B \neq \varnothing$, $a \varrho b$ and $b \varrho a$, for any $a, b \in B$, and for $a \in B$, $b \in A$, $a \varrho b$ and $b \varrho a$ imply that $b \in B$. Let P be the set of all blocks and set $B_1 \leq B_2$ ($B_1, B_2 \in P$) iff $b_1 \varrho b_2$, for some (thus for all) $b_1 \in B_1$ and $b_2 \in B_2$. Prove that $\langle P; \leq \rangle$ is a poset.

29. Let A be a set of sets. Let $a \varrho b$ mean that there is a one-to-one map from a into b. What is $\langle P; \leq \rangle$? (Notation is that of Exercise 28.)

30. Let the binary operation $\circ$ on the set A be associative. Give a rigorous proof of the statement that any meaningful bracketing of $a_0 \circ \cdots \circ a_{n-1}$ will yield the same element.

31. Suppose that in a poset $b \vee c$, $a \vee (b \vee c)$, and $a \vee b$ exist. Prove that $(a \vee b) \vee c$ exists and that $a \vee (b \vee c) = (a \vee b) \vee c$.

32. Prove that if $a \wedge b$ exists in a poset, so does $a \vee (a \wedge b)$.

33. Let H and K be subsets of a poset. Suppose that $\sup H$, $\sup K$, and $\sup(H \cup K)$ exist. ($H \cup K$ is the set union of H and K.) Under these conditions verify that $(\sup H) \vee (\sup K)$ exists and equals $\sup(H \cup K)$.

34. In a poset P define the *comparability relation* γ: For $a, b \in P$, $a \gamma b$ iff $a \leq b$ or $b \leq a$. Take a sequence $a_1, \ldots, a_k$ of elements of P satisfying the following conditions:

    (i) $a_i \neq a_{i+1}$, $a_i \gamma a_{i+1}$, for $i = 1, \ldots, k - 1$, and $a_k \neq a_1$, $a_k \gamma a_1$.

    (ii) For no elements $a, b \in P$, and $i, j < k$, $i \neq j$ is $a = a_i = a_j$, $b = a_{i+1} = a_{j+1}$, or $a = a_i = a_k$, $b = a_{i+1} = a_1$.

    (iii) For no $1 \leq i \leq k - 2$ is $a_i \gamma a_{i+2}$, and neither $a_{k-1} \gamma a_1$ nor $a_k \gamma a_2$ holds.

    Prove that k is even.

*35. Prove that a binary relation γ on a set A is the comparability relation of some poset $\langle A; \leq \rangle$ iff γ satisfies the condition of Exercise 34 (A. Ghouilà-Houri [1962]; see also P. C. Gilmore and A. J. Hofman [1964] and M. Aigner [1969]).

2. How to Describe Lattices

To illustrate results and to refute conjectures, we shall have to describe a large number of examples of lattices. This can be done by basing the examples on known mathematical structures, as illustrated in the exercises of Section 1. In this section we list a few other methods.

A finite lattice can always be described by a *meet-table* and a *join-table*. For example, the following two tables describe a lattice on the set $L = \{0, a, b, 1\}$:

$\wedge$	0	a	b	1
0	0	0	0	0
a	0	a	0	a
b	0	0	b	b
1	0	a	b	1

$\vee$	0	a	b	1
0	0	a	b	1
a	a	a	1	1
b	b	1	b	1
1	1	1	1	1

We see that much of the information provided by the tables is redundant. Since both operations are commutative, the tables are symmetric with respect to the diagonal. Furthermore, $x \wedge x = x$ and $x \vee x = x$; thus the diagonals themselves do not provide information. Therefore, the two tables can be condensed into one:

$\vee\wedge$	0	a	b	1
0		0	0	0
a	a		0	a
b	b	1		b
1	1	1	1	

It should be emphasized that the part above the diagonal determines the part below the diagonal since either determines the partial ordering. To show that this table defines a lattice, we have only to check the associative and absorption identities.

An alternative way is to describe the partial ordering, that is, all pairs $\langle x, y \rangle$ with $x \leq y$. In the preceding example, we get

$$\leq = \{\langle 0,0 \rangle, \langle 0, a \rangle, \langle 0, b \rangle, \langle 0, 1 \rangle, \langle a, a \rangle, \langle a, 1 \rangle, \langle b, b \rangle, \langle b, 1 \rangle, \langle 1, 1 \rangle\}.$$

Obviously, all pairs of the form $\langle x, x \rangle$ can be omitted from the list, since we know that $x \leq x$. Also, if $x \leq y$ and $y \leq z$, then $x \leq z$. For instance, if we know that $0 \leq a$ and $a \leq 1$, we do not have to be told that $0 \leq 1$. To make this idea more precise, let us say that in the poset $\langle P; \leq \rangle$, a *covers* b or b *is covered by* a (in notation, $a \succ b$ or $b \prec a$) iff $b < a$ and, for no x, $b < x < a$. The covering relation of the preceding examples is simply:

$$\prec = \{\langle 0, a \rangle, \langle 0, b \rangle, \langle a, 1 \rangle, \langle b, 1 \rangle\}.$$

Does the covering relation determine the partial ordering? The following lemma shows that in the finite case it does.

Lemma 1. *Let $\langle P; \leq \rangle$ be a finite poset. Then $a \leq b$ iff $a = b$ or there exists a finite sequence of elements $x_0, \ldots, x_{n-1}$ such that $x_0 = a$, $x_{n-1} = b$, and $x_i \prec x_{i+1}$, for $0 \leq i < n - 1$.*

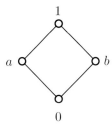

$$1$$

$$a \qquad b$$

$$0$$

Figure 1

Proof. If there is such a finite sequence, then $a = x_0 \le x_1 \le \cdots \le x_{n-1} = b$, and a trivial induction on n yields $a \le b$. Thus it suffices to prove that if $a < b$, then there is such a sequence. Fix $a, b \in P$, $a < b$, and take all subsets H of P such that H is a chain in P, a is the smallest element of H, and b is the largest element of H. There are such subsets: $\{a, b\}$, for example. Choose such an H with the largest possible number of elements, say with m elements. (P is finite.) Then $H = \{x_0, \ldots, x_{m-1}\}$, and we can assume that $x_0 < x_1 < \cdots < x_{m-1}$. We claim that in $\langle P; \le \rangle$, we have $a = x_0 \prec x_1 \prec \cdots \prec x_{m-1}$. Indeed, $x_i < x_{i+1}$ by assumption. Thus if $x_i \prec x_{i+1}$ does not hold, then $x_i < x < x_{i+1}$, for some $x \in P$, and $H \cup \{x\}$ will be a chain of $m+1$ elements between a and b, contrary to the maximality of the number of elements of H. □

The *diagram* of a poset $\langle P; \le \rangle$ represents the elements with small circles ∘; the circles representing two elements x, y are connected by a straight line iff one covers the other; if x covers y, then the circle representing x is higher than the circle representing y. The diagram of the lattice discussed above is shown in Figure 1.

In a diagram the intersection of two lines does not indicate an element. A diagram is *planar* if no two lines intersect. A diagram is *optimal*, if the number of pairs of intersecting lines is minimal. Figures 1, 2, and 3 show planar diagrams; Figure 4 is an optimal, but not planar, diagram; Figure 6 is not optimal. As a rule, optimal diagrams are the most practical to use.

Sometimes the "diagram" of an infinite poset is drawn; see, for instance, Figure VI.1.6. Such diagrams are always accompanied by explanations in the text.

We shall describe lattices by combining the methods described above. The lattice N_5 in Figure 2 has five elements: o, a, b, c, i, and $b < a$, $b \vee c = i$, $a \wedge c = 0$. This description is complete—that is, all the relations follow from the ones given. M_3 has five elements: o, a, b, c, i, and

$$a \wedge b = a \wedge c = b \wedge c = o,$$
$$a \vee b = a \vee c = b \vee c = i.$$

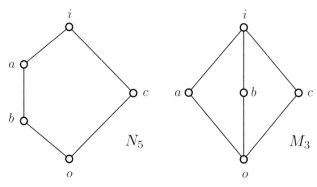

Figure 2

In contrast, we can start with some elements (say a, b, c), with some relations (say, $b < a$), and ask for the "most general" lattice that can be formed *without* specifying the elements to be used. (The exact meaning of "most general" will be given in Section 5.) In this case, we continue to form meets and joins until we get a lattice. A meet (or join) formed is identified with an element that we already have only if this identification is forced by the lattice axioms or by the given relations. The lattice we get from a, b, c and satisfying $b < a$ is shown in Figure 3.

To illustrate these ideas, we give a part of the computation that goes into the construction of the most general lattice L generated by a, b, c with $b < a$. We start by constructing the meets and joins $a \wedge c$, $b \wedge c$, $a \vee c$, $b \vee c$; note that $b \wedge (a \wedge c) = (a \wedge b) \wedge c = b \wedge c$, since $a \wedge b = b$; similarly, $a \vee (b \vee c) = a \vee c$. Next

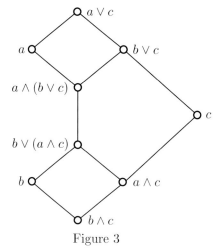

Figure 3

we have to show that the seven elements

$$a, \ b, \ c, \ a \wedge c, \ b \wedge c, \ a \vee c, \ b \vee c$$

that we already have are all distinct. Remember that two were equal if such equality would follow from the relation $(b < a)$ and the lattice axioms. Therefore, to show a pair of them distinct, it is enough to find a lattice K with $a, b, c \in K$, $b < a$, where the pair of elements is distinct. For instance, to show $a \neq a \vee c$, take the lattice $\{0, 1, 2\}$, with $0 < 1 < 2$, and $b = 0$, $a = 1$, $c = 2$.

The next step is to form a further join and meet: $b \vee (a \wedge c)$, $a \wedge (b \vee c)$ and claim that the nine elements

$$a, \ b, \ c, \ a \vee c, \ b \vee c, \ a \wedge c, \ b \wedge c, \ b \vee (a \wedge c), \ a \wedge (b \vee c)$$

form a lattice. We have to prove that by joins and meets we cannot get anything new. The thirty-six joins and thirty-six meets we have to check are all trivial. For instance,

$$b \wedge (a \wedge (b \vee c)) = b \wedge a \wedge (b \vee c) = b,$$

by the absorption identity, since $a \wedge b = b$; also

$$c \wedge (a \wedge (b \vee c)) = (c \wedge a) \wedge (b \vee c) = a \wedge c,$$

since $c \wedge a \leq b \vee c$.

Exercises

1. Give the meet- and join-table of the lattice in Exercise 1.9(i) for one-, two-, and three-element sets X.

2. Give the set $\leq$ for the lattices in Exercise 1. Which is simpler: the meet- and join-table or $\leq$?

3. Describe a practical method of checking associativity in a join- (meet-) table.

4. Relate Lemma 1 to Exercise 1.27.

5. Let $\langle P; \leq \rangle$ be a poset, $a, b \in P$, $a < b$, and let C denote the set of all chains H in P with smallest element a, largest element b. Let $H_0 \leq H_1$ mean $H_0 \subseteq H_1$, for $H_0, H_1 \in C$. Show that $\langle C; \leq \rangle$ is a poset with zero $\{a, b\}$.

6. The poset $\langle Q; \leq \rangle$ is said to satisfy the *Ascending Chain Condition* iff all increasing chains terminate; that is, if $x_i \in Q$, $i = 0, 1, 2, \ldots$, and $x_0 \leq x_1 \leq \cdots \leq x_i \leq \cdots$, then for some m, we have $x_m = x_{m+1} = \cdots$. The element x of Q is *maximal* if $x \leq y$ $(y \in Q)$ implies that $x = y$. Show that the Ascending Chain Condition implies the existence of maximal elements and that, in fact, every element is included in a maximal element (but not conversely).

7. Dualize Exercise 6. (The dual of maximal is *minimal* and the dual of ascending is *descending*.)

8. If $\langle Q; \leq \rangle$ is a lattice and x is a maximal element, then x is the unit. Show that this statement is not, in general, true in a poset.

9. Give examples of posets without maximal elements and of posets with maximal elements in which not every element is included in a maximal element.

10. Let $\langle P; \leq \rangle$ be a poset with the property that for every a, $b \in P$, $a < b$, all chains in P with smallest element a and the largest element b are finite. Show that the poset $\langle C; \leq \rangle$ formed from $\langle P; \leq \rangle$ in Exercise 5 satisfies the Ascending Chain Condition (see Exercise 6).

11. Extend Lemma 1 to posets satisfying the condition of Exercise 10 (combine Exercises 6 and 10).

12. Could the result of Exercise 11 be proved using the reasoning of Lemma 1?

13. Is the result of Exercise 11 the best possible? (No.)

14. Describe a method of finding the meet- and join-table of a lattice given by a diagram.

15. Are the posets of Figures 4 and 5 lattices?

16. Show that Figures 6 and 7 represent the same lattice.

17. Simplify Figure 8.

18. Simplify Figure 9. What is the number of pairs of intersecting lines in an optimal diagram? (Zero.)

19. Draw the diagrams of P(X) (Exercise 1.9(i)) for $|X| = 3$ and $|X| = 4$. ($|X|$ is the cardinality of X.) Does P(X) have a planar diagram for $|X| = 3$?

20. Draw the diagrams of the lattice of binary relations on X (partially ordered by set inclusion) for $|X| \leq 4$.

21. Describe the most general lattice generated by a, b, c such that $a > b$, $a \vee c = b \vee c$, and $a \wedge c = b \wedge c$.

*22. Describe the most general lattice generated by a, b, c, d such that $a > b > c$. (Hint: see Figure VI.1.2.)

*23. Show that the most general lattice generated by a, b, c, d such that $a > b$ and $c > d$ is infinite. (Hint: see Figure VI.1.6.)

24. Let N be the set of positive integers, $L = \{ \langle n, i \rangle \mid n \in N, \; i = 0, \; 1 \}$. Set $\langle n, i \rangle \leq \langle m, j \rangle$ iff $n \leq m$ and $i \leq j$. Show that L is a lattice and draw the "generalized diagram" of L.

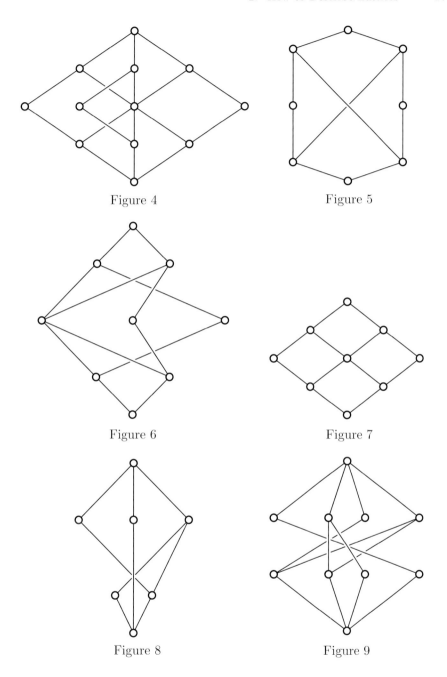

Figure 4

Figure 5

Figure 6

Figure 7

Figure 8

Figure 9

25. Draw the diagrams of all lattices with at most six elements.

To dispel the impression that may have been created by Sections 1 and 2, namely, that it is always easy to prove that a poset is a lattice, we present Exercises 26–36 showing that the poset T_n defined in Exercise 34 is a lattice. This is a result of D. Tamari [1951], first published in H. Friedman and D. Tamari [1967]. The present proof is based on S. Huang and D. Tamari [1972]. See also D. Huguet [1975].

26. Let T_n denote the set of all possible *binary bracketings* of $x_0 x_1 \cdots x_n$; for instance,

$$T_0 = \{x_0\},$$
$$T_1 = \{(x_0 x_1)\},$$
$$T_2 = \{(x_0 x_1) x_2), (x_0 (x_1 x_2))\},$$
$$T_3 = \{(x_0(x_1(x_2 x_3))), (x_0((x_1 x_2) x_3)), ((x_0 x_1)(x_2 x_3)), ((x_0(x_1 x_2)) x_3),$$
$$(((x_0 x_1) x_2) x_3)\}.$$

Give a formal (inductive) definition of T_n.

27. Replacing consistently all occurrences of (AB) by $A(B)$ in a binary bracketing, we get the *right bracketing* of the expression. For instance, the right bracketing of $((x_0(x_1 x_2)) x_3)$ is $x_0(x_1(x_2))(x_3)$ and of $((x_0 x_1)(x_2 x_3))$ is $x_0(x_1)(x_2(x_3))$. Give a formal (inductive) definition of right bracketing and prove that there is a one-to-one correspondence between binary and right bracketings.

28. Show that in a right bracketing of $x_0 x_1 \cdots x_n$, there is one and only one opening bracket preceding any x_i, $1 \le i \le n$.

29. Associate with a right bracketing of $x_0 x_1 \cdots x_n$ a *bracketing function*

$$E \colon \{1, \dots, n\} \to \{1, \dots, n\}$$

defined as follows: For $1 \le i \le n$, there is, by Exercise 28, an opening bracket before x_i; let this bracket close following x_j; set $E(i) = j$. Show that E has the following properties:

(i) $i \le E(i)$, for $1 \le i \le n$;

(ii) $i \le j \le E(i)$ imply that $E(j) \le E(i)$, for $1 \le i \le j \le n$.

*30. Show that (i) and (ii) of Exercise 29 characterize bracketing functions.

31. Let E_n denote the set of all bracketing functions defined on $\{1, \dots, n\}$. For E, $F \in E_n$, set $E \le F$ iff $E(i) \le F(i)$, for all i, $1 \le i \le n$. Show that $\le$ is a partial ordering, E_n as a poset is a lattice, and $(E \wedge F)(i) = \inf\{E(i), F(i)\}$.

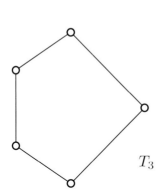
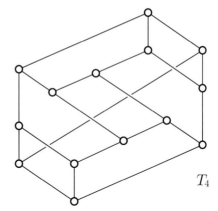

T_3 T_4

Figure 10

32. The *semiassociative identity* applied at the place i is a map $\sigma_i \colon T_n \to T_n$ defined as follows: If $E = \cdots (A(BC)) \cdots$, where the first variable in B and C is x_i and x_j, respectively, then $E\sigma_i = \cdots ((AB)C) \cdots$; if E is not of such form, then $E\sigma_i = E$. Let X and Y denote the bracketing functions associated with E and $E\sigma_i$, respectively. Show that $X(k) = Y(k)$, for $k \neq i$, $i \leq k \leq n$, and $Y(i) = j - 1$. Conclude that $X > Y$.

33. Show the converse of Exercise 32.

34. For $E, F \in T_n$, define $E < F$ to mean the existence of a sequence $E = X_0$, $X_1, \ldots, X_k = F$, $X_i \in T_n$, $0 \leq i \leq k$, such that X_{i+1} can be obtained from X_i by some application of the semiassociative law, for $0 \leq i < k$. Let $E \leq F$ mean $E = F$ or $E < F$. Show that $\leq$ is a partial ordering. Verify that Figure 10 is a diagram of T_3 and T_4. Is the diagram of T_4 optimal? (No.)

*35. Let $X, Y \in T_n$ and $X \succ Y$. Let E and F be the binary bracketings associated with X and Y, respectively. Show that $F = E\sigma_i$, for some i.

36. Show that T_n is a lattice, for each $n \geq 0$. (In fact, $T_n \cong E_n$.)

3. Some Algebraic Concepts

The purpose of any algebraic theory is the investigation of algebras up to isomorphism. We can introduce two concepts of isomorphism for lattices.

The lattices $\mathfrak{L}_0 = \langle L_0; \leq \rangle$ and $\mathfrak{L}_1 = \langle L_1; \leq \rangle$ *are isomorphic* (in symbol, $\mathfrak{L}_0 \cong \mathfrak{L}_1$), and the map $\varphi \colon L_0 \to L_1$ is an *isomorphism* iff φ is one-to-one and onto and

$$a \leq b \text{ in } \mathfrak{L}_0 \quad \text{iff} \quad a\varphi \leq b\varphi \text{ in } \mathfrak{L}_1.$$

The lattices $\mathfrak{L}_0 = \langle L_0; \wedge, \vee \rangle$ and $\mathfrak{L}_1 = \langle L_1; \wedge, \vee \rangle$ are *isomorphic* (in symbol, $\mathfrak{L}_0 \cong \mathfrak{L}_1$), and the map $\varphi \colon L_0 \to L_1$ is an *isomorphism* iff φ is one-to-one and onto and

$$(a \wedge b)\varphi = a\varphi \wedge b\varphi,$$
$$(a \vee b)\varphi = a\varphi \vee b\varphi.$$

An isomorphism of a lattice with itself is called an *automorphism*.

It is easy to see that the two isomorphism concepts coincide under the equivalence of Theorem 1.1. However, when we generalize these to homomorphism concepts, we get various new nonequivalent notions. In order to avoid confusion, they will be given different names.

From now on, we shall abandon the precise notation $\langle L; \wedge, \vee \rangle$ and $\langle L; \leq \rangle$ for lattices and posets; we shall simply write italic capitals, indicating the underlying sets, unless for some reason we want to be more exact.

Note that the first definition of isomorphism can be applied to any two posets L_0 and L_1, thus yielding an isomorphism concept for arbitrary posets. Having this concept of isomorphism, we can restate the content of Lemma 2.1: *The diagram of a finite poset determines the poset up to isomorphism.*

Let C_n denote the set $\{0, \dots, n-1\}$ ordered by $0 < 1 < 2 < \cdots < n-1$. Then C_n is an n-element chain. Observe that $l(C_n) = n - 1$. If $C = \{x_0, \dots, x_{n-1}\}$ is an n-element chain, $x_0 < x_1 < \cdots < x_{n-1}$, then $\varphi \colon i \mapsto x_i$ is an isomorphism between C_n and C. Therefore, the n-element chain is unique up to isomorphism.

The isomorphism of posets generalizes as follows.

The map $\varphi \colon P_0 \to P_1$ is an *isotone map* (also called *monotone map* or *order-preserving map*) of the poset P_0 into the poset P_1 iff $a \leq b$ in P_0 implies that $a\varphi \leq b\varphi$ in P_1.

Let us define a *homomorphism* of the semilattice $\langle S_0; \cdot \rangle$ into the semilattice $\langle S_1; \cdot \rangle$ as a map $\varphi \colon S_0 \to S_1$ satisfying $(a \cdot b)\varphi = a\varphi \cdot b\varphi$. Since a lattice $\mathfrak{L} = \langle L; \wedge, \vee \rangle$ is a semilattice both under $\wedge$ and under $\vee$, we get two homomorphism concepts, *meet-homomorphism* ($\wedge$-homomorphism) and *join-homomorphism* ($\vee$-homomorphism). A *homomorphism* is a map that is both a meet-homomorphism and a join-homomorphism. Thus a homomorphism φ of the lattice L_0 into the lattice L_1 is a map of L_0 into L_1 satisfying both

$$(a \wedge b)\varphi = a\varphi \wedge b\varphi,$$
$$(a \vee b)\varphi = a\varphi \vee b\varphi.$$

A homomorphism of a lattice into itself is called an *endomorphism*. A one-to-one homomorphism will also be called an *embedding*. (The list of homomorphism concepts will be further extended in Section 6.)

Note that meet-homomorphisms, join-homomorphisms, and (lattice) homomorphisms are all isotone. Let us prove this statement for meet-homomorphisms. If $\varphi \colon L_0 \to L_1$, $(a \wedge b)\varphi = a\varphi \wedge b\varphi$, for all a, $b \in L_0$, and if x, $y \in L_0$, $x \leq y$

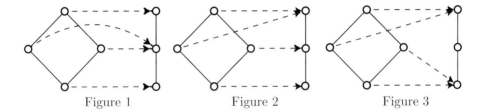

Figure 1   Figure 2   Figure 3

in L_0, then $x = x \wedge y$; thus $x\varphi = (x \wedge y)\varphi = x\varphi \wedge y\varphi$, and $x\varphi \leq y\varphi$ in L_1. Note that the converse does not hold, nor is there any connection between meet- and join-homomorphisms.

Figures 1–3 show three maps of the four-element lattice L of Figure 2.1 into the three-element chain C_3. The map of Figure 1 is isotone but is neither a meet-nor a join-homomorphism. The map of Figure 2 is a join-homomorphism but is not a meet-homomorphism, thus not a homomorphism. The map of Figure 3 is a homomorphism.

The second basic algebraic concept is that of a subalgebra:

A *sublattice* $\mathfrak{K} = \langle K; \wedge, \vee \rangle$ of the lattice $\mathfrak{L} = \langle L; \wedge, \vee \rangle$ is defined on a nonempty subset K of L with the property that $a, b \in K$ implies that $a \wedge b$, $a \vee b \in K$ (the operations $\wedge$, $\vee$ are taken in $\mathfrak{K}$), and the $\wedge$ and the $\vee$ of $\mathfrak{K}$ are restrictions to K of the $\wedge$ and the $\vee$ of $\mathfrak{L}$.

To put this in simpler language, we take a nonempty subset K of a lattice L such that K is closed under $\wedge$ and $\vee$. Under the same $\wedge$ and $\vee$, K is a lattice; this is a sublattice of L.

The concept of a lattice as a poset would suggest the following sublattice concept: Take a nonempty subset K of the lattice L; if the subposet K is a lattice, call K a sublattice of L. This concept is different from the previous one and we shall not use it at all; recently, some use was made of this in D. Kelly and I. Rival [1975a].

Let A_λ, $\lambda \in \Lambda$, be sublattices of L. Then $\bigcap(A_\lambda \mid \lambda \in \Lambda)$ (the set theoretic intersection of A_λ, $\lambda \in \Lambda$) is also closed under $\wedge$ and $\vee$; thus, for every $H \subseteq L$, $H \neq \varnothing$, there is a smallest $[H] \subseteq L$ containing H and closed under $\wedge$ and $\vee$. The sublattice $[H]$ is called the *sublattice of L generated by H*, and H is a *generating set* of $[H]$. If $H = \{a, b, c, \dots\}$, then we write $[a, b, c, \dots]$ for $[H]$.

The subset K of the lattice L is called *convex* iff $a, b \in K$, $c \in L$, and $a \leq c \leq b$ imply that $c \in K$. For $a, b \in L$, $a \leq b$, the *interval*

$$[a, b] = \{ x \mid a \leq x \leq b \}$$

is an important example of a *convex sublattice*. For a chain C, $a, b \in C$, $a \leq b$, we can also define the *half-open intervals*:

$$(a, b] = \{ x \mid a < x \leq b \},$$
$$[a, b) = \{ x \mid a \leq x < b \},$$

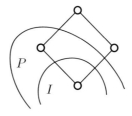

Figure 4

and the *open interval*:

$$(a,b) = \{\, x \mid a < x < b \,\}.$$

These are also, whenever nonempty, examples of convex sublattices. A sublattice I of L is an *ideal* iff $i \in I$ and $a \in L$ imply that $a \wedge i \in I$. An ideal I of L is *proper* iff $I \neq L$. A proper ideal I of L is *prime* iff $a, b \in L$ and $a \wedge b \in I$ imply that $a \in I$ or $b \in I$. In Figure 4, I is an ideal and P is a prime ideal; note that I is not prime.

   The concept of a convex sublattice is a typical example of the interplay between the algebraic and order theoretic concepts. We shall now examine these concepts more closely.

   Since the intersection of any number of convex sublattices (ideals) is a convex sublattice (ideal) unless void, we can define the *convex sublattice generated by a subset H*, and the *ideal generated by a subset H* of the lattice L, provided that $H \neq \varnothing$. The ideal generated by a subset H will be denoted by $(H]$, and if $H = \{a\}$, we write $(a]$ for $(\{a\}]$; we shall call $(a]$ a *principal ideal*.

Lemma 1. *Let L be a lattice and let H and I be nonempty subsets of L.*

   (i) *I is an ideal iff the following two conditions hold:*

      (i_1) *$a, b \in I$ implies that $a \vee b \in I$,*

      (i_2) *$a \in I$, $x \in L$, and $x \leq a$ imply that $x \in I$.*

   (ii) *$I = (H]$ iff the following three conditions hold:*

      (i_1) *I is an ideal,*

      (i_2) *$H \subseteq I$,*

      (ii_3) *for all $i \in I$, there exists an integer $n \geq 1$ and there exist elements $h_0, \ldots, h_{n-1} \in H$ such that $i \leq h_0 \vee \cdots \vee h_{n-1}$.*

   (iii) *For $a \in L$,*

$$(a] = \{\, x \mid x \leq a \,\} = \{\, x \wedge a \mid x \in L \,\}.$$

Proof.

(i) Let I be an ideal. Then a, $b \in I$ implies that $a \vee b \in I$, since I is a sublattice, verifying (i_1). If $x \leq a \in I$, then $x = x \wedge a \in I$, and (i_2) is verified. Conversely, let I satisfy (i_1) and (i_2). Let a, $b \in I$. Then $a \vee b \in I$ by (i_1), and, since $a \wedge b \leq a \in I$, we also have $a \wedge b \in I$ by (i_2); thus I is a sublattice. Finally, if $x \in L$ and $a \in I$, then $a \wedge x \leq a \in I$, thus $a \wedge x \in I$ by (i_2), proving that I is an ideal.

(ii) Set

$$I_0 = \{\, i \mid i \leq h_0 \vee \cdots \vee h_{n-1},$$
$$\text{for some } h_0, \ldots, h_{n-1} \in H \text{ and for some integer } n \geq 1 \,\}.$$

Using (i), it is clear that I_0 is an ideal, and obviously $H \subseteq I_0$. Finally, if $H \subseteq J$ and J is an ideal, then $I_0 \subseteq J$, and thus I_0 is the smallest ideal containing H; that is, $I = I_0$.

(iii) This proof is obvious directly, or by application of (ii). $\square$

Let $\mathrm{Id}\, L$ denote the set of all ideals of L and let $\mathrm{Id}_0\, L = \mathrm{Id}\, L \cup \{\varnothing\}$. We call $\mathrm{Id}\, L$ the *ideal lattice* and $\mathrm{Id}_0\, L$ the *augmented ideal lattice* of L. Many papers still use the older notation $I(L)$ for $\mathrm{Id}\, L$.

Corollary 2. $\mathrm{Id}\, L$ *and* $\mathrm{Id}_0\, L$ *are posets under set inclusion, and as posets they are lattices.*

In fact, $I \vee J = (I \cup J]$, if we agree that $(\varnothing] = \varnothing$. From (ii) of Lemma 1, we see that for I, $J \in \mathrm{Id}\, L$, $x \in I \vee J$ iff $x \leq i \vee j$, for some $i \in I$, $j \in J$. It does not matter that we consider only two ideals in this formula. Using the notation $(I_\lambda \mid \lambda \in \Lambda)$ for a *family* (of ideals) indexed by Λ, the operations $\bigcap$, $\bigcup$ for families of sets and $\bigwedge$, $\bigvee$ for families of elements of a poset, we obtain, in general,

$$\bigwedge (I_\lambda \mid \lambda \in \Lambda) = \bigcap (I_\lambda \mid \lambda \in \Lambda),$$
$$\bigvee (I_\lambda \mid \lambda \in \Lambda) = \left(\bigcup (I_\lambda \mid \lambda \in \Lambda) \right];$$

so any nonempty subset of $\mathrm{Id}\, L$ has a supremum. Combining this formula with Lemma 1(ii), we obtain:

Corollary 3. *Let* I_λ, $\lambda \in \Lambda$, *be ideals and let* $I = \bigvee (I_\lambda \mid \lambda \in \Lambda)$. *Then* $i \in I$ *iff* $i \leq j_{\lambda_0} \vee \ldots \vee j_{\lambda_{n-1}}$, *for some integer* $n \geq 1$ *and for some* $\lambda_0, \ldots, \lambda_{n-1} \in \Lambda$, $j_{\lambda_i} \in I_{\lambda_i}$.

Now observe the formulas:

$$(a] \wedge (b] = (a \wedge b],$$
$$(a] \vee (b] = (a \vee b].$$

Since $a \neq b$ implies that $(a] \neq (b]$, these yield:

Corollary 4. *L can be embedded in* Id *L (and also in* Id_0 *L), and* $a \mapsto (a]$ *is an embedding.*

Let us connect homomorphisms and ideals (recall that C_2 denotes the two-element chain with elements 0 and 1).

Lemma 5.

(i) *I is a proper ideal of L iff there is a join-homomorphism φ of L onto C_2 such that $I = 0\varphi^{-1}$ (the complete inverse image of 0, that is, $I = \{\, x \mid x\varphi = 0 \,\}$).*

(ii) *I is a prime ideal of L iff there is a homomorphism φ of L onto C_2 with $I = 0\varphi^{-1}$.*

Proof.

(i) Let I be a proper ideal and define φ by

$$x\varphi = \begin{cases} 0, & \text{if } x \in I; \\ 1, & \text{if } x \notin I; \end{cases}$$

obviously, this φ is a join-homomorphism.

Conversely, let φ be a join-homomorphism of L onto C_2 and $I = 0\varphi^{-1}$. Then, for $a, b \in I$, we have $a\varphi = b\varphi = 0$; thus $(a \vee b)\varphi = a\varphi \vee b\varphi = 0 \vee 0 = 0$, that is, $a \vee b \in I$. If $a \in I$, $x \in L$, $x \leq a$, then $x\varphi \leq a\varphi = 0$, that is, $x\varphi = 0$; thus $x \in I$. Finally, φ is onto, therefore $I \neq L$.

(ii) If I is prime, take the φ constructed in the proof of (i) and note that φ can violate the property of being a homomorphism only with $a, b \notin I$. However, since I is prime, $a \wedge b \notin I$; consequently, $(a \wedge b)\varphi = 1 = a\varphi \wedge b\varphi$, and so φ is a homomorphism. Conversely, let φ be a homomorphism of L onto C_2 and let $I = 0\varphi^{-1}$. If $a, b \notin I$, then $a\varphi = b\varphi = 1$, thus $(a \wedge b)\varphi = a\varphi \wedge b\varphi = 1$, and therefore $a \wedge b \notin I$; I is prime. $\qquad\square$

By dualizing, we get the concepts of *dual ideal* (also called *filter*), *principal dual ideal*, $[a)$ *(principal filter)*, *dual ideal* $[H)$ *generated by* H, *proper dual ideal*, *prime dual ideal (prime filter* or *ultra filter)*, the lattice Du L of dual ideals ordered by set inclusion, and $\mathrm{Du}_0\, L = (\mathrm{Du}\, L) \cup \{\varnothing\}$ ordered by set inclusion. Note that in Du L (and in $\mathrm{Du}_0\, L$) the largest element is L; if L has 0 and 1, then $L = [0)$ is the largest and $\{1\} = [1)$ is the smallest element of Du L. Furthermore, for $a, b \in L$, we have

$$[a) \wedge [b) = [a \vee b),$$
$$[a) \vee [b) = [a \wedge b).$$

(In the literature, filter means one of the following four concepts: (i) dual ideal; (ii) proper dual ideal; (iii) dual ideal of the lattice of all subsets of a set; (iv) proper dual ideal of the lattice of all subsets a set. Further variants allow the empty set as a filter under (i) or (ii).)

Lemma 6. *Let I be an ideal and let D be a dual ideal. If $I \cap D \neq \varnothing$, then $I \cap D$ is a convex sublattice, and every convex sublattice can be expressed in this form in one and only one way.*

Proof. The first statement is obvious. To prove the second, let C be a convex sublattice and set $I = (C]$ and $D = [C)$. Then $C \subseteq I \cap D$. If $t \in I \cap D$, then $t \in I$, and thus by (ii) of Lemma 1, $t \leq c$, for some $c \in C$; also, $t \in D$; therefore, by the dual of (ii) of Lemma 1, $t \geq d$, for some $d \in C$. This implies that $t \in C$ since C is convex, and so $C = I \cap D$.

Suppose now that C has another representation, $C = I_1 \cap D_1$. Since $C \subseteq I_1$, we have $(C] \subseteq I_1$. Let $a \in I_1$ and let c be an arbitrary element of C. Then $a \vee c \in I_1$ and $a \vee c \geq c \in D_1$, so $a \vee c \in D_1$, thus $a \vee c \in I_1 \cap D_1 = C$. Finally, $a \leq a \vee c \in C$; therefore, $a \in (C]$. This shows that $I_1 = (C]$. The dual argument shows that $D_1 = [C)$. Hence the uniqueness of such representations. $\square$

An *equivalence relation* Θ (that is, a reflexive, symmetric, and transitive binary relation) on a lattice L is called a *congruence relation* of L iff

$$a_0 \equiv b_0 \pmod{\Theta},$$
$$a_1 \equiv b_1 \pmod{\Theta}$$

imply that

$$a_0 \wedge a_1 \equiv b_0 \wedge b_1 \pmod{\Theta},$$
$$a_0 \vee a_1 \equiv b_0 \vee b_1 \pmod{\Theta}$$

(*Substitution Property*). Trivial examples are ω, ι, defined by

$$x \equiv y \quad (\omega) \qquad \text{iff } x = y;$$
$$x \equiv y \quad (\iota), \qquad \text{for all } x, y \in L.$$

(ω is the Greek o and stands for 0; ι is the Greek i and stands for identity and 1.) For $a \in L$, we write $[a]\Theta$ for the *congruence class* containing a, that is,

$$[a]\Theta = \{\, x \mid x \equiv a \ (\Theta) \,\}.$$

Lemma 7. *Let Θ be a congruence relation of L. Then for every $a \in L$, $[a]\Theta$ is a convex sublattice.*

Proof. Let $x, y \in [a]\Theta$; then $x \equiv a \ (\Theta)$ and $y \equiv a \ (\Theta)$. Therefore,

$$x \wedge y \equiv a \wedge a = a \quad (\Theta),$$
$$x \vee y \equiv a \vee a = a \quad (\Theta),$$

proving that $[a]\Theta$ is a sublattice. If $x \leq t \leq y$ and $x, y \in [a]\Theta$, then $x \equiv a \ (\Theta)$ and $y \equiv a \ (\Theta)$. Therefore,

$$t = t \wedge y \equiv t \wedge a \quad (\Theta),$$

and so

$$t = t \vee x \equiv (t \wedge a) \vee x \equiv (t \wedge a) \vee a = a \quad (\Theta),$$

proving that $[a]\Theta$ is convex.   □

Sometimes a long computation is required to prove that a given binary relation is a congruence relation. Such computations are often facilitated by the following lemma (G. Grätzer and E. T. Schmidt [1958d] and F. Maeda [1958]):

Lemma 8. *A reflexive binary relation Θ on a lattice L is a congruence relation iff the following three properties are satisfied, for x, y, z, $t \in L$:*

(i) $x \equiv y \ (\Theta)$ *iff* $x \wedge y \equiv x \vee y \ (\Theta)$.

(ii) $x \leq y \leq z$, $x \equiv y \ (\Theta)$, *and* $y \equiv z \ (\Theta)$ *imply that* $x \equiv z \ (\Theta)$.

(iii) $x \leq y$ *and* $x \equiv y \ (\Theta)$ *imply that* $x \wedge t \equiv y \wedge t \ (\Theta)$ *and* $x \vee t \equiv y \vee t \ (\Theta)$.

Proof. The "only if" part being trivial, assume now that a reflexive binary relation Θ satisfies conditions (i)–(iii). Let b, $c \in [a, d]$ and $a \equiv d \ (\Theta)$; we claim that $b \equiv c \ (\Theta)$. Indeed, $a \equiv d \ (\Theta)$ and $a \leq d$ imply by (iii) that

$$b \wedge c = a \vee (b \wedge c) \equiv d \vee (b \wedge c) = d \quad (\Theta).$$

Now $b \wedge c \leq d$ and (iii) imply that

$$b \wedge c = (b \wedge c) \wedge (b \vee c) \equiv d \wedge (b \vee c) = b \vee c \quad (\Theta);$$

thus by (i), $b \equiv c \ (\Theta)$.

To prove that Θ is transitive, let $x \equiv y \ (\Theta)$ and $y \equiv z \ (\Theta)$. Then by (i), $x \wedge y \equiv x \vee y \ (\Theta)$, and by (iii),

$$y \vee z = (y \vee z) \vee (x \wedge y) \equiv (y \vee z) \vee (x \vee y) = x \vee y \vee z \quad (\Theta),$$

and similarly, $x \wedge y \wedge z \equiv y \wedge z \ (\Theta)$. Therefore,

$$x \wedge y \wedge z \equiv y \wedge z \equiv y \vee z \equiv x \vee y \vee z \quad (\Theta)$$

and

$$x \wedge y \wedge z \leq y \wedge z \leq y \vee z \leq x \vee y \vee z.$$

Thus, applying (ii) twice, we get $x \wedge y \wedge z \equiv x \vee y \vee z \ (\Theta)$. Now we apply the statement of the previous paragraph with $a = x \wedge y \wedge z$, $b = x$, $c = z$, $d = x \vee y \vee z$, to conclude that $x \equiv z \ (\Theta)$.

Let $x \equiv y \ (\Theta)$; we claim that $x \wedge t \equiv y \wedge t \ (\Theta)$. Indeed, $x \wedge y \equiv x \vee y \ (\Theta)$ by (i); thus by (iii), $x \wedge y \wedge t \equiv (x \vee y) \wedge t \ (\Theta)$. Since

$$x \wedge t, \ y \wedge t \in [x \wedge y \wedge t, (x \vee y) \wedge t],$$

we conclude, by the first paragraph of the proof, that $x \wedge t \equiv y \wedge t$ (Θ).

To prove the Substitution Property for $\wedge$, let $x_0 \equiv y_0$ (Θ) and $x_1 \equiv y_1$ (Θ). Then $x_0 \wedge x_1 \equiv y_0 \wedge y_1 \equiv x_0 \wedge y_1$ (Θ), implying that $x_0 \wedge x_1 \equiv y_0 \wedge y_1$ (Θ), since Θ is transitive. The Substitution Property for $\vee$ is similarly proved. $\square$

Let $\mathrm{Con}\, L$ denote the set of all congruence relations on L partially ordered by set inclusion (remember that every $\Theta \in \mathrm{Con}\, L$ is a subset of L^2). As a first application of Lemma 8, we prove

---

Theorem 9. $\mathrm{Con}\, L$ is a lattice. For Θ, $\Phi \in \mathrm{Con}\, L$,

$$\Theta \wedge \Phi = \Theta \cap \Phi.$$

The join, $\Theta \vee \Phi$, can be described as follows:

> $x \equiv y$ ($\Theta \vee \Phi$) iff there is a sequence $z_0 = x \wedge y$, z_1, $\ldots$, $z_{n-1} = x \vee y$ of elements of L such that $z_0 \leq z_1 \leq \cdots \leq z_{n-1}$ and for each i, $0 \leq i < n-1$, $z_i \equiv z_{i+1}$ (Θ) or $z_i \equiv z_{i+1}$ (Φ).   —

Remark. $\mathrm{Con}\, L$ is called the *congruence lattice* of L; it was denoted by $\Theta(L)$ and $C(L)$ in earlier papers. Observe that $\mathrm{Con}\, L$ is a sublattice of $\mathrm{Part}\, L$ (Exercises 45, 46); that is, the join and meet of congruence relations as congruence relations and as equivalence relations (partitions) coincide.

Proof. $\Theta \wedge \Phi = \Theta \cap \Phi$ is obvious. To prove the statement for the join, let Ψ be the binary relation described in Theorem 9. Then $\Theta \subseteq \Psi$ and $\Phi \subseteq \Psi$ are obvious. If Γ is a congruence relation, $\Theta \subseteq \Gamma$, $\Phi \subseteq \Gamma$, and $x \equiv y$ (Ψ), then for each i, either $z_i \equiv z_{i+1}$ (Θ) or $z_i \equiv z_{i+1}$ (Φ); thus $z_i \equiv z_{i+1}$ (Γ), for all i. By the transitivity of Γ, $x \wedge y \equiv x \vee y$ (Γ); thus $x \equiv y$ (Γ). Therefore, $\Psi \subseteq \Gamma$. This shows that if Ψ is a congruence relation, then $\Psi = \Theta \vee \Phi$.

Ψ is obviously reflexive and satisfies Lemma 8(i). If $x \leq y \leq z$, $x \equiv y$ (Ψ), and $y \equiv z$ (Ψ), then $x \equiv z$ (Ψ) is established by putting together the sequences showing $x \equiv y$ (Ψ) and $y \equiv z$ (Ψ); this verifies Lemma 8(ii). To show Lemma 8(iii), let $x \equiv y$ (Ψ), $x \leq y$, with z_0, $\ldots$, z_{n-1} establishing this, and $t \in L$. Then $x \wedge t \equiv y \wedge t$ (Ψ) and $x \vee t \equiv y \vee t$ (Ψ) is demonstrated by the sequences $z_0 \wedge t$, $\ldots$, $z_{n-1} \wedge t$ and $z_0 \vee t$, $\ldots$, $z_{n-1} \vee t$, respectively. Thus the hypotheses of Lemma 8 hold for Ψ, and we conclude that Ψ is a congruence relation. $\square$

Homomorphisms and congruence relations express two sides of the same phenomenon. To establish this fact, we first define quotient lattices (also called *factor lattices*). Let L be a lattice and let Θ be a congruence relation on L. Let L/Θ denote the set of blocks of the partition of L induced by Θ, that is,

$$L/\Theta = \{\, [a]\Theta \mid a \in L \,\}.$$

Set

$$[a]\Theta \wedge [b]\Theta = [a \wedge b]\Theta,$$
$$[a]\Theta \vee [b]\Theta = [a \vee b]\Theta.$$

This defines $\wedge$ and $\vee$ on L/Θ. Indeed, if $[a]\Theta = [a_1]\Theta$ and $[b]\Theta = [b_1]\Theta$, then $a \equiv a_1$ (Θ) and $b \equiv b_1$ (Θ); therefore, $a \wedge b \equiv a_1 \wedge b_1$ (Θ), that is $[a \wedge b]\Theta = [a_1 \wedge b_1]\Theta$. Thus $\wedge$, and dually $\vee$, are well defined on L/Θ. The lattice axioms are easily verified. The lattice L/Θ is the *quotient lattice* of L modulo Θ.

Lemma 10. *The map*

$$\varphi_\Theta : x \mapsto [x]\Theta, \qquad \textit{for } x \in L,$$

is a homomorphism of L onto L/Θ.

Remark. The lattice K is a *homomorphic image* of the lattice L iff there is a homomorphism of L *onto* K. Lemma 10 states that any quotient lattice is a homomorphic image.

Proof. The proof is trivial.   □

Theorem 11 (The Homomorphism Theorem).   Let L be a lattice. Any homomorphic image of L is isomorphic to a suitable quotient lattice of L. In fact, if $\varphi \colon L \to L_1$ is a homomorphism of L onto L_1 and if Θ is the congruence relation of L defined by $x \equiv y$ (Θ) iff $x\varphi = y\varphi$, then

$$L/\Theta \cong L_1;$$

an isomorphism (see Figure 5) is given by

$$\psi \colon [x]\Theta \mapsto x\varphi, \qquad x \in L. \qquad \text{—}$$

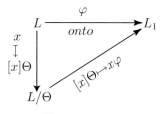

Figure 5

Proof. It is easy to check that Θ is a congruence relation. To prove that ψ is an isomorphism we have to check that ψ (i) is well defined, (ii) is one-to-one, (iii) is onto, and (iv) preserves the operations.

(i) Let $[x]\Theta = [y]\Theta$. Then $x \equiv y$ (Θ); thus $x\varphi = y\varphi$, that is, $([x]\Theta)\psi = ([y]\Theta)\psi$.

(ii) Let $([x]\Theta)\psi = ([y]\Theta)\psi$, that is $x\varphi = y\varphi$. Then $x \equiv y$ (Θ); and so $[x]\Theta = [y]\Theta$.

(iii) Let $a \in L_1$. Since φ is onto, there is an $x \in L$ with $x\varphi = a$. Thus $([x]\Theta)\psi = a$.

(iv)

$$([x]\Theta \wedge [y]\Theta)\psi = ([x \wedge y]\Theta)\psi = (x \wedge y)\varphi = x\varphi \wedge y\varphi = ([x]\Theta)\psi \wedge ([y]\Theta)\psi.$$

The computation for $\vee$ is identical.   □

For a homomorphism $\varphi\colon L \to L_1$ (not necessarily onto), the relation Θ described in Theorem 11 is called the *congruence kernel* of the homomorphism φ; it will be denoted by Ker φ. If L_1 has a zero, 0, then $0\varphi^{-1}$ is an ideal of L, called the *ideal kernel* of the homomorphism φ.

Let L be a lattice and let Θ be a congruence relation of L. If L/Θ has a zero, $[a]\Theta$, then $[a]\Theta$ as a subset of L is an ideal, called the *ideal kernel* of the congruence relation Θ.

In contrast with group or ring theory, all three kernel concepts are useful in lattice theory. Note the obvious connections; for instance, if I is the ideal kernel of φ, then it is the ideal kernel of Ker φ.

The final algebraic concept introduced in this section is that of direct product. Let L, K be lattices and form the set $L \times K$ of all ordered pairs $\langle a, b \rangle$ with $a \in L$, $b \in K$. Define $\wedge$ and $\vee$ in $L \times K$ "componentwise":

$$\langle a_0, b_0 \rangle \wedge \langle a_1, b_1 \rangle = \langle a_0 \wedge a_1, b_0 \wedge b_1 \rangle,$$
$$\langle a_0, b_0 \rangle \vee \langle a_1, b_1 \rangle = \langle a_0 \vee a_1, b_0 \vee b_1 \rangle.$$

This makes $L \times K$ into a lattice, called the *direct product* of L and K (for an example, see Figure 6).

Lemma 12. *Let L, L_1, K, K_1 be lattices, $L \cong L_1$, $K \cong K_1$. Then*

$$L \times K \cong L_1 \times K_1 \cong K_1 \times L_1.$$

Remark. This means that $L \times K$ is determined up to isomorphism if we know L and K up to isomorphism, and the direct product is determined up to isomorphism by the factors; the order in which they are given is irrelevant.

Proof. Let $\varphi\colon L \to L_1$ and $\psi\colon K \to K_1$ be isomorphisms; for $a \in L$ and $b \in K$, define $\langle a, b \rangle \chi = \langle a\varphi, b\psi \rangle$. Then $\chi\colon L \times K \to L_1 \times K_1$ is an isomorphism. Of course, $L_1 \times K_1 \cong K_1 \times L_1$ is proved by showing that $\langle a, b \rangle \mapsto \langle b, a \rangle$ $(a \in L_1, b \in K_1)$ is an isomorphism.   □

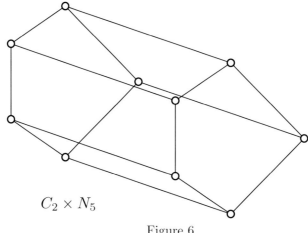

$C_2 \times N_5$

Figure 6

If L_i, $i \in I$, is a family of lattices, we first form the *Cartesian product* $\prod(L_i \mid i \in I)$ of the sets, which is defined as the set of all functions

$$f : I \to \bigcup(L_i \mid \in I)$$

such that $f(i) \in L_i$, for all $i \in I$. We then define $\wedge$ and $\vee$ "componentwise", that is, $f \wedge g = h$, $f \vee g = k$ means:

$$f(i) \wedge g(i) = h(i),$$
$$f(i) \vee g(i) = k(i),$$

for all $i \in I$. The resulting lattice is the *direct product* $\prod(L_i \mid \in I)$. If $L_i = L$, for all $i \in I$, we get the *direct power* L^I. Letting n denote the set $\{0, \ldots, n-1\}$, L^n is

$$\underbrace{(\cdots (L \times L) \times \cdots) \times L}_{n\text{-times}}$$

(at least, up to isomorphism). In particular, if we identify $f \colon 2 \to L$ with $\langle f(0), f(1) \rangle$, then we get $L^2 = L \times L$.

We shall use the convention that a one-element lattice (respectively, algebra) is the direct product of the empty family of lattices (respectively, algebras).

A very important property of direct products is:

Theorem 13.   Let L and K be lattices, let Θ be a congruence relation of L, and let Φ be a congruence relation of K. Define the relation $\Theta \times \Phi$ on $L \times K$ by

$$\langle a, b \rangle \equiv \langle c, d \rangle \quad (\Theta \times \Phi) \qquad \text{iff} \qquad a \equiv c \quad (\Theta) \quad \text{and} \quad b \equiv d \quad (\Phi).$$

Then $\Theta \times \Phi$ is a congruence relation on $L \times K$. Conversely, every congruence relation of $L \times K$ is of this form.   —

Proof. The first statement is obvious. Now let Ψ be a congruence relation on $L \times K$. For a, $b \in L$, define $a \equiv b$ (Θ) iff $\langle a, c \rangle \equiv \langle b, c \rangle$ (Ψ), for some $c \in K$. Let $d \in K$. Joining both sides with $\langle a \wedge b, d \rangle$ and then meeting with $\langle a \vee b, d \rangle$, we get $\langle a, d \rangle \equiv \langle b, d \rangle$ (Ψ); thus $\langle a, c \rangle \equiv \langle b, c \rangle$ (Ψ), for *some* $c \in K$, is equivalent to $\langle a, c \rangle \equiv \langle b, c \rangle$ (Ψ), for *all* $c \in K$.

Similarly, define for a, $b \in K$, $a \equiv b$ (Φ) iff $\langle c, a \rangle \equiv \langle c, b \rangle$ (Ψ) for any/for all $c \in L$. It is easily seen that Θ and Φ are congruences. Let $\langle a, b \rangle \equiv \langle c, d \rangle$ ($\Theta \times \Phi$); then $\langle a, x \rangle \equiv \langle c, x \rangle$ (Ψ), $\langle y, b \rangle \equiv \langle y, d \rangle$ (Ψ), for all $x \in K$ and $y \in L$. Joining the two congruences, with $y = a \wedge c$ and $x = b \wedge d$, we get $\langle a, b \rangle \equiv \langle c, d \rangle$ (Ψ). Finally, let $\langle a, b \rangle \equiv \langle c, d \rangle$ (Ψ). Meeting with $\langle a \vee c, b \wedge d \rangle$, we get $\langle a, b \wedge d \rangle \equiv \langle c, b \wedge d \rangle$ (Ψ); therefore, $a \equiv c$ (Θ). Similarly, $b \equiv d$ (Φ), and so $\langle a, b \rangle \equiv \langle c, d \rangle$ ($\Theta \times \Phi$), proving that $\Psi = \Theta \times \Phi$. $\square$

In conclusion, we introduce a nonalgebraic concept to balance the (false) impression that might have been created in this section that all lattice theoretic concepts are algebraic.

A lattice L is called *complete* if $\bigwedge H$ and $\bigvee H$ exist, for *any subset* $H \subseteq L$. The concept is self-dual, and half of the hypothesis is redundant.

Lemma 14. *Let P be a poset in which $\bigwedge H$ exists, for all $H \subseteq P$. Then P is a complete lattice.*

Proof. For $H \subseteq P$, let K be the set of all upper bounds of H. By hypothesis, $\bigwedge K$ exists; set $a = \bigwedge K$. If $h \in H$, then $h \leq k$, for all $k \in K$; therefore $h \leq a$ and $a \in K$. Thus a is the smallest member of K, that is, $a = \bigvee H$. $\square$

Lemma 14 can be applied to $\mathrm{Id}_0\, L$ and $\mathrm{Con}\, L$. For a further application, let $\mathrm{Sub}\, L$ denote the set of all subsets A of L closed under $\wedge$ and $\vee$ partially ordered under set inclusion. In other words, if $A \in \mathrm{Sub}\, L$ and $A \neq \varnothing$, then A is a sublattice of L. Obviously, $\mathrm{Sub}\, L$ is closed under arbitrary intersections.

Corollary 15. $\mathrm{Id}_0\, L$ *and* $\mathrm{Con}\, L$ *are complete lattices. If L has a smallest element, $\mathrm{Id}\, L$ is a complete lattice. The lattice $\mathrm{Sub}\, L$ is a complete lattice.*

Exercises

1. Formalize and prove the equivalence of the two isomorphism concepts.

2. Let $\varphi \colon L_0 \to L_1$, $\psi \colon L_1 \to L_0$ be (lattice) homomorphisms. Show that if $\varphi\psi$ is the identity map on L_0 and $\psi\varphi$ is the identity map on L_1, then φ is an isomorphism and $\psi = \varphi^{-1}$ (the *inverse map*, $\varphi^{-1} \colon L_1 \to L_0$). Furthermore, prove that if $\varphi \colon L_0 \to L_1$, $\psi \colon L_1 \to L_2$ are isomorphisms, then so are φ^{-1} and $\varphi\psi$.

3. A one-to-one and onto homomorphism is an isomorphism. Is a one-to-one and onto isotone map an isomorphism?

4. Find a general construction of meet- (join-) homomorphisms that are not homomorphisms.

5. Find a subset H of a lattice L such that H is not a sublattice of L but H is a lattice under the partial ordering of L restricted to H.

6. Show that a lattice L is a chain iff every nonempty subset of L is a sublattice.

7. Prove that a sublattice generated by two distinct elements has two or four elements.

*8. Find an infinite lattice generated by three elements.

9. Verify that a nonempty subset I of a lattice L is an ideal iff, for $a, b \in L$, $a \vee b \in I$ is equivalent to $a, b \in I$.

10. Prove that if L is finite, then L and $\operatorname{Id} L$ are isomorphic. How about $\operatorname{Id}_0 L$?

*11. Is there an infinite lattice L such that $L \cong \operatorname{Id} L$, but not every ideal is principal? (There is no such lattice; see D. Higgs [1971].)

12. Prove the completeness of $\operatorname{Id}_0 L$ without using Lemma 14.

13. Prove that $\operatorname{Con} L$ is complete by showing the following description of infinite joins:

> Let $H \subseteq \operatorname{Con} L$, $\Phi = \bigvee H$. Then $x \equiv y \ (\Phi)$ iff there exists a finite sequence $x \wedge y = z_0, z_1, \ldots, z_{n-1} = x \vee y$, such that $z_0 \le z_1 \le \cdots \le z_{n-1}$, and $z_i \equiv z_{i+1} \ (\Theta_i)$, for $0 \le i < n - 1$ and for some $\Theta_i \in H$.

14. Let $\varphi \colon L \to K$ be an onto homomorphism, let I be an ideal of L, and let J be an ideal of K. Show that $I\varphi$ is an ideal of K, and $J\varphi^{-1} = \{\, a \mid a \in L,\ a\varphi \in J \,\}$ is an ideal of L.

15. Is the image $P\varphi$ of a prime ideal under a homomorphism φ prime again?

16. Show that the complete inverse image $P\varphi^{-1}$ of a prime ideal P under an onto lattice homomorphism φ is prime again.

17. Show that an ideal P is a prime ideal of L iff $L - P$ is a dual ideal—in fact, a prime dual ideal.

18. Let H be a nonempty subset of the lattice L such that $a, b \in H$ implies that $a \vee b \in H$. Then

$$(H] = \{\, t \mid t \leq h, \text{ for some } h \in H \,\};$$

that is,

$$(H] = \bigcup (\, (h] \mid h \in H \,).$$

19. Show that if K is a sublattice of L, then K is isomorphic to a sublattice of Id L.

*20. Verify that the converse of Exercise 19 is false even for some finite K.

21. Prove that if N_5 (see Figure 2.2) is isomorphic to a sublattice of Id L, then N_5 is isomorphic to a sublattice of L.

22. Find a lattice L and a convex sublattice C of L that cannot be represented as $[a]\Theta$, for any congruence relation Θ of L.

23. State and prove an analogue of Lemma 8 for *join-congruence relations*, that is, for equivalence relations on a lattice satisfying the Substitution Property for joins.

24. Describe $\Theta \vee \Phi$ for join-congruences.

*25. Find a lattice L such that $L \cong L/\Theta$, for all $\Theta \neq \iota$, and there are infinitely many such Θ.

26. Describe the congruence lattice of N_5.

27. Describe the congruence lattice of an n-element chain.

28. Describe the congruence lattice of the lattice of Figure 2.3, and list all quotient lattices.

29. Construct a lattice that has exactly three congruence relations.

30. Construct infinitely many lattices L such that each lattice is isomorphic to its congruence lattice.

*31. Can an L in Exercise 30 be infinite?

32. Generalize Lemma 12 to the direct product of more than two lattices.

33. Show that $N_5 \cong L \times K$ implies that L or K has only one element.

34. Show that Id L is *conditionally complete*: Every nonempty set H with an upper bound has a supremum, and dually, Id L is complete iff L has a zero.

35. Show that the ideal kernel of a homomorphism is an ideal of L.

36. Find an ideal that is the ideal kernel of no homomorphism.

37. Find an ideal that is the kernel of more than one (infinitely many) homomorphisms (congruence relations).

38. Prove that every ideal of a lattice L is prime iff L is a chain.

39. Under what conditions does $L \times K$ have a planar diagram?

40. Let L and K be lattices and let $\varphi \colon L \to K$ be one-to-one and onto satisfying

$$\{(a \wedge b)\varphi, (a \vee b)\varphi\} = \{a\varphi \wedge b\varphi, a\varphi \vee b\varphi\},$$

for all $a, b \in L$. Let $A \in \mathrm{Sub}(L)$. Show that $A\varphi \in \mathrm{Sub}\,K$; in fact, $A \mapsto A\varphi$ is a (lattice) isomorphism between $\mathrm{Sub}\,L$ and $\mathrm{Sub}\,K$.

41. Prove the converse of Exercise 40.

42. Generalize Theorem 13 to finitely many lattices.

43. Show that the first part of Theorem 13 holds for any number of lattices but that the second part does not.

44. Show that the second statement of Theorem 13 fails for (Abelian) groups.

45. For a set X, let Part X denote the set of all equivalence relations on X partially ordered under set inclusion. Show that Part X is a (complete) lattice. (Equivalence relations and partitions can be identified—as explained in Section IV.4—and there we use the notation Part X for the set of all partitions on X.)

46. Show that Con L is a sublattice of Part L.

47. **The First Isomorphism Theorem.** *Let L be a lattice, let Θ be a congruence relation on L, and let L_1 be a sublattice of L. If, for every $a \in L$, there exists exactly one $b \in L_1$ satisfying $a \equiv b \ (\Theta)$, then*

$$L/\Theta \cong L_1.$$

48. Let L_0 and L_1 be lattices and let $\pi_i \colon \langle x_0, x_1 \rangle \mapsto x_i$ be the projection map of $L_0 \times L_1$ onto L_i, $i = 0, 1$. Prove that if A is a lattice and φ_i is a homomorphism of A into L_i, $i = 0, 1$, then there is a unique homomorphism $\psi \colon A \to L_0 \times L_1$ satisfying $\psi\pi_i = \varphi_i$, $i = 0, 1$.

49. In what way does Exercise 48 characterize $L_0 \times L_1$?

50. Characterize $\prod(\,L_i \mid i \in i\,)$ by projection maps and homomorphisms.

4.   Polynomials, Identities, and Inequalities

From variables $x_0, x_1, \dots, x_n, \dots$, we can form polynomials in the usual manner using $\wedge$, $\vee$, and, of course, parentheses. Examples of polynomials are:

$$x_0,$$
$$x_3,$$
$$x_0 \vee x_0,$$
$$(x_0 \wedge x_2) \vee (x_3 \wedge x_0),$$
$$(x_0 \wedge x_1) \vee ((x_0 \vee x_2) \wedge (x_1 \vee x_2)).$$

A formal definition is:

Definition 1. The set $\mathbf{P}^{(n)}$ of *n-ary lattice polynomials* is the smallest set satisfying (i) and (ii):

(i) $x_i \in \mathbf{P}^{(n)}$, $0 \le i < n$.

(ii) If $p, q \in \mathbf{P}^{(n)}$, then $(p \wedge q)$, $(p \vee q) \in \mathbf{P}^{(n)}$.

Remark. We shall omit the outside parentheses and write

$$p_1 \wedge p_2 \wedge \cdots \wedge p_n \quad \text{for} \quad (\cdots (p_1 \wedge p_2) \wedge \cdots \wedge p_n),$$

and the same for $\vee$. Thus we write $x_0 \wedge x_1$ for $(x_0 \wedge x_1)$ and $x_0 \wedge x_1 \wedge x_2$ for $((x_0 \wedge x_1) \wedge x_2)$. Note that if $n < m$, then $\mathbf{P}^{(n)} \subset \mathbf{P}^{(m)}$.

It is easy to define formally what it means to substitute in a polynomial p all occurrences of x_i by x_j; for instance, substituting x_4 for x_1, $p = (x_1 \vee x_2) \wedge (x_1 \wedge x_3)$ becomes $q = (x_4 \vee x_2) \wedge (x_4 \wedge x_3)$; observe that $p \in \mathbf{P}^{(4)}$ but $q \notin \mathbf{P}^{(4)}$. In general, for $p = p(x_0, \dots, x_{n-1}) \in \mathbf{P}^{(n)}$, we have $p(x_{i_0}, \dots, x_{i_{n-1}}) \in \mathbf{P}^{(m)}$, for any $m > i_0, \dots, i_{n-1}$.

Remark. In many current papers, polynomials and polynomial functions are called *terms* and *term functions*, respectively.

By Definition 1, a polynomial is just a sequence of symbols. It is defined because in terms of such a sequence of symbols we can define a function on any lattice:

Definition 2. An n-ary polynomial p defines a function in n variables (a *polynomial function*, or simply, a *polynomial*) on a lattice L by the following rules $(a_0, \dots, a_{n-1} \in L)$:

(i) If $p = x_i$, then $p(a_0, \dots, a_{n-1}) = a_i$, $0 \le i < n$.

(ii) If $p(a_0, \dots, a_{n-1}) = a$, $q(a_0, \dots, a_{n-1}) = b$, and $p \wedge q = r$, $p \vee q = t$, then $r(a_0, \dots, a_{n-1}) = a \wedge b$ and $t(a_0, \dots, a_{n-1}) = a \vee b$.

Thus if $p = (x_0 \wedge x_1) \vee (x_2 \vee x_1)$, then $p(a,b,c) = (a \wedge b) \vee (c \vee b) = b \vee c$. Definitions 1 and 2 are quite formal but their meaning is very simple.

A *polynomial* is an n-ary polynomial, for some n. We shall also use the variables x, y, z, ... instead of the x_i.

Note that if p is a *unary* $(n = 1)$ lattice polynomial, then $p(a) = a$, for any $a \in L$. If p is *binary*, then, for all a, $b \in L$, one of the following four possibilities holds:

$$p(a,b) = \begin{cases} a; \\ b; \\ a \wedge b; \\ a \vee b. \end{cases}$$

We shall prove statements on polynomials by induction on the *rank*. The rank of x_i is 1; that of $p \wedge q$ and of $p \vee q$ is the sum of the ranks of p and q.

Now we are in a position to describe $[H]$, the sublattice generated by $H \neq \varnothing$:

Lemma 3. $a \in [H]$ *iff* $a = p(h_0, \dots, h_{n-1})$, *for some integer* $n \geq 1$, *for some* n-ary polynomial p, and for some h_0, ... , $h_{n-1} \in H$.

Proof. First we must show that if $a = p(h_0, \dots, h_{n-1})$, h_0, ... , $h_{n-1} \in H$, then $a \in [H]$, which can be easily accomplished by induction on the rank of p. Then we form the set

$$\{ a \mid a = p(h_0, \dots, h_{n-1}), \ n \geq 1, \ h_0, \ \dots, \ h_{n-1} \in H, \ p \in \mathbf{P}^{(n)} \};$$

observe that this set contains H and that it is closed under $\wedge$ and $\vee$. Since it is contained in $[H]$, it has to equal $[H]$. □

Corollary 4. $|[H]| \leq |H| + \aleph_0$.

Proof. By Lemma 3, every element of $[H]$ can be associated with a finite sequence of elements of $H \cup \{(,),\wedge,\vee\}$, and there are no more than $|H| + \aleph_0$ such sequences. □

Definition 5. A *lattice identity (resp., inequality)* is an expression of the form $p = q$ (resp., $p \leq q$), where p and q are polynomials. An *identity,* $p = q$ (resp., *inequality,* $p \leq q$) *holds in the lattice* L iff $p(a_0, \dots, a_{n-1}) = q(a_0, \dots, a_{n-1})$ (resp., $p(a_0, \dots, a_{n-1}) \leq q(a_0, \dots, a_{n-1})$) holds, for any a_0, ... , $a_{n-1} \in L$.

An identity $p = q$ is equivalent to the two inequalities $p \leq q$ and $q \leq p$, and the inequality $p \leq q$ is equivalent to the identity $p \wedge q = p$ (and to $p \vee q = q$). Frequently, the validity of identities is shown by verifying that the two inequalities hold.

One of the most basic properties of polynomials is:

Lemma 6. *A polynomial (function) p is* isotone; *that is, if $a_0 \leq b_0, \ldots, a_{n-1} \leq b_{n-1}$, then $p(a_0, \ldots, a_{n-1}) \leq p(b_0, \ldots, b_{n-1})$. Furthermore,*

$$x_0 \wedge \cdots \wedge x_{n-1} \leq p(x_0, \ldots, x_{n-1}) \leq x_0 \vee \cdots \vee x_{n-1}.$$

Proof. We prove the first statement by induction on the rank of p. The first statement is certainly true for $p = x_i$. Suppose that it is true for q and r and that

$$p(x_0, \ldots, x_{n-1}) = q(x_0, \ldots, x_{n-1}) \wedge r(x_0, \ldots, x_{n-1}).$$

Then

$$p(a_0, \ldots, a_{n-1}) \wedge p(b_0, \ldots, b_{n-1})$$
$$= (q(a_0, \ldots, a_{n-1}) \wedge r(a_0, \ldots, a_{n-1})) \wedge (q(b_0, \ldots, b_{n-1}) \wedge r(b_0, \ldots, b_{n-1}))$$
$$= (q(a_0, \ldots, a_{n-1}) \wedge q(b_0, \ldots, b_{n-1})) \wedge (r(a_0, \ldots, a_{n-1}) \wedge r(b_0, \ldots, b_{n-1}))$$
$$= q(a_0, \ldots, a_{n-1}) \wedge r(a_0, \ldots, a_{nx-1}) = p(a_0, \ldots, a_{n-1});$$

thus $p(a_0, \ldots, a_{n-1}) \leq p(b_0, \ldots, b_{n-1})$. The proof is similar for $p = q \vee r$.

Since

$$x_0 \wedge \cdots \wedge x_{n-1} \leq x_i \leq x_0 \vee \cdots \vee x_{n-1},$$

for $0 \leq i \leq n-1$, using the idempotency of $\wedge$ and $\vee$, we obtain:

$$x_0 \wedge \cdots \wedge x_{n-1} = p(x_0 \wedge \cdots \wedge x_{n-1}, \ldots, x_0 \wedge \cdots \wedge x_{n-1}) \leq p(x_0, x_1, \ldots, x_{n-1})$$
$$\leq p(x_0 \vee \cdots \vee x_{n-1}, \ldots, x_0 \vee \cdots \vee x_{n-1}) = x_0 \vee \cdots \vee x_{n-1},$$

proving the second statement. ☐

A simple application is:

Lemma 7. *Let $p_i = q_i$, $0 \leq i < n$, be lattice identities. Then there is a single identity $p = q$ such that all $p_i = q_i$, $0 \leq i < n$, hold in a lattice L iff $p = q$ holds in L.*

Proof. Let us take two identities, $p_0 = q_0$ and $p_1 = q_1$. Suppose that all polynomials are n-ary and consider the identity (we use now $2n$-ary polynomials formed by substitution, see Remark to Definition 1):

(N) $$p_0(x_0, \ldots, x_{n-1}) \wedge p_1(x_n, \ldots, x_{2n-1})$$
$$= q_0(x_0, \ldots, x_{n-1}) \wedge q_1(x_n, \ldots, x_{2n-1}).$$

It is obvious that if $p_0 = q_0$ and $p_1 = q_1$ hold in L, then (N) holds in L. Now let (N) hold in L and let $a_0, \ldots, a_{n-1} \in L$. Substitute $x_0 = a_0, \ldots, x_{n-1} = a_{n-1}$, $x_n = \cdots = x_{2n-1} = a_0 \vee \cdots \vee a_{n-1} = a$. By Lemma 6,

$$p_0(a_0, \ldots, a_{n-1}) \leq p_0(a, \ldots, a) = a = p_1(a, \ldots, a),$$

whence $p_0(a_0, \ldots, a_{n-1}) = p_0(a_0, \ldots, a_{n-1}) \wedge p_1(a, \ldots, a)$, and similarly for q_0, q_1; thus (N) yields $p_0(a_0, \ldots, a_{n-1}) = q_0(a_0, \ldots, a_{n-1})$. The second identity is derived similarly from (N). The general proof is similar.   □

The most important (and, in fact, characteristic) properties of identities are given by

Lemma 8. *Identities are preserved under the formation of sublattices, homomorphic images, direct products, and ideal lattices.*

Proof. Let the polynomials p and q both be n-ary and let $p = q$ hold in L. If L_1 is a sublattice of L, then $p = q$ obviously holds in L_1. Let $\varphi \colon L \to K$ be an onto homomorphism. A simple induction shows that

$$p(a_0, \ldots, a_{n-1})\varphi = p(a_0\varphi, \ldots, a_{n-1}\varphi),$$

and the similar formula for q. Therefore,

$$p(a_0\varphi, \ldots, a_{n-1}\varphi) = p(a_0, \ldots, a_{n-1})\varphi = q(a_0, \ldots, a_{n-1})\varphi = q(a_0\varphi, \ldots, a_{n-1}\varphi),$$

and so $p = q$ holds in K. The statement for direct products is also obvious.

The last statement is an easy corollary of the following formula. Let p be an n-ary polynomial and let $I_0, \ldots, I_{n-1}$ be ideals of L. Then $I_0, \ldots, I_{n-1} \in \mathrm{Id}\, L$; thus we can substitute the I_j into p: $p(I_0, \ldots, I_{n-1})$ is also in $\mathrm{Id}\, L$, that is, an ideal of L. This ideal can be described by a simple formula:

$$p(I_0, \ldots, I_{n-1}) = \{\, x \mid x \leq p(i_0, \ldots, i_{n-1}),\ \text{for some } i_0 \in I_0,\ \ldots,\ i_{n-1} \in I_{n-1} \,\}.$$

This follows easily from Lemma 6 and the formula in Section 3 describing $I \vee J$, by induction on the rank of p.   □

Now we list a few important inequalities:

Lemma 9. *The following inequalities hold in any lattice:*

(i) $$(x \wedge y) \vee (x \wedge z) \leq x \wedge (y \vee z),$$
(ii) $$x \vee (y \wedge z) \leq (x \vee y) \wedge (x \vee z),$$
(iii) $$(x \wedge y) \vee (y \wedge z) \vee (z \wedge x) \leq (x \vee y) \wedge (y \vee z) \wedge (z \vee x),$$
(iv) $$(x \wedge y) \vee (x \wedge z) \leq x \wedge (y \vee (x \wedge z)).$$

Remark. (i)–(iii) are called *distributive inequalities*, and (iv) is the *modular inequality*.

Proof. We prove (iv) as an example and leave the rest to the reader. Since $x \wedge y \leq x$ and $x \wedge z \leq x$, we get

$$(x \wedge y) \vee (x \wedge z) \leq x.$$

Moreover, $x \wedge y \le y \le y \vee (x \wedge z)$ and $x \wedge z \le y \vee (x \wedge z)$, therefore,

$$(x \wedge y) \vee (x \wedge z) \le y \vee (x \wedge z).$$

Meeting the two displayed inequalities, we obtain (iv).   □

Lemma 10. *Consider the following two identities and inequality:*

(i) $x \wedge (y \vee z) = (x \wedge y) \vee (x \wedge z),$

(ii) $x \vee (y \wedge z) = (x \vee y) \wedge (x \vee z),$

(iii) $(x \vee y) \wedge z \le x \vee (y \wedge z).$

Then (i), (ii), *and* (iii) *are equivalent in any lattice L.*

Remark. A lattice satisfying identities (i) or (ii) is called *distributive*. Note that (i) and (ii) are *not* equivalent for fixed elements; that is, (i) can hold for three elements a, b, c of a lattice L, whereas (ii) does not.

Proof. Let (i) hold in L, and let a, b, $c \in L$; then, using (i) with $x = a \vee b$, $y = a$, $z = c$,

$$(a \vee b) \wedge (a \vee c) = ((a \vee b) \wedge a) \vee ((a \vee b) \wedge c)$$

(since $a = (a \vee b) \wedge a$ and using (i) with $x = c$, $y = a$, $z = b$)

$$= a \vee (a \wedge c) \vee (b \wedge c) = a \vee (b \wedge c),$$

verifying (ii).

   The proof that (ii) implies (i) follows by duality.
   Let (ii) hold in L; then

$$x \vee (y \wedge z) = (x \vee y) \wedge (x \vee z) \ge (x \vee y) \wedge z,$$

since $x \vee z \ge z$, verifying (iii).
   Let (iii) hold in L. Then, with $x = a$, $y = b$, $z = a \vee c$ in (iii),

$$(a \vee b) \wedge (a \vee c) \le a \vee (b \wedge (a \vee c)) = a \vee ((a \vee c) \wedge b)$$

(and with $x = a$, $y = c$, $z = b$ in (iii))

$$\le a \vee (a \vee (c \wedge b)) = a \vee (c \wedge b).$$

This, combined with the dual of Lemma 9(i), gives (ii).   □

Corollary 11. *The dual of a distributive lattice is distributive.*

Lemma 12. *The identity*

$$(x \wedge y) \vee (x \wedge z) = x \wedge (y \vee (x \wedge z))$$

is equivalent to the condition:

$$x \geq z \text{ implies that } (x \wedge y) \vee z = x \wedge (y \vee z).$$

Remark. A lattice satisfying either condition is called *modular*.

Proof. If $x \geq z$, then $z = x \wedge z$; thus the implication follows from the identity. Conversely, if the implication holds, then since $x \geq x \wedge z$, we have

$$(x \wedge y) \vee (x \wedge z) = x \wedge (y \vee (x \wedge z)).$$

Exercises

1. Give a formal definition of substituting all occurrences of x_i by x_j in a polynomial p. Prove formally the statement of the last sentence of the Remark to Definition 1 and the statement of the first sentence of the proof of Lemma 3.

2. Let $H = \{h_0, \ldots, h_{n-1}\}$. Prove that $a \in [H]$ iff $a = p(h_0, \ldots, h_{n-1})$, for some n-ary polynomial p.

3. Show that the upper bound in Corollary 4 is best possible if $|H| \geq 3$. Give the best estimates for $|H| \leq 2$.

4. Give a formal proof of Lemma 8.

5. Prove (without reference to Lemma 8) that if L is distributive (modular), then so is Id L.

6. Show that the dual of a modular lattice is modular.

7. Prove that L is distributive iff the identity

    $$(x \wedge y) \vee (y \wedge z) \vee (z \wedge x) = (x \vee y) \wedge (y \vee z) \wedge (z \vee x)$$

    holds in L.

8. Prove that every distributive lattice is modular, but not conversely. Find the smallest modular but nondistributive lattice.

9. Find an identity $p = q$ characterizing distributive lattices such that neither $p \leq q$ nor $q \leq p$ holds in a general lattice.

10. Show that in any lattice

    $$\bigvee(\bigwedge(x_{ij} \mid i < m) \mid j < n) \leq \bigwedge(\bigvee(x_{ij} \mid j < n) \mid i < m).$$

11. Prove that the following identity holds in a distributive lattice:

$$\bigvee(\bigwedge(x_{ij} \mid j < n) \mid i < m) = \bigwedge(\bigvee(x_{if(i)} \mid i < m) \mid f \in F),$$

where F is the set of all functions

$$f: \{0, 1, \dots, m - 1\} \rightarrow \{0, 1, \dots, n - 1\}.$$

12. Derive (i)–(iii) of Lemma 9 from Exercise 10.

13. Verify that any chain is a distributive lattice.

14. Let L be a lattice with more than one element, let $L' = L \cup \{0\}$, and let $0 < x$, for all $x \in L$. Show that L' is then a lattice and that an identity $p = q$ holds in L iff it holds in L'.

15. Show that the identity $x_0 = p(x_0, \dots, x_{n-1})$ holds in the two-element lattice, then it holds in every lattice.

16. Prove that $\mathrm{P}(X)$ is a distributive lattice.

17. Show that Part X is distributive iff $|X| \leq 2$ and modular iff $|X| \leq 3$. (See Exercise 3.45.)

18. Find an identity that holds in C_2 but not in N_5.

19. Find an identity that holds in M_3 but not in N_5.

20. Let K and L be modular lattices, let D be a dual ideal of K, and let I be an ideal of L such that there exists an isomorphism φ of D with I. Set $A = K \cup (L - I)$ (disjoint union) and define $x \leq y$, for $x, y \in A$, as follows:

    (i)   For $x, y \in K$ or $x, y \in L$, let $x \leq y$ retain its meaning;

    (ii) for $x \in K$, $y \in L - I$, let $x \leq y$ mean the existence of a $z \in D$ such that $x \leq z$ in K, $z\varphi \leq y$ in L;

    (iii) $x \leq y$, for no $x \in L - I$, $y \in K$.

    Is $\langle A; \leq \rangle$ then a modular lattice? (P. M. Whitman [1943] and R. P. Dilworth and M. Hall [1944].)

21. If K and L of Exercise 20 are distributive, is A distributive?

22. In Exercise 20, put $I = D = \{a\}$ and show that any identity holding in K and L holds in A as well.

23. Examine the statements of Lemma 8 for properties of the form, "If $p_0 = q_0$, then $p_1 = q_1$."

*24.   Show that $\langle L; \wedge, \vee \rangle$ is a lattice satisfying the identity $p = q$ iff it satisfies the identities

$$(w \wedge x) \vee x = x,$$
$$(((x \wedge p) \wedge z) \vee u) \vee v = (((q \wedge z) \wedge x) \vee v) \vee ((t \vee u) \wedge u),$$

where x, z, u, v, and w are variables that do not occur in p or q (R. Padmanabhan [1968]).

*25.   Show that the result of Exercise 24 is best possible: If $p = q$ is an identity *not* satisfied by some lattice, then the two identities of Exercise 24 cannot be replaced by one (R. N. McKenzie [1970]).

26.   Show that the lattice L is modular iff the inequality

$$x \wedge (y \vee z) \leq y \vee ((x \vee y) \wedge z)$$

holds in L.

5.   Free Lattices

Though it is quite easy to develop a feeling for the most general lattice (generated by a set of elements and satisfying some relation) of Section 2 by way of some examples, a general definition seems hard to accept. So we ask the reader to withhold judgment on whether Definition 2 expresses his intuitive feelings until the theory is developed and further examples are presented.

The most general lattice will be called *free*. Since we might be interested, for instance, in the most general distributive lattice generated by a, b, c satisfying $b < a$, it seems desirable to define freeness with respect to a class **K** of lattices.

Definition 1. Let $p_i = q_i$ be identities, for $i \in I$. The class **K** of all lattices satisfying all identities $p_i = q_i$, $i \in I$, is called a *variety of lattices*. A variety is *trivial* iff it contains one-element lattices only.

Remark. Varieties were called *equational classes* in many earlier papers.

The class **L** of all lattices, the class **D** of all distributive lattices, and the class **M** of all modular lattices are examples of varieties of lattices.

Next we have to agree on what kinds of relations to allow in a generating set. Can we prescribe only relations of the form $b \leq a$, or do we allow relations of the form $a \wedge b = c$ or $d \vee e = f$? Lemma 9 (below) can be rephrased to furnish an example in which the four generators a, b, c, d are required to satisfy $a \vee b = a \vee c = b \vee c = d$, showing the usefulness of relations of the form $a \vee b = d$. Let us therefore agree that for a generating set we take a poset P, and for relations we take all $a \leq b$ that hold in P, all $a \wedge b = c$, where $\inf\{a, b\} = c$ in P, and all $a \vee b = c$, where $\sup\{a, b\} = c$ in P. (A more liberal approach will be presented later in this section.)

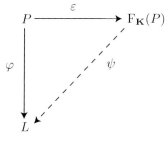

Figure 1

Now we are ready to formulate our basic concept.

Definition 2. Let P be a poset and let $\mathbf{K}$ be a variety of lattices. A lattice $F_{\mathbf{K}}(P)$ is called a *free lattice* over $\mathbf{K}$ *generated by* P iff the following conditions are satisfied:

(i) $F_{\mathbf{K}}(P) \in \mathbf{K}$.

(ii) $P \subseteq F_{\mathbf{K}}(P)$, and for a, b, $c \in P$, $\inf\{a,b\} = c$ in P iff $a \wedge b = c$ in $F_{\mathbf{K}}(P)$, and $\sup\{a,b\} = c$ in P iff $a \vee b = c$ in $F_{\mathbf{K}}(P)$.

(iii) $[P] = F_{\mathbf{K}}(P)$.

(iv) Let $L \in \mathbf{K}$ and let $\varphi\colon P \to L$ be an isotone map with the following two properties, for a, b, $c \in P$:

(iv$_1$) if $\inf\{a,b\} = c$ in P, then $a\varphi \wedge b\varphi = c\varphi$ in L,
(iv$_2$) if $\sup\{a,b\} = c$ in P, then $a\varphi \vee b\varphi = c\varphi$ in L.

Then there exists a (lattice) homomorphism $\psi\colon F_{\mathbf{K}}(P) \to L$ extending φ (that is, satisfying $a\varphi = a\psi$, for all $a \in P$).

Let ε denote the identity map on P; then the crucial condition (iv) can be expressed by Figure 1. In that and in all such similar *commutative diagrams*, the capital letters represent lattices or posets, and the arrows indicate homomorphisms, or maps with certain properties, so that the maps compose as indicated; in this case, $\varepsilon\psi = \varphi$, which is (iv). Note that (ii) is also included in the diagram, if the arrows are supposed to represent maps as required in (iv). In fact, we could have required in (ii) that the identity map on P be an embedding of P into $F_{\mathbf{K}}(P)$ in the sense of (iv).

If P is an unordered set, $|P| = \mathfrak{m}$, we shall write $F_{\mathbf{K}}(\mathfrak{m})$ for $F_{\mathbf{K}}(P)$ and call it a *free lattice on* $\mathfrak{m}$ *generators over* $\mathbf{K}$. In case $\mathbf{K} = \mathbf{L}$, we may omit "over $\mathbf{L}$"; thus "*free lattice generated by* P" shall mean "free lattice over $\mathbf{L}$ generated by P", or, in notation, $F(P)$.

It should be noted that if $b \in F_{\mathbf{K}}(P)$, then by (iii) and by Lemma 4.3, $b = p(a_0, \dots, a_{n-1})$, where p is a polynomial and $a_0, \dots, a_{n-1} \in P$. Thus if the

ψ of (iv) exists (in fact, if φ is any map of P into L and if $\psi\colon \mathrm{F_K}(P) \to L$ is any homomorphism extending φ), then we must have

$$b\psi = p(a_0, \dots, a_{n-1})\psi$$

(and since ψ is a homomorphism)

$$= p(a_0\psi, \dots, a_{n-1}\psi) = p(a_0\varphi, \dots, a_{n-1}\varphi),$$

since $a_i\psi = a_i(\varepsilon\psi) = a_i\varphi$. From this we conclude that there is at most one homomorphism $\psi\colon \mathrm{F_K}(P) \to L$ extending φ, whence:

Corollary 3. *The homomorphism ψ in* (iv) *is unique.*

This corollary is used to prove:

Corollary 4. *Let both $\mathrm{F_K}(P)$ and $\mathrm{F_K^*}(P)$ satisfy the conditions of Definition 2. Then there exists an isomorphism $\chi\colon \mathrm{F_K}(P) \to \mathrm{F_K^*}(P)$, and χ can be chosen so that $a\chi = a$, for all $a \in P$. In other words, free lattices (over $\mathbf{K}$ generated by P) are unique up to isomorphism.*

Proof. Let us use Figure 1 with $L = \mathrm{F_K}(P)$ and $\varphi = \varepsilon$. Then there exist $\psi_1\colon \mathrm{F_K}(P) \to \mathrm{F_K^*}(P)$ and $\psi_2\colon \mathrm{F_K^*}(P) \to \mathrm{F_K}(P)$ such that $\varepsilon\psi_1 = \varepsilon$ and $\varepsilon\psi_2 = \varepsilon$. Thus $\psi_1\psi_2\colon \mathrm{F_K}(P) \to \mathrm{F_K}(P)$ is the identity map ε on P. By the statement preceding Corollary 3, ε has a unique extension to a homomorphism $\mathrm{F_K}(P) \to \mathrm{F_K}(P)$; the identity map on $\mathrm{F_K}(P)$ is one such extension. Therefore, $\psi_1\psi_2$ is the identity map on $\mathrm{F_K}(P)$. Similarly, $\psi_2\psi_1$ is the identity map on $\mathrm{F_K^*}(P)$, and so (Exercise 3.2) ψ_1 is the required isomorphism. $\qquad\square$

This settles the uniqueness, but how about existence? Naturally, $\mathrm{F_K}(P)$ need not exist. For instance, $\mathrm{F_D}(N_5)$ should be N_5, since $\mathrm{F_K}(P) = P$, if P is a lattice, by (ii) and (iii) of Definition 2, but $N_5 \notin \mathbf{D}$, so (i) is violated.

Theorem 5.   Let P be a poset and let $\mathbf{K}$ be a variety of lattices. Then $\mathrm{F_K}(P)$ exists iff the following condition is satisfied:

(E) There exists a lattice L in $\mathbf{K}$ such that $P \subseteq L$ and, for $a, b, c \in P$,

$$\inf\{a, b\} = c \text{ in } P \text{ iff } a \wedge b = c \text{ in } L,$$
$$\sup\{a, b\} = c \text{ in } P \text{ iff } a \vee b = c \text{ in } L. \qquad\qquad —$$

Proof. Condition (E) is obviously necessary for the existence of $\mathrm{F_K}(P)$; indeed, if $\mathrm{F_K}(P)$ exists, (E) can always be satisfied with $L = \mathrm{F_K}(P)$ by (i) and (ii) of Definition 2.

Now assume that (E) is satisfied. In this proof, a map $\varphi\colon P \to N$ ($N \in \mathbf{K}$) will be called a *homomorphism* iff it satisfies the conditions set forth in Definition 2(iv).

Obviously, in Definition 2(iv), it suffices to consider L with $L = [P\varphi]$. Let $\langle N, \varphi \rangle$ denote this situation—that is, $N \in \mathbf{K}$, $\varphi \colon P \to N$ is a homomorphism, and $N = [P\varphi]$. Then $\mathbf{F_K}(P)$ or, more precisely $\langle \mathbf{F_K}(P), \varepsilon \rangle$, has the property that, for any $\langle L, \varphi \rangle$, there exists a $\psi \colon \mathbf{F_K}(P) \to L$ with $\varphi = \varepsilon \psi$. To construct $\mathbf{F_K}(P)$, we have to construct a lattice having this property for *all* $\langle L, \varphi \rangle$.

How would we construct such a lattice for two?

Let $\langle L_1, \varphi_1 \rangle$ and $\langle L_2, \varphi_2 \rangle$ be given. Form $L_1 \times L_2$ and define a map $\varphi \colon P \to L_1 \times L_2$ by $p\varphi = \langle p\varphi_1, p\varphi_2 \rangle$; set $N = [P\varphi]$. The fact that φ is a homomorphism is easy to check. A simple example is illustrated in Figures 2–4. Now we define $\psi_i \colon \langle x_1, x_2 \rangle \mapsto x_i$. Obviously, for $p \in P$, $p\varphi\psi_i = p\varphi_i$ and $\psi_i \colon N \to L_i$.

If we are given any number of $\langle L_i, \varphi_i \rangle$, $i \in I$, we can proceed as before and get $\langle N, \varphi \rangle$; if one of the $\langle L_i, \varphi_i \rangle$ is the $\langle L, \varepsilon \rangle$ given by (E), then (ii) of Definition 2 will also be satisfied. There is only one problem: All the pairs $\langle L_i, \varphi_i \rangle$ do not form a set, so their direct product cannot be formed. The $\langle L_i, \varphi_i \rangle$ do not form a set because a lattice and all its isomorphic copies do not form a set; therefore, if we can somehow restrict taking too many isomorphic copies, the previous procedure can be followed. Observe that, by Corollary 4.4, in every pair $\langle L_i, \varphi_i \rangle$ we have

$$|L_i| \le |P\varphi_i| + \aleph_0 \le |P| + \aleph_0.$$

Thus, by choosing a large enough set S and taking only those $\langle L_i, \varphi_i \rangle$ that satisfy $L_i \subseteq S$, we can solve our problem.

Now we are ready to proceed with the formal proof. Choose a set S satisfying $|P| + \aleph_0 = |S|$. Let Q be the set of all pairs $\langle M, \psi \rangle$, where $M \subseteq S$. Form

$$A = \prod (M \mid \langle M, \psi \rangle \in Q),$$

and, for each $p \in P$, let $f_p \in A$ be defined by

$$f_p(\langle M, \psi \rangle) = p\psi.$$

Finally, set

$$N = [\{ f_p \mid p \in P \}].$$

We claim that if, for all $p \in P$, we identify p with f_p, then N satisfies (i)–(iv) of Definition 2, and thus $N = \mathbf{F_K}(P)$.

(i) N is constructed from members of $\mathbf{K}$ by forming a direct product and by taking a sublattice. By Lemma 4.8, $N \in \mathbf{K}$, since $\mathbf{K}$ is a variety.

(ii) Let $\inf\{a, b\} = c$ in P. Then for every $\langle M, \psi \rangle \in Q$, $a\psi \wedge b\psi = c\psi$, so

$$f_a(\langle M, \psi \rangle) \wedge f_b(\langle M, \psi \rangle) = f_c(\langle M, \psi \rangle);$$

that is, $f_a \wedge f_b = f_c$. Since p is identified with f_p, we conclude that $a \wedge b = c$ in N.

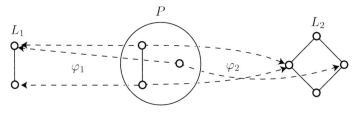

Figure 2

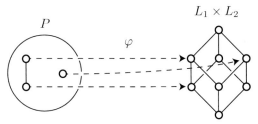

Figure 3

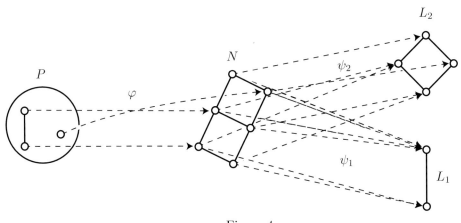

Figure 4

Conversely, let $a \wedge b = c$ in N, that is, $f_a \wedge f_b = f_c$. Let L be a lattice given by (E) and let ε be the identity map on P. We can assume that $L = [P]$; thus we can form $\langle L, \varepsilon \rangle$. By Corollary 4.4, $|L| \leq |S|$, so there is a one-to-one map $\alpha \colon L \to S$. Let $L_1 = L_\alpha$ and make L_1 into a lattice by defining

$$a\alpha \wedge b\alpha = (a \wedge b)\alpha,$$
$$a\alpha \vee b\alpha = (a \vee b)\alpha.$$

Then $L \cong L_1$ and we can form the pair $\langle L_1, \alpha_1 \rangle$, where α_1 is the restriction of α

to P $(\subseteq L)$. Since $L_1 \subseteq S$, $\langle L_1, \alpha_1 \rangle \in Q$. Now $f_a \wedge f_b = f_c$ yields

$$f_a(\langle L_1, \alpha_1 \rangle) \wedge f_b(\langle L_1, \alpha_1 \rangle) = f_c(\langle L_1, \alpha_1 \rangle),$$

that is, $a\alpha \wedge b\alpha = c\alpha$, which in turn gives $a \wedge b = c$, since α is an isomorphism. By (E), $a \wedge b = c$ in L implies that $\inf\{a, b\} = c$. The second part of (ii) follows by duality.

(iii) This part of the proof is obvious by the definition of N.

(iv) Take $\langle L, \varphi \rangle$; we have to find a homomorphism $\psi \colon N \to L$ satisfying $a\varphi = a\psi$, for $a \in P$. Using $|L| \leq |S|$, the argument given in (ii) can be repeated to find $\langle L_1, \varphi_1 \rangle$, an isomorphism $\alpha \colon L \to L_1$ such that $a\varphi\alpha = a\varphi_1$, for all $a \in P$ and $L_1 \subseteq S$. Therefore $\langle L_1, \varphi_1 \rangle \in Q$. Set

$$\psi_1 \colon f \mapsto f(\langle L_1, \varphi_1 \rangle), \quad f \in N.$$

Then, for $a \in P$,

$$a\psi_1 = f_a\psi_1 = f_a(\langle L_1, \varphi_1 \rangle) = a\varphi_1 = a\varphi\alpha.$$

Thus the homomorphism $\psi = \psi_1\alpha^{-1} \colon N \to L$ will satisfy the requirement of (iv). $\square$

Two consequences of this theorem are very important:

Corollary 6. *For any nontrivial variety* $\mathbf{K}$ *and for any cardinal* $\mathfrak{m}$, *a free lattice over* $\mathbf{K}$ *with* $\mathfrak{m}$ *generators,* $F_{\mathbf{K}}(\mathfrak{m})$, *exists.*

Proof. It suffices to find an $L \in \mathbf{K}$, $X \subseteq L$, such that $|X| = \mathfrak{m}$, and, for x, $y \in X$, $x \neq y$, x and y are incomparable. This is easily done. Since $\mathbf{K}$ is nontrivial, there exists an $N \in \mathbf{K}$, $|N| > 1$; thus, C_2 is a sublattice of N. By Lemma 4.8, $C_2 \in \mathbf{K}$; by Lemma 4.8, $C_2^I \in \mathbf{K}$, for any set I. Let $|I| = \mathfrak{m}$, let $L = C_2^I$; for $i \in I$, define $f_i \in L$ by $f_i(i) = 1$, $f_i(j) = 0$, for $i \neq j$, and set $X = \{ f_i \mid i \in I \}$. Obviously, X satisfies the condition. $\square$

Corollary 7. *For any poset* P, *a free lattice (over* $\mathbf{L}$) *generated by the poset* P *exists.*

Proof. Take a poset P and define $\mathrm{Id}_0\, P$ to be the set of all subsets I of P satisfying the condition: $\sup\{a, b\} \in I$ iff $a, b \in I$. Partially ordering $\mathrm{Id}_0\, P$ by set inclusion makes $\mathrm{Id}_0\, P$ a lattice. Identifying a with $\{ x \mid x \leq a \}$, we see that $\mathrm{Id}_0\, P$ contains P and satisfies (E) of Theorem 5. The detailed computation is almost the same as that for Theorem 20, so it will be omitted. $\square$

The argument proving Corollary 3 shows that whenever L is generated by P, any homomorphism φ of P has at most one extension to L, and if there is one, it is given by

$$\psi \colon p(a_0, \ldots, a_{n-1}) \mapsto p(a_0\varphi, \ldots, a_{n-1}\varphi).$$

This formula gives a homomorphism iff ψ is well defined; in other words, iff

$$p(a_0, \ldots, a_{n-1}) = q(b_0, \ldots, b_{m-1})$$

implies that

$$p(a_0\varphi, \ldots, a_{n-1}\varphi) = q(b_0\varphi, \ldots, b_{m-1}\varphi),$$

for any $a_0, \ldots, a_{n-1}, b_0, \ldots, b_{m-1} \in P$ and $\varphi \colon P \to N \in \mathbf{K}$.

This yields a very practical method of finding free lattices and verifying their freeness.

Theorem 8.   In the definition of $\mathrm{F}_\mathbf{K}(P)$, (iv) can be replaced by the following condition:

> If $b \in \mathrm{F}_\mathbf{K}(P)$ has two representations, $b = p(a_0, \ldots, a_{n-1})$ and $b = q(b_0, \ldots, b_{m-1})$ $(a_0, \ldots, a_{n-1}, b_0, \ldots, b_{m-1} \in P)$, then
>
> $$p(a_0, \ldots, a_{n-1}) = q(b_0, \ldots, b_{m-1})$$
>
> can be derived from the identities defining $\mathbf{K}$ and the relations of P of the form $a \wedge b = c$ and $a \vee b = c$.   —

Remark. Thus, in proving $p = q$, we can use only the meet- and join-table of P, but we cannot use $a \neq b$ or $a \neq b \vee c$, and so on.

We illustrate Theorem 8 first by determining $\mathrm{F}_\mathbf{D}(3)$. The following simple observation will be useful:

Lemma 9. *Let x, y, and z be elements of a lattice L and let $x \vee y$, $y \vee z$, $z \vee x$ be pairwise incomparable. Then $\{x \vee y, y \vee z, z \vee x\}$ generates a sublattice of L isomorphic to C_2^3 (see Figure 5).*

Proof. Almost all the meets and joins are obvious; by symmetry, the nonobvious ones are typified by the following two:

$$((x \vee y) \wedge (y \vee z)) \vee ((x \vee y) \wedge (z \vee x)) = x \vee y,$$
$$((x \vee y) \wedge (y \vee z)) \vee (z \vee x) = x \vee y \vee z.$$

Since $y \leq (x \vee y) \wedge (y \vee z)$ and $x \leq (x \vee y) \wedge (z \vee x)$, we get

$$x \vee y \leq ((x \vee y) \wedge (y \vee z)) \vee ((x \vee y) \wedge (z \vee x)),$$

and $\geq$ is trivial. The second equality follows from $y \leq (x \vee y) \wedge (y \vee z)$.

Note that, for example,

$$(x \vee y) \wedge (y \vee z) = (x \vee y) \wedge (y \vee z) \wedge (z \vee x)$$

would imply, by joining both sides with $z \vee x$, that $x \vee y \vee z = z \vee x$; thus $x \vee y \leq z \vee x$, a contradiction. Therefore, all eight elements are distinct.   □

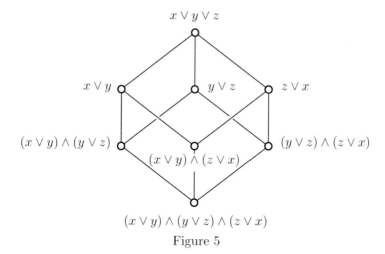

$$x \lor y \lor z$$

$$x \lor y \qquad y \lor z \qquad z \lor x$$

$$(x \lor y) \land (y \lor z) \qquad\qquad (y \lor z) \land (z \lor x)$$

$$(x \lor y) \land (z \lor x)$$

$$(x \lor y) \land (y \lor z) \land (z \lor x)$$

Figure 5

Theorem 10.   A free distributive lattice on three generators, $F_D(3)$, has eighteen elements (see Figure 6).   —

Proof. Let x, y, and z be the free generators. The top eight and the bottom eight elements form sublattices by Lemma 9 and its dual; note that

$$(x \land y) \lor (y \land z) \lor (z \land x) = (x \lor y) \land (y \lor z) \land (z \lor x)$$

by Exercise 4.7.

According to Theorem 8, we have only to verify that the lattice L of Figure 6 is a distributive lattice, and that if p, q, r are polynomials representing elements of L and $p \land q = r$ in L, then $p \land q = r$ in every distributive lattice and similarly for $\lor$. The first statement is easily proved by representing L by sets (see Exercise 13). The second statement requires a complete listing of all triples p, q, r with $p \land q = r$. If p, q, r belong to the top eight or bottom eight elements, the statement follows from Lemma 9. The remaining cases are all trivial except when p or q is x, y, or z. By symmetry, only $p = x$, $q = y \lor z$, $r = (x \land y) \lor (x \land z)$ is left to discuss, but then $p \land q = r$ is the distributive law.   □

Theorem 11 (R. Dedekind [1900]).   A free modular lattice on three generators, $F_M(3)$, has twenty-eight elements (see Figure 7).   —

Proof. Let x, y, and z be the free generators. Again, modularity is proved by a representation (see Exercise 16). Theorem 10 takes care of most meets and joins not involving x_1, y_1, z_1. Of the rest, only one relation (and the symmetric and the dual cases) is nontrivial to prove: $x_1 \wedge y_1 = u$. This we do now, leaving the rest to the reader.

Compute:

$$x_1 \wedge y_1 = ((x \wedge v) \vee u) \wedge ((y \wedge v) \vee u)$$

(since $u \leq (y \wedge v) \vee u$)

$$= ((x \wedge v) \wedge ((y \wedge v) \vee u)) \vee u$$

(by modularity)

$$= ((x \wedge v) \wedge (y \vee u) \wedge v) \vee u$$

$$a = (x \vee y) \wedge (x \vee z) \wedge (y \vee z)$$
$$= (x \wedge y) \vee (x \wedge z) \vee (y \wedge z)$$
$$b = (x \vee y) \wedge (y \vee z)$$
$$c = (x \wedge y) \vee (y \wedge z)$$

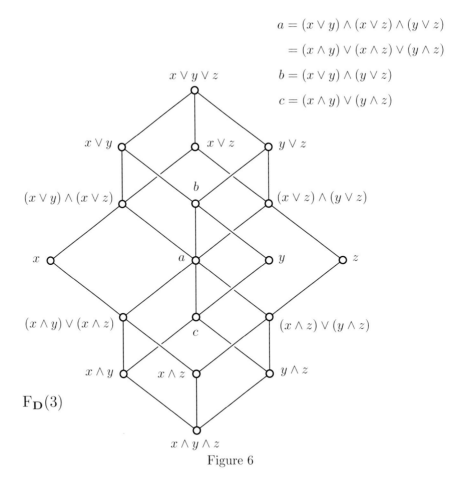

Figure 6

(substitute u and v)

$$= (x \wedge (y \vee z) \wedge (y \vee (x \wedge z))) \vee u$$
$$= (x \wedge y) \vee (x \wedge z) \vee u$$
$$= u. \qquad \qquad \Box$$

Consider the lattice represented by Figure 8. We would like to say that it is freely generated by $\{0, a, b, 1\} = P$, but this is clearly not the case according to Definition 2, since $\sup\{a, b\} = 1$ in P, whereas in the lattice, $a \vee b < 1$. So to get

$u = (x \wedge y) \vee (y \wedge z) \vee (x \wedge z)$

$v = (x \vee y) \wedge (y \vee z) \wedge (x \vee z)$

$x_1 = (x \wedge v) \vee u$

$y_1 = (y \wedge v) \vee u$

$z_1 = (z \wedge v) \vee u$

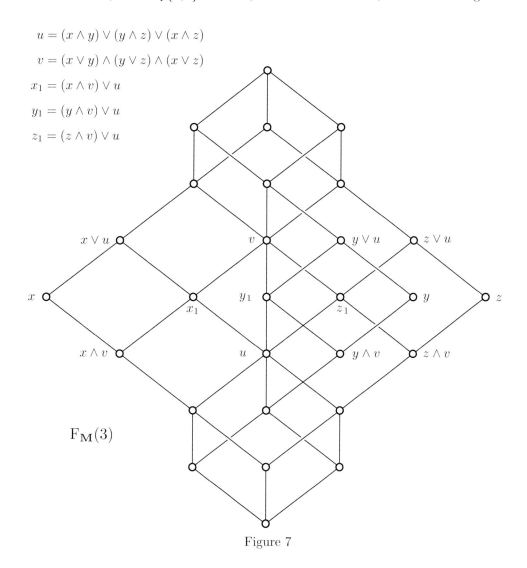

$\mathrm{F_M}(3)$

Figure 7

$$\{x, y\} = \{a, c\} \text{ and } z = f,$$

$$\{x, y\} = \{b, d\} \text{ and } z = g,$$

$$\{x, y\} = \{f, g\} \text{ and } z = 1.$$

Then $\langle H; \wedge, \vee \rangle$ is a weak partial lattice (check the axioms). Now suppose that there exists a lattice L, $H \subseteq L$, such that $\langle H; \wedge, \vee \rangle$ is a relative sublattice of L. Then $1 = (a \vee c) \vee (b \vee d)$ in L, and thus $1 = \sup\{a, b, c, d\}$. Since $e \geq a$, b, and $h \geq c$, d in L, and $1 \geq e$, h, we get $1 = \sup\{e, h\}$ in L. The fact that e, h, $1 \in H$ implies that $e \vee h$ is defined in H (and equals 1), contrary to the definition of $\langle H; \wedge, \vee \rangle$. (Compare this with Lemma 19.)

To avoid such anomalies we shall introduce two further conditions. To prepare for them, we prove:

Lemma 16. *Let $\langle H; \wedge, \vee \rangle$ be a weak partial lattice. Then we define a partial ordering relation $\leq$ on H by*

$$a \leq b \text{ iff } a \wedge b \text{ exists and } a \wedge b = a.$$

If $a \vee b$ exists, then $a \vee b = \sup\{a, b\}$. If $a \wedge b$ exists, then $a \wedge b = \inf\{a, b\}$. Also, $a \leq b$ iff $a \vee b = b$.

Proof. This proof is the same as the proof of the corresponding parts of Theorem 1.1, except that the arguments are a bit longer. $\square$

Note that, in a partial lattice, $\sup\{a, b\}$ may exist but $a \vee b$ does not. For instance, let L be the lattice of Figure 8, $H = \{0, a, b, 1\}$. Then $\sup\{a, b\} = 1$ but $a \vee b$ is not defined in H because $a \vee b \notin H$.

Definition 17. An *ideal* of a weak partial lattice H is a nonempty subset I of H satisfying the following conditions:

   (i) if a, $b \in I$ and $a \vee b$ exists, then $a \vee b \in I$,

   (ii) $x \leq a \in I$ implies that $x \in I$.

Again, we set

$$(a] = \{\, x \mid x \leq a \,\}.$$

Dual ideal and $[a)$ are defined dually. $\mathrm{Id}_0\, H$ is the lattice consisting of $\varnothing$ and all ideals of H (partial ordering is $\subseteq$), $\mathrm{Du}_0\, H$ is the lattice consisting of $\varnothing$ and all dual ideals of H (partial ordering is $\subseteq$). For $K \subseteq H$, $(K]$ is the ideal and $[K)$ is the dual ideal generated by K.

Corollary 18. *Let H and L be given as in Definition 12. Let I be an ideal of L. Then $I \cap H$ is an ideal of H provided that $I \cap H \neq \varnothing$.*

Lemma 19. *Any partial lattice H satisfies the following condition:*

(I) *If $(a] \vee (b] = (c]$ in $\mathrm{Id}_0\, H$, then $a \vee b$ exists in H and equals c.*

Proof. Let H and L be given as in Definition 12, let a, b, $c \in H$, and let $(a] \vee (b] = (c]$ in $\mathrm{Id}_0\, H$. Set $I = (a \vee b]_L$. Then $(a]_H$, $(b]_H \subseteq I \cap H$, thus

$$(c]_H = (a]_H \vee (b]_H \subseteq (a \vee b]_L;$$

that is, $c \leq a \vee b$. Since $a \leq c$ and $b \leq c$, we conclude that $a \vee b = c$. $\square$

Let (D) denote the condition dual to (I), namely:

(D) *If $[a) \vee [b) = [c)$ in $\mathrm{Du}_0\, H$, then $a \wedge b$ exists in H and equals c.*

Theorem 20 (N. Funayama [1953]).   A partial lattice can be characterized as a weak partial lattice satisfying conditions (I) and (D).   —

Proof. Corollary 15, Lemma 19, and its dual prove that a partial lattice is a weak partial lattice satisfying (I) and (D). Conversely, let $\langle H; \wedge, \vee \rangle$ be a weak partial lattice satisfying (I) and (D). Consider the map

$$\varphi \colon x \mapsto \langle (x], [x) \rangle,$$

sending H into $\mathrm{Id}_0\, H \times \widetilde{\mathrm{Du}_0\, H}$, where $\widetilde{\mathrm{Du}_0\, H}$ is the dual of $\mathrm{Du}_0\, H$. This map φ is one-to-one. If $x \vee y = z$, then $(x] \vee (y]$ in $\mathrm{Id}_0\, H$ and $[x) \vee [y) = [z)$ in $\widetilde{\mathrm{Du}_0\, H}$, thus $x\varphi \vee y\varphi = (x \vee y)\varphi$. Conversely, if $x\varphi \vee y\varphi = z\varphi$, then $(x] \vee (y] = (z]$ in $\mathrm{Id}_0\, H$. Therefore, by (I), $x \vee y$ exists and equals z, so $x\varphi \vee y\varphi = z\varphi$ implies that $x \vee y = z$. A similar argument shows that $x \wedge y = z$ iff $x\varphi \wedge y\varphi = z\varphi$. Thus we can identify x with $x\varphi$, getting $H \subseteq L = \mathrm{Id}_0\, H \times \widetilde{\mathrm{Du}_0\, H}$. We have just proved that $\langle H; \wedge, \vee \rangle$ is a relative sublattice of L. $\square$

Let $\langle P; \leq \rangle$ be a poset. We make P into a partial lattice as follows: $a \wedge b$ is defined iff $\inf\{a, b\}$ exists and $a \wedge b = \inf\{a, b\}$, and similarly for $a \vee b$.

Lemma 21. $\langle P; \wedge, \vee \rangle$ *is a partial lattice.*

Proof. It is easy to verify that (I), (D), (i)–(iv) of Lemma 13, and (i')–(iv') of Lemma 13' hold. $\square$

We need some further definitions.

Definition 22. Let $\langle A; \wedge, \vee \rangle$, $\langle B; \wedge, \vee \rangle$ be weak partial lattices, $\varphi \colon A \to B$. We call φ a *homomorphism* iff whenever $a \wedge b$ exists, for a, $b \in A$, then $a\varphi \wedge b\varphi$ exists and $(a \wedge b)\varphi = a\varphi \wedge b\varphi$, and the dual condition holds for $\vee$. A one-to-one homomorphism φ is an *embedding* provided that $a \wedge b$ exists iff $a\varphi \wedge b\varphi$ exists, and the dual condition holds for $\vee$. If φ is onto and φ is an embedding, then φ is an *isomorphism*.

Now we are ready again to define the most general lattices of Section 2.

Definition 23. Let $\mathfrak{A} = \langle A; \wedge, \vee \rangle$ be a partial lattice and let $\mathbf{K}$ be a variety of lattices. The lattice $F_{\mathbf{K}}(\mathfrak{A})$ (or simply, $F_{\mathbf{K}}(A)$) is a *free lattice over* $\mathbf{K}$ *generated by* $\mathfrak{A}$ iff the following conditions are satisfied:

(i) $F_{\mathbf{K}}(A) \in \mathbf{K}$.

(ii) $A \subseteq F_{\mathbf{K}}(A)$, and A is a relative sublattice of $F_{\mathbf{K}}(A)$.

(iii) $[A] = F_{\mathbf{K}}(A)$.

(iv) If $L \in \mathbf{K}$ and $\varphi\colon A \to L$ is a homomorphism, then there exists a homomorphism $\psi\colon F_{\mathbf{K}}(A) \to L$ extending φ (that is, $a\varphi = a\psi$, for $a \in A$).

If P is a poset, then $F_{\mathbf{K}}(P)$ is a free lattice over $\mathbf{K}$ generated by P, where P is considered a partial lattice as in Lemma 21; thus Definition 23 contains Definition 2 as a special case. The general theory can be developed exactly as it was in the first part of this section. The final result is:

Theorem 24.   Let $\mathfrak{A} = \langle A; \wedge, \vee \rangle$ be a partial lattice and let $\mathbf{K}$ be a variety. Then $F_{\mathbf{K}}(\mathfrak{A})$ exists iff there exists a lattice L in $\mathbf{K}$ such that $\mathfrak{A}$ is a relative sublattice of L.   —

As an application, we prove the existence of the lattice completely freely generated by a poset. Let P be a poset; we define a partial lattice P^m on P as follows:

$$x \wedge y = z \text{ in } P^m \text{ iff } x \text{ and } y \text{ are comparable and } z = \inf\{x, y\};$$

$$x \vee y = z \text{ in } P^m \text{ iff } x \text{ and } y \text{ are comparable and } z = \sup\{x, y\}.$$

Definition 25. $F(P^m)$ $(= F_{\mathbf{L}}(P^m))$ is called a lattice *completely freely generated by* P.

Theorem 26.   For any poset P, a lattice completely freely generated by P exists.   —

Proof. A subset $A \subseteq P$ is called *hereditary* iff $x \in A$ and $y \leq x$ imply that $y \in A$. Let $H(P)$ be the set of all hereditary subsets of P partially ordered under set inclusion. Let $P_1 = H(P)$ and $\widetilde{P}_2 = H(\widetilde{P}_1)$, where $\widetilde{P}_i$ is the dual of P_i. Identifying $p \in P$ with $(p]$, we get $P \subseteq P_1$; identifying $p \in P_1$ with $[p)$, we get $P_1 \subseteq P_2$, thus $P \subseteq P_2$. Obviously, P_2 is a lattice. Let $a, b, c \in P$ and $a \vee b = c$ in P_2. If a and b are incomparable, then $(a] \cup (b]$ $(\neq (c])$ is an upper bound for a and b, thus $a \vee b < c$ in P_2. A similar argument works for $a \wedge b = c$. Therefore, $P_2 \subseteq P$ satisfies the condition of Theorem 24.   □

CF(P) is the usual notation for $F(P^m)$ in the literature.

Exercises

1. Show that a variety is closed under the formation of sublattices, homomorphic images, and direct products.

2. Show that the class $\mathbf{T}$ of all one-element lattices is a variety. For any variety $\mathbf{K}$, prove that $\mathbf{K} \supseteq \mathbf{T}$.

3. Let $\mathbf{K}_i$, $i \in I$, be varieties. Show that $\bigcap(\mathbf{K}_i \mid i \in I)$ is again a variety.

4. Define a variety $\mathbf{A}$ by

$$(x \vee y) \wedge (x \vee z) \wedge (x \vee u)$$
$$= x \vee ((x \vee y) \wedge z \wedge u) \vee ((x \vee z) \wedge y \wedge u) \vee ((x \vee u) \wedge y \wedge z).$$

   Prove that $\mathbf{D} \subset \mathbf{A} \subset \mathbf{M}$.

5. Let $\mathbf{K}$ be a nontrivial variety. Show that $\mathbf{K}$ contains arbitrarily large lattices.

6. Let $\mathbf{P} = \{0, a, b\}$, $\inf\{a, b\} = 0$, and let $\mathbf{K}$ be a nontrivial variety. Show that $F_{\mathbf{K}}(P) \cong F(2)$.

7. Find an example of an automorphism φ of $F_{\mathbf{K}}(P)$ that is *not* the identity map on P.

8. Let P be a poset, let $\mathbf{K}$, $\mathbf{N}$ be varieties, and assume that $F_{\mathbf{K}}(P)$ and $F_{\mathbf{N}}(P)$ exist. Prove that if $\mathbf{K} \supseteq \mathbf{N}$, then there exists a homomorphism φ from $F_{\mathbf{K}}(P)$ onto $F_{\mathbf{N}}(P)$ such that φ is the identity map on P.

9. Work out a proof of the existence of $F(3)$ without any reference to Theorem 5.

10. Prove Theorem 8.

11. Formulate and prove the form of Theorem 8 that is used in the proofs of Theorems 10 and 11.

12. Prove that $F_{\mathbf{M}}(4)$ is infinite (G. Birkhoff [1933]). (Hint: Let R be the set of real numbers and let L be the lattice of vector subspaces of R^3. Set

$$a = \{\langle x, 0, x \rangle \mid x \in R\}, \qquad b = \{\langle 0, x, x \rangle \mid x \in R\},$$
$$c = \{\langle 0, 0, x \rangle \mid x \in R\}, \qquad d = \{\langle x, x, x \rangle \mid x \in R\}.$$

   Then $[\{a, b, c, d\}]$ is infinite.)

13. Let A and B be disjoint three-element sets. Let L be the set of the following subsets of $A \cup B$: all $X \subseteq A$, all $A \cup Y$, $Y \subseteq B$, all three-element sets Z with $|Z \cap A| = 2$. Prove that $\langle L; \subseteq \rangle$ is a lattice, that $\wedge$ and $\vee$ are intersection and union, and that thus L is distributive. Figure 6 is the diagram of L (A. D. Campbell [1943]).

14. Represent the lattice of Figure 6 as a sublattice of C_2^6.

15. Prove that the exponent six is best possible in Exercise 14.

16. Represent the lattice of Figure 7 as a sublattice of $L \times M_3$, where L is the lattice of Exercise 13 and M_3 is given in Figure 2.2.

17. Show that the condition (E) of Theorem 5 is equivalent to the following:

> For any a, b, $c \in P$ not satisfying $\inf\{a, b\} = c$, there exist a lattice L in **K** and a homomorphism $\varphi\colon P \to L$ with $a\varphi \wedge b\varphi \neq c\varphi$, and dually.

18. The statement "$\mathfrak{A} = \langle A; \wedge, \vee \rangle$ is a *partial algebra*" means that A is a nonempty set and that $\wedge$, $\vee$ are partial binary operations on A. For an n-ary polynomial p, a_0, $\ldots$, $a_{n-1} \in A$, interpret $p(a_0, \ldots, a_{n-1})$. (When is it defined and what is its value?)

19. In the weak partial lattice $\langle H; \wedge, \vee \rangle$ (derived from the lattice $\mathfrak{H}$ of Figure 9), $p = x \wedge ((x \vee y) \vee (x \vee z))$ is not defined for $x = 0$, $y = e$, $z = h$. Verify that in every lattice $p = q$, where $q = x$ and q is defined in every partial lattice.

20. An identity $p = q$ *holds in the partial algebra* $\mathfrak{A} = \langle A; \wedge, \vee \rangle$ iff the following three conditions are satisfied:

   (i) If $p(a_0, \ldots, a_{n-1})$, $q(a_0, \ldots, a_{n-1})$ are defined (a_0, $\ldots$, $a_{n-1} \in A$), then $p(a_0, \ldots, a_{n-1}) = q(a_0, \ldots, a_{n-1})$.

   (ii) If $p(a_0, \ldots, a_{n-1})$ is defined, $q = q_0 * q_1$, where $*$ is $\wedge$ or $\vee$, and both $q_0(a_0, \ldots, a_{n-1})$ and $q_1(a_0, \ldots, a_{n-1})$ are defined, then

$$q_0(a_0, \ldots, a_{n-1}) * q_1(a_0, \ldots, a_{n-1})$$

   is defined.

   (iii) This condition is the same as (ii) with p and q interchanged.

   Check that the Lemmas 13 and 13$'$ give this interpretation to the lattice axioms.

21. Let

$$p = (((x \vee z) \vee (y \vee u)) \vee v) \vee w,$$
$$q = ((v \vee x) \vee (v \vee y)) \vee ((w \vee z) \vee (w \vee u)).$$

Show that $p = q$ in any lattice. Show that $p = q$ does not hold in the weak partial lattice defined in connection with Figure 9.

22. Let I_0 and I_1 be ideals of a weak partial lattice. Set

$$J_0 = I_0 \cup I_1,$$
$$J_n = \{\, x \mid x \leq y \vee z,\ y,\ z \in J_{n-1} \,\}, \qquad n = 1,\ 2, \dots,$$
$$J = \bigcup (\, J_i \mid i = 0,\ 1,\ 2,\ \dots).$$

Show that $J = I_0 \vee I_1$.

23. Let the weak partial lattice L violate (I) (of Lemma 19); that is, $(a] \vee (b] = (c]$, but $a \vee b$ is undefined. Let $I_0 = (a]$, $I_1 = (b]$, and $c \in J_n$ (see Exercise 22). Generalizing Exercise 21, find an identity $p = q$ that holds in any lattice but not in L. (Exercise 21 is the special case $n = 2$.)

24. Prove that a partial algebra $\mathfrak{A} = \langle A; \wedge, \vee \rangle$ is a partial lattice iff every identity $p = q$ holding in any lattice also holds in $\mathfrak{A}$.

25. A *homomorphism* of a partial algebra $\mathfrak{A} = \langle A; \wedge, \vee \rangle$ into a lattice L is a map $\varphi \colon A \to L$ such that $(a \wedge b)\varphi = a\varphi \wedge b\varphi$, whenever $a \wedge b$ exists, and the same for $\vee$. Prove that there exists a one-to-one homomorphism of $\mathfrak{A}$ into some lattice L iff there exists a partial ordering $\leq$ on A satisfying $a \wedge b = \inf\{a, b\}$, whenever $a \wedge b$ is defined in $\mathfrak{A}$, and $a \vee b = \sup\{a, b\}$, whenever $a \vee b$ is defined in $\mathfrak{A}$.

26. Show that every weak partial lattice satisfies the condition of Exercise 25, but not conversely.

27. Let $A = \{0, a, b, c, 1\}$, $0 \leq a, b, c \leq 1$. For $x \leq y$, define

$$x \wedge y = y \wedge x = x,$$
$$x \vee y = y \vee x = y,$$

and define

$$a \wedge b = b \wedge a = 0,$$
$$a \vee b = b \vee a = 1.$$

Show that $\langle A; \wedge, \vee \rangle$ is a partial lattice.

28. Let $A = \{0, a, b, c, d, 1\}$, $0 \le a$, b, c, $d \le 1$. For $x \le y$, define

$$x \wedge y = y \wedge x = x,$$
$$x \vee y = y \vee x = y,$$

and define

$$a \wedge b = b \wedge a = c \wedge d = d \wedge c = 0,$$
$$a \vee b = b \vee a = c \vee d = d \vee c = 1.$$

Show that $\langle A; \wedge, \vee \rangle$ is a partial lattice.

29. Let L and K be lattices and let $L \cap K$ be a sublattice of L and of K. For x, $y \in L \cup K$ define $x \wedge y = z$ iff x, y, $z \in L$ and $x \wedge y = z$ in L, or x, y, $z \in K$ and $x \wedge y = z$ in K; define $x \vee y$ similarly. Is $\langle L \cup K; \wedge, \vee \rangle$ a partial lattice?

30. Are the eight axioms of weak partial lattice independent?

31. Are the ten axioms of a partial lattice independent?

32. Define weak partial semilattice and partial semilattice; prove the analogue of Theorem 20 for partial semilattices.

33. Let A be a weak partial lattice in which $a \wedge b$ exists for all a, $b \in A$. Then A is a partial lattice iff $a \mapsto (a]$ is an embedding of A into $\mathrm{Id}_0\, A$.

34. Let $\mathbf{T}$ be the variety of all one-element lattices. Show that $\mathbf{F_T}(A)$ exists iff $|A| = 1$.

35. Let $P = C_2^2$. Show that for any nontrivial variety $\mathbf{K}$, $\mathbf{F_K}(P)$ and $\mathbf{F_K}(P^m)$ exist and that always $\mathbf{F_K}(P) \not\cong \mathbf{F_K}(P^m)$.

36. Determine $\mathbf{F}(P)$, $\mathbf{F_M}(P)$, and $\mathbf{F_D}(P)$, where $P = \{a, b, c\}$ and $a < b$.

37. Discuss the set of all weak partial lattices on A inducing a given partial ordering on A.

38. Repeat Exercise 37 for partial lattices.

39. Show that a one-to-one homomorphism of weak partial lattices need not be an embedding.

40. Show that in Theorem 24, the condition "there exists a lattice L in $\mathbf{K}$ such that $\mathfrak{A}$ is a relative sublattice of L" can be replaced by the following condition:

For all $a, b, c \in A$ for which $a \wedge b = c$ does not hold, there exists a lattice L in $\mathbf{K}$, and a homomorphism φ of A into L such that $a\varphi \wedge b\varphi \neq c\varphi$, and the same condition for $\vee$.

41. Let $\langle A; \wedge, \vee \rangle$ be a partial algebra, let $\mathbf{K}$ be a variety of lattices, let $L \in \mathbf{K}$, and let M be a relative sublattice of L. Then M is called a *maximal homomorphic image of A in* $\mathbf{K}$ iff there is a homomorphism φ of A onto M such that whenever ψ is a homomorphism of A into $N \in \mathbf{K}$, then there is a homomorphism $\alpha\colon M \to N$ such that $\varphi\alpha = \psi$. Prove that a maximal homomorphic image is unique up to isomorphism, provided it exists.

42. Starting with an arbitrary partial algebra $\langle A; \wedge, \vee \rangle$, carry out the construction of Theorem 5 (Theorem 24).

43. Are there posets $P_0 \subset P$ such that $\mathrm{F}(P_0^m) = \mathrm{F}(P^m)$?

44. After M. M. Gluhov [1960], a finite partial lattice $\mathfrak{A}$ is a *basis* of a lattice L iff $L = \mathrm{F}(\mathfrak{A})$, but $L = \mathrm{F}(\mathfrak{A}_0)$, for no $A_0 \subset A$. Show that the lattice $L = \mathrm{F}(\mathfrak{A})$ has more than one basis, where $\mathfrak{A}$ is defined as follows: Let $\langle A; \leq \rangle$ be the poset given by Figure 10; for $x \leq y$, let $x \wedge y = y \wedge x = x$; furthermore, the join of any two elements is defined in $\{0, a, b, c, d, e, f, 1\}$ as supremum, and $1 \vee g = g \vee 1 = 1$. (This example, which is due to C. Herrmann [1975], contradicts M. M. Gluhov's result.)

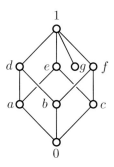

Figure 10

45. Let L and L_1 be lattices, let $L = [A]$, and let φ be a map of A into L_1. Then φ can be extended to a homomorphism ψ of L into L_1 iff

$$p(a_0, \ldots, a_{n-1}) = q(a_0, \ldots, a_{n-1})$$

implies that

$$p(a_0\varphi, \ldots, a_{n-1}\varphi) = q(a_0\varphi, \ldots, a_{n-1}\varphi)$$

holds for any $a_0, \ldots, a_{n-1} \in A$, for any integer n, and pair p, q of n-ary polynomials.

46.   Under the hypotheses of Exercise 45 describe ψ.

47.   Characterize **K**-free lattices using Exercise 45.

48.   Under the hypotheses of Exercise 45, show that if A is finite and φ is isotone, then φ can always be extended to an isotone map of L into L_1.

6.   Special Elements

A *bounded* poset is one that has both 0 and 1. A $\{0,1\}$-*homomorphism* (of a bounded lattice into another one) is a homomorphism taking zero into zero and unit into unit. A $\{0,1\}$-*sublattice* of a bounded lattice L is a sublattice containing the 0 and 1 of L. Similarly, we can define $\{0\}$-*homomorphism*, and so on, for lattices and semilattices.

In a bounded lattice L, a is a *complement* of b iff

$$a \wedge b = 0,$$
$$a \vee b = 1.$$

Lemma 1. *In a bounded distributive lattice, an element can have only one complement.*

Proof. If b_0 and b_1 are both complements of a, then

$$b_0 = b_0 \wedge 1 = b_0 \wedge (a \vee b_1) = (b_0 \wedge a) \vee (b_0 \wedge b_1) = 0 \vee (b_0 \wedge b_1) = b_0 \wedge b_1;$$

similarly, $b_1 = b_0 \wedge b_1$, thus $b_0 = b_1$.   □

Let $a \in [b, c]$; x is a *relative complement* of a in $[b, c]$ iff

$$a \wedge x = b,$$
$$a \vee x = c.$$

Lemma 2. *In a bounded distributive lattice, if a has a complement, then it also has a relative complement in any interval containing it.*

Proof. Let d be the complement of a. Then $x = (d \vee b) \wedge c$ is the relative complement of a in $[b, c]$, provided that $b \leq a \leq c$. Indeed,

$$a \wedge x = a \wedge (d \vee b) \wedge c = ((a \wedge d) \vee (a \wedge b)) \wedge c = (0 \vee b) \wedge c = b,$$

and $a \vee x = c$, by duality.   □

Lemma 3 (De Morgan's Identities). *In a bounded distributive lattice, if a and b have complements, a' and b', respectively, then $a \wedge b$ and $a \vee b$ have complements, $(a \wedge b)'$ and $(a \vee b)'$, respectively, and*

$$(a \wedge b)' = a' \vee b',$$
$$(a \vee b)' = a' \wedge b'.$$

Proof. By Lemma 1, it suffices to prove that

$$(a \wedge b) \wedge (a' \vee b') = 0,$$
$$(a \wedge b) \vee (a' \vee b') = 1$$

to verify the first identity; the second is dual. Compute:

$$(a \wedge b) \wedge (a' \vee b') = (a \wedge b \wedge a') \vee (a \wedge b \wedge b') = 0 \vee 0 = 0$$

and

$$(a \wedge b) \vee (a' \vee b') = (a \vee a' \vee b') \wedge (b \vee a' \vee b') = 1 \wedge 1 = 1. \qquad \square$$

A *complemented lattice* is a bounded lattice in which every element has a complement. A *relatively complemented lattice* is a lattice in which every element has a relative complement in any interval containing it. A *Boolean lattice* is a complemented distributive lattice. Thus, in a Boolean lattice B, every element a has a unique complement, and B is also relatively complemented.

A *Boolean algebra* is a Boolean lattice in which 0, 1, and $'$ (complementation) are also considered to be operations. Thus a Boolean algebra is a system: $\langle B; \wedge, \vee, ', 0, 1 \rangle$, where $\wedge$, $\vee$ are binary operations, $'$ is a unary operation, and 0, 1 are nullary operations. (A nullary operation on B picks out an element of B.) A *homomorphism* φ of a Boolean algebra preserves 0, 1 and $'$; that is, it is a $\{0, 1\}$-homomorphism satisfying $(x\varphi)' = x'\varphi$. A *subalgebra* of a Boolean algebra is a $\{0, 1\}$-sublattice closed under complementation. B_2 will denote the two-element Boolean algebra.

Note that in a bounded distributive lattice L, if b is a complement of a, then b is the largest element x of L with $a \wedge x = 0$. More generally, let L be a lattice with 0; an element a^* is a *pseudocomplement* of a ($\in L$) iff $a \wedge a^* = 0$, and $a \wedge x = 0$ implies that $x \leq a^*$. An element can have at most one pseudocomplement. A *pseudocomplemented lattice* is one in which every element has a pseudocomplement.

The concept of pseudocomplement involves only the meet operation. Thus we can also define *pseudocomplemented semilattices*.

Theorem 4.   Let L be a pseudocomplemented meet-semilattice,

$$S(L) = \{\, a^* \mid a \in L \,\}.$$

Then the partial ordering of L partially orders $S(L)$ and makes $S(L)$ into a Boolean lattice. For a, $b \in S(L)$, we have $a \wedge b \in S(L)$, and the join in $S(L)$ is described by

$$a \vee b = (a^* \wedge b^*)^*. \qquad\qquad —$$

Remark. This result was proved for complete distributive lattices by V. Glivenko [1929] and published in its full generality for the first time by O. Frink [1962]. Both proofs used special axiomatizations of Boolean algebras to get around the difficulty of proving distributivity. The present proof is direct. Note that even if L is a lattice, the join in L need not be the same as the join in $S(L)$.

Proof. We start with the following observations:

(1) $$a \le a^{**}.$$
(2) $$a \le b \quad \text{implies that} \quad a^* \ge b^*.$$
(3) $$a^* = a^{***}.$$
(4) $$a \in S(L) \quad \text{iff} \quad a = a^{**}.$$
(5) $$a, \ b \in S(L) \quad \text{implies that} \quad a \wedge b \in S(L).$$
(6) $$\text{For } a, \ b \in S(L), \quad \sup_{S(L)}\{a, b\} = (a^* \wedge b^*)^*.$$

Formulas (1) and (2) follow from the definitions.

Formulas (1) and (2) yield $a^* \ge a^{***}$, and by (1) $a^* \le a^{***}$, thus (3).

If $a \in S(L)$, then $a = b^*$; therefore, by (3), $a^{**} = b^{***} = b^* = a$. Conversely, if $a = a^{**}$, then $a = b^*$ with $b = a^*$; thus $a \in S(L)$, proving (4).

If $a, b \in S(L)$, then $a = a^{**}$ and $b = b^{**}$, and so $a \ge (a \wedge b)^{**}$ and $b \ge (a \wedge b)^{**}$, thus $a \wedge b \ge (a \wedge b)^{**}$; by (1), $a \wedge b = (a \wedge b)^{**}$, thus $a \wedge b \in S(L)$. If $x \in S(L)$, $x \le a$, and $x \le b$, then $x \le a \wedge b$; therefore $a \wedge b = \inf_{S(L)}\{a, b\}$, proving (5).

$a^* \ge a^* \wedge b^*$, thus by (2) and (4), $a \le (a^* \wedge b^*)^*$. Similarly, $b \le (a^* \wedge b^*)^*$. If $a \le x$ and $b \le x$ ($x \in S(L)$), then $a^* \ge x^*$ and $b^* \ge x^*$ by (2); thus by (2) and (4), $(a^* \wedge b^*)^* \le x$, proving (6).

For $a, b \in S(L)$, define

$$a \vee b = (a^* \wedge b^*)^*.$$

By formulas (5) and (6), $\langle S(L); \wedge, \vee \rangle$ is a bounded lattice. $S(L)$ is a complemented lattice since

$$a \vee a^* = (a^* \wedge a^{**})^* = 0^* = 1,$$
$$a \wedge a^* = 0,$$

for $a \in S(L)$. Now we need only prove that $S(L)$ is distributive. For x, y, $z \in S(L)$,

$$x \wedge z \le x \vee (y \wedge z),$$
$$y \wedge z \le x \vee (y \wedge z);$$

therefore,

$$x \wedge z \wedge (x \vee (y \wedge z))^* = 0,$$
$$y \wedge z \wedge (x \vee (y \wedge z))^* = 0.$$

Thus $z \wedge (x \vee (y \wedge z))^* \leq x^*, y^*$, and so $z \wedge (x \vee (y \wedge z))^* \leq x^* \wedge y^*$. Consequently, $z \wedge (x \vee (y \wedge z))^* \wedge (x^* \wedge y^*)^* = 0$, which implies that

$$z \wedge (x^* \wedge y^*)^* \leq (x \vee (y \wedge z))^{**}.$$

Now the left-hand side is $z \wedge (x \vee y)$, by formula (6), and the right-hand side is $x \vee (y \wedge z)$, by formula (4). Thus we get

$$z \wedge (x \vee y) \leq x \vee (y \wedge z),$$

which is distributivity, by Lemma 4.10.   □

Other types of special elements: An element a is an *atom*, if $a \succ 0$ and a *dual atom*, if $a \prec 1$; it is *join-irreducible*, if $a = b \vee c$ implies that $a = b$ or $a = c$; it is *meet-irreducible*, if $a = b \wedge c$ implies that $a = b$ or $a = c$. An element which is both join- and meet-irreducible is called *doubly irreducible*. Examples are given in the following exercises.

Exercises

1. Find a homomorphism of bounded lattices that is not a $\{0, 1\}$-homomorphism, and a sublattice that is not a $\{0, 1\}$-sublattice.

2. Find a modular lattice in which every element $x \neq 0$, 1 has exactly m complements.

3. Let L be a distributive lattice, $a, b \in L$. Prove that if $a \wedge b$ and $a \vee b$ have complements, so do a and b.

4. Show that Lemma 2 holds in any modular lattice.

5. In a bounded lattice L, let x be a relative complement of a in $[b, c]$; let y be a relative complement of c in $[x, 1]$; let z be a relative complement of b in $[0, x]$; and let t be a relative complement of x in $[z, y]$. Verify that t is a complement of a.

6. Let B_0, B_1 be Boolean algebras and let φ be a $\{0, 1\}$-(lattice) homomorphism of B_0 into B_1. Show that φ is a homomorphism of the Boolean algebras.

7. Let L be a distributive lattice with 0. Show that $\mathrm{Id}\, L$ is pseudocomplemented.

8. Is the converse of Exercise 7 true?

9. Prove that if L is a lattice in Theorem 4, then $a \vee b$ in $S(L)$ can be described by $a \vee b = (a \vee b)^{**}$.

10. Let L be a pseudocomplemented lattice. Show that

$$a^{**} \vee b^{**} = (a \vee b)^{**}.$$

11. Find arbitrarily large pseudocomplemented lattices in which

$$S(L) = \{0, 1\}.$$

12. Prove that in a Boolean lattice, $x \neq 0$ is join-irreducible iff x is an atom.

13. Show that in a finite lattice every element is the join of join-irreducible elements.

14. Verify that "finite lattice" in Exercise 13 can be replaced by "lattice satisfying the Descending Chain Condition." (A lattice L or, in general, a poset, satisfies the *Descending Chain Condition* iff $x_0, x_1, x_2, \ldots \in L$ and $x_0 \geq x_1 \geq x_2 \geq \cdots$ imply that $x_n = x_{n+1} = \cdots$, for some n; see Exercise 2.7.)

15. Show that some form of the Axiom of Choice must be used to verify Exercise 14.

16. Prove that if a lattice satisfies the Descending Chain Condition, then every nonzero element contains an atom.

17. The dual of the Descending Chain Condition is the Ascending Chain Condition (see Exercise 2.6). Dualize Exercises 13–16.

18. Show that a lattice satisfies the Ascending Chain Condition and the Descending Chain Condition iff all chains are finite.

19. Find a lattice in which all chains are finite but the lattice contains a chain of n elements, for every natural number n.

20. Find a lattice in which there are no join- or meet-irreducible elements.

21. Prove that the Ascending Chain Condition (Descending Chain Condition) holds in a lattice L iff every ideal (dual ideal) of L is principal.

22. Let L be a lattice and let C be a chain with $|C| \leq \aleph_0$. Prove that if there is a homomorphism of L onto C, then L contains an isomorphic copy of C. How does this result extend to countable lattices C? Does this result hold for the chain of real numbers? (No).

23. Let $\wedge$ be a binary operation on L, let * be a unary operation on L (that is, for every $a \in L$, $a^* \in L$), and let 0 be a nullary operation (that is, $0 \in L$). Let us assume that the following hold, for all a, b, $c \in L$:

$$a \wedge b = b \wedge a,$$
$$(a \wedge b) \wedge c = a \wedge (b \wedge c),$$
$$a \wedge a = a,$$
$$0 \wedge a = 0,$$
$$a \wedge (a \wedge b)^* = a \wedge b^*,$$
$$a \wedge 0^* = a,$$
$$(0^*)^* = 0.$$

Show that $\langle L; \wedge \rangle$ is a meet-semilattice with 0 as zero, and for all $a \in L$, a^* is the pseudocomplement of a (R. Balbes and A. Horn [1970]).

24. Let L be a pseudocomplemented meet-semilattice and let a, $b \in L$. Verify the formula

$$(a \wedge b)^* = (a^{**} \wedge b)^* = (a^{**} \wedge b^{**})^*.$$

25. Let L be a meet-semilattice and let a, $b \in L$. The *pseudocomplement of a relative to b* is an element $a * b$ of L satisfying $a \wedge x \leq b$ iff $x \leq a * b$. Show that $a * b$ is unique if it exists; show that $a * a$ exists iff L has a unit.

26. Let L be a *relatively pseudocomplemented meet-semilattice* (that is, L is a meet-semilattice and $a * b$ exists, for all a, $b \in L$). Show that L has 1; for all a, b, $c \in L$:

$$a * (b * c) = (a \wedge b) * c,$$
$$a * (b * c) = (a * b) * (a * c).$$

Furthermore, if L is a lattice, then L is distributive.

27. Let L be a pseudocomplemented distributive lattice. Prove that for each $a \in L$, $(a]$ is a pseudocomplemented distributive lattice; in fact, the pseudocomplement of $x \in (a]$ in $(a]$ is $x^* \wedge a$.

*28. Using the notation of Exercise 27, let $S(a)$ denote the elements of the form $x^* \wedge a$, $x \leq a$. Then $S(a)$ is a Boolean algebra by Theorem 4. Let $\vee_a$ denote the join in $S(a)$. Show that if x, $y \in S(a)$ and x, $y \in S(b)$ (a, $b \in L$), then $x \vee_a y = x \vee_b y$.

*29. Let $b \in S(a)$. Prove that $S(b) \subseteq S(a)$. (The results of Exercises 28 and 29 first appeared in G. Grätzer [1971].)

30. Show that T_n (see Exercise 2.36) is complemented.

31. Show that every interval is pseudocomplemented in T_n. (Exercises 30 and 31 are due to H. Lakser.)

32. For a finite lattice L, let $\operatorname{Irr} L$ denote the set of all doubly irreducible elements. Prove that $|L| \geq 2(l(L) + 1) - |\operatorname{Irr} L|$. (Exercises 32–35 are based on I. Rival [1974a].)

33. Show that for a finite lattice L,

$$l(\operatorname{Sub} L) = |\operatorname{Irr} L| + l(\operatorname{Sub}(L - \operatorname{Irr} L)).$$

34. A finite lattice L of n elements is *dismantlable* iff there is a chain $L_1 \subset L_2 \subset \cdots \subset L_n = L$ of sublattices satisfying $|L_i| = i$. Show that every lattice with at most seven elements is dismantlable. (Hint: use Exercise 32.)

35. Show that, for every integer $n \geq 8$, there is a lattice of n elements which is not dismantlable.

36. A finite lattice is *planar* iff it has a planar diagram (see Section 2). Show that every finite planar lattice is dismantlable.

37. For every integer $n \geq 9$, construct an n-element dismantlable lattice which is not planar. (For Exercises 36–37, see K. A. Baker, P. C. Fishburn, and F. S. Roberts [1970].)

38. If a dismantlable lattice L is not a chain, then it contains two incomparable doubly irreducible elements.

39. Prove that every sublattice and homomorphic image of a dismantlable lattice is dismantlable. Is this also true for planar lattices? (Exercises 38 and 39 are from D. Kelly and I. Rival [1974].)

*   *   *

Let P and Q be posets and let $\alpha\colon P \to Q$ and $\beta\colon Q \to P$ satisfy the following conditions.

(G$_1$) For x, $y \in P$, if $x \leq y$, then $x\alpha \leq y\alpha$.

(G$_2$) For x, $y \in Q$, if $x \leq y$, then $x\beta \leq y\beta$.

(G$_3$) For $x \in P$, $x\alpha\beta \geq x$.

(G$_4$) For $x \in Q$, $x\beta\alpha \geq x$.

Then $\langle \alpha, \beta \rangle$ is a *Galois connection* between P and Q.

This concept was introduced in G. Birkhoff [1940]. The results presented in the following exercises are based on G. Birkhoff [1940] and O. Ore [1944]. They are modeled after the correspondence between subgroups of the Galois group and subfields of a separable field extension, which is the subject of the classical Fundamental Theorem of Galois Theory.

40. Let L be a pseudocomplemented meet-semilattice. Let $P = Q = L$ and, for $x \in L$, let $x\alpha = x\beta = x^*$. Show that $\langle \alpha, \beta \rangle$ is a Galois connection.

41. Let X_0 and X_1 be sets and $r \subseteq X_0 \times X_1$. For $A \subseteq X_0$, set

$$A^1 = \{\, x_1 \mid x_1 \in X_1 \text{ and } \langle x_0, x_1 \rangle \in r, \text{ for some } x_0 \in A \,\};$$

for $B \subseteq X_1$, set

$$B^0 = \{\, x_0 \mid x_0 \in X_0 \text{ and } \langle x_0, x_1 \rangle \in r, \text{ for some } x_1 \in B \,\}.$$

$A \mapsto A^1$ and $B \mapsto B^0$ set up a Galois connection between $\mathrm{P}(X_0)$ and $\mathrm{P}(X_1)$.

42. Let $\langle \alpha, \beta \rangle$ be a Galois connection between the posets P and Q. Show that $\alpha\beta$ is a closure map on P and $\beta\alpha$ is a closure map on Q. (A *closure map* χ on a poset is an isotone map of the poset into itself satisfying $x \leq x\chi$, $x\chi\chi = x\chi$. If $x\chi = x$, then x is called *closed*.)

43. Let P_C and Q_C be the set of closed elements in P and Q, respectively (P, Q, α, β as in Exercise 42). Prove that if P and Q are complete lattices, then so are P_C and Q_C, α is an isomorphism of P_C with the dual of Q_C, and β is the inverse of α.

44. How much of Theorem 4 can be derived from Exercises 40–43?

Further Topics and References

Many of the concepts and results discussed in Chapter 1 are special cases of universal algebraic concepts and results. To see this, the reader needs the definition of a universal algebra. An *n-ary operation* f on a nonempty set A is a map from A^n into A; in other words, if $a_1, \dots, a_n \in A$, then $f(a_1, \dots, a_n) \in A$. If $n = 1$, f is called *unary*; if $n = 2$, f is called *binary*. Since $A^0 = \{\varnothing\}$, a *nullary operation* ($n = 0$) is determined by $f(\varnothing) \in A$, and f is sometimes identified with $f(\varnothing)$. A *universal algebra*, or simply *algebra*, consists of a nonempty set A and a set F of operations; each $f \in F$ is an n-ary operation, for some n (depending on f). We denote this algebra by $\mathfrak{A}$ or $\langle A; F \rangle$. Many of the results of Sections 3–5 can be formulated and proved for arbitrary universal algebras. For more details, see Chapters 1–4 of the author's book [1968].

In every poset, we can introduce (as suggested by the real line) a ternary relation r called *betweenness*: $r(a, b, c)$ iff $a \leq b \leq c$ or $c \leq b \leq a$. M. Altwegg [1950] proves that the partial ordering can be defined in terms of betweenness (naturally, up to duality); see also J. M. Cibulskis [1969]. In lattices, several betweenness relations are known, see E. Pitcher and M. F. Smiley [1942]. For applications of these relations, see M. F. Smiley and W. R. Transcue [1943] and R. Padmanabhan [1966a].

Lattices and Boolean algebras can be defined by identities in innumerable ways. Of the eight identities we used to define lattices ((L1)–(L4) of Section 1), two (the identities in (L1)) can be dropped. Ju. I. Sorkin [1951] reduces six to four, R. Padmanabhan [1969] to two, and finally, in R. N. McKenzie [1970], a single identity is found characterizing lattices. Ju. I. Sorkin's identities use only three variables; the others use more. More recently, R. Padmanabhan [1972] has found two identities in three variables characterizing lattices. It is easy to see that two variables would not suffice. Take the lattice of Figure 5.8 and redefine the join of the two atoms to be 1; otherwise keep all the joins and meets. The resulting algebra is not a lattice, but every subalgebra generated by two elements is a lattice. Therefore, lattices cannot be defined by identities in two variables. By means of a more complicated construction, A. H. Diamond and J. C. C. McKinsey [1947] derive the same conclusion for Boolean algebras. Finite algebras in which every two-generated subalgebra is Boolean have been investigated in R. W. Quackenbush [1974].

A result of A. Tarski [1968] states that, given any integer n, there exists a set of n identities defining lattices such that no identity can be dropped from the set.

By Exercise 4.24, modular and distributive lattices can be defined by two identities. Nicer sets of identities for these cases can be found in M. Kolibiar [1956]; for instance,

$$(a \vee (b \wedge b)) \wedge b = b,$$
$$((a \wedge b) \wedge c) \vee (a \wedge d) = ((d \wedge a) \vee (c \wedge b)) \wedge a$$

characterize modular lattices. By M. Sholander [1951],

$$a \wedge (a \vee b) = a,$$
$$a \wedge (b \vee c) = (c \wedge a) \vee (b \wedge a)$$

characterize distributive lattices. See also J. Riečan [1958].

E. V. Huntington [1904] provides one of the most useful axiomatizations of Boolean algebras: *A Boolean algebra is a complemented lattice in which the complementation is pseudocomplementation* (that is, $a \wedge x = 0$ implies that $x \leq a'$). Contrast this with Corollary VI.3.8. Observe that a proof of Huntington's result is implicit in the proof of Theorem 6.4.

One of the briefest axiom systems of Boolean algebras in terms of $\wedge$ and $'$ is due to L. Byrne [1946]:

$$a \wedge b = b \wedge a,$$
$$a \wedge (b \wedge c) = (a \wedge b) \wedge c,$$
$$a \wedge b' = c \wedge c' \quad \text{iff} \quad a \wedge b = a.$$

Characterization by identities is usually longer. Independently, R. N. McKenzie, A. Tarski, and the author observed that Boolean algebras can be defined by a single identity (see A. Tarski [1968], and G. Grätzer and R. N. McKenzie [1967]). A thorough survey of the axiom systems of Boolean algebras is given in S. Rudeanu [1963]; see also F. M. Sioson [1964] and S. Rudeanu [1974]. The only known irredundant selfdual axiom system can be found in R. Padmanabhan [1983].

Boolean algebras originated as an algebraic formalization of propositional logic. Most introductory logic texts give satisfactory expositions of these ideas; see also P. R. Halmos [1963]. There is a similar relationship between Boolean algebras and the theory of switching circuits; see M. A. Harrison [1965].

P. Ribenboim [1949] first pointed out that pseudocomplementation can be described by identities (involving *). A. Monteiro [1955] accomplished the same for relative pseudocomplementation; see also R. Balbes and A. Horn [1970a] and R. Balbes and P. Dwinger [1974]. This fact is applied in G. Grätzer [1969].

The examples of lattice identities we have seen so far seem to suggest that all such identities are selfdual. The identity

$$a = (a \wedge b) \vee (a \wedge c),$$

where

$$a = x \wedge ((x \wedge y) \vee (y \wedge z) \vee (z \wedge x)),$$
$$b = (x \wedge y) \vee (y \wedge z),$$
$$c = (x \wedge z) \vee (y \wedge z),$$

is an example of a nonselfdual identity. This identity holds in a lattice L iff L does not have a sublattice isomorphic to the dual of the lattice of Figure V.2.7; see H. F. Löwig [1943].

A diagram of a finite lattice is a graph; however, very little work has been done in lattice theory from a graph theoretic point of view. It is known that a finite distributive lattice is planar iff no element is covered by more than two elements. Planar modular lattices are characterized in R. Wille [1974]. It is pointed out in K. A. Baker, F. C. Fishburn, and F. S. Roberts [1970] that a finite lattice is planar iff it has order dimension one or two (a poset $\langle P; \leq \rangle$ is of *order dimension* n iff the relation $\leq$ is the set intersection of n, but not of fewer, linear orders on P), a result they attribute to J. Zilber; they also point out that the corresponding result for posets does not hold. A graph theoretic characterization of planar lattices is given in C. R. Platt [1976]. A characterization by exclusion of an infinite list of lattices as subposets can be found in D. Kelly and I. Rival [1975a]. (D. Kelly and I. Rival [1974] investigate dismantlable lattices, a generalization of planar lattices, see Exercises 6.34–6.39.) For a general review of planar lattices see R. W. Quackenbush [1973]. Pairs of lattices whose diagrams are isomorphic as graphs are investigated by J. Jakubik [1954], [1954a]. See also J. Jakubik and M. Kolibiar [1954].

Automorphism groups of planar graphs are described in L. Babai [1972] and [1975]. L. Babai used these results to characterize automorphism groups of planar lattices.

Sublattices suggest a number of problems: For a variety $\mathbf{K}$ of lattices, let $f_{\mathbf{K}}(k, n)$ denote the smallest integer such that any lattice in $\mathbf{K}$ having at least $f_{\mathbf{K}}(k, n)$ elements has at least k sublattices of exactly n elements; write $f(k, n)$ for $f_{\mathbf{L}}(k, n)$. The function $f(1, n)$ has been investigated:

(i) $f(1, t) = t$, for $t \leq 6$ (I. Kaplansky);

(ii) $f(1, 7) = 9$, $f(1, 8) = 8$, $f(1, 9) \geq 17$, $f(1, n) \geq n + 2$, for $n \geq 10$ (I. Rival [1974a] and R. Nowakowski and I. Rival [1977]);

(iii) $f(1, n) \leq n^{3n}$ (G. Havas and M. Ward [1969]).

See also M. Curzio [1953]. Related questions concerning the *spectrum* of a finite lattice L, that is, the set

$$\{\, k \mid L \text{ has a } k \text{ element sublattice} \,\},$$

the size of maximal sublattices, dismantlability, etc. have been investigated in a number of papers; see I. Rival [1974a], D. Kelly and I. Rival [1974] and [1975a].

Infinite lattices have, as a rule, very many large sublattices. T. P. Whaley [1969] proves that if $|L|$ is an infinite regular cardinal, then the lattice L has an infinite chain of sublattices of cardinality $|L|$.

The intersection $\Phi(L)$ of all maximal proper sublattices of a lattice is called the *Frattini sublattice* (a misnomer, since $\Phi(L) = \varnothing$ is possible). If the lattice L satisfies the Ascending or the Descending Chain Condition, then $\Phi(L) = \varnothing$ iff L is a chain; furthermore, every lattice K can be represented as $\Phi(L)$, for a suitable lattice K; see K. M. Koh [1971]. For planar modular lattices and distributive lattices, the Frattini sublattices are investigated in K. M. Koh [1973a] and C. C. Chen, K. M. Koh, and S. K. Tan [1973], [1975]. It is proved in M. E. Adams [1973] that every distributive lattice is the Frattini sublattice of a suitable distributive lattice.

The lattice of sublattices Sub L and the lattice of subsemilattices of a semilattice are investigated in N. D. Filippov [1966] and L. N. Ševrin [1964], respectively. They give necessary and sufficient conditions for Sub $L \cong$ Sub L_1. Filippov's results show that if L is distributive (Boolean) and Sub $L \cong$ Sub L_1, then L_1 is distributive (Boolean); see also I. Rival [1972]. The lattice of subalgebras of a Boolean algebra has been characterized in D. Sachs [1962]; see also G. Grätzer, K. M. Koh, and M. Makkai [1972].

Sub L has few of the properties discussed in this book. Sub L is modular iff L is a chain (K. M. Koh [1973]) and it is semimodular (see Section IV.2) iff L does not contain a sublattice isomorphic to $C_2 \times C_3$ (H. Lakser [1973b]).

A new area of study is the investigation of finite sublattices of lattices to prove that many properties can be characterized by the existence of finite sublattices of a certain type. Of course, results of this sort are well known for identities; see Theorem II.1.1 and the discussion of splitting lattices in Section V.2. Recently, similar results were found for properties of many different kinds: three-generated lattices (B. A. Davey and I. Rival [1976]), $(SD_\wedge)$ and $(SD_\vee)$ of Section VI.1 (B. A. Davey, W. Poguntke, and I. Rival [1975]), (W) of Section VI.1 in the presence of $(SD_\wedge)$ and $(SD_\vee)$ (R. Antonius and I. Rival [1974]). A free lattice also has very many finite sublattices; see H. S. Gaskill [1978]. Apparently, an infinite lattice has many more typical finite sublattices than has been suspected before.

By Lemma I.4.8, identities are preserved when passing from L to Id L. In G. Grätzer [1970] the problem is posed: which first-order properties (properties that can be formulated as a first order sentence in the sense of Exercises V.1.23 and V.1.24) are preserved when passing from L to Id L, or conversely? The theory of transferable and sharply transferable lattices and semilattices that grew out of this problem will be discussed in the Retrospective. See also K. A. Baker and A. W. Hales [1974] and I. Rival and B. Sands [1975].

A very important property of complete lattices is the following:

Fixed-Point Theorem.   Any isotone map f of a complete lattice L into itself has a fixed point (that is, $f(a) = a$, for some $a \in L$); in fact, $f(a) = a$ for

$$a = \bigvee (\, b \mid b \in L, \ b \leq f(b)\,).$$ —

See A. Tarski [1955] and B. Knaster [1928]. A. C. Davis [1955] proved that if this theorem holds in a lattice L, then L is complete. Various generalizations are given in A. Tarski [1955], E. S. Wolk [1957], A. Pelczar [1961] and [1962], S. Abian and A. B. Brown [1961], V. Devidé [1963], S. R. Kogalovskiĭ[1964], R. Demarr [1964], and B. R.-Salinas [1969]. Posets of length one in which every isotone map has a fixed point are characterized in I. Rival [1976].

The poset of all partial order relations on a set X has an interesting property recently discovered by B. M. Schein [1972a]: every lattice can be embedded in such a poset as a sublattice. Compare this with Whitman's result that every lattice can be embedded in a partition lattice (a stronger form of which is Theorem IV.4.4).

It is hard to overemphasize the importance of free lattices; they provide one of the most important research tools of lattice theory. Two typical applications to modular lattices can be found in G. Grätzer and E. T. Schmidt [1961] and G. Grätzer [1966]. We discuss free lattices in great detail in Section VI.2 and we use the method of constructing special lattices $F_{\mathbf{K}}(P)$ throughout the book.

Few of the topics in this chapter have been investigated for semilattices. Congruences are one exception (see G. Zacher [1952], D. Papert [1964], R. A. Dean and R. H. Oehmke [1964], R. Permutti [1964], J. C. Varlet [1965], E. T. Schmidt [1969], R. Freese and J. B. Nation [1973]). Congruences of partial lattices are considered in G. Grätzer and H. Lakser [1968].

Distributivity (see also Section II.5) and modularity of semilattices is investigated in J. B. Rhodes [1975] and B. M. Schein [1972], see also I. Fleischer [1975]. For some other topics on semilattices, see G. Grätzer and H. Lakser [1969b], G. Bruns and H. Lakser [1970], A. Horn and N. Kimura [1971], H. S. Gaskill [1972], K. A. Baker [1973], and K. M. Koh [1971a].

Problems

1. Characterize the comparability relations of semilattices and lattices. (See Exercises 1.34 and 1.35. R. N. McKenzie pointed out that neither of these two problems has a first-order solution.)

2. Characterize the lattices T_n and finite lattices that can be embedded in some T_n. (See Exercises 2.26–2.36, 6.30–6.31.)

3. Characterize the lattice $\operatorname{Sub} L$ of all sublattices of a lattice L.

4. Find conditions under which $\operatorname{Sub} L$ determines L up to isomorphism.

5. For what varieties $\mathbf{K}$ does $L \in \mathbf{K}$ and $\operatorname{Sub} L \cong \operatorname{Sub} L_1$ imply that $L_1 \in \mathbf{K}$? (This is known for $\mathbf{K} = \mathbf{L}$, $\mathbf{M}$, $\mathbf{D}$, and $\mathbf{T}$.)

6. For a variety $\mathbf{K}$, let $\operatorname{Sub} \mathbf{K}$ denote the variety generated by

$$\{\, \operatorname{Sub} L \mid L \in \mathbf{K} \,\}.$$

   Determine all varieties $\mathbf{K}$, for which $\operatorname{Sub} \mathbf{K} = \mathbf{K}$.

7. Determine all varieties $\mathbf{K}$, for which $\operatorname{Sub} \mathbf{K} = \mathbf{L}$.

8. Which first-order properties of lattices are preserved under projectivities of lattices, that is, under $\operatorname{Sub} L \cong \operatorname{Sub} L_1$?

9. Which first-order properties are preserved when passing from L to Sub L, and conversely?

10. Let Csub L be the lattice of all convex sublattices (including $\varnothing$) of the lattice L. Solve Problems 3–9 with Csub L replacing Sub L. (In Problem 8, projectivity, that is, Sub $L \cong$ Sub L_1, has to be replaced by Csub $L \cong$ Csub L_1. Csub L is investigated in K. M. Koh [1972], C. C. Chen and K. M. Koh [1972], and J. W. de Baker [1967]. See also W. D. Duthie [1942].)

11. Relate the automorphism groups of L, Sub L, and Csub L.

12. Characterize planar posets.

13. Determine the order dimensions of planar posets.

14. Is there an analogue of the Kelly-Rival theorem for finite lattices of order dimension n (of order dimension $\leq n$)?

15. Relate the automorphism group of a planar lattice L to the automorphism group of a planar diagram of L (as a graph).

16. Define the *complexity* of a finite lattice as the number of intersections of lines in an optimal diagram. Investigate this concept.

17. For a variety **K** of lattices, determine the set of all natural numbers n such that n occurs as the complexity of a finite $L \in$ **K**.

18. Determine the complexity of $P(A) = C_2^n$ and Part A, for $|A| = n$.

19. Compute the complexity of $A \times B$.

20. Is there a connection between complexity and order dimension?

21. Investigate first-order properties which are preserved when passing from L to Id L; and conversely. (See G. Grätzer [1970] and K. A. Baker and A. W. Hales [1974].)

22. A (semi) lattice **K** is called *transferable* iff whenever, for a (semi) lattice L, K has an embedding φ in Id L, then K has an embedding ψ in L. Characterize transferability.

23. Is a sublattice or the dual of a finite transferable lattice transferable?

24. If in Problem 22 the embedding ψ can be chosen so as to satisfy the condition: $x\psi \in y\varphi$ iff $x \leq y$, then K is called *sharply transferable*. Characterize sharply transferable lattices.

25.   Find a direct proof that a sublattice and the dual of a finite sharply transferable lattice are sharply transferable.

26.   Is a finite transferable lattice sharply transferable? The same question for finite semilattices.

27.   Impose conditions on a class **K** of partial lattices such that **K**-transferable and sharply **K**-transferable become meaningful and include the corresponding concepts for lattices and semilattices. Investigate these new concepts, especially in the light of Problems 22–26.

28.   For a variety **K** of lattices, investigate **K**-transferability and sharp **K**-transferability.

29.   Is there a variety **K** of lattices, **K** $\neq$ **L**, such that any finite lattice L which is sharply **K**-transferable is projective in **K**?

30.   Describe the variety of lattices with a single identity in three variables.

31.   Find the shortest single identity characterizing lattices.

32.   Define and investigate weak partial lattices and partial lattices satisfying an identity.

33.   Which lattice identities are equivalent to one of the form $p \leq q$, where p and q are lattice polynomials and no variable occurs twice in p. (See Lemma IV.5.11.)

34.   For a partial lattice A, define a congruence relation as follows: A binary relation Θ on A is a *congruence* iff there exists a lattice L and a congruence relation Φ on L such that (i) A is a relative sublattice of L and (ii) Φ restricted to A is Θ. Give an intrinsic characterization of this concept of congruence relation. (This is known for universal algebras in general; see G. Grätzer [1968], Theorem 13.3.)

35.   Characterize the lattice of all congruence relations of a semilattice.

36.   Let L be a meet-semilattice with 0 for which $[0, a]$ is pseudocomplemented, for each $a \in L$; let S(a) denote the pseudocomplements in $[0, a]$. Characterize the family of Boolean algebras $\{\, S(a) \mid a \in L \,\}$. (See Exercises 6.28 and 6.29.)

37.   What is the order of magnitude of $f(k, n)$? (Note that $f(k, n)$ is defined on page 72.)

38.   For what varieties **K**, do all $f_{\mathbf{K}}(k, n)$ exist?

39. Investigate the lattice of join-endomorphisms of a finite lattice. (See G. Grätzer and E. T. Schmidt [1958].)

40. Determine those varieties **K** of lattices for which every $K \in \mathbf{K}$ is the Frattini sublattice of a suitable $L \in \mathbf{K}$.

41. For a variety **K**, a partial **K**-lattice is a relative sublattice of a lattice in **K**. Examine special cases when characterizations along the lines of Theorem 5.20 can be found.

42. Find a class **K** of lattices such that, for all pairs of ordinals $\alpha < \beta$, $\varinjlim{}^{\alpha}\mathbf{K} \subset \varinjlim{}^{\beta}\mathbf{K}$, and the same problem for inverse limits. Can this **K** (or the **K** for which $\varinjlim{}^{\alpha}\mathbf{K} \neq \varinjlim{}^{\alpha+1}\mathbf{K}$) be constructed to satisfy some nontrivial identities? ($\varinjlim{}^{\alpha}\mathbf{K}$ and $\varprojlim{}^{\alpha}\mathbf{K}$ is obtained from **K** by applying direct limits and inverse limits, respectively, α times, see Problem 26 of G. Grätzer [1968]. For fixed α and β, C. R. Platt [1974] constructed a class **K** of lattices satisfying $\varinjlim{}^{\alpha}\mathbf{K} \neq \varinjlim{}^{\alpha+1}\mathbf{K}$ and $\varprojlim{}^{\beta}\mathbf{K} \neq \varprojlim{}^{\beta+1}\mathbf{K}$.)

Distributive Lattices

1. Characterization Theorems and Representation Theorems

The two typical examples of nondistributive lattices are N_5 and M_3, whose diagrams are given in Figure 1. We characterize distributivity by the absence of these lattices as sublattices. We introduce special names and notation for them. A subset A of a lattice L is called a *pentagon* or *diamond* iff A is a sublattice isomorphic to N_5 or M_3, respectively. If we say that $A = \{x_0, x_1, x_2, x_3, x_4\}$ is a pentagon (respectively, a diamond), we also assume that $x_0 \mapsto 0$, $x_1 \mapsto a$, $x_2 \mapsto b$, $x_3 \mapsto c$, $x_4 \mapsto i$ is an isomorphism of A and with N_5 (respectively, with M_3).

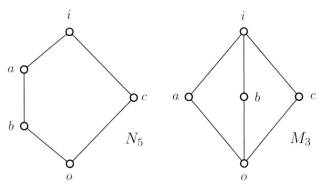

Figure 1

We state the characterization theorem in two forms. Theorem 1 is a striking and useful characterization of distributive lattices; Theorem 2 is a more detailed version of Theorem 1.

Theorem 1.   A lattice L is distributive iff L does not contain a pentagon or a diamond.   —

Theorem 2.

(i) A lattice L is modular iff it does not contain a pentagon.

(ii) A modular lattice L is distributive iff it does not contain a diamond.   —

Proof.

(i) If L is modular, then every sublattice of L is also modular; N_5 is not modular, thus it cannot be isomorphic to a sublattice of L. Conversely, let L be nonmodular, let a, b, $c \in L$, $a \geq b$ and let $(a \wedge c) \vee b \neq a \wedge (c \vee b)$. The free lattice generated by a, b, and c with $a \geq b$ is shown in Figure I.2.3. Therefore, the sublattice of L generated by a, b, and c must be a homomorphic image of the lattice of Figure I.2.3. Observe that if any two of the five elements $a \wedge c$, $(a \wedge c) \vee b$, $a \wedge (b \vee c)$, $b \vee c$, c are identified under a homomorphism, then so are $(a \wedge c) \vee b$ and $a \wedge (b \vee c)$. Consequently, these five elements are distinct in L, and they form a pentagon.

(ii) Let L be modular, but nondistributive, and choose x, y, $z \in L$ such that $x \wedge (y \vee z) \neq (x \wedge y) \vee (x \wedge z)$. The free modular lattice generated by x, y, z is shown in Figure I.5.7. By inspecting the diagram we see that u, x_1, y_1, z_1, v form a diamond. Thus in *any* modular lattice, they form a sublattice isomorphic to a quotient lattice of M_3. But M_3 has only two quotient lattices: M_3 and the one-element lattice. In the former case we have finished the proof. In the latter case, note that if u and v collapse, then so do $x \wedge (y \vee z)$ and $(x \wedge y) \vee (x \wedge z)$, contrary to our assumption.   □

Naturally, Theorems 1 and 2 could be proved without any reference to free lattices. A routine proof of (ii) runs as follows: Take x, y, z in a modular lattice L such that $x \wedge (y \vee z) \neq (x \wedge y) \vee (x \wedge z)$ and define u, x_1, y_1, z_1, v as the corresponding polynomials of Figure I.5.7. Then a direct computation shows that u, x_1, y_1, z_1, v form a diamond. There are some very natural objections to such a proof. How are the appropriate polynomials found? How is it possible to guess the result? And there is only one answer: by working out the free lattice.

Corollary 3. *A lattice L is distributive iff every element has at most one relative complement in any interval.*

Proof. The "only if" part was proved in Section I.6. If L is nondistributive, then, by Theorem 2, it contains a pentagon or a diamond, and each has an element with two relative complements in some interval.   □

Corollary 4. *A lattice L is distributive iff, for any two ideals I, J of L:*

$$I \vee J = \{\, i \vee j \mid i \in I, \ j \in J \,\}.$$

Proof. Let L be distributive. By Lemma I.3.1(ii), if $t \in I \vee J$, then $t \leq i \vee j$, for some $i \in I$ and $j \in J$. Therefore, $t = (t \wedge i) \vee (t \wedge j)$ and $t \wedge i \in I$, $t \wedge j \in J$. Conversely, if L is nondistributive, then L contains elements a, b, c as in Figure 1. Let $I = (b]$ and $J = (c]$; observe that $a \in I \vee J$, since $a \leq b \vee c$. However, a has no representation as required by Corollary 4, since if $a = i \vee j$, $i \in I$, $j \in J$, then $j \leq a$ and $j \leq c$. Therefore, $j \leq a \wedge c < b$; thus $j \in I$, and so $a = i \vee j \in I$, a contradiction. $\qquad\square$

Another important property of ideals of a distributive lattice is:

Lemma 5. *Let I and J be ideals of a distributive lattice L. If $I \wedge J$ and $I \vee J$ are principal, then so are I and J.*

Proof. Let $I \wedge J = (x]$ and $I \vee J = (y]$. Then $y = i \vee j$, for some $i \in I$ and $j \in J$. Set $c = x \vee i$ and $b = x \vee j$; note that $c \in I$ and $b \in J$ (since $x \in I \wedge J = I \cap J$). We claim that $I = (c]$ and $J = (b]$. Indeed, if, for instance, $J \neq (b]$, then there is an $a > b$, $a \in J$, and $\{x, a, b, c, y\}$ form a pentagon. $\qquad\square$

Theorem 6.   Let L be a distributive lattice and let $a \in L$. Then the map

$$\varphi \colon x \mapsto \langle x \wedge a, x \vee a \rangle, \quad x \in L,$$

is an embedding of L into $(a] \times [a)$; it is an isomorphism if a has a complement.

Proof. φ is one-to-one, since if $x\varphi = y\varphi$, then x and y are both relative complements of a in the same interval; thus $x = y$ by Corollary 3. Distributivity also implies that φ is a homomorphism.

   If a has a complement b and $\langle u, v \rangle \in (a] \times [a)$, then for $x = (u \vee b) \wedge v$, $x\varphi = \langle u, v \rangle$; therefore, φ is an isomorphism. $\qquad\square$

We start the detailed investigation of the structure of distributive lattices with the finite case.

Definition 7. For a distributive lattice L, let $J(L)$ denote the set of all nonzero join-irreducible elements, regarded as a poset under the partial ordering of L. For $a \in L$, set

$$r(a) = \{\, x \mid x \leq a, \ x \in J(L) \,\} = (a] \cap J(L).$$

Definition 8. For a poset P, call $A \subseteq P$ *hereditary* iff $x \in A$ and $y \leq x$ imply that $y \in A$. Let $H(P)$ denote the set of all hereditary subsets partially ordered by set inclusion.

Note that $\mathrm{H}(P)$ is a lattice in which meet and join are intersection and union, respectively, and thus $\mathrm{H}(P)$ is distributive. The structure of finite distributive lattices is revealed by the following result:

Theorem 9.   Let L be a finite distributive lattice. Then the map

$$\varphi \colon a \mapsto r(a)$$

is an isomorphism between L and $\mathrm{H}(\mathrm{J}(L))$.   —

Proof. Since L is finite, every element is the join of nonzero join-irreducible elements; thus

$$a = \bigvee r(a),$$

showing that φ is one-to-one. Obviously, $r(a) \cap r(b) = r(a \wedge b)$, and so $(a \wedge b)\varphi = a\varphi \wedge b\varphi$. The formula $(a \vee b)\varphi = a\varphi \vee b\varphi$ is equivalent to

$$r(a \vee b) = r(a) \cup r(b).$$

To verify this formula, note that $r(a) \cup r(b) \subseteq r(a \vee b)$ is trivial. Now let $x \in r(a \vee b)$. Then

$$x = x \wedge (a \vee b) = (x \wedge a) \vee (x \wedge b);$$

therefore, $x = x \wedge a$ or $x = x \wedge b$, since x is join-irreducible. Thus $x \in r(a)$ or $x \in r(b)$, that is, $x \in r(a) \cup r(b)$.

Finally, we have to show that if $A \in \mathrm{H}(\mathrm{J}(L))$, then $a\varphi = A$. Set $a = \bigvee A$. Then $r(a) \supseteq A$ is obvious. Let $x \in r(a)$; then

$$x = x \wedge a = x \wedge \bigvee A = \bigvee(x \wedge y \mid y \in A).$$

So $x = x \wedge y$, for some $y \in A$, implying that $x \in A$, since A is hereditary.   □

Corollary 10. *The correspondence $L \mapsto \mathrm{J}(L)$ makes the class of all finite distributive lattices with more than one element correspond to the class of all finite posets; isomorphic lattices correspond to isomorphic posets, and vice versa.*

Proof. This is obvious from $\mathrm{J}(\mathrm{H}(P)) \cong P$ and $\mathrm{H}(\mathrm{J}(L)) \cong L$.   □

A subset S of $\mathrm{P}(A)$ is called a *ring of sets* if X, $Y \in S$ implies that $X \cap Y$, $X \cup Y \in S$. Since $\mathrm{H}(\mathrm{J}(L))$ is a ring of sets, we obtain:

Corollary 11. *A finite lattice is distributive iff it is isomorphic to a ring of sets.*

If the poset Q is unordered, $\mathrm{H}(Q) = \mathrm{P}(Q)$; if B is Boolean, $\mathrm{J}(B)$ is the set of all atoms or $\varnothing$ and therefore $\mathrm{J}(B)$ is unordered. Thus we get:

Corollary 12. *A finite lattice is Boolean iff it is isomorphic to the Boolean lattice of all subsets of a finite set.*

A representation $a = x_0 \vee \cdots \vee x_{n-1}$ is *redundant* iff

$$a = x_0 \vee \cdots \vee x_{i-1} \vee x_{i+1} \vee \cdots \vee x_{n-1},$$

for some $0 \leq i < n$; otherwise it is *irredundant*.

Corollary 13. *Every element of a finite distributive lattice has a unique irredundant representation as a join of join-irreducible elements.*

Proof. The existence of such a representation is obvious. If

$$a = x_0 \vee \cdots \vee x_{n-1}$$

is an irredundant representation, then

$$r(a) = \bigcup (r(x_i) \mid 0 \leq i < n).$$

Thus x occurs in such a representation iff x is a maximal element of $r(a)$; hence the uniqueness. $\square$

Corollary 14. *Every maximal chain C of the finite distributive lattice L is of length $|J(L)|$.*

Proof. For $a \in J(L)$, let $m(a)$ be the smallest member of C containing a. Then $a \mapsto m(a)$ is a one-to-one map of $J(L)$ onto the nonzero elements of C. Indeed, if $m(a) = m(b)$, $m(a) \succ x$, and $x \in C$, then $x \vee a = x \vee b$; therefore, $a = (a \wedge x) \vee (a \wedge b)$, implying that $a \leq x$ or $a \leq b$. But $a \leq x$ implies that $m(a) \leq x < m(a)$, a contradiction. Consequently, $a \leq b$; similarly, $b \leq a$; thus $a = b$. Let $y \in C$, $y \succ z$, $z \in C$. Then $r(y) \supset r(z)$, and so $y = m(a)$, for any $a \in r(y) - r(z)$. $\square$

Let $M(L)$ denote the set of meet-irreducible elements of L excluding 1. Corollary 14 and its dual yield:

$$|J(L)| = |M(L)|.$$

For the modular case, see Exercise IV.5.26.

The crucial Theorem 9 and its most important consequence, Corollary 11, depend on the existence of sufficiently many join-irreducible elements. In an infinite distributive lattice, there may be no join-irreducible element. Note that in a distributive lattice L, the element $a \neq 0$ is join-irreducible iff $L - [a)$ is a prime ideal. In the infinite case, the role of join-irreducible elements is taken over by prime ideals. The crucial result is the existence of sufficiently many prime ideals (as illustrated in Figure 2).

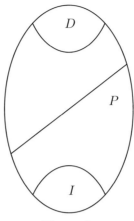

Figure 2

Theorem 15 (M. H. Stone [1936]).   Let L be a distributive lattice, let I be an ideal, let D be a dual ideal of L, and let $I \cap D = \varnothing$. Then there exists a prime ideal P of L such that $P \supseteq I$ and $P \cap D = \varnothing$.   —

Proof. Some form of the Axiom of Choice is needed to prove this statement. The most convenient form for this proof is:

Zorn's Lemma. *Let A be a set and let $\mathfrak{X}$ be a nonempty subset of $\mathrm{P}(A)$. Let us assume that $\mathfrak{X}$ has the following property: If C is a chain in $\langle \mathfrak{X}; \subseteq \rangle$, then $\bigcup C \in \mathfrak{X}$. Then $\mathfrak{X}$ has a maximal member.*

Let $\mathfrak{X}$ be the set of all ideals of L that contain I and are disjoint from D. We verify that $\mathfrak{X}$ satisfies the hypothesis of Zorn's Lemma. Since $I \in \mathfrak{X}$, we conclude that $\mathfrak{X}$ is nonempty. Let C be a chain in $\mathfrak{X}$ and let $M = \bigcup C$. If $a, b \in M$, then $a \in X$ and $b \in Y$, for some $X, Y \in C$; since C is a chain, either $X \subseteq Y$ or $Y \subseteq X$; if, say, $X \subseteq Y$, then $a, b \in Y$, and so $a \vee b \in Y \subseteq M$, since Y is an ideal. Also, if $b \leq a \in M$, then $a \in X \in C$; since X is an ideal, $b \in X \subseteq M$. Thus M is an ideal. It is obvious that $M \supseteq I$ and $M \cap D = \varnothing$, verifying that $M \in \mathfrak{X}$. Therefore, by Zorn's Lemma, $\mathfrak{X}$ has a maximal element P. We claim that P is a prime ideal. Indeed, if P is not prime, then there exist $a, b \in L$ such that a, $b \notin P$ but $a \wedge b \in P$. The maximality of P yields that $(P \vee (a]) \cap D \neq \varnothing$ and $(P \vee (b]) \cap D \neq \varnothing$. Thus there are $p, q \in P$ such that $p \vee a \in D$ and $q \vee b \in D$. Then $x = (p \vee a) \wedge (q \vee b) \in D$, since D is a dual ideal. Also,

$$x = (p \wedge q) \vee (p \wedge b) \vee (a \wedge q) \vee (a \wedge b) \in P;$$

thus $P \cap D \neq \varnothing$, a contradiction.   □

Corollary 16. *Let L be a distributive lattice, let I be an ideal of L, and let $a \in L$ and $a \notin I$. Then there is a prime ideal P such that $P \supseteq I$ and $a \notin P$.*

Proof. Apply Theorem 15 to I and $D = [a)$. □

Corollary 17. *Let L be a distributive lattice, $a,\ b \in L$ and $a \neq b$. Then there is a prime ideal containing exactly one of a and b.*

Proof. Either $(a] \cap [b) = \varnothing$ or $[a) \cap (b] = \varnothing$. □

Corollary 18. *Every ideal I of a distributive lattice is the intersection of all prime ideals containing it.*

Proof. Let

$$I_1 = \bigcap (\, P \mid P \supseteq I,\ P \text{ is a prime ideal of } L\,).$$

If $I \neq I_1$, then there is an $a \in I_1 - I$, and so, by Corollary 16, there is a prime ideal P, with $P \supseteq I$ and $a \notin P$. But then $a \notin P \supseteq I_1$ is a contradiction. □

Theorem 19 (G. Birkhoff [1933] and M. H. Stone [1936]).   A lattice is distributive iff it is isomorphic to a ring of sets. ―

Proof. Let L be a distributive lattice and let X be the set of all prime ideals of L. For $a \in L$, set

$$r(a) = \{\, P \mid a \notin P,\ P \in X \,\}.$$

Then the $r(a)$ form a ring of sets, and $a \mapsto r(a)$ is an isomorphism. The details are similar to the proof of Theorem 9, except for the first step, which now uses Corollary 17. □

This result has a very useful corollary.

Corollary 20. *Let L be a distributive lattice with more than one element. An identity holds in L iff it holds in the two-element chain, C_2.*

Proof. Let $p = q$ hold in L. Since $|L| > 1$, C_2 is a sublattice of L, and so $p = q$ holds in C_2. Conversely, let $p = q$ hold in C_2. Note that $C_2 = \mathrm{P}(X)$, with $|X| = 1$, and that $\mathrm{P}(A)$ is isomorphic to the direct power $\mathrm{P}(X)^{|A|}$. Therefore, $p = q$ holds in any $\mathrm{P}(A)$. By Theorem 19, L is a sublattice of some $\mathrm{P}(A)$; thus $p = q$ holds in L. □

We can reformulate Theorem 19 using the concept of a *field of sets*: a ring of sets closed under set complementation.

Corollary 21 (M. H. Stone [1936]). *A lattice is Boolean iff it is isomorphic to a field of sets.*

Proof. Use the representation of Theorem 19. Obviously, $r(a') = X - r(a)$, and thus complements are also preserved. $\square$

For a distributive lattice L with more than one element, let $\mathcal{P}(L)$ denote the set of all prime ideals of L, regarded as a poset under $\subseteq$. The importance of $\mathcal{P}(L)$ should be clear from the previous results. A topology on $\mathcal{P}(L)$ will be discussed in Section 5.

Some interesting properties of L are reflected in $\mathcal{P}(L)$. An important result of this type is the following theorem (see also L. Rieger [1949]):

Theorem 22 (L. Nachbin [1947]).   Let L be a bounded distributive lattice with $0 \neq 1$. Then L is a Boolean lattice iff $\mathcal{P}(L)$ is unordered.   —

Proof. Let L be Boolean, $P, Q \in \mathcal{P}(L)$, and $P \subset Q$. Choose $a \in Q - P$. Since $a \in Q$, we have $a' \notin Q$, and thus, $a' \notin P$. Thus $a, a' \notin P$, but $a \wedge a' = 0 \in P$, a contradiction, showing that $\mathcal{P}(L)$ is unordered. This proof, in fact, verifies that in a Boolean algebra every prime ideal is maximal.

Now let $\mathcal{P}(L)$ be unordered and $a \in L$, and let us assume that a has no complement. Set

$$D = \{\, x \mid a \vee x = 1 \,\}.$$

Then D is a dual ideal. Take

$$D_1 = D \vee [a) = \{\, x \mid x \geq d \wedge a, \text{ for some } d \in D \,\}.$$

The dual ideal D_1 does not contain 0, since $0 = d \wedge a$, $a \vee d = 1$ would mean that d is a complement of a. Thus there exists a prime ideal P disjoint to D_1. Note that $1 \notin (a] \vee P$, otherwise $1 = a \vee p$, for some $p \in P$, contradicting $P \cap D = \varnothing$. Thus some prime ideal Q contains $(a] \vee P$; and so $P \subset Q$, which is impossible since $\mathcal{P}(L)$ is unordered. $\square$

According to Corollary 18, every ideal is an intersection of prime ideals. When is this representation unique?

Theorem 23 (J. Hashimoto [1952]).   Let L be a bounded distributive lattice with $0 \neq 1$. Every ideal has a unique representation as an intersection of prime ideals iff L is a finite Boolean lattice.   —

Proof. If L is a finite Boolean lattice, then P is a prime ideal iff $P = (a]$, where a is a dual atom; the uniqueness follows from Corollary 13 (or is obvious by direct computation).

Now let every ideal of L have a unique representation as a meet of prime ideals. We claim that $\mathrm{Id}\,L$ is Boolean. Let $I \in \mathrm{Id}\,L$; define

$$J = \bigcap(\, P \mid P \in \mathcal{P}(L),\ P \not\supseteq I \,).$$

Then

$$I \wedge J = \bigcap (P \mid P \in \mathcal{P}(L)) = (0].$$

If $L \neq I \vee J$, then there is a prime ideal $P_0 \supseteq I \vee J$, and consequently J has two representations:

$$\bigcap (P \mid P \not\supseteq I) = P_0 \cap \bigcap (P \mid P \not\supseteq I).$$

Thus $L = I \vee J$ and J is a complement of I in $\operatorname{Id} L$.

By Lemma 5, every ideal of L is principal; therefore $L \cong \operatorname{Id} L$, and so L is Boolean. By Exercise I.6.21, L satisfies the Ascending Chain Condition; thus every element of L other than the unit is contained in a dual atom. Since the complement of a dual atom is an atom, by taking complements we find that every nonzero element of L contains an atom.

If $p_0, p_1, \ldots, p_n, \ldots$ are distinct atoms in L, then the ascending chain

$$p_0, \; p_0 \vee p_1, \; \ldots, \; p_0 \vee p_1 \vee \cdots \vee p_n, \; \ldots$$

does not terminate, and thus L has only finitely many atoms, $p_0, \ldots, p_{n-1}$. Let $a = p_0 \vee \ldots \vee p_{n-1}$. If $a' \neq 0$, then a' has to contain an atom, which is impossible. Therefore, $a' = 0$, $a = 1$, and $L \cong \mathrm{P}(X)$, with $|X| = n$. $\qquad \square$

Exercises

1. Work out a direct proof of Theorem 2(i).

2. Work out a direct proof of Theorem 2(ii).

3. Let K be a five-element distributive lattice. Is there an identity $p = q$ such that $p = q$ holds in a lattice L iff L has no sublattice isomorphic to K?

4. Does the property stated in Lemma 5 characterize distributive lattices?

5. Let L be a distributive lattice with 0 and 1. Prove that the direct decompositions $L_0 \times L_1$ of L are in one-to-one correspondence with the complemented elements of L.

6. Prove that the complemented elements of a distributive lattice form a sublattice.

7. Let L be a distributive lattice with 0 and 1. Let

$$L \cong L_0 \times L_1 \cong K_0 \times K_1.$$

Show that there is a direct decomposition

$$L \cong A_0 \times A_1 \times A_2 \times A_3$$

such that

$$A_0 \times A_1 \cong L_0,$$
$$A_2 \times A_3 \cong L_1,$$
$$A_0 \times A_2 \cong K_0,$$
$$A_1 \times A_3 \cong K_1.$$

8. Extend Theorem 9 to distributive lattices satisfying the Descending Chain Condition (see Exercise I.6.14).

9. Extend Corollary 10 to distributive lattices satisfying the Descending Chain Condition.

10. Can Exercises 8 and 9 be further sharpened?

11. Let L be a distributive lattice with 0 and 1. Let C_0 and C_1 be finite chains in L. Show that there exist chains $D_0 \supseteq C_0$ and $D_1 \supseteq C_1$ such that $|D_0| = |D_1|$.

12. Derive from Exercise 11 the result that all maximal chains of a finite distributive lattice have the same length.

13. Find examples showing that Exercise 11 is not valid if "finite" is omitted.

14. Prove the theorem "L is modular iff Id L is modular" by showing that "L contains a pentagon iff Id L contains a pentagon."

*15. Is the second statement of Exercise 14 true for the diamond rather than for the pentagon?

16. Let L be a distributive lattice, a, b, $c \in L$, and $a \le b$. Is it true that $[a, b]$ is Boolean iff $[a \wedge c, b \wedge c]$ and $[a \vee c, b \vee c]$ are Boolean?

17. For a poset P, let $\mathrm{H_F}(P)$ denote the lattice of all subsets of the form $(a_0] \cup (a_1] \cup \cdots \cup (a_{n-1}]$, where n is an arbitrary natural number. Show that Theorem 9 holds for $\mathrm{H_F}(P)$.

18. Prove that if P is a finite poset, then $\mathrm{H_F}(P) = \mathrm{H}(P)$.

19. Is it true that if Corollary 13 holds for a lattice L with more than one element, then $L \cong \mathrm{H_F}(P)$, for some poset P?

20. Show that the Ascending Chain Condition implies the Descending Chain Condition for Boolean lattices.

21. Show that Exercise 20 fails to hold for *generalized Boolean lattices* (that is, relatively complemented distributive lattices with 0).

22. Use Exercise 20 to simplify the proof of Theorem 23.

23. Let L be a lattice, let P be a prime ideal of L, and let a, b, $c \in L$. Prove that if $a \vee (b \wedge c) \in P$, then $(a \vee b) \wedge (a \vee c) \in P$.

24. Using Exercise 23, show that the lattice L is distributive iff, for all x, $y \in L$, $x < y$, there exists a prime ideal P with $x \in P$ and $y \notin P$.

25. Verify the statement of Exercise 24 using Theorem 6.

26. Verify the statement of Exercise 24 using Theorem 1.

27. Let L be a distributive lattice. Then L is relatively complemented iff $\mathcal{P}(L)$ is unordered.

28. Prove Theorem 15 by well-ordering L, $L = \{\, a_\gamma \mid \gamma < \alpha \,\}$, and deciding one by one for each a_γ whether $a_\gamma \in P$ or $a_\gamma \notin P$.

29. Let L be a distributive lattice with 1. Show that every prime ideal P is contained in a *maximal prime ideal* Q (that is, $P \subseteq Q$, and for any prime ideal X of L, $Q \subseteq X$ implies that $Q = X$).

30. Let L be a distributive lattice with 0. Verify that every prime ideal P contains a *minimal prime ideal* Q (that is, $P \supseteq Q$, and for any prime ideal X of L, $Q \supseteq X$ implies that $Q = X$).

31. Find a distributive lattice L with no minimal and no maximal prime ideals.

32. Investigate the connections among the Ascending Chain Condition (and Descending Chain Condition) for a distributive lattice L, for Id L, and for $\mathcal{P}(L)$.

33. Let L be a distributive lattice with 0 and let $I \in$ Id L. Show that the pseudocomplement of I is $I^* = \{\, x \mid (x] \wedge I = (0] \,\}$.

34. Prove that $I = I^{**}$, for any $I \in$ Id L of a distributive lattice with 0, iff L is a generalized Boolean lattice satisfying the Descending Chain Condition.

35. The congruence relations Θ and Φ *permute* iff

$$a \equiv b \quad (\Theta) \qquad \text{and} \qquad b \equiv c \quad (\Phi)$$

imply that

$$a \equiv d \quad (\Phi) \qquad \text{and} \qquad d \equiv c \quad (\Theta),$$

for some d. Show that the congruences of a relatively complemented lattice permute.

36. Prove the converse of Exercise 35 for distributive lattices.

37. Generalize Theorem 23 to distributive lattices without 0 and 1.

*38. Let L be a distributive lattice, let $a \in L$, let S be a sublattice of L, and let $a \notin S$. Show that there exists a prime ideal P and a prime dual ideal Q such that $a \notin P \cup Q \supseteq S$, provided that a is not the 0 or 1 of L (J. Hashimoto [1952]).

39. Let L be a relatively complemented distributive lattice. A sublattice K of L is *proper* iff $K \neq L$. Show that every proper sublattice of L can be extended to a maximal proper sublattice of L (K. Takeuchi [1951]; see also J. Hashimoto [1952] and G. Grätzer and E. T. Schmidt [1958c]).

40. Show that the statement of Exercise 39 is not valid in general if L is not relatively complemented (K. Takeuchi [1951]). (See also M. E. Adams [1973].)

41. Generalize Corollary 13 to infinite distributive lattices, claiming the unique irredundant representation of certain ideals as a meet of prime ideals.

42. If P is a prime ideal of L, then $(P]$ is a principal prime ideal of Id L. Is the converse true?

43. Show that Corollary 17 characterizes distributivity.

44. A chain C in a poset P is called *maximal* iff, for any chain D in P, $C \subseteq D$ implies that $C = D$. Using Zorn's Lemma, show that every chain is contained in a maximal chain.

45. Prove that a finite distributive lattice is planar iff no element is covered by three elements.

46. Show that a finite distributive lattice is planar iff it is dismantlable.

47. Let S be a sublattice of the finite lattice L. Then S can be represented in the form

$$L - \bigcup ([a, b] \mid a \text{ is join-irreducible and } b \text{ is meet-irreducible})).$$

(Exercises 47 and 48 are from I. Rival [1973].)

48. Prove the converse of Exercise 47 for distributive lattices.

2. Polynomials and Freeness

We can introduce an equivalence relation $\equiv$ for lattice polynomials: $p \equiv q$ iff p and q define the same functions in the class $\mathbf{D}$ of distributive lattices. More formally, if p and q are lattice polynomials (see Section I.4), then $p \equiv q$ iff, for any distributive lattice L and $a_1, \ldots, a_n \in L$, we have $p(a_1, \ldots, a_n) = q(a_1, \ldots, a_n)$ (see Definitions I.4.1 and I.4.2). For an n-ary lattice polynomial p, let $[p]\mathbf{D}$ denote the set of all n-ary lattice polynomials q satisfying $p \equiv q$ and let $P_{\mathbf{D}}(n)$ denote the set of all these equivalence classes, that is,

$$P_{\mathbf{D}}(n) = \{\, [p]\mathbf{D} \mid p \in \mathbf{P}^{(n)} \,\}.$$

Observe that, for $p, p_1, q, q_1 \in \mathbf{P}^{(n)}$, if $p \equiv p_1$ and $q \equiv q_1$, then $p \wedge q \equiv p_1 \wedge q_1$ and $p \vee q \equiv p_1 \vee q_1$. Thus

$$[p]\mathbf{D} \wedge [q]\mathbf{D} = [p \wedge q]\mathbf{D},$$
$$[p]\mathbf{D} \vee [q]\mathbf{D} = [p \vee q]\mathbf{D},$$

define $\wedge$ and $\vee$ on $P_{\mathbf{D}}(n)$. It is easily seen that $P_{\mathbf{D}}(n)$ is a distributive lattice and $[p]\mathbf{D} \leq [q]\mathbf{D}$ iff the inequality $p \leq q$ holds in the class $\mathbf{D}$.

To describe the structure of $P_{\mathbf{D}}(n)$, let $Q(n)$ denote the dual of the poset of all proper nonempty subsets of $\{0, 1, \ldots, n-1\}$. Define $\mathrm{H}(\varnothing) = C_1$.

Theorem 1.

(i) $P_{\mathbf{D}}(n)$ is isomorphic with $\mathrm{H}(Q(n))$.

(ii) $P_{\mathbf{D}}(n)$ is a free distributive lattice on n generators.

(iii) $2n - 2 \leq |P_{\mathbf{D}}(n)| \leq 2^{2^n - 2}$.

(iv) A finitely generated distributive lattice is finite.   —

Proof.

(i) A lattice polynomial p is called a *meet-polynomial* iff it is of the form $x_{i_0} \wedge \cdots \wedge x_{i_{k-1}}$. For $J \subseteq \{0, \ldots, n-1\}$, $J \neq \varnothing$, set

$$p_J = \bigwedge (\, x_i \mid i \in J \,).$$

We claim that $[p_J]\mathbf{D} \leq [p_K]\mathbf{D}$ $(J, K \subseteq \{0, \ldots, n-1\})$ iff $J \supseteq K$. The "if" part is obvious. Now let $J \not\supseteq K$; then there exists an $i \in K$ such that $i \notin J$. Consider the two-element chain C_2 and substitute $x_i = 0$ and $x_j = 1$, for $j \neq i$. Obviously, $p_J = 1$ and $p_K = 0$; thus the inequality $p_J \leq p_K$ fails in C_2, and therefore in $\mathbf{D}$.

Every lattice polynomial is equivalent to one of the form $\bigvee p_J$; because every x_i is of this form, the join of two such polynomials is of this form, and the same holds for the meet in view of

$$\bigvee p_{J_i} \wedge \bigvee p_{K_j} \equiv \bigvee (p_{J_i} \wedge p_{K_j})$$

and

$$p_{J_i} \wedge p_{K_j} \equiv p_{J_i \cup K_j}.$$

Next we claim that $[p]\mathbf{D} \in J(P_\mathbf{D}(n))$ iff it is a $[p_J]\mathbf{D}$. Since every $[p]\mathbf{D} \in P_\mathbf{D}(n)$ is a join of some $[p_J]\mathbf{D}$, it suffices to prove that a $[p_J]\mathbf{D}$ is join-irreducible. Let

$$p_J \equiv \bigvee (p_{J_i} \mid i \in K, \ J_i \subseteq \{0, \dots, n-1\});$$

$J \subseteq J_i$ follows from $[p_J]\mathbf{D} \geq [p_{J_i}]\mathbf{D}$. Now if $[p_J]\mathbf{D} > [p_{J_i}]\mathbf{D}$, for all i, then we have $J \subset J_i$. Choose $j_i \in J_i$, $j_i \notin J$, for all $i \in K$. In C_2, put $x_k = 0$, for all $k = j_i$ and $x_k = 1$, otherwise. Then $p_J = 1$, and $\bigvee (p_{J_i} \mid i \in K) = 0$, which is a contradiction.

A reference to Theorem 1.9 completes the proof of (i).

(ii) Let L be a distributive lattice, $a_0, \dots, a_{n-1} \in L$. Then the map $x_i \mapsto a_i$ can be extended to the homomorphism

$$[p]\mathbf{D} \mapsto p(a_0, \dots, a_{n-1}),$$

proving (ii).

(iii) This proof is obvious from (i).

(iv) This proof is obvious from (iii).   □

Figure I.5.6 is a diagram of $P_\mathbf{D}(3)$.

Boolean polynomials are defined exactly as lattice polynomials except that all five operations $\wedge$, $\vee$, $'$, 0, 1 are used in the formation of polynomials. A formal definition is the same as Definition I.4.1 with two clauses added: If p is a Boolean polynomial, so is p'; 0 and 1 are Boolean polynomials. An n-ary Boolean polynomial p defines a function in n variables on any Boolean algebra B; $p(a_0, \dots, a_{n-1})$ can be defined imitating Definition I.4.2.

For the Boolean polynomials p and q, set $p \equiv q$ iff, for any Boolean algebra B and $a_0, \dots, a_{n-1} \in B$, we have $p(a_0, \dots, a_{n-1}) = q(a_0, \dots, a_{n-1})$. Let $[p]\mathbf{B}$ denote the equivalence class containing p. Observe that $p \equiv q$ is equivalent to the identity $p = q$ holding in the class $\mathbf{B}$ of all Boolean algebras.

Let $P_\mathbf{B}(n)$ denote the set of all $[p]\mathbf{B}$, where p is an n-ary Boolean polynomial. It is easily seen that

$$[p]\mathbf{B} \wedge [q]\mathbf{B} = [p \wedge q]\mathbf{B},$$
$$[p]\mathbf{B} \vee [q]\mathbf{B} = [p \vee q]\mathbf{B},$$
$$([p]\mathbf{B})' = [p']\mathbf{B},$$
$$0 = [0]\mathbf{B},$$
$$1 = [1]\mathbf{B}$$

define the Boolean operations on $P_\mathbf{B}(n)$, and thus $P_\mathbf{B}(n)$ is a Boolean algebra.

Theorem 2.

(i) $P_{\mathbf{B}}(n)$ is isomorphic to $B_2^{2^n}$.

(ii) $P_{\mathbf{B}}(n)$ is a free Boolean algebra on n generators.

(iii) $P_{\mathbf{B}}(n) = 2^{2^n}$.

(iv) A finitely generated Boolean algebra is finite.   —

Proof. A Boolean polynomial is called *atomic* iff it is of the form

$$x_0^{i_0} \wedge \cdots \wedge x_{n-1}^{i_{n-1}},$$

where $i_j = 0$ or 1, x^0 denotes x and x^1 denotes x'. For every $J \subseteq \{0, \ldots, n-1\}$, there is an atomic polynomial p_J, for which $i_j = 0$ iff $j \in J$. The crucial statement is:

$$[p_{J_0}]\mathbf{B} \leq [p_{J_1}]\mathbf{B} \quad \text{iff} \quad J_0 = J_1.$$

Indeed, let $J_0 \neq J_1$. We make the following substitution in B_2: $x_i = 1$ if $i \in J_0$, $x_i = 0$ if $i \notin J_0$; this makes $p_{J_0} = 1$, $p_{J_1} = 0$, contradicting that $[p_{J_0}]\mathbf{B} \leq [p_{J_1}]\mathbf{B}$.

Let $B(n)$ be the set of all Boolean polynomials that are equivalent to one of the form $\bigvee(p_{J_i} \mid i \in K)$. Then $B(n)$ is closed under $\vee$ and $\wedge$, since

$$\bigvee p_{J_i} \wedge \bigvee p_{I_k} \equiv \bigvee (p_{J_i} \wedge p_{I_k})$$

and $p_{J_i} \wedge p_{I_k} \equiv p_{J_k}$, if $J_i = I_k$, and $p_{J_i} \wedge p_{I_k} \equiv 0$, otherwise.

Now we prove by induction on n that $x_i, x_i' \in B(n)$, for $i < n$. For $n = 1$, x_0, x_0' are atomic polynomials, so $x_0, x_0' \in B(1)$. By induction, $x_0 \equiv \bigvee(p_{J_i} \mid i \in K)$, where the p_{J_i} are atomic $(n-1)$-ary polynomials; then

$$x_0 \equiv x_0 \wedge (x_{n-1} \vee x_{n-1}') \equiv (x_0 \wedge x_{n-1}) \vee (x_0 \wedge x_{n-1}')$$

$$\equiv \bigvee(p_{J_i} \wedge x_{n-1} \mid i \in K) \vee \bigvee(p_{J_i} \wedge x_{n-1}' \mid i \in K),$$

and similarly for x_0'. Thus $x_0, x_0' \in B(n)$, and, by symmetry, $x_i, x_i' \in B(n)$, for all $i < n$. Since

$$(p_J)' \equiv \bigvee(x_i' \mid i \in J) \vee \bigvee(x_i \mid i \notin J),$$

we conclude that $p_J' \in B(n)$; therefore $B(n)$ is closed under $'$. Thus $B(n)$ is closed under $\wedge$, $\vee$, $'$, 0, 1. Since $B(n)$ includes all x_i, $i < n$, it is the set of all n-ary Boolean polynomials.

Consequently, every $[p]\mathbf{B}$ is a join of atomic ones, the $[p]\mathbf{B}$ for p atomic polynomials are unordered and 2^n in number, implying (i) and (iii). The proof of (ii) is routine (same proof as that of Theorem 1(ii)), and (iv) follows trivially from (iii). □

We can use Theorems 1 and 2 to characterize free distributive lattices and free Boolean algebras, respectively.

Theorem 3.   Let L be a distributive lattice generated by $\{\, a_i \mid i \in I \,\}$. L is freely generated by the a_i iff the validity in L of a relation of the form

$$\bigwedge(\, a_i \mid i \in I_0 \,) \leq \bigvee(\, a_i \mid i \in I_1 \,)$$

implies that $I_0 \cap I_1 \neq \varnothing$, for finite nonempty subsets I_0, I_1 of I.   —

Proof. The "only if" part can be easily verified by using substitutions in C_2. For the converse, let F be the distributive lattice freely generated by x_i, $i \in I$, and let φ be the homomorphism of F into (in fact, onto) L satisfying $x_i\varphi = a_i$, for $i \in I$. It suffices to prove that for the lattice polynomials p, q, $p\varphi \leq q\varphi$ implies that $[p]\mathbf{D} \leq [q]\mathbf{D}$. (We think of the elements of F as equivalence classes of polynomials in the x_i, $i \in I$.)
    Let

$$p \equiv \bigvee(\,\bigwedge(x_i \mid i \in I_j\,) \mid j \in J\,),$$
$$q \equiv \bigwedge(\,\bigvee(x_i \mid i \in K_t\,) \mid t \in T\,).$$

Then $p\varphi \leq q\varphi$ takes the form

$$\bigvee(\,\bigwedge(\, a_i \mid i \in I_j\,) \mid j \in J\,) \leq \bigwedge(\,\bigvee(\, a_i \mid i \in K_t\,) \mid t \in T\,),$$

which is equivalent to

$$\bigwedge(\, a_i \mid i \in I_j\,) \leq \bigvee(\, a_i \mid i \in K_t\,),$$

for all $j \in J$, $t \in T$. By assumption, this implies that $I_j \cap K_t \neq \varnothing$, for all $j \in J$, $t \in T$; thus

$$\bigwedge(\, x_i \mid i \in J_j\,) \leq \bigvee(\, x_i \mid i \in K_t\,),$$

for all $j \in J$, $t \in T$; implying that $[p]\mathbf{D} \leq [q]\mathbf{D}$.   □

Theorem 4.   Let B be a Boolean algebra generated by $\{\, a_i \mid i \in I \,\}$. Then B is freely generated by $\{\, a_i \mid i \in I \,\}$ iff, whenever I_0, I_1, J_0, J_1 are finite subsets of I with $I_0 \cup I_1 = J_0 \cup J_1$ and $I_0 \cap I_1 = \varnothing$, then

$$\bigwedge(\, a_i \mid i \in I_0\,) \wedge \bigwedge(\, a_i' \mid i \in I_1\,) \leq \bigwedge(\, a_i \mid i \in J_0\,) \wedge \bigwedge(\, a_i' \mid i \in J_1\,)$$

implies that $I_0 = J_0$ and $I_1 = J_1$.   —

Proof. Again, the "only if" part is by substitution into B_2. On the other hand, clearly B is freely generated by $\{\, a_i \mid i \in I \,\}$ iff, for every finite subset K of I, the subalgebra $[\{\, a_i \mid i \in K \,\}]$ is freely generated by $\{\, a_i \mid i \in K \,\}$. By Theorem 2, the latter holds iff $[\{\, a_i \mid i \in K \,\}]$ has $2^{2^{|K|}}$ elements, which, in turn, is equivalent to $[\{\, a_i \mid i \in K \,\}]$ having $2^{|K|}$ atoms. Using the proof of Theorem 2 and the present hypothesis for $I_0 \cup I_1 = K$, we can see that the elements of the form

$$\bigwedge(\, a_i \mid i \in I_0\,) \wedge \bigwedge(\, a_i' \mid i \in I_1\,),$$

where $I_0 \cup I_1 = K$ and $I_0 \cap I_1 = \emptyset$, are distinct atoms in $[\{\, a_i \mid i \in K \,\}]$, thus completing the proof. $\square$

For a simpler variant of Theorem 4, see Exercise 3.43.

Now we turn our attention to an important application of polynomials: finding homomorphisms.

Theorem 5.   Let the Boolean algebra B be generated by the subalgebra D_1 and the element a. Let D_2 be a Boolean algebra and let φ be a homomorphism of D_1 into D_2. The extensions of φ to homomorphisms of B into D_2 are in one-to-one correspondence with the elements p of D_2 satisfying the following conditions:

 (i) If $x \in D_1$ and $x \leq a$, then $x\varphi \leq p$.

 (ii) If $x \in D_1$ and $x \geq a$, then $x\varphi \geq p$.   —

To prepare for the proof of this theorem we verify a simple lemma, in which $+$ denotes the *symmetric difference*; that is,

$$x + y = (x' \wedge y) \vee (x \wedge y').$$

Lemma 6. *Let the Boolean algebra B be generated by the subalgebra D_1 and the element a. Then every element x of B can be represented in the form*

$$x = (a \wedge x_0) \vee (a' \wedge x_1), \quad x_0, \ x_1 \in D_1.$$

This representation is not unique. In fact,

$$(a \wedge x_0) \vee (a' \wedge x_1) = (a \wedge y_0) \vee (a' \wedge y_1), \quad x_0, \ x_1, \ y_0, \ y_1 \in D_1,$$

iff

$$a \leq (x_0 + y_0)' \quad and \quad x_1 + y_1 \leq a.$$

Proof. Let D_0 denote the set of all elements of B having such a representation. If $x \in D_1$, then $x = (a \wedge x) \vee (a' \wedge x)$; thus $D_1 \subseteq D_0$. Also $a = (a \wedge 1) \vee (a' \wedge 0)$, and so $a \in D_0$. Therefore, to show that $D_0 = B$, it suffices to verify that D_0 is

13. Let L and L_1 be distributive lattices, let $L = [A]$, and let φ be a map of
     A into L_1. Show that there is a homomorphism of L into L_1 extending φ
     iff, for any pair of finite nonempty subsets A_1 and A_2 of A,

$$\bigwedge A_1 \leq \bigvee A_2 \quad \text{implies that} \quad \bigwedge A_1\varphi \leq \bigvee A_2\varphi.$$

(Compare this with Exercise I.5.45.)

14. State and prove Exercise 13 for Boolean algebras.

15. Interpret Lemma 6 using Exercise 14.

16. Extend the last statement of Corollary 7 to the case in which D_1 is gen-
     erated by B and $a_0, \ldots, a_{n-1} \in D_1$, $n > 1$.

17. Let p and q be lattice polynomials. Since p and q can also be regarded as
     Boolean polynomials, $p \equiv q$ was defined in two ways in this section: with
     respect to $\mathbf{D}$ and with respect to $\mathbf{B}$. Show that the two definitions are
     equivalent for lattice polynomials.

18. Define $\equiv$ for lattice polynomials with respect to a class $\mathbf{K}$ of lattices.
     Show that $P_{\mathbf{K}}(n) \in \mathbf{K}$ iff the free lattice over $\mathbf{K}$ with n generators exists,
     in which case $P_{\mathbf{K}}(n)$ is a free lattice with n generators.

3.   Congruence Relations

Let L be a lattice and let $H \subseteq L^2$. We denote by $\Theta(H)$ the smallest congruence
relation such that $a \equiv b$, for all $\langle a, b \rangle \in H$, and call it the *congruence relation
generated by H*.

Lemma 1. *For any $H \subseteq L^2$, $\Theta(H)$ exists.*

Proof. Let

$$\Phi = \bigwedge (\Theta \mid a \equiv b \; (\Theta), \text{ for all } \langle a, b \rangle \in H).$$

Since the meet is intersection in the lattice $\mathrm{Con}\, L$, it is obvious that $a \equiv b \; (\Theta)$,
for all $\langle a, b \rangle \in H$; thus $\Phi = \Theta(H)$.   □

We shall use special notation in two cases: If $H = \{\langle a, b \rangle\}$, we write $\Theta(a, b)$ for
$\Theta(H)$. If $H = I^2$, where I is an ideal, we write $\Theta[I]$ for $\Theta(H)$. The congruence
relation $\Theta(a, b)$ is called *principal*; its importance is revealed by the following
formula.

Lemma 2. $\Theta(H) = \bigvee (\Theta(a, b) \mid \langle a, b \rangle \in H)$.

Proof. The proof is obvious.   □

Note that $\Theta(a, b)$ is the smallest congruence relation under which $a \equiv b$, whereas $\Theta[I]$ is the smallest congruence relation under which I is contained in a single class.

In general lattices, not much can be said about $\Theta(H)$. In distributive lattices, the following description of $\Theta(a, b)$ is important (G. Grätzer and E. T. Schmidt [1958d]):

Theorem 3.   Let L be a distributive lattice, a, b, x, $y \in L$, and let $a \leq b$. Then

$$x \equiv y \quad (\Theta(a, b)) \quad \text{iff} \quad x \wedge a = y \wedge a \text{ and } x \vee b = y \vee b. \qquad \text{---}$$

Remark. This situation is illustrated in Figure 1.

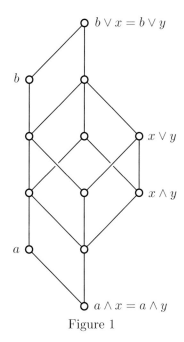

Figure 1

Proof. Let Φ denote the binary relation under which $x \equiv y$ (Φ) iff $x \wedge a = y \wedge a$ and $x \vee b = y \vee b$. Φ is obviously an equivalence relation. If $x \equiv y$ (Φ) and $z \in L$, then

$$(x \vee z) \wedge a = (x \wedge a) \vee (z \wedge a) = (y \wedge a) \vee (z \wedge a) = (y \vee z) \wedge a,$$

and

$$(x \vee z) \vee b = z \vee (x \vee b) = z \vee (y \vee b) = (y \vee z) \vee b;$$

thus $x \vee z \equiv y \vee z$ (Φ). Similarly, $x \wedge z \equiv y \wedge z$ (Φ), and so Φ is a congruence relation. That $a \equiv b$ (Φ) is obvious. Finally, let Θ be any congruence relation such that $a \equiv b$ (Θ) and let $x \equiv y$ (Φ). Then

$$x \wedge a = y \wedge a,$$
$$x \vee b = y \vee b,$$
$$x \vee a \equiv x \vee b \quad (\Theta),$$
$$x \wedge b \equiv x \wedge a \quad (\Theta).$$

Computing modulo Θ, we obtain

$$x = x \vee (x \wedge a) = x \vee (y \wedge a) = (x \vee y) \wedge (x \vee a) \equiv (x \vee y) \wedge (x \vee b)$$
$$= (x \vee y) \wedge (y \vee b) = y \vee (x \wedge b) \equiv y \vee (x \wedge a) = y \vee (y \wedge a) = y,$$

that is, $x \equiv y$ (Θ), proving that $\Phi \leq \Theta$.   □

Explanation. Since $a \equiv b$ implies that $(a \vee p) \wedge q \equiv (b \vee p) \wedge q$, we must have $x \equiv y$ ($\Theta(a,b)$) if

$$x \wedge y = (a \vee p) \wedge q,$$
$$x \vee y = (b \vee p) \wedge q.$$

It is easy to check that the x, y satisfying the conditions of Theorem 3 are exactly the same as those for which such p, q exist. Thus Theorem 3 can be interpreted as follows: We get all pairs x, y with $x \equiv y$ ($\Theta(a,b)$) and $x \leq y$, by applying the Substitution Property "twice". No further application of the Substitution Property or transitivity is needed. An equivalent form of Theorem 3 is to require

$$x \vee y \leq b \vee (x \wedge y),$$
$$(a \vee (x \wedge y)) \wedge (x \vee y) = x \wedge y.$$

Some applications of Theorem 3 follow.

Corollary 4. *Let I be an ideal of the distributive lattice L. Then $x \equiv y$ ($\Theta[I]$) iff $x \vee y = (x \wedge y) \vee i$, for some $i \in I$. Therefore, I is a congruence class modulo $\Theta[I]$.*

Remark. This situation is illustrated in Figure 2, in which the dotted line indicates congruence modulo $\Theta[I]$.

Proof. If $x \vee y = (x \wedge y) \vee i$, then $x \equiv y$ ($\Theta(x \wedge y \wedge i, i)$), $x \wedge y \wedge i$, $i \in I$, and so $x \equiv y$ ($\Theta[I]$). Conversely, $\Theta[I] = \bigvee (\Theta(u, v)] \mid u,\ v \in I$) by Lemma 2. However,

$$\Theta(u, v) \vee \Theta(u_1, v_1) \leq \Theta(u \wedge v \wedge u_1 \wedge v_1, u \vee v \vee u_1 \vee v_1);$$

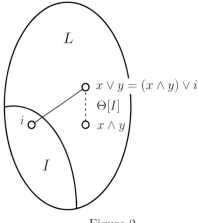

Figure 2

therefore, $\Theta[I] = \bigcup(\Theta(u,v) \mid u, \ v \in I)$. If $x \equiv y \ (\Theta(u,v))$, u, $v \in I$, $u \le v$, then $x \vee v = y \vee v$, and so $(x \wedge y) \vee (v \wedge (x \vee y)) = x \vee y$; thus the condition of Corollary 4 is satisfied with $i = v \wedge (x \vee y) \in I$. Finally, if $a \in I$ and $a \equiv b$ $(\Theta[I])$, then $a \vee b = (a \wedge b) \vee i$, $i \in I$, and so $a \vee b \in I$ and $b \in I$, showing that I is a full congruence class.   □

Corollary 5. *Let L be a distributive lattice, x, y, a, $b \in L$, and let $x \le y \le a \le b$ or $a \le b \le x \le y$. Then $x \equiv y \ (\Theta(a,b))$ implies that $x = y$.*

A very important congruence relation has already been used in the proof of Lemma I.3.5(ii): Given a prime ideal P of the lattice L, we can construct a congruence relation that has exactly two congruence classes, P and $L - P$. This statement can be generalized as follows: Let A be a set of prime ideals of a lattice L and let us call two elements x and y congruent modulo A iff, for every $P \in A$, either x, $y \in P$ or x, $y \in L - P$; this describes a congruence relation on L. For instance, if $A = \{P, Q, R\}$, $Q \subset P$, $R \subset P$, then we get five congruence classes as shown in Figure 3; the quotient lattice is shown in Figure I.5.8.

This principle will be used often. An interesting application is:

Theorem 6.   Let K be a sublattice of a distributive lattice L. Any congruence relation Θ of K can be extended to L; that is, there exists a congruence relation Φ on L such that $x \equiv y \ (\Phi)$ iff $x \equiv y \ (\Theta)$, for x, $y \in K$.   ―

Proof. Let $\varphi \colon x \mapsto [x]\Theta$ be the natural homomorphism of K onto K/Θ; then, for every prime ideal P of K/Θ, $P\varphi^{-1}$ is a prime ideal of K. Therefore, $(P\varphi^{-1}]$ is an ideal of L and $[K - P\varphi^{-1})$ is a dual ideal of L; thus by Theorem 1.15, we can choose a prime ideal P_1 of L such that $P_1 \supseteq P\varphi^{-1}$ and $P_1 \cap (K - P\varphi^{-1}) = \varnothing$. For every prime ideal P of K, we choose such a prime ideal P_1 of L; let A denote

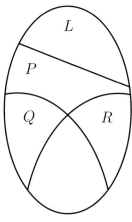

Figure 3

the collection of all such prime ideals. Let Φ be the congruence relation associated with A, as previously described. Now for x, $y \in K$, the condition $x \equiv y$ (Θ) is equivalent to $x\varphi = y\varphi$, and so, for every $P_1 \in A$, either x, $y \in P_1$ or x, $y \notin P_1$; thus $x \equiv y$ (Φ). Conversely, if $x \equiv y$ (Φ), then, for every $P_1 \in A$, either x, $y \in P_1$ or x, $y \notin P_1$, and so either $x\varphi$, $y\varphi \in P$ or $x\varphi$, $y\varphi \notin P$. Since every pair of distinct elements of K/Θ is separated by a prime ideal (Corollary 1.17), we conclude that $x\varphi = y\varphi$ and thus $x \equiv y$ (Θ). $\qquad \square$

It is well known that in rings, ideals are in a one-to-one correspondence with congruence relations. In one class of lattices the situation is exactly the same as in the class of rings.

Theorem 7.   Let L be a Boolean lattice. Then $\Theta \mapsto [0]\Theta$ is a one-to-one correspondence between congruence relations and ideals of L. $-$

Proof. By Corollary 4, the map is onto; therefore, we have only to prove that it is one-to-one, that is, that $[0]\Theta$ determines Θ. This fact, however, is obvious, since $a \equiv b$ (Θ) iff $a \wedge b \equiv a \vee b$ (Θ), which, in turn, is equivalent to $c \equiv 0$ (Θ), where c is the relative complement of $a \wedge b$ in $[0, a \vee b]$ (see Figure 4). Thus $a \equiv b$ (Θ) iff $c \in [0]\Theta$. $\square$

This proof does not make full use of the hypothesis that L is a complemented distributive lattice. In fact, all we need to make the proof work is that L has a zero and is relatively complemented. Such a distributive lattice is called a *generalized Boolean lattice*.

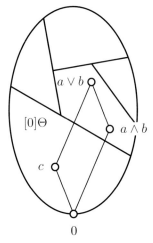

Figure 4

Theorem 8 (J. Hashimoto [1952]).   Let L be a lattice. There is a one-to-one correspondence between ideals and congruence relations of L under which the ideal corresponding to a congruence relation Θ is a whole congruence class under Θ iff L is a generalized Boolean lattice.   —

Proof. (G. Grätzer and E. T. Schmidt [1958d].) The "if" part is Theorem 7. We proceed with the "only if" part. The ideal corresponding to ω has to be $(0]$, and thus L has a 0. If L contains a diamond, $\{0, a, b, c, i\}$, then $(a]$ cannot be a congruence class, because $a \equiv 0$ implies that

$$i = a \vee c \equiv o \vee c = c,$$
$$b = b \wedge i \equiv b \wedge c = o.$$

But $o \in (a]$, and thus any congruence class containing $(a]$ contains b, and $b \notin (a]$. Similarly, if L contains a pentagon, $\{o, a, b, c, i\}$, and a congruence class contains $(b]$, then $b \equiv o$; thus

$$i = b \vee c \equiv o \vee c = c,$$

and so

$$a = a \wedge i \equiv a \wedge c = o.$$

Therefore, this congruence class has to contain a, and $a \notin (b]$. Thus, by Theorem 1.1, L is distributive. Let $a < b$ and $I = [0]\Theta(a, b)$. By Corollary 4, $\Theta[I]$ is also a congruence relation of L having I as a whole congruence class;

consequently, $\Theta[I] = \Theta(a, b)$, and so $a \equiv b \ (\Theta[I])$. Thus, again by Corollary 4, $b = a \vee i$, for some $i \in I$, and $i \equiv 0 \ (\Theta(a, b))$. The latter is equivalent to $i \vee b = 0 \vee b$ and $i \wedge a = 0 \wedge a$. Thus $a \vee i = b$ and $a \wedge i = 0$, and so i is the relative complement of a in $[0, b]$. $\qquad\qquad\qquad\qquad\qquad\qquad\qquad$ $\square$

It is no coincidence that, in the class of generalized Boolean lattices, congruences and ideals behave as they do in rings. Indeed, generalized Boolean lattices are rings in disguise.

Theorem 9 (M. H. Stone [1936]).

(i) Let $\mathfrak{B} = \langle B; \wedge, \vee \rangle$ be a generalized Boolean lattice. Define the binary operations $\cdot$ and $+$ on B by setting

$$x \cdot y = x \wedge y$$

and by defining $x + y$ as the relative complement of $x \wedge y$ in $[0, x \vee y]$ (see Figure 5). Then $\mathfrak{B}^R = \langle B; +, \cdot \rangle$ is a Boolean ring—that is, an (associative) ring satisfying $x^2 = x$, for all $x \in B$ (and, consequently, satisfying $xy = yx$ and $x + x = 0$, for $x, y \in B$).

(ii) Let $\mathfrak{B} = \langle B; +, \cdot \rangle$ be a Boolean ring. Define the binary operations $\wedge$ and $\vee$ in B by

$$x \wedge y = x \cdot y,$$
$$x \vee y = x + y + x \cdot y.$$

Then $\mathfrak{B}^L = \langle B; \wedge, \vee \rangle$ is a generalized Boolean lattice.

(iii) Let $\mathfrak{B}$ be a generalized Boolean lattice. Then $(\mathfrak{B}^R)^L = \mathfrak{B}$.

(iv) Let $\mathfrak{B}$ be a Boolean ring. Then $(\mathfrak{B}^L)^R = \mathfrak{B}$. $\qquad\qquad$ —

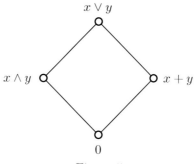

Figure 5

The proof of this theorem is purely computational. Some steps will be given in the exercises.

The correspondence between Boolean rings and generalized Boolean lattices preserves many algebraic properties.

Theorem 10.   Let $\mathfrak{B}_0$ and $\mathfrak{B}_1$ be generalized Boolean lattices.

(i) Let $I \subseteq B_0$. Then I is an ideal of $\mathfrak{B}_0$ iff I is an ideal of $\mathfrak{B}_0^R$.

(ii) Let $\varphi \colon B_0 \to B_1$. Then φ is a $\{0\}$-homomorphism of $\mathfrak{B}_0$ into $\mathfrak{B}_1$ iff φ is a homomorphism of $\mathfrak{B}_0^R$ into $\mathfrak{B}_1^R$.

(iii) $\mathfrak{B}_0$ is a $\{0\}$-sublattice of $\mathfrak{B}_1$ iff $\mathfrak{B}_0^R$ is a subring of $\mathfrak{B}_1^R$.   ▬

The proof is again left to the reader.

Congruence relations on an arbitrary lattice have an interesting connection with distributive lattices:

Theorem 11 (N. Funayama and T. Nakayama [1942]).   Let L be an arbitrary lattice. Then $\operatorname{Con} L$, the lattice of all congruence relations of L, is distributive.   ▬

Proof. Let Θ, Φ, $\Psi \in \operatorname{Con} L$. Since

$$\Theta \wedge (\Phi \vee \Psi) \geq (\Theta \wedge \Phi) \vee (\Theta \wedge \Psi),$$

it suffices to prove that

$$a \equiv b \quad (\Theta \wedge (\Phi \vee \Psi)) \qquad \text{implies that} \qquad a \equiv b \quad ((\Theta \wedge \Phi) \vee (\Theta \wedge \Psi)).$$

So let $a \equiv b\ (\Theta \wedge (\Phi \vee \Psi))$; that is, $a \equiv b\ (\Theta)$ and $a \equiv b\ (\Phi \vee \Psi)$. By Theorem I.3.9, there exists a sequence

$$a \wedge b = z_0 \leq \cdots \leq z_n = a \vee b$$

such that

$$z_i \equiv z_{i+1} \quad (\Phi) \qquad \text{or} \qquad z_i \equiv z_{i+1} \quad (\Psi),$$

for every $0 \leq i < n$. Since $a \equiv b\ (\Theta)$, we also have $a \wedge b \equiv a \vee b\ (\Theta)$, and so $z_i \equiv z_{i+1}\ (\Theta)$, for every $0 \leq i < n$. Thus, for every $0 \leq i < n$,

$$z_i \equiv z_{i+1} \quad (\Theta \wedge \Phi) \qquad \text{or} \qquad z_i \equiv z_{i+1} \quad (\Theta \wedge \Psi),$$

implying that

$$a \equiv b \quad ((\Theta \wedge \Phi) \vee (\Theta \wedge \Psi)).$$ □

Another property of congruence lattices is given in the following definition.

Definition 12.

    (i) Let L be a complete lattice and let a be an element of L. Then a is called *compact* iff $a \leq \bigvee X$, for some $X \subseteq L$, implies that $a \leq \bigvee X_1$, for some finite $X_1 \subseteq X$.

    (ii) A complete lattice is called *algebraic* iff every element is the join of compact elements.

The name, algebraic lattice, is due to G. Birkhoff [1967], however Birkhoff does not assume completeness. In the literature, algebraic lattices are also called *compactly generated lattices*.

Just as for lattices, a nonempty subset I of a join-semilattice F is an *ideal* iff, for $a, b \in F$, we have $a \vee b \in I$ exactly if a and $b \in I$. Again, $\mathrm{Id}\, F$ is the poset (not necessarily a lattice) of all ideals of F partially ordered under set inclusion. If F has a zero, then $\mathrm{Id}\, F$ is a lattice.

Using $\mathrm{Id}\, F$, we give a useful characterization of algebraic lattices:

Theorem 13.   A lattice L is algebraic iff it is isomorphic to the lattice of all ideals of a join-semilattice with 0.   —

Proof. Let F be a join-semilattice with 0; we prove that $\mathrm{Id}\, F$ is algebraic. We know that $\mathrm{Id}\, F$ is complete. We claim that, for $a \in F$, $(a]$ is a compact element of $\mathrm{Id}\, F$. Let $X \subseteq \mathrm{Id}\, F$ and let

$$(a] \subseteq \bigvee X.$$

Just as in the proof of Corollary I.3.2,

$$\bigvee X = \{\, x \mid x \leq t_0 \vee \cdots \vee t_{n-1},\ t_i \in I_i,\ I_i \in X \,\}.$$

Therefore, $a \leq t_0 \vee \cdots \vee t_{n-1}$, $t_i \in I_i$, $I_i \in X$. Thus with $X_1 = \{I_0, \ldots, I_{n-1}\}$,

$$(a] \subseteq \bigvee X_1.$$

Since, for any $I \in \mathrm{Id}\, F$, we have

$$I = \bigvee (\, (a] \mid a \in I\,),$$

we see that $\mathrm{Id}\, F$ is algebraic.

Now let L be an algebraic lattice and let F be the set of compact elements of L. Obviously, $0 \in F$. Let $a, b \in F$ and $a \vee b \leq \bigvee X$, for some $X \subseteq L$. Then $a \leq a \vee b \leq \bigvee X$, and so $a \leq \bigvee X_0$, for some finite $X_0 \subseteq X$. Similarly, $b \leq \bigvee X_1$, for some finite $X_1 \subseteq X$. Thus $a \vee b \leq \bigvee(X_0 \cup X_1)$, and $X_0 \cup X_1$ is a finite subset of X. So $a \vee b \in F$.

Therefore, $\langle F; \vee \rangle$ is a join-semilattice with 0. Consider the map:

$$\varphi: a \mapsto \{\, x \mid x \in F,\ x \leq a \,\}, \quad a \in L.$$

Obviously, φ maps L into $\operatorname{Id} F$. By the definition of an algebraic lattice, $a = \bigvee a\varphi$, and thus φ is one-to-one. To prove that φ is onto, let $I \in \operatorname{Id} F$, $a = \bigvee I$. Then $a\varphi \supseteq I$. Let $x \in a\varphi$. Then $x \leq \bigvee I$, so that by the compactness of x, $x \leq \bigvee I_1$, for some finite $I_1 \subseteq I$. Therefore $x \in I$, proving that $a\varphi \subseteq I$. Consequently, $a\varphi = I$, and so φ is onto. Thus φ is an isomorphism. □

Now we connect the foregoing with congruence lattices.

Lemma 14. *Every principal congruence relation is compact.*

Proof. Let L be a lattice, $a, b \in L$, $X \subseteq \operatorname{Con} L$, and

$$\Theta(a, b) \leq \bigvee X.$$

Then $a \equiv b\ (\bigvee X)$, and thus (just as in Theorem I.3.9) there exists a sequence $a = x_0, x_1, \ldots, x_n = b$, in which, for all i with $0 \leq i < n$, $x_i \equiv x_{i+1}\ (\Theta_i)$, for some $\Theta_i \in X$. Therefore, $a \equiv b\ (\bigvee X_0)$, where $X_0 = \{\Theta_0, \ldots, \Theta_{n-1}\}$, and so $\Theta(a, b) \leq \bigvee X_0$, where X_0 is a finite subset of X. □

Theorem 15.   Let L be an arbitrary lattice. Then $\operatorname{Con} L$ is an algebraic lattice.   —

Proof. For every $\Theta \in \operatorname{Con} L$,

$$\Theta = \bigvee (\, \Theta(a, b) \mid a \equiv b\,(\Theta)\,).$$

Consequently, this theorem follows from Lemma 14 and Corollary I.3.15.   □

Combining Theorems 11 and 15 we get:

Corollary 16. *Let L be an arbitrary lattice. Then $\operatorname{Con} L$ is a distributive algebraic lattice.*

The converse of Corollary 16 is a long-standing conjecture of lattice theory. We shall verify the conjecture in the finite case. This was first established by R. P. Dilworth. The present proof combines a construction of G. Grätzer and E. T. Schmidt [1962] (Lemma 18) with a result of G. Grätzer and H. Lakser [1968] (Lemma 19). The reader may want to page over the proof of Theorem 17 at the first reading of this book.

Theorem 17.   Let K be a finite distributive lattice. Then there exists a finite lattice L such that K is isomorphic to $\operatorname{Con} L$.   —

To prepare for the proof of this result, we introduce some new concepts and prove two lemmas.

Let M be a poset satisfying the following two condition:

(i) $\inf\{a, b\}$ exists in M, for any $a, b \in M$;

(ii) $\sup\{a, b\}$ exists for any $a, b \in M$ having a common upper bound in M.

We define in M:

$$a \wedge b = \inf\{a, b\},$$
$$a \vee b = \sup\{a, b\},$$

whenever $\sup\{a, b\}$ exists in M. This makes M into a partial lattice, called a *chopped lattice*. Observe that, for a finite poset, the first condition implies the second, hence every finite meet-semilattice can be regarded as a chopped lattice.

We now define congruences and ideals of chopped lattices, as it was done for lattices.

An equivalence relation Θ of a chopped lattice M is a *congruence relation* iff $a_0 \equiv b_0$ (Θ) and $a_1 \equiv b_1$ (Θ) imply that $a_0 \wedge a_1 \equiv b_0 \wedge b_1$ (Θ) and, whenever $a_0 \vee a_1$ and $b_0 \vee b_1$ exist, $a_0 \vee a_1 \equiv b_0 \vee b_1$ (Θ). The set $\operatorname{Con} M$ of all congruence relations of M partially ordered by set inclusion is again a lattice.

A subset I of the chopped lattice M is an *ideal* if

(i) $i \in I$ and $a \in M$ imply that $a \wedge i \in I$;

(ii) $i, j \in I$ implies that $i \vee j \in I$, provided that $i \vee j$ exists in M.

The set $\operatorname{Id} M$ of all ideals of M partially ordered by set inclusion is a lattice.

Lemma 18. *Let K be a finite distributive lattice. Then there exists a finite chopped lattice M such that $\operatorname{Con} M$ is isomorphic to K.*

Proof. Take the finite set $M_0 = J(K) \cup \{0\}$ and make it a meet-semilattice by defining $\inf\{a, b\} = 0$ if $a \neq b$ ($J(K)$ is the set of nonzero join-irreducible elements of K; see Section 1), as illustrated in Figure 6. Note that in the finite chopped lattice M_0, $a \equiv b$ (Θ) and $a \neq b$ imply that $a \equiv 0$ (Θ) and $b \equiv 0$ (Θ); therefore, congruence relations of M_0 are in one-to-one correspondence with

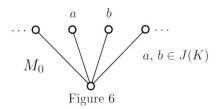

Figure 6

subsets of $J(K)$. Thus $\text{Con}\, M_0$ is a Boolean lattice whose atoms are associated with elements of $J(K)$; the congruence Φ_a associated with $a \in J(K)$ is: $x \equiv y$ (Φ_a) if $\{x, y\} = \{a, 0\}$, and if $\{x, y\} \neq \{a, 0\}$, then $x \equiv y$ (Φ_a) implies that $x = y$.

If $J(K)$ is unordered in K, then we are ready. However, if, say, $a, b \in J(K)$, $a > b$ in K, then we must force $\Phi_a > \Phi_b$. The simplest way to make this happen is to use the lattice $M(a, b)$ of Figure 7. Note that $M(a, b)$ has three congruence relations, namely, ω, ι, and Θ, where Θ is the congruence relation with congruence classes $\{0, b_1, b_2, b\}$, $\{a_1, a(b)\}$. Thus $\Theta(a_1, 0) = \iota$. In other words, $a_1 \equiv 0$ "implies" that $b_1 \equiv 0$, but $b_1 \equiv 0$ "does not imply" that $a_1 \equiv 0$.

We construct the finite chopped lattice M by "inserting" $M(a, b)$ in M_0, whenever $a > b$ in $J(K)$. Figure 8 gives M for the three-element chain.

More precisely, M consists of four kinds of elements: (i) 0; (ii) all maximal join-irreducible elements of K (that is, all $a \in J(K)$ such that there is no $x \in J(K)$ with $a < x$ in K); (iii) for any nonmaximal join-irreducible element a of K, three elements: a, a_1, a_2; (iv) for each pair $a, b \in J(K)$ with $a > b$, a new element, $a(b)$. To simplify the notation, for each maximal join-irreducible element a, we write $a = a_1 = a_2$. For $a, b \in J(K)$ with $a > b$, we set $M(a, b) = \{0, a_1, b, b_1, b_2, a(b)\}$.

Observe that

$$M(a, b) \cap M(c, d) = \begin{cases} M(a, b), & \text{if } a = c \text{ and } b = d; \\ \{0, b, b_1, b_2\}, & \text{if } a \neq c \text{ and } b = d; \\ \{0, a_1\}, & \text{if } a = c \text{ and } b \neq d; \\ \{0, b_1\} & \text{if } b = c; \\ \{0\}, & \text{otherwise.} \end{cases}$$

For $x, y \in M$, let us define $x \leq y$ to mean that, for some $a, b \in J(K)$ with $a > b$, we have $x, y \in M(a, b)$ and $x \leq y$ in the lattice $M(a, b)$, as illustrated in Figure 7. It is easily seen that $x \leq y$ does not depend on the choice of a and b,

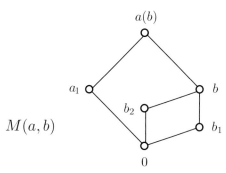

$M(a, b)$

Figure 7

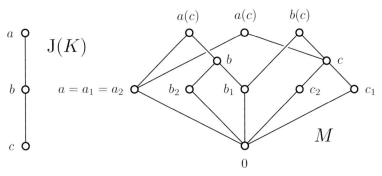

Figure 8

and that $\leq$ is a partial ordering relation. Since, under this partial ordering, all $M(a,b)$ and $M(a,b) \cap M(c,d)$ are lattices and x, $y \in M$, $x \in M(a,b)$, and $y \leq x$ imply that $y \in M(a,b)$, we conclude that $\inf\{u,v\}$ exists, for all u, $v \in M$. We regard M as a chopped lattice.

Now we describe Con M. Let $H \in \mathrm{H}(\mathrm{J}(K))$ (notation of Definition 1.8). We define a binary relation Θ_H on M by $x \equiv y$ (Θ_H) iff one of the following hold:

x, $y \in \bigcup(\, M(a,b) \mid a,\ b \in H, a > b\,) \cup \bigcup(\, \{0, a_1, a_2, a\} \mid a \in H\,)$;

x, $y \in \{a_1, a(b), a(c)\}$,   where $a > b$, $a > c$, $a \notin H$, b, $c \in H$;

$x = y$.

In other words, $[0]\Theta_H$ contains all a_1, a_2, a with $a \in H$; and if $a > b$, a, $b \in H$, then it also contains $a(b)$. Outside this class, the only nontrivial congruences are $a(b) \equiv a_1 \equiv a(c)$, where $a \notin H$, and b, $c \in H$, $a > b$, $a > c$.

Θ_H is obviously an equivalence relation. The fact that Θ_H restricted to any $M(a,b)$ is a congruence relation easily implies that Θ_H is a congruence relation of the chopped lattice M. Given a Θ_H, we get

$$H = \{\, a \mid a_1 \equiv 0 \ (\Theta_H)\,\};$$

thus the map

$$\varphi \colon H \mapsto \Theta_H$$

is a one-to-one order-preserving map of $\mathrm{H}(\mathrm{J}(K))$ into Con M. To show that φ is an isomorphism, we have to show that φ is onto. So let Θ be a congruence relation of M, and

$$H = \{\, a \mid a_1 \equiv 0 \ (\Theta)\,\}.$$

Since in $M(a,b)$ every congruence Θ is determined by the atoms in $[0]\Theta$, the same holds in M. Therefore, $\Theta = \Theta_H$. Thus $\mathrm{H}(\mathrm{J}(K)) \cong$ Con M. By Theorem 1.9, $K \cong \mathrm{H}(\mathrm{J}(K))$, and so $K \cong$ Con M. $\qquad\square$

Lemma 19. *Let M be a finite chopped lattice. Then, for every congruence relation Θ, there exists exactly one congruence relation $\overline{\Theta}$ of $\mathrm{Id}\, M$ such that, for $a, b \in M$,*

$$(a] \equiv (b] \quad (\overline{\Theta}) \qquad \textit{iff} \qquad a \equiv b \quad (\Theta).$$

Proof. Since arbitrary meets exist in M, $(m]$ is a (finite) lattice, for every element $m \in M$. So if $\{x, y\}$ has an upper bound, then $x \vee y$ exists.

Let Θ be a congruence relation of M. For $X \subseteq M$, set

$$[X]\Theta = \bigcup(\,[x]\Theta \mid x \in X\,);$$

that is,

$$[X]\Theta = \{\,y \mid x \equiv y\ (\Theta),\ \text{for some } x \in X\,\}.$$

If $I, J \in \mathrm{Id}\, M$, define $I \equiv J\ (\overline{\Theta})$ iff $[I]\Theta = [J]\Theta$. Obviously, $\overline{\Theta}$ is an equivalence relation. Let $I \equiv J\ (\overline{\Theta})$, $N \in \mathrm{Id}\, M$, and $x \in I \cap N$. Then $x \equiv y\ (\Theta)$, for some $y \in J$, and so $x \equiv x \wedge y\ (\Theta)$ and $x \wedge y \in J \cap N$. This shows that $[I \cap N]\Theta \subseteq [J \cap N]\Theta$. Similarly, $[J \cap N]\Theta \subseteq [I \cap N]\Theta$, so $I \cap N \equiv J \cap N\ (\overline{\Theta})$.

To show that $I \vee N \equiv J \vee N\ (\overline{\Theta})$, recall the description of $I \vee N$ given in Exercise I.5.22: set

$$A_0 = I \cup N,$$
$$A_n = \{\,x \mid x \leq t_0 \vee t_1, \quad t_0,\ t_1 \in A_{n-1}\,\}, \quad \text{for } 0 < n < \omega;$$

then $I \vee N = \bigcup(\,A_n \mid n < \omega\,)$. By induction on n, we prove that $A_n \subseteq [J \vee N]\Theta$. For $n = 0$,

$$A_0 = I \cup N \subseteq [J]\Theta \cup N \subseteq [J \vee N]\Theta.$$

Suppose that $A_{n-1} \subseteq [J \vee N]\Theta$ and let $x \in A_n$. Then $x \leq t_0 \vee t_1$, for some t_0, $t_1 \in A_{n-1}$. Thus $t_0 \equiv u_0\ (\Theta)$ and $t_1 \equiv u\ (\Theta)$, for some u_0, $u_1 \in J \vee N$, and so $t_0 \equiv t_0 \wedge u_0\ (\Theta)$ and $t_1 \equiv t_1 \wedge u_1\ (\Theta)$. Observe that $t_0 \vee t_1$ is an upper bound for $\{t_0 \wedge u_0, t_1 \wedge u_1\}$; consequently, $(t_0 \wedge u_0) \vee (t_1 \wedge u_1)$ exists. Therefore,

$$t_0 \vee t_1 \equiv (t_0 \wedge u_0) \vee (t_1 \wedge u_1) \quad (\Theta).$$

Finally,

$$x = x \wedge (t_0 \vee t_1) \equiv x \wedge ((t_0 \wedge u_0) \vee (t_1 \wedge u_1)) \quad (\mathrm{mod}\ \Theta),$$
$$x \wedge ((t_0 \wedge u_0) \vee (t_1 \wedge u_1)) \in J \vee N.$$

Thus $x \in [J \vee N]\Theta$. Since $I \vee N$ is $\bigcup(\,A_n \mid n < \omega\,)$, we conclude that $I \vee N \subseteq [J \vee N]\Theta$. Similarly, $J \vee N \subseteq [I \vee N]\Theta$, proving that $I \vee N \equiv J \vee N\ (\overline{\Theta})$. This completes the verification that $\overline{\Theta}$ is a congruence relation of $\mathrm{Id}\, M$.

17. Prove that the verification of Theorem 9(i) can be reduced to the Boolean lattice case and that in this case

$$x + y = (x \wedge y') \vee (x' \wedge y).$$

18. Let B be a Boolean lattice. Verify that

$$x + y = (x \vee y) \wedge (x' \vee y').$$

19. Let B be a Boolean lattice. Verify that

$$(x + y) + z = (x \wedge y' \wedge z') \vee (x' \wedge y \wedge z') \vee (x' \wedge y' \wedge z)$$

and conclude that $+$ is associative.

20. Prove that $x(y + z) = xy + xz$.

21. Prove Theorem 9(i).

22. Prove Theorem 9(ii).

23. Let $\mathfrak{B}$ be a generalized Boolean lattice. Observe that, for $x, y \in B$, $x \wedge y$ is the same in $\mathfrak{B}$ as in $(\mathfrak{B}^R)^L$ (namely, $x \cdot y$); conclude that $\mathfrak{B} = (\mathfrak{B}^R)^L$.

24. Verify Theorem 9(iv).

25. Verify Theorem 10.

26. Show that, using the concept of a distributive semilattice (see Section 5), Corollary 16 can be reformulated as follows: Let L be an arbitrary lattice. Prove that there exists a distributive join-semilattice F with 0 such that $\mathrm{Con}\, L$ is isomorphic to $\mathrm{Id}\, F$.

27. Characterize the lattice of all ideals of a lattice using the concept of an algebraic lattice.

28. Characterize the lattice of all ideals of a Boolean lattice.

29. Find a result for chopped lattices analogous to Lemma I.3.8.

30. Generalize Theorem I.3.9 to chopped lattices.

31. For a chopped lattice M, is $\mathrm{Con}\, M$ necessarily distributive?

32. Let M be a chopped lattice satisfying the following conditions:

    (i)   There exists $H \subseteq M$ such that, for $h \in H$, $(h]$ is a lattice.

(ii) For each ideal I that belongs to the sublattice $\mathrm{Id}_\mathrm{p}\, M$ of $\mathrm{Id}\, M$ generated by the principal ideals, there exist finite $\{h_1,\dots,h_n\} \subseteq H$ and $\{i_1,\dots,i_n\} \subseteq I$ with $i_j \le h_j$ such that $I = (i_1] \cup \cdots \cup (i_n]$.

Under these conditions, prove Lemma 19 for M, replacing $\mathrm{Id}\, M$ with $\mathrm{Id}_\mathrm{p}\, M$ (G. Grätzer and H. Lakser [1968]).

Exercises 33–36 are from G. Grätzer and E. T. Schmidt [1962].

*33. Let K be a finite distributive lattice and let L be the lattice of Theorem 17 as constructed in Lemmas 18 and 19. Show that L is sectionally complemented.

34. Let L be given as in Exercise 33. Show that the congruences of L permute.

35. Let n be the length of the longest chain in K. Show that the L of Theorem 17 can be constructed so that L is of length $2n - 1$ (define $a(b)$ only for $a \succ b$).

36. Let P be a poset. Generalize Theorem 17 for $K = \mathrm{Id}_\mathrm{p}\, P$ (notation as in Exercise 1.17).

Exercises 37 and 38 are from G. Grätzer and E. T. Schmidt [1958b].

*37. Show that a chain C is the congruence lattice of a lattice iff C is algebraic.

38. Prove that a Boolean lattice B is the congruence lattice of a lattice iff B is algebraic.

39. Let L be a distributive lattice. Show that $a \mapsto \Theta[(a]]$ embeds L into $\mathrm{Con}\, L$.

40. Let L be a bounded distributive lattice. Show that, for a, $b \in L$, $a \le b$, $\Theta(a,b)$ has a complement in $\mathrm{Con}\, L$, namely, $\Theta(0,a) \vee \Theta(b,1)$.

41. Let L be a bounded distributive lattice. Show that the compact elements of $\mathrm{Con}\, L$ form a Boolean lattice (J. Hashimoto [1952], G. Grätzer and E. T. Schmidt [1958e]).

42. Let B be a Boolean algebra generated by X and let L be the sublattice of B generated by X. Show that B is freely generated by X iff L is freely generated by X in $\mathbf{D}$ (see Exercise 2.12).

43. Let the Boolean algebra B be generated by X. Show that B is freely generated by X iff $\bigwedge X_0 \le \bigvee X_1$ implies that $X_0 \cap X_1 \ne \varnothing$, for any finite nonempty X_0, $X_1 \subseteq X$ (compare this with Theorems 2.3 and 2.4).

4. Boolean Algebras R-generated by Distributive Lattices

Theorem 1.   Every distributive lattice can be embedded in a Boolean lattice.

Proof. By Theorem 1.19, every distributive lattice L is isomorphic to a ring of subsets of some set X. Obviously, L can be embedded into $P(X)$. $\square$

Definition 2. Let L be a $\{0\}$-sublattice of the generalized Boolean lattice B. Then L is said to *R-generate* B iff L generates B as a ring, and if L has a unit element, then the same element is the unit element of B.

The last condition is equivalent to the following: If $\bigvee L$ exists, then $\bigvee B$ exists and $\bigvee L = \bigvee B$.

Our goal is to show the uniqueness of the generalized Boolean lattice R-generated by L. The first result is essentially due to H. M. MacNeille [1939]:

Lemma 3. *Let B be R-generated by L. Then every $a \in B$ can be expressed in the form*

$$a_0 + a_1 + \cdots + a_{n-1}, \qquad a_0 \leq a_1 \leq \cdots \leq a_{n-1}, \qquad a_0, \ldots, a_{n-1} \in L.$$

Example. Let B be the Boolean lattice shown in Figure 1, $L = \{0, a_0, a_1, a_2\}$. Then L R-generates B.

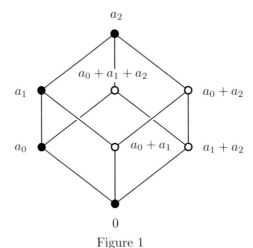

Figure 1

Proof. Let B_1 denote the set of all elements that can be represented in the form $a_0 + \cdots + a_{n-1}$, $a_0, \ldots, a_{n-1} \in L$. Then $L \subseteq B_1$, and B_1 is closed under $+$

and $-$ (since $x - y = x + y$). Furthermore,

$$(a_0 + \cdots + a_{n-1})(b_0 + \cdots + b_{m-1}) = \sum a_i b_j,$$

and each term $a_i b_j = a_i \wedge b_j \in L$, so B_1 is closed under multiplication. We conclude that $B_1 = B$.

Note that L is a sublattice of B; therefore, for $a, b \in L$, $a \vee b$ in L is the same as $a \vee b$ in B. Thus $a \vee b = a + b + ab$, and so

$$a + b = ab + (a \vee b) = (a \wedge b) + (a \vee b).$$

Take $a_0 + \cdots + a_{n-1} \in B$. We prove by induction on n that the summands can be made to form an increasing sequence. For $n = 1$, this is obvious. Let us assume that $a_1 \leq \cdots \leq a_{n-1}$. Then

$$a_0 + a_1 + \cdots + a_{n-1}$$
$$= (a_0 \wedge a_1) + (a_0 \vee a_1) + a_2 + \cdots + a_{n-1}$$
$$= (a_0 \wedge a_1) + ((a_0 \vee a_1) \wedge a_2) + (a_0 \vee a_2) + a_3 + \cdots + a_{n-1}$$
$$= (a_0 \wedge a_1) + ((a_0 \vee a_1) \wedge a_2) + ((a_0 \vee a_2) \wedge a_3) + (a_0 \vee a_3) + \cdots + a_{n-1}$$
$$\cdots$$
$$= (a_0 \wedge a_1) + ((a_0 \vee a_1) \wedge a_2) + \cdots + ((a_0 \vee a_{n-2}) \wedge a_{n-1}) + (a_0 \vee a_{n-1}),$$

and

$$a_0 \wedge a_1 \leq (a_0 \vee a_1) \wedge a_2 \leq \cdots \leq (a_0 \vee a_{n-2}) \wedge a_{n-1} \leq a_0 \vee a_{n-1}. \qquad \square$$

Lemma 4. *Let L be a distributive lattice with 0. Then there exists a generalized Boolean lattice B freely R-generated by L—that is, a generalized Boolean lattice B with the following properties:*

(i) *B is R-generated by L.*

(ii) *If B_1 is R-generated by L, then there is a homomorphism φ of B onto B_1 that is the identity map on L.*

Proof. The existence of B can be proved by copying the proof of Theorem I.5.5 (Theorem I.5.24), *mutatis mutandis*. $\qquad \square$

Lemma 5 (J. Hashimoto [1952]). *Let B be a generalized Boolean lattice R-generated by the distributive lattice L with 0. Then every congruence relation of L has one and only one extension to B.*

Proof. The existence of an extension was proved in Theorem 3.6. By Theorems 3.7 and 3.10(i), the following statement implies the uniqueness of the extension:

If I and J are (ring) ideals of B with I ⊂ J, then there are elements a, b ∈ L, a ≠ b, such that a ≡ b (mod J) and a ≢ b (mod I).

Indeed, let $x \in J - I$. By Lemma 3, x can be represented in the form

$$x = x_0 + \cdots + x_{n-1}, \qquad x_0 \leq \cdots \leq x_{n-1}, \qquad x_0, \ldots, x_{n-1} \in L.$$

If n is odd, then $x_0 = x \cdot x_0 \leq x \in J$, and thus $x_0 \in J$; also, $x_0 + x_1 + x_2 = x \cdot x_2 \in J$, therefore

$$x_1 + x_2 = x_0 + (x_0 + x_1 + x_2) \in J.$$

Similarly,

$$x_3 + x_4, \ x_5 + x_6, \ \ldots \in J.$$

Since

$$x_0 + (x_1 + x_2) + (x_3 + x_4) + \cdots \in J - I,$$

we conclude that either $x_0 \in J - I$, or $x_{2i} + x_{2i+1} \in J - I$, for some $2i < n$. If n is even, then we obtain $x_0 + x_1, x_2 + x_3, \ldots \in J$ (by multiplying x by $x_1, x_3, \ldots$), and we conclude that, for some $2i < n$, $x_{2i} + x_{2i+1} \in J - I$.

Now if $x_{2i} + x_{2i+1} \in J - I$, then $x_{2i} \equiv x_{2i+1}$ (mod J), but $x_{2i} \not\equiv x_{2i+1}$ (mod I), $x_{2i}, x_{2i+1} \in L$. Finally, if $x_0 \in J - I$, then $x_0 \equiv 0$ (mod J) and $x_0 \not\equiv 0$ (mod I). □

Theorem 6.   If D_1 and D_2 are generalized Boolean lattices R-generated by a distributive lattice L with 0, then D_1 and D_2 are isomorphic.   —

Remark. For a distributive lattice L with 0, we shall denote by $B(L)$ a generalized Boolean lattice R-generated by L.

Proof. Let B be a free generalized Boolean lattice R-generated by L (as defined in Lemma 4). Let φ be a homomorphism of B onto D_1 such that φ is the identity on L (see Lemma 4(ii)). We want to show that φ is an isomorphism. Indeed, if φ is not an isomorphism, then the ideal kernel I of φ is not 0. Thus by Lemma 5, $a \equiv b$ (mod I), for some $a, b \in L$, $a \neq b$. This means that $a\varphi = b\varphi$, contrary to our assumptions. Similarly, there is an isomorphism ψ between B and D_2, Obviously, $\varphi^{-1}\psi$ is an isomorphism between D_1 and D_2. □

Corollary 7. *Let L_0 and L_1 be distributive lattices with 0 and let φ be a $\{0\}$-homomorphism of L_0 onto L_1. Then φ can be extended to a homomorphism of $B(L_0)$ onto $B(L_1)$.*

Proof. Let Θ be the congruence kernel of φ, and let $\overline{\Theta}$ be the extension of Θ to $B(L_0)$ (Lemma 5). Then $B(L_0)/\overline{\Theta}$ is a generalized Boolean lattice R-generated by $L_0/\Theta \cong L_1$. Thus

$$B(L_0)/\overline{\Theta} \cong B(L_1)$$

by Theorem 6, and using this, the proof of Corollary 7 becomes trivial. $\qquad\square$

Corollary 8. *Let L_0 be a $\{0\}$-sublattice of the distributive lattice L_1 with 0, and assume that if L_1 has 1, then L_0 is a $\{0,1\}$-sublattice of L_1. Let B denote the subalgebra of $B(L_1)$ R-generated by L_0. Then $B(L_0) \cong B$.*

Proof. The proof is trivial. $\qquad\square$

Let L_0 and L_1 be given as in Corollary 8. It is natural to ask under what conditions does L_0 R-generate $B(L_1)$. For $H \subseteq B(L_1)$, let $[H]_R$ denote the generalized Boolean sublattice of $B(L_1)$ R-generated by H. We can answer our query by determining $[L_0]_R \cap L_1$.

Lemma 9. *Let L_0 and L_1 be given as in Corollary 8. Then $L_1 \cap [L_0]_R$ is the smallest sublattice of L_1 containing L_0 that is closed under taking relative complements in L_1. Therefore L_0 R-generates $B(L_1)$ iff the smallest sublattice of L_1 containing L_0 and closed under relative complementation in L_1 is L_1 itself.*

Proof. It is obvious that $L_0 \subseteq L_1 \cap [L_0]_R$. If a, b, $c \in L_1 \cap [L_0]_R$, $d \in L_1$, and d is a relative complement of b in $[a, c]$, then $d = a + b + c \in L_1 \cap [L_0]_R$, since (see Figure 2) d is a relative complement of $a + b$ in the interval $[0, c]$. Thus $d \in L_1 \cap [L_0]_R$. Now suppose that L is a sublattice of L_1 containing L_0 and closed under relative complementation in L_1. If $x \in L_1 \cap [L_0]_R$, then, by Lemma 3, we can represent x as

$$x = a_0 + \cdots + a_{n-1}, \quad a_0, \dots, a_{n-1} \in L_0, \quad a_0 \leq \cdots \leq a_{n-1}.$$

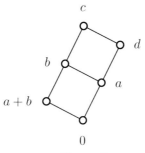

Figure 2

We prove $x \in L$ by induction on n. If $n = 1$, then $x = a_0 \in L_0 \subseteq L$. If $n = 2$, then x is the relative complement of a_0 in $[0, a_1]$, 0, a_0, $a_1 \in L_0$, thus $x \in L$.

If $n = 3$, then (see Figure 1) $x = a_0 + a_1 + a_2$ is the relative complement of a_1 in $[a_0, a_2]$, and so $x \in L$. Now let $n > 3$, and let $y \in L$ be proved for all $y = b_0 + \cdots + b_{k-1}$, $b_0, \ldots, b_{k-1} \in L_0$, $b_0 \leq \cdots \leq b_{k-1}$, and $k < n$. Note that $x \in L_1$ and $a_{n-3} \in L_0$ imply that

$$x a_{n-3} = a_0 + \cdots + a_{n-3} + a_{n-1} + a_{n-3} + a_{n-3} = a_0 + \cdots + a_{n-3} \in L_1$$

and

$$x \vee a_{n-3} = x + a_{n-3} + x a_{n-3}$$
$$= a_0 + \cdots + a_{n-1} + a_{n-3} + a_0 + \cdots + a_{n-3}$$
$$= a_{n-3} + a_{n-2} + a_{n-1} \in L_1.$$

By the induction hypothesis,

$$a_0 + \cdots + a_{n-3} \in L \text{ and } a_{n-3} + a_{n-2} + a_{n-1} \in L;$$

therefore, x is the relative complement in L_1 of an element (namely, of a_{n-3}) of L in an interval (namely, in $[a_0, + \cdots + a_{n-3}, a_{n-3} + a_{n-2} + a_{n-1}]$) in L, and so, by assumption, $x \in L$. Thus $L_1 \cap [L_0]_R \subseteq L$. □

In Theorem 1, we embedded L into $\mathrm{P}(X)$, which is a complete Boolean lattice. The question arises whether we can require this embedding to be complete, that is, to preserve arbitrary meets and joins, if they exist in L.

It is easy to see that not every complete distributive lattice has a complete embedding into a complete Boolean lattice.

Lemma 10 (J. von Neumann [1936]). *Let B be a complete Boolean lattice. Then B satisfies the Join Infinite Distributive Identity*

(JID) $$x \wedge \bigvee(x_i \mid i \in I) = \bigvee(x \wedge x_i \mid i \in I),$$

and its dual, the Meet Infinite Distributive Identity (MID).

Of course, (JID) is not an identity in the sense of Section I.4 but is only an infinitary analogue of an identity.

Proof. $x \wedge x_j \leq x$ and $x \wedge x_j \leq \bigvee(x_i \mid i \in I)$, for any $j \in I$; therefore, $x \wedge \bigvee(x_i \mid i \in I)$ is an upper bound for $\{ x \wedge x_i \mid i \in I \}$. Now let u be any upper bound, that is, $x \wedge x_i \leq u$, for all $i \in I$. Then

$$x_i = x_i \wedge (x \vee x') = (x_i \wedge x) \vee (x_i \wedge x') \leq u \vee x'.$$

Thus

$$x \wedge \bigvee(x_i \mid i \in I) \leq x \wedge (u \vee x') = (x \wedge u) \vee (x \wedge x') = x \wedge u \leq u,$$

showing that $x \wedge \bigvee(x_i \mid i \in I)$ is the least upper bound for $\{ x \wedge x_i \mid i \in I \}$. (MID) follows by duality. □

Corollary 11. *Any complete distributive lattice that has a complete embedding into a complete Boolean lattice satisfies both* (JID) *and* (MID).

Easy examples show that (JID) and (MID) need not hold in a complete distributive lattice.

Our task now is to show the converse of Corollary 11 (N. Funayama [1959]). The construction depends on a property of $B(L)$ and on Theorem I.6.4.

Lemma 12 (V. Glivenko [1929]). *Let L be a distributive lattice with 0. Then* Id L *is a pseudocomplemented lattice in which*

$$I^* = \{\, x \mid x \wedge i = 0, \ for \ all \ i \in I \,\}.$$

Let

$$\mathrm{S}(\mathrm{Id}\,L) = \{\, I^* \mid I \in \mathrm{Id}\,L \,\}.$$

If L is a Boolean lattice, then $\mathrm{S}(\mathrm{Id}\,L)$ *is a complete Boolean lattice and the map* $a \mapsto (a]$ *embeds L into* $\mathrm{S}(\mathrm{Id}\,L)$; *this embedding preserves all existing meets and joins.*

Proof. The first statement is trivial. Now let L be Boolean. It follows from Theorem I.6.4 that $\mathrm{S}(\mathrm{Id}\,L)$ is a Boolean lattice. Furthermore, it is easily seen that for any $X \subseteq \mathrm{Id}\,L$, the inf and sup of X in $\mathrm{S}(\mathrm{Id}\,L)$ are $\bigwedge X$ and $(\bigvee X)^{**}$, respectively, where $\bigwedge$ and $\bigvee$ are the meet and join of X in $\mathrm{Id}\,L$, respectively. Since

$$\bigwedge(\,(x] \mid x \in X\,) = (\inf X],$$

whenever $\inf X$ exists in L, the map $a \mapsto (a]$ preserves all existing meets in L. Observe that, for $x, a \in L$, $x \wedge a' = 0$ iff $x \leq a$, and so $(a] = (a']^* \in \mathrm{S}(\mathrm{Id}\,L)$. Now let $a = \sup X$ in L and set $I = (X] \ (= \bigvee(\,(x] \mid x \in X\,))$. To show that $x \mapsto (x]$ is join-preserving, we have to verify that $I^{**} = (a]$, or, equivalently, that $I^* = (a']$. Indeed, if $b \in I^*$, then $b \wedge x = 0$, for all $x \in I$, and thus $x \leq b'$. Therefore, $a = \sup X \leq b'$, proving $a' \geq b$, that is, $b \in (a']$. Conversely, let $b \in (a']$. Then $b' \geq a$; therefore, $b' \geq a = \sup X \geq x$, for all $x \in X$, and so $b \wedge x = 0$, for all $x \in X$. This shows that $b \in I^*$, proving that $I^* = (a']$. □

Lemma 13. *Let L be a complete lattice satisfying* (JID) *and* (MID). *Then the identity map is a complete embedding of L into $B(L)$.*

Proof. Let us write $a \in B(L)$ in the form

$$a = a_0 + \cdots + a_{n-1}, \quad a_0 \leq \cdots \leq a_{n-1}, \quad a_0, \ldots, a_{n-1} \in L.$$

If n is even, let us replace a_0 by $0 + a_0$; thus we can assume that n is odd. We claim that, for $x \in L$ and $a \in B(L)$, we have $x \leq a$ iff

$$x \wedge a_0 = x \wedge a_1,$$
$$x \wedge a_2 = x \wedge a_3,$$
$$\dots$$
$$x \wedge a_{n-3} = x \wedge a_{n-2},$$
$$x \leq a_{n-1}.$$

Indeed, let $x \leq a$. Then

$$xa_1 = xa_1(a_0 + \dots + a_{n-1}) = x(a_0 + a_1 + a_1 + \dots + a_1) = xa_0;$$

therefore, $x \wedge a_0 = x \wedge a_1$. Thus

$$x(a_2 + \dots + a_{n-1}) = (xa_0 + xa_1) + x(a_2 + \dots + a_{n-1}) = xa = x,$$

and so $x \leq a_2 + \dots + a_{n-1}$. Conversely, if $x \wedge a_0 = x \wedge a_1$ and $x \leq a_2 + \dots + a_{n-1}$, then

$$xa = xa_0 + xa_1 + x(a_2 + \dots + a_{n-1}) = x,$$

proving that $x \leq a$. A simple induction completes the proof of the claim.

Let $X \subseteq L$, $y = \sup X$ in L, and $a \in B(L)$. If $x \leq a$, for all $x \in X$, then the formulas of the preceding claim hold for all x and a and, by (JID), for y and a, proving that $y \leq a$. Thus $y = \sup X$ in $B(L)$. The dual argument, using (MID), completes the proof.   □

Theorem 14 (N. Funayama [1959]).   A complete lattice L has a complete embedding into a complete Boolean lattice iff L satisfies (MID) and (JID).   —

Proof. Combine Lemmas 10–13.   □

The representation for $a \in B(L)$ given in Lemma 3 is not unique in general; the only exception is when L is a chain. Since this case is of special interest, we shall investigate it in detail.

Repeating the definition, a Boolean lattice B is R-*generated by a chain* C with 0 iff $B = B(C)$. This concept is due to A. Mostowski and A. Tarski [1939] and can be extended to distributive lattices as follows.

A distributive lattice L with 0 is R-*generated by a chain* C ($\subseteq L$) with 0 iff C R-generates $B(L)$.

Lemma 15. *Let L be a distributive lattice with 0 and let C be a chain in L, $0 \in C$. Then C R-generates L iff L is the smallest sublattice of itself containing C and closed under formation of relative complements.*

Proof. Apply Lemma 9 to C. $\square$

An explicit representation of $B(C)$ is given as follows: for a chain C with 0, let $B[C]$ be the set of all subsets of C of the form

$$(a_0] + (a_1] + \cdots + (a_{n-1}], \quad 0 < a_0 < a_1 \leq \cdots \leq a_{n-1}, \quad a_0, \ldots, a_{n-1} \in C,$$

where $+$ is the symmetric difference. We consider $B[C]$ as a poset, partially ordered by $\subseteq$. We identify $a \in C$ with $(a]$, for $a \neq 0$, and 0 with $\varnothing$. Thus $C \subseteq B[C]$.

Lemma 16. $B[C]$ *is the generalized Boolean lattice R-generated by* C.

Proof. The proof is obvious, by construction and by Theorem 6. $\square$

Note that every nonempty element a of $B[C]$ can be represented in the form

$$a = (a_0] \cup (b_1, a_1] \cup \cdots \cup (b_{n-1}, a_{n-1}], \quad 0 < a_0 < b_1 < a_1 < \cdots < b_{n-1} < a_{n-1},$$

where the union is disjoint union and the first term $(a_0]$ may be missing. An element of the form $(x, y]$ is nothing but $x + y$. Thus,

$$a = a_0 + b_1 + a_1 + \cdots + b_{n-1} + a_{n-1},$$

and so we conclude:

Corollary 17. *In* $B(C)$, *every nonzero element* a *has a unique representation in the form*

$$a = a_0 + a_1 + \cdots + a_{n-1}, \quad 0 < a_0 < a_1 < \cdots < a_{n-1}, \quad a_0, \ldots, a_{n-1} \in C.$$

The following results show that many distributive lattices can be R-generated by chains.

Lemma 18. *Every finite Boolean lattice* B *can be R-generated by a chain; in fact,* $B = [C]_R$, *for any maximal chain* C *of* B.

Proof. Let B_1 be the subalgebra of B R-generated by C. Using the notation of Corollary 1.14, the length of C equals $|\mathrm{J}(B)|$; also, the length of C equals $|\mathrm{J}(B_1)|$; thus $|\mathrm{J}(B)| = |\mathrm{J}(B_1)| = n$. We conclude that both B and B_1 have 2^n elements, proving that $B = B_1$. $\square$

Corollary 19. *Every finite distributive lattice* L *can be R-generated by a chain, in fact, by any maximal chain of* L.

Proof. Let C be a maximal chain in L and let $B = B(L)$. Then $|\mathrm{J}(L)| = |\mathrm{J}(B)|$. By Corollary 1.14, C is maximal in B. Thus, $B = B(C) \supseteq L$. $\square$

Theorem 20.   Let L be a countable distributive lattice with 0. Then L can be R-generated by a chain.   —

Proof. Let $L = \{a_0, a_1, a_2, \dots, a_n, \dots\}$, $a_0 = 0$, and let L_n be the sublattice of L generated by $a_0, \dots, a_n$. Let A_0 be a maximal chain of L_0, and, inductively, let A_n be a maximal chain of L_n containing A_{n-1}. Set $A = \bigcup(A_i \mid i < \omega)$. Obviously, $0 \in A$. We claim that A R-generates L. Take $a \in B(L)$;

$$a = x_0 + \cdots + x_{m-1}, \quad x_0, \ \dots, \ x_{m-1} \in L.$$

$L = \bigcup(L_i \mid i < \omega)$; thus, for some n, $x_0, \ \dots, \ x_{m-1} \in L_n$, and so $a \in B(L_n)$. Since L_n is finite, we get $B(L_n) = B(A_n)$; therefore $a \in B(A_n) \subseteq B(A)$, proving that $L \subseteq B(A)$.   □

Corollary 21. *The correspondence $C \mapsto B(C)$ maps the class of countable chains with 0 onto the class of countable generalized Boolean lattices. Under this correspondence, $\{0\}$-subchains and $\{0\}$-homomorphic images correspond to $\{0\}$-subalgebras and $\{0\}$-homomorphic images.*

Note, however, that $C \cong C'$ is *not* implied by $B(C) \cong B(C')$.

Much is known about countable chains. Utilizing the previous results, such information can be used to prove results on countable generalized Boolean lattices. To help distinguish an important class of chains, we introduce the concept of a prime interval. An interval $[a, b]$ is *prime* iff $a \prec b$.

Lemma 22. *Every countable chain C can be embedded in the chain $\mathbb{Q}$ of rational numbers. Every countable chain not containing any prime interval is isomorphic to one of the intervals $(0, 1)$, $[0, 1)$, $(0, 1]$, and $[0, 1]$ of $\mathbb{Q}$.*

Proof. Let $C = \{x_0, x_1, \dots, x_{n-1}, \dots\}$. We define the map φ inductively as follows: Pick an arbitrary $r_0 \in \mathbb{Q}$ and set $x_0\varphi = r_0$. If $x_0\varphi, \ \dots, \ x_{n-1}\varphi$ have already been defined, we define $x_n\varphi$ as follows: Let

$$L_n = \bigcup((x_i\varphi] \mid x_i < x_n, \ i < n),$$
$$U_n = \bigcup([x_i\varphi) \mid x_i > x_n, \ i < n)$$

($L_n = \varnothing$ or $U_n = \varnothing$ is possible). Note that if $L_n \neq \varnothing$, then it has a greatest element l_n, and if $U_n \neq \varnothing$, then it has a smallest element u_n. If both are nonempty, then $l_n < u_n$. In any case, we can choose an $r_n \in \mathbb{Q}$ satisfying $r_n \notin L_n \cup U_n$. We set $x_n\varphi = r_n$. Obviously, φ is an embedding.

The second part of the proof reduces to the following statement:

Let C and D be bounded countable chains with no prime intervals. Then $C \cong D$.

To prove this, let $C = \{c_0, c_1, \dots\}$ and $D = \{d_0, d_1, \dots\}$. We define two maps: $\varphi \colon C \to D$ and $\psi \colon D \to C$. Let us assume that $c_0 = 0$, $c_1 = 1$, and that $d_0 = 0$, $d_1 = 1$. For each $n < \omega$, we shall define inductively finite chains $C^{(n)}$, $D^{(n)}$ ($C^{(n)} \subseteq C$, $D^{(n)} \subseteq D$) and an isomorphism $\varphi_n \colon C^{(n)} \to D^{(n)}$ with inverse $\psi_n \colon D^{(n)} \to C^{(n)}$. Set $C^{(0)} = \{c_0, c_1\} = \{0, 1\}$, $D^{(0)} = \{d_0, d_1\} = \{0, 1\}$, and $i\varphi_0 = i$, $i\psi_0 = i$, for $i = 0$, 1. Given $C^{(n)}$, $D^{(n)}$, φ_n, ψ_n, and n even, let k be the smallest integer with $c_k \notin C^{(n)}$. Define $u_k = \bigwedge([c_k) \cap C^{(n)})$ and $l_k = \bigvee((c_k] \cup C^{(n)})$. Then $l_k < c_k < u_k$, and so $l_k\varphi_n < u_k\varphi_n$. Since D contains no prime intervals, we can choose a $d \in D$ satisfying $l_k\varphi_n < d < u_k\varphi_n$. Since ψ_n is isotone, $d \notin D^{(n)}$. Define $C^{(n+1)} = C^{(n)} \cup \{c_k\}$, $D^{(n+1)} = D^{(n)} \cup \{d\}$, φ_{n+1} restricted to $C^{(n)}$ to be φ_n, and $c_k\varphi_{n+1} = d$, ψ_{n+1} restricted to $D^{(n)}$ to be ψ_n and $d\psi_{n+1} = c_k$. If n is odd, then we proceed in a similar way, but we interchange the role of C and D, $C^{(n)}$ and $D^{(n)}$, φ_n and ψ_n, respectively.

Finally, put $\varphi = \bigcup(\varphi_n \mid n < \omega)$. Clearly,

$$C = \bigcup(C^{(n)} \mid n < \omega),$$
$$D = \bigcup(D^{(n)} \mid n < \omega),$$

and φ is the required isomorphism. $\qquad\qquad\qquad\qquad\qquad\qquad\square$

Corollary 23. *Up to isomorphism, there is exactly one countable Boolean lattice with no atoms and exactly one countable generalized Boolean lattice with no atoms and no unit element, $B(\mathbb{Q})$.*

Proof. Take the rational intervals $[0, 1]$ and $(0, 1)$. The generalized Boolean lattices in question are $B([0, 1])$ and $B((0, 1))$. This follows from the observation that $[a, b]$ is a prime interval in C iff $a + b$ is an atom in $B(C)$. The results follow from Lemmas 16 and 22 and Theorem 20. $\qquad\qquad\qquad\qquad\qquad\square$

Theorem 24.   Let B be a countable Boolean algebra. Then B has either $\aleph_0$ or $2^{\aleph_0}$ prime ideals. $\qquad\qquad\qquad\qquad\qquad\qquad\qquad\qquad\quad$ —

Remark. This is obvious if we assume the Continuum Hypothesis. Interestingly, we can give a proof without it.

Proof. For a Boolean algebra B and an ideal I of B, we shall write B/I for $B/\Theta[I]$. If J is an ideal of B with $J \supseteq I$, then $J/I = \{\,[x]\Theta[I] \mid x \in J\,\}$ is an ideal of B/I. (J/I is the usual notation in ring theory.) Let B be a Boolean algebra. We define the ideals I_γ by transfinite induction. $I_0 = (0]$; I_1 is the ideal generated by the atoms of B; given I_γ, let I be the ideal of B/I_γ generated by the atoms of B/I_γ and let $\varphi\colon x \mapsto x + I_\gamma$ be the homomorphism of B onto B/I_γ; we set $I_{\gamma+1} = I\varphi^{-1}$. Finally, if γ is a limit ordinal, set $I_\gamma = \bigcup(I_\delta \mid \delta < \gamma)$. The *rank* of B is defined to be the smallest ordinal α such that $I_\alpha = I_{\alpha+1}$. Obviously, the cardinality of α is at most $|B|$.

Claim 1. *Let B be countable. If $I_\alpha \neq B$, then $|\mathcal{P}(B)| = 2^{\aleph_0}$.*

Indeed, if $I_\alpha \neq B$, then B/I_α has no atoms, and thus $B/I_\alpha \cong B(C)$, where C is the rational interval $[0, 1]$. By Lemma 5 (see Exercise 27), $|\mathcal{P}(B/I_\alpha)| = |\mathcal{P}(C)| = |\operatorname{Id} C| = 2^{\aleph_0}$.

Claim 2. *Let B be countable. If $I_\alpha = B$, then $|\mathcal{P}(B)| = \aleph_0$.*

Indeed, for $\gamma < \alpha$, let $\mathcal{P}_\gamma(B)$ be the set of prime ideals P of B for which $I_\gamma \subseteq P$, $I_{\gamma+1} \nsubseteq P$. Since α is finite or countable, it suffices to show that $|\mathcal{P}_\gamma(B)| = \aleph_0$. If $P \in \mathcal{P}_\gamma(B)$, then, by Corollary 1.16 and Theorem 1.22, we have $P \vee I_{\gamma+1} = B$. It follows that for $P, Q \in \mathcal{P}_\gamma(B)$, $P \neq Q$, we have $P \cap I_{\gamma+1} \neq Q \cap I_{\gamma+1}$. Thus $P \mapsto (P \cap [I_{\gamma+1}]_R)/I_\gamma$ is a one-to-one correspondence of $\mathcal{P}_\gamma(B)$ into (in fact, onto) $\mathcal{P}([I_{\gamma+1}]_R/I_\gamma)$; but $[I_{\gamma+1}]_R/I_\gamma$ is just the generalized Boolean lattice of all finite subsets of a countable set. Therefore, $|\mathcal{P}_\gamma(B)| = \aleph_0$. $\square$

In order to avoid giving the impression that most Boolean algebras can be R-generated by chains, we state:

Lemma 25. *Let B be a complete Boolean algebra R-generated by a chain C with 0. Then B is finite.*

Proof. Let $B = [C]_R$ and let the chain C be infinite. We can assume that C contains a subchain

$$\{x_0, x_1, \dots, x_n, \dots\}, \quad 0 < x_0 < x_1 < \cdots < x_n < \cdots$$

(or dually, in which case the complements form an increasing sequence). Then we define

$$a_n = x_0 + x_1 + \cdots + x_{2n} + x_{2n+1},$$

for each $n < \omega$. We claim that $\bigvee(\, a_i \mid i < \omega\,)$ does not exist. Indeed, let a be an upper bound for $\{\, a_i \mid i < \omega\,\}$. By the remarks immediately following Lemma 16, we can represent each a_n by a set

$$(x_0, x_1] \cup (x_2, x_3] \cup \cdots \cup (x_{2n}, x_{2n+1}],$$

and we can represent a in the form

$$a = (b_0] \cup (b_1, b_2] \cup \cdots \cup (b_{m-2}, b_{m-1}],$$

where $0 < b_0 < b_1 < \cdots < b_{m-1}$, $b_i \in C$, for $i < m$, and the first term, $(b_0]$, may be missing. Since a contains each a_n, there must exist an n and a $j < m$ such that both $(x_{2n}, x_{2n+1}]$ and $(x_{2n+2}, x_{2n+3}]$ are contained in $(b_{j-1}, b_j]$ or in $(b_0]$. Therefore, the interval $(x_{2n+1}, x_{2n+2}]$ can be deleted from a, and it will still contain all the a_i, that is, $a + x_{2n+1} + x_{2n+2}$ is an upper bound for $\{\, a_i \mid i < \omega\,\}$, and $a + x_{2n+1} + x_{2n+2} < a$. We conclude that $\{\, a_i \mid i < \omega\,\}$ does not have a least upper bound. $\square$

Next we consider which chains with 0 can be R-generating chains of a given distributive lattice.

Definition 26. Let L be a distributive lattice with 0 and let C be a chain in L, $0 \in C$. The chain C is called *strongly maximal in* L iff, for any homomorphism φ of L onto a distributive lattice L_1, the chain $C\varphi$ is maximal in L_1.

Lemma 27. *Let L be a distributive lattice with 0 and let C be a chain in L, $0 \in C$. If L is R-generated by C, then C is maximal in L.*

Proof. If C is not maximal in L, then we can find $a \in L$, $a \notin C$, such that $C \cup \{a\}$ is a chain. Write

$$a = a_0 + a_1 + \cdots + a_{n-1},$$

with $0 < a_0 < a_1 < \cdots < a_{n-1}$ and $a_i \in C$, for $i < n$. Since $a \notin C$, $n > 1$. Now

$$a \wedge a_0 = a_0 + a_0 + \cdots + a_0,$$

which is a_0 if n is odd and 0 if n is even. But $a_0 \neq a$ and $0 \neq a$, so, since a and a_0 are comparable, $a \wedge a_0 = a_0$ and n is odd. Then

$$a \wedge a_1 = a_0 + a_1 + \cdots + a_1 = a_0,$$

contradicting the comparability of a and a_1. $\qquad\square$

The converse of Lemma 27 is false. The following theorem settles the matter.

Theorem 28.   Let L be a distributive lattice with 0 and let C be a chain in L, $0 \in C$. Then C R-generates L iff C is strongly maximal in L.   —

Proof. If C R-generates L, then $C\varphi$ R-generates $L\varphi$, for any onto homomorphism φ. By Lemma 27, $C\varphi$ is maximal in $L\varphi$, so C is strongly maximal in L.

Next assume that C is strongly maximal in L but does not R-generate L. Without any loss of generality, we can assume that L and C have a greatest element. (Otherwise, add one. Then $C \cup \{1\}$ is strongly maximal in $L \cup \{1\}$ but does not R-generate $L \cup \{1\}$.) Let $B_1 = B(L)$ and let $B_0 = [C]_R$. By hypothesis, $B_0 \neq B_1$, so there exists an $a \in B_1 - B_0$. We claim that there exist prime ideals $P_1 \neq P_2$ of B_1 with $B_0 \cap P_1 = B_0 \cap P_2$. With $I = ((a] \cap B_0]$ and $D = [a)$, we have $I \cap D = \varnothing$, so, by Theorem 1.15, there is a prime ideal P_1 such that $I \subset P_1$ and $P_1 \cap D = \varnothing$. Then let $I_1 = (a]$ and $D_1 = [B_0 - P_1)$. Since $(a] \cap B_0 \subseteq P_1$, it follows that $I_1 \cap D_1 = \varnothing$. Let P_2 be a prime ideal with $I_1 \subseteq P_2$, $P_2 \cap D_1 = \varnothing$. Then $a \in P_2 - P_1$, so $P_1 \neq P_2$. Because $P_2 \cap (B_0 - P_1) = \varnothing$, $P_2 \cap B_0 \subseteq P_1 \cap B_0$. Then by Theorem 1.22 (prime ideals of a Boolean lattice are unordered), $P_1 \cap B_0 = P_2 \cap B_0$, proving our claim.

Now (again by using Theorem 1.22) we can map B_1 onto C_2^2 by a homomorphism ψ: For $x \in P_1 \cap P_2$, $x\psi = \langle 0, 0 \rangle$; for $x \in P_1 - P_2$, $x\psi = \langle 0, 1 \rangle$; for $x \in P_2 - P_1$, $x\psi = \langle 1, 0 \rangle$; for $x \notin P_1 \cup P_2$, $x\psi = \langle 1, 1 \rangle$. Since $C\psi = \{\langle 0, 0 \rangle, \langle 1, 1 \rangle\}$ is not maximal, we conclude that C is not strongly maximal in L. $\qquad\square$

Corollary 29. *Let C and D be strongly maximal chains of the distributive lattice L with 0. Then $|C| = |D|$ and $|\operatorname{Id} C| = |\operatorname{Id} D|$.*

Proof. If L is finite, these conclusions follow from Corollary 1.14. If $|L|$ is infinite, then $[C] = [D] = B(L)$, and so $|C| = |D| = |L|$. Also, by Lemma 5, $|\mathcal{P}(C)| = |\mathcal{P}(B(L))| = |\mathcal{P}(D)|$, and $\mathcal{P}(C) = \operatorname{Id} C$, $\mathcal{P}(D) = \operatorname{Id} D$; hence the second statement. □

Corollary 29 is the strongest known extension of Corollary 1.14 to the infinite case. The second part of Corollary 29 is from G. Grätzer and E. T. Schmidt [1957].

Boolean algebras generated by chains were first investigated by A. Mostowski and A. Tarski [1939]. Theorem 20 for Boolean lattices and Theorem 24 were communicated to the author by J. R. Büchi. These results have been known for some time in topology (via the Stone topological representation theorem, see Section 5). Some of the other results appeared first in G. Grätzer [1971].

Exercises

1. Give a detailed proof of Lemma 4.

2. Try to describe the most general situation to which the idea of the proof of Theorem I.5.5 (Theorem I.5.24) could be applied.

3. Show that Lemma 5 does not remain valid if the word "generalized" is omitted.

4. Find necessary and sufficient conditions on a distributive lattice L in order that L have a Boolean extension B to which every congruence of L has exactly one extension.

5. Let L be a distributive lattice with 0 and define L_1 to be the lattice L if L has a unit element; let L_1 be L with a unit added if L does not have a unit element. The *Boolean lattice $B[L]$ R-generated by L* is defined as $B(L_1)$. Show that if B is any Boolean lattice, containing L as a sublattice, and B is generated by L under $\wedge$, $\vee$, and $'$, then B is isomorphic to the Boolean lattice R-generated by L.

6. Work out Corollaries 7 and 8 for the Boolean lattice R-generated by L.

*7. The *Complete Infinite Distributive Identity* is (for I, $J \neq \varnothing$):

$$\bigwedge(\bigvee(a_{ij} \mid j \in J) \mid i \in I) = \bigvee(\bigwedge(a_{i\,i\varphi} \mid i \in I) \mid \varphi \colon I \to J).$$

Show that this holds in a complete Boolean lattice B iff it is atomic (A. Tarski [1930]). (Hint: apply the identity to $\bigwedge(a \vee a' \mid a \in B) = 1$.)

8. Prove that the Complete Infinite Distributive Identity is selfdual for Boolean lattices.

9. For a subset A of a lattice L, set

$$A^u = \{\, x \mid x \in L,\ x \text{ is an upper bound of } A \,\},$$
$$A^l = \{\, x \mid x \in L,\ x \text{ is a lower bound of } A \,\}.$$

Prove that this sets up a Galois connection and therefore $(A^u)^l \supseteq A$, $(A^l)^u \supseteq A$, $A^u = ((A^u)^l)^u$, and $A^l = ((A^l)^u)^l$.

10. Call an ideal I of lattice L *normal* iff $I = (I^u)^l$. Show that every principal ideal is normal.

11. Let $\mathrm{Id}_N L$ denote the set of all normal ideals of L. Show that $\mathrm{Id}_N L$ is a complete lattice but that it is not necessarily a sublattice of $\mathrm{Id}_0 L$.

12. Show that the map: $x \mapsto (x]$ is an embedding of L into $\mathrm{Id}_N L$, preserving all meets and joins that exist in L. ($\mathrm{Id}_N L$ is called the *MacNeille completion* of L; see H. M. MacNeille [1937].)

13. Let B be a Boolean lattice and let I be an ideal of B. Show that I is normal iff $I = I^{**}$ (for the concepts, see Exercise 10 and Lemma 12).

14. Prove that the Boolean lattice $S(\mathrm{Id}\, L)$ of Lemma 12 is the MacNeille completion of the Boolean lattice L.

*15. Show that the MacNeille completion of a distributive lattice need not even be modular.

16. Let L be a distributive algebraic lattice. Show that L satisfies the Join Infinite Distributive Identity. (Thus, for any lattice K, $\mathrm{Con}\, K$ satisfies (JID).)

17. Let L be a distributive lattice, $a_i, b_i \in L$, for $i < \omega$, and

$$[a_0, b_0] \supset [a_1, b_1] \supset \cdots$$

Define

$$\Theta = \bigvee(\,\Theta(a_0, a_i) \vee \Theta(b_0, b_i) \mid i < \omega\,).$$

Show that

$$\Theta \vee \bigwedge(\,\Theta(a_i, b_i) \mid i < \omega\,) \neq \bigwedge(\,\Theta \vee \Theta(a_i, b_i) \mid i < \omega\,).$$

18. Use Exercise 17 to show that, for a distributive lattice L, the Meet Infinite Distributive Identity holds in $\mathrm{Con}\, L$ iff every interval in L is finite (G. Grätzer and E. T. Schmidt [1958b]).

19. For a chain C, introduce and describe $B(C)$ using Corollary 17.

*20. Prove the converse of Lemma 18: If every maximal chain R-generates the Boolean lattice B, then B is finite.

21. Why is it not possible to use transfinite induction to extend Theorem 20 to the uncountable case?

22. Let C be a chain with 0 and 1 and let $a \in C$. Define a new order on C: For x, $y < a$, and for $a \le x$, y, let $x \le y$ retain its meaning; for $x < a \le y$, let $y < x$; let C' be the set C with the new order. Then C' is a chain, and $B(C) \cong B(C')$, but, in general, $C \cong C'$ does not hold.

23. Describe a countable family of pairwise nonisomorphic countable Boolean algebras.

24. Give an example of a bounded distributive lattice L with a maximal chain C such that C is not maximal in $B(L)$ (G. W. Day [1970]).

25. Let L_0 be the $[0, 1]$ rational interval and let L_1 be the $[0, 1]$ real interval. Let $C = \{\, \langle x, x \rangle \mid 0 \le x \le 1,\ x \text{ rational} \,\}$. Then C is a maximal chain in $L_0 \times L_1$. Show that C is not strongly maximal (G. Grätzer and E. T. Schmidt [1957]).

26. In $L_0 \times L_1$ of Exercise 25, find maximal chains of cardinality $\aleph_0$ and $2^{\aleph_0}$. What are the cardinalities of strongly maximal chains?

27. Let L be a distributive lattice with 0 and let $B = B(L)$. Show that

$$P \mapsto P \cap L, \text{ for } P \in \mathcal{P}(B)$$

is a one-to-one correspondence between the prime ideals of L and B.

*28. Let A be an infinite set, $B = \mathrm{P}(A)$. Prove that B has maximal chains of cardinality $|A|$ and $2^{|A|}$.

29. Construct an example in which the sequence of ideals I_γ of Theorem 24 does not terminate in finitely many steps.

30. Let C be the $[0, 1]$ interval of the rational numbers. Show that $B(C)$ is $\mathrm{F_B}(\aleph_0)$.

5.   Topological Representation

The poset $\mathcal{P}(L)$ of prime ideals does give a great deal of information about the distributive lattice L, but obviously it does not characterize L. For instance, for a countably infinite Boolean algebra L, $\mathcal{P}(L)$ is an unordered set of cardinality $\aleph_0$ or $2^{\aleph_0}$, whereas there are surely more that two such Boolean algebras up to isomorphism.

Therefore, it is necessary to endow $\mathcal{P}(L)$ with more structure if we want it to characterize L. M. H. Stone [1937] endowed $\mathcal{P}(L)$ with a topology (see also L. Rieger [1949]). In this section, we shall discuss his approach in a slightly more general but, in our opinion, more natural framework.

A join-semilattice L is called *distributive* iff $a \le b_0 \vee b_1$ (a, b_0, $b_1 \in L$) implies the existence of a_0, $a_1 \in L$, with $a_0 \le b_0$, $a_1 \le b_1$, and $a = a_0 \vee a_1$ (see Figure 1). Note that a_0 and a_1 need not be unique.

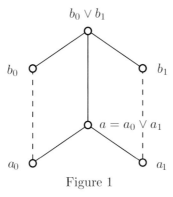

Figure 1

Some elementary properties of a distributive join-semilattice are as follows:

Lemma 1.

(i) *If $\langle L; \wedge, \vee \rangle$ is a lattice, then the join-semilattice $\langle L; \vee \rangle$ is distributive iff the lattice $\langle L; \wedge, \vee \rangle$ is distributive.*

(ii) *If a join-semilattice L is distributive, then, for any a, $b \in L$, there is a $d \in L$ with $d \le a$ and $d \le b$. Consequently, $\operatorname{Id} L$ is a lattice.*

(iii) *A join-semilattice L is distributive iff $\operatorname{Id} L$, as a lattice, is distributive.*

Proof.

(i) If $\langle L; \wedge, \vee \rangle$ is distributive, and $a \le b_0 \vee b_1$, then with $a_0 = a \wedge b_0$, $a_1 = a \wedge b_1$, we obtain that $a = a_0 \vee a_1$. Conversely, if $\langle L; \vee \rangle$ is distributive, and the lattice L contains a diamond or a pentagon $\{o, a, b, c, i\}$, then $a \le b \vee c$, but a cannot be represented as $a = a_0 \vee a_1$ with $a_0 \le b$ and $a_1 \le c$, a contradiction.

(ii) $a \le a \vee b$, thus $a = a_0 \vee b_0$, where $a_0 \le a$, $b_0 \le b$. Since, in addition, $b_0 \le a$, b_0 is a lower bound for a and b.

(iii) First we observe that, for I, $J \in \text{Id }L$,

$$I \vee J = \{ i \vee j \mid i \in I, \; j \in J \}$$

follows from the assumption that the join-semilattice L is distributive. Therefore, the distributivity of $\text{Id }L$ can be easily proved. Conversely, if $\text{Id }L$ is distributive and $a \leq b_0 \vee b_1$, then

$$(a] = (a] \wedge ((b_0] \vee (b_1]) = ((a] \wedge (b_0]) \vee ((a] \wedge (b_1]),$$

and so $a = a_0 \vee a_1$, $a_0 \in (b_0]$, $a_1 \in (b_1]$, which is distributivity for L. $\square$

A subset D of a join-semilattice L is called a *dual ideal* iff $a \in D$ and $x \geq a$ imply that $x \in D$, and a, $b \in D$ implies that there exists a lower bound $d \in D$ of a and b. An ideal I of L is *prime* iff $I \neq L$ and $L - I$ is a dual ideal. Again, let $\mathcal{P}(L)$ denote the set of all prime ideals of L.

Lemma 2. *Let I be an ideal and let D be a nonempty dual ideal of a distributive join-semilattice L. If $I \cap D = \varnothing$, then there exists a prime ideal P of L with $P \supseteq I$ and $P \cap D = \varnothing$.*

Proof. The proof is a routine modification of the proof of Theorem 1.15. $\square$

In the rest of this section, unless stated otherwise, let L stand for a distributive join-semilattice with zero.

In $\mathcal{P}(L)$, sets of the form

$$r(a) = \{ P \mid a \notin P \}$$

represent the elements of L. We will make all these sets open.

Let $\mathcal{S}(L)$ denote the topological space defined on $\mathcal{P}(L)$ by postulating that the sets of the form $r(a)$ be a subbase for the open sets; we shall call $\mathcal{S}(L)$ the *Stone space* of L. (Exercises 1–22 review all the topological concepts used in this section.)

Lemma 3. *For an ideal I of L, define*

$$r(I) = \{ P \mid P \in \mathcal{S}(L), \; P \not\supseteq I \}.$$

Then $r(I)$ is open in $\mathcal{S}(L)$. Conversely, every open set U of $\mathcal{S}(L)$ can be uniquely represented as $r(I)$, for some ideal I of L.

Proof. We simply observe that

$$r(I) \cap r(J) = r(I \wedge J),$$

$$r\left(\bigvee (I_j \mid j \in K) \right) = \bigcup (r(I_j) \mid j \in K),$$

and $r(a]) = r(a)$, from which it follows that the $r(I)$ form the smallest collection of sets closed under finite intersection and arbitrary union containing all the $r(a)$. Observe that $a \in I$ iff $r(a) \subseteq r(I)$. Thus $r(I) = r(J)$ iff $a \in I$ is equivalent to $a \in J$, that is, iff $I = J$. $\square$

Lemma 4. *The subsets of $\mathcal{S}(L)$ of the form $r(a)$ can be characterized as compact open sets.*

Proof. Indeed, if a family of open sets $\{\, r(I_k) \mid k \in K \,\}$ is a cover for $r(a)$, that is,

$$r(a) \subseteq \bigcup (\, r(I_k) \mid k \in K\,) = r(\bigvee (\, I_k \mid k \in K\,)),$$

then $a \in \bigvee (\, I_k \mid k \in K\,)$. This implies that $a \in \bigvee (\, I_k \mid k \in K_0\,)$, for some finite $K_0 \subseteq K$, proving that $r(a) \subseteq \bigcup (\, r(I_k) \mid k \in K_0\,)$. Thus $r(a)$ is compact. Conversely, if I is not principal, then

$$r(I) \subseteq \bigcup (\, r(a) \mid a \in I\,),$$

but

$$r(I) \nsubseteq \bigcup (\, r(a) \mid a \in I_0\,),$$

for any finite $I_0 \subseteq I$.   □

From Lemma 4, we immediately conclude:

Theorem 5.   The Stone space $\mathcal{S}(L)$ determines L up to isomorphism.   —

If we want to use Stone spaces to construct new ones in order that new distributive lattices can be constructed from given ones, then we have to know what Stone spaces look like. Stone spaces are characterized in Theorem 8. To prepare for the proof of Theorem 8, we prove Lemma 6.

Let P be a prime ideal of L. Then P is represented as an element of $\mathcal{S}(L)$ and also by $r(P)$. The connection between P and $r(P)$ is given in Lemma 6 and is illustrated by Figure 2.

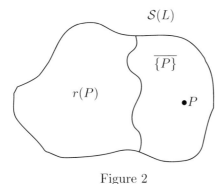

Figure 2

Lemma 6. *For every prime ideal P of L,*

$$\overline{\{P\}} = \mathcal{S}(L) - r(P),$$

where $\overline{\{P\}}$ is the topological closure of $\{P\}$.

Proof. By the definition of closure,

$$\overline{\{P\}} = \{\, Q \mid Q \in r(a) \text{ implies that } P \in r(a) \,\} = \{\, Q \mid Q \supseteq P \,\}$$
$$= \mathcal{S}(L) - \{\, Q \mid Q \not\supseteq P \,\} = \mathcal{S}(L) - r(P). \qquad \square$$

Corollary 7. *If $P \neq Q$, then $\overline{\{P\}} \neq \overline{\{Q\}}$; in other words, $\mathcal{S}(L)$ is a T_0-space.*

Proof. Combine Lemmas 3 and 6. $\qquad \square$

Lemma 6 also shows that if P is a prime ideal, then $\mathcal{S}(L) - r(P)$ must be the closure of a singleton. In other words:

(C) If U is an open set with the property that, for the compact open sets U_0 and U_1, $U_0 \cap U_1 \subseteq U$ implies that $U_0 \subseteq U$ or $U_1 \subseteq U$, then $\mathcal{S}(L) - U = \overline{\{P\}}$, for some element P.

Now we can state the characterization theorem.

Theorem 8.   The Stone space $\mathcal{S}$ of a distributive join-semilattice with zero can be characterized (up to homeomorphism) by the following two properties:

(S1) $\mathcal{S}$ is a T_0-space in which the compact open sets form a base for the open sets.

(S2) If F is a closed set in $\mathcal{S}$, $\{\, U_k \mid k \in K \,\}$ is a dually directed family (that is, $K \neq \varnothing$ and, for any $k, l \in K$, there exists a $t \in K$ such that $U_t \subseteq U_k \cap U_l$) of compact open sets of $\mathcal{S}$, and $U_k \cap F \neq \varnothing$, then $\bigcap (\, U_k \mid k \in K \,) \cap F \neq \varnothing$.

Remark. The meaning of condition (S1) is clear. Condition (S2) is a complicated way of ensuring that (C) holds and that Lemma 2 holds for the join-semilattice of compact open sets of $\mathcal{S}(L)$.

Proof. To show that (S1) holds for $\mathcal{S}(L)$, we have to verify that the $r(a)$, $a \in L$, form a base (not only a subbase) for the opens sets of $\mathcal{S}(L)$. In other words, for $a, b \in L$ and $P \in r(a) \cap r(b)$, we have to find a $c \in L$ with $P \in r(c)$ and $r(c) \subseteq r(a) \cap r(b)$. By assumption, $a \notin P$ and $b \notin P$; thus, since P is prime, there exists a $c \in L$, $c \notin P$, $c \leq a$, $c \leq b$. Then $P \in r(c)$, $r(c) \subseteq r(a)$, and $r(c) \subseteq r(b)$, as required. To verify (S2) for $\mathcal{S}(L)$, let $F = \mathcal{S}(L) - r(I)$ and $U_k = r(a_k)$. Thus $F = \{\, P \mid P \supseteq I \,\}$ and $U_k = \{\, P \mid a_k \notin P \,\}$. The assumptions on the a_k mean that $D = \{\, x \mid x \geq a_k, \text{ for some } k \in K \,\}$ is a dual ideal; since $U_k \cap F \neq \varnothing$, we have $r(a_k) \nsubseteq r(I)$; that is, $a_k \notin I$, showing that $D \cap I = \emptyset$. Therefore, by Lemma 2, there exists a prime ideal P with $P \supseteq I$ and $P \cap D = \emptyset$. Then $a_k \notin P$, and so $P \in r(a_k)$, for all $k \in K$. Also $P \supseteq I$, thus $P \notin r(I)$, and so $P \in F$, proving that $P \in F \cap \bigcap(U_k \mid k \in K)$, verifying (S2).

Conversely, let $\mathcal{S}$ be a topological space satisfying conditions (S1) and (S2) of the theorem. Let L be the set of compact open sets of $\mathcal{S}$. Obviously, $\emptyset \in L$ and if A, $B \in L$, then $A \cup B \in L$, and thus L is a join-semilattice with zero. Let

$$A \subseteq B_0 \cup B_1, \quad \text{with } A,\ B_0,\ B_1 \in L.$$

Then $A \cap B_i$ is open, and therefore

$$A \cap B_i = \bigcup(A_j^i \mid j \in J_i), \quad i = 0,\ 1,$$

where the A_j^i are compact open sets. Since

$$A = (A \cap B_0) \cup (A \cap B_1) \subseteq \bigcup(A_j^i \mid j \in J_0 \cup J_1,\ i = 0,\ 1),$$

by the compactness of A, we get

$$A \subseteq \bigcup(A_j^i \mid j \in J_0^* \text{ or } j \in J_1^*),$$

where J_i^* is a finite subset of J_i, $i = 0$, 1. Set

$$A_i = \bigcup(A_j^i \mid j \in J_i^*), \quad i = 0,\ 1.$$

Then A_0, $A_1 \in L$, $A = A_0 \cup A_1$, and $A_0 \subseteq B_0$, $A_1 \subseteq B_1$, showing that L is distributive.

It follows from (S1) that the open sets of $\mathcal{S}$ are uniquely associated with ideals of L: for an ideal I of L, let

$$U(I) = \bigcup(a \mid a \in I)$$

(keep in mind that an $a \in L$ is a subset of $\mathcal{S}$, as illustrated in Figure 3). Note that, for $a \in L$, $a \in I$ iff $a \subseteq U(I)$.

Now let P be a prime ideal of L, $F = \mathcal{S} - U(P)$, and let $\{U_k \mid k \in K\}$ be the set of all compact open sets of $\mathcal{S}$ that have nonempty intersections with F. Thus, the U_k are exactly those elements of L that are not in P. Therefore, by the definition of a prime ideal, given k, $l \in K$, there exists $t \in K$ with $U_t \subseteq U_k$, $U_t \subseteq U_l$, proving that F and $\{U_k \mid k \in K\}$ satisfy the hypothesis of condition (S2). By (S2) we conclude that there exists a $p \in F \cap \bigcap(U_k \mid k \in K)$. If $q \in F$, then for every compact open set U with $q \in U$, we have $U \cap F \neq \emptyset$; thus $p \in U$, proving that $\overline{\{p\}} = F$. Note that $\mathcal{S}$ is T_0; therefore p is unique.

Conversely, if $p \in \mathcal{S}$, let $I = \{a \mid a \in L,\ a \subseteq \mathcal{S} - \overline{\{p\}}\}$. Then I is an ideal of L, and $\mathcal{S} - \overline{\{p\}} = U(I)$. We claim that I is prime. Indeed, if U, $V \in L$, $U \notin I$,

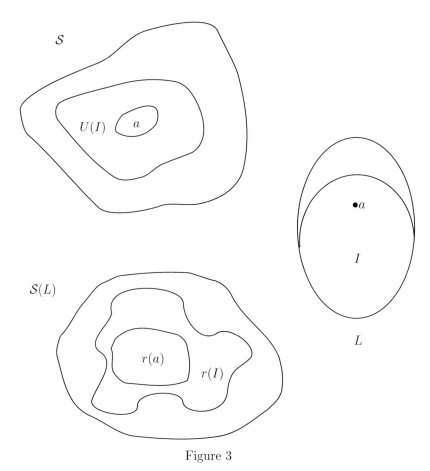

Figure 3

$V \notin I$, then $U \cap \overline{\{p\}} \neq \varnothing$, $V \cap \overline{\{p\}} \neq \varnothing$, and therefore $p \in U$, $p \in V$. Thus, $p \in U \cap V$ and so $U \cap V \not\subseteq U(I)$. By (S1), there exists a $W \in L$ with $W \subseteq U \cap V$ and $W \not\subseteq U(I)$. Therefore, $W \notin I$, and so I is prime.

Summing up, the map $\varphi \colon P \mapsto p$ is one-to-one and onto between $\mathcal{S}(L)$ and $\mathcal{S}$. To show that φ is a homeomorphism, it suffices to show that U is open in $\mathcal{S}(L)$ iff $U\varphi$ is open in $\mathcal{S}$. Since a typical open set in $\mathcal{S}(L)$ is of the form $r(I)$ ($I \in \operatorname{Id} L$), and an open set of $\mathcal{S}$ is of the form $U(I)$, we need only prove that $r(I)\varphi = U(I)$ and $U(I)\varphi^{-1} = r(I)$—in other words, that $P \in r(I)$ iff $(P\varphi =) p \in U(I)$. Indeed, $P \in r(I)$ means that $P \not\supseteq I$, which is equivalent to $U(P) \not\supseteq U(I)$; this, in turn, is the same as

$$U(I) \cap (\mathcal{S} - U(P)) \neq \varnothing.$$

Since $\mathcal{S}(P) = \overline{\{p\}}$, with $p = P\varphi$, the last condition means that

$$U(I) \cap \overline{\{p\}} \neq \varnothing,$$

which holds iff $p \in U(I)$. (Indeed, if $p \notin U(I)$, then $U(I) \subseteq U(P)$, and so $U(I) \cap \overline{\{p\}} = \varnothing$.) □

Corollary 9 (M. H. Stone [1937]). *The Stone space of a distributive lattice is characterized by* (S1), (S2), *and*

(S3) *The intersection of two compact open sets is compact.*

Corollary 10 (M. H. Stone [1936]). *The Stone space $\mathcal{S}$ of a Boolean lattice (called a* Boolean space*) can be characterized as a compact Hausdorff space in which the closed open (clopen) sets form a base for open sets. (In other words, $\mathcal{S}$ is* totally disconnected.*)*

Proof. Let $\mathcal{S} = \mathcal{S}(B)$, where B is a Boolean lattice. Then $\mathcal{S} = r(1)$, and thus $\mathcal{S}$ is compact. Let P, $Q \in \mathcal{S}$ and $P \neq Q$ and take $a \in P - Q$. Then $Q \in r(a)$ and $P \in r(a')$; therefore, every pair of elements of $\mathcal{S}$ can be separated by clopen sets, verifying that $\mathcal{S}$ is Hausdorff. Obviously, $\mathcal{S}$ is totally disconnected. Conversely, let $\mathcal{S}$ be compact, Hausdorff, and totally disconnected. Then (S1) is obvious. (S2) follows from the observation that F and U_i, $i \in I$, are now closed sets having the finite intersection property; therefore, by compactness, they have an element in common. The compact open sets of $\mathcal{S}$ form a Boolean lattice B, and thus $\mathcal{S}$ is homeomorphic to $\mathcal{S}(B)$. □

As an interesting application, we prove:

Theorem 11.   Let B be an infinite *Boolean lattice*. Then $|\mathcal{P}(B)| \geq |B|$.   —

Proof. Let $\mathcal{S}$ be a totally disconnected compact Hausdorff space. For a, $b \in \mathcal{S}$ with $a \neq b$, fix a pair of clopen sets $U_{a,b}$ and $U_{b,a}$ such that $a \in U_{a,b}$, $b \in U_{b,a}$, and $U_{a,b} \cap U_{b,a} = \varnothing$. Now let U be clopen and $a \in U$. Then

$$\mathcal{S} - U \subseteq \bigcup (U_{b,a} \mid b \in \mathcal{S} - U),$$

and so, by the compactness of $\mathcal{S} - U$,

$$\mathcal{S} - U \subseteq \bigcup (U_{b,a} \mid b \in X),$$

for some finite $X \subseteq \mathcal{S} - U$. Then $V_a = \bigcap (V_a \mid b \in X)$ is open and $a \in V_a \subseteq U$. Thus, $U \subseteq \bigcup (V_a \mid a \in U)$, so by the compactness of U, $U \subseteq \bigcup (V_a \mid a \in A)$, for some finite $A \subseteq U$. Therefore, $U = \bigcup (V_a \mid a \in A)$. Thus every clopen set is a finite union of finite intersections of $U_{a,b}$, and so there are no more clopen sets than there are finite sequences of elements of $\mathcal{S}$; this cardinality is $|\mathcal{S}|$, provided that $|\mathcal{S}|$ is infinite. □

It might be illuminating to compare this to an algebraic proof; see Exercise 36.

Theorem 8 and its corollaries provide topological representations for distributive join-semilattices, distributive lattices, and Boolean lattices, respectively. It is possible to give a topological representation for homomorphisms. We do it here only for $\{0, 1\}$-homomorphisms of bounded distributive lattices.

Lemma 12. *Let L_0 and L_1 be bounded distributive lattices and let φ be a $\{0, 1\}$-homomorphism of L_0 into L_1. Then*

$$S(\varphi)\colon P \mapsto P\varphi^{-1}$$

maps $S(L_1)$ into $S(L_0)$; $S(\varphi)$ is a continuous function with the property that if U is compact open in $S(L_0)$, then $U(S(\varphi))^{-1}$ is compact in $S(L_1)$. Conversely, if $\psi\colon S(L_1) \to S(L_0)$ has these properties, then $\psi = S(\varphi)$, for exactly one $\varphi\colon L_0 \to L_1$.

Proof. If $U = r(a)$, $a \in L_0$, then

$$
\begin{aligned}
US(\varphi)^{-1} &= \{\, P \mid P \in S(L_1),\ P\varphi^{-1} \in r(a) \,\} \\
&= \{\, P \mid P \in S(L_1),\ a \notin P\varphi^{-1} \,\} \\
&= \{\, P \mid P \in S(L_1),\ a\varphi \notin P \,\} \\
&= r(a\varphi),
\end{aligned}
$$

and so $S(\varphi)$ is continuous, having the desired property.

Conversely, if such a map ψ is given and $U = r(a)$, $a \in L_0$, then $U\psi^{-1}$ is compact open, and so $U\psi^{-1} = r(b)$, for a unique $b \in L_1$. The map $\varphi\colon a \mapsto b$ is a $\{0, 1\}$-homomorphism, and $\psi = S(\varphi)$. $\qquad\square$

The following interpretation of (S2) will be useful. Let S be a topological space. The *Booleanization of S* is a topological space S^B on S that has the compact open sets of S and their complements as a subbase for open sets.

Lemma 13. *A compact topological space S satisfies (S1), (S2), and (S3) iff S^B is a Boolean space.*

Proof. Let S satisfy (S1), (S2), and (S3). Then S^B is obviously Hausdorff and totally disconnected. To verify the compactness of S^B, let $\mathcal{F}_0$ be a collection of compact open sets of S, and let $\mathcal{F}_1$ be a collection of complements of compact open sets of S such that in $\mathcal{F} = \mathcal{F}_0 \cup \mathcal{F}_1$ no finite intersection is void. Because of (S3), we can assume that $\mathcal{F}_0$ is closed under finite intersection. Since members of $\mathcal{F}_1$ are closed in S and S is compact, $\bigcap(\, X \mid X \in \mathcal{F}_1\,) = F$ is nonempty. Also, for any $U \in \mathcal{F}_0$ and $X \in \mathcal{F}_1$, $U \cap X$ is closed in U, and thus

$$U \cap F = \bigcap(\, U \cap X \mid X \in \mathcal{F}_1 \,) \neq \varnothing.$$

Applying (S2) to F and $\mathcal{F}_0$, we conclude that $\bigcap \mathcal{F} \neq \varnothing$, which, by Alexander's Theorem (see Exercise 14), proves compactness.

Conversely, if $\mathcal{S}^B$ is Boolean, then the compact opens sets of $\mathcal{S}^B$ form a Boolean lattice L. We can easily verify that the compact open sets of $\mathcal{S}$ form a sublattice L_1 of L. Thus L_1 is a distributive lattice, and so, by Corollary 9, $\mathcal{S} = \mathcal{S}(L_1)$ satisfies (S1), (S2), and (S3).   □

We close this section with an interesting application.

Let L_i, $i \in I$, be pairwise disjoint distributive lattices. Then

$$Q = \bigcup (L_i \mid i \in I)$$

is a partial lattice. A free lattice generated by Q over the class $\mathbf{D}$ of all distributive lattices is called a *free distributive product* of the L_i, $i \in I$. To prove the existence of free distributive products, it suffices by Theorem I.5.24 to show that there exists a distributive lattice L containing Q as a relative sublattice. This is easily done: Let L be the direct product of the $L_i \cup \{0\}$, $i \in I$, where 0 is a new zero element of L_i. Identify $x \in L_i$ with $f \in L$ defined by $f(i) = x$, $f(j) = 0$, for $j \neq i$. Then Q becomes a relative sublattice of L.

An equivalent definition is:

Definition 14. Let $\mathbf{K}$ be a class of lattices and let L_i, $i \in I$, be lattices in $\mathbf{K}$. A lattice L in $\mathbf{K}$ is called a *free $\mathbf{K}$-product* of the L_i, $i \in I$, iff every L_i has an embedding ε_i into L such that:

(i) L is generated by $\bigcup (L_i \varepsilon_i \mid i \in I)$.

(ii) If K is any lattice in $\mathbf{K}$ and φ_i is a homomorphism of L_i into K, for $i \in I$, then there exists a homomorphism φ of L into K satisfying $\varphi_i = \varepsilon_i \varphi$, for all $i \in I$ (see Figure 4).

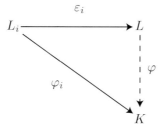

Figure 4

For distributive lattices, this is equivalent to the first definition (see Exercise 55). In most cases, we will assume that each L_i is a sublattice of L and that ε_i is the inclusion map; then (ii) will simply state that the φ_i have a common extension. Note that in all cases we shall consider, (i) can be replaced by the requirement that the φ in (ii) be unique.

If, in Definition 14, **K** is a class of bounded lattices and all homomorphisms are assumed to be $\{0, 1\}$-homomorphisms, we get the concept of a *free* **K** $\{0, 1\}$-*product*. In particular, if **K** = **L**, we get the concept of *free* $\{0, 1\}$-*product*, see Section VI.3, and if **K** = **D**, we obtain the concept of *free* $\{0, 1\}$-*distributive product*.

Our final result is the existence and description of a free $\{0, 1\}$-distributive product of a family of bounded distributive lattices.

Theorem 15 (A. Nerode [1959]).   Let L_i, $i \in I$, be distributive lattices with 0 and 1. Let $\mathcal{S} = \prod(\mathcal{S}(L_i) \mid i \in I)$ (see Exercise 15). Then $\mathcal{S}$ is a Stone space, and thus $\mathcal{S} \cong \mathcal{S}(L)$, for some distributive lattice L. Any such lattice L is a free $\{0, 1\}$-distributive product of the L_i, $i \in I$.   —

The proof of Theorem 15 will be preceded by two lemmas. In these two lemmas a *Stone space* is a topological space satisfying (S1), (S2), and (S3).

Lemma 16. *Let $\mathcal{S}_i$, $i \in I$, be compact Stone spaces. Then*

$$\prod(\mathcal{S}_i^B \mid i \in I) = (\prod(\mathcal{S}_i \mid i \in I))^B.$$

Proof. For $U \subseteq \mathcal{S}_j$, let

$$E(U) = \{\, f \mid f \in \prod \mathcal{S}_i, \ f(j) \in U \,\}$$

(see Exercise 15). The compact open sets form a base for open sets in $\mathcal{S}_j$; therefore,

$$\{\, E(U) \mid U \text{ compact open in some } \mathcal{S}_j \,\}$$

is a subbase for open sets in $\prod(\mathcal{S}_i \mid i \in I)$. Note that all the sets $E(U)$ are compact open in $\prod \mathcal{S}_i$; therefore, $V \subseteq \prod \mathcal{S}_i$ is compact open iff it is a finite union of finite intersections of some of the $E(U)$. Consequently, declaring the complements of compact open sets to be open (in forming $(\prod \mathcal{S}_i)^B$) is equivalent to making the complements of the sets $E(U)$ open. But the complement of $E(U)$ is $E(\mathcal{S}_i - U)$, and $\mathcal{S}_i - U$ is a typical open set of $\mathcal{S}_i^B$. Thus $\prod \mathcal{S}_i^B$ and $(\prod \mathcal{S}_i)^B$ have the same topology.   □

Lemma 17. *The product of compact Stone spaces is again a compact Stone space.*

Proof. Let $\mathcal{S}_i$, $i \in I$, be Stone spaces. Then $\mathcal{S} = \prod \mathcal{S}_i$ is T_0 and compact (Exercises 16 and 21). Since $\mathcal{S}_j^B$ is Boolean (Lemma 13), so is $\prod \mathcal{S}_i^B$ (Exercises 20–22). By Lemma 16, $\mathcal{S}^B = \prod \mathcal{S}_i^B$, and thus $\mathcal{S}^B$ is Boolean. Therefore, by Lemma 13, $\mathcal{S}$ is a Stone space.   □

Proof of Theorem 15. Let e_i be the ith projection $(e_i\colon \mathcal{S}(L) \to \mathcal{S}(L_i)$, $fe_i = f(i))$. By Lemma 12, there is a unique $\{0,1\}$-homomorphism $\varepsilon_i\colon L_i \to L$ satisfying $\mathcal{S}(\varepsilon_i) = e_i$. It is easy to visualize ε_i; think of the elements of L_i as compact open sets of $\mathcal{S}_i$; then $U\varepsilon_i = E(U) = Ue_i^{-1}$. It is obvious from this that ε_i is an embedding.

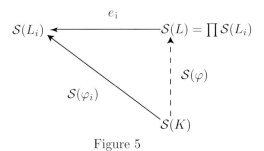

Figure 5

By applying $\mathcal{S}$ to Figure 4, we obtain Figure 5. Thus the method of defining $\mathcal{S}(\varphi)$ is clear. For $x \in \mathcal{S}(K)$, $x\mathcal{S}(\varphi)$ is a member of $\prod \mathcal{S}(L_i)$, and $x\mathcal{S}(\varphi)(i) = x\mathcal{S}(\varphi_i)$, for $i \in I$.

To show that this correspondence is indeed of the form $\mathcal{S}(\varphi)$, for some homomorphism $\varphi\colon L \to K$, we have to verify that (a) $\mathcal{S}(\varphi)$ is continuous (this statement follows from Exercise 18), and that (b) if V is compact open in $\mathcal{S}(L)$ then $V\mathcal{S}(\varphi)^{-1}$ is compact open in $\mathcal{S}(K)$. It is enough to verify (b) for $V = E(U)$, where U is compact open in some $\mathcal{S}(L_i)$. Then

$$V\mathcal{S}(\varphi)^{-1} = E(U)\mathcal{S}(\varphi)^{-1} = Ue_i^{-1}\mathcal{S}(\varphi)^{-1} = U(\mathcal{S}(\varphi)e_i)^{-1} = U\mathcal{S}(\varphi_i)^{-1},$$

and therefore $V\mathcal{S}(\varphi)^{-1}$ is compact open since $\mathcal{S}(\varphi_i)$ satisfies the condition of Lemma 12. $\square$

Exercises

The first 22 exercises review the basic concepts and theorems of topology that are utilized in this section.

1. A *topological space* is a set A and a collection T of subsets of A, with $A \in T$, closed under finite intersections and arbitrary unions; a member of T is called an *open set*. Call a set *closed* iff its complement is open. Characterize the family of closed sets.

2. A family of nonempty sets $B \subseteq T$ is a *base* for open sets iff every open set is a union of members of B. Show that for a set A, a collection B of subsets of A is a base of open sets of some topological space defined on A iff $\bigcup B = A$, and for $X, Y \in B$ and $p \in X \cap Y$, there exists a $Z \in B$ with $p \in Z$, $Z \subseteq X$, and $Z \subseteq Y$.

3. A family of nonempty sets $C \subseteq P(A)$ is a *subbase* for open sets iff the finite intersections of members of C form a base for open sets. Show that $C \subseteq P(A)$ is a subbase of some topology defined on A iff $\bigcup C = A$.

4. Let A be a topological space and let $X \subseteq A$. Then there exists a smallest closed set $\overline{X}$ containing X; $\overline{X}$ is the *closure* of X. Show that $\overline{\varnothing} = \varnothing$ and that, for $X, Y \subseteq A$,

   (i) $X \subseteq Y$ implies that $\overline{X} \subseteq \overline{Y}$,

   (ii) $X \subseteq \overline{X}$,

   (iii) $\overline{X \cup Y} = \overline{X} \cup \overline{Y}$,

   (iv) $\overline{\overline{X}} = \overline{X}$.

5. Prove that the properties of $\overline{X}$ given in Exercise 4 characterize it.

6. Show that $a \in \overline{X}$ iff every open set (in a given base) containing a has a nonempty intersection with X.

7. A space A is a T_0-space iff $\overline{\{x\}} = \overline{\{y\}}$ implies that $x = y$. Show that A is a T_0-space iff, for every $x, y \in A$, $x \neq y$, there exists an open set (in a given base) containing exactly one of x and y.

8. A is a T_1-space iff $\overline{\{x\}} = \{x\}$, for all $x \in A$. A T_1-space is a T_0-space. Show that A is a T_1-space iff, for $x, y \in A$, $x \neq y$, there exists an open set (in a given base) containing x but not y.

9. Let A and B be topological spaces and $f \colon A \to B$. Then f is called *continuous* iff, for every open set U of B, $f^{-1}(U)$ is open in A. f is a *homeomorphism* iff f is one-to-one and onto and if both f and f^{-1} are continuous. Show that continuity can be checked by considering only those $f^{-1}(U)$, where U belongs to a given subbase.

10. Show that $f \colon A \to B$ is continuous iff $f(\overline{X}) \subseteq \overline{f(X)}$, for all $X \subseteq A$.

11. A subset X of a topological space A is *compact* iff

$$X \subseteq \bigcup (U_i \mid U_i \text{ open}, i \in I)$$

implies that

$$X \subseteq \bigcup (U_i \mid i \in I_1),$$

for some finite $I_1 \subseteq I$. The space A is *compact* iff $X = A$ is compact. Show that A is compact, iff, for every family F of closed sets, if $\bigcap F_1 \neq \varnothing$, for all finite $F_1 \subseteq F$, then $\bigcap F \neq \varnothing$.

12. Let A be a compact topological space and let X be a closed set in A. Show that X is compact.

13. Prove that a space A is compact iff, in the lattice of closed sets of A, every maximal dual ideal is principal.

*14. Show that a T_1-space A is compact iff it has a subbase C of closed sets (that is, $\{A - X \mid X \in C\}$ is a subbase for open sets) with the property: If $\bigcap D = \varnothing$, for some $D \subseteq C$, then $\bigcap D_1 = \varnothing$, for some finite $D_1 \subseteq D$ (J. W. Alexander [1939]).

15. Let A_i, $i \in I$, be topological spaces and set $A = \prod(A_i \mid i \in I)$. For $U \subseteq A_i$, set $E(U) = \{f \mid f \in A, \ f(i) \in U\}$. The *product topology* on A is the topology determined by taking all the sets $E(U)$ as a subbase for open sets, where U ranges over all open sets of A_i, for all $i \in I$. Show that the projection map $e_i \colon f \mapsto f(i)$ is a continuous map of A onto A_i.

16. Show that if the A_i are T_0-spaces (T_1-spaces), so is $A = \prod(A_i \mid i \in I)$.

17. A map $f \colon A \to B$ is *open* iff $f(U)$ is open in B for every open $U \subseteq A$. Show that the projection maps (see Exercise 15) are open.

18. Prove that a function $f \colon B \to \prod A_i$ is continuous iff, for each $i \in I$, $e_i f \colon B \to A_i$ is continuous.

19. A space A is a *Hausdorff space* (T_2-space) iff, for x, $y \in A$ with $x \neq y$, there exist open sets U, V such that $x \in U$, $y \in V$, $U \cap V = \varnothing$. Show that:

    (i) A is Hausdorff iff $\Delta = \{\langle x, x \rangle \mid x \in A\}$ is closed in $A \times A$.

    (ii) A compact subset of a T_2-space is closed.

20. Prove that a product of Hausdorff spaces is a Hausdorff space.

21. Show that a product of compact spaces is compact (Tihonov's Theorem). (Hint: use Exercise 14.)

22. A space A is *totally disconnected* iff, for x, $y \in A$, $x \neq y$, there exists a closed open set U with $x \in U$, $y \notin U$. Show that the product of totally disconnected sets is totally disconnected.

$$* \qquad * \qquad *$$

23. Let I and J be ideals of a join-semilattice. Verify that

$$I \vee J = \{t \mid t \leq i \vee j, \ i \in I, \ j \in J\}.$$

24. Show that for a join-semilattice L, Id L is a lattice iff any two elements of L have a common lower bound.

25. Give a detailed proof of Lemma 2.

26. Prove that every join-semilattice can be embedded in a Boolean lattice (considered as a join-semilattice).

27. Show that a finite distributive join-semilattice is a distributive lattice.

28. Let L be a join-semilattice and let Θ be a *join-congruence*, that is, an equivalence relation on L having the Substitution Property for join. Then L/Θ is also a join-semilattice. Show that the distributivity of L does not imply the distributivity of L/Θ.

29. Let F be a free join-semilattice; let F_0 be F with a new 0 added. Show that F_0 is a distributive join-semilattice.

30. Let φ be a join-homomorphism of the join-semilattice F_0 onto the join-semilattice F_1. Show that, for distributive join-semilattices F_0, F_1, the proper homomorphism concept is the one requiring that if P is a prime ideal of F_1, then $P\varphi^{-1}$ is a prime ideal of F_0.

31. Show that there is no "free distributive join-semilattice" with the homomorphism concept of Exercise 30.

32. Does Theorem 1.22 generalize to distributive join-semilattices?

33. Characterize the Stone spaces of finite Boolean lattices and of finite chains.

34. Let S_0 and S_1 be disjoint topological spaces; let $S = S_0 \cup S_1$ and call $U \subseteq S$ open iff $U \cap S_0$ and $U \cap S_1$ are open. Show that if S_0 and S_1 are Stone spaces, then so is S.

35. If, in Exercise 34, $S_i = S(L_i)$, $i = 0, 1$, then $S = S(L_0 \times L_1)$.

36. To prove Theorem 11, pick an element $a(P, Q) \in P - Q$, for all $P, Q \in P(B)$ with $P \neq Q$. Show that the $a(P, Q)$ R-generate all of B.

37. In Lemma 12, characterize $S(\varphi)$ for one-to-one φ and for onto φ.

38. Determine the connection between the Stone space of a lattice and the Stone space of a sublattice.

39. Call the Stone space of a generalized Boolean lattice a *generalized Boolean space*; characterize it. (Compactness of S should be replaced by *local compactness*: For every $p \in S$, there exists an open set U with $p \in U$ and a set V with $U \subseteq V$ such that V is compact.)

40. Show that the product of (generalized) Boolean spaces is (generalized) Boolean.

41. Call the join-semilattice L *modular* iff, for a, b, $c \in L$, $a \leq b$ and $b \leq a \vee c$ imply the existence of $c_1 \in L$ with $c_1 \leq c$ and $b = a \vee c_1$. Show that a distributive join-semilattice is modular.

42. Show that Lemma 1 remains valid if all occurrences of the word "distributive" are replaced by the word "modular."

43. Show that the set of all finitely generated normal subgroups of a group (and also the finitely generated ideals of a ring) form a modular join-semilattice.

44. The lattice of congruence relations of a join-semilattice L is distributive iff any pair of elements with a lower bound are comparable (D. Papert [1964], R. A. Dean and R. H. Oehmke [1964]).

$$* \qquad * \qquad *$$

We have seen, in this section, how Theorem I.5.24 can be used to show the existence of free distributive products. The same method, however, does not apply to distributive lattices with 0 and 1. Nevertheless, the idea of the proofs of Theorems I.5.5 and I.5.24 can be used to get a much stronger result on the existence of free products; it is easiest to formulate this result (G. Grätzer [1968], Theorem 29.2) for universal algebras. To do so, we have to introduce some concepts.

A *type* τ of algebras is a sequence $\langle n_0, n_1, \ldots, n_\gamma, \ldots \rangle$ of nonnegative integers, $\gamma < o(\tau)$, where $o(\tau)$ is an ordinal called the *order* of τ. An algebra $\mathfrak{A}$ of *type* τ is an ordered pair $\langle A; F \rangle$, where A is a nonempty set and F is a sequence $\langle f_0, \ldots, f_\gamma, \ldots \rangle$, $\gamma < o(\tau)$, where f_γ is an n_γ-ary operation on A. If $o(\tau)$ is finite, $o(\tau) = n$, then we write $\langle A; f_0, \ldots, f_{n-1} \rangle$ for $\langle A; F \rangle$.

45. Define the concepts of subalgebra, polynomial, identity, and variety for algebras of a given type τ. Show that if $\mathbf{K}$ is an algebra in a variety, $\mathfrak{A}$ is an algebra in $\mathbf{K}$, and $\mathfrak{B}$ is a subalgebra of $\mathfrak{A}$, then $\mathfrak{B}$ is in $\mathbf{K}$.

46. Define the concepts of homomorphism, homomorphic image, and direct product for algebras of a given type. Show that a variety is closed under the formation of homomorphic images and direct products.

47. Let $\mathfrak{A} = \langle A; F \rangle$ be an algebra, let $H \subseteq A$, and let $H \neq \varnothing$. Show that there exists a smallest subset $[H]$ of A, $[H] \supseteq H$ such that $\langle [H]; F \rangle$ is a subalgebra of $\mathfrak{A}$. (This subalgebra is said to be *generated by* H.)

48. Show that $\|[H]\| \leq |H| + |F| + \aleph_0$.

49. Modify Definition 14 for algebras. Show that the φ in (ii) is unique.

50. Let $\mathfrak{B}$ and $\mathfrak{C}$ be free **K**-products of $\mathfrak{A}_i$, $i \in I$, with embeddings ε_i, $i \in I$, and χ_i, $i \in I$, respectively. Show that there exists an isomorphism $\alpha \colon B \to C$ such that $\varepsilon_i \alpha = \chi_i$ and $\chi_i \alpha^{-1} = \varepsilon_i$, for all $i \in I$.

51. Let **K** be a variety of algebras and let $\mathfrak{A}_i \in$ **K**, for $i \in I$. Choose a set S satisfying

$$|S| \geq \sum |A_i| + |F| + \aleph_0.$$

Let Q be the set of all pairs $\langle \mathfrak{B}, (\varphi_i \mid i \in I) \rangle$ such that $B \subseteq S$, φ_i is a homomorphism of $\mathfrak{A}_i$ into $\mathfrak{B}$, and $B = [\bigcup(A_i\varphi \mid i \in I)]$. Form

$$\mathfrak{A} = \prod(\mathfrak{B} \mid \langle \mathfrak{B}, (\varphi_i \mid i \in I) \rangle \in Q)$$

(direct product), and, for $a \in A_i$, define $f_a \in A$ by

$$f_a(\langle \mathfrak{B}, (\varphi_i \mid i \in I) \rangle) = a\varphi_i.$$

Finally, let $\mathfrak{N}$ be the subalgebra generated by the f_a, $a \in A_i$, $i \in I$. Show that $\mathfrak{N} \in$ **K**, $a \mapsto f_a$ is a homomorphism ε_i of $\mathfrak{A}_i$ into $\mathfrak{N}$, for $i \in I$, and that $\mathfrak{N}$ is generated by $\bigcup(A_i\varepsilon_i \mid i \in I)$.

52. Show that ε_i is one-to-one iff, for $i \in I$, a, $b \in A_i$, $a \neq b$, there exists an algebra $\mathfrak{C} \in$ **K** and homomorphisms $\psi_j \colon \mathfrak{A}_j \to \mathfrak{C}$, for all $j \in I$, such that $a\psi_i \neq b\psi_i$.

53. Combine the previous exercises to prove the following result.

Existence Theorem for Free Products. *Let* **K** *be a variety of algebras, $\mathfrak{A}_i \in$ **K**, for $i \in I$. A free* **K**-*product of $\mathfrak{A}_i$, $i \in I$, exists iff, for $i \in I$, a, $b \in A_i$, $a \neq b$, there exist a $\mathfrak{C} \in$ **K**, homomorphisms $\psi_j \colon \mathfrak{A}_j \to \mathfrak{C}$, for all $j \in I$, such that $a\psi_i \neq b\psi_i$.*

54. Show that in proving the existence of free distributive products and free $\{0,1\}$-distributive products, we can always choose $\mathfrak{C} = C_2$, the two-element chain, in applying Exercise 53.

55. Show that the two definitions of free **K**-product are equivalent for any class **K** of lattices.

56. Show that the free Boolean algebra on $\mathfrak{m}$ generators is a free $\{0,1\}$-distributive product of $\mathfrak{m}$ copies of the free Boolean algebra on one generator.

57. Prove that the free Boolean algebra on $\mathfrak{m}$ generators can be represented by the clopen subsets of $\{0,1\}^{\mathfrak{m}}$, where $\{0,1\}$ is the two-element discrete topological space.

58. Find a topological representation for the free distributive lattice on $\mathfrak{m}$ generators (G. Ja. Areškin [1953a]).

6. Distributive Lattices with Pseudocomplementation

In this section, we shall deal exclusively with pseudocomplemented distributive lattices. There are two concepts that we should be able to distinguish: a lattice, $\langle L; \wedge, \vee \rangle$, in which every element has a pseudocomplement; and an algebra $\langle L; \wedge, \vee, {}^*, 0, 1 \rangle$, where $\langle L; \wedge, \vee, 0, 1 \rangle$ is a bounded lattice and where, for every $a \in L$, the element a^* is the pseudocomplement of a. We shall call the former a *pseudocomplemented lattice* and the latter a *lattice with pseudocomplementation* (as an operation)—the same kind of distinction we make between Boolean lattices and Boolean algebras. As defined in the exercises of Section 5, a pseudocomplemented lattice is an algebra of type $\langle 2, 2 \rangle$, whereas a lattice with pseudocomplementation is an algebra of type $\langle 2, 2, 1, 0, 0 \rangle$. To see the difference in viewpoint, consider the lattice of Figure 1. As a distributive lattice, it has twenty-five sublattices and eight congruences; as a lattice with pseudocomplementation, it has three subalgebras and five congruences.

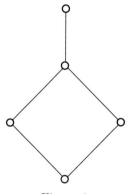

Figure 1

Thus, for a lattice with pseudocomplementation L, a *subalgebra* L_1 is a $\{0, 1\}$-sublattice of L closed under * (that is, $a \in L_1$ implies that $a^* \in L_1$). A *homomorphism* φ is a $\{0, 1\}$-homomorphism that also satisfies $(x\varphi)^* = x^*\varphi$. Similarly, a *congruence relation* Θ shall have the Substitution Property also for *, that is, $a \equiv b \ (\Theta)$ implies that $a^* \equiv b^* \ (\Theta)$.

A wide class of examples is provided by

Theorem 1.   Any complete lattice that satisfies the Join Infinite Distributive Identity (JID) is a pseudocomplemented distributive lattice.   —

Proof. Let L be such a lattice. For $a \in L$, set

$$a^* = \bigvee (x \mid x \in L, \ a \wedge x = 0).$$

Then, by (JID),

$$a \wedge a^* = a \wedge \bigvee(x \mid a \wedge x = 0) = \bigvee(a \wedge x \mid a \wedge x = 0) = \bigvee 0 = 0.$$

Furthermore, if $a \wedge x = 0$, then $x \leq a^*$ by the definition of a^*; thus a^* is indeed the pseudocomplement of a.   □

Corollary 2. *Every distributive algebraic lattice is pseudocomplemented.*

Proof. Let L be a distributive algebraic lattice. We can assume that $L = \operatorname{Id} S$, where S is a distributive join-semilattice with 0, by Theorem 3.13 and Lemma 5.1. Let I, $I_j \in \operatorname{Id} S$, for $j \in J$. Then

$$I \wedge I_k \subseteq I \wedge \bigvee(I_j \mid j \in J),$$

for any $k \in J$, and thus

$$\bigvee(I \wedge I_j \mid j \in J) \subseteq I \wedge \bigvee(I_j \mid j \in J).$$

To prove the reverse inclusion, let

$$a \in I \wedge \bigvee(I_j \mid j \in J),$$

that is, $a \in I$ and $a \in \bigvee(I_j \mid j \in J)$. The latter implies that

$$a \leq t_1 \vee \cdots \vee t_n, \text{ where } t_1 \in I_{j_1}, \ \ldots, \ t_n \in I_{j_n}, \ j_1, \ \ldots, \ j_n \in J.$$

Thus $a \in I_{j_1} \vee \cdots \vee I_{j_n}$ and so, using the distributivity of $\operatorname{Id} L$, we obtain

$$a \in I \wedge (I_{j_1} \vee \cdots \vee I_{j_n}) = (I \wedge I_{j_1}) \vee \cdots \vee (I \wedge I_{j_n}) \subseteq \bigvee(I \wedge I_j \mid j \in J),$$

completing the proof of (JID). The statement now follows from Theorem 1. □

Thus, the lattice of all congruence relations of an arbitrary lattice and the lattice of all ideals of a distributive (semi)-lattice with zero are examples of pseudocomplemented distributive lattices. Note that, for $I \in \operatorname{Id} K$,

$$I^* = \{ x \mid x \in K, \ x \wedge i = 0, \text{ for all } i \in I \}.$$

Also, any finite distributive lattice is pseudocomplemented. Therefore, our investigations include all finite distributive lattices.

The first class of distributive lattices with pseudocomplementation, other than the class of Boolean algebras, to be examined in detail was the class of Stone algebras (so named in G. Grätzer and E. T. Schmidt [1957a]). A distributive lattice with pseudocomplementation L is called a *Stone algebra* iff it satisfies the *Stone identity*:

$$a^* \vee a^{**} = 1.$$

The corresponding pseudocomplemented lattice is called a *Stone lattice*. To understand the meaning of this identity, define the *skeleton* of L:

$$S(L) = \{\, a^* \mid a \in L \,\}.$$

The elements of $S(L)$ are called *skeletal*. By Theorem I.6.4, $\langle S(L); \wedge, \vee, ^*, 0, 1 \rangle$ is a Boolean algebra. For a Stone algebra L, $S(L)$ is a subalgebra of L:

Lemma 3. *For a distributive lattice with pseudocomplementation L, the following conditions are equivalent:*

(i) *L is a Stone algebra.*

(ii) *$(a \wedge b)^* = a^* \vee b^*$, for $a, b \in L$.*

(iii) *$a, b \in S(L)$ implies that $a \vee b \in S(L)$.*

(iv) *$S(L)$ is a subalgebra of L.*

Proof. The proofs that (ii) implies (iii), that (iii) implies (iv), and that (iv) implies (i) are trivial. Now let L be a Stone algebra; we show that $a^* \vee b^*$ is the pseudocomplement of $a \wedge b$. Indeed,

$$(a \wedge b) \wedge (a^* \vee b^*) = (a \wedge b \wedge a^*) \vee (a \wedge b \wedge b^*) = 0 \vee 0 = 0.$$

If $(a \wedge b) \wedge x = 0$, then $(b \wedge x) \wedge a = 0$, and so $b \wedge x \leq a^*$. Meeting both sides by a^{**} yields

$$b \wedge x \wedge a^{**} \leq a^* \wedge a^{**} = 0;$$

that is, $x \wedge a^{**} \wedge b = 0$, implying that $x \wedge a^{**} \leq b^*$. By the Stone identity, $a^* \vee a^{**} = 1$, and thus

$$x = x \wedge 1 = x \wedge (a^* \vee a^{**}) = (x \wedge a^*) \vee (x \wedge a^{**}) \leq a^* \vee b^*. \qquad \square$$

This is already enough to yield the structure theorem for finite Stone algebras (G. Grätzer and E. T. Schmidt [1957a]):

Corollary 4. *A finite distributive lattice is a Stone lattice iff it is the direct product of finite distributive dense lattices, that is, finite distributive lattices with only one atom.*

Proof. By Lemma 3, a Stone lattice L has a complemented element $a \notin \{0, 1\}$ iff $S(L) \neq \{0, 1\}$; thus the decomposition of Theorem 1.6 can be repeated until each factor L_i satisfies $S(L_i) = \{0, 1\}$. In a direct product, * is formed componentwise; therefore, all the L_i are Stone lattices. For a finite distributive lattice K with $S(K) = \{0, 1\}$, the condition that K has exactly one atom is equivalent to K being a Stone lattice.   □

Another significant subset of a Stone algebra is the *dense set*:

$$D(L) = \{\, a \mid a^* = 0 \,\}.$$

The elements of $D(L)$ are called *dense*.

We can easily check that $D(L)$ is a dual ideal of L and that $1 \in D(L)$; thus $D(L)$ is a distributive lattice with 1. Since $a \vee a^* \in D(L)$, for every $a \in L$, we can interpret the identity

$$a = a^{**} \wedge (a \vee a^*)$$

to mean that every $a \in L$ can be represented in the form

$$a = b \wedge c, \quad b \in S(L), \; c \in D(L).$$

Such an interpretation correctly suggests that if we know $S(L)$ and $D(L)$ and the relationships between elements of $S(L)$ and $D(L)$, then we can describe L. The relationship is expressed by the homomorphism $\varphi(L) \colon S(L) \to \operatorname{Du} D(L)$ defined by

$$\varphi(L) \colon a \mapsto \{\, x \mid x \in D(L), \; x \geq a^* \,\}.$$

Theorem 5 (C. C. Chen and G. Grätzer [1969a]).   Let L be a Stone algebra. Then $S(L)$ is a Boolean algebra, $D(L)$ is a distributive lattice with 1, and $\varphi(L)$ is a $\{0, 1\}$-homomorphism of $S(L)$ into $\operatorname{Du} D(L)$. The triple

$$\langle S(L), D(L), \varphi(L) \rangle$$

characterizes L up to isomorphism.   —

Proof. The first statement is easily verified. For $a \in S(L)$, set

$$F_a = \{\, x \mid x^{**} = a \,\}.$$

The sets $\{\, F_a \mid a \in S(L) \,\}$ form a partition of L; for a simple example, see Figure 2. Obviously, $F_0 = \{0\}$ and $F_1 = D(L)$. The map $x \mapsto x \vee a^*$ sends F_a into $F_1 = D(L)$; in fact, the map is an isomorphism between F_a and $a\varphi(L) \subseteq D(L)$. Thus $x \in F_a$ is completely determined by a and $x \vee a^* \in a\varphi(L)$—that is, by a

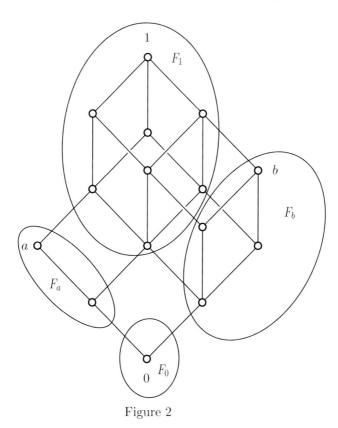

Figure 2

pair $\langle a, z \rangle$, where $a \in S(L)$, $z \in a\varphi(L)$—and every such pair determines one and only one element of L. To complete our proof, we have to show how the partial ordering on L can be determined by such pairs.

Let $x \in F_a$ and $y \in F_b$. Then $x \leq y$ implies that $x^{**} \leq y^{**}$, that is, $a \leq b$. Since $x \leq y$ iff

$$a \vee x \leq a \vee y \text{ and } x \vee a^* \leq y \vee a^*,$$

and since the first of these two conditions is trivial, we obtain:

$$x \leq y \quad \text{iff} \quad a \leq b \text{ and } x \vee a^* \leq y \vee a^*.$$

Identifying x with $\langle x \vee a^*, a \rangle$ and y with $\langle y \vee b^*, b \rangle$, we see that the preceding conditions are stated in terms of the components of the ordered pairs, except that $y \vee a^*$ will have to be expressed by the triple.

Because $\varphi(L)$ is a $\{0, 1\}$-homomorphism and a is the complement of a^*, we conclude that $a\varphi(L)$ and $a^*\varphi(L)$ are complementary dual ideals of $D(L)$. Thus every $z \in D(L)$ can be written in a unique fashion in the form $z = z\varrho_a \wedge z_1$, where $z\varrho_a \in a\varphi(L)$ and $z_1 \in a^*\varphi(L)$. Observe that ϱ_a is expressed in terms of the triple. Finally,

$$y \vee a^* = y \vee b^* \vee a^* = (y \vee b^*)\varrho_a.$$

Thus for $u \in a\varphi(L)$ and $v \in b\varphi(L)$, we have

$$\langle u, a \rangle \leq \langle v, b \rangle \quad \text{iff} \quad a \leq b \text{ and } u \leq v\varrho_a. \qquad \square$$

This theorem shows that the behavior of the skeleton and the dense set is decisive for Stone algebras. This conclusion leads us to formulate the goal of research for Stone algebras and *for all* distributive lattices with pseudocomplementation as follows:

> A problem for distributive lattices with pseudocomplementation is considered solved if it can be reduced to two problems: one for Boolean algebras and one for distributive lattices with unit.

By applying Zorn's Lemma to prime dual ideals of a lattice with zero, we obtain that every prime dual ideal is contained in a maximal prime dual ideal, or, equivalently, we get that every prime ideal contains a *minimal prime ideal P*, that is, a prime ideal P such that $Q \subset P$ for no prime ideal Q (see Exercise 1.30). Minimal prime ideals play an important role in the theory of distributive lattices with pseudocomplementation, as illustrated by the following result:

Theorem 6 (G. Grätzer and E. T. Schmidt [1957a]).   Let L be a distributive lattice with pseudocomplementation. Then L is a Stone algebra iff

$$P \vee Q = L,$$

for any two distinct minimal prime ideals P and Q.   —

Proof. Let L be a Stone algebra and let P and Q be distinct minimal prime ideals. Choose $a \in P - Q$ (note that $P \not\subset Q$, since Q is minimal, and $Q \neq P$, hence $P - Q \neq \varnothing$); since $a \wedge a^* = 0 \in Q$, $a \notin Q$, and Q is prime, we obtain that $a^* \in Q$.

$L - P$ is a maximal dual prime ideal, hence by the dual of Corollary 1.18, it is a maximal dual ideal of L. Thus $(L - P) \vee [a) = L$ and so $0 = a \wedge x$, for some $x \in L - P$. Therefore, $a^* \geq x \in L - P$ and so $a^* \notin P$. Hence $a^* \in Q - P$. Similarly, $a^{**} \in P - Q$, which implies that

$$1 = a^* \vee a^{**} \in P \vee Q,$$

yielding $P \vee Q = L$.

To prove the converse (for this proof, see J. C. Varlet [1966]), let us assume that L is not a Stone algebra and let $a \in L$ such that $a^* \vee a^{**} \neq 1$. Let R be a prime ideal (see Corollary 1.17) such that $a^* \vee a^{**} \in R$.

We claim that $(L - R) \vee [a^*) \neq L$. Indeed, if $(L - R) \vee [a^*) = L$, then there exists an $x \in L - R$ such that $x \wedge a^* = 0$. Then $a^{**} \geq x \in L - R$, hence $a^{**} \in L - R$, a contradiction. Let F be a maximal dual prime ideal containing $(L - R) \vee [a^*)$ and, similarly, let G be a maximal dual prime ideal containing $(L - R) \vee [a^{**})$. We set $P = L - F$ and $Q = L - G$. Then P and Q are minimal prime ideals. Moreover, $P \neq Q$, because $a^* \in F = L - P$ and hence $a^* \notin P$; thus $a^{**} \in P$, while $a^{**} \notin Q$. Finally, $P, Q \subseteq R$, hence $P \vee Q \neq L$.   □

We conclude this section by proving two representation theorems for Stone algebras that correspond to the two representation theorems for distributive lattices given in Section 1. The proofs we present use the Subdirect Product Representation Theorem of G. Birkhoff [1944]. Direct proofs are possible but we shall present a proof that can be generalized to other varieties of distributive lattices with pseudocomplementation.

In the remainder of this section "algebra" means universal algebra. For the purpose of this book, the reader can substitute "lattice" or "lattice with pseudocomplementation" for "algebra". Just as for posets and lattices, we write A for the algebra $\mathfrak{A} = \langle A; F \rangle$ if there is no danger of confusion.

Definition 7. An algebra A is called *subdirectly irreducible* iff there exist elements $u, v \in A$ such that $u \neq v$ and $u \equiv v \; (\Theta)$, for all congruences $\Theta > \omega$.

In other words, A has at least two elements and

$$\text{Con } A = \{\omega\} \cup [\Theta(u, v)).$$

An equivalent form is:

Corollary 8. *The algebra A is subdirectly irreducible iff $\omega = \bigwedge(\Theta_i \mid i \in I)$ ($\Theta_i \in \text{Con } A$, for $i \in I$) implies that $\Theta_i = \omega$, for some $i \in I$.*

Example 9. A distributive lattice L is subdirectly irreducible iff $|L| = 2$.

Proof. If $|L| = 1$, then L is not subdirectly irreducible by definition. If $|L| = 2$, then obviously L is subdirectly irreducible. Let $|L| > 2$. Then there exist a, b, $c \in L$, $a < b < c$. We claim that $\Theta(a, b) \wedge \Theta(b, c) = \omega$, which by Corollary 2 shows that L is not subdirectly irreducible. Let $x \equiv y \; (\Theta(a, b) \wedge \Theta(b, c))$. By Theorem 3.3, this implies that $x \vee b = y \vee b$ and $x \wedge b = y \wedge b$; thus $x = y$ by Corollary 1.3.   □

Example 10. B_2 is the only subdirectly irreducible Boolean algebra.

Proof. Let B be Boolean. The statement is obvious for $|B| \leq 2$. If $|B| > 2$, then B has a direct product representation, $B = D_1 \times D_2$, $|D_1|, |D_2| \geq 2$ (use Exercise 1.5); thus B cannot be subdirectly irreducible.   □

We shall need a simple universal algebraic lemma:

Lemma 11 (The Second Isomorphism Theorem). *Let A be an algebra and let Θ be a congruence relation of A. For any congruence Φ of A such that $\Phi \geq \Theta$, define the relation Φ/Θ on A/Θ by*

$$[x]\Theta \equiv [y]\Theta \quad (\Phi/\Theta) \qquad iff \qquad x \equiv y \quad (\Phi).$$

Then Φ/Θ is a congruence of A/Θ. Conversely, every congruence Ψ of A/Θ can be (uniquely) represented in the form $\Psi = \Phi/\Theta$, for some congruence $\Phi \geq \Theta$. In particular, the congruence lattice of A/Θ is isomorphic with the dual ideal $[\Theta)$ of the congruence lattice of A.

Proof. We have to prove that Φ/Θ is well-defined, it is an equivalence relation, and it has the Substitution Property. To represent Ψ, define Φ by

$$x \equiv y \quad (\Phi) \qquad iff \qquad [x]\Theta \equiv [y]\Theta \quad (\Psi).$$

Again, we have to verify that Φ is a congruence. $\Phi/\Theta = \Psi$ follows from the definition of Φ. The details are trivial and left to the reader. $\qquad\square$

Varieties of universal algebras can be introduced by defining polynomials and identities, just as in the case of lattices. However, in the next theorem (see G. Birkhoff [1944]), the reader can avoid the use of this terminology by substituting for "variety" the phrase "class closed under the formation of subalgebras, homomorphic images, and direct products". (This does not make the result more general, see Theorem V.1.3.)

Theorem 12 (Birkhoff's Subdirect Representation Theorem).   Let **K** be a variety of algebras. Every algebra A in **K** can be embedded in a direct product of subdirectly irreducible algebras in **K**.   —

Proof. For a, $b \in A$, $a \neq b$, let $\mathfrak{X}$ denote the set of all congruences Θ of A satisfying $a \not\equiv b\,(\Theta)$. $\mathfrak{X}$ is not empty since $\omega \in \mathfrak{X}$. Let C be a chain in $\mathfrak{X}$. Then $\Theta = \bigcup C$ is a congruence, $a \not\equiv b\,(\Theta)$, and thus every chain in $\mathfrak{X}$ has an upper bound. By Zorn's Lemma, $\mathfrak{X}$ has a maximal element $\Psi(a, b)$. We claim that $A/\Psi(a, b)$ is subdirectly irreducible; in fact, $u = [a]\Psi(a, b)$ and $v = [b]\Psi(a, b)$ satisfy the condition of Definition 7. Indeed, if Θ is any congruence of $A/\Psi(a, b)$, $\Theta \neq \omega$, then by Lemma 11, $\Theta = \Phi/\Psi(a, b)$. Since $\Theta \neq \omega$, we obtain that $\Phi > \Psi(a, b)$, and so $a \equiv b\ (\Phi)$. Thus $u \equiv v\ (\Theta)$, as claimed.

Let $B = \prod(\,A/\Psi(a, b) \mid a, \ b \in A, \ a \neq b\,)$; then B is a direct product of subdirectly irreducible algebras. We embed A into B by $\varphi \colon x \mapsto f_x$, where f_x takes on the value $[x]\Psi(a, b)$ in the algebra $A/\Psi(a, b)$. Clearly, φ is a homomorphism. To show that φ is one-to-one, assume that $f_x = f_y$. Then $x \equiv y\ (\Psi(a, b))$, for all a, $b \in A$, $a \neq b$. Therefore,

$$x \equiv y \quad (\bigwedge(\,\Psi(a, b) \mid a, \ b \in A, \ a \neq b\,)),$$

and so $x = y$. $\qquad\square$

We got a little bit more than claimed. If we pick $w \in A/\Psi(a,b)$, then $w = [x]\Psi(a,b)$, for some $x \in A$. Thus there is an element in the representation of A whose component in $A/\Psi(a,b)$ is w; such a representation is called *subdirect*.

Corollary 13. *In a variety* **K**, *every algebra can be represented as a subdirect product of subdirectly irreducible algebras in* **K**.

Observe how strong Theorem 12 is. If combined with Example 9, it yields Theorem 1.19; when combined with Example 10, we obtain Corollary 1.21.

The readers should note that subdirect representations of an algebra A are in one-to-one correspondence with families $(\Theta_i \mid i \in I)$ of congruence relations of A satisfying

$$\bigwedge(\Theta_i \mid i \in I) = \omega.$$

A subdirect representation by subdirectly irreducible algebras corresponds to such families $(\Theta_i \mid i \in I)$ that satisfy, in addition, the property that all Θ_i are completely meet-irreducible (an element Θ is *completely meet-irreducible* iff $\Theta = \bigwedge(\Phi_j \mid j \in J)$ implies that $\Theta = \Phi_j$, for some $j \in J$). Thus Lemma 11 and Theorem 12 combine to yield

Corollary 14. *Every congruence relation of an algebra is a meet of completely meet-irreducible elements.*

Let S_1 denote the three-element chain $\{0, e, 1\}$ $(0 < e < 1)$ as a distributive lattice with pseudocomplementation.

Theorem 15.   Up to isomorphism, B_2 and S_1 are the only subdirectly irreducible Stone algebras.   —

Proof. B_2 and S_1 are obviously subdirectly irreducible (the congruence lattice of S_1 is a three-element chain).

Now let L be a subdirectly irreducible Stone algebra. By Lemma 3, $S(L)$ is a subalgebra of L. By definition, $|L| > 1$. If $|S(L)| > 2$, then $S(L)$ is directly decomposable and therefore so is L. Thus $|S(L)| = 2$, that is, $S(L) = \{0, 1\}$. If $|D(L)| > 2$, then there exist congruences Θ and Φ on $D(L)$ such that $\Theta \wedge \Phi = \omega$ on $D(L)$ (by Example 9). Extend Θ and Φ to L by defining $\{0\}$ as the only additional congruence class. We conclude that L is subdirectly reducible.

Thus $S(L) = \{0, 1\}$ and so $L = D(L) \cup \{0\}$ and $|D(L)| \leq 2$, yielding that $L \cong B_2$ or $L \cong S_1$. $\square$

Corollary 16 (G. Grätzer [1969]). *Every Stone algebra can be embedded in a direct product of two- and three-element chains (regarded as Stone algebras).*

Proof. Combine Corollary 13 and Theorem 15. $\square$

Every distributive lattice can be embedded in some $P(X)$. O. Frink [1962] asked whether every Stone algebra can be embedded in some Id $P(X)$?

Theorem 17 (G. Grätzer [1963]).   A distributive lattice with pseudocomplementation L is a Stone algebra iff it can be embedded into some Id $P(X)$. ___

Proof. Id $P(X)$, and therefore any of its subalgebras, is a Stone algebra by Corollary 2.

It is obvious that the class of Stone algebras that can be embedded into some Id $P(X)$ is closed under the formation of direct products and subalgebras. Hence by Corollary 16, it is sufficient to prove that B_2 and S_1 can be so embedded. For B_2 this is obvious. To embed S_1, take an infinite set X and embed S_1 into Id $P(X)$ as follows:

$$0 \mapsto \{\varnothing\},$$
$$e \mapsto \{\, A \mid A \subseteq X \text{ and } A \text{ is finite} \,\},$$
$$1 \mapsto P(X).$$

It is obvious that this is an embedding. □

Exercises

1. Show that every bounded chain is a pseudocomplemented distributive lattice.

2. Let L be a lattice with 1. Adjoin a new zero 0 to L: $L_1 = L \cup \{0\}$, $0 < x$, for all $x \in L$. Show that L_1 is a pseudocomplemented lattice.

3. Call a lattice with 0 *dense* iff 0 is meet-irreducible. Show that every bounded dense lattice K is pseudocomplemented and that every such lattice can be constructed by the method of Exercise 2 with $L = D(K)$.

4. Find an example of a complete distributive lattice L that is not pseudocomplemented.

5. Prove that if L is a complete Stone lattice, then so is Id L. (Hint: $I^* = (a]$, where $a = \bigwedge (\, x^* \mid x \in I \,)$.)

6. Show that a distributive pseudocomplemented lattice is a Stone lattice iff

$$(a \vee b)^{**} = a^{**} \vee b^{**},$$

    for all $a, b \in L$.

7. Find a small set of identities characterizing Stone algebras.

8. Let L be a Stone algebra. Show that $S(L)$ is a retract of L.

9. Let L be a Stone algebra, a, $b \in S(L)$, and $a \leq b$. Prove that

$$x \mapsto (x \vee a^*) \wedge b$$

embeds F_a into F_b.

10. Let B be a Boolean algebra. Define $B^{[2]} \subseteq B^2$ by $\langle a, b \rangle \in B^{[2]}$ iff $a \leq b$. Verify that $B^{[2]}$ is a sublattice of B^2 but not a subalgebra of B^2. Show that $B^{[2]}$ is a Stone lattice.

11. Let L be a pseudocomplemented distributive lattice. Show that, for a, $b \in L$,

$$(a \vee b)^* = a^* \wedge b^*,$$
$$(a \wedge b)^{**} = a^{**} \wedge b^{**}.$$

12. Prove that a prime ideal P of a Stone algebra L is minimal iff P as an ideal of L is generated by $P \cap S(L)$.

13. Show that a distributive lattice with pseudocomplementation is a Stone algebra iff every prime ideal contains exactly one minimal prime ideal (G. Grätzer and E. T. Schmidt [1957a]).

*14. Prove that a poset Q is isomorphic to the poset of all prime ideals of a Stone algebra iff

  (i) every element of Q contains exactly one minimal element;

  (ii) for every minimal element m of Q, the poset $\{ x \mid x > m, \; x \in Q \}$ is isomorphic to the poset of all prime ideals of some distributive lattice with 1.

(See C. C. Chen and G. Grätzer [1969b].)

15. Give a detailed proof of the Second Isomorphism Theorem.

16. Prove Corollary 16 directly.

Exercises 17–31 are from C. C. Chen and G. Grätzer [1969a] and [1969b].

Let B be a Boolean algebra, let D be a distributive lattice with 1, and let φ be a $\{0, 1\}$-homomorphism of B into $\mathrm{Du}\, D$. Set

$$L = \{ \langle x, a \rangle \mid a \in B, \; x \in a\varphi \},$$

and define

$$\langle x, a \rangle \leq \langle y, b \rangle \quad \text{iff} \quad a \leq b \text{ and } x \leq y\varrho_a,$$

where $[y\varrho_a) = a\varphi \wedge [y)$.

17. Verify the following formulas:

   (i)   If $a \in B$ and $d \in D$, then $d\varrho_a = d$ iff $d \in a\varphi$.

   (ii) $d\varrho_a \geq d$, for $a \in B$, $d \in D$.

   (iii) $d\varrho_a \wedge d\varrho_{a'} = d$, for $a \in B$, $d \in D$ (where a' is the complement of a in B).

   (iv) $\varrho_a \varrho_b = \varrho_{a \wedge b}$, for $a, b \in C$.

18. Prove that:

   (i) $d\varrho_a \wedge d\varrho_b = d\varrho_{a \vee b}$, for $a, b \in B$, $d \in D$.

   (ii) $d\varrho_{a \wedge b} = d\varrho_a \vee d\varrho_b$, for $a, b \in B$, $d \in D$.

19. Show that L is a poset under the given partial ordering.

20. For $\langle x, a \rangle$, $\langle y, b \rangle \in L$, verify that

$$\langle x, a \rangle \wedge \langle y, b \rangle = \langle x\varrho_b \wedge y\varrho_a, a \wedge b \rangle.$$

*21. Show that

$$\langle x, a \rangle \vee \langle y, b \rangle = \langle (x\varrho_{b'} \wedge y) \vee (x \wedge y\varrho_{a'}), a \vee b \rangle.$$

22. For $\langle x, a \rangle$, $\langle y, b \rangle$, $\langle z, c \rangle \in L$, let

$$U = (\langle x, a \rangle \wedge \langle y, b \rangle) \vee \langle z, c \rangle,$$
$$V = (\langle x, a \rangle \vee \langle z, c \rangle) \wedge (\langle y, b \rangle \vee \langle z, c \rangle).$$

Compute U; show that

$$V = \langle d, (a \vee c) \wedge (b \vee c) \rangle,$$

where

$$d = d_0 \vee d_1 \vee d_2 \vee d_3,$$
$$d_0 = x\varrho_{b \wedge c'} \wedge y\varrho_{a \wedge c'} \wedge z,$$
$$d_1 = x\varrho_{b \wedge c'} \wedge y\varrho_{a \wedge c} \wedge z,$$
$$d_2 = x\varrho_{b \vee c} \wedge y\varrho_{a \wedge c'} \wedge z,$$
$$d_3 = x\varrho_{b \vee c} \wedge y\varrho_{b \vee c} \wedge z\varrho_{a' \vee b'}.$$

23. Show that $d_0 \geq d_1$ and $d_0 \geq d_2$; therefore, $d = d_0 \vee d_3$.

24. Show that L is distributive.

25. Show that L is a Stone lattice.

26. Identify $b \in B$ with $\langle 1, b \rangle$ and $d \in D$ with $\langle d, 1 \rangle$. Verify that $S(L) = B$, $D(L) = D$, and $\varphi(L) = \varphi$. In other words, we have proved the following theorem of C. C. Chen and G. Grätzer [1969a]:

Construction Theorem. *Given a Boolean algebra B, a distributive lattice D with 1, and a $\{0, 1\}$-homomorphism $\varphi \colon B \to \mathrm{Du}\, D$, there exists a Stone algebra L whose triple is $\langle B, D, \varphi \rangle$.*

27. Describe isomorphisms and homomorphisms of Stone algebras in terms of triples.

28. Describe subalgebras of Stone algebras in terms of triples.

29. For a given Boolean algebra B with more than one element and distributive lattice D with 1, construct a Stone algebra with $S(L) \cong B$ and $D(L) \cong D$. ($S(L)$ and $D(L)$ are independent.)

30. Show that a Stone algebra L is complete if $S(L)$ and $D(L)$ are complete.

*31. Characterize the completeness of Stone algebras in terms of triples.

*32. Show that a distributive lattice with pseudocomplementation L has the *Congruence Extension Property*: for a subalgebra K of L and for a congruence relation Θ of K, there exists a congruence relation Φ of L extending Θ, that is, for $a, b \in K$, $a \equiv b \ (\Theta)$ iff $a \equiv b \ (\Phi)$ (G. Grätzer and H. Lakser [1971]).

33. Let L be a lattice or a lattice with additional operations. If $a, b \in L$ and $[b, a]$ is simple, then $\Psi(a, b)$ (defined in the proof of Theorem 12) is unique.

Further Topics and References

A number of papers are devoted to the subject of guaranteeing the distributivity of a lattice by imposing equations on a fixed generating set. For instance, if L is generated by a, b, and c, then L is distributive iff

$$x \wedge (y \vee z) = (x \wedge y) \vee (x \wedge z)$$

and its dual holds for any permutation x, y, z of a, b, c and

$$(a \wedge b) \vee (b \wedge c) \vee (c \wedge a) = (a \vee b) \wedge (b \vee c) \wedge (c \vee a),$$

by a result of O. Ore [1940]. These seven conditions (there are only seven by commutativity) are independent but if L is assumed to be modular, all seven are equivalent, as it is evident from Figure I.5.7. Generalizations can be found in M. Kolibiar [1972] and S. Tamura [1971], [1971a], and for the modular case in B. Jónsson [1955], R. Musti and E. Buttafuoco [1956], and R. Balbes [1969].

For some special classes of lattices, Theorem 1.1 and 1.2 have various stronger forms that claim the existence of very large or very small pentagons and diamonds. For instance, a bounded relatively complemented nonmodular lattice always contains a pentagon as a $\{0, 1\}$-sublattice. The same is true of the diamond in certain complemented modular lattices; such results are implicit in J. von Neumann [1936]. If the lattice is finite, modular, and nondistributive, then it contains a diamond in which a, b, c cover o, and i covers a, b, c. If L is finite and nonmodular, then the pentagon it contains can be required to satisfy $a \succ b$.

It seems hard to generalize the uniqueness of an irredundant join-representation of an element of a finite distributive lattice. The best generalization is that of R. P. Dilworth and P. Crawley [1960] to distributive algebraic lattices in which every proper interval has an atom. See the survey article by R. P. Dilworth [1961] and S. Kinugawa and J. Hashimoto [1966]. Some results on, and references to, the modular and semimodular cases can be found in Chapter IV.

By Theorem 1.19, every distributive lattice is a subdirect product of copies of C_2. Distributive lattices that are subdirect products of copies of C_n are characterized in F. W. Anderson and R. L. Blair [1961].

For a finite distributive lattice L, what is the smallest k such that L is a subdirect product of k chains? For $a \in L$, let n_a be the number of elements of L covering a. Then $k = \max\{n_a \mid a \in L\}$. This is an easy application of a well-known combinatorial result of R. P. Dilworth [1950] (first discovered by T. Gallai in 1936), according to which a poset P is a union of at most k chains iff P is of width $\leq k$. See also H. Tverberg [1967].

The size of maximal sublattices of a finite distributive lattice is investigated in I. Rival [1973] and [1974]; maximal $\{0, 1\}$-sublattices of finite Boolean lattices are considered in H. [1968] and D. Steven [1968]. $\{0, 1\}$-sublattices of finite Boolean lattices and finite topologies are closely related. See also R. P. Stanley [1973].

Theorems 1.9 and 1.19 can be generalized by using ideals other than prime ideals. In this connection see Theorem IV.5.17 and G. Birkhoff and O. Frink [1948].

The problem of determining $|\mathbf{F_D}(n)|$ goes back to R. Dedekind [1900]. For some older and some recent contributions (including the tabulation of $|\mathbf{F_D}(n)|$, for $n \leq 7$) see R. Church [1940], E. N. Gilbert [1954], G. Hansel [1967], D. Kleitman [1969], R. Church [1965] (this gives the correct value for $|\mathbf{F_D}(7)|$), and F. Lunnon [1971].

Infinitary Boolean polynomials are considered in H. Gaifman [1964] and A. W. Hale [1964]; they prove that free complete Boolean algebras do not exist.

Free distributive lattices have many interesting properties. All chains are finite or countable (the proof of this is similar to that of Theorem VI.2.10). If a

and H are such that $x \wedge y = a$, for all x, $y \in H$, $x \neq y$, call H a-*disjoint*. In a free distributive lattice all a-disjoint sets are finite, see R. Balbes [1967].

I. Reznikoff [1963] and A. Horn [1968] prove that all chains of a free Boolean algebra are also finite or countable.

A more general form of Theorem 2.5 can be found in R. Sikorski [1964]; for the universal algebraic background, see Theorem 12.2 in G. Grätzer [1968].

Some properties of $\Theta[I]$ can be generalized to certain ideals of a general lattice, see Chapter III.

Let L be a lattice and let f be an n-ary function on L, that is, $f \colon L^n \to L$. We say that f has the *Congruence Substitution Property* iff, for every congruence relation Θ of L and a_i, $b_i \in L$, $1 \leq n$, $a_i \equiv b_i$ (Θ), $1 \leq i \leq n$, imply that $f(a_1, \ldots, a_n) \equiv f(b_1, \ldots, b_n)$ (Θ). On a Boolean algebra B, a function has the Congruence Substitution Property iff it is a Boolean algebraic function, that is, a Boolean polynomial in which elements of B are substituted for some variables. Functions satisfying the Congruence Substitution Property on a bounded distributive lattice were described in G. Grätzer [1964].

The method given in Theorem 3.9 is not the only one used to introduce ring operations in a generalized Boolean lattice. G. Grätzer and E. T. Schmidt [1958d] prove that ring operations $+$, $\cdot$ can be introduced on a distributive lattice L such that $+$ and $\cdot$ satisfy the Congruence Substitution Property iff L is relatively complemented. Furthermore, $+$ and $\cdot$ are uniquely determined by the zero of the ring, which can be an arbitrary element of L.

Whether every distributive algebraic lattice is isomorphic to the congruence lattice of some lattice is one of the longest-standing problems of lattice theory. The method used in Section 3 for the finite case can be easily extended to infinite algebraic lattices in which every element is a finite join of join-irreducible elements. A further extension of this result is in E. T. Schmidt [1962] and [1968]. (In reading the two papers, the reader should disregard the Theorem and Lemmas 9 and 10 of the first paper.) See also E. T. Schmidt [1969], [1974], [1975] and J. Berman [1972].

Congruence lattices of distributive lattices are not considered in this text because their characterization problem is trivial by Lemma 4.5: A lattice is isomorphic to the congruence lattice of a distributive lattice iff it is isomorphic to Id B, where B is a generalized Boolean lattice; Id B, by Theorem 3.13, is characterized as a distributive algebraic lattice in which the compact elements form a relatively complemented sublattice.

Algebraic lattices originated in A. Komatu [1943], G. Birkhoff and O. Frink [1948], L. Nachbin [1949], and J. R. Büchi [1952]. The original definition in G. Birkhoff and O. Frink [1948] is as follows:

   (i) L is complete;

  (ii) every element in L is the join of join-inaccessible elements;

 (iii) L is join-continuous.

In this definition, an element a of L is *join-accessible* iff there is a nonempty subset H of L such that H is *directed* (for $x, y \in H$, there exists an upper bound $z \in H$), $\bigvee H = a$, and $a \notin H$; otherwise, a is *join-inaccessible*; L is *join-continuous* iff, for any $a \in L$ and directed $H \subseteq L$, we have $a \wedge \bigvee H = \bigvee(a \wedge h \mid h \in H)$.

Interestingly, it is sufficient to formulate conditions (i) and (iii) of the previous paragraph for chains only. In other words, a lattice L is complete iff $\bigwedge C$ and $\bigvee C$ exist for any chain C of L; and a (complete) lattice L is join-continuous iff $a \wedge \bigvee C = \bigvee(a \wedge c \mid \in C)$, for any chain C of L. These statements are immediate consequences of the following result of T. Iwamura [1944]. Let H be an infinite directed set. Then H has a decomposition $H = \bigcup(H_\gamma \mid \gamma < \alpha)$, where each H_γ is directed; for $\gamma < \delta < \alpha$ we have $H_\gamma \subseteq H_\delta$, and for $\gamma < \alpha$, we have $|H_\gamma| < H$. Extensions of lattices preserving (iii) are considered in D. A. Kappos and F. Papangelou [1966].

The results of Section 4 were presented in G. Grätzer [1971] without the assumption that L has a 0. The necessity of this assumption is pointed out in R. D. Byrd and R. A. Mena [1975] and R. D. Byrd, R. A. Mena, and L. A. Troy [1975]. These papers give a detailed analysis of what happens if this assumption is dropped.

W. Hanf [1976] proves that there is no algorithmic way to find a generating chain in all countable Boolean algebras. There are, however, always generating chains of a rather special order type; see R. S. Pierce [1973].

The problem of completions of a lattice has been extensively studied. The standard method is the MacNeille completion (see Exercises 4.9–4.14, which are from H. M. MacNeille [1937]). This method is described in detail in G. Bruns [1962]; see also Y. Sampei [1953]. One of the important shortcomings of this method is that it does not preserve identities, not even distributivity; see M. Cotlar [1944] and N. Funayama [1944]. See also P. Crawley [1962], R. P. Dilworth and J. E. McLaughlin [1952], and W. H. Cornish [1974]. What is preserved for complemented modular lattices is examined in J. E. McLaughlin [1961].

If we define a *completion* $\hat{L}$ of a lattice L as any complete lattice $\hat{L}$ containing L as a sublattice such that all infinite meets and joins that exist in L are preserved in $\hat{L}$, then we can ask whether there is *any* distributive completion of a distributive lattice. The answer, in general, is in the negative; see P. Crawley [1962].

For Boolean algebras, set representations preserving all joins and meets, or joins and meets of certain types, play an important role, especially in relationship with infinite distributive identities. The larger part of the book by R. Sikorski [1964] is devoted to this problem. For distributive lattices, these questions become even more complicated—see, for instance, G. Bruns [1959], [1961], and [1962a]; C. C. Chang and A. Horn [1962]; R. P. Dilworth and J. E. McLaughlin [1952]; A. Horn [1962]; J. Jakubik [1972]; and G. N. Raney [1952], [1953]; see also R. Balbes and P. Dwinger [1974].

J. Jakubik [1957], [1958] and G. W. Day [1970] consider cardinalities of maximal chains in Boolean algebras. See also Exercise 4.28.

R. S. Pierce [1972] gave a deep analysis of the structure of countable Boolean algebras. A generalization of Boolean algebras generated by chains can be found in R. W. Quackenbush and H. C. Reichel [1975].

The method used in establishing Lemma 4.22 was employed by B. Jónsson [1956] to construct "universal" lattices, that is, lattices containing a large class of lattices as sublattices (see also B. Jónsson [1960] and M. D. Morley and R. L. Vaught [1962]).

For a distributive lattice L, $\mathcal{P}(L)$ is a poset and, by Section 5, a topological space. These two approaches are married in H. A. Priestley [1972] where $\mathcal{P}(L)$ is considered, and characterized, as an ordered topological space. This approach proved very convenient in applications; see H. A. Priestley [1972], [1974], and [1975]; see also M. E. Adams [1974], [1975], [1976], B. A. Davey [1973], [1974], T. P. Speed [1972a]. T. P. Speed [1972] characterizes $\mathcal{P}(L)$ for bounded L by the property that $\mathcal{P}(L) \cup \{0, 1\}$ is a profinite poset. See also M. E. Adams [1975], R. Balbes [1971], B. A. Davey [1973], T. P. Speed [1972a] and [1974].

A solution to the word problem of free $\{0, 1\}$-distributive products is given in G. Grätzer and H. Lakser [1969]. In the same paper, it is proved that if $\mathfrak{m}$ is a regular cardinal and L_i, $i \in I$, are bounded distributive lattices satisfying $|C| < \mathfrak{m}$ for any chain C in any L_i, then the same holds in the free $\{0, 1\}$-distributive product. In particular, if all L_i satisfy the Countable Chain Condition, so does the free $\{0, 1\}$-distributive product; see also Section 12 of G. Grätzer [1971]. M. E. Adams and D. Kelly [1977] extended this result to free products of lattices. See also B. Jónsson [1971]. The analogous problem for a-disjoint sets is considered in H. Lakser [1973a] and M. E. Adams and D. Kelly [1977]. An interesting description of free $\{0, 1\}$-distributive products can be found in R. W. Quackenbush [1972a].

A weaker form of Corollary 6.16 can be found in T. P. Speed [1969].

Subdirect products of lattices have an interesting property due to G. M. Bergman (see K. A. Baker and A. F. Pixley [1975]): a subdirect product L of L_1, ..., L_n is determined up to isomorphism by its projections onto $L_i \times L_j$, for all i and j with $1 \leq i < j \leq n$. Subdirect products of ideal lattices are investigated in R. Freese [1975].

The interdependence of the various "finiteness conditions" for distributive lattices is investigated in S. P. Avann [1964] and S. Rudeanu [1964].

Boolean algebras with only the trivial automorphisms are constructed in M. Katětov [1951], B. Jónsson [1951], and L. Reiger [1951]; Katětov's example has $2^{\aleph_0}$ elements, the others are very large.

Some categorical concepts have been studied for distributive lattices. For a class $\mathbf{K}$ of algebras, $P \in \mathbf{K}$ is called *projective in* $\mathbf{K}$ iff, for any A, $B \in \mathbf{K}$, homomorphisms $\gamma \colon A \to B$ and $\beta \colon P \to B$, if γ is onto, then there is a homomorphism $\alpha \colon P \to A$ satisfying $\alpha\gamma = \beta$. Free algebras are projective.

R. Balbes [1967] proves that a finite distributive lattice L is projective in $\mathbf{D}$ iff the join of any two meet-irreducible elements is meet-irreducible. See also R. Balbes and A. Horn [1970b], G. Grätzer and B. Wolk [1970]. R. Engelking

[1965] has shown that a subalgebra of a free Boolean algebra need not be projective, but every countable Boolean algebra is projective, see P. R. Halmos [1961].

I is *injective in* $\mathbf{K}$ iff, for any A, $B \in \mathbf{K}$, embedding $\gamma\colon B \to A$, and homomorphism $\beta\colon B \to I$, there is a homomorphism $\alpha\colon A \to I$ satisfying $\gamma\alpha = \beta$. Injective Boolean algebras are the complete ones, see P. R. Halmos [1961]; and injective distributive lattices are the complete Boolean lattices. For more details on categorical concepts for distributive lattices, see Section 13 of G. Grätzer [1971].

R. Dedekind found the distributive identity by investigating ideals of number fields. Rings with a distributive lattice of ideals have been investigated by E. Noether [1927], L. Fuchs [1949] (who names such rings *arithmetical rings*), I. S. Cohen [1950], and C. U. Jensen [1963]. Varieties of rings with distributive ideal lattices were considered in G. Michler and R. Wille [1970] and in H. Werner and R. Wille [1970]. E. A. Behrens [1960] and [1961] considered rings in which one-sided ideals form a distributive lattice. Rings with a distributive lattice of subrings were classified in P. A. Freĭdman [1967]. In this context, G. M. Bergman's work on the distributive-divisor-lattice of free algebras should be mentioned, see Chapter 4 of P. M. Cohn [1971].

For groups, the corresponding problem is much simpler: The subgroup lattice of a group G is distributive iff G is locally cyclic (see O. Ore [1937] and [1938]). H. L. Silcock [1977] has proved that every finite distributive lattice is isomorphic to the lattice of normal subgroups of a suitable group.

The distributivity of congruence lattices of lattices has a number of important consequences. B. Jónsson [1967] discovered that many of these results hold for arbitrary universal algebras with distributive congruence lattices. His results have found applications that go far beyond lattice theory—they have already been applied to lattice-ordered algebras, closure algebras, nonassociative lattices, cylindric algebras, monadic algebras, lattices with pseudocomplementation, primal algebras, and multi-valued logics.

The foregoing examples show the central role played by distributive lattices in applications of the lattice concept.

Except for V. Glivenko's early work [1929], the study of pseudocomplemented distributive lattices started only in 1956 with a solution of Problem 70 of G. Birkhoff [1948] giving characterization of Stone lattices by minimal prime ideals (G. Grätzer and E. T. Schmidt [1957a]; for a simplified proof, see J. C. Varlet [1966]), see Theorem 6.6. The idea of a triple was conceived by the author as a tool to prove Frink's conjecture (see O. Frink [1962] and Theorem 6.17). This attempt failed and as a result triples were not utilized until C. C. Chen and G. Grätzer [1969a] and [1969b]. Frink's conjecture was solved using the Compactness Theorem in G. Grätzer [1963] (see Theorem 6.17) and the proof was simplified in G. Bruns [1965] and G. Grätzer [1969]. An interesting generalization can be found in H. Lakser [1971]. Projective and injective Stone algebras are investigated in R. Balbes and G. Grätzer [1971]; see also R. Beazer

[1975]. See also Sections 16 and 17 of G. Grätzer [1971].

Let $\mathbf{B}_n$ denote the variety of distributive lattices with pseudocomplementation satisfying the identity

$$(\mathrm{L}_n) \qquad (x_1 \wedge \cdots \wedge x_n)^* \vee (x_1^* \wedge \cdots \wedge x_n)^* \vee \cdots \vee (x_1 \wedge \cdots \wedge x_n^*)^* = 1,$$

for $n \geq 1$. Then $\mathbf{B}_1$ is the class of Stone algebras. K. B. Lee [1970] has proved that $\mathbf{B}_n$, $-1 \leq n \leq \omega$, is a complete list of varieties of distributive lattices with pseudocomplementation, where $\mathbf{B}_{-1}$ is the trivial class, $\mathbf{B}_0$ is the class of Boolean algebras, and $\mathbf{B}_\omega$ is the class of all distributive lattices with pseudocomplementation. Moreover,

$$\mathbf{B}_{-1} \subset \mathbf{B}_0 \subset \mathbf{B}_1 \subset \cdots \subset \mathbf{B}_n \subset \cdots \subset \mathbf{B}_\omega.$$

In H. Lakser [1971] and G. Grätzer and H. Lakser [1971] and [1972], most of the structure theorems known for Stone algebras have been generalized to the classes $\mathbf{B}_n$. In these papers the amalgamation class of $\mathbf{B}_n$ (in the sense of Section V.4) is also considered. See also Chapter 3 of G. Grätzer [1971], where many of these results are presented in detail.

In more than 30 papers, T. Katriňák investigates various generalizations and specializations of distributive lattices with pseudocomplementation and establishes triple constructions for many classes. These classes include relative Stone algebras, generalized Stone algebras, pseudocomplemented posets, generalized Stone join-semi-lattices, distributive pseudocomplemented meet-semilattices, modular pseudocomplemented lattices, modular pseudocomplemented lattices satisfying $x^* \vee x^{**} = 1$, double Stone algebras, and so on. Free algebras, varieties, injective hulls and many other questions are considered. The reader is referred to the papers of T. Katriňák listed in the Bibliography and to R. Balbes and A. Horn [1970], R. Balbes and P. Dwinger [1974], G. Bruns and G. Kalmbach [1971] and [1972], A. Day [1970a], H. Lakser [1970] and [1973], J. C. Varlet [1963], [1966], and [1968], P. V. R. Murty and V. V. R. Rao [1974], U. M. Swamy and V. V. R. Rao [1975].

Relatively pseudocomplemented distributive lattices (often called *Heyting algebras*) arise from nonclassical logic and were first investigated by T. Skolem about 1920.

For a detailed development, see H. B. Curry [1963], which contains all the important rules of computation with $a * b$. Injectives and projectives were investigated in the work of R. Balbes and A. Horn [1970a] and A. Day [1970a]; see also A. Horn [1969] and [1969a], W. C. Nemitz [1965]).

In connection with nonclassical logic, many algebras emerged that are distributive lattices endowed with some additional structure. The best known of these, the Post algebras, happen to be Stone lattices. In fact, a *Post algebra* is a free $\{0, 1\}$-distributive product of a finite chain and a Boolean algebra, in which the elements of the chain are regarded as nullary operations; see E. L. Post

[1921], P. C. Rosenbloom [1942], G. Epstein [1960], T. Traczyk [1963] and [1967], C. C. Chang and A. Horn [1961], G. Rousseau [1970], R. Balbes and P. Dwinger [1971] and [1971a], and M. Mandelker [1970].

For some other types of distributive lattices with additional operations, see N. D. Belnap and J. H. Spencer [1966], H. M. Dunn and N. D. Belnap [1968], H. Rasiowa and R. Sikorski [1963], and J. C. Varlet [1968].

Two books have appeared recently dealing with distributive lattices with additional operations. R. Balbes and P. Dwinger [1974] deals with them from a lattice theoretical viewpoint while H. Rasiowa [1974] investigates their usefulness to logic.

Some of the results for pseudocomplemented distributive lattices referred to above have been extended to ideals of general distributive lattices; see R. Beazer [1975a], W. H. Cornish [1972], [1974a], B. A. Davey [1974a], M. Mandelker [1970].

Some results do not require the distributivity of the whole lattice; for instance, to prove that $\mathrm{Id}\, L$ is pseudocomplemented, it is sufficient that L be 0-*distributive*, that is, $a \wedge b = a \wedge c = 0$ implies that $a \wedge (b \vee c) = 0$, for $a,\ b,\ c \in L$. This and some related concepts have been in the literature for 20–30 years. Here are some recent references: W. H. Cornish [1974b], M. F. Janowitz [1975], J. C. Varlet [1968a].

Problems

1. Find short one-identity axioms characterizing Boolean algebras.

2. Characterize the automorphism groups of Boolean algebras. (An unpublished result of B. Jónsson states that the finite automorphism groups of Boolean algebras are exactly the finite symmetric groups. See also R. N. McKenzie and J. D. Monk [1973].)

3. Let B and C be infinite Boolean algebras and let $\mathfrak{m}$ be a cardinal number satisfying $\mathfrak{m} \geq |B|$ and $|C|$. Does there exist a Boolean algebra D containing B as a subalgebra and such that the automorphism groups of C and D are isomorphic?

4. Investigate further the poset $\mathcal{P}(L)$ of prime ideals of a distributive lattice.

5. Characterize $\mathcal{P}(L)$ under the additional assumption that L has a unit and/or a zero.

6. For which classes $\mathbf{K}$ of finite lattices can we claim that if $L \in \mathbf{K}$ and L is nonmodular, then L contains a "minimal" pentagon?

7. Characterize the congruence lattices of lattices. (Conjecture: distributive algebraic lattices).

8. Characterize the congruence lattices of sectionally complemented lattices. (Conjecture: distributive algebraic lattices).

9. Determine $f_{\mathbf{D}}(k, n)$.

10. Investigate the cardinalities of maximal chains in complete Boolean algebras.

11. What can be proved about the cardinalities of maximal chains in a complete and atomic Boolean algebra without the Generalized Continuum Hypothesis? For instance, do they form a convex set?

12. In a free $\{0, 1\}$-distributive product there are two normal forms for an element: normal $\vee$-representation and normal $\wedge$-representation. (See G. Grätzer and H. Lakser [1969] and also Section 12 of G. Grätzer [1971].) Compare the number and form of the terms in a normal $\vee$-representation and normal $\wedge$-representation of an element in a free $\{0, 1\}$-distributive product.

13. For a bounded distributive lattice L, let $\mathrm{F}(L)$ and $\mathrm{F}_n(L)$ denote the lattice of all functions and all n-ary functions on L, respectively, with the Congruence Substitution Property. To what extent do $\mathrm{F}(L)$ and $\mathrm{F}_n(L)$ determine the structure of L?

14. Characterize $\mathrm{F}(L)$ and $\mathrm{F}_n(L)$.

15. Describe the functions with the Congruence Substitution Property on unbounded distributive lattices.

16. Study Problems 13 and 14 in the unbounded case.

17. For a natural number n, let $g(n)$ denote the smallest integer such that, for every distributive lattice D of length n, there is a lattice L of length at most $g(n)$ satisfying $\mathrm{Con}\, L \cong D$. Determine $g(n)$. ($g(n) \leq 2n - 1$ by Exercise 3.35. See also J. Berman [1972] and E. T. Schmidt [1975].)

18. Let K be a lattice and let G be a group. Does there exist a lattice L satisfying $\mathrm{Con}\, L \cong \mathrm{Con}\, K$ and $\mathrm{Aut}\, L \cong G$? ($\mathrm{Aut}\, L$ is the automorphism group of L.)

19. In 18, if K is finite, can L be chosen to be finite?

20. Is there a fixed type of algebras such that every algebraic lattice is isomorphic to the congruence lattice of an algebra of that type? Can this type be $\langle 2 \rangle$?

21. Is the statement "There exist two bounded distributive lattices in which every a-disjoint set is countable but there exists an uncountable a-disjoint set in their free $\{0, 1\}$-distributive product" equivalent to the Souslin Hypothesis?

22. Prove that, for every infinite distributive lattice A, there exist distributive lattices B and C such that the free distributive products of A, B and A, C are elementarily equivalent but B and C are not elementarily equivalent. (Two lattices are *elementarily equivalent* iff a first-order sentence holding for one, also holds for the other. For A finite, such B and C do not exist, see B. Jónsson and P. Olin [1976].)

23. Prove that a distributive lattice L is equationally compact iff L is complete and satisfies (JID) and (MID).

24. Let $\mathbf{K}$ be a variety of lattices such that the word problem of $F_{\mathbf{K}}(\aleph_0)$ can be solved. Give a solution to the word problem of (bounded) free $\mathbf{K}$-products.

25. Describe the projective Stone algebras.

26. Characterize free $\mathbf{B}_n$-products $(n > 1)$.

27. Describe Amal $\mathbf{B}_n$, for $n > 2$.

28. Determine the projectives in $\mathbf{B}_n$, $n > 1$.

29. Let $n \geq 0$. Determine all *injective structures* in the sense of J.-M. Maranda [1964] in the category $\mathbf{B}_n$.

30. For every identity I for distributive lattices with pseudocomplementation, there exists a first-order sentence $\Phi(I)$ such that I holds for L iff $\Phi(I)$ holds for $\mathcal{P}(L)$. (In K. B. Lee [1970], the sentence for $(\mathbf{L}_n)$ is given as follows: "Every element contains at most n minimal elements." For $n = 0$, use Theorem 1.22.) Is there a natural class of first-order sentences properly containing all identities for which the same statement can be made?

31. Investigate the lattice of implicational classes of distributive lattices with pseudocomplementation.

32. For a distributive lattice L, define the number $n(L)$ as follows: Let $n(L)$ be the smallest integer n such that $\operatorname{Id} L \in \mathbf{B}_n$ if L has a zero and $\operatorname{Id}_0 L \in \mathbf{B}_n$ if L does not have a zero. Classify and investigate distributive lattices according to the value of $n(L)$. (The case $n = 1$ was considered in T. Katriňák [1966] and [1968].)

33. Let $\mathbf{K}$ be a variety of algebras satisfying $\mathbf{HS}(\mathbf{K}_1) = \mathbf{ISH}(\mathbf{K}_1)$, for all $\mathbf{K}_1 \subseteq \mathbf{K}$. Does $\mathbf{K}$ satisfy the Congruence Extension Property? Does the assumption that the congruence lattices of algebras in $\mathbf{K}$ are distributive make any difference?

34. Find maximal classes $\mathbf{K} \subset \mathbf{D}$ such that, for $L_1, L_2 \in \mathbf{K}$, $\mathcal{P}(L_1) \cong \mathcal{P}(L_2)$ implies that $L_1 \cong L_2$.

Congruences and Ideals

1. Weak Projectivity and Congruences

Let a, b, c, and d be elements of a lattice L; if, for any congruence relation Θ of L, $a \equiv b \ (\Theta)$ implies that $c \equiv d \ (\Theta)$, then we can say that "$a \equiv b$ forces $c \equiv d$". It is necessary to understand "forcing" in order to study the structure of congruence relations of lattices; this will be accomplished in the present section.

In Section I.3, we proved that $a \equiv b \ (\Theta)$ iff $a \wedge b \equiv a \vee b \ (\Theta)$ and so it is enough to deal with pairs of comparable elements. To simplify our notation, let a/b denote an ordered pair of elements a, b of a lattice L satisfying $b \leq a$; a/b is called a *quotient* of L. (This notation obviously imitates quotient groups: G/H.) c/d is a *subquotient* of a/b iff $b \leq d \leq c \leq a$; we call a/b a *proper quotient* iff $b < a$. If $b \prec a$, then a/b is called a *prime quotient*. We write $a/b = c/d$ iff $a = c$ and $b = d$.

Now take a look at Figures 1 and 2, permitting the degenerate cases: $a = c$ and $b = d$. In both cases, $a \equiv b \ (\Theta)$ iff $c \equiv d \ (\Theta)$, because $x \wedge y \equiv x \ (\Theta)$ iff $y \equiv x \vee y \ (\Theta)$. In either case, we shall write $a/b \sim c/d$, and say that a/b is *perspective* to c/d. If we want to show whether the perspectivity is "up" or "down", we shall write $a/b \nearrow c/d$ in the first case and $a/b \searrow c/d$ in the second case.

If, for some natural number n, there exist

$$a/b = e_0/f_0, \ e_1/f_1, \ \ldots, \ e_n/f_n = c/d$$

such that

$$e_i/f_i \sim e_{i+1}/f_{i+1}, \quad i = 0, \ \ldots, \ n-1,$$

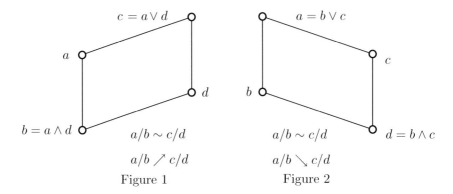

$c = a \vee d$

a

d

$b = a \wedge d$

$a/b \sim c/d$

$a/b \nearrow c/d$

Figure 1

$a = b \vee c$

c

b

$a/b \sim c/d$

$a/b \searrow c/d$

$d = b \wedge c$

Figure 2

then we shall say that a/b is *projective* to c/d and write $a/b \approx c/d$. (Note that $a/b \approx a/b$, with $n = 1$, and $a/b \sim c/d$ implies that $a/b \approx c/d$.) Projectivity is the transitive extension of perspectivity. Observe that $a/b \approx c/d$ and $a = b$ imply that $c = d$.

The concept of projectivity is sufficient for the study of "forcing" in many large classes of lattices (for instance, in the class of modular lattices, see Section IV.1). In general, however, we have to introduce somewhat more cumbersome notions: weak perspectivity and weak projectivity.

Consider Figures 3 and 4, permitting the degenerate cases $a_1 = c$ and $b_1 = d$. If $a \equiv b \ (\Theta)$, then $a_1 \equiv b_1 \ (\Theta)$ by Lemma I.3.7; since $a_1/b_1 \sim c/d$, $a \equiv b \ (\Theta)$ implies that $c \equiv d \ (\Theta)$. Let us say that c/d is *weakly projective into* a/b iff we can get from c/d into a/b in finitely many steps as described in Figures 3 and 4. Figures 3 and 4 then should describe the concept: c/d is *weakly perspective into* a/b. It turns out, however, that a special case of Figures 3 and 4 suffices to describe the same.

As illustrated in Figures 5 and 6, we write $c/d \searrow_w a/b$ iff $b \le d$ and $c = a \vee d$; similarly, $c/d \nearrow_w a/b$ iff $c \le a$ and $d = b \wedge c$. If $c/d \nearrow_w a/b$ or $c/d \searrow_w a/b$, then c/d is *weakly perspective into* a/b, in symbols, $c/d \sim_w a/b$. If, for some natural

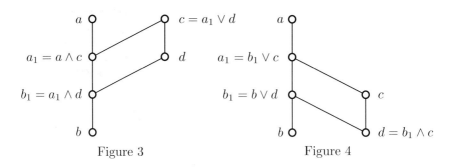

a

$a_1 = a \wedge c$

$b_1 = a_1 \wedge d$

b

Figure 3

$c = a_1 \vee d$

d

a

$a_1 = b_1 \vee c$

$b_1 = b \vee d$

b

Figure 4

c

$d = b_1 \wedge c$

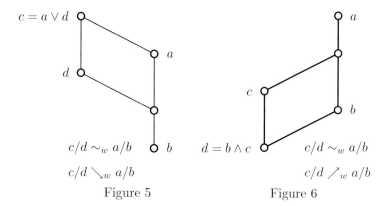

$$c/d \sim_w a/b$$
$$c/d \searrow_w a/b$$

Figure 5

$$d = b \wedge c$$

$$c/d \sim_w a/b$$
$$c/d \nearrow_w a/b$$

Figure 6

number n and

$$c/d = e_0/f_0, \ e_1/f_1, \ \ldots, \ e_n/f_n = a/b,$$

we have

$$e_i/f_i \sim_w e_{i+1}/f_{i+1}, \quad i = 0, \ \ldots, \ n-1,$$

then c/d is *weakly projective into* a/b, in notation, $c/d \approx_w a/b$. (Note that c/d is weakly projective into c/d with $n = 1$.) Weak projectivity is the transitive extension of weak perspectivity. Observe that neither weak perspectivity nor weak projectivity is a symmetric relation. The concept of weak perspectivity is due to R. P. Dilworth [1950a].

The notational system we use is geometrically motivated. Many papers (starting with G. Grätzer and E. T. Schmidt [1958d]) use the notation $[b, a] \rightarrow [d, c]$ or $a/b \rightarrow c/d$ for $c/d \approx_w a/b$; these notations appear to be more natural from a universal algebraic point of view and have the advantage of emphasizing the nonsymmetric nature of weak projectivity.

The following elementary observation shows the equivalence of the various forms of the definition of weak projectivity.

Lemma 1. *Let L be a lattice, a, b, c, $d \in L$, $b \leq a$, and $d \leq c$. Then the following conditions are equivalent:*

(i) *c/d is weakly projective into a/b.*

(ii) *There is an integer m and there are elements $e_0, \ \ldots, \ e_{m-1} \in L$ such that*

$$p_m(a, e_0, \cdots, e_{m-1}) = c,$$
$$p_m(b, e_0, \cdots, e_{m-1}) = d,$$

where the polynomial p_m is defined by

$$p_m(x, y_0, \ldots, y_{m-1}) = \cdots (((x \wedge y_0) \vee y_1) \wedge y_2) \vee \cdots.$$

(iii) *There is an integer n and there are quotients*

$$c/d = e_0/f_0, \ e_0'/f_0', \ e_1/f_1, \ e_1'/f_1', \ \ldots, \ e_n/f_n = a/b$$

such that $e_i/f_i \sim e_i'/f_i'$ *and* e_i'/f_i' *is a subquotient of* e_{i+1}/f_{i+1}, *for* $i = 0$, $1, \ldots, n-1$.

Remark. Condition (iii) is the definition of weak projectivity using Figures 3 and 4.

Proof. Figures 5 and 6 are special cases of Figures 3 and 4, respectively; hence (i) implies (iii). Now let (iii) hold. Then, for each $i = 0, 1, \ldots, n-1$, either $e_i/f_i \nearrow e_i'/f_i'$ or $e_i/f_i \searrow e_i'/f_i'$. In both cases,

$$e_i = p_4(e_{i+1}, e_i', f_i', e_i, f_i),$$
$$f_i = p_4(f_{i+1}, e_i', f_i', e_i, f_i).$$

Repeating these steps n times, we get (ii) with $m \leq 4n$. Thus (iii) implies (ii). Finally, let (ii) hold. Then

$$a \wedge e_0/b \wedge e_0 \nearrow_w a/b,$$
$$(a \wedge e_0) \vee e_1/(b \wedge e_0) \vee e_1 \searrow_w a \wedge e_0/b \wedge e_0,$$

and so on; in m steps, this yields that $c/d \approx_w a/b$. □

We shall write $c/d \overset{k}{\approx}_w a/b$ iff Lemma 1 (ii) holds with $m = k$. This is slightly artificial. It corresponds to requiring that the series of k weak perspectivities end with $\nearrow_w$ and that $\nearrow_w$ and $\searrow_w$ alternate throughout.

Intuitively, "$a \equiv b$ forces $c \equiv d$" iff c/d is put together from pieces each weakly projective into a/b. To state this more precisely, we describe $\Theta(a, b)$, the smallest congruence relation under which $a \equiv b$ (see Section II.3).

Theorem 2 (R. P. Dilworth [1950a]).   Let L be a lattice, $a, b, c, d \in L$, $b \leq a$, $d \leq c$. Then $c \equiv d \ (\Theta(a, b))$ iff, for some sequence

$$c = e_0 \geq e_1 \geq \cdots \geq e_m = d,$$

we have

$$e_j/e_{j+1} \approx_w a/b, \qquad \text{for} \quad j = 0, \ldots, m-1.$$

_____

Proof. Let Φ denote the following relation on L: $x \equiv y \ (\Phi)$ iff $x \vee y = c$ and $x \wedge y = d$ satisfy the condition of the theorem.

We first prove that Φ is a congruence relation by verifying the conditions of Lemma I.3.8. Φ is reflexive since, for any $c \in L$, we get

$$c/c \nearrow_w a \vee c/b \vee c \searrow_w a/b.$$

It is also obvious that if $a_1 \geq a_2 \geq a_3$ and $a_1 \equiv a_2$ (Φ), $a_2 \equiv a_3$ (Φ), then $a_1 \equiv a_3$ (Φ). Indeed, take the sequences establishing the two congruences; putting the two sequences together we get a sequence establishing $a_1 \equiv a_3$ (Φ). Now let $c \equiv d$ (Φ), $c \geq d$, and $f \in L$. Let

$$c = e_0 \geq e_1 \geq \cdots \geq e_m = d$$

be the sequence establishing $c \equiv d$ (Φ), that is, $e_i/e_{i+1} \approx_w a/b$, for $i = 0, \ldots,$ $m - 1$. Then

$$c \wedge f = e_0 \wedge f \geq e_1 \wedge f \geq \cdots \geq e_m \wedge f = d \wedge f,$$
$$e_i \wedge f/e_{i+1} \wedge f \nearrow_w e_i/e_{i+1} \approx_w a/b,$$

hence $e_i \wedge f/e_{i+1} \wedge f \approx_w a/b$, for $i = 0, \ldots, n-1$; this proves that $c \wedge f \equiv d \wedge f (\Phi)$. Similarly, $c \vee f \equiv d \vee f$ (Φ). Thus by Lemma I.3.8, Φ is a congruence relation.

$a \equiv b$ (Φ) and so Φ is a congruence relation under which $a \equiv b$.

Now let Θ be any congruence relation satisfying $a \equiv b$ (Θ). It is easy to see that for $x \geq y$ and $u \geq v$, $x \equiv y$ (Θ) and $u/v \sim_w x/y$ imply that $u \equiv v$ (Θ). By a trivial induction, $x \equiv y$ (Θ) and $u/v \approx_w x/y$ imply that $u \equiv v$ (Θ). So finally, let $c \equiv d$ (Φ), established by

$$c \vee d = e_0 \geq e_1 \geq \cdots \geq e_m = c \wedge d.$$

Since $e_i/e_{i+1} \approx_w a/b$, we conclude that $e_i \equiv e_{i+1}$ (Θ), for $i = 0, \ldots, m - 1$. Therefore, by the transitivity of Θ, we obtain that $c \equiv d$ (Θ). This proves that Φ is the smallest congruence relation under which $a \equiv b$, and so $\Phi = \Theta(a,b)$. $\square$

Let L be a lattice and $H \subseteq L^2$. To compute $\Theta(H)$, the smallest congruence relation Θ under which $a \equiv b$ (Θ), for all $\langle a, b \rangle \in H$, we use the formula (Lemma II.3.2)

$$\Theta(H) = \bigvee (\Theta(a,b) \mid \langle a, b \rangle \in H),$$

and we need a formula for joins:

Lemma 3. *Let L be a lattice and let Θ_i, $i \in I$, be congruence relations of L. Then $a \equiv b$ $(\bigvee(\Theta_i \mid i \in I))$ iff there is a sequence*

$$z_0 = a \wedge b \leq z_1 \leq \cdots \leq z_n = a \vee b$$

such that, for each j with $0 \leq j < n$, there is an $i_j \in I$ satisfying $z_j \equiv z_{j+1}$ (Θ_{i_j}).

The proof of Lemma 3 is the same as that of Theorem I.3.9, namely, a direct application of Lemma I.3.8.

By combining Theorem 2 and Lemma 3, we get:

Corollary 4. *Let L be a lattice, let $H \subseteq L^2$, and let a, $b \in L$ with $b \leq a$. Then $a \equiv b \ (\Theta(H))$ iff, for some integer n, there exists a sequence*

$$a = c_0 \geq c_1 \geq \cdots \geq c_n = b$$

such that, for each i with $0 \leq i < n$, there exists a $\langle d, e \rangle \in H$ satisfying

$$c_i/c_{i+1} \approx_w d \vee e/d \wedge e.$$

Corollary 5. *Let L be a lattice, let I be an ideal of L, and let a, $b \in L$, $b \leq a$. Then $a \equiv b \ (\Theta[I])$ iff, for some integer n, there exists a sequence*

$$a = c_0 \geq c_1 \geq \cdots \geq c_n = b$$

such that, for each i with $0 \leq i < n$, there exist d, $e \in I$ with $e \leq d$ satisfying $c_i/c_{i+1} \approx_w d/e$.

Recall that an ideal I is called the ideal kernel of a congruence relation Θ iff I is a congruence class modulo Θ (see Section I.3).

Corollary 6. *Let L be a lattice and let I be an ideal of L. Then I is a kernel of a congruence relation iff*

$$a/b \approx_w c/d, \ a \in L, \ and \ b, \ c, \ d \in I \quad imply \ that \quad a \in I.$$

Proof. Combine Corollary 5 with the observation that I is an ideal kernel of some congruence relation iff it is the kernel of $\Theta[I]$.   □

In distributive lattices, every ideal is the kernel of some congruence relation, in fact, this property characterizes distributivity. In general lattices, we shall introduce various classes of ideals that are kernels and for which $\Theta[I]$ can be nicely described. This will be done and applied in Sections 2–4.

In a sense, weak projectivity describes the structure of congruence relations of a lattice. It is not surprising, therefore, that many important classes of lattices can be described by weak projectivities. We give two examples.

To introduce the first class, the class of weakly modular lattices, we need a lemma for motivation.

Lemma 7. *Let L be a lattice, a, b, c, $d \in L$, $b < a$, $d < c$, and let $a/b \approx_w c/d$. If L is modular or if L is relatively complemented, then there exists a proper subquotient c'/d' of c/d such that $c'/d' \approx_w a/b$.*

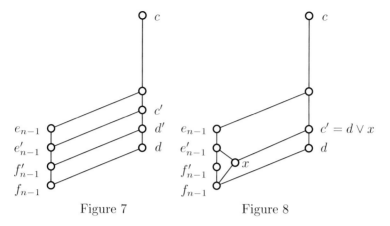

Figure 7   Figure 8

Proof. Let

$$a/b = e_0/f_0 \sim_w e_1/f_1 \sim_w \cdots \sim_w e_n/f_n = c/d;$$

we prove the statement by induction on n. By duality, we can assume that $e_{n-1}/f_{n-1} \nearrow_w c/d$ and, by the induction hypothesis, there exist $e'_{n-1} > f'_{n-1}$ such that $e'_{n-1}/f'_{n-1} \approx_w a/b$ and $f_{n-1} \leq f'_{n-1} < e'_{n-1} \leq e_{n-1}$.

Let L be modular. As in Figure 7, define $d' = d \vee f'_{n-1}$ and $c' = d \vee e'_{n-1}$. Then

$$d' \vee e'_{n-1} = (d \vee f'_{n-1}) \vee e'_{n-1} = d \vee (f'_{n-1} \vee e'_{n-1}) = d \vee e'_{n-1} = c'.$$

By modularity,

$$d' \wedge e'_{n-1} = (d \vee f'_{n-1}) \wedge e_{n-1}$$

(since $f'_{n-1} \leq e'_{n-1}$)

$$= (d \wedge e'_{n-1}) \vee f'_{n-1} = f_{n-1} \vee f'_{n-1} = f'_{n-1}.$$

Hence $c'/d' \sim e'_{n-1}/f'_{n-1}$. Thus $d' < c'$ and $c'/d' \sim e'_{n-1}/f'_{n-1} \approx_w a/b$; that is, $c'/d' \approx_w a/b$.

Next, let L be relatively complemented. Let x be the relative complements of f'_{n-1} in $[f_{n-1}, e'_{n-1}]$ and define $c' = d \vee x$, $d' = d$ (see Figure 8). Then $e'_{n-1}/f'_{n-1} \searrow x/f_{n-1} \nearrow c'/d'$ and so $c'/d' \approx e'_{n-1}/f'_{n-1}$. We conclude, as before, that $d' < c'$ and that $c'/d' \approx_w a/b$. $\square$

Definition 8 (G. Grätzer and E. T. Schmidt [1958 d]). *Let us call a lattice L weakly modular iff a, b, c, $d \in L$, $b < a$, $d < c$, $a/b \approx_w c/d$ imply the existence of a proper subquotient c'/d' of c/d satisfying $c'd' \approx_w a/b$.*

Corollary 9. *Every modular and every relatively complemented lattice is weakly modular.*

The importance of the class of weakly modular lattices will be illustrated in Sections 2 and 4.

To introduce the second class of lattices, we again need a lemma.

Lemma 10. *Let L be a lattice and let a_i, $b_i \in L$, $b_i < a_i$, for $i = 1, \ldots, n$. Then*

$$\bigwedge(\Theta(a_i, b_i) \mid i = 1, 2, \ldots, n) \neq \omega$$

iff there exists a proper quotient a/b of L such that

$$a/b \approx_w a_i/b_i, \quad for \quad i = 1, 2, \ldots, n.$$

Proof. If such a, b exist, then $a \equiv b$ $(\Theta(a_i, b_i))$, for $i = 1, 2, \ldots, n$. Hence

$$a \equiv b \quad (\bigwedge(\Theta(a_i, b_i) \mid i = 1, 2, \ldots, n))$$

and so

$$\bigwedge(\Theta(a_i, b_i) \mid i = 1, 2, \ldots, n) \neq \omega.$$

We prove the converse, by induction on n, in a somewhat stronger form: if $v < u$ and

$$u \equiv v \quad (\bigwedge(\Theta(a_i, b_i) \mid i = 1, 2, \ldots, n)),$$

then there exists a proper subquotient a/b of u/v such that $a/b \approx_w a_i/b_i$, for $i = 1, 2, \ldots, n$. For $n = 1$, apply Theorem 2. Assuming the statement proved for $n - 1$, we get a'/b' a proper subquotient of u/v satisfying $a'/b' \approx_w a_i/b_i$, for $i = 1, \ldots, n - 1$. Since $u \equiv v$ $(\Theta(a_n, b_n))$, we get $a' \equiv b'$ $(\Theta(a_n, b_n))$. Hence, by Theorem 2, we obtain a proper subquotient a/b of a'/b' with $a/b \approx_w a_n/b_n$. Since $a/b \approx_w a'/b'$, we also have $a/b \approx_w a_i/b_i$, for $i = 1, 2, \ldots, n - 1$. $\square$

ω is meet-irreducible in a subdirectly irreducible lattice, therefore, we conclude:

Corollary 11. *Let L be a subdirectly irreducible lattice and let a_i/b_i be proper quotients of L, for $i = 1, \ldots, n$. Then there exists a proper quotient a/b of L satisfying $a/b \approx_w a_i/b_i$, $i = 1, \ldots, n$.*

In fact, Corollary 11 holds for any lattice in which ω is meet-irreducible.

Definition 12 (R. Wille [1972]). *Let L be a lattice and let P be a finite nonempty subset of L. Then P is called* primitive *iff*

$$\bigwedge(\Theta(x, x \vee y) \mid x, y \in P, x \neq x \vee y) \neq \omega.$$

Combining Lemma 10 and Definition 12, we obtain

Corollary 13. *P is primitive iff there exists a proper quotient a_p/b_p of L such that $a_p/b_p \approx_w c \vee d/c$, for all c, $d \in P$, $c \neq c \vee d$.*

Observe also that, by Corollary 11, every finite nonempty subset of a subdirectly irreducible lattice is primitive.

It will be pointed out in Chapter V that important classes of lattices can be defined by the property of not containing a primitive subset (poset) isomorphic to a member of a given collection of posets.

We study weak projectivities to enable us to obtain descriptions of quotient lattices such as $L/\Theta(a,b)$. Sometimes, however, L/Θ can be identified very simply within L: if there is a sublattice L_1 of L having one and only one element in every congruence class, then $L/\Theta \cong L_1$. Such congruence relations are called *representable*. It happens much more often that we can get a meet-subsemilattice or a join-subsemilattice L_1 of L having one and only one element in every congruence class; in such cases we call Θ *meet-representable* or *join-representable*, respectively.

Lemma 14. *Let L be a lattice and let Θ be a congruence relation of L. If every congruence class of Θ has a minimal element, then Θ is join-representable.*

Proof. Let L_1 be the set of minimal elements of congruence classes of Θ; then L_1 contains exactly one element of each congruence class. Now let a, $b \in L_1$. We show that $a \vee b$ is the smallest element of $[a \vee b]\Theta = [a]\Theta \vee [b]\Theta$. Indeed, if $c < a \vee b$ and $c \equiv a \vee b$ (Θ), then $a \wedge c \equiv a$ (Θ) and $b \wedge c \equiv b$ (Θ). Since $c < a \vee b$, we get that $a \wedge c < a$ or $b \wedge c < b$, say $a \wedge c < a$. Then $a \wedge c < a$ and $a \wedge c \in [a]\Theta$ contradict that $a \in L_1$. $\qquad\square$

Thus if L satisfies the Descending Chain Condition, then every congruence relation is join-representable, and dually.

Exercises

1. For all pairs of quotients x/y, u/v of N_5, investigate when "$x \equiv y$ forces $u \equiv v$". Repeat the investigation for M_3.

2. Show that if L is *sectionally complemented*, that is, L has a zero and, for every $a \in L$, $[0, a]$ is complemented, then in order to learn the congruence structure of L it is sufficient to consider "$a \equiv b$ forces $c \equiv d$" in the special case $b = d = 0$.

3. Prove that if a lattice is modular or relatively complemented, then $a/b \approx_w c/d$ iff $a/b \approx c'/d'$, for some subquotient c'/d' of c/d.

4. Show, by an example, that the conclusion of Exercise 3 is false for lattices in general.

5. Let L be a distributive lattice and let $a/b \approx_w c/d$. Prove that there exists a quotient e/f and a subquotient c'/d' of c/d such that

$$a/b \nearrow e/f \searrow c'/d'.$$

6. Let $p = p(x_0, \dots, x_{n-1})$ be a lattice polynomial, let L be a lattice, and let $a_1, \dots, a_{n-1} \in L$. Let $q(x) = p(x, a_1, \dots, a_{n-1})$ be an algebraic function. Under what conditions on p is it true that, for all quotients a/b of L,

$$q(a)/q(b) \approx_w a/b.$$

(This is true for $p = \cdots((x_0 \wedge x_1) \vee x_2) \wedge \cdots$ by Lemma 1.)

7. Referring to the proof of Lemma 1, axiomatize those properties of p_2 which make it possible to define p_m from p_2 to obtain generalizations of Lemma 1 to some varieties of algebras.

8. Show that $p_m(x, y_0, \dots, y_i, y_i, y_{i+2}, \dots, y_{m-1})$ does not depend on $x, y_0, \dots, y_{i-1}$ (that is, the value of $p_m(a, e_0, \dots, e_i, e_i, e_{i+2}, \dots, e_{m-1})$ is independent of $a, e_0, \dots, e_{i-1}$).

9. Prove that $p_m(x, c, d, c, d, \dots) = x$, for any $d \leq x \leq c$.

10. Rephrase and prove Lemma 1 using $q_m = \cdots((x \vee y_0) \wedge y_1) \vee \cdots$. Show that $\overset{m}{\approx}_w$, defined in terms of q_m, may differ from $\overset{m}{\approx}_w$ defined by p_m.

11. Show that in any sequence of weak perspectivities, any number of steps $\nearrow_w$ and $\searrow_w$ can be put in. Observe that in any nonredundant sequence of weak perspectivities, $\nearrow_w$ and $\searrow_w$ have to alternate.

12. Find examples to show that m cannot be bounded in Theorem 2. (Make all your examples planar modular lattices.)

13. Find examples to show that in Theorem 2, $\approx_w$ cannot be replaced by $\overset{n}{\approx}_w$, for any natural number n. (Make all your examples planar modular lattices.)

14. Prove that an ideal I is a kernel iff I is the kernel of $\Theta[I]$.

15. Let L be a lattice, and let I and J be ideals of L with $J \subseteq I$. Assume that J is a kernel in the lattice I and that I is a kernel in the lattice L. Is J a kernel in L?

16. Show that the ideal kernels of the lattice L form a sublattice $\mathrm{Id}_C L$ of $\mathrm{Id}\, L$, the lattice of all ideals of L. In fact, $\mathrm{Id}_C L$ is closed under arbitrary joins and under all meets existing in $\mathrm{Id}\, L$.

17. Let L be a distributive lattice, let I be an ideal of L, and let a/b be a quotient of L. Then $a \equiv b \ (\Theta[I])$ iff $a/b \sim a'/b'$, for some quotient a'/b' of I.

18. Show that Exercise 5 characterizes distributivity.

19. Show that Exercise 17 characterizes distributivity.

In Exercises 20–23 (which are due to R. Wille [1972]), let L and L' be lattices and let φ be a homomorphism of L onto L'.

20. Let a'/b' be a quotient of L' and let c/d be a quotient of L with $a'/b' \approx_w c\varphi/d\varphi$. Prove that there is a quotient a/b of L with $a/b \approx_w c/d$, $a\varphi = a'$, and $b\varphi = b'$.

21. Let $c_1/d_1, \ldots, c_n/d_n$ be quotients of L such that $c_1\varphi = c_2\varphi = \cdots = c_n\varphi$ and $d_1\varphi = d_2\varphi = \cdots = d_n\varphi$. Show that there is a quotient a/b of L satisfying $a/b \approx_w c_i/d_i$ and $a\varphi/b\varphi = c_i\varphi/d_i\varphi$, for $i = 1, 2, \ldots, n$.

22. Let P' be a primitive subset of L', and let P be a subset of L such that φ restricted to P is a (poset) isomorphism of P with P'. Show that P is primitive in L.

23. Let P be a primitive subset of L, let a_P, b_P be given as in Corollary 13, and let $a_P\varphi \neq b_P\varphi$. Show that $P\varphi$ is primitive in L' and that φ restricted to P is a poset isomorphism of P with $P\varphi$.

24. Let L be a lattice and let Ψ be a congruence relation of L. Verify the formula:

$$\Theta([a]\Psi, [b]\Psi) = (\Theta(a,b) \vee \Psi)/\Psi.$$

25. Let a_i, $b_i \in L$, for $1 \leq i \leq n$. Show that

$$\bigwedge(\Theta([a_i]\Psi, [b_i]\Psi) \mid 1 \leq i \leq n) = \left(\bigwedge(\Theta(a_i, b_i) \mid 1 \leq i \leq n) \vee \Psi\right)/\Psi.$$

26. Use Exercises 24 and 25 to verify Exercises 22 and 23.

27. Show that, for ideals I and J of a lattice L,

$$\Theta[I] \vee \Theta[J] = \Theta[I \vee J].$$

28. Let L be a lattice and let J_i, $i \in I$, be ideals of L. Prove that

$$\bigvee(\Theta[J_i] \mid i \in I) = \Theta\left[\bigvee(J_i \mid i \in I)\right].$$

29. Show that $\Theta[I] \wedge \Theta[J] = \Theta[I \wedge J]$ does not hold in general.

30. Verify

$$\Theta[I] \wedge \Theta[J] = \Theta[I \wedge J],$$

for ideals I and J of a distributive lattice.

31. Show that the formula of Exercise 30 holds also under the condition that every ideal is the kernel of at most one congruence relation.

32. Show that Corollary 11 does not hold for infinitely many a_i/b_i.

33. Show that in a finite lattice (or in a lattice in which all chains are finite) every congruence relation is meet- and join-representable.

34. Find a congruence relation Θ of the lattice of Figure 9 which is not representable. (Observe that Θ is both meet- and join-representable.)

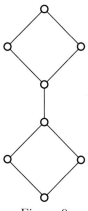

Figure 9

35. Let L be a finite lattice and let $L/\Theta \cong M_3$. Show that Θ is representable.

36. Give a formal proof of the statement that if L_1 represents the congruence relation Θ of L, then $L/\Theta \cong L_1$.

37. A congruence relation Θ of a lattice L is *order-representable* iff there exists a subset $H \subseteq L$ satisfying $[H]\Theta = L$ and

$$a \leq b \text{ in } L \quad \text{iff} \quad [a]\Theta \leq [b]\Theta \text{ in } L/\Theta, \quad \text{for } a, \ b \in H.$$

Show that H as a subposet of L is a lattice and $H \cong L/\Theta$.

38. Prove that meet-representability implies order-representability for congruence relations.

39. Let L be a lattice and let Θ be a congruence relation of L. Verify that if L/Θ is finite or countable, then Θ is order-representable.

40. Find a chain C, a lattice L, and a congruence relation Θ of L such that $L/\Theta \cong C$ but Θ is not order-representable.

2. Distributive, Standard, and Neutral Elements

The three types of elements of a lattice mentioned in the title of this section were discovered by O. Ore [1935], G. Grätzer [1959], and G. Birkhoff [1940a], respectively. It turned out later that all three can be defined using distributive equations only. This may be a coincidence, but it could be considered as a confirmation of the principle that distributive lattice theory provides the foundation of general lattice theory.

Definition 1. *Let L be a lattice and let a be an element of L.*

(i) *The element a is called* distributive *iff*

$$a \vee (x \wedge y) = (a \vee x) \wedge (a \vee y),$$

for all x, $y \in L$.

(ii) *The element a is called* standard *iff*

$$x \wedge (a \vee y) = (x \wedge a) \vee (x \wedge y),$$

for all x, $y \in L$.

(iii) *The element a is called* neutral *iff*

$$(a \wedge x) \vee (x \wedge y) \vee (y \wedge a) = (a \vee x) \wedge (x \vee y) \wedge (y \vee a),$$

for all x, $y \in L$.

For instance, in N_5, o, i, a, and c are distributive; o, i, and a are standard but c is not standard; only o and i are neutral. In M_3, only o and i are distributive; they are also standard and neutral. Of course, every element of a distributive lattice is distributive, standard, and neutral.

We can also dualize these definitions and define *dually distributive elements* and *dually standard elements*. Observe that the concept of neutrality is selfdual.

Various useful equivalent forms of these definitions are given in the three theorems that follow.

Theorem 2 (O. Ore [1935]).   Let L be a lattice and let a be an element of L. The following conditions on the element a are equivalent:

(i) a is distributive.

(ii) The map

$$\varphi \colon x \mapsto a \vee x \quad (x \in L)$$

is a homomorphism of L onto $[a)$.

(iii) The binary relation Θ_a on L defined by

$$x \equiv y \ \ (\Theta_a) \qquad \text{iff} \qquad a \vee x = a \vee y$$

is a congruence relation.   —

Remark. (i) and (ii) are equivalent since φ is a homomorphism iff it preserves meets, while (ii) and (iii) are equivalent because $\Theta_a = \operatorname{Ker}\varphi$. A more formal proof follows.

Proof.
    (i) *implies* (ii). φ maps L into $[a)$, for any element a of L; in fact, φ is always onto since $b\varphi = b$, for all $b \geq a$. The map φ is always a join-homomorphism since

$$x\varphi \vee y\varphi = (a \vee x) \vee (a \vee y) = a \vee (x \vee y) = (x \vee y)\varphi.$$

In view of Definition 1(i), if a is distributive, we also have

$$x\varphi \wedge y\varphi = (a \vee x) \wedge (a \vee y) = a \vee (x \wedge y) = (x \wedge y)\varphi,$$

and so φ is a homomorphism.
    (ii) *implies* (iii). Θ_a is the kernel of the homomorphism φ and therefore Θ_a is a congruence relation.
    (iii) *implies* (i). Since $x \vee a = (a \vee x) \vee a$, it follows that $x \equiv a \vee x \ (\Theta_a)$. Similarly, $y \equiv a \vee y \ (\Theta_a)$. Therefore,

$$z \wedge y \equiv (a \vee x) \wedge (a \vee y) \ \ (\Theta_a).$$

By the definition of Θ_a, we get

$$a \vee (x \wedge y) = a \vee ((a \vee x) \wedge (a \vee y)) = (a \vee x) \wedge (a \vee y),$$

proving (i). $\square$

Theorem 3 (G. Grätzer and E. T. Schmidt [1961]).   Let L be a lattice and let a be an element of L. The following conditions are equivalent:

(i) a is standard.

(ii) The binary relation Θ_a on L defined by

$$x \equiv y \quad (\Theta_a) \qquad \text{iff} \qquad (x \wedge y) \vee a_1 = x \vee y, \quad \text{for some } a_1 \le a,$$

is a congruence relation.

(iii) a is distributive and, for $x, y \in L$,

$$a \wedge x = a \wedge y \text{ and } a \vee x = a \vee y \qquad \text{imply that} \qquad x = y.$$

Proof.

(i) *implies* (ii). Let a be standard and let Θ_a be defined as in (ii). We use Lemma I.3.8 to verify that Θ_a is a congruence relation. By definition, Θ_a is reflexive and $x \equiv y$ (Θ_a) iff $x \wedge y \equiv x \vee y$ (Θ_a). If

$$x \le y \le z,$$
$$x \equiv y \quad (\Theta_a),$$
$$y \equiv z \quad (\Theta_a),$$

then

$$x \vee a_1 = y,$$
$$y \vee a_2 = z,$$

for some $a_1, a_2 \le a$. Hence $x \vee (a_1 \vee a_2) = z$, showing that $x \equiv z$ (Θ_a) since $a_1 \vee a_2 \le a$. Finally, let $x \le y$ and $x \equiv y$ (Θ_a), that is, $x \vee a_1 = y$ with $a_1 \le a$. For any $t \in L$, $(x \vee t) \vee a_1 = y \vee t$, hence $x \vee t \equiv y \vee t$ (Θ_a). To show the Substitution Property for meet, observe that

$$y \wedge t \le y = x \vee a_1 \le x \vee a,$$

and so, using the fact that a is standard we get

$$y \wedge t = (y \wedge t) \wedge (x \vee a) = ((y \wedge t) \wedge x) \vee ((y \wedge t) \wedge a) = (x \wedge t) \vee a_2,$$

where $a_2 = y \wedge t \wedge a \le a$. Thus the conditions of Lemma I.3.8 are verified, and Θ_a is a congruence relation.

(ii) *implies* (iii). Let us assume that the Θ_a defined in (ii) is a congruence relation. We can show that a is distributive just as in Theorem 2. Now let

$$a \wedge x = a \wedge y,$$
$$a \vee x = a \vee y.$$

Since $y \equiv a \vee y$ (Θ_a), meeting both sides with x and using $a \vee y = a \vee x$, we obtain

$$x \wedge y \equiv x \wedge (a \vee y) = x \wedge (a \vee x) = x \quad (\Theta_a).$$

Thus $x = (x \wedge y) \vee a_1$, for some $a_1 \leq a$. Also, $a_1 \leq x$, hence $a_1 \leq a \wedge x = a \wedge y$, and so $a_1 \leq x \wedge y$. We conclude that $x = x \wedge y$. Similarly, $y = x \wedge y$, and so $x = y$.

(iii) *implies* (i). Let us assume (iii), let x, $y \in L$, and define

$$b = x \wedge (a \vee y),$$
$$c = (x \wedge a) \vee (x \wedge y).$$

In order to show $b = c$, by (iii), it will be sufficient to prove that

$$a \wedge b = a \wedge c,$$
$$a \vee b = a \vee c.$$

To prove the first,

$$a \wedge x \leq a \wedge c$$

(using $c \leq b$)

$$\leq a \wedge b = a \wedge x \wedge (a \vee y) = a \wedge x,$$

and hence $a \wedge c = a \wedge b$. To prove the second, we compute, using the fact that a is distributive:

$$a \vee b = a \vee ((x \wedge (a \vee y))) = (a \vee x) \wedge (a \vee y)$$
$$= a \vee (x \wedge y) = a \vee (x \wedge a) \vee (x \wedge y) = a \vee c. \qquad \square$$

Theorem 4.   Let L be a lattice and let a be an element of L. The following conditions are equivalent:

(i) a is neutral.

(ii) a is distributive, a is dually distributive, and

$$a \wedge x = a \wedge y,$$
$$a \vee x = a \vee y$$

imply that $x = y$, for any x, $y \in L$.

(iii) There is an embedding φ of L into a direct product $A \times B$ where A has 1 and B has a 0 and $a\varphi = \langle 1, 0 \rangle$.

(iv) For any x, $y \in L$, the sublattice generated by a, x, and y is distributive.

—

Remark. The equivalence of (ii)–(iv) is due to G. Birkhoff [1940a]; in fact, G. Birkhoff used (iv) as a definition of neutrality. The equivalence of (i) to (ii)–(iv) was conjectured in G. Grätzer and E. T. Schmidt [1961]. This was proved in G. Grätzer [1962], J. Hashimoto and S. Kinugawa [1963], and Iqbalunissa [1964].

Proof.

(i) *implies* (ii). Let a be neutral. Then

(*) $$a \vee (x \wedge y) = x \wedge (a \vee y), \quad \text{for } x \geq a.$$

Indeed,

$$a \vee (x \wedge y) = (a \wedge x) \vee (x \wedge y) \vee (y \wedge a)$$

(by Definition 1 (iii))

$$= (a \vee x) \wedge (x \vee y) \wedge (y \vee a) = x \wedge (a \vee y).$$

To show that a is distributive, compute:

$$a \vee (x \wedge y) = a \vee (a \wedge x) \vee (x \wedge y) \vee (y \wedge a)$$

(by Definition 1 (iii)

$$= a \vee ((a \vee x) \wedge (x \vee y) \wedge (y \vee a))$$

(apply (*) to a, $a \vee x$, and $(x \vee y) \wedge (y \vee a)$)

$$= (a \vee x) \wedge (a \vee ((x \vee y) \wedge (y \vee a)))$$

(apply (*) to a, $y \vee a$, and $x \vee y$)

$$= (a \vee x) \wedge (y \vee a) \wedge (a \vee x \vee y) = (a \vee x) \wedge (a \vee y),$$

as claimed. By duality we get that a is dually distributive.

Finally, let $a \wedge x = a \wedge y$ and $a \vee x = a \vee y$. Then

$$x = x \wedge (a \vee x) \wedge (a \vee y) \wedge (x \vee y)$$
$$= x \wedge ((a \wedge x) \vee (x \wedge y) \vee (a \wedge y))$$
$$= x \wedge ((a \wedge x) \vee (x \wedge y))$$
$$= (a \wedge x) \vee (x \wedge y)$$
$$= (a \wedge x) \vee (a \wedge y) \vee (x \wedge y).$$

Since, the right-hand side is symmetric in x and y, we conclude that $x = y$.

(ii) *implies* (iii). Let (ii) hold for a and define $A = (a]$ and $B = [a)$. Let

$$\varphi : x \mapsto \langle x \wedge a, x \vee a \rangle.$$

Since a is distributive and dually distributive, φ is a homomorphism of L into $A \times B$ (by Theorem 1(ii) and its dual). The map φ is one-to-one, since if $x\varphi = y\varphi$, for $x, y \in L$, then

$$\langle x \wedge a, x \vee a \rangle = \langle y \wedge a, y \vee a \rangle,$$

that is,

$$x \wedge a = y \wedge a,$$
$$x \vee a = y \vee a;$$

thus $x = y$ by (ii). So φ is an embedding, $a\varphi = \langle a, a \rangle$, and a is the unit element of A and the zero of B.

(iii) *implies* (iv). The following three statements are obvious:

(iv) holds for the zero and the unit in any lattice;

(iv) holds for $\langle a_0, a_1 \rangle$ in $A_0 \times A_1$ iff it holds for a_0 in A_i and a_1 in A_1;

if a is an element of L_0, L_0 is a sublattice of L_1, (iv) holds for a in L_1, then (iv) holds for a in L_0.

(iii) along with these three statements proves (iv).

(iv) *implies* (i). Obvious by Exercise I.4.7.   □

The results stated in Theorems 2–4 make it possible to verify the most important properties of distributive, standard, and neutral elements.

Theorem 5.

(i) Every neutral element is standard.

(ii) Every standard element is distributive.

(iii) Every standard and dually standard (dually distributive) element is neutral.

(iv) If a is distributive or standard, then the Θ_a in Theorem 2(iii) and Theorem 3(ii), respectively, agrees with $\Theta[(a]]$.   ——

Proof.

(i) and (iii). By Theorem 4(ii) and Theorem 3(iii).

(ii) By Theorem 3(iii).

(iv) If $u \equiv a$ (Θ), for any $u \leq a$ and any congruence relation Θ, then $a \vee x = a \vee y$ implies that

$$x = x \vee (a \wedge x) \equiv x \vee a = y \vee a \equiv y \vee (a \wedge y) = y \quad (\Theta),$$

hence for the relation Θ_a given by Theorem 2(iii), we have $\Theta_a \leq \Theta$. Similarly, if $(x \wedge y) \vee a_1 = x \vee y$, for some $a_1 \leq a$, then

$$x \wedge y = (x \wedge y) \vee (a \wedge x \wedge y) \equiv (x \wedge y) \vee a_1 = x \vee y \quad (\Theta),$$

and so $x \equiv y$ (Θ); hence, for the relation Θ_a of Theorem 3(ii), we obtain that $\Theta_a \leq \Theta$.   □

For a principal ideal $(a] = I$, let $\Theta[I]$ be denoted by Θ_a. Then by Theorem 5(iv), the Θ_a of Theorem 2(iii) and the Θ_a of Theorem 3(ii) are indeed the Θ_a just defined in case a is distributive or standard. Hence 2(iii) and 3(ii) are definitions of distributive and standard elements, respectively, in terms of the properties of Θ_a.

As we have already seen, the converse of Theorem 5(i), as well as that of Theorem 5(ii), is false. Two wide classes of lattices in which the converse holds are the class of modular lattices and the class of relatively complemented lattices.

Theorem 6 (G. Grätzer and E. T. Schmidt [1961]).   In a weakly modular lattice, an element is distributive iff it is neutral.   —

Before proving the theorem, we verify a lemma connecting the distributivity of an element with weak projectivity.

Lemma 7. *Let L be a lattice and let a be an element of L. The element a is distributive iff $u \leq z \leq a \leq y \leq x$ and $x/y \approx_w z/u$ imply that $x = y$.*

Proof. Let a be a distributive element, $u \leq z \leq a \leq y \leq x$, and $x/y \approx_w z/u$. Since $u \leq z \leq a$, we get $u \equiv z$ (Θ_a), hence $x \equiv y$ (Θ_a). By Theorem 2(iii), $x = x \vee a = y \vee a = y$.

Let us now assume that a is not distributive. Then there exist $b, c \in L$ such that

$$a \vee (b \wedge c) \neq (a \vee b) \wedge (a \vee c).$$

Since $a \equiv a \wedge b \wedge c$ (Θ_a), we obtain

$$(a \vee b) \wedge (a \vee c) = a \vee ((a \vee b) \wedge (a \vee c))$$
$$\equiv a \vee (((a \wedge b \wedge c) \vee b) \wedge ((a \wedge b \wedge c) \vee c))$$
$$= a \vee (b \wedge c) \quad (\Theta_a),$$

so, by Corollary 1.5, there exist $x, y, u \in L$ satisfying, with $z = a$,

$$u < z \leq a \leq (a \vee b) \wedge (a \vee c) \leq y < x \leq a \vee (b \wedge c),$$
$$x/y \approx_w z/u.$$ □

Proof of Theorem 6. Let L be a weakly modular lattice and let a be a distributive element. If a is not dually distributive, then by Lemma 7, there exist $u < z \leq a \leq y \leq x$ satisfying $z/u \approx_w x/y$. By weak modularity, there exists a proper subquotient x_1/y_1 of x/y satisfying $x_1/y_1 \approx_w z/u$, contradicting, by Lemma 7, that a is distributive.

Now let

$$a \wedge x = a \wedge y,$$
$$a \vee x = a \vee y,$$

and $x \neq y$, for some x, $y \in L$. Proceeding just as in the step "(ii) implies (iii)" in the proof of Theorem 3, we conclude that

$$x \equiv y \quad (\Theta(a, a \wedge x)),$$

and dually

$$x \equiv y \quad (\Theta(a, a \vee x)).$$

Applying Theorem 1.2 to $x \equiv y$ $(\Theta(a, a \wedge x))$, we get a proper subquotient x_1/y_1 of $x \vee y/x \wedge y$ satisfying $x_1/y_1 \approx_w a/a \wedge x$. Since $x \equiv y$ $(\Theta(a, a \vee x))$, we also have $x_1 \equiv y_1$ $(\Theta(a, a \vee x))$, and again, by Theorem 1.2, we get a proper subquotient x_2/y_2 of x_1/y_1 such that $x_2/y_2 \approx_w a \vee x/a$. Using the weak modularity of L, we obtain a proper subquotient x_3/y_3 of $a \vee x/a$ satisfying $x_3/y_3 \approx_w x_2/y_2$. Hence

$$x_3/y_3 \approx_w x_2/y_2 \approx_w x_1/y_1 \approx_w a/a \wedge x,$$

contradicting Lemma 7. By Theorem 4(ii), a is a neutral. $\qquad\square$

Corollary 8. *In a weakly modular lattice, every standard element is neutral.*

Theorem 9.   Let L be a lattice.

(i) Let D denote the set of all distributive elements of L. Then a and $b \in D$ imply that $a \vee b \in D$.

(ii) Let S denote the set of all standard elements of L. Then a and $b \in S$ imply that $a \wedge b$ and $a \vee b \in S$.

(iii) Let N denote the set of all neutral elements of L. Then a and $b \in N$ imply that $a \wedge b$ and $a \vee b \in N$.   ——

Proof.
    (i) Let a, $b \in D$ and compute:

$$(a \vee b) \vee (x \wedge y) = a \vee b \vee (x \wedge y)$$
$$= a \vee ((b \vee x) \wedge (b \vee y))$$
$$= (a \vee b \vee x) \wedge (a \vee b \vee y),$$

so $a \vee b \in D$.

(ii) Let $a, b \in S$. First we do the join:

$$x \wedge (a \vee b \vee y) = (x \wedge a) \vee (x \wedge (b \vee y))$$
$$= (x \wedge a) \vee (x \wedge b) \vee (x \wedge y)$$
$$= (x \wedge (a \vee b)) \vee (x \wedge y),$$

proving that $a \vee b \in S$. Now we verify the formula

$$\Theta_a \wedge \Theta_b = \Theta_{a \wedge b},$$

where Θ_a, Θ_b, and $\Theta_{a \wedge b}$ are the relations described by Theorem 3(ii). Since $\Theta_a \wedge \Theta_b \geq \Theta_{a \wedge b}$ is trivial, let $x \equiv y$ ($\Theta_a \wedge \Theta_b$). Then $x \equiv y$ (Θ_a), and so $(x \wedge y) \vee a_1 = x \vee y$, for some $a_1 \leq a$. We also have $x \equiv y$ (Θ_b), and therefore

$$a_1 = a_1 \wedge (x \vee y) \equiv a_1 \wedge x \wedge y \quad (\Theta_b).$$

Thus, for some $b_1 \leq b$, we have $a_1 = (a_1 \wedge x \wedge y) \vee b_1$. Now

$$(x \wedge y) \vee b_1 = (x \wedge y) \vee (a_1 \wedge x \wedge y) \vee b_1 = (x \wedge y) \vee a_1 = x \vee y;$$

since $b_1 \leq b$ and $b_1 \leq a_1 \leq a$, we obtain that $b_1 \leq a \wedge b$; these verify that $x \equiv y$ ($\Theta_{a \wedge b}$).

This formula shows that if $a, b \in S$, then the relation $\Theta_{a \wedge b}$ of Theorem 3(ii) is a congruence relation, hence $a \wedge b \in S$ by Theorem 3.

(iii) Let $a, b \in N$. By Theorem 5, $a, b \in S$. Hence by (ii), $a \wedge b \in S$. By Theorems 3 and 4, to show that $a \wedge b \in N$, we have to prove only that $a \wedge b$ is dually distributive. Since a and b are dually distributive, they are distributive elements of the dual of L, hence by (i), $a \vee b$ in the dual of L is distributive and so $a \wedge b$ in L is dually distributive. □

Figure 1 shows that $a, b \in D$ does not imply that $a \wedge b \in D$.

Exercises

1. Let L be a bounded lattice. Show that 0 and 1 are distributive, standard, and neutral.

2. Let L be the lattice of Figure 2. Then $(a]$ is the kernel of some congruence relation but a is not distributive.

3. If a is a distributive element of L, then $L/\Theta_a \cong [a)$. To what extent does $L/\Theta_a \cong [a)$ characterize the distributivity of a?

4. Find distributive elements that are not dually distributive.

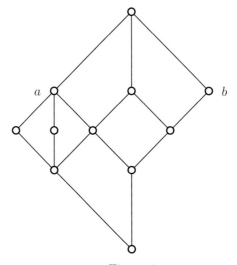

Figure 1

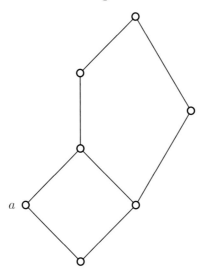

Figure 2

5. Investigate the relation Θ_a of Theorem 2(iii) as a congruence relation of the join-semilattice.

6. Let L be a lattice, let φ be a homomorphism of L onto L', and let L_1 be a sublattice of L'. Let a be an element of L satisfying $a\varphi \in L_1$. Show that if a is a distributive in L, then $a\varphi$ is distributive in L_1.

7. Prove the analogue of Exercise 6 for standard and neutral elements.

8. Let a be a distributive element of L. Show that $(a]$ is a distributive element of $\operatorname{Id} L$.

9. Prove the analogue of Exercise 8 for standard and neutral elements.

10. Show that in Theorem 3(iii) (and in Theorem 4(ii)) the condition can be weakened by assuming that $x \geq y$.

11. Verify that the element a of a lattice L is standard iff $x \leq a \vee y$ implies that $x = (x \wedge a) \vee (x \wedge y)$ (G. Grätzer and E. T. Schmidt [1961]).

12. Prove that the join of two distributive elements is again distributive by verifying the formula $\Theta_a \vee \Theta_b = \Theta_{a \vee b}$.

13. Find classes of lattices in which the meet of two distributive elements is distributive.

14. Show that the map $a \mapsto \Theta_a$ for standard elements is an embedding of the sublattice of standard elements into the congruence lattice.

15. Let L be a lattice, and let a, b, $c \in L$. Show that a, b, and c generate a distributive sublattice iff, for any permutation x, y, z of a, b, c, we have

$$x \wedge (y \vee z) = (x \wedge y) \vee (x \wedge z),$$
$$x \vee (y \wedge z) = (x \vee y) \wedge (x \vee z),$$
$$(x \wedge y) \vee (y \wedge z) \vee (z \wedge x) = (x \vee y) \wedge (y \vee z) \wedge (z \vee x)$$

(O. Ore [1940]).

16. Let L be a lattice and let a, b be standard elements of L. Show that, for any $c \in L$, a, b, and c generate a distributive sublattice.

17. Do three distributive elements generate a distributive sublattice?

18. Verify directly that in a modular lattice, distributive element = standard element = neutral element.

19. Verify the conclusion of Exercise 18 in a relatively complemented lattice.

20. An element of a modular lattice is neutral iff it has at most one relative complement in any interval containing it (G. Grätzer and E. T. Schmidt [1961]).

21. Let L be a modular lattice and let $a \in L$. Then a is neutral iff a is neutral in every interval of the form $[a \wedge x, a \vee x]$, for $x \in L$.

22. Show that the conclusion of Exercise 21 fails in weakly modular lattices. (See Figure 3.)

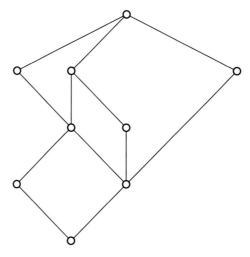

Figure 3

23. Show that Exercise 21 also fails for weakly modular lattices and standard elements. (See Figure 4.)

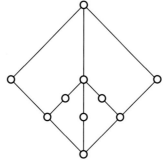

Figure 4

24. Let L be a bounded relatively complemented lattice. Then an element a of L is neutral iff it has a unique complement (J. Hashimoto and S. Kinugawa [1963]).

25. Show that the N of Theorem 9(iii) is the intersection of the maximal distributive sublattices of L (G. Birkhoff [1940a]).

26. Prove Theorem 9(iii) using Theorem 4(iii).

3. Distributive, Standard, and Neutral Ideals

The three types of ideals mentioned in the title of this section derive naturally from the concepts introduced in Section 2.

Definition 1. An ideal I of a lattice L is called *distributive, standard* or *neutral*, respectively, iff I is distributive, standard, or neutral, respectively, as an element of $\operatorname{Id} L$, the lattice of all ideals of L.

To establish a connection between the type of elements and the type of ideals they generate, we need a lemma:

Lemma 2. *Let L be a lattice, let I be an ideal of L, and let p and q be n-ary polynomials. Let us assume that, for all $a \in I$, there is an element $b \in I$ such that $a \le b$ and*

$$p(b, c_1, \dots, c_{n-1}) = q(b, c_1, \dots, c_{n-1}),$$

for all $c_1, \dots, c_{n-1} \in L$. Then

$$p(I, J_1, \dots, J_{n-1}) = q(I, J_1, \dots, J_{n-1})$$

holds in $\operatorname{Id} L$, for all $J_1, \dots, J_{n-1} \in \operatorname{Id} L$.

Proof. In Section I.4 (in the proof of Lemma I.4.8), we proved the formula:

$$p(I_0, \dots, I_{n-1}) = \{x \mid x \le p(i_0, \dots, i_{n-1}),$$
$$\text{for some } i_0 \in I_0, \ \dots, \ i_{n-1} \in I_{n-1}\}.$$

Thus we obtain:

$$
\begin{aligned}
p(I, &J_1, \dots, J_{n-1}) \\
&= \{x \mid x \le p(a, j_1, \dots, j_{n-1}), \\
&\qquad \text{for some } a \in I, \ j_1 \in J_1, \ \dots, \ j_{n-1} \in J_{n-1}\} \\
&= \{x \mid x \le p(b, j_1, \dots, j_{n-1}), \\
&\qquad \text{where } b \in I, \ j_1 \in J_1, \ \dots, \ j_{n-1} \in J_{n-1}, \text{ and } b \text{ satisfies} \\
&\qquad p(b, c_1, \dots, c_{n-1}) = q(b, c_1, \dots, c_{n-1}), \\
&\qquad \text{for all } c_1, \ \dots, \ c_{n-1} \in L\} \\
&= \{x \mid x \le q(b, j_1, \dots, j_{n-1}), \\
&\qquad \text{for } j_i \in J_1, \ \dots, \ j_{n-1} \in J_{n-1}\} \\
&= q(I, J_1, \dots, J_{n-1}).
\end{aligned}
$$

$\square$

Corollary 3. *Let L be a lattice, and let $a \in L$. Then a is distributive, standard, or neutral, respectively, iff $(a]$, as an ideal, is distributive, standard, or neutral, respectively.*

The main characterization theorems can be proved similarly to the proofs of Theorems 2.2–2.4.

Theorem 4.   Let L be a lattice, and let I be an ideal of L. The following conditions on I are equivalent:

(i) I is distributive.

(ii) $\Theta[I]$ can be described as follows:

$$x \equiv y \quad (\Theta[I]) \qquad \text{iff} \qquad x \vee i = y \vee i, \quad \text{for some } i \in I. \qquad —$$

Proof. If in (ii) we put $(x] \vee I = (y] \vee I$ in place of $x \vee i = y \vee i$, then the equivalence can be proved as in Section 2 since

$$x \equiv y \quad (\Theta[I])$$

in L iff

$$(x] \equiv (y] \quad (\Theta_I)$$

in Id L. Now, if $x \vee i = y \vee i$, for some $i \in I$, then $(x] \vee I = (y] \vee I$ is obvious. Conversely, if $(x] \vee I = (y] \vee I$, then $x \leq y \vee i_0$ and $y \leq x \vee i_1$, for some $i_0, i_1 \in I$. Therefore, $x \vee i = y \vee i$, with $i_0 \vee i_1 = i \in I$.   □

Theorem 5.   Let L be a lattice, and let I be an ideal of L. The following conditions on I are equivalent:

(i) I is standard.

(ii) The equality

$$(a] \wedge (I \vee (b]) = ((a] \wedge I) \vee (a \wedge b]$$

holds, for all $a, b \in L$.

(iii) For any ideal J of L,

$$I \vee J = \{ i \vee j \mid i \in I \text{ and } j \in J \}.$$

(iv) $\Theta[I]$ can be described by

$$x \equiv y \quad (\Theta[I]) \qquad \text{iff} \qquad (x \wedge y) \vee i = x \vee y, \quad \text{for some } i \in I.$$

(v) I is a distributive ideal and, for all $J, K \in \text{Id } L$,

$$I \wedge J = I \wedge K \text{ and } I \vee J = I \vee K \qquad \text{imply that} \qquad J = K. \qquad —$$

Remark. This result, and all the other unreferenced results in this section, are based on G. Grätzer [1959] and G. Grätzer and E. T. Schmidt [1961].

Proof. The equivalence of the five conditions can be verified as in Section 2. Only (iii) is new. But (iii) follows from (ii): if I satisfies (ii) and $a \in I \vee J$, then $a \leq i \vee j$, for some $i \in I$ and $j \in J$; hence $(a] = (a] \wedge (I \vee (j]) =$ (by (ii)) $=$ $((a] \wedge I) \vee (a \wedge j]$. Therefore, $a \leq i_1 \vee j_1$, where $i_1 \in (a] \wedge I$ and $j_1 \leq a \wedge j$. Consequently, $a = i_1 \vee j_1$. Finally, observe that when proving the analogue of "(i) implies (iv)" it is sufficient to use (iii). $\qquad\square$

Theorem 6.   Let L be a lattice, and let I be an ideal of L. The following conditions on I are equivalent:

(i) I is neutral.

(ii) For all j, $k \in L$,

$$(I \wedge (j]) \vee ((j] \wedge (k]) \vee ((k] \wedge I) = (I \vee (j]) \wedge ((j] \vee (k]) \wedge ((k] \vee I).$$

(iii) For all J, $K \in \mathrm{Id}\, L$, I, J, and K generate a distributive sublattice of $\mathrm{Id}\, L$.

(iv) I is distributive and dually distributive, and, for all J, $K \in \mathrm{Id}\, L$,

$$I \wedge J = I \wedge K \text{ and } I \vee J = I \vee K \qquad \text{imply that} \qquad J = K.$$

Proof. We can verify that (i) is equivalent to (ii) by using the argument of Lemma 2. The rest of the proof is the same as in Section 2. $\qquad\square$

Observe that every distributive ideal I of a lattice L is the kernel of $\Theta[I]$. Indeed, if $i \in I$, $a \in L$, and $i \equiv a \ (\Theta[I])$, then $i \vee j = a \vee j$, for some $j \in I$, thus $a \leq i \vee j \in I$, and so $a \in I$. This explains why $L/\Theta[I]$ can always be described. This description is especially simple if I is principal, $I = (a]$. Then $\Theta[(a]] = \Theta_a$ is a representable congruence relation and $(a]$ represents Θ_a, hence

$$L/\Theta_a \cong (a].$$

If I is not principal, then we describe $\mathrm{Id}(L/\Theta[I])$ as follows:

Theorem 7.   Let I be a distributive ideal of a lattice L. Then $\mathrm{Id}(L/\Theta[I])$ is isomorphic with the lattice of all ideals of L containing I, that is, with the interval $[I, L]$ in $\mathrm{Id}\, L$.

Proof. Let φ be the homomorphism $x \mapsto [x]\Theta$ of L onto L/Θ. Then the map $\psi \colon K \to K\varphi^{-1}$ maps $\mathrm{Id}(L/\Theta)$ into $[I, L]$. To show that this map is onto, it is sufficient to see that $[J]\Theta = J$, for all $J \supseteq I$. Indeed, if $j \in J$, $a \in L$ and $j \equiv a$

$(\Theta[I])$, then $j \vee i = a \vee i$, for some $i \in I$, and so $a \leq i \vee j \in J$ and $a \in J$, as claimed. Since ψ is obviously isotone and one-to-one, we conclude that it is an isomorphism. $\qquad\square$

Corollary 8. $L/\Theta[I]$ *is determined by the interval* $[I, L]$ *of* Id L, *provided that* I *is a distributive ideal of* L.

Proof. We know (Section II.3) that $\mathrm{Id}(L/\Theta[I])$ determines $L/\Theta[I]$ up to isomorphism. $\qquad\square$

Call a congruence relation Θ *distributive, standard, neutral*, respectively, iff $\Theta = \Theta[I]$, where I is a distributive, standard, or neutral ideal, respectively. The following result was first established by J. Hashimoto [1952], for neutral congruences, and by G. Grätzer and E. T. Schmidt [1961], in its present form.

The congruences Φ and Ψ of a lattice L are said to *permute* (or are *permutable*) iff $\Phi \vee \Psi = \Phi \cdot \Psi$, where $\Phi \cdot \Psi$ is the binary relation defined by

$$a \equiv b \quad (\Phi \cdot \Psi) \qquad \text{iff} \qquad \text{there exists a } c \in L \text{ with } a \equiv c \quad (\Phi) \text{ and } c \equiv b \quad (\Psi).$$

An equivalent definition is that Φ and Ψ permute iff $\Phi \cdot \Psi = \Psi \cdot \Phi$, which is, in turn, equivalent to $\Phi \cdot \Psi$ being a congruence relation.

Theorem 9.   Any two standard congruences of a lattice permute.   —

Proof. Let Φ and Ψ be standard congruences of the lattice L, that is, $\Phi = \Theta[I]$ and $\Psi = \Theta[J]$, where I and J are standard ideals of L. Let

$$a \equiv b \quad (\Phi) \qquad \text{and} \qquad b \equiv c \quad (\Psi), \quad a, \, b, \, c \in L.$$

We want to show that there exists a $d \in L$ satisfying

$$a \equiv d \quad (\Psi) \qquad \text{and} \qquad d \equiv c \quad (\Phi).$$

If

$$x \leq y \leq z, \qquad x \equiv y \quad (\Phi), \qquad \text{and } y \equiv z \quad (\Psi),$$

then $y = x \vee i$, for some $i \in I$, and $z = y \vee j$, for some $j \in J$. With $u = x \vee j$, we have $z = u \vee i$, and hence $x \equiv u$ (Ψ) and $u \equiv z$ (Φ).

Applying this observation to

$$a \equiv a \vee b \quad (\Phi), \qquad a \vee b \equiv a \vee b \vee c \quad (\Psi),$$

and to

$$c \equiv b \vee c \quad (\Psi), \qquad b \vee c \equiv a \vee b \vee c \quad (\Phi),$$

we obtain elements e, $f \in L$ satisfying

$$a \equiv e \ (\Psi), \qquad e \equiv a \vee b \vee c \ (\Phi), \qquad a \leq e \leq a \vee b \vee c,$$
$$c \equiv f \ (\Phi), \qquad f \equiv a \vee b \vee c \ (\Psi), \qquad c \leq f \leq a \vee b \vee c.$$

Set $d = e \wedge f$. Then

$$a \equiv e = e \wedge (a \vee b \vee c) \equiv e \wedge f = d \ (\Psi),$$
$$c \equiv f = f \wedge (a \vee b \vee c) \equiv f \wedge e = d \ (\Phi). \qquad \square$$

Theorem 9 is significant because in a wide class of lattices all congruences are standard. Recall that a lattice L is *sectionally complemented* iff L has a zero, 0, and all intervals $[0, a]$ of L are complemented.

Theorem 10.   In a sectionally complemented lattice L all congruence relations are standard.   —

Proof. Let Θ be a congruence relation of L and let $I = [0]\Theta$ be the ideal kernel of Θ. Let $a, b \in L$ and $a \equiv b \ (\Theta)$. Let c be the complement of $a \wedge b$ in $[0, a \vee b]$. Then

$$c = c \wedge (a \vee b) \equiv c \wedge (a \wedge b) = 0 \quad (\Theta),$$

hence $c \in I$. Thus $a \equiv b \ (\Theta)$ iff $(a \wedge b) \vee i = a \vee b$, for some $i \in I$. By Theorem 5(iv), this shows that I is standard and $\Theta = \Theta[I]$. $\qquad \square$

Corollary 11. *Let L be a weakly modular sectionally complemented lattice. If L satisfies the Ascending Chain Condition, then all congruences of L are neutral, in fact, of the form Θ_a, where a is a neutral element.*

Proof. Let Θ be a congruence relation of L. By Theorem 10, $\Theta = \Theta[I]$, where I is a standard ideal of L. By the Ascending Chain Condition, $I = (a]$. By Corollary 3, a is standard. Since L is weakly modular, by Theorem 2.6, a is neutral. Hence $\Theta = \Theta_a$, with a neutral. $\qquad \square$

Exercises

1. Prove that an ideal generated by a set of distributive elements is distributive.

2. Show that an ideal generated by a set of standard elements is standard.

3. Does the analogue of Exercises 1 and 2 hold for neutral ideals?

4. Verify that the converse of Exercise 2 does not hold. (Hint: Consider the lattice of Figure 1.)

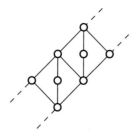

Figure 1

5. Prove Corollary 3 directly.

6. Consider the lattice L as a sublattice of Id L under the natural embedding $x \mapsto (x]$. Show that every congruence relation of L can be extended to Id L.

7. Characterize those congruence relations of Id L that are extensions of congruences of L.

8. For any ideal I of a lattice L, relate the congruence relation $\Theta[I]$ of L with the congruence relation Θ_I of Id L.

9. Characterize a standard ideal I of L in terms of the congruence relation Θ_I on Id L.

10. Show that in Theorem 5 it is sufficient to assume that condition (iii) holds for principal ideals.

*11. Construct a lattice L and an ideal I of L such that Theorem 5(v) holds for all principal ideals J and K, yet I is not standard (Iqbalunissa [1965]).

12. Show that we can assume that $J \subseteq K$ in Theorem 5(v) and Theorem 6(iv).

13. Show that the congruence relation Θ and Φ of the lattice L permute iff, for all $a, b, c \in L$ with

$$a \le b \le c, \qquad a \equiv b \ (\Theta), \qquad b \equiv c \ (\Phi),$$

there exists a $d \in L$ satisfying

$$a \le d \le c, \qquad a \equiv d \ (\Phi), \qquad d \equiv c \ (\Theta).$$

14. Use Exercise 13 to reprove Theorem 9.

15. Show that $I \mapsto \Theta[I]$ is an isomorphism between the lattice of standard ideals and the lattice of standard congruence relations.

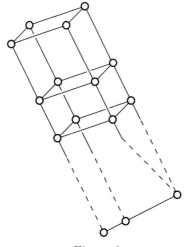

Figure 2

16. Construct a lattice L and standard ideals I_j, $j \in J$, of L such that $I = \bigwedge(I_j \mid j \in J) \neq \emptyset$ and I is not a standard ideal. (Hint: Consider the lattice of Figure 2.)

17. Let L be a lattice and let I and J be ideals of L. If I is standard and $I \wedge J$, $I \vee J$ are principal, then J is principal.

18. For a lattice L and standard ideal I of L, let L/I denote the quotient lattice $L/\Theta[I]$. Verify the First Isomorphism Theorem for Standard Ideals: For any ideal J of L, $I \wedge J$ is a standard ideal of I and
$$I \vee J/I \cong J/I \wedge J.$$

19. Prove the Second Isomorphism Theorem for Standard Ideals: Let L be a lattice, let I and J be ideals of L, $J \subseteq I$, and let J be standard. Then I is standard iff I/J is standard in L/J, and in this case
$$L/I \cong (L/J)/(I/J).$$

20. State and verify the Second Isomorphism Theorem for Neutral Ideals (J. Hashimoto [1952]).

4. Structure Theorems

Let A and B be bounded lattices and $L = A \times B$. Then $a = \langle 1, 0 \rangle$ and $b = \langle 0, 1 \rangle$ are neutral elements, and they are complementary. Conversely, if L is a bounded

such that

$$A_i \cong C_{0,i} \times C_{1,i} \times \cdots \times C_{m-1,i}, \quad \text{for } 0 \le i < n,$$

and

$$B_j \cong C_{j,0} \times C_{j,1} \times \cdots \times C_{j,n-1}, \quad \text{for } 0 \le j < m. \qquad \underline{}$$

Proof. Let $\langle I_0, I_1, \ldots, I_{n-1} \rangle$ and $\langle J_0, J_1, \ldots, J_{m-1} \rangle$ be the neutral ideals associated with the two decompositions as in Theorem 2. Then

$$\langle I_0 \wedge J_0, I_1 \wedge J_0, \ldots, I_{n-1} \wedge J_0, \ldots, I_0 \wedge J_{m-1}, I_1 \wedge J_{m-1}, \ldots, I_{n-1} \wedge J_{m-1} \rangle$$

is again a sequence of neutral ideals satisfying the conditions of Theorem 2. (We only need Theorem 2.9 to help verify this.) The direct decomposition associated with this sequence will yield the decomposition required. $\qquad \square$

A lattice L is called *directly indecomposable* iff L has no representation in the form $L = A \times B$, where both A and B have more than one element.

Corollary 4. *Let L be a lattice with 0. If L has a representation*

$$L = A_0 \times \cdots \times A_{n-1},$$

where each A_i, $0 \le i < n$, is directly indecomposable, then for any other decomposition

$$L = B_0 \times \cdots \times B_{m-1}$$

of L into directly indecomposable factors, we have that $n = m$ and that there exists a permutation α of $\{0, \ldots, n-1\}$ such that $A_i \cong B_{i\alpha}$, for all $0 \le i < n$.

Results analogous to Corollaries 3 and 4 also hold for general lattices, see the exercises.

In trying to sharpen Corollary 4 to a structure theorem, there are two difficulties we have to overcome. A lattice need not have a decomposition into directly indecomposable factors. Directly indecomposable lattices are hard to accept as "building blocks"; one would rather have *simple lattices*, that is, lattices with only the two trivial congruences, ω and ι.

The first difficulty is easy to overcome by chain conditions. Observe that if A and B are lattices with 0, $C = \{0, a_1, a_2, \ldots, a_n\}$ is a chain in A, and $D = \{0, b_1, \ldots, b_m\}$ is a chain in B, then

$$E = \{ \langle 0, 0 \rangle, \langle a_1, 0 \rangle, \ldots, \langle a_n, 0 \rangle, \langle a_n, b_1 \rangle, \ldots, \langle a_n, b_m \rangle \}$$

is a chain in $A \times B$ of length $n+m$. This easily implies that $l(A \times B) = l(A)+l(B)$, for lattices A and B of finite length.

Lemma 5. *Let L be a bounded lattice. If L is of finite length, then L is isomorphic to a direct product of directly indecomposable lattices.*

The passage from "directly indecomposable" to "simple" requires stronger hypotheses.

Theorem 6.   Let L be a sectionally complemented, weakly modular lattice. If L is of finite length, then L can be represented as a direct product of simple lattices.
                                                                                    ——

Proof. First observe that if $L \cong A \times B$ and L satisfies the hypotheses of this theorem, then so do A and B; only the statement about weak modularity has to be verified and it follows from the observation that

$$\langle a_1, a_2 \rangle / \langle b_1, b_2 \rangle \approx_w \langle c_1, c_2 \rangle / \langle d_1, d_2 \rangle \quad \text{implies that} \quad a_1/b_1 \approx_w c_1/d_1.$$

By Lemma 5, $L \cong L_1 \times \cdots \times L_k$, where all L_i are directly indecomposable. By the observation above, all L_i satisfy the hypotheses of this theorem. So it is sufficient to prove that if L satisfies the hypotheses of the theorem and L is directly indecomposable, then L is simple. Indeed, if Θ is a congruence relation of L, then $\Theta = \Theta[I]$, where I is a standard ideal by Theorem 3.10. By the chain condition, $I = (a]$, and a is standard by Corollary 3.3. By Theorem 2.6, a is neutral. Since L is complemented, by Theorem 1, $(a]$ is a direct factor of L. But L is directly indecomposable, hence $a = 0$ or $a = 1$, yielding $\Theta = \omega$ or $\Theta = \iota$, verifying that L is simple.   □

Since modular lattices and relatively complemented lattices are special cases of weakly modular lattices, we obtain two famous structure theorems as special cases (G. Birkhoff [1935b] and K. Menger [1936]):

Corollary 7 (The Birkhoff-Menger Theorem). *Let L be a complemented modular lattice. If L is of finite length, then L is isomorphic to a direct product of simple lattices.*

We shall see in Chapter IV that these simple lattices are exactly C_2 and the nondegenerate projective geometries of finite dimension.

Corollary 8 (R. P. Dilworth [1950]). *Let L be a relatively complemented lattice. If L is of finite length, then L is isomorphic to a direct product of simple lattices.*

If L is a direct product of finitely many simple lattices $L_1, \ldots, L_k$, then by Theorem I.3.13

$$\operatorname{Con} L \cong \operatorname{Con} L_1 \times \cdots \times \operatorname{Con} L_k \cong C_2^k,$$

that is, Con L is a Boolean lattice. This led G. Birkhoff to suggest (in G. Birkhoff [1948]) the study of lattices L for which Con L is Boolean. We are going to present a solution to this problem.

Let us call a congruence relation Θ of a lattice L *separable* iff, for all a, $b \in L$, $a < b$, there exists a sequence $a = x_0 < x_1 < \cdots < x_n = b$ such that, for each $0 \leq i < n$, $x_i \equiv x_{i+1}$ (Θ) or $u \equiv v$ (Θ) for no proper subquotient u/v of x_{i+1}/x_i.

Theorem 9 (G. Grätzer and E. T. Schmidt [1958d]).   Let L be a lattice. Then Con L is Boolean iff L is weakly modular and all congruences of L are separable.   —

Proof. Let Con L be Boolean. Let

$$a/b \approx_w c/d, \quad a,\ b,\ c,\ d \in L, \quad a \neq b;$$

we wish to find a proper subquotient c'/d' of c/d satisfying $c'/d' \approx_w a/b$. Let $\Theta = \Theta(a,b)$ and let Θ' be the complement of Θ. Then $\Theta \vee \Theta' = \iota$ and so $c \equiv d$ $(\Theta \vee \Theta')$. By Theorem I.3.9, there is a sequence $d = x_0 < x_1 < \cdots < x_n = c$ such that, for each $0 \leq i < n$,

$$x_i \equiv x_{i+1} \quad (\Theta) \quad \text{or} \quad x_i \equiv x_{i+1} \quad (\Theta').$$

If, for all $0 \leq i < n$, $x_i \equiv x_{i+1}$ (Θ'), then $c \equiv d$ (Θ'); since $a/b \approx_w c/d$ and $a \equiv b$ (Θ'). But $\Theta = \Theta(a,b)$, and so $a \equiv b$ (Θ). We conclude that $a \equiv b$ $(\Theta \wedge \Theta')$, that is, $a \equiv b$ (ω), contradicting $a \neq b$. Therefore $x_i \equiv x_{i+1}$ (Θ), for some $0 \leq i < n$. Applying Theorem 1.2 to $x_i \equiv x_{i+1}$ $(\Theta(a,b))$, we obtain a proper subquotient c'/d' of x_{i+1}/x_i satisfying $c'/d' \approx_w a/b$, proving that L is weakly modular.

Now let Θ be a congruence relation of L and let a, $b \in L$, $a < b$. Then $a \equiv b$ $(\Theta \vee \Theta')$, where Θ' is the complement of Θ, and so, by Theorem I.3.9, there exists a sequence $a = x_o < x_1 < \cdots < x_n = b$ such that, for each $0 \leq i < n$,

$$x_i \equiv x_{i+1} \quad (\Theta) \quad \text{or} \quad x_i \equiv x_{i+1} \quad (\Theta').$$

We claim that the same sequence establishes the separability of Θ. Indeed, if, for some i, $x_i \not\equiv x_{i+1}(\Theta)$, and u, $v \in [x_i, x_{i+1}]$, $u \equiv v$ (Θ), then $x_i \equiv x_{i+1}$ (Θ') and so $u \equiv v$ (Θ'), implying that $u \equiv v$ $(\Theta \wedge \Theta')$, that is $u = v$.

To prove the converse, let us start out with a weakly modular lattice L and a congruence relation Θ of L. We claim:

The binary relation Φ *defined on* L *by*

$$a \equiv b\ (\Phi) \text{ iff } u \equiv v\ (\Theta) \text{ for no proper subquotient } u/v \text{ of } a \vee b/a \wedge b$$

is a congruence relation; in fact, Φ *is the pseudocomplement of* Θ.

We prove the first part of this claim by verifying that Φ satisfies conditions (i)–(iii) of Lemma I.3.8. Condition (i) is clear by the definition of Φ. To verify (ii), let

$$a,\ b,\ c \in L, \qquad a \leq b \leq c, \qquad a \equiv b \ (\Phi), \qquad \text{and } b \equiv c \ (\Phi);$$

we wish to show that $a \equiv c \ (\Phi)$. Assume to the contrary that $a \not\equiv c \ (\Phi)$; then $u \equiv v \ (\Theta)$, for some proper subquotient u/v of c/a. Since

$$u \equiv v \quad (\Theta(a,c)),$$
$$\Theta(a,c) = \Theta(a,b) \vee \Theta(b,c),$$

by Theorem 1.2 and Lemma 1.3 there exists a proper subquotient x/y of u/v satisfying $x/y \approx_w c/b$ or $x/y \approx_w b/a$, say, $x/y \approx_w c/b$. Therefore, $b < c$. By weak modularity, there exists a proper subquotient c_1/b_1 of c/b satisfying $c_1/b_1 \approx_w x/y$. But $x,\ y \in [v,u]$ and $u \equiv v \ (\Theta)$, so $x \equiv y \ (\Theta)$. Hence $c_1 \equiv b_1$ (Θ), contradicting that $c_1,\ b_1 \in [b,c]$ and $b \equiv c \ (\Phi)$. This shows that (ii) holds.

To verify (iii), assume to the contrary that

$$a,\ b,\ c \in L, \qquad a \leq b, \qquad a \equiv b \ (\Phi), \qquad \text{but } a \wedge c \not\equiv b \wedge c(\Phi).$$

Then there exists a proper subquotient u/v of $b \wedge c/a \wedge c$ satisfying $v \equiv u \ (\Theta)$. It follows that

$$u/v \approx_w b \wedge c/a \wedge c \approx_w b/a,$$

hence $u/v \approx_w b/a$; so $a < b$ and, by weak modularity, $b_1/a_1 \approx_w u/v$, where b_1/a_1 is a proper subquotient of b/a. Since $u \equiv v \ (\Theta)$, we obtain that $b_1 \equiv a_1$ (Θ), contradicting $a \equiv b \ (\Phi)$. By duality, $a \vee c \equiv b \vee c \ (\Phi)$.

This shows that Φ is a congruence relation. $\Theta \wedge \Phi = \omega$ is trivial. Now if

$$\Theta \wedge \Psi = \omega, \qquad a \equiv b \ (\Psi), \qquad u,\ v \in [a \wedge b, a \vee b], \qquad \text{and } u \equiv v \quad (\Theta),$$

then also $u \equiv v \ (\Psi)$, hence $u = v$, showing that $\Psi \subseteq \Phi$. This completes the proof of the claim.

Now to complete the proof of the theorem, let L be weakly modular and let the congruences of L be separable. Let Θ be a congruence of L. Let Φ be the congruence constructed in the claim. For $a, b \in L$, $a < b$, let

$$a = x_0 < x_1 < \cdots < x_n = b$$

be the chain establishing the separability of Θ. Then, for each $0 \leq i < n$, either $x_i \equiv x_{i+1} \ (\Theta)$, or, by the definitions of separability and of Φ, $x_i \equiv x_{i+1} \ (\Phi)$. Therefore, $a \equiv b \ (\Theta \vee \Phi)$, and so $\Theta \vee \Phi = \iota$. This shows that Φ is a complement of Θ. Since $\mathrm{Con}\,L$ is distributive (Theorem II.3.11) this proves that $\mathrm{Con}\,L$ is Boolean. $\square$

We get a large number of corollaries from Theorem 9 yielding that $\mathrm{Con}\, L$ is Boolean for special classes of lattices. Some of these will be discussed in the exercises.

The *center* of a bounded lattice L is the sublattice $\mathrm{Cen}\, L$ of complemented neutral elements. It is obvious that $0,\, 1 \in \mathrm{Cen}\, L$ and $\mathrm{Cen}\, L$ is a Boolean lattice. In some chapters of lattice theory, it is important to know when $\mathrm{Cen}\, L$ is a complete lattice. (If this is the case, L can often be "coordinatized" over $\mathrm{Cen}\, L$.) This was established for continuous geometries by J. von Neumann [1936].

Let L be a complete lattice and $K \subseteq L$. Then K is a *complete sublattice* of L iff $\bigwedge X$ and $\bigvee X \in K$, for all $X \subseteq K$ ($\bigwedge X$ and $\bigvee X$ are formed in L).

Theorem 10.   Let L be a complete, sectionally complemented, dually sectionally complemented, weakly modular lattice. Then the center of L is a complete sublattice of L.   —

Proof. Let $X \subseteq \mathrm{Cen}\, L$. Set $a = \bigwedge X$. Then $(a]$ is the kernel of the congruence relation $\bigwedge(\Theta_x \mid x \in X)$. Hence, by Theorem 3.10, $(a]$ is standard; by Corollary 3.3, a is standard. Now Corollary 2.8 tells us that a is neutral. Since a is complemented, $a \in \mathrm{Cen}\, L$.

By duality, we obtain that $\bigvee X \in \mathrm{Cen}\, L$.   □

Corollary 11 (M. F. Janowitz [1967]). *The center of a complete, relatively complemented lattice is a complete sublattice.*

This result has some interesting implications concerning direct decompositions of complete relatively complemented lattices; see S. Maeda [1966].

Exercises

1. Show that the representation of a lattice L with 0 as a direct product of two lattices are (up to isomorphism) in one-to-one correspondence with pairs of ideals $\langle I, J \rangle$ satisfying $I \cap J = \{0\}$ such that every element a of L has exactly one representation of the form $a = i \vee j,\, i \in I,\, j \in J$.

2. Let $L = A_1 \times A_2$. Define the binary relation Θ_i on L by

    $$\langle a_1, a_2 \rangle \equiv \langle b_1, b_2 \rangle \quad (\Theta_i)$$

    iff $a_i = b_i$ $(i = 1,\, 2)$. Show that Θ_1 and Θ_2 are congruence relations of L,

    $$\Theta_1 \wedge \Theta_2 = \omega,$$
    $$\Theta_1 \vee \Theta_2 = \iota.$$

3. Show that the congruence relations Θ_1 and Θ_2 of Exercise 2 are permutable.

4. Prove that the representations of a lattice L as a direct product of two lattices are (up to isomorphism) in a one-to-one correspondence with pairs of congruence relations $\langle \Theta_1, \Theta_2 \rangle$ that are complementary and permutable.

5. Show that if, in Exercise 4, we pass from two to n factors, then we get an n-tuple of congruences $\langle \Theta_1, \ldots, \Theta_n \rangle$ such that

$$\Theta_1 \wedge \cdots \wedge \Theta_n = \omega,$$
$$(\Theta_1 \wedge \cdots \wedge \Theta_{i-1}) \vee \Theta_i = \iota,$$
$$\Theta_1 \wedge \cdots \wedge \Theta_{i-1} \text{ and } \Theta_i \text{ permute,}$$

for $i = 2, \ldots, n$.

6. Can we replace, in Exercise 5, the condition "$\Theta_1 \wedge \cdots \wedge \Theta_{i-1}$ and Θ_i permute" by "any two Θ_i and Θ_j permute"?

7. Use Exercise 5 to verify Corollary 3 for arbitrary lattices.

8. Verify Corollary 4 for arbitrary lattices.

9. Relate Exercise 5 to Theorem 2.

10. Let B be the Boolean lattice R-generated by the rational interval $[0, 1]$. Show that, for any natural number n, B has a representation as a direct product of n lattices $L_1, \ldots, L_n$ with $|L_i| > 1$, for $i = 1, \ldots, n$, but B has no representation as a direct product of directly indecomposable lattices.

11. Construct a lattice which is not of finite length but every chain in the lattice is finite.

12. Statements 5–8 of this section deal with lattices of finite length. Which of these statements remain valid for lattices in which every chain is finite?

Exercises 13–19 are based on G. Grätzer and E. T. Schmidt [1958d].

13. Prove that every congruence relation of a finite lattice or of a lattice of finite length is separable.

14. Verify that if in a lattice L, for every a, $b \in L$, $a < b$, there is a finite maximal chain in $[a, b]$, then all congruence relations of L are separable. This holds, in particular, if the lattice is *locally finite*, that is, if all intervals are finite.

15. Let L be a distributive lattice, a, b, a_1, a_2, $\ldots \in L$, $a = a_1 < a_2 < a_3 < \cdots < b$. Then $\Theta = \bigvee(\Theta(a_{2i-1}, a_{2i}) \mid i = 1, 2, \ldots)$ is not a separable congruence relation.

16. For a distributive lattice L, $\operatorname{Con} L$ is a Boolean iff L is locally finite. (Use Exercises 14 and 15. J. Hashimoto [1952].)

17. The separable congruences of a lattice L form a sublattice of $\operatorname{Con} L$.

18. Let L be a lattice with 1. A neutral congruence relation $\Theta[I]$ is separable iff I is principal.

19. Let L be a complemented modular lattice. Then $\operatorname{Con} L$ is Boolean iff all neutral ideals of L are principal (Shih-chiang Wang [1953]).

20. Find a complete lattice L and a sublattice K of L such that K is a complete lattice but not a complete sublattice of L.

Further Topics and References

The properties of weak projectivity and of the congruences $\Theta[I]$ are discussed in greater detail in G. Grätzer and E. T. Schmidt [1958d]. For universal algebras, A. I. Mal'cev introduced similar concepts; the difference is that while for lattices it is sufficient to consider "unary algebraic functions" of a special form (namely, $\cdots(((x \wedge a_0) \vee a_1) \wedge a_2) \cdots)$, for universal algebras, in general, we have to consider arbitrary unary algebraic functions (see, for instance, G. Grätzer [1968]). The polynomial p_2 is also of special interest; identities of p_2 that hold for lattices imply that the congruence lattices of lattices are distributive. A general condition for the distributivity of congruence lattices of algebras in a given of algebras can be found in B. Jónsson [1967].

    Weak modularity is a rather complicated condition. It can be somewhat simplified for finite lattices; a finite lattice is weakly modular iff $a/b \approx_w c/d$ and $a > b \geq c > d$ imply the existence of a proper subquotient c'/d' of c/d satisfying $c'/d' \approx_w a/b$ (G. Grätzer [1963a]).

    In general, if in a lattice L any interval contains a finite maximal chain, then it is sufficient to consider weak projectivities of prime quotients, see, for instance, N. Funayama [1942], J. Jakubik [1955 a], and D. T. Finkbeiner [1960]. Under this condition, L is simple iff any pair of prime quotients are projective. In general, no bound can be put on the number of perspectivities. However, if L is a relatively complemented lattice of length n, a, $b \succ 0$ and $a/0 \approx b/0$, then $a/0 \overset{k}{\approx} b/0$, where $k \leq 2[\frac{1}{2}(n+1)]$, see J. E. McLaughlin [1951] and [1953].

    The investigation of the number of projectivities becomes important in the study of varieties of lattices, see Section V.3 and C. Herrmann [1973].

    Local conditions implying the projectivity of any pair of prime quotients can be found in F. A. Smith [1974].

    Modular lattices and relatively complemented lattices share many properties stronger than weak modularity. For instance, in P. Crawley and R. P. Dilworth [1973], a lattice is said to have the *Projectivity Property* iff whenever a/b is weakly projective into c/d, then a/b is projective to some subquotient c'/d' of c/d.

Weak modularity seems to be a very natural concept. It appears quite surprisingly in some results. The following result of Iqbalunissa [1966] is a good illustration of this point: if every congruence relation is neutral in the lattice L, then L is weakly modular.

A lot more information on standard elements and ideals can be found in G. Grätzer [1959] and G. Grätzer and E. T. Schmidt [1961]. Among the topics discussed in these papers are the Isomorphism Theorems, the Zassenhaus Lemma, the Jordan-Hölder Theorem for standard ideals, and the Schreier extension problem. It is also shown that in a finite modular lattice exactly the neutral ideals satisfy the First Isomorphism Theorem. Some of these results were inspired by J. Hashimoto [1952].

In the proof of Theorem 4.9, we describe the pseudocomplement Φ of a congruence relation Θ of a weakly modular lattice. Iqbalunissa [1966] observes that the converse also holds: if in a lattice L, the relation Φ given by any congruence relation Θ (as described in the proof of Theorem 4.9) is always a congruence relation, then L is weakly modular.

For some additional results on standard ideals, see M. F. Janowitz [1964], [1964a], and [1965], in which some types of ideals, more general than standard ideals, are discussed. Congruences of relatively complemented lattices, in particular, a description of $\Theta(a,b)$, is given in M. F. Janowitz [1968]. Some counterexamples are given in Iqbalunissa [1965] and [1965a].

Lattices whose congruences form a Boolean lattice are discussed also in T. Tanaka [1951], J. Hashimoto [1957], P. Crawley [1960]. See also D. T. Finkbeiner [1960] and J. Jakubik [1955]. Lattices whose congruences form a Stone lattice are studied in M. F. Janowitz [1968] and [1968a], and Iqbalunissa [1971].

Standard elements in the lattice of all subsemigroups of a group are studied in S. G. Ivanov [1966].

It is pointed out in S. Maeda [1974] that if L is the dual of the lattice of all T_1-topologies on an infinite set, then L has infinitely many standard elements, but only the elements 0 and 1 are neutral.

The concept of a standard ideal has recently been extended to convex sublattices in E. Fried and E. T. Schmidt [1975].

Distributivity of a pair of lattice elements is investigated in P. G. Kontorovič, S. G. Ivanov, and G. P. Kondrašov [1965]; see also F. and S. Maeda [1970]. We shall consider modular pairs of elements in Section IV.2.

A great deal of useful information on congruence relations of lattices, in particular, about regularity and permutability, can be found in J. Hashimoto [1963].

There is a connection between projectivity and representability. If L/Θ is projective (in a class containing L), then Θ is representable. A. Day [1973] considers a weaker concept implying the representability of Θ for finite L.

Problems

1. Generalize the concepts of distributive and neutral ideals to convex sub-lattices.

2. Let $L = I_0 \subseteq I_1 \subseteq \cdots \subseteq I_{n+1} = I$ be a descending sequence of ideals. If I_{j+1} is a standard ideal of I_j, for $j = 0, \ldots, n$, then I is a *standard ideal of order n of L*. Investigate standard ideals of order 2 (order n). Under what conditions do they form a sublattice?

3. An ideal I of a lattice L is said to *satisfy the First Isomorphism Theorem* (G. Grätzer and E. T. Schmidt [1958d]) iff

   $$(I \vee J)/\Theta[I] \cong J/\Theta[I \cap J],$$

   for any ideal J of L (under the natural isomorphism). Investigate this concept and relate it to standard ideals.

4. Under what conditions do the ideals of a weakly modular lattice form a weakly modular lattice?

5. Investigate lattices whose congruences form a Stone lattice.

6. For $n > 1$, investigate lattices whose congruence lattices, as distributive lattices with pseudocomplementation, belong to $\mathbf{B}_n$.

7. Develop structure theorems for lattices all of whose congruences are standard (distributive, neutral).

Modular and Semimodular Lattices

1. Modular Lattices

The modular identity is unquestionably the most important identity apart from the distributive identity. In this section, we examine the most important consequences of modularity.

Theorem 1. For a lattice L, the following conditions are equivalent:

(i) L is modular, that is,

$$x \geq z \quad \text{implies that} \quad x \wedge (y \vee z) = (x \wedge y) \vee z.$$

(ii) L satisfies the *shearing identity*:

$$x \wedge (y \vee z) = x \wedge ((y \wedge (x \vee z)) \vee z).$$

(iii) L does not contain a pentagon. ——

Remark. In Section II.1, we have already proved the equivalence of (i) and (iii). The importance, or convenience, of the shearing identity (named by I. Halperin) is that it can be applied to any expressions of the form $x \wedge (y \vee z)$ without any assumptions. Observe also the dual of the shearing identity:

$$x \vee (y \wedge z) = x \vee ((y \vee (x \wedge z)) \wedge z).$$

Proof.

(i) *implies* (ii). Since $x \vee z \geq z$, by modularity,

$$(y \wedge (x \vee z)) \vee z = (y \vee z) \wedge (x \vee z),$$

and so

$$x \wedge ((y \wedge (x \vee z)) \vee z) = x \wedge ((y \vee z) \wedge (x \vee z)) = x \wedge (y \vee z).$$

(ii) *implies* (iii). In N_5, (ii) fails; indeed,

$$a \wedge (c \vee b) = a \wedge i = a,$$
$$a \wedge ((c \wedge (a \vee b)) \vee b) = a \wedge ((c \wedge a) \vee b) = a \wedge b = b.$$

Thus (ii) implies (iii).

(iii) *implies* (i). As in Section II.1.   □

The most important form of modularity is the following:

Theorem 2 (The Isomorphism Theorem).   Let L be a modular lattice and let $a, b \in L$. Then

$$\varphi_b \colon x \mapsto x \wedge b, \quad x \in [a, a \vee b],$$

is an isomorphism of $[a, a \vee b]$ and $[a \wedge b, b]$. The inverse isomorphism is

$$\psi_a \colon y \mapsto x \vee a, \quad y \in [a \wedge b, a].$$ —

(See Figure 1.)

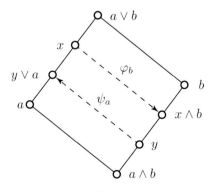

Figure 1

Proof. It is sufficient to show that $x\varphi_b\psi_a = x$, for $x \in [a, a \vee b]$. Indeed, if this is true, then by duality, $y\psi_a\varphi_b = y$, for all $y \in [a \wedge b, b]$. The isotone maps φ_b and ψ_a, thus, compose into the identity maps, hence they are isomorphisms, as claimed.

So let $x \in [a, a \vee b]$. Thus $x\varphi_b\psi_a = (x \wedge b) \vee a$. Since $x \in [a, a \vee b]$, we have $a \leq x$, and so modularity applies:

$$x\varphi_b\psi_a = (x \wedge b) \vee a = x \wedge (b \vee a) = x,$$

because $x \leq a \vee b$. □

Corollary 3. *Let L be a modular lattice and let a, $b \in L$. Then, for x, $y \in [a \wedge b, b]$, we have*

$$a \vee (x \wedge y) = (a \vee x) \wedge (a \vee y).$$

Most applications of the Isomorphism Theorem are like Corollary 3: the special form of the isomorphism is used. There are two important applications where the form of the isomorphism plays no role.

Let us write $a \preceq b$ (or $b \succeq a$) for $a \prec b$ or $a = b$. A lattice L is said to satisfy the *Upper Covering Condition* iff $a \preceq b$ implies that $a \vee c \preceq b \vee c$, for a, b, $c \in L$. The *Lower Covering Condition* is the dual.

Theorem 4.   A modular lattice satisfies both the Upper Covering Condition and the Lower Covering Condition.   —

Proof. Let L be a modular lattice. Let a, b, $c \in L$ and $b \prec a$. If $a \vee c = b \vee c$, we have nothing to prove. If $a \vee c \neq b \vee c$, then $a \nleq b \vee c$, and so $a \wedge (b \vee c) = b$. Applying the Isomorphism Theorem to a and $b \vee c$, we obtain

$$[b, a] \cong [b \vee c, a \vee c].$$

Since $[b, a]$ is a prime interval, so is $[b \vee c, a \vee c]$, that is, $b \vee c \prec a \vee c$, proving the Upper Covering Condition. By duality, we get the Lower Covering Condition. □

The second application deals with representations of elements. In Corollary II.1.13, we proved that in a finite distributive lattice the irredundant representation of an element as a join of join-irreducible elements is unique. This is obviously false in M_3. But we have the following results (A. G. Kuroš [1935], O. Ore [1936]):

Theorem 5 (The Kuroš-Ore Theorem).   Let L be a modular lattice and let $a \in L$. If $a = x_0 \vee \cdots \vee x_{n-1}$ and $a = y_0 \vee \cdots \vee y_{m-1}$ are irredundant

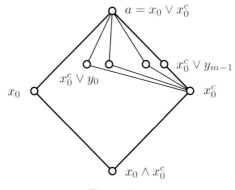

Figure 2

representations of a as joins of join-irreducible elements, then for every x_i, there is a y_j such that

$$a = x_0 \vee \cdots \vee x_{i-1} \vee y_j \vee x_{i+1} \vee \cdots \vee x_{n-1}$$

and $n = m$.   —

Proof. Let us prove the first statement, say, for x_0. Let $x_0^c = x_1 \vee \cdots \vee x_{n-1}$. (See Figure 2.) Since $y_0 \vee \cdots \vee y_{m-1} = a$, we obtain that

$$(x_0^c \vee y_0) \vee (x_0^c \vee y_1) \vee \cdots \vee (x_0^c \vee y_{m-1}) = a,$$

where $x_0^c \vee y_0, \dots, x_0^c \vee y_{m-1} \in [x_0^c, a]$. By the Isomorphism Theorem,

$$[x_0^c, a] \cong [x_0 \wedge x_0^c, x_0],$$

and the image of a under any such isomorphism is x_0. But x_0 is join-irreducible in L, and, therefore, in $[x_0 \wedge x_0^c, x_0]$; thus a is join-irreducible in $[x_0^c, a]$. Hence $x_0^c \vee y_j^c = a$, for some j, proving the first statement.

Now, let $a = z_0 \vee \cdots \vee z_{k-1}$ be an irredundant representation of a as a join of join-irreducibles, with k minimal. Applying the statement we have just proved to $a = z_0 \vee \cdots \vee z_{k-1}$, z_0, and $a = x_0 \vee \cdots \vee x_{n-1}$, we obtain that $a = x_{j_0} \vee z_1 \vee \cdots \vee z_{k-1}$. This representation is irredundant, otherwise, the minimality of k would be contradicted. Repeating this, we eventually obtain that $a = x_{j_0} \vee \cdots \vee x_{j_{k-1}}$. However, $a = x_0 \vee \cdots \vee x_{n-1}$ is an irredundant representation and so $\{j_0, \dots, j_{k-1}\} = \{0, \dots, n-1\}$. This shows that $n \leq k$. Thus $k = n \ (= m)$. □

Weak projectivities in modular lattices can be described rather precisely. Let us call a sequence of perspectivities:

$$x_0/y_0 \sim x_1/y_1 \sim \cdots \sim x_n/y_n$$

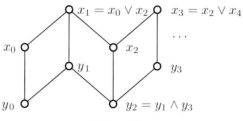

Figure 3

alternating iff, for each i with $0 < i < n$, either $x_{i-1}/y_{i-1} \nearrow x_i/y_i \searrow x_{i+1}/y_{i+1}$ or $x_{i-1}/y_{i-1} \searrow x_i/y_i \nearrow x_{i+1}/y_{i+1}$. If, in addition, $x_i = x_{i-1} \vee x_{i+1}$, in the first case, and $y_i = y_{i-1} \wedge y_{i+1}$, in the second, we call the sequence *normal*. (See Figure 3.)

Theorem 6 (G. Grätzer [1966]).   Let L be a modular lattice, and let a/b and c/d be quotients of L. Then $a/b \approx_w c/d$ iff there exists a normal sequence of perspectivities $a/b = x_0/y_0 \sim x_1/y_1 \sim \cdots \sim x_n/y_n$, where x_n/y_n is a subquotient of c/d.   ——

Proof. It follows from the Isomorphism Theorem that if $x/y \nearrow u/v$ and x_1/y_1 is a subquotient of x/y, then $x_1/y_1 \nearrow u_1/v_1$, where $u_1 = y_1 \vee v$ and $v_1 = x_1 \vee v$. This, the dual statement, and a simple induction show that if $a/b \approx_w c/d$, then there is an alternating sequence of perspectivities

$$a/b = x_0/y_0 \sim \cdots \sim x_n/y_n,$$

where x_n/y_n is a subquotient of c/d. Starting with this sequence, we show, by induction on n, that there is a normal sequence of perspectivities

$$a/b = u_0/v_0 \sim u_1/v_1 \sim \cdots \sim u_n/v_n = x_n/y_n.$$

This statement is obvious, for $n \leq 1$. We can assume that $u_{n-1}/v_{n-1} \searrow u_n/v_n$, by duality. By the induction hypothesis, there is a normal sequence

$$a/b = u_0/v_0 \sim \cdots \sim u_{n-1}/v_{n-1} = x_{n-1}/y_{n-1}.$$

We define a new sequence as follows:

$$a/b = u_0/v_0 \sim u_1/v_1 \sim \cdots$$
$$\searrow u_{n-2}/v_{n-2} \nearrow u_{n-2} \vee x_n/v_{n-1} \wedge (u_{n-2} \vee x_n) \searrow x_n/y_n.$$

We claim that this is a normal sequence of perspectivities. The following formulas have to be checked; (i) and (ii) are the perspectivities, and (iii) is the normality:
(i) $u_{n-2}/v_{n-2} \nearrow u_{n-2} \vee x_n/v_{n-1} \wedge (u_{n-2} \vee x_n)$;
(ii) $u_{n-2} \vee x_n/v_{n-1} \wedge (u_{n-2} \vee x_n) \searrow x_n/y_n$;

(iii) $v_{n-2} = v_{n-3} \wedge (v_{n-1} \wedge (u_{n-2} \vee x_n))$.
Since $v_{n-2} \leq v_{n-1} \wedge (u_{n-2} \vee x_n) \leq v_{n-1}$,

$$u_{n-2} \wedge (v_{n-1} \wedge (u_{n-2} \vee x_n)) = v_{n-2}$$

is obvious. Now compute:

$$u_{n-2} \vee (v_{n-1} \wedge (u_{n-2} \vee x_n))$$

(by modularity and $u_{n-2} \leq u_{n-2} \vee x_n$)

$$= (u_{n-2} \vee v_{n-1}) \wedge (u_{n-2} \vee x_n)$$
$$= u_{n-2} \vee x_n,$$

verifying (i). The proof of (ii) is similar. For (iii), observe that $v_{n-2} \leq u_{n-2} \vee x_n$ and $v_{n-3} \wedge v_{n-1} = v_{n-2}$, hence

$$v_{n-3} \wedge (v_{n-1} \wedge (u_{n-2} \vee x_n)) = (v_{n-3} \wedge v_{n-1}) \wedge (u_{n-2} \vee x_n) = v_{n-2}.$$

This completes the induction step.   □

We can do even more than this: we can determine the sublattices generated by three consecutive quotients.

Theorem 7 (G. Grätzer [1966]).   Let L be a modular lattice and let

$$x_0/y_0 \nearrow x_1/y_1 \searrow x_2/y_2$$

be a normal sequence of quotients. Let A be the sublattice of L generated by x_0, x_1, x_2, y_0, y_1, and y_2. Then either A is distributive, in which case A is the lattice of Figure 4 or some quotient thereof and we also have

$$x_0/y_0 \searrow (x_0 \wedge x_2)/(y_0 \wedge y_2) \nearrow x_2/y_2,$$

or A is not distributive and A is the lattice of Figure 5 or some quotient not collapsing the diamond in A.   —

Proof. A is generated by the elements x_0, y_1, and x_2, hence A is isomorphic to a quotient lattice of $F_M(3)$, see Figure I.5.7. The generators satisfy the relation $x_0 \vee y_1 = y_1 \vee x_2 = x_0 \vee x_2$ (the last one comes in because of normality). The corresponding quotient lattice is given in Figure 5. A quotient of this is distributive iff the diamond is collapsed, yielding Figure 4.   □

Corollary 8. *Let L be a modular lattice and let $a/b \approx c/d$ in L. There is a shortest normal sequence of quotients $a/b = x_0/y_0 \sim x_1/y_1 \sim x_n/y_n = c/d$. Then either $n \leq 2$, or, for each i, $0 < i < n$, the sublattice generated by x_{i-1}, x_i, x_{i+1}, y_{i-1}, y_i, y_{i+1} is isomorphic to one of the lattices of Figure 5, Figure 6, or their duals.*

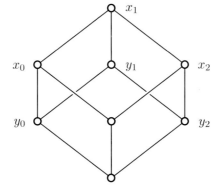

Figure 4

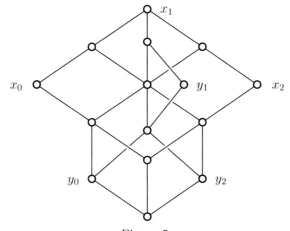

Figure 5

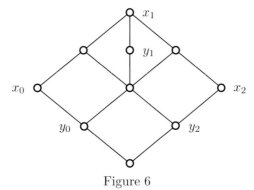

Figure 6

Proof. If $n > 2$, then the sublattice generated by x_{i-1}, x_i, x_{i+1}, y_{i-1}, y_i, y_{i+1} cannot be distributive, otherwise we could turn the down arrow up and the up arrow down and get a sequence of length $n-1$. Thus these elements generate the lattice of Figure 5, or its dual, or a quotient lattice of Figure 5, or its dual. The only nontrivial quotient lattice of the lattice of Figure 5 is given by Figure 6.   □

As a typical example of the advantages of using the shearing identity we consider independence in the sense of J. von Neumann [1936]. This plays a very important role in the applications of lattice theory to direct decompositions of groups and rings and also in continuous geometries.

Definition 9. *Let L be a lattice with 0. A subset I of $L - \{0\}$ is called independent iff, for any two finite subsets X, Y of I, we have*

$$\bigvee X \wedge \bigvee Y = \bigvee (X \cap Y).$$

Corollary 10. *A subset I of a lattice L is independent iff*

$$\varphi \colon X \mapsto \bigvee X$$

is an isomorphism between $[I]$ and the generalized Boolean lattice of all finite subsets of I.

Proof. $\bigvee X \vee \bigvee Y = \bigvee (X \cup Y)$ and so if I is independent, then φ is a homomorphism. If φ is not one-to-one, then $\bigvee X = \bigvee Y$, for some $X \neq Y$, $X, Y \subseteq I$, and X, Y finite. Let, say, $X \not\subseteq Y$, and let $a \in X - Y$. Then $a \leq \bigvee Y$, $a \notin Y$. Therefore,

$$a = a \wedge \bigvee Y = \bigvee \{a\} \wedge \bigvee Y = \bigvee (\{a\} \cap Y) = \bigvee \varnothing = 0,$$

a contradiction. Thus φ is an isomorphism. The converse is obvious.   □

A singleton $\{a\}$ ($a \in L - \{0\}$) is always independent. $\{a, b\}$ is independent iff $a \wedge b = 0$ ($a, b \in L - \{0\}$ and $a \neq b$). For $\{a, b, c\}$ to be independent (a, b, $c \in L - \{0\}$ and a, b, c all distinct), we have to require that

$$a \wedge (b \vee c) = 0,$$
$$b \wedge (a \vee c) = 0,$$
$$c \wedge (a \vee b) = 0,$$
$$(a \vee b) \wedge (a \vee c) = a,$$
$$(b \vee c) \wedge (b \vee a) = b,$$
$$(c \vee a) \wedge (c \vee b) = c.$$

For modular lattices, fewer relations will do:

Theorem 11.   Let L be a modular lattice with 0. Then an n element set $\{a_1, \ldots, a_n\} \subseteq L - \{0\}$ is independent iff

$$(a_1 \vee \cdots \vee a_i) \wedge a_{i+1} = 0, \quad \text{for } i = 1, 2, \ldots, n - 1. \qquad —$$

Proof. We obtain the necessity of the condition by setting $X = \{a_1, \ldots, a_i\}$, $Y = \{a_{i+1}\}$. Now assume that $\{a_1, \ldots, a_n\}$ satisfies the condition. Let X, $Y \subseteq \{a_1, \ldots, a_n\}$, $X \cap Y = \varnothing$. We claim that

$$\bigvee X \wedge \bigvee Y = 0.$$

Indeed, let $a_k \in X \cup Y$ with k maximal. Let, say, $a_k \in Y$. Apply the shearing identity to $\bigvee X \wedge \bigvee Y$ with $x = \bigvee X$, $y = a_k$, and $z = \bigvee(Y - \{a_k\})$:

$$\bigvee X \wedge \bigvee Y = \bigvee X \wedge (a_k \vee \bigvee(Y - \{a_k\}))$$
$$= \bigvee X \wedge ((a_k \wedge (\bigvee X \vee \bigvee(Y - \{a_k\}))) \vee \bigvee(Y - \{a_k\}))$$
$$= \bigvee X \wedge \bigvee(Y - \{a_k\}),$$

since $a_k \wedge (\bigvee X \vee \bigvee(Y - \{a_k\})) \leq (a_1 \vee \cdots \vee a_{k-1}) \wedge a_k = 0$.

Proceeding thus, we can eliminate all the a_i belonging to $X \cup Y$, getting $\bigvee X \wedge \bigvee Y = \bigvee \varnothing = 0$. Now, in the general case,

$$\bigvee X \wedge \bigvee Y = \bigvee X \wedge (\bigvee(X \cap Y) \vee \bigvee(Y - X))$$

(by modularity)

$$= \bigvee(X \cap Y) \vee (\bigvee X \wedge \bigvee(Y - X))$$

(by $X \cap (Y - X) = \varnothing$)

$$= \bigvee(X \cap Y). \qquad \square$$

We conclude this section with three important "sublattice theorems".

Theorem 12 (J. von Neumann [1936]).   Let L be a modular lattice and let a, b, $c \in L$. The sublattice of L generated by a, b, and c is distributive iff $a \wedge (b \vee c) = (a \wedge b) \vee (a \wedge c)$. $\qquad —$

Proof. By inspection of the diagram of $\mathsf{F_M}(3)$ (see Figure I.5.7). If a, b, and c are the generators and $a \wedge (b \vee c) \equiv (a \wedge b) \vee (a \wedge c)$ (Θ), where Θ is a congruence relation, then Θ collapses the only diamond and so $\mathsf{F_M}(3)/\Theta$ is distributive. $\qquad \square$

One can view the definition of modularity as requiring that any sublattice generated by three elements, two of which are comparable, has to be distributive. This is true in general for any two chains:

Theorem 13 (G. Birkhoff [1940]).   Let L be a modular lattice. Let us assume that C and C' be chains in L. Then the sublattice of L generated by $C \cup C'$ is distributive.   —

Proof. Since a lattice is distributive iff every finitely generated sublattice is distributive, it is sufficient to verify this result for finite C and C'. Let

$$C = \{a_0, \ldots, a_{m-1}\}, \quad a_0 < \cdots < a_{m-1},$$
$$C' = \{b_0, \ldots, b_{n-1}\}, \quad b_0 < \cdots < b_{n-1}.$$

To simplify the notation, we shall write $a_m = b_n = a_{m-1} \vee b_{n-1}$. Let us define

$$x(r, \alpha, \beta) = (a_{\alpha(1)} \wedge b_{\beta(1)}) \vee \cdots \vee (a_{\alpha(r)} \wedge b_{\beta(r)}),$$

where

$$\alpha = \langle \alpha(1), \ldots, \alpha(r) \rangle, \quad m \geq \alpha(1) > \alpha(2) > \cdots > \alpha(r) \geq 1,$$
$$\beta = \langle \beta(1), \ldots, \beta(r) \rangle, \quad 1 \leq \beta(1) < \beta(2) < \cdots < \beta(r) \leq n.$$

Let A denote the set of all elements of L of the form $x(r, \alpha, \beta)$. Then $C, C' \subseteq A$, since $a_i = a_i \wedge b_n$ and $b_i = a_m \wedge b_i$. It is also easy to see that A is closed under join: if we form $x(r, \alpha, \beta) \vee x(s, \alpha', \beta')$, then if we have both $a_i \wedge b_j$ and $a_i \wedge b_k$, we can eliminate one by absorption; if $a_i < a_j$, $b_k < b_t$ and we have $a_i \wedge b_k$ and $a_j \wedge b_t$, the former is eliminated by absorption; what is left can be written in the form $x(r', \alpha'', \beta'')$.

Now we prove, by induction, the formula

$$x(r, \alpha, \beta) = a_{\alpha(1)} \wedge (b_{\beta(1)} \vee a_{\alpha(2)}) \wedge \cdots \wedge (b_{\beta(r-1)} \vee a_{\alpha(r)}) \wedge b_{\beta(r)}$$

and its dual. The case $r = 1$ is obvious. Let us assume that this formula and its dual have been verified for $r - 1$. Now compute:

$$x(r, \alpha, \beta) = (a_{\alpha(1)} \wedge b_{\beta(1)}) \vee \cdots \vee (a_{\alpha(r)} \wedge b_{\beta(r)})$$

(observe that $a_{\alpha(1)} \geq (a_{\alpha(2)} \wedge b_{\beta(2)}) \vee \cdots \vee (a_{\alpha(r)} \wedge b_{\beta(r)})$ and apply modularity)

$$= a_{\alpha(1)} \wedge (b_{\beta(1)} \vee (a_{\alpha(2)} \wedge b_{\beta(2)}) \vee \cdots \vee (a_{\alpha(r)} \wedge b_{\beta(r)}))$$

(observe that $b_{\beta(r)} \geq b_{\beta(1)} \vee \cdots \vee (a_{\alpha(r-1)} \wedge b_{\beta(r-1)})$ and apply modularity)

$$= a_{\alpha(1)} \wedge (b_{\beta(1)} \vee (a_{\alpha(2)} \wedge b_{\beta(2)}) \vee \cdots \vee (a_{\alpha(r-1)} \wedge b_{\beta(r-1)}) \vee a_{\alpha(r)})$$
$$\wedge b_{\beta(r)}$$

(apply to the expression in parentheses the dual of the formula for $r - 1$)

$$= a_{\alpha(1)} \wedge (b_{\beta(1)} \vee a_{\alpha(2)}) \wedge (b_{\beta(2)} \vee a_{\alpha(3)}) \wedge \cdots \wedge (b_{\beta(r-1)} \vee a_{\alpha(r)})$$
$$\wedge b_{\beta(r)},$$

completing the proof.

Now we easily see that A is a sublattice. Indeed, $x(r, \alpha, \beta) \wedge x(s, \alpha', \beta')$ can be rewritten by the above formula as a meet of $b_i \vee a_j$; but by the dual of the above formula, the result can again be put in the form $x(r, \alpha, \beta)$.

So we conclude that A is the sublattice generated by $C \cup C'$ and therefore $|A| \leq$ the number of expressions of the form $x(r, \alpha, \beta)$.

Now consider the set $X = [0, n + 1] \times [0, m + 1]$, let F be the sublattice of $P(X)$ generated by

$$a_i = \{ \langle x, y \rangle \mid y \leq i + 1 \}, \quad \text{for } i = 0, 1, \ldots, m - 1,$$
$$b_i = \{ \langle x, y \rangle \mid x \leq i + 1 \}, \quad \text{for } i = 0, 1, \ldots, n - 1,$$

and let $C = \{a_0, \ldots, a_{m-1}\}$ and $C' = \{b_0, \ldots, b_{n-1}\}$. In this lattice

$$x(r, \alpha, \beta) = \bigcup (\{ \langle x, y \rangle \mid x \leq \alpha(i) + 1, \ y \leq \beta(i) + 1) \} \mid i = 1, 2, \ldots, r),$$

and so all these represent different elements. Consequently, F is the free modular lattice generated by $C \cup C'$. Since F is distributive, any modular lattice generated by $C \cup C'$ is distributive. $\square$

Finally, we generalize Theorem II.1.6 to modular lattices:

Theorem 14. Let L be a modular lattice and let $a, b \in L$. Then the sublattice of L generated by $[a \wedge b, a] \cup [a \wedge b, b]$ is isomorphic to $[a \wedge b, a] \times [a \wedge b, b]$. —

Remark. In the distributive case, the sublattice generated by $[a \wedge b, a] \cup [a \wedge b, b]$ is $[a \wedge b, a \vee b]$; this does not hold for modular lattices, as exemplified by M_3.

Proof. The isomorphism we set up is

$$\varphi \colon \langle x, y \rangle \mapsto x \vee y, \quad \text{for } x \in [a \wedge b, a] \text{ and } y \in [a \wedge b, b].$$

Using the formula for the dual of $x(r, \alpha, \beta)$ in the proof of Theorem 13, we obtain:

$$x \vee y = (a \vee y) \wedge (b \vee x).$$

Hence

$$a \wedge (x \vee y) = a \wedge (a \vee y) \wedge (b \vee x) = a \wedge (b \vee x) = (a \wedge b) \vee x = x,$$
$$b \wedge (x \vee y) = y,$$

proving that φ is one-to-one. φ is obviously onto and preserves join. Let x, $x_1 \in [a \wedge b, a]$ and $y, y_1 \in [a \wedge b, b]$. Then

$$(x \vee y) \wedge (x_1 \vee y_1) = (a \vee y) \wedge (b \vee x) \wedge (a \vee y_1) \wedge (b \vee x_1)$$
$$= (a \vee y) \wedge (a \vee y_1) \wedge (b \vee x) \wedge (b \vee x_1)$$

(use Corollary 3)

$$= (a \vee (y \wedge y_1)) \wedge (b \vee (x \wedge x_1))$$
$$= (x \wedge x_1) \vee (y \wedge y_1),$$

proving that φ is an isomorphism.   □

Exercises

1.   Show that a lattice L is modular iff it satisfies the identity

$$(x \vee (y \wedge z)) \wedge (y \vee z) = (x \wedge (y \vee z)) \vee (y \wedge z).$$

2.   Let

$$p = (x \vee y) \wedge (y \vee z) \wedge (z \vee x),$$
$$d_0 = a \vee b \vee c,$$

and define, for $n \geq 0$,

$$d_{n+1} = p(a \wedge d_n, b \wedge d_n, c \wedge d_n).$$

Show that the identity $d_1 = d_2$ is equivalent to modularity but $d_2 = d_3$ is not (B. Wolk).

3.   A finite lattice L is modular iff it does not contain a pentagon $\{0, a, b, c, i\}$ satisfying $a \succ b$.

4.   Can the numbers of covering pairs in Exercise 3 be increased?

5.   Let L be a lattice and $a, b \in L$. Then $\varphi_b \colon x \mapsto x \wedge b$ maps $[a, a \vee b]$ into $[a \wedge b, b]$, and $\psi_a \colon x \mapsto x \vee a$ maps $[a \wedge b, b]$ into $[a, a \vee b]$. Show that $\varphi_b \psi_a \varphi_b = \varphi_b$ and $\psi_a \varphi_b \psi_a = \psi_a$.

6.   Using the notation of Exercise 5, prove that

$$\{\, x \mid x \in [a \wedge b, b], \ x\psi_a\varphi_b = x \,\}$$

is isomorphic to

$$\{\, x \mid x \in [a, a \vee b], \ x\varphi_b\psi_a = x \,\}.$$

(Exercises 5 and 6 are due to W. Schwan [1948], [1948a], [1949], and [1949a].)

7. Does Corollary 3 characterize the modularity of a lattice?

8. Find a nonmodular lattice satisfying the Upper Covering Condition and the Lower Covering Condition.

9. Show that the Kuroš-Ore Theorem fails in the first lattice of Figure 2.1.

10. Prove that the representation

$$a = x_0 \vee \cdots \vee x_{i-1} \vee y_j \vee x_{i+1} \vee \cdots \vee x_{n-1}$$

is always irredundant in the Kuroš-Ore Theorem.

11. Using the notation of the Kuroš-Ore Theorem, show that there is a permutation π of $\{0, \dots, m-1\}$ such that

$$a = x_0 \vee \cdots \vee x_{i-1} \vee y_{i\pi} \vee x_{i+1} \vee \cdots \vee x_{n-1}$$

holds, for all i, $0 \leq i < n$ (R. P. Dilworth [1946]).

12. Can the Kuroš-Ore Theorem be sharpened so as to require that all x_i can be simultaneously replaced by y_j? (No, inspect Figure 3.4b.)

13. Does Theorem 6 characterize modularity?

14. Prove Theorem 7 directly, that is, without reference to $\mathbf{F_M}(3)$.

15. Show that, for finite modular lattices, it is sufficient to consider projectivities of prime intervals. Simplify Theorems 6 and 7 for prime intervals.

16. Extend Exercise 15 to locally finite lattices.

17. Show that Exercise 15 does not generalize to nonmodular lattices.

18. Show that to define independence in modular lattices, it is sufficient to assume that $\bigvee X \wedge \bigvee Y = \varnothing$, for $X \cap Y = \varnothing$, $X, Y \subseteq I$, X and Y finite.

19. Let L be a complete lattice and let I be a set of compact elements of L. Then I is independent iff $X \mapsto \bigvee X$ is an isomorphism between the Boolean lattice of all subsets of I and the complete sublattice generated by I.

20. Let L be a complemented modular lattice and let $a_0 = 0 < a_1 < \cdots < a_n = 1$ be elements of L. Let $b_1 = a_1$, let b_2 be a relative complement of a_1 in $[0, a_2]$, $\dots$, let b_n be a relative complement of a_{n-1} in $[0, a_n]$. Then $B = \{b_1, \dots, b_n\}$ is an independent set.

21. Prove the converse of Exercise 20.

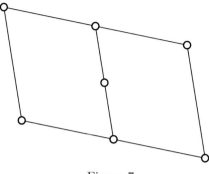

Figure 7

*22. Let L be a modular lattice and let C, C', and C'' be chains in L. The sublattice generated by $C \cup C' \cup C''$ is distributive iff $a \wedge (b \vee c) = (a \wedge b) \vee (a \wedge c)$, for all $a \in C$, $b \in C'$, and $c \in C''$ (B. Jónsson [1955]).

23. Let $\mathbf{K}$ be an variety of lattices and let P be a poset such that $F_{\mathbf{K}}(P)$ exists in the sense of Definition I.5.2. Let A be the class of all *finite* lattices L such that L is generated by some homomorphic image of P (in the sense defined in the proof of Theorem I.5.5). Let us assume that there is an $L \in A$ such that $|L| \geq |L'|$, for any $L' \in A$. Is it true that $F_{\mathbf{K}}(P)$ exists and is finite, in fact, is it true we can take $F_{\mathbf{K}}(P) = L$?

24. Prove that in a modular lattice L, the sublattice generated by $[a \wedge b, a] \cup [a \wedge b, b]$ equals $[a \wedge b, a \vee b]$, for any a, $b \in L$ iff L is distributive.

25. Find equations to replace "$a \wedge (b \vee c) = (a \wedge b) \vee (a \wedge c)$" in Theorem 12.

26. Find a direct proof of the statement that, in a modular lattice, every distributive element is neutral.

27. Show that, in a distributive lattice, $a/b \approx c/d$ and c/d a subquotient of a/b imply that $a/b = c/d$.

28. Let D be a bounded distributive lattice. Define a subset M of D^3 by

$$\langle x, y, z \rangle \in M \quad \text{iff} \quad x \wedge y = y \wedge z = z \wedge x.$$

Then M as a subposet of D^3 is a lattice. For α, $\beta \in M$, $\alpha \wedge \beta$ in M is the same as $\alpha \wedge \beta$ in D^3. Compute $\alpha \vee \beta$.

29. Show that the lattice M of Exercise 28 is a modular lattice.

30. Show that in M of Exercise 28, $\langle 1, 0, 0 \rangle / \langle 0, 0, 0 \rangle \cong \langle 0, 1, 0 \rangle / \langle 0, 0, 0 \rangle \cong D$. (Exercises 28–30 are due to E. T. Schmidt [1962].)

31. Show that the identity $d_2 = d_3$ of Exercise 2 fails in the lattice of Figure 7 (B. Wolk).

32. Show that a finite modular lattice is dismantlable iff it has breadth two or less (D. Kelly and I. Rival [1974]).

33. Verify that if L is a finite modular lattice satisfying

$$|L| \le \frac{1}{3}(5l(L) + 7),$$

then L has an m element sublattice, for all $m \le |L|$. (I. Rival [1975]. Hint: Use Exercise I.6.32.)

34. Does the fact that φ_b of Theorem 2 is one-to-one, for all $a, b \in L$, characterize the modularity of L?

2. Semimodular Lattices

A lattice L is called *semimodular* iff it satisfies the *Upper Covering Condition*, that is, for $a, b \in L$,

$$a \prec b \quad \text{implies that} \quad a \vee c \prec b \vee c \text{ or } a \vee c = b \vee c.$$

Examples of semimodular lattices include finite modular lattices and some important lattices from geometry (see Section 3). Two further examples are given in Figure 1.

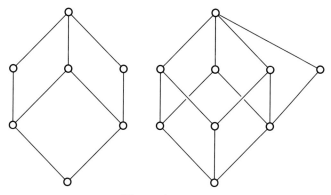

Figure 1

Let L be a lattice with 0. We define the *height function*: for $a \in L$, let $h(a)$ denote the length of a longest maximal chain in $[0, a]$, if there is a finite longest maximal chain; otherwise, put $h(a) = \infty$. If L is of finite length, $0 \le h(a) < \omega$, for all $a \in L$. We first show that, for semimodular lattices of finite length, $h(a)$ is the length of *any* maximal chain in $[0, a]$.

Theorem 1 (The Jordan-Hölder Chain Condition. O. Ore [1943]).
Let L be a lattice of finite length. If L is semimodular, then any two maximal chains of L are of the same length.   —

Proof. Let $C = \{a_0, \ldots, a_n\}$, $0 = a_0 < \cdots < a_n = 1$, be a maximal chain of length n. We prove that any other maximal chain is of length n, by induction on n. If $n \leq 1$, the statement is trivial. Let us assume the statement to hold for length $< n$. Let $C' = \{b_0, b_1, \ldots, b_m\}$, $0 = b_0 < b_1 < \cdots < b_m = 1$ be another maximal chain in L. If $a_1 = b_1$, then, in the semimodular lattice $[a_1)$, the maximal chain $C - \{a_0\}$ is of length $n - 1$, so the maximal chain $C' - \{b_0\}$ has to be of length $n - 1$, therefore $n = m$. If $a_1 \neq b_1$ (see Figure 2), then let C'' be a maximal chain in $[a_1 \vee b_1)$ and let k be the length of C''. Because of semimodularity, $a_1 \vee b_1 \succ a_1$ and $a_1 \vee b_1 \succ b_1$. Hence $C'' \cup \{a_1\}$ is a maximal chain of length $k + 1$ and $C - \{a_0\}$ is a maximal chain of length $n - 1$ in $[a_1)$; thus $k + 1 = n - 1$. Similarly, $k + 1 = m - 1$, hence $n = m$. ☐

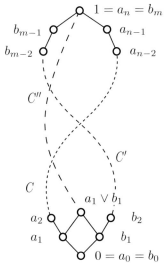

Figure 2

Theorem 1 suggests that semimodularity can be characterized in terms of the height function (the equivalence of (i)–(iii), and that they imply (iv) can be found in G. Birkhoff [1933], see also G. Birkhoff [1967]):

Theorem 2. Let L be a lattice of finite length. The following conditions on L are equivalent, for all a, b, $c \in L$:

(i) L is semimodular.

(ii) If $a \neq b$, a and b cover $a \wedge b$, then $a \vee b$ covers a and b.

(iii) If $a \leq b$ and C is a maximal chain in $[a, b]$, then $\{\, x \vee c \mid x \in C \,\}$ is a maximal chain in $[a \vee c, b \vee c]$.

(iv) $h(a) + h(b) \geq h(a \wedge b) + h(a \vee b)$.   —

Proof.

(ii) *implies* (i). Let $a \prec b$. If $c \leq a$ or $a \vee c \geq b$, then $b \vee c \succeq a \vee c$. If $c \nleq a$ and $a \vee c \ngeq b$, then $b \wedge (a \vee c) = a$. Let $a = a_0 < a_1 < \cdots < a_n = a \vee c$ be a maximal chain in $[a, a \vee c]$. Since b and a_1 cover a and $b \neq a_1$, $b \vee a_1$ covers a_1. An easy induction shows that $b \vee a_i$ covers a_i, for $i = 1, 2, \ldots, n$. Thus $b \vee a_n$ covers a_n, that is, $b \vee c$ covers $a \vee c$.

(i) *implies* (iii). Obvious, since if $x \prec y$ in C, then $c \vee x \preceq c \vee y$.

(iii) *implies* (iv). (iii) obviously implies semimodularity and so, by Theorem 1, we can assume that the Jordan-Hölder Chain Condition holds. Let C be a maximal chain in $[a \wedge b, b]$. By the Jordan-Hölder Chain Condition, the length of C is $h(b) - h(a \wedge b)$. By (iii), $D = \{\, a \vee x \mid x \in C \,\}$ is a maximal chain in $[a, a \vee b]$. The length of D is at most the length of C, that is, $h(b) - h(a \wedge b)$; on the other hand, the length of D is $h(a \vee b) - h(a)$, by the Jordan-Hölder Chain Condition, and so

$$h(b) - h(a \wedge b) \geq h(a \vee b) - h(a),$$

which was to be proved.

(iv) *implies* (ii). Certainly, (ii) holds in $(0]$. By induction on $h(x)$, let $x \in L$ and let (ii) hold in $(y]$, for all $y < x$. Let $a, b \in (x]$, $a, b \succ a \wedge b$, and $a \neq b$. Then $a < x$, so (ii) holds in $(a]$ and (ii) implies (i), hence $(a]$ is semimodular. Thus $h(a) - h(a \wedge b) = 1$. Hence, by (iv),

$$h(a \vee b) \leq h(a) + h(b) - h(a \wedge b) = h(b) + 1,$$

and so $a \vee b \succ b$. Similarly, $a \vee b \succ a$.   □

Using Theorem 2, it is easy to check that some constructions yield semimodular lattices. We give one example. Take a semimodular lattice L of length n and pick a $k < n$. Let L_k be the set of all $x \in L$ with $h(x) \leq k$ (a chopped lattice, see Section II.3) along with 1. It is easy to check that L_k is a semimodular lattice of length $k + 1$. The result of this construction with $L = C_2^4$ and $k = 2$ is shown in Figure 3.

Another application of Theorem 2 is

Corollary 3. *Let L be a lattice of finite length. The following conditions on L are equivalent:*

(i) *L is modular.*

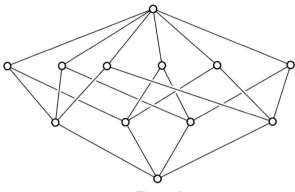

Figure 3

(ii) *L satisfies the Upper and the Lower Covering Conditions.*

(iii) $h(a) + h(b) = h(a \wedge b) + h(a \vee b)$.

Proof. We know, from Section 1, that (i) implies (ii). Theorem 2 and its dual (to be more precise, the dual of 2(iii)) yield (iii). Now assume (iii). If L is not modular, then L contains a pentagon $\{o, a, b, c, i\}$. Thus

$$h(i) = h(b) + h(c) - h(o),$$
$$h(i) = h(a) + h(c) - h(o),$$

implying that $h(a) = h(b)$, a contradiction.   □

Independence in semimodular lattices can be easily handled only for atoms.

Theorem 4.   Let $I = \{a_1, \ldots, a_n\}$ be a set of n atoms of a semimodular lattice. Then the following conditions on I are equivalent:

(i) I is independent.

(ii) $(a_1 \vee \cdots \vee a_i) \wedge a_{i+1} = 0$,   for $i = 1, 2, \ldots, n - 1$.

(iii) $h(a_1 \vee \cdots \vee a_n) = n$.   —

Proof.
    (i) *implies* (ii) is obvious.
    (ii) *implies* (iii). By induction on i, we prove that $h(a_1 \vee \cdots \vee a_i) = i$. This is true for $i = 1$. If $h(a_1 \vee \cdots \vee a_i) = i$, then by (ii) and semimodularity,

$$a_1 \vee \cdots \vee a_i \prec a_1 \vee \cdots \vee a_i \vee a_{i+1},$$

and so

$$h(a_1 \vee \cdots \vee a_i \vee a_{i+1}) = h(a_1 \vee \cdots \vee a_i) + 1 = i + 1.$$

(iii) *implies* (i). We show that the condition of Corollary 1.10 holds for I. It follows from (iii) that $h(\bigvee X) = |X|$, for any $X \subseteq I$, so the map $\varphi \colon X \mapsto \bigvee X$ is one-to-one. Obviously, φ is join preserving.

Now take $X, Y \subseteq I$, and let $a = \bigvee X \wedge \bigvee Y$ and $b = \bigvee(X \cap Y)$. Then $a \geq b$ and by Theorem 2(iv)

$$h(\bigvee X) + h(\bigvee Y) \geq h(a) + h(\bigvee(X \cup Y)),$$

so by (iii),

$$|X| + |Y| \geq h(a) + |X \cup Y|.$$

We conclude that

$$h(a) \leq |X| + |Y| - |X \cup Y| = |X \cap Y|.$$

On the other hand,

$$h(a) \geq h(b) = |X \cap Y|,$$

hence $h(a) = h(b)$ and $a = b$. $\qquad\square$

Let A be a set of atoms. Then $G \subseteq A$ *spans* A iff, for every $a \in A$, there is a finite $G_1 \subseteq G$ such that $a \leq \bigvee G_1$. The following generalizes a result familiar from any first course on vector spaces.

Theorem 5.   Let A be a set of atoms of a semimodular lattice L, let I be an independent subset of A, and let $G \supseteq I$ span A. Then there is an independent subset J of A such that $G \supseteq J \supseteq I$ and J spans A.   —

Proof. Let $\mathfrak{X}$ be the set of all independent subsets X of A with $I \subseteq X \subseteq G$. If $\mathcal{C} \subseteq \mathfrak{X}$ and $\mathcal{C}$ is a chain, then $\bigcup \mathcal{C}$ is again independent, since independence is tested with the finite subsets of $\bigcup \mathcal{C}$ and every finite subset of $\bigcup \mathcal{C}$ is also a finite subset of some $X \in \mathcal{C}$. Hence, by Zorn's Lemma (Section II.1), there is a maximal independent subset J of A with $I \subseteq J \subseteq G$. We wish to show that J spans A. It is sufficient to show that J spans G. Indeed, let $g \in G$. If $g \in J$, there is nothing to prove. If $g \in G - J$, then $J \cup \{g\}$ is not independent, and so there is a finite $J_1 \subseteq J$ such that $J_1 \cup \{g\}$ is not independent, that is, by Theorem 4(iii) and by semimodularity,

$$h(\bigvee(J_1 \cup \{g\})) < |J_1| + 1.$$

Since $h(\bigvee J_1) = |J_1|$, we obtain that $\bigvee(J_1 \cup \{g\}) = \bigvee J_1$, that is, $g \leq \bigvee J_1$, proving that A is spanned by J. $\qquad\square$

The definition of semimodularity was given for arbitrary lattices but it is obvious that it is not very useful for lattices without many prime intervals since lattices without prime intervals are always semimodular. Various attempts have been made to rectify this situation, that is, to come up with a definition agreeing with semimodularity for lattices of finite length and that also selects an interesting class of lattices which are not of finite length.

Definition 6 (L. R. Wilcox [1939] and S. Maeda [1965]). *Let L be a lattice. A pair of elements $\langle a, b \rangle$ of L is called* modular *(see Figure 4), in notation, $a \, M \, b$, iff*

$$x \leq b \quad \text{implies that} \quad x \vee (a \wedge b) = (x \vee a) \wedge b.$$

The lattice L is called M-symmetric *iff $a \, M \, b$ implies that $b \, M \, a$, for any $a, b \in L$.*

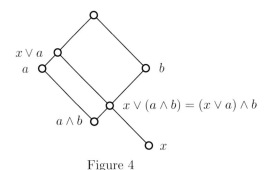

Figure 4

Now it is obvious that L is modular iff $a \, M \, b$, for all $a, b \in L$. In the proof of $[a \wedge b, b] \cong [a, a \vee b]$ in the Isomorphism Theorem, we only used that $a \, M \, b$ in L and $b \, M \, a$ in the dual of L, hence

Corollary 7. *Let L be a lattice and $a, b \in L$. If $a \, M \, b$ in L and $b \, M \, a$ in the dual of L, then $[a \wedge b, b] \cong [a, a \vee b]$. This isomorphism is given by ψ_a, whose inverse is φ_b.*

Remark. The notation φ_b and ψ_a are from Section 1 (see Figure 1.1).

If only $a \, M \, b$ is assumed, half of this conclusion still holds.

Lemma 8. *Let L be a lattice and let $a, b \in L$. The following conditions are equivalent:*

(i) *$a \, M \, b$;*

(ii) *φ_b is onto;*

(iii) *ψ_a is one-to-one.*

Proof.

(i) *implies* (ii). If $x \in [a \wedge b, b]$, then $x \leq b$, and so

$$(a \vee x)\varphi_b = (a \vee x) \wedge b = x \vee (a \wedge b) = x.$$

(ii) *implies* (iii). Let $x, y \in [a \wedge b, b]$, $x\psi_a = y\psi_a$, and $x \neq y$. Then $(x \vee y)\psi_a = x\psi_a = y\psi_a$ and x or $y < x \vee y$, say $x < x \vee y$. There exists a $z \in [a, a \vee b]$ such that $z\varphi_b = x$. We have $z \geq x$ and $z \geq a$, hence $z \geq x \vee a = x\psi_a = (x \vee y)\psi_a$. Thus $z \geq x \vee y$ and so $z\varphi_b \geq x \vee y > x$, a contradiction.

(iii) *implies* (i). Let $x \leq b$. Then

$$(x \vee (a \wedge b))\psi_a = x \vee (a \wedge b) \vee a \quad = x \vee a,$$
$$((x \vee a) \wedge b)\psi_a = ((x \vee a) \wedge b) \vee a = x \vee a;$$

hence, using that ψ_a is one-to-one, we obtain that $a \, M \, b$. To verify that

$$((x \vee a) \wedge b) \vee a = x \vee a,$$

observe that $x \vee a \geq$ and $x \vee a \geq (x \vee a) \wedge b$, hence $x \vee a \geq ((x \vee a) \wedge b) \vee a$. Moreover, if $t \geq (x \vee a) \wedge b$ and $t \geq a$, then $t \geq (x \vee a) \wedge b \geq x$ (since $b \geq x$), and so $t \geq a \vee x$. □

The following result is implicit in L. R. Wilcox [1939].

Theorem 9.   Let L be a lattice of finite length. Then L is semimodular iff L is M-symmetric.   —

Proof. Let L be a semimodular lattice of finite length. We shall prove that

$$a \, M \, b \quad \text{iff} \quad h(a) + h(b) = h(a \wedge b) + h(a \vee b),$$

from which M-symmetry trivially follows.

Using the notation of "(iii) implies (iv)" in the proof of Theorem 2, the length of C is $h(b) - h(a \wedge b)$ and the length of D is $h(a \vee b) - h(a)$. So if $a \, M \, b$, then ψ_a is one-to-one; $|C| = |D|$, and $h(b) - h(a \wedge b) = h(a \vee b) - h(a)$. Conversely, if $a \, M \, b$ fails, then ψ_a is not one-to-one, and C can be chosen so as to include $x, y \in [a \wedge b, b]$, $x\psi_a = y\psi_a$. Then $|D| < |C|$ and we obtain that $h(b) - h(a \wedge b) > h(a \vee b) - h(a)$.

To prove the converse, we do not have to assume that L is of finite length. So let L be an M-symmetric lattice, let $a, b, c \in L$, and let $b \succ a$. If $b \vee c = a \vee c$, then we have nothing to prove. If $b \vee c > a \vee c$, then put $d = a \vee c$ and we have $b \wedge d = a$, $b \vee d = b \vee c$. We have to prove that $b \vee d \succ d$. Indeed, let $b \vee d > x \geq d$. Then $x \not\geq b$ and so $b \wedge x = a$ and $b \vee x = b \vee d$. Since $b \succ b \wedge x$, φ_b as a map from $[x, x \vee b]$ into $[x \wedge b, b]$ is an onto map and so, by Lemma 8, we obtain that $x \, M \, b$. By M-symmetry, $b \, M \, x$, which means, by definition, that, for any $y \leq x$,

$$y \vee (b \wedge x) = (y \vee b) \wedge x.$$

Let $y = d$; then we obtain

$$d = d \vee (b \wedge x) = (d \vee b) \wedge x = x,$$

that is, $b \vee c \succ a \vee c$. $\square$

Examples of M-symmetric lattices not of finite length include the lattice of all closed subspaces of a Banach space and the projection lattice of a von Neumann algebra.

Exercises

1. Show that a lattice L is semimodular iff $x \succ x \wedge y$ implies that $x \vee y \succ y$.

2. Modify the proof of the Jordan-Hölder Theorem. Assume only that C' is a chain and $n < m$, and derive a contradiction. What conclusion can be drawn from this proof?

3. Let A, B, and C be sets of atoms of a semimodular lattice. Show that if A spans B and B spans C, then A spans C.

4. Show that (i)–(iii) of Lemma 8 are equivalent to

    (iv) $\psi_a \varphi_b$ is the identity map.

5. Let L be a semimodular lattice. Prove that if p and q are atoms of L, $a \in L$, and $a < a \vee q \le a \vee p$, then $a \vee p = a \vee q$ (*Steinitz-MacLane Exchange Axiom*).

6. Let L be a lattice. Show that all sublattices of L are semimodular iff L is modular.

7. Show that direct products and convex sublattices of semimodular lattices are again semimodular.

8. Prove that a homomorphic image of a semimodular lattice of finite length is again semimodular.

9. Investigate the statements of Exercises 6–8 for M-symmetric lattices.

10. Show that Part A, the lattice of all partitions on a set A, is a semimodular lattice.

11. Show that the lattice of all congruence relations of a semilattice is a semi-modular lattice (R. Freese and J. B. Nation [1973]).

12. Let L be a modular lattice and let I be an ideal of L. Then $L' = (L - I) \cup \{0\}$ is a lattice. Under what conditions is L' semimodular?

13. Is the lattice L' of Exercise 12 always M-symmetric?

14. Let L be a lattice. Prove that $\text{Sub}\, L$ is semimodular iff L is a chain (K. M. Koh [1973]).

15. A poset P is called *graded* iff an integer-valued function h can be defined on P satisfying, for $x, y \in P$:

$$x \le y \text{ and } h(x) + 1 = h(y) \quad \text{iff} \quad x \prec y.$$

Show that P is graded iff every interval of P is of finite length and satisfies the Jordan-Hölder Chain Condition.

16. Let L be a lattice. Show that $\text{Sub}\, L$ is graded iff the dual of $\text{Sub}\, L$ is semimodular, which, in turn, is equivalent to the condition that L has no sublattice isomorphic to $C_2 \times C_3$ (H. Lakser [1973b]).

17. Prove that a lattice of finite length is semimodular iff it is graded and satisfies $h(x) + h(y) \ge h(x \wedge y) + h(x \vee y)$ (G. Birkhoff [1933]).

18. The *Infinite Jordan-Hölder Chain Condition* holds for a lattice L iff, for any two maximal chains C and C' of L, we have $|C| = |C'|$. Show that this fails for the lattice $L = A \times B$, where A is the $[0, 1]$ real interval and B is the $[0, 1]$ rational interval.

19. Show that the Infinite Jordan-Hölder Chain Condition holds for the Boolean lattice of all subsets of a set A iff A is finite.

20. A lattice L is said to satisfy the *MacLane Condition* iff, for a, b, $c \in L$ satisfying $a \parallel b$ and $a \wedge b < c \le a$, there exists a $d \in L$ satisfying $a \wedge (b \vee d) = c$. Show that the MacLane Condition implies the Upper Covering Condition.

21. Prove that a lattice of finite length is semimodular iff it satisfies the MacLane Condition. (Exercises 20 and 21 are from S. MacLane [1938].)

22. Find M-symmetric lattices that fail to satisfy the MacLane Condition.

23. Find lattices satisfying the MacLane Condition that are not M-symmetric.

3. Geometric Lattices

Just as most lattices arising out of algebraic examples are algebraic (Section II.3, see also Concluding Remarks), most lattices arising out of geometry are geometric in the following sense:

Definition 1. *A lattice L is called* geometric *iff L is semimodular, L is algebraic, and the compact elements of L are exactly the finite joins of atoms of L.*

Equivalently, L is complete, L is *atomistic* (that is, every element of L is a join of atoms), all atoms are compact, and L is semimodular.

For lattices of finite length, this concept was introduced by G. Birkhoff [1935a], influenced by H. Whitney [1935]. Geometric lattices with no restriction on length were introduced and investigated by S. MacLane [1938] under the name *exchange lattices*. The name, geometric lattice, is due to M. L. Dubreil-Jacotin, L. Lesieur, and R. Croisot [1953]. G. Birkhoff [1967] and some others retain in their definition the requirement that the lattice be of finite length. Many authors call these lattices *matroid lattices*. (For combinatorial definitions of matroids, see the Further Topics and References for this chapter.)

Later in this section, we shall investigate the connection between geometric lattices and geometries. We start out, however, by investigating the lattice theoretic consequences of this definition.

Corollary 2.

  (i) *Let L be a geometric lattice. Then the set F of elements of finite height is an ideal of L. The lattice F is semimodular and every element of F is a finite join of atoms. L is isomorphic to $\operatorname{Id} F$, the lattice of all ideals of F.*

 (ii) *An interval of a geometric lattice is again a geometric lattice.*

Proof.

  (i) Let a, $b \in F$, $c \in L$, and $c \leq a$. Then $h(c) \leq h(a)$, hence $c \in F$. By Theorem 2.2,

$$h(a \vee b) \leq h(a) + h(b) - h(a \wedge b),$$

and so $a \vee b \in F$. Thus F is an ideal. The second statement of (i) is obvious. The third statement follows from the proof of Theorem II.3.13.

  (ii) An interval $[a, b]$ of an algebraic lattice L is again algebraic. Moreover,

$$\{\, a \vee p \mid p \text{ an atom of } L,\ p \nleq a,\ \text{and } p \leq b \,\}$$

is, by semimodularity, the set of all atoms of $[a, b]$. The rest is easy.   □

Now we utilize the theory of independence.

Lemma 3. *Let L be a geometric lattice. Every element a of L is a join of an independent set of atoms, in fact, of any maximal independent set of atoms of L contained in $(a]$.*

Proof. Let A be a maximal independent set of atoms of L contained in $(a]$ (A exists by Theorem 2.5 with $I = \varnothing$ and G is the set of all atoms in $(a]$). By Theorem 2.5 and the compactness of atoms, A spans all the atoms in $(a]$. Since L is atomistic, $a = \bigvee A$.   □

Theorem 4 (G. Birkhoff [1935a] and S. MacLane [1938]). Any geometric lattice L is complemented; in fact, it is relatively complemented. —

Proof. Let $a \in L$, let A be a maximal independent set of atoms in $(a]$, and let K be the set of all atoms not contained in $(a]$. Then by Theorem 2.5, there is a maximal independent set I of atoms of L satisfying $A \subseteq I \subseteq A \cup K$. Set $b = \bigvee (I - A)$. Since $\bigvee I = 1$ by Theorem 2.5, we obtain that $a \vee b = 1$. Let $c = a \wedge b$. If $c \neq 0$, then there is an atom $p \leq c$. Since $p \leq \bigvee I$ and $p \leq \bigvee (I - A)$, by the compactness of p there exist finite $I_1 \subseteq I$ and $I_2 \subseteq I - A$ such that $p \leq \bigvee I_1$ and $p \leq \bigvee I_2$. By the definition of independence,

$$p \leq \bigvee I_1 \wedge \bigvee I_2 = \bigvee (I_1 \cap I_2) = \bigvee \varnothing = 0,$$

a contradiction. Hence $a \wedge b = 0$, proving that L is complemented. The second statement follows from the first and from Corollary 2. □

Our next task is to prove the structure theorem of geometric lattices due to F. Maeda [1951] and in its sharper form due to U. Sasaki and S. Fujiwara [1952].

Theorem 5. Every geometric lattice is isomorphic to a direct product of directly indecomposable geometric lattices. —

This theorem is augmented by a characterization theorem of direct indecomposability. Let us call the elements a and b *perspective*, in symbol, $a \sim b$, iff they have a common complement, that is,

$$a \vee x = b \vee x = 1,$$
$$a \wedge x = b \wedge x = 0,$$

for some element x. The reader should not confuse the perspectivity of elements with perspectivity of quotients. (See Exercise 13.)

Theorem 6. A geometric lattice is directly indecomposable iff any two atoms are perspective. —

For the proof of Theorem 5, we need a lemma.

Lemma 7. *Let L be a geometric lattice. If a, b, $x \in L$,*

(*)
$$a \wedge x = b \wedge x = 0,$$
$$a \vee x = b \vee x,$$

then $a \sim b$. If a, $b \in L$, $a \sim b$, and $h(a)$, $h(b) < \infty$, then there is an $x \in L$ satisfying (*) and $h(x) < \infty$.

Proof. Let $(*)$ hold. Let y be the relative complement (which exists by Theorem 4) of $a \vee x = b \vee x$ in $[x, 1]$. Then

$$a \wedge y = a \wedge (a \vee x) \wedge y = a \wedge x = 0,$$
$$a \vee y = a \vee x \vee y = 1;$$

similarly, $b \wedge y = 0$ and $b \vee y = 1$, that is, $a \sim b$. To prove the second statement, let

$$a \wedge y = b \wedge y = 0,$$
$$a \vee y = b \vee y = 1,$$

and observe that $a \leq b \vee y$ and $h(a), h(b) < \infty$, hence, by the compactness of a, there is an $x_1 \leq y$, $h(x_1) < \infty$ satisfying $a \leq b \vee x_1$. Similarly, choose $x_2 \leq y$, $h(x_2) < \infty$ satisfying $b \leq a \vee x_2$. Then $x = x_1 \vee x_2$ will satisfy the lemma.   □

Proof of Theorem 5. For $a, b \in L$, let us write $a \approx b$ and call a and b *projective* iff $a = x_0 \sim x_1 \sim \cdots \sim x_n = b$, for some $x_1, \ldots, x_{n-1} \in L$. For an atom p of L, we define

$$R(p) = \{\, q \mid q \text{ is an atom of } L \text{ and } p \approx q \,\}$$

and

$$Z = \{\, R(p) \mid p \text{ is an atom of } L \,\}.$$

For $x \in Z$, $x = R(p)$, we set $c(x) = c(p) = \bigvee R(p)$.

We are going to show that L is isomorphic to $\prod(\, [0, c(x)] \mid x \in Z \,)$.

Every element a of L is a join of atoms, hence if we set

$$a_{R(p)} = \bigvee((a] \wedge R(p)),$$

then

$$a = \bigvee(\, a_{R(p)} \mid R(p) \in Z \,),$$

where $a_{R(p)} \leq c(p)$.

To establish that

$$a \mapsto \langle a_{R(p)} \mid R(p) \in Z \rangle$$

is the required isomorphism, it is sufficient by Exercise 21 to show that every $a \in L$ has a unique representation in the form

$$a = \bigvee(\, a_{R(p)} \mid R(p) \in Z \,)$$

with $a_{R(p)} \leq c(p)$. Indeed, let a have two such distinct representations:

$$a = \bigvee (\, x_{R(p)} \mid R(p) \in Z \,) = \bigvee (\, y_{R(p)} \mid R(p) \in Z \,),$$

$x_{R(p)}, y_{R(p)} \leq c(p)$, and for at least one atom, say q, $x_{R(q)} \neq y_{R(q)}$. By taking the join of the two representations, we can assume that $x_{R(p)} \leq y_{R(p)}$ and so $x_{R(q)} < y_{R(q)}$.

We now need the following simple statement:

If $A \cup \{p\}$ is a set of atoms of L and $p \leq \bigvee A$, then $p \sim r$, for some $r \in A$.

Indeed, by the compactness of p, $p \leq \bigvee A_1$, for some finite $A_1 \subseteq A$. Choose A_1 minimal. Then $p \sim r$, for any $r \in A_1$, because if $x = \bigvee (A_1 - \{r\})$, then $r \vee x = \bigvee A_1$; $p \not\leq x$ (since A_1 is minimal), hence $h(p \vee x) = h(x) + 1 = h(\bigvee A_1)$ and so $p \vee x = \bigvee A_1$. By minimality, A_1 is independent, hence $r \wedge x = 0$ and $p \not\leq x$, hence $p \wedge x = 0$. Thus $p \sim r$, by Lemma 7.

It follows from this statement that there is an atom $t \in R(q)$ such that $t \leq y_{R(q)}$ and $t \not\leq x_{R(q)}$. It also follows that if we set

$$a_1 = x_{R(q)},$$
$$a_2 = \bigvee (x_{R(p)} \mid p \not\sim q),$$

then $a = a_1 \vee a_2$ and $a_1 \wedge a_2 = y_{R(q)} \wedge a_2 = 0$. So we obtain elements t, a_1, and a_2 satisfying

$$t \wedge a_1 = 0,$$
$$a_1 \wedge a_2 = 0,$$
$$t \leq a_1 \vee a_2$$

(see Figure 1). Let x_1 be the relative complement of $t \vee a_1$ in $[a_1, a_1 \vee a_2]$ and x_2 the relative complement of $a_2 \wedge x_1$ in $[0, a_2]$. We claim that

$$t \wedge x_1 = x_2 \wedge x_1 = 0,$$
$$t \vee x_1 = x_2 \vee x_1,$$

and therefore $t \sim x_2$ by Lemma 7. Indeed,

$$t \wedge x_1 = t \wedge (t \vee a_1) \wedge x_1 = t \wedge a_1 = 0,$$
$$x_2 \wedge x_1 = a_2 \wedge x_2 \wedge x_1 = x_2 \wedge (a_2 \wedge x_1) = 0,$$
$$t \vee x_1 = t \vee a_1 \vee x_1 = a_1 \vee a_2,$$
$$x_2 \vee x_1 = x_2 \vee (a_2 \wedge x_1) \vee x_1 = a_2 \vee x_1 = a_1 \vee a_2.$$

Now observe that $x_1 \prec t \vee x_1 = x_1 \vee x_2$. Therefore, for any atom $u \leq x_2$, we have $x_1 \vee u = t \vee x_1$ and, of course, $x_1 \wedge u = 0$. Hence $t \sim u$ by Lemma 7. Now recall

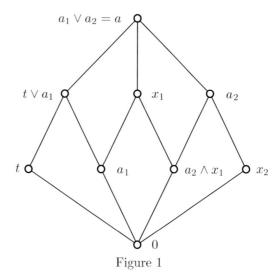

Figure 1

that $t \in R(q)$, so $t \sim u$ implies that $u \in R(q)$. But $u \leq a_2 = \bigvee(\, x_{R(p)} \mid p \not\approx q\,)$, hence $u \in R(p)$, for some $p \neq q$, a contradiction.

Finally, we have to show that $[0, c(p)]$ is directly indecomposable. If it is not, then, by Theorem III.4.1, it has a pair a, b of complemented neutral elements, a, $b \notin \{0, c(p)\}$. So we can choose atoms u, v of $[0, c(p)]$ such that $u \leq a$ and $v \leq b$. Since $u \approx v$ in L, $u \approx v$ in $[0, c(p)]$ (since $[0, c(p)]$ is a direct factor). Taking the homomorphism $\varphi\colon x \mapsto x \wedge a$ of $[0, c(p)]$ onto $[0, a]$, we obtain that $u\varphi \approx v\varphi$. But $u = u\varphi$, $v\varphi = 0$, and $u \approx 0$ imply that $u = 0$, a contradiction.   □

It is clear, from the proof of Theorem 5, that indecomposability is equivalent with the projectivity of atoms. Therefore, to prove Theorem 6, it suffices to prove the following statement:

Theorem 6'.   In a geometric lattice, perspectivity of atoms is transitive.   —

Proof. Let p, q, and r be atoms of L, $p \sim q$, $q \sim r$; by Lemma 7, we can choose x and y such that

$$p \wedge x = q \wedge x = 0,$$
$$p \vee x = q \vee x,$$
$$h(x) = m,$$
$$q \wedge y = r \wedge y = 0,$$
$$q \vee y = r \vee h,$$
$$h(y) = n.$$

Choose x so as to minimize $h(x)$. Let $\{s_1, \ldots, s_m\}$ be an independent set of atoms in $(x]$, and $\{t_1, \ldots, t_n\}$ in $(y]$, and let

$$x_i = s_1 \vee \cdots \vee s_{i-1} \vee s_{i+1} \vee \cdots \vee s_m,$$

for $1 \le i \le m$. We claim that $p \not\le x_i \vee q$. Otherwise, $p \le x_i \vee q$ and so $x_i \vee p \le x_i \vee q$. However, $h(x_i) = m - 1$ and, therefore, $h(x_i \vee p) = h(x_i \vee q) = m$, implying that $x_i \vee p = x_i \vee q$ and then x_i could replace x, contradicting the minimality of $h(x)$. So $h(p \vee q \vee x_i) = 2 + m - 1 = m + 1$, and

$$p \vee q \vee x_i = p \vee x = q \vee x,$$

for $1 \le i \le m$. Now by Theorem 2.4 $\{s_1, \ldots, s_m, p\}$ is an independent set of atoms so, by Theorem 2.5, we can choose a subset $\{t'_1, \ldots, t'_k\}$ of $\{t_1, \ldots, t_n\}$ such that

$$\{p, s_1, \ldots, s_m, t'_1, \ldots, t'_k\}$$

is independent. Set $y' = t'_1 \vee \cdots \vee t'_k$. Then $\{p, s_1, \ldots, s_m, y'\}$ is independent, and so is $\{q, s_1, \ldots, s_m, y'\}$. Therefore, using the definition of independence, we obtain that

$$y' = (x \vee y') \wedge (q \vee x_1 \vee y') \wedge \cdots \wedge (q \vee x_m \vee y').$$

Since $r \wedge y = 0$, we have $r \not\le y$; hence $r \not\le y'$ and, therefore, either $r \not\le x \vee y'$ or $r \not\le q \vee x_i \vee y'$, for some $1 \le i \le m$. Set $a = x \vee y'$, if $r \not\le x \vee y'$, or $a = q \vee x_i \vee y'$, if $r \not\le q \vee x_i \vee y'$. Then

$$r \wedge a = 0,$$
$$r \le p \vee a$$

(the latter because $r \le q \vee y \le p \vee x \vee y = p \vee x \vee y' = p \vee q \vee x_i \vee y'$). Finally, since $r \not\le a$, we obtain that $p \not\le a$, that is,

$$p \wedge a = 0,$$

thus $a \le a \vee r \le a \vee p$ and $a \prec a \vee p$, and so

$$a \vee r = a \vee p,$$

proving that $p \sim r$. $\square$

If the geometric lattice L is of finite length, then L is said to be *finite dimensional*. In this case, Theorem 5 follows from Lemma III.4.5. With some effort, the general case could be reduced to the finite dimensional case. However, without the present proof of Theorem 5, it would be more difficult to prove Theorem 6.

Theorem III.4.6 could be used to prove the result that every finite dimensional geometric lattice is isomorphic to a direct product of simple geometric lattices. The factors we obtain from Theorems 5 and 6 are such that any two atoms are perspective. Now if L is a geometric lattice of finite length in which any two atoms are perspective, then L is simple. Indeed, if Θ is a congruence relation of L and $\Theta \neq \omega$, then there exist a, $b \in L$, $a < b$, such that $a \equiv b$ (Θ). Choose an atom p such that $p \leq b$ and $p \not\leq a$. Then $p \equiv 0$ (Θ), since $p = p \wedge b \equiv p \wedge a = 0$ (Θ). Now let q be any atom of L. Since $p \sim q$, p and q have a common complement x. Then

$$p/0 \nearrow 1/x \searrow q/0,$$

and so $q \equiv 0$ (Θ). Since $h(1) < \infty$, $1 = q_1 \vee \cdots \vee q_n$, for some finite set of atoms $\{q_1, \ldots, q_n\}$ and $q_i \equiv 0$ (Θ), $i = 1, \ldots, n$, hence $1 \equiv 0$ (Θ). Thus $\Theta = \iota$ and L is simple.

Exactly the same proof yields that if L is a geometric lattice in which any two atoms are perspective, then L is subdirectly irreducible; the only atom Φ of $\operatorname{Con} L$ contained in every $\Theta \neq \omega$ is the congruence $\Phi = \Theta(p, 0)$, where p is any atom. If $a \in L$ and $h(a) < \infty$, then $a \equiv 0$ (Φ). In general, however, L is not simple.

By a *geometry* we mean a set A certain subsets of which are called *subspaces* (or *flats*) such that certain basic properties hold; we shall list these properties in the next definition.

Definition 8. *A geometry $\langle A, ^- \rangle$ is a set A and a function $X \mapsto \overline{X}$ of $\mathrm{P}(A)$ into itself satisfying the following conditions:*

(i) $^-$ *is a closure relation, that is,*

    (i$_1$) $X \subseteq \overline{X}$;
    (i$_2$) *if $X \subseteq Y$, then $\overline{X} \subseteq \overline{Y}$;*
    (i$_3$) $\overline{\overline{X}} = \overline{X}$.

(ii) $\overline{\varnothing} = \varnothing$, *and $\overline{\{x\}} = \{x\}$, for $x \in A$.*

(iii) *If $x \in \overline{X \cup \{y\}}$, but $x \notin \overline{X}$, then $y \in \overline{X \cup \{x\}}$.*

(iv) *If $x \in \overline{X}$, then $x \in \overline{X_1}$, for some finite $X_1 \subseteq X$.*

(i) means that $\langle A, ^- \rangle$ is a *closure space*. A subset X of a closure space is called *closed* iff $\overline{X} = X$. It follows simply form (i$_1$)–(i$_3$), that $\overline{X}$ is the smallest closed set containing X.

Lemma 9. *Let $\langle A, ^- \rangle$ be a closure space. Then $L\langle A, ^- \rangle = \{ \overline{X} \mid X \subseteq A \}$ is a complete lattice with respect to set inclusion, and $\bigwedge Y = \bigcap Y$, for $Y \subseteq L\langle A, ^- \rangle$. Conversely, if $L \subseteq \mathrm{P}(A)$ is closed under arbitrary intersection, then*

$$\overline{X} = \bigcap (Z \mid Z \supseteq X \text{ and } Z \in L)$$

defines a closure space $\langle A, ^- \rangle$ such that a set is closed iff it belongs to L.

Proof. Let $Y \subseteq L\langle A, ^- \rangle$ and let $U = \bigcap Y$. Then $U \subseteq X$, for all $X \in Y$, so $\overline{U} \subseteq \overline{X} = X$, and therefore $\overline{U} \subseteq \bigcap Y = U$. Since $U \subseteq \overline{U}$ always holds, $U = \overline{U}$. The remaining steps in the proof of Lemma 9 are very easy and will be left to the reader. $\square$

Thus a closure space is completely determined by the lattice L of closed sets. In case of a geometry, we have $\overline{\{x\}} = \{x\}$, meaning that the elements of A can be identified as atoms of L, hence L determines the geometry even if L is known only up to isomorphism.

For geometries, a closed set is called a *subspace* and $\overline{X}$ is called the *subspace spanned by* X. Thus (iv) means that if x belongs to the subspace spanned by X, then x belongs to the subspace spanned by some finite $X_1 \subseteq X$. A closure space satisfying (iv) is usually called *algebraic*.

Lemma 10. *A lattice L is algebraic iff L is isomorphic to the lattice of closed sets of an algebraic closure space.*

Proof. Let L be an algebraic lattice. By Theorem II.3.13, $L \cong \operatorname{Id} F$, where F is a join-semilattice with 0. Take $A = F$ and, for $X \subseteq A$, define $\overline{X} = (X]$, the ideal generated by X. The verification of (i) and (iv) for $\langle A, ^- \rangle$ is contained in the proof of Theorem II.3.13. Then $\operatorname{Id} F$ is the set of closed sets and so $L \cong \operatorname{Id} F$ by assumption.

Conversely, if $\langle A, ^- \rangle$ is an algebraic closure space and L is the lattice of closed sets, then it is easily seen that $X \in L$ is compact iff $X = \overline{X}_1$, for some finite $X_1 \subseteq X$, from which it follows immediately that L is algebraic. $\square$

Condition (iii) makes it possible to define the *rank* or *dimension* of a subspace. Let $r(\varnothing) = 0$ and $r(\{a\}) = 1$. If X is a subspace, $r(X) = n$, and $y \notin X$, then $r(\overline{X \cup \{y\}}) = n + 1$. (iii) implies that this defines r uniquely.

Finally, (ii) states that $\varnothing$ is a subspace and so is every singleton $\{x\}$. If we drop this last condition, we get what is known as a *pregeometry*, from which a geometry can be obtained by identifying any two elements, x, y of A, for which $\overline{\{x\}} = \overline{\{y\}}$.

Theorem 11.   Let $\langle A, ^- \rangle$ be a geometry. Then $L = L\langle A, ^- \rangle$ is a geometric lattice. Conversely, if L is a geometric lattice, A is the set of atoms of L, and, for $X \subseteq A$, $\overline{X}$ is the set of atoms spanned by X (in the sense defined in Section 2), then $\langle A, ^- \rangle$ is a geometry and $L \cong L\langle A, ^- \rangle$.   —

Proof. Let $\langle A, ^- \rangle$ be a geometry. Then, by Lemma 9, $L = L\langle A, ^- \rangle$ is algebraic. Since X is compact iff $X = \overline{X}_1$, for some finite $X_1 \subseteq X$, we see that X is compact iff X is a finite join of atoms (for $x \in A$, $\{x\}$ is an atom by 8(ii)). It remains to show that L is semimodular. Let $X, Y \in L$, let $Y = \overline{X \cup \{x\}}$, and let $x \notin X$. We claim that $X \prec Y$. Indeed, if $Z \in L$ and $X \subset Z \subseteq Y$, then there is a $z \in Z - X$ and $z \in Z \subseteq Y = \overline{X \cup \{x\}}$, and so by 8(iii), $x \in \overline{X \cup \{z\}} \subseteq Z$.

Thus, $Y = \overline{X \cup \{x\}} \subseteq Z$; we conclude that $Y = Z$, proving $X \prec Y$. Now let $U \in L$. Then

$$Y \vee U = \overline{Y \cup U} = \overline{X \cup U \cup \{x\}},$$
$$X \vee U = \overline{X \cup U},$$

hence either $x \in \overline{X \cup U}$ and so $X \vee U = Y \vee U$ or $x \notin \overline{X \cup U}$ and in this case $X \vee U \prec Y \vee U$.

Conversely, let L be a geometric lattice, let A be the set of atoms of L, and let $\overline{X}$ be the set of atoms spanned by X, for $X \subseteq A$. It is immediate that $\langle A, ^- \rangle$ is a closure space. 8(ii) and 8(iv) are also clear by definition. To verify 8(iii), let $x \in \overline{X \cup \{y\}}$ and $x \notin \overline{X}$. Since $\{y\}$ is an atom, by semimodularity,

$$\overline{X \cup \{y\}} = \overline{X} \vee \{y\} \succ \overline{X};$$

thus

$$\overline{X} \subset \overline{X \cup \{x\}} \subseteq \overline{X \cup \{y\}}$$

implies that

$$\overline{X \cup \{x\}} = \overline{X \cup \{y\}};$$

so $y \in \overline{X \cup \{x\}}$.

Now let φ be the map $X \mapsto \bigvee X$, for $X \subseteq A$, $X \in L\langle A, ^- \rangle$. Then φ maps $L\langle A, ^- \rangle$ into L. Since every element of L is a join of atoms, $X \subseteq Y$ iff $\bigvee X \leq \bigvee Y$. Therefore, φ is onto, one-to-one and both φ and φ^{-1} are isotone. Hence φ is an isomorphism.   □

The lattice of subspaces thus completely determines the geometry and vice versa. The diagrams of geometric lattices are, as a rule, too complicated to be of any use. Many times, however, we can draw a picture of the geometry associated with the lattice.

Let us call an element of height 2 a *line*. A line is the join of any two distinct points in the line. Some geometries (for instance, projective geometries, see Section 5) are completely determined by their lines, some are not. When drawing the picture of a geometry, we try to represent lines by straight lines or by curves. Lines having exactly two points are not drawn as a rule. Figures 2a, 3a, and 4a are pictures of geometries and Figures 2b, 3b, and 4b are the corresponding lattice diagrams. Observe the simplicity of geometries compared to the intricate diagrams of the lattices.

Closure spaces satisfying the additional requirements 8(ii)–8(iv) can be found in large numbers, especially in algebra, geometry, and combinatorics. Here is an example from combinatorics.

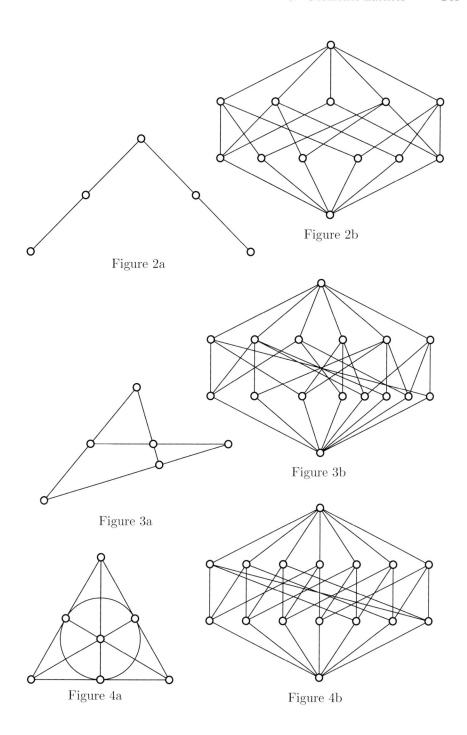

Figure 2a

Figure 2b

Figure 3a

Figure 3b

Figure 4a

Figure 4b

A *graph* is a set G with a fixed set E (the *edges*) of two-element subsets of G (unoriented graph without loops). Figure 5 shows a diagram of a graph. Two points a and b (elements of G) are connected with a straight line segment iff $\{a, b\} \in E$.

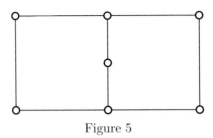

Figure 5

Let a, $b \in G$ and $A \subseteq E$. We shall call a and b *A-connected* iff there is a sequence of edges $\{x_0, x_1\}$, $\{x_1, x_2\}$, ... , $\{x_{n-1}, x_n\} \in A$ such that $a = x_0$ and $x_n = b$.

Now we define a geometry, called the *edge geometry*, on E: for $\{a, b\} \in E$ and $A \subseteq E$, let $\{a, b\} \in \overline{A}$ iff a and b are A-connected.

Theorem 12. $\langle E, ^- \rangle$ is a geometry.   —

Proof. All the properties of a geometry obviously hold except perhaps 8(iii). Let a and b be A-connected; then there is a connecting sequence

$$\{x_0, x_1\}, \ \ldots, \ \{x_{n-1}, x_n\},$$

for which n is minimal. We claim that no edge can occur twice in this sequence. Indeed, let $0 \leq i < j \leq n - 1$ and $\{x_i, x_{i+1}\} = \{x_j, x_{j+1}\}$. Now if $x_i = x_j$ and $x_{i+1} = x_{j+1}$, then, by dropping $\{x_{i+1}, x_{i+2}\}$, ... , $\{x_j, x_{j+1}\}$ from the sequence, we get a shorter connecting sequence. If $x_i = x_{j+1}$ and $x_{i+1} = x_j$, then we can drop all of $\{x_i, x_{i+1}\}$, ... , $\{x_j, x_{j+1}\}$.

Now we prove 8(iii). Let $x \in \overline{X \cup \{y\}}$ but $x \notin \overline{X}$, where $x = \{a, b\}$ and $y = \{c, d\}$. Let e_0, ... , $e_{n-1} \in X \cup \{y\}$ be a shortest sequence connecting a and b. Since $x \notin \overline{X}$, one of the e_0, ... , e_{n-1} must be y. By the observation above, exactly one of e_0, ... , e_{n-1}, say e_i, equals y. But then e_{i+1}, ... , e_{n-1}, x, e_0, ... , e_{i-1} will connect c and d, hence $y \in \overline{X \cup \{x\}}$.   □

For instance, if we take the graph of Figure 6, then Figure 2a shows the edge geometry and Figure 2b the lattice, called the *edge lattice* associated with it.

To conclude this section we discuss two results of a combinatorial nature.

Let L be a finite geometric lattice. Let $E(i)$ be the set of elements of L of height i and let us define the *Whitney number*:

$$W_i = |E(i)|.$$

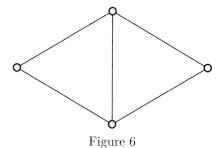

Figure 6

Observe that each $E(i)$ is an antichain of L.

An antichain E of L is a *maximum sized antichain* iff $|X| \leq |E|$, for any antichain X of L. The following two theorems establish the size of a maximum sized antichain, for some classes of geometric lattices. The first is credited to R. P. Dilworth and L. H. Harper in K. A. Baker [1969].

Theorem 13.   Let the finite geometric lattice L have the property that every element x of height i is covered by a_i and covers b_i elements (a_i and b_i depend only on i). Then a maximum sized antichain of L has $\max(W_i \mid 0 \leq i \leq l(L))$ elements.   ___

Proof. Let $n = l(L)$ and let $\max(W_i \mid 0 \leq i \leq n) = W_k$; then $E(k)$ is an antichain with W_k elements. We have to prove that if S is an antichain, then $|S| \leq W_k$.

For $x \in L$, let $s(x)$ denote the number of maximal chains going through x. It is obvious, that if $h(x) = i$, then $s(x) = b_1 \cdots b_i \cdot a_i \cdots a_{n-1}$, hence $s(x)$ depends only on i. Since every maximal chain has exactly one element of order i, we obtain that $s(x) \cdot W_i = s$, where s is the number of maximal chains in L. Using $W_i \leq W_k$, we obtain

$$s(x) = \frac{s}{W_i} \geq \frac{s}{W_k}.$$

Finally, a maximal chain goes through at most one element of the antichain S. Therefore,

$$s \geq \sum (s(x) \mid x \in S) \geq |S| \cdot \frac{s}{W_k},$$

that is, $|\, S \,| \leq W_k$.   □

Theorem 13, as well as the next result, was motivated by *Sperner's Lemma* (E. Sperner [1928]), stating the conclusions of Theorem 13 for $P(X)$.

A finite geometric *lattice* L of length n is *unimodal* iff for some integer k we have that

$$W_1 \leq \cdots \leq W_{k-1} \leq W_k \geq W_{k+1} \geq \cdots \geq W_{n-1}.$$

If X and Y are subsets of L we say that there is a *matching* between X and Y iff there is a one-to-one map $\varphi\colon X \mapsto Y$ (or $\varphi\colon Y \mapsto X$, if $|X| \geq |Y|$) such that x is comparable with $x\varphi$, for all $x \in X$. The following is G.-C. Rota's approach to the problem of finding maximum sized antichains:

Theorem 14.   Let L be a finite geometric lattice of length n. If L is unimodal $(W_1 \leq \cdots \leq W_k \geq \cdots \geq W_{n-1})$ and there is a matching between $E(i)$ and $E(i+1)$, for all $i < n$, then a maximum sized antichain of L has W_k elements. $\quad\overline{}$

Proof. For a subset S of L, let

$$t(S) = \max(h(x) - h(y) \mid x,\ y \in S)$$

be the *thickness* of S. Let S be an antichain of L. We shall prove $|S| \leq W_k$, by induction on $t(S)$.

If $t(S) = 0$, then $S \subseteq E(i)$, for some i, hence $|S| \leq W_i \leq W_k$. Now let $t(S) > 0$ and assume the inequality for antichains with a smaller thickness. Since $t(S) > 0$, there is an $x \in S$ with $h(x) \neq k$, say $h(x) < k$. Let $i = \min(h(x) \mid x \in S)$, and set $S = S_0 \cup S_1$, where $S_1 = S \cap E(i)$ and $S_0 = S - S_1$. By assumption, $i < k$, and so $|E(i)| \leq |E(i+1)|$, hence there is a matching $\varphi\colon E(i) \to E(i+1)$. Define

$$S' = S_0 \cup S_1\varphi.$$

S_0 and $S_1\varphi$ are disjoint, because if $x \in S_0$ and $x \in S_1\varphi$, that is, $x = y\varphi$ with $y \in S_1$, then $x,\ y \in S$, $x \neq y$, and x and y are comparable, contradicting that S is an antichain. Thus $|S'| = |S|$. Moreover, S' is an antichain; indeed, if x, $y \in S'$, $x \neq y$, x and y are comparable, then $x \in S_0$ and $y \in S_1\varphi$, that is, $y = z\varphi$, for some $z \in S_1$. Since $h(y) = i + 1$, we have $h(y) \leq h(x)$, therefore $x < y$ is impossible. Thus $y < x$, contradicting that $z \leq y$ and $z \parallel x$.

Thus S' is an antichain, $t(S') = t(S) - 1$, and so

$$|S| = |S'| \leq W_k.$$

$\quad\square$

Exercises

1.   Let F be a semimodular lattice with 0 in which every element is a finite join of atoms. Show that $\operatorname{Id} F$ is a geometric lattice.

2.   Is the converse of Exercise 1 true?

3.   Call a lattice L *atomic* iff L has a 0 and, for every $a \in L$, $a \neq 0$, there is an atom $p \leq a$. A subset D of L is *directed* iff, for all $x,\ y \in D$, there is a $z \in D$ satisfying $x \leq z$ and $y \leq z$. A lattice L is (*upper*) *continuous* iff

$$a \wedge \bigvee D = \bigvee(a \wedge x \mid x \in D),$$

for any (upper) directed subset D of L. Prove that a lattice L is geometric iff L is atomic, relatively complemented, continuous, and semimodular (G. Birkhoff [1948] and F. Maeda [1951]).

4. Show that in Exercise 3 "semimodular" can be replaced by the condition that if a and b cover $a \wedge b$ and $a \neq b$, then $a \vee b$ covers a and b.

5. Prove that in Exercise 3, "semimodularity" can be replaced by "M-symmetry" (F. and S. Maeda [1970]).

6. Let F be a semilattice and let $\operatorname{Con} F$ be the lattice of all congruence relations of F. Prove that $\operatorname{Con} F$ is geometric iff $\operatorname{Con} F$ is atomistic.

7. Let K be a subfield of the field K'. Let L be the lattice of all relatively algebraically closed subfields $M \subseteq K$. (M is *relatively algebraically closed* iff a polynomial with coefficients in K that factors in K' also factors in M.) Prove that L is a geometric lattice.

8. Prove that the lattice L of Exercise 7 is modular iff the transcendence degree of K' over K is at most two (S. MacLane [1938]).

9. Let G be an abelian group. Let L denote the lattice of all subgroups H with the property that G/H (the quotient group) has no element of finite order excepting zero. Prove that L is a geometric lattice.

10. Consider the concepts of p-independence and p-basis in fields of characteristic p (see, for instance, B. L. van der Waerden [1931]). Can these be viewed as independence in a suitable geometric lattice?

11. Let L be a geometric lattice. Show that $a \sim b$ in L iff

$$a \wedge x = b \wedge x,$$
$$a \vee x = b \vee x,$$

for some $x \in L$.

12. Let a and b be atoms in the geometric lattice L. Show that $a \sim b$ iff there exists a finite independent set of atoms I such that $b \in I$ and, for $I_1 \subseteq I$, $a \leq \bigvee I_1$ iff $I_1 = I$.

13. Show that if $a \sim b$, then $a/0 \nearrow 1/x \searrow b/0$, for some x.

14. Prove that, in a geometric lattice, $a \sim b$ iff $a/0 \sim x/y \sim b/0$, for some quotient x/y.

15. Relate $a \approx b$ and $a/0 \approx b/0$ in a general lattice, in a geometric lattice, and in a modular geometric lattice.

16. Let L be a modular geometric lattice. Show that L is simple iff L is a directly indecomposable lattice of finite length.

17. Let I and J be maximal independent sets of atoms of a geometric lattice. Prove that $|I| = |J|$.

18. Show that, in a geometric lattice L, for every element a, there is a smallest element $c(a)$ of the center of L, satisfying $a \leq c(a)$.

19. Prove that, in a geometric lattice, $a \sim b$ iff $c(a) = c(b)$ (with $c(x)$ as in Exercise 18).

20. A lattice L is called *left-complemented* iff L is bounded and, for $a, b \in L$, there exists $b_1 \leq b$ such that

$$a \vee b_1 = a \vee b,$$
$$a \wedge b_1 = 0,$$

and $a \, M \, b_1$ (L. R. Wilcox [1942]). Is every geometric lattice left-complemented?

21. Let L be a complete lattice and let $X \subseteq L$. Then $L \cong \prod(\,(x] \mid x \in X\,)$ iff every element $a \in L$ has one and only one expression of the form

$$a = \bigvee(\, a_x \mid x \in X\,),$$

where $a_x \leq x$, for all $x \in X$.

22. Define the isomorphism of two geometries. Prove that two geometries are isomorphic iff the associated geometric lattices are isomorphic.

23. Define a *circuit* of a graph as a set of edges $\{x_0, x_1\}, \ldots, \{x_{n-1}, x_n\}$, $\{x_n, x_0\}$. Define the subspaces of the geometry on the edges of the graph in terms of the circuits. What role is played by the minimal circuits?

24. Draw the lattice associated with the graph of Figure 5.

25. Does the lattice associated with a graph determine the graph (up to isomorphism)?

26. Generalize Theorem 13 to posets.

27. Prove Sperner's Lemma using Theorem 13.

28. Prove Sperner's Lemma using Theorem 14.

29. Does Theorem 13 or Theorem 14 apply to finite projective geometries?

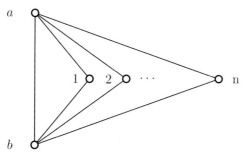

Figure 7

30. Consider the graph G_n of Figure 7. Let L_n be the edge lattice of G_n. If $A \in L_n$ and $\{a, b\} \in A$, then A is determined by

$$\{\, i \mid \{a, i\} \text{ and } \{b, i\} \in A \,\}.$$

If $\{a, b\} \notin A$, then A is determined by

$$\{\, i \mid \{a, i\} \text{ or } \{b, i\} \in A \,\}$$

and a choice function selecting one of $\{a, i\}$ and $\{b, i\}$. Conclude from this, that

$$W_k = 2^k \binom{n}{k} + \binom{n}{k-1}.$$

31. Show that in L_{10} (as defined in Exercise 30) there is an antichain contained in $E(6) \cup E(7)$ that has more elements than any $E(k)$. (Exercises 30 and 31 are due to R. P. Dilworth and C. Greene [1971].)

Let P be a finite poset. The *Möbius function* μ is an integer valued function defined on $P \times P$ by the formulas:

$$\mu(x, y) = \begin{cases} 1, & \text{for } x = y \in P; \\ 0, & \text{if } x \not\leq y; \\ -\sum(\mu(x, z) \mid x \leq z < y), & \text{if } x < y. \end{cases}$$

(See A. F. Möbius [1832] and L. Weisner [1935].)

32. Establish the Möbius inversion formula: if f and g are real valued functions on a finite partially ordered set P and $g(x) = \sum(f(y) \mid y \leq x)$, for every $x \in P$, then

$$f(x) = \sum(g(y)\mu(y, x) \mid y \leq x).$$

33. Let P and Q be finite poses with Möbius functions μ_P and μ_Q, respectively. Then the Möbius function $\mu_{P \times Q}$ of $P \times Q$ satisfies

$$\mu_{P \times Q}(\langle x, y \rangle, \langle u, v \rangle) = \mu_P(x, u)\mu_Q(y, v),$$

for x, $u \in P$ and y, $v \in Q$.

34. Show that the Möbius function μ of the Boolean lattice B is given by

$$\mu(x, y) = (-1)^{h(y) - h(x)},$$

for $x \leq y$, where $h(x)$ is the height of x in B.

35. Let μ be the Möbius function of a finite lattice L and let x, y, $z \in L$.

(i) Let us assume that $x \leq y$ and y is not the join of elements covering x; then $\mu(x, y) = 0$ (P. Hall [1934]).

(ii) If $x \leq z \leq y$, then

$$\sum (\mu(x, t) \mid z \vee t = y) = \begin{cases} \mu(x, y), & \text{if } z = x; \\ 0, & \text{if } z \neq x. \end{cases}$$

(See L. Weisner [1935].)

36. Show that the Möbius function μ of a finite distributive lattice L is given by:

$$\mu(x, y) = \begin{cases} 0, & \text{if } y \text{ is not the join of elements covering } x; \\ (-1)^n, & \text{if } y \text{ is the join of } n \text{ distinct elements covering } x. \end{cases}$$

(Hint: Use Exercises 34 and 35(i).)

37. Let μ be the Möbius function of a finite geometric lattice L and x, $y \in L$, $x \leq y$. Then $\mu(x, y) \neq 0$; $\mu(x, y)$ is positive if $h(y) - h(x)$ is even; $\mu(x, y)$ is negative, if $h(y) - h(x)$ is odd (G.-C. Rota [1964]). (Hint: Apply Exercise 35(ii).)

4. Partition Lattices

A *partition* of a set A is a set π of nonempty pairwise disjoint subsets of A whose union is A. The members of π are called the *blocks* of π. A singleton as a block is called *trivial*. If a and b (a, $b \in A$) belong to the same block we write $a \equiv b$ (π) or $a \pi b$.

Given a partition π, we can define an equivalence relation ε by $\langle x, y \rangle \in \varepsilon$ iff $x \equiv y$ (π). Conversely, if ε is an equivalence relation, then $\pi = \{ [a]\varepsilon \mid a \in A \}$ is a partition of A. Thus we set up a one-to-one correspondence between partitions and equivalence relations. Partitions are easier to visualize, so we shall use them; if the reader prefers equivalence relations, he can use them throughout.

Part A will denote the set of all partitions of A partially ordered by

$$\pi_0 \leq \pi_1 \quad \text{iff} \quad x \equiv y \;\; (\pi_0) \text{ implies that } x \equiv y \;\; (\pi_1).$$

Part A with this partial ordering (which corresponds to set inclusion for the corresponding equivalence relations) forms a complete lattice, called the *partition lattice* (or *equivalence lattice*) of A.

We draw a picture of a partition by drawing the boundary lines of the blocks. Then $\pi_0 \leq \pi_1$ iff the boundary lines of π_1 are also boundary lines of π_0 (but π_0 may have some more boundary lines). Equivalently, the blocks of π_1 are unions of blocks of π_0. (See Figure 1.)

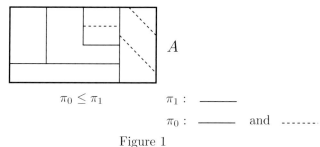

$$\pi_0 \leq \pi_1$$

$\pi_1 :$ ———

$\pi_0 :$ ——— and ·······

Figure 1

Lemma 1.

(i) Part A *is a complete lattice and*

$$x \equiv y \quad \Big(\bigwedge(\pi_i \mid i \in I)\Big)$$

iff $x \equiv y \;\; (\pi_i)$, *for all* $i \in I$;

$$x \equiv y \quad \Big(\bigvee(\pi_i \mid i \in I)\Big)$$

iff there is a natural number n, $i_0, \dots, i_n \in I$, *and* $x_0, \dots, x_{n+1} \in A$ *such that* $x = x_0$, $y = x_{n+1}$, *and* $x_j \equiv x_{j+1} \;\; (\pi_{i_j})$, *for* $0 \leq j \leq n$.

The zero of Part A *has only trivial blocks and the unit has only one block, namely,* A.

(ii) *The atoms of* Part A *are the partitions with exactly one nontrivial block and this block has exactly two elements. The dual atoms of* Part A *are the partitions with exactly two blocks.*

(iii) $\pi_0 \prec \pi_1$ *in* Part A *iff* π_1 *is the result of replacing two blocks of* π_0 *by their union.*

(iv) *In Part A, $[\pi)$ is isomorphic with the partition lattice of the set π.*

(v) *In Part A, $(\pi]$ is isomorphic with the direct product of all Part X, where X ranges over the nontrivial blocks of π.*

Proof.

(i) This is implicit in the proof of Theorem I.3.9.

(ii) and (iii) are trivial.

(iv) If $\pi \le \xi$, then each block of ξ is a union of blocks of π. Thus ξ defines a partition of the blocks of π, that is, of π. This sets up the required isomorphism.

(v) If $\xi \le \pi$, then for each nontrivial block X of π, ξ defines a partition ξ_X of X. Since a block of ξ that intersects X is completely within X, the map

$$\xi \mapsto \langle \xi_X \mid X \text{ is a nontrivial block of } \pi \rangle$$

sets up the required isomorphism.   □

From Lemma 1 we easily conclude:

Theorem 2 (O. Ore [1942]).   Part A is a simple geometric lattice.   —

Proof. By Lemma 1(i), Part A is complete. It follows from 1(ii) that every element is a join of atoms and, by the formula for join, the atoms are compact. If π_0 and π_1 satisfy the condition of 1(iii), then so do $\pi_0 \vee \xi$ and $\pi_1 \vee \xi$ unless $\pi_0 \vee \xi = \pi_1 \vee \xi$. Therefore, Part A is semimodular. This shows that Part A is geometric.

Let π and ξ be atoms. By 1(ii), π and ξ have only one nontrivial block each, say $\{a,b\}$ and $\{c,d\}$, respectively. Let τ be a partition with two blocks as follows: if $\{a,b\} \cap \{c,d\} = \varnothing$, the blocks of τ are $A - \{a,c\}$ and $\{a,c\}$; if $\{a,b\} \cap \{c,d\} = \{e\}$, then the blocks of τ are $A - \{e\}$ and $\{e\}$. In either case, τ is a complement of π and ξ and so $\pi \sim \xi$. By the discussion in Section 3, Part A is simple, if A is finite.

Let A be infinite and let Θ be a congruence relation of Part A, $\Theta \ne \omega$. Then all atoms are congruent modulo Θ to the zero of Part A. Since A is infinite, we can split A up into two disjoint sets A_0 and A_1 of the same cardinality. Let φ be a one-to-one map of A_0 onto A_1. Now we define some partitions of A. Let π have the two blocks, A_0 and A_1. Let ξ_i have exactly one nontrivial block, A_i ($i = 0$, 1). The blocks of τ are all the sets of the form $\{x, x\varphi\}$, where $x \in A_0$. Then these partitions form a sublattice of Part A; a diagram of this sublattice is given in Figure 2, where π_0 and π_1 denote the zero and unit of Part A, respectively.

Let α be any atom $\le \tau$. Then $\alpha \equiv \pi_0$ (Θ), hence $\pi \equiv \pi \vee \alpha$ (Θ). Since $\pi \prec \pi_1$, we obtain that $\pi \vee \alpha = \pi_1$ and so $\pi_1 \equiv \pi$ (Θ). Taking the meets of both sides with τ and joining with ξ_i, we get $\xi_i \equiv \pi_1$ (Θ), $i = 0$, 1, hence $\pi_0 = \xi_0 \wedge \xi_1 \equiv \pi_1$ (Θ). Thus $\Theta = \iota$ and Part A is simple.   □

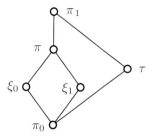

Figure 2

Call a partition π *finite* iff all blocks of π are finite and only finitely many blocks are nontrivial. Then it is clear that π is finite iff it is a finite join of atoms (recall 1(ii)). Since Part A is geometric, this is, furthermore, equivalent to π being compact.

Let $\mathrm{Part}_{\text{fin}}\, A$ be the set of all finite partitions. This is obviously an ideal of Part A. By Corollary 3.2 we obtain:

Corollary 3. $\mathrm{Part}_{\text{fin}}\, A$ *is an ideal of* Part A *and the following isomorphism holds:*

$$\mathrm{Id}\,\mathrm{Part}_{\text{fin}}\, A \cong \mathrm{Part}\, A.$$

The importance of partition lattices lies in the fact that they are as universal for lattices as permutation groups are for groups. P. M. Whitman [1946] proved that every lattice L can be embedded in some partition lattice. One drawback of this result is that the joins of the partitions representing elements of L have to be computed by 1(i), which can be rather complicated. This was corrected by B. Jónsson [1953]. To state Whitman's theorem as improved by Jónsson, we need some concepts.

A *representation* of a lattice L is an embedding α of L into some Part A; α is called *of type* 3 iff, for all $a, b \in L$ and $x, y \in A$,

$$x \equiv y \quad ((a \vee b)\alpha)$$

iff there exist $z_1, z_2, z_3 \in A$ such that

$$x \equiv z_1 \quad (a\alpha),$$
$$z_1 \equiv z_2 \quad (b\alpha),$$
$$z_2 \equiv z_3 \quad (a\alpha),$$
$$z_3 \equiv y \quad (b\alpha).$$

In other words, type 3 means that for the partitions $a\alpha$, $a \in L$, the join formula in 1(i) always holds with $n \leq 3$. The number 3 appears arbitrary but we shall see in Theorem 5 that 3 cannot be reduced to 2 unless L is modular.

Theorem 4.   Every lattice has a type 3 representation.   —

Definition 9. *Let* x_0, x_1, x_2, y_0, y_1, y_2 *be variables. We define some polynomials:*

$$z_{ij} = (x_i \vee x_j) \wedge (y_i \vee y_j), \quad 0 \le i < j < 3,$$
$$z = z_{01} \wedge (z_{02} \vee z_{12}).$$

The Arguesian identity *is*

$$(x_0 \vee y_0) \wedge (x_1 \vee y_1) \wedge (x_2 \vee y_2) \le ((z \vee x_1) \wedge x_0) \vee ((z \vee y_1) \wedge y_0).$$

A lattice satisfying this identity is called Arguesian.

Remark. This identity is a lattice theoretic form of Desargues' Theorem; see B. Jónsson [1953]; see also M. Schützenberger [1945]. To understand the meaning of the identity, the reader should read the discussion of Desargues' Theorem and its connection with the Arguesian identity in Section 5.

Theorem 10 (B. Jónsson [1953]).   Any lattice having a type 1 representation is Arguesian.   —

Proof. Let $\alpha\colon L \to \operatorname{Part} A$ be a type 1 representation of the lattice L. We shall show that $L\alpha$ is Arguesian (this is sufficient to prove since $L \cong L\alpha$). Let a_0, a_1, a_2, b_0, b_1, $b_2 \in L\alpha$; thus a_0, a_1, a_2, b_0, b_1, b_2 are partitions of the set A. If x, $y \in A$ satisfy

$$x \equiv y \quad ((a_0 \vee b_0) \wedge (a_1 \vee b_1) \wedge (a_2 \vee b_2)),$$

then there exist u_0, u_1, $u_2 \in A$ (see Figure 5) such that

$$x \equiv u_i \quad (a_i),$$
$$u_i \equiv y \quad (b_i),$$

for $i = 1$, 2, 3. Let

$$c_{ij} = (a_i \vee a_j) \wedge (b_i \vee b_j),$$

for $0 \le i < j < 3$. Then, for all such i and j, $u_i \equiv u_j$ (c_{ij}). Set

$$c = c_{01} \wedge (c_{02} \vee c_{12});$$

then $u_0 \equiv u_1$ (c). So

$$x \equiv u_0 \quad ((c \vee a_1) \wedge a_0),$$
$$u_0 \equiv y \quad ((c \vee b_1) \wedge b_0),$$

therefore

$$x \equiv y \quad (((c \vee a_1) \wedge a_0) \vee ((c \vee b_1) \wedge b_0)),$$

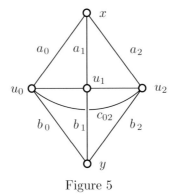

Figure 5

proving that

$$(a_0 \vee b_0) \wedge (a_1 \vee b_1) \wedge (a_2 \vee b_2) \leq ((c \vee a_1) \wedge a_0) \vee ((c \vee b_1) \wedge b_0),$$

so $L\alpha$ is arguesian. $\square$

Every arguesian lattice is modular. Indeed, to see this, it is sufficient to show that N_5 is not Arguesian. If we take N_5, and substitute

$$x_0 = y_1 = c,$$
$$x_1 = x_2 = y_0 = b,$$
$$y_2 = a,$$

then $z = a$, however, the identity demands that $a \leq b$, which is false.

It follows from the results of Section 5 that there are modular lattices that are not Arguesian.

Exercises

1. Compute $|\operatorname{Part} A|$, for $|A| \leq 5$.

2. Show that $\operatorname{Part} A$ is modular iff $|A| \leq 3$.

3. Let $A \subset B$. Find embeddings of $\operatorname{Part} A$ into $\operatorname{Part} B$.

4. Let A_i, $i \in I$, be pairwise disjoint sets and let α_i be a partition of A_i, for $i \in I$. Define a partition α on $A = \bigcup(A_i \mid i \in I)$ by

   $$u \equiv v \ (\alpha) \quad \text{iff} \quad u, \ v \in A_i, \text{ for some } i \in I \text{ and } u \equiv v \ (\alpha_i).$$

   Show that this defines an embedding of $\prod(\operatorname{Part} A_i \mid i \in I)$ into $\operatorname{Part} A$.

5. With A_i, $i \in I$, and A as in Exercise 4, is there a $\{0, 1\}$-embedding of $\prod(\operatorname{Part} A_i \mid i \in I)$ into $\operatorname{Part} A$?

6. Is there an embedding of $\prod(\operatorname{Part} A_i \mid i \in I)$ into $\operatorname{Part} \prod(A_i \mid i \in I)$?

7. Let A be a set and let $B = A \cup \{z\}$ $(z \notin A)$. For $X \subseteq A$, define a partition $\alpha(X)$ on B by

$$x \equiv y\ (\alpha(X)) \quad \text{iff} \quad x,\ y \in X \cup \{z\}.$$

Show that $X \mapsto \alpha(X)$ is an embedding of $P(A)$ into $\operatorname{Part} B$.

8. Using Exercise 7, show that every distributive lattice has an embedding into a partition lattice.

9. Show that every finite distributive lattice can be embedded in a finite partition lattice.

10. Let $\mathfrak{A}$ be a finite (universal) algebra. Define the concept of a congruence relation of an algebra. The congruence relations of $\mathfrak{A}$ form a lattice, Con $\mathfrak{A}$. Verify that the congruence lattice of a finite algebra can be embedded in a finite partition lattice.

11. Following O. Ore [1942], for $\alpha,\ \beta \in \operatorname{Part} A$, we define a graph $G(\alpha, \beta)$ on the set $\alpha \cup \beta$; an edge $\{X, Y\}$ of $G(\alpha, \beta)$ is a block X of α and a block Y of β such that $X \cap Y \neq \varnothing$. Show that the blocks of $\alpha \vee \beta$ are maximal connected subgraphs of $G(\alpha, \beta)$.

12. Let $\alpha,\ \beta \in \operatorname{Part} A$. Prove that the modular equation,

$$\alpha \wedge (\beta \vee \gamma) = (\alpha \wedge \beta) \vee \gamma,$$

holds for all $\gamma \leq \alpha$ $(\gamma \in \operatorname{Part} A)$ iff $G(\alpha, \beta)$ is a *tree* (that is, a graph without cycles) (O. Ore [1942]).

13. Let $\beta,\ \gamma \in \operatorname{Part} A$. Show that

$$\alpha \wedge (\beta \vee \gamma) = (\alpha \wedge \beta) \vee (\alpha \wedge \gamma)$$

holds for all $\alpha \in \operatorname{Part} A$ iff β and γ are permutable (O. Ore [1942]).

14. Let $\alpha,\ \gamma \in \operatorname{Part} A$. Prove that

$$\alpha \vee (\beta \wedge \gamma) = (\alpha \vee \beta) \wedge (\alpha \vee \gamma)$$

holds for all $\beta \in \operatorname{Part} A$ iff α is the zero of $\operatorname{Part} A$, or $\alpha \geq \gamma$, or γ has only one nontrivial block and $\gamma \geq \alpha$ (O. Ore [1942]).

15. Show that the zero and the unit are the only distributive elements of $\operatorname{Part} A$.

16. Let D be a distributive ideal of Part A. Show that every element of D is contained in another element of D with only one nontrivial block. Also show that $\bigvee D$ is a distributive element of Part A.

17. Show that Part A has no distributive ideal.

18. Use Theorem III.3.10 and Exercise 17 to show that Part A is simple (O. Ore [1942]).

19. Define type n representations. Find a representation which is not of type n, for any $n < \omega$.

20. Find a representation which is not associated with a distance function.

21. Disprove the converse of Corollary 7.

*22. Disprove the converse of Theorem 10 (B. Jónsson).

5. Complemented Modular Lattices

The key to the investigation of complemented modular lattices is provided by modular geometric lattices and projective spaces. We begin with some properties of modular geometric lattices.

Lemma 1. *Let L be a modular geometric lattice and let $\langle A, ^- \rangle$ be the associated geometry.*

(i) *Let $p, q \in A$ and $p \neq q$. Then $p \sim q$ iff there exists an $r \in A$ such that $r \neq p, q$ and $r \leq p \vee q$ (thus $\{0, p, q, r, p \vee q\}$ is a diamond in L).*

(ii) *Let $X \subseteq A$. Then X is a subspace iff $p, q \in X$ implies that $r \in X$, for all $r \leq p \vee q$.*

Proof.
(i) Let x be a complement of p and q and define $r = (p \vee q) \wedge x$. Then

$$p \vee r = p \vee ((p \vee q) \wedge x)$$

(by modularity and $p \leq p \vee q$)

$$= (p \vee x) \wedge (p \vee q) = 1 \wedge (p \vee q) = p \vee q.$$

Similarly, $q \vee r = p \vee q$. Obviously, $p \wedge r = q \wedge r = 0$. Hence

$$h(r) = h(p \vee q) + h(0) - h(p) = 1.$$

Thus $r \in A$. Since $r \neq 0$ and $r \wedge p = r \wedge q = 0$, we have $r \neq p, q$.

(ii) Let X $(\subseteq A)$ satisfy the condition that $r \leq p \vee q$ and $p, q \in X$ imply that $r \in X$. We prove by induction on n that

$$r \in A, \ r \leq p_1 \vee \cdots \vee p_n, \ \text{and} \ p_1, \ \ldots, \ p_n \in X \quad \text{imply that} \quad r \in X.$$

This is obvious for $n = 1$ and, by assumption, for $n = 2$. Let us assume this statement for $n - 1$ and let $r, p_1, \ldots, p_n$ be given satisfying $r \leq p_1 \vee \cdots \vee p_n$. We can assume that $\{p_1, \ldots, p_n\}$ is independent, since otherwise $r \leq p_{i_1} \vee \cdots \vee p_{i_k}$, for some $\{i_1, \ldots, i_k\} \subset \{1, \ldots, n\}$, and so $r \in X$, by induction. With $a = p_2 \vee \cdots \vee p_n$, we have then $r \vee a \leq p_1 \vee a$ and $h(r \vee a) = h(p_1 \vee a) = n$, hence $r \vee a = p_1 \vee a$. If $r \leq a$, then $r \in X$ by the induction hypothesis, so we can assume that $r \not\leq a$. Thus $r \wedge a = p_1 \wedge a = 0$. By the proof of (i), $t = (p_1 \vee r) \wedge a$ is an atom and, by the induction hypothesis, $t \in X$. Since t, $p_1 \in X$ and $r \leq t \vee p_1$, we conclude that $r \in X$. This completes the proof of the "if" part of (ii); the "only if" part is trivial. $\qquad\square$

Lemma 2. *Under the conditions of Lemma 1, the geometry $\langle A, \overline{} \rangle$ satisfies the following property (see Figure 1):*

The Pasch Axiom. If p, q, r, x, y are points, $x \leq p \vee q$, $y \leq q \vee r$, and $x \neq y$, then there is a point z such that $z \leq (p \vee r) \wedge (x \vee y)$.

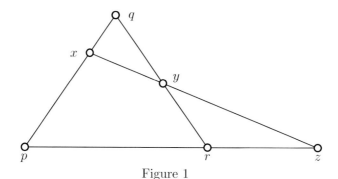

Figure 1

Remark. Apart from degenerate cases, the Pasch Axiom requires that if a line $x \vee y$ intersects two sides, $p \vee q$ and $q \vee r$, of the triangle $\{p, q, r\}$, then it intersects the third side, $p \vee r$. If $x \neq p$, q and $y \neq q$, r, then we shall also have $z \neq p$, r.

Proof. If $|\{p, q, r, x, y\}| < 5$, then we can always choose a $z \in \{p, q, r, x, y\}$ satisfying $z \leq (p \vee r) \wedge (x \vee y)$. If $r \leq p \vee q$, again one can choose $z = x$. So let $|\{p, q, r, x, y\}| = 5$, and let $r \not\leq p \vee q$. Then $h(p \vee r) = h(x \vee y) = 2$ and $h(p \vee q \vee r) = 3$.

Thus

$$h((p \vee r) \wedge (x \vee y)) = h(p \vee r) + h(x \vee y) - h(p \vee r \vee x \vee y).$$

Now observe that $p \vee q \vee x \vee y = p \vee q \vee r$, hence

$$h((p \vee r) \wedge (x \vee y)) = 2 + 2 - 3 = 1,$$

and so $z = (p \vee r) \wedge (x \vee y)$ is the required atom.   □

We can see, from Lemmas 1 and 2, that many important properties of a geometry associated with a modular geometric lattice are formulated only with points and lines. This pattern is followed in the next definition:

Definition 3. *Let A be a set and let L be a collection of subsets of A. $\langle A, L \rangle$ is called a* projective space *iff the following properties hold:*

(i) *Every $l \in L$ has at least two elements.*

(ii) *For any two distinct p, $q \in A$, there is exactly one $l \in L$ satisfying p, $q \in l$.*

(iii) *For p, q, r, x, $y \in A$ and l_1, $l_2 \in L$ satisfying p, q, $x \in l_1$ and q, r, $y \in l_2$, there exist $z \in A$ and l_3, $l_4 \in L$ satisfying p, r, $z \in l_3$ and x, y, $z \in l_4$.*

Let us call the members of A *points*, and those of L, *lines*. For p, $q \in A$, $p \neq q$, let $p + q$ denote the (unique) line containing p and q; if $p = q$, set $p + q = \{p\}$.

A set $X \subseteq A$ is called a *linear subspace* iff p and $q \in X$ imply that $p + q \subseteq X$. If X and Y are linear subspaces define

$$X + Y = \bigcup (x + y \mid x \in X \text{ and } y \in Y).$$

Lemma 4. *If X and Y are linear subspaces of a projective space, then so is $X + Y$.*

Proof. Take the points p, q, r such that $r \in p + q$ and p, $q \in X + Y$. We wish to show that $r \in X + Y$. Now if p, $q \in X$ or p, $q \in Y$, then $r \in X \cup Y \subseteq X + Y$, since X and Y are linear subspaces. If $p \in X$ and $q \in Y$ (or the other way around), then $r \in X + Y$, by the definition of $X + Y$. Therefore, we can assume that $p \notin X \cup Y$, and so there exist $p_x \in X$ and $p_y \in Y$ such that $p \in p_x + p_y$. Now we distinguish two cases. Firstly, if $q \in X \cup Y$, say $q \in X$, then consider $q + p_x$, $q + p$, $p + p_x$, and $r + p_y$ (see Figure 2). We can assume that q, p, p_x, p_y, and r are all distinct, and so $r + p_y$ has a point in common with $q + p$ and $p + p_y$; hence, by 3(iii), there is a $t \in X$ such that $t \in q + p_x$ and $t \in r + p_y$. If $t = p_y$, then $p \in X$, a contradiction. Hence $t \neq p_y$ and so $r \in t + p_y \in X + Y$. Secondly, let $q \notin X \cup Y$. Again we apply 3(iii), this time to $p + p_x$, $p + q$, $p_x + q$, and $p_y + r$ (see Figure 3). Hence there is a point $t \in p_x + q$ such that $r \in t + p_y$. By the first case, $t \in X + Y$. Again by the first case, $r \in X + Y$.   □

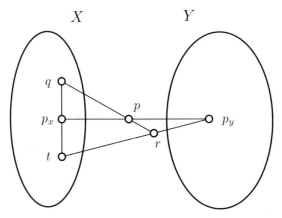

Figure 2

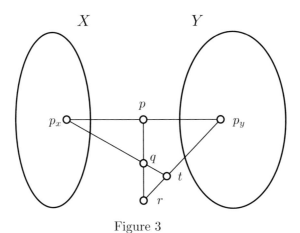

Figure 3

Now we come to the crucial step:

Theorem 5.   The linear subspaces of a projective space form a modular geometric lattice.   —

Proof. Since the intersection of any number of linear subspaces is a linear subspace again, we have a closure space $\langle A,^- \rangle$. For $X \subseteq A$, the closure $\overline{X}$ can be described as follows: set

$$X_0 = X,$$

$$\cdots$$

$$X_n = X_{n-1} + X_{n-1},$$

$$\cdots$$

Then

$$\overline{X} = \bigcup (\,X_i \mid i = 0,\ 1,\ 2,\ \dots\,).$$

It follows immediately that $\langle A, ^- \rangle$ is an algebraic closure space, and so the linear subspaces form an algebraic lattice and, for the linear subspaces X and Y, the formula $X \vee Y = \overline{X \cup Y}$ holds.

If X, Y, and Z are linear subspaces and $X \supseteq Z$, then $X \wedge (Y \vee Z) \supseteq (X \wedge Y) \vee Z$ obviously holds.

Now let $p \in X \wedge (Y \vee Z)$, that is, $p \in X$ and $p \in Y \vee Z$. Since $p \in Y \vee Z = Y + Z$, there exist $p_y \in Y$ and $p_z \in Z$ such that $p \in p_y + p_z$. From $X \supseteq Z$, it follows that p and $p_z \in X$. If $p = p_z$, then $p \in Z$ and so $p \in (X \wedge Y) \vee Z$. If $p \neq p_z$, then $p_y \in p + p_z \subseteq X$. Thus $p_y \in X \wedge Y$ and $p_z \in Z$; therefore, $p \in (X \wedge Y) \vee Z$. This proves that the lattice is modular.

3.8(i), (ii), (iv) have been verified or are obvious and (iii) follows from modularity. ☐

We have proved somewhat more than stated. We also obtained that the linear subspaces define a geometry whose subspaces are exactly the linear subspaces.

Call a geometry *projective*, if the associated geometric lattice is modular.

Theorem 6.   There is a one-to-one correspondence between projective spaces (defined by points and lines) and projective geometries (defined as geometries with modular subspace lattices). Under this correspondence, linear subspaces of projective spaces correspond to subspaces of projective geometries.   —

Applying the results of Section 3, we obtain that every modular geometric lattice is a direct product of indecomposable modular geometric lattices, which, by Lemma 1(i), are characterized by the property that in the associated projective space *every line has at least three points*; such projective spaces (and projective geometries) are called *nondegenerate*.

Corollary 7. *Every modular geometric lattice can be represented as a direct product of modular geometric lattices that are associated with nondegenerate projective geometries.*

The most important example of a modular geometric lattice associated with a nondegenerate projective geometry is the following.

Let D be a *division ring* (that is, an associative ring with unit, in which a^{-1} exists, for any $a \in D$, $a \neq 0$) and let $\mathfrak{m} > 0$ be a cardinal number (finite or infinite). We take a set I with $|I| = \mathfrak{m}$ and we construct $V = V(D, \mathfrak{m})$, an $\mathfrak{m}$-dimensional vector space over D, as the set of all functions $f \colon I \to D$ such that $f(i) = 0$, for all but a finite number of $i \in I$; we define $h = f + g$ and $k = af$ $(a \in D)$ by $h(i) = f(i) + g(i)$ and $k(i) = af(i)$, respectively. U is a *submodule* of V iff U is nonempty and $f, g \in U$ imply that $f + ag \in U$, for all $a \in D$.

The submodules of V form a lattice denoted by $L(D, \mathfrak{m})$. It is easily seen that $L(D, \mathfrak{m})$ is a modular geometric lattice. The atoms of $L(D, \mathfrak{m})$ are exactly the submodules $\{\, af \mid a \in D \,\}$ where f is any element of V that is not the identically zero function, f_0.

The submodules $\{\, af \mid a \in D \,\}$ ($f \neq f_0$) form the points of the associated projective geometry. If $\{\, af \mid a \in D \,\}$ and $\{\, ag \mid a \in D \,\}$ are distinct points, then the points of the line through these two points are

$$\{\, a(xf + yg) \mid a \in D \,\}, \quad x, \, y \in D, \; xy \neq 0.$$

f ($\neq f_0$) will be called a *representative* of the point $\{\, af \mid a \in D \,\}$.

This geometry is nondegenerate: if f and g are representatives of two distinct points, then $f - g$ represents a third point of the line.

This projective geometry is typical of projective geometries in which Desargues' Theorem holds.

Let us call a set of points X *collinear* iff $X \subseteq l$, for some line l. A triple $\langle a_0, a_1, a_2 \rangle$ of noncollinear points is a *triangle*. Two triangles $\langle a_0, a_1, a_2 \rangle$ and $\langle b_0, b_1, b_2 \rangle$ are *perspective with respect to the point* p iff $a_i \neq b_i$, $a_i + a_j \neq b_i + b_j$, for $0 \leq i, \, j < 3$, and the points p, a_i, b_i are collinear, for $i = 0, \, 1, \, 2$. They are *perspective with respect to a line* l iff c_{01}, c_{12}, $c_{20} \subseteq l$, where c_{ij} is the intersection of $a_i + a_j$ and $b_i + b_j$. The classical *Desargues' Theorem* states that if two triangles are perspective with respect to a point, then they are perspective with respect to a line.

Now we shall prove that the Arguesian identity, of the last section, is a lattice theoretic formulation of Desargues' Theorem. See B. Jónsson [1953], [1954], [1959b] and M. Schützenberger [1945].

Theorem 8.   Let L be a modular geometric lattice. Then L satisfies the Arguesian identity iff Desargues' Theorem holds in the associated projective geometry.

—

The proof of Theorem 8 will follow immediately from Lemmas 9–11.

Lemma 9. *Let L be a modular geometric lattice. Then the Arguesian identity holds for the atoms of L iff Desargues' Theorem holds in the associated projective geometry.*

Proof. Let us substitute the atoms a_0, a_1, b_0, b_1, b_2 in the Arguesian identity. Let us form the c_{ij}, $0 \leq i < j \leq 2$, and let

$$c = c_{01} \wedge (c_{02} \vee c_{12}),$$
$$p = (a_0 \vee b_0) \wedge (a_1 \vee b_1) \wedge (a_2 \vee b_2);$$

see Figure 4. We have to verify that

$$p \leq ((c \vee a_1) \wedge a_0) \vee ((c \vee b_1) \wedge b_0).$$

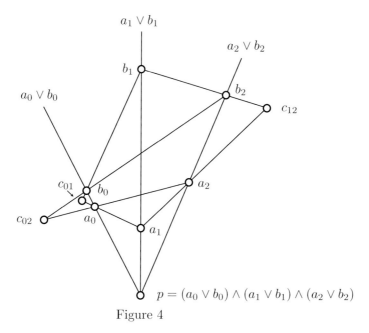

$$p = (a_0 \vee b_0) \wedge (a_1 \vee b_1) \wedge (a_2 \vee b_2)$$

Figure 4

Let us assume first that a_0, a_1, a_2 and b_0, b_1, b_2 are not collinear, $a_i \neq b_i$, $i = 0$, 1, 2, and that the triangles $\langle a_0, a_1, a_2 \rangle$ and $\langle b_0, b_1, b_2 \rangle$ are perspective with respect to the point p. Then the c_{ij} are all distinct atoms and they are collinear iff $c = c_{01} \wedge (c_{12} \vee c_{02})$ is an atom, in fact, $c = c_{01}$, otherwise, $c = 0$. If $c = c_{01}$, then $((c \vee a_1) \wedge a_0) \vee ((c \vee b_1) \wedge b_0)$ is the line $a_0 \vee b_0$, and so

$$p \leq ((c \vee a_1) \wedge a_0) \vee ((c \vee b_1) \wedge b_0);$$

otherwise, $c = 0$ and $(c \vee a_1) \wedge a_0 = (c \vee b_1) \wedge b_0 = 0$ and $p \not\leq 0$. Thus, under the conditions stated, the identity holds iff the geometry satisfies Desargues' Theorem.

Now observe that if the above conditions are not satisfied, then

$$p \leq ((c \vee a_1) \wedge a_0) \vee ((c \vee b_1) \wedge a_1)$$

always holds. There are several cases to consider ($p = 0$, p is a line, a_0, a_1, a_2 are collinear, a_1, b_1, b_2 are collinear, $a_i = b_i$, for some i, and so on), but all are trivial. □

Lemma 10. *Let L be a modular geometric lattice. Let p be an n-ary polynomial in which each variable x_i, $0 \leq i < n$, occurs at most once. Then, for an atom a and elements b_0, ... , $b_{n-1} \in L$,*

$$a \leq p(b_0, \dots, b_{n-1})$$

holds iff there exist a_0, ... , $a_{n-1} \in L$ such that

(i) $a \leq p(a_0, \ldots, a_{n-1})$,

(ii) $a_i \leq b_i$, for all $0 \leq i < n$,

(iii) a_i is an atom, if $b_i \neq 0$.

Proof. We use induction on the rank of p. Let $p = x_i$. Then $a \leq b_i$, and so we can take $a_i = a$. Now let us assume the statement to hold for

$$p_0 = p_0(x_0, \ldots, x_{n-1}),$$
$$p_1 = p_1(y_0, \ldots, y_{m-1}),$$

where $\{x_0, \ldots, x_{n-1}\} \cap \{y_0, \ldots, y_{m-1}\} = \varnothing$.

First, let $p = p_0 \wedge p_1$. Now if

$$a \leq p(b_0, \ldots, b_{n-1}, b'_0, \ldots, b'_{m-1}) = p_0(b_0, \ldots, b_{n-1}) \wedge p_1(b'_0, \ldots, b'_{m-1}),$$

then

$$a \leq p_0(b_0, \ldots, b_{n-1}),$$
$$a \leq p_1(b'_0, \ldots, b'_{m-1}).$$

So by the induction hypothesis, we can choose $a_i \leq b_i$ and $a'_j \leq b'_j$ such that a_i is an atom, if $b_i \neq 0$, and a'_j is an atom, if $b'_j \neq 0$, and

$$a \leq p_0(a_0, \ldots, a_{n-1}) \wedge p_1(a'_0, \ldots, a'_{m-1}) = p(a_0, \ldots, a_{n-1}, a'_0, \ldots, a'_{m-1}).$$

(Observe that if $x_i = y_j$, then we would not know how to choose $a_i = a'_j$.)

Second, let $p = p_0 \vee p_1$. Then

$$a \leq p_0(b_0, \ldots, b_{n-1}) \vee p_1(b'_0, \ldots, b'_{m-1}).$$

If $p_0(b_0, \ldots, b_{n-1}) = 0$, then $a \leq p_1(b'_0, \ldots, b'_{m-1})$, hence we can choose the a'_i by the induction hypothesis. We choose a_i as an arbitrary atom $\leq b_i$, if $b_i \neq 0$, and let $a_i = 0$, if $b_i = 0$. If $p_1(b'_0, \ldots, b'_{m-1}) = 0$, we proceed similarly. Now if

$$p_0(b_0, \ldots, b_{n-1}) \neq 0,$$
$$p_1(b'_0, \ldots, b'_{m-1}) \neq 0,$$

then Lemma 4 and Theorem 6 imply that there exist atoms

$$a' \leq p_0(b_0, \ldots, b_{n-1}),$$
$$a'' \leq p_1(b'_0, \ldots, b'_{m-1})$$

such that $a \leq a' \vee a''$. Applying the induction hypothesis twice, we obtain a_0, $\ldots$, a_{n-1}, a'_0, $\ldots$, a'_{m-1} such that $a_i \leq b_i$ $(0 \leq i < n)$, $a'_i \leq b'_i$ $(0 \leq i < m)$, a_i is an atom, if $b_i \neq 0$, a'_i is an atom, if $b'_i \neq 0$, and

$$a' \leq p_0(a_0, \ldots, a_{n-1}),$$
$$a'' \leq p_1(a'_0, \ldots, a_{m-1}).$$

Thus

$$a \leq a' \vee a'' \leq p_0(a_0, \dots, a_{n-1}) \vee p_1(a_0', \dots, a_{m-1}'),$$

completing the induction. □

Lemma 11 (G. Grätzer and H. Lakser [1973a]). *Let us assume that the lattice identity ε can be written in the form $p \leq q$, where p and q are lattice polynomials and each variable occurs in p at most once. If ε holds for the atoms and the zero of a modular geometric lattice L, then ε holds for L.*

Proof. Let $b_0, \dots, b_{n-1} \in L$; we want to verify that

$$p(b_0, \dots, b_{n-1}) \leq q(b_0, \dots, b_{n-1}).$$

It is sufficient to show that $a \leq q(b_0, \dots, b_{n-1})$, for any atom $a \leq p(b_0, \dots, b_{n-1})$. So let a be an atom and $a \leq p(b_0, \dots, b_{n-1})$. By Lemma 10, we can choose $a_i \leq b_i$ such that the a_i are atoms or zero, $a_i \leq b_i$, for $0 \leq i < n$, and

$$a \leq p(a_0, \dots, a_{n-1}).$$

By assumption, ε holds for atoms and zero, hence

$$p(a_0, \dots, a_{n-1}) \leq q(a_0, \dots, a_{n-1}).$$

Using these two inequalities and the fact that q is isotone, we obtain that

$$a \leq p(a_0, \dots, a_{n-1}) \leq q(a_0, \dots, a_{n-1}) \leq q(b_0, \dots, b_{n-1}). \qquad \square$$

Proof of Theorem 8. If Desargues' Theorem fails in the projective geometry, then, by Lemma 9, the Arguesian identity ε fails in L. Conversely, if Desargues' Theorem holds in the projective geometry, then, by Lemma 9, ε holds for the atoms of L. A trivial step extends this to the statement that ε holds for the atoms and the zero of L. Observe that ε is of the form required in Lemma 11, hence ε holds in L. □

Corollary 12. *Desargues' Theorem holds in the projective space associated with the lattice $L(D, \mathfrak{m})$.*

Proof. With a submodule U, we associate an equivalence relation $\alpha(U)$ on $V = V(D, \mathfrak{m})$, defined by $f \, \alpha(U) \, g$ iff $f - g \in U$. Then obviously $U \mapsto \alpha(U)$ is a type 1 representation of $L(D, \mathfrak{m})$, hence, by Theorem 4.10, $L(D, \mathfrak{m})$ is Arguesian. Thus, by Theorem 8, Desargues' Theorem holds in the projective space associated with $L(D, \mathfrak{m})$. □

The next result is a well-known theorem of geometry; it is the classical proof of Desargues' Theorem in a three-dimensional space.

Theorem 13.   Let L be a modular geometric lattice. Let us assume that the projective geometry associated with L is nondegenerate. If the length of L is at least 4, then L is Arguesian.   —

Proof. Let the triangles $\langle a_0, a_1, a_2 \rangle$ and $\langle b_0, b_1, b_2 \rangle$ be perspective with respect to the point p. A *plane* α is an element of height 3. $\alpha = a_0 \vee a_1 \vee a_2$ and $\beta = b_0 \vee b_1 \vee b_2$ are planes.

Let us assume that $\alpha \neq \beta$. Since

$$\alpha \vee \beta = p \vee \alpha = p \vee \beta,$$
$$h(\alpha \vee \beta) = 4,$$

it follows that

$$h(\alpha \wedge \beta) = h(\alpha) + h(\beta) - h(\alpha \vee \beta) = 3 + 3 - 4 = 2;$$

thus $\alpha \wedge \beta$ is a line. The lines $a_i \vee a_j$ and $b_i \vee b_j$ are in the plane $p \vee a_i \vee a_j$, hence

$$h(c_{ij}) = h(a_i \vee a_j) + h(b_i \vee b_j) - h(p \vee a_i \vee a_j) = 1,$$

and so c_{ij} is a point. Since $a_i \vee a_j$ is in the plane α and $b_i \vee b_j$ is in β, c_{ij} is in $\alpha \wedge \beta$. Therefore, c_{01}, c_{02}, and c_{12} are collinear.

Now let us assume that $\alpha = \beta$. Since the length of L is at least 4, α is not the unit element. Take a $\pi \in L$, $\pi \succ \alpha$. Let l be a relative complement of α in $[p, \pi]$. Then

$$h(l) = h(\pi) + h(p) - h(\alpha) = 4 + 1 - 3 = 2,$$

so l is a line. The projective geometry associated with L is nondegenerate, hence we can choose two distinct points p' and p'' on l, $p' \neq p$, $p'' \neq p$.

Now define

$$d_i = (p' \vee a_i) \wedge (p'' \vee b_i),$$

for $i = 0, 1, 2$. It is easily seen that $\langle d_0, d_1, d_2 \rangle$ is a triangle and the plane $\delta = d_0 \vee d_1 \vee d_2 \neq \alpha, \beta$. Furthermore, $\langle a_0, a_1, a_2 \rangle$ and $\langle d_0, d_1, d_2 \rangle$ are perspective with respect to p', and $\langle b_0, b_1, b_2 \rangle$ and $\langle d_0, d_1, d_2 \rangle$ are perspective with respect to p''. Thus we can apply the first case to conclude that $(a_0 \vee a_1) \wedge (d_0 \vee d_1)$ and also $(b_0 \vee b_1) \wedge (d_0 \vee d_1)$ are in $l' = \alpha \wedge \delta = \beta \wedge \delta$; but this implies that $(a_0 \vee a_1) \wedge (b_0 \vee b_1) \in l'$, hence $c_{01} \in l'$. Similarly, c_{02} and $c_{12} \in l'$. This shows that the two triangles are perspective with respect to the line l'.   □

Before we come to the classical Coordinatization Theorem, we prove one more lemma.

Lemma 14. *A projective space satisfying Desargues' Theorem also satisfies the converse (or "dual") statement:*

If two triangles are perspective with respect to a line, then they are perspective with respect to a point.

Proof. Let us use the notation of Figure 4. We now assume that c_{01}, c_{02}, and c_{12} are on a line l and we do not have the point p. Define p as the intersection of the lines $a_1 + b_1$ and $a_2 + b_2$. To show the two triangles perspective with respect to p, we have to verify that a_0, b_0, and p are collinear.

Consider the triangles $\langle a_2, b_2, c_{02} \rangle$ and $\langle a_1, b_1, c_{01} \rangle$; they are perspective with respect to the point c_{12}. By Desargues' Theorem, the intersections of the corresponding sides, a_0, b_0, and p are collinear. $\qquad\square$

Now we are ready to state the classical Coordinatization Theorem of Projective Geometry.

Theorem 15. Let L be a directly irreducible Arguesian geometric lattice of length at least three. Then there exist a division ring D, unique up to isomorphism, and a unique cardinal number $\mathfrak{m}$ such that $L \cong L(D, \mathfrak{m})$. $\qquad$ —

Remark. For lattices of finite length, this was already included in O. Veblen and W. H. Young [1910] (of course, in a form appropriate for projective spaces). Some authors claim that this proof is not complete and that no complete proof appeared until J. von Neumann [1936]. The condition of finite dimensionality was eliminated in O. Frink [1946].

The method described below is "von Staudt's algebra of throws". A more modern (maybe less intuitive) proof can be found in R. Baer [1952]; see also E. Artin [1940].

Sketch of proof. A complete proof of this result is too long and of not enough interest for a lattice theory book. However, the mere statement that there is such an isomorphism $\varphi \colon L \to L(D, \mathfrak{m})$ is insufficient for workers in lattice theory. So we choose a middle course: we are going to construct D and φ but will not provide detailed proofs.

Let l be an arbitrary line in L; we fix three distinct points of l and call them 0, 1, and ∞. We set $D = l - \{\infty\}$, and define addition and multiplication on D.

Let a, $b \in D$. Fix two distinct points p and q not in L such that 0, p, and q are collinear (see Figure 5). Then form the points

$$r = (a \vee p) \wedge (q \vee \infty),$$
$$s = (p \vee \infty) \wedge (b \vee q),$$

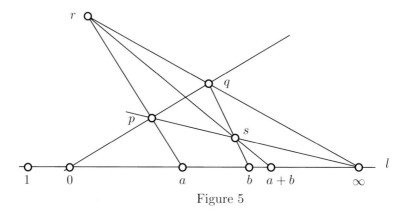

Figure 5

and set

$$a + b = (r \lor s) \land l.$$

It is easy to see that $a + b$ does not depend on the choice of p and q. Indeed, let p' and q' be another pair of distinct points not in l such that 0, p', and q' are collinear and let us form r' and s', as before. Then $\langle p, q, r \rangle$ and $\langle p', q', r' \rangle$ and also $\langle p, q, s \rangle$ and $\langle p', q', s' \rangle$ are perspective with respect to the line l; so by Lemma 14, they are perspective with respect to a point, in fact, the same point u, that is

$$u, \ p, \ p'; \quad u, \ q, \ q'; \quad u, \ r, \ r'; \quad u, \ s, \ s'$$

are all collinear. Hence $\langle r, s, q \rangle$ and $\langle r', s', q' \rangle$ are perspective with respect to u; therefore by Desargues' Theorem, they are perspective with respect to a line l'. But

$$(s \lor q) \land (s' \lor q') = b \in l,$$
$$(r \lor q) \land (r' \lor q') = \infty \in l,$$

hence $l = l'$. Therefore, $(r \lor s) \land (r' \lor s') \in l$, that is,

$$(r \lor s) \land l = (r' \lor s') \land l,$$

showing that $a + b$ is independent of the choice of p and q.

To define ab, choose two distinct points p and q not in l such that p, q, and ∞ are collinear (see Figure 6). Then we form the points

$$r = (1 \lor p) \land (q \lor b),$$
$$s = (0 \lor r) \land (p \lor a),$$

and we define

$$ab = (q \lor s) \land l.$$

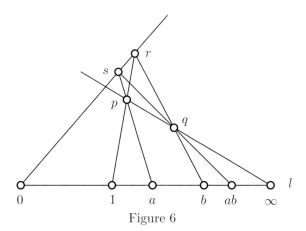

Figure 6

Then D is a division ring with 0 as a zero and 1 as the unit element.

Now we choose a maximal independent set I of atoms of L. Let $|I| = \mathfrak{m}$ and we consider the lattice $L(D, \mathfrak{m})$, where each atom is represented by a function $f \colon I \to D$, which is 0 at all but a finite number of elements; we wish to associate with each r of L such a function $d_r \colon I \to D$.

Fix $p, t \in I$, $p \neq t$, and set $l = p \vee t$. We choose a point t_1 on l, $t_1 \neq p$ and $t_1 \neq t$. We define a division ring D on l with p as a zero, t_1 as the unit, and t as infinity.

Now, for any $q \in I$, $q \neq p$, and $q \neq t$, fix a point q_1 of the line $p \vee q$, $q_1 \neq p$ and $q_1 \neq q$. The map

$$\psi_q \colon x \mapsto (x \vee ((t \vee q) \wedge (t_1 \vee q_1))) \wedge l$$

is one-to-one and onto between $p \vee q$ and l and it fixes p and takes q_1 into p_1.

We define $d_r \colon I \to D$, for any atom r of L. First, let $r \not\leq \bigvee(I - \{p\})$. Let $r \leq \bigvee I_1$, where I_1 is a minimal (finite) subset of I satisfying this. By assumption, $p \in I_1$. We define

$$d_r(x) = \begin{cases} 1 \text{ (the unit of } D), & \text{for } x = p; \\ 0 \text{ (the zero of } D), & \text{for } x \notin I_1; \\ ((r \vee \bigvee(I_1 - \{x\})) \wedge (p \vee x))\psi_x \in D, & \text{for } x \in I_1 - \{p\}. \end{cases}$$

Second, if $r \leq \bigvee(I - \{p\})$, then $r \leq \bigvee I_1$, where I_1 is a minimal (finite) subset of $I - \{p\}$ with this property. By assumption, $p \notin I_1$. Let s be any point on $p \vee r$, $s \neq p$, $s \neq r$. Then $s \not\leq \bigvee(I - \{p\})$, hence d_s has already been defined. Now we define

$$d_r(x) = \begin{cases} d_s(x), & \text{for } x \neq p; \\ 0, & \text{for } x = p. \end{cases}$$

The claim is then that $r \mapsto d_r$ maps the atoms of L onto the set of functions representing the atoms of $L(D, \mathfrak{m})$ in such a manner that collinearity is preserved. Hence this map can be extended to an isomorphism φ of L and $L(D, \mathfrak{m})$.   □

Combining Theorems 13 and 15, we obtain

Corollary 16. *Let L be a directly indecomposable modular geometric lattice of length at least 4. Then $L \cong L(D, \mathfrak{m})$, with a suitable division ring D and cardinal number $\mathfrak{m}$.*

The following embedding theorem is the most important embedding result for complemented modular lattices:

Theorem 17 (O. Frink [1946]).   Let L be a complemented modular lattice. Then L can be embedded in a modular geometric lattice K. This embedding can be chosen to be a $\{0, 1\}$-embedding and K can be chosen to satisfy all the identities of L.
                                                                          —

Remark. The last observation was made in B. Jónsson [1954].

Proof. By Zorn's Lemma, every proper dual ideal of L can be extended to a maximal dual ideal. (A maximal dual ideal of L is proper ($\neq L$) by definition.) Let M denote the set of maximal dual ideals of L. For $x \in L$, we set

$$R(x) = \{ D \mid x \in D \text{ and } D \in M \}.$$

Take the dual of $\mathrm{Du}\, L$ and let A be the sublattice generated by M. Then $K = \mathrm{Id}\, A$ is a modular geometric lattice and, if we identify $D \in M$ with the principal ideal it generates, M is the set of atoms of K. Thus there is a projective space $\langle M, ^- \rangle$ associated with K. In fact, we can define $\langle M, ^- \rangle$ directly as a projective space:

$$D \in D_1 + D_2 \quad \text{iff} \quad D \supseteq D_1 \cap D_2.$$

We claim that $x \mapsto R(x)$ is a lattice embedding of L into the lattice of subspaces of $\langle M, ^- \rangle$, that is, into K.

This claim proves the theorem, since this is a $\{0, 1\}$-embedding; furthermore, by Lemma I.4.8 and its dual, any identity that holds in L holds in the dual of $\mathrm{Du}\, L$, and therefore in A and $K = \mathrm{Id}\, A$.

$R(x)$ *is a subspace.* If D_0, $D_1 \in R(x)$ and $D \supseteq D_0 \cap D_1$, then $x \in D_0 \cap D_1$, hence $x \in D$, and so $D \in R(x)$.

$R(x \wedge y) = R(x) \cap R(y)$. Indeed, $D \in R(x) \cap R(y)$ iff x, $y \in D$, which is equivalent to $x \wedge y \in D$, that is, $D \in R(x \wedge y)$.

If $x \neq y$, then $R(x) \neq R(y)$. Let $y \not\geq x$, and let x_1 be a relative complement of $x \wedge y$ in $[0, x]$. Let D be a maximal dual ideal containing x_1. (D exists by Zorn's Lemma, since L has a 0). Then $D \in R(x)$ but $D \notin R(y)$, otherwise y, $x_1 \in D$, which contradicts that $x \wedge y_1 = 0$.

$R(x \vee y) = R(x) \vee R(y)$. We already know that R is isotone, so it is sufficient to prove that if $D \in R(x \vee y)$, then $D \in R(x) \vee R(y)$. This is trivial if x or y is 0. So let x, $y \neq 0$. By Lemma 1(ii), we have to prove the following:

If x, $y \in L$, x, $y \neq 0$, and D is a maximal dual ideal satisfying $x \vee y \in D$, then there exist maximal dual ideals E and F satisfying $x \in E$, $y \in F$, and $E \cap F \subseteq D$.

This is obvious if x or $y \in D$ (if $x \in D$, choose $E = D$ and let F be any maximal dual ideal F containing y). So let x, $y \notin D$.

First, let us assume that $x \wedge y = 0$.

Let $D_1 = D \cap [y)$. Since $y \notin D_1$, we have $[y) \supset D_1$. Computing in $\operatorname{Du} L$, we obtain

$$[x) \wedge [y) = [x) \wedge D_1 \ (= [x \vee y)),$$

since in $\operatorname{Du} L$ the $\wedge$ is set intersection; and so by modularity

$$[x) \vee [y) \supset [x) \vee D_1.$$

Therefore, the dual ideal generated by x and D_1 is proper. Thus there exists a maximal dual ideal $E \supseteq [x) \vee D_1$; in particular, $x \in E$. We repeat this trick. Let $E_1 = E \cap D$. Then $D \supset E_1$ (since E and D are both maximal) and

$$[y) \wedge E_1 = [y) \cap E \cap D$$

(use $E \supseteq [y) \cap D = D_1$)

$$= [y) \cap D.$$

Hence, by modularity,

$$[y) \vee D \supset [y) \vee E_1.$$

Therefore, there exists a maximal dual ideal $F \supseteq [y) \vee E_1$; in particular, $y \in F$.

Now since $F \supseteq E_1 = D \cap E$, we conclude that F is on the line spanned by D and E and since D, E, F are all distinct ($E \neq F$ since $x \in E$, $y \in F$, and $x \wedge y = 0$), D is on the line spanned by E and F, that is, $D \supseteq E \cap F$, which was to be proved.

Second, if $x \wedge y \neq 0$, then let y_1 be a relative complement of $x \wedge y$ in $[0, y]$. Then $x \wedge y_1 = 0$ and $x \vee y_1 = x \vee y$, hence there exist maximal dual ideals E and F satisfying $x \in E$, $y_1 \in F$, and $D \supseteq E \cap F$. Obviously, $y \in F$ so E and F will do for x and y. □

We can combine the previous results to obtain the following embedding theorems:

Corollary 18. *Every complemented modular lattice can be embedded in a direct product of lattices $L(D_i, \mathfrak{m}_i)$ and subspace lattices of projective planes that do not satisfy Desargues' Theorem.*

Proof. This is immediate from Theorem 17, Corollary 7, and Theorems 13 and 15.   □

Corollary 19. *A complemented modular lattice L can be embedded in a direct product of lattices $L(D_i, \mathfrak{m}_i)$ iff L is Arguesian.*

Proof. If L has such an embedding, then it is arguesian by Corollary 12. If L is Arguesian, then, by Theorem 17, L can be embedded in an Arguesian geometric lattice. The direct factorization of Corollary 7 gives us Arguesian directly indecomposable geometric lattices, which, by Theorem 15, are of the form $L(D_i, \mathfrak{m}_i)$.   □

Summarizing these, we obtain a result of B. Jónsson [1960a]:

Theorem 20.   Let L be a complemented modular lattice. The following conditions on L are equivalent:

   (i) L can be embedded in a direct product of $L(D_i, \mathfrak{m}_i)$.

   (ii) L can be embedded in an Arguesian geometric lattice.

   (iii) L can be embedded in the lattice of all subgroups of an abelian group.

   (iv) L has a representation of type 1.

   (v) L is Arguesian.   —

Proof. (i) is equivalent to (v) by Corollary 19. The equivalence of (i) and (ii) was proved in the proof of Corollary 19. (i) implies (iii), since the elements of L are represented by subgroups of the additive group of $\prod D_i$. (iii) implies (iv) was observed in the proof of Corollary 12. (iv) implies (v) by Theorem 4.10.   □

Theorem 20 explains the significance of the Arguesian identity for complemented modular lattices. It is interesting that a weaker version of the Arguesian identity holds for all complemented modular lattices.
   Let

$$\varepsilon \colon p(x_1, \ldots, x_6) \leq q(x_1, \ldots, x_6)$$

be the Arguesian identity. Choose three new variables, x, y, z, and define

$$u = (x \vee y) \wedge (y \vee z) \wedge (z \vee x),$$
$$v = (x \wedge y) \vee (y \wedge z) \vee (z \wedge x),$$
$$\overline{x}_i = x_i \wedge ((u \wedge x) \vee v), \qquad\qquad i = 1, 2, \ldots, 6.$$

Theorem 21 (G. Grätzer and H. Lakser [1973a]).   The identity

$$\bar{\varepsilon}: p(\bar{x}_1, \dots, \bar{x}_6) \vee v \le q(\bar{x}_1, \dots, \bar{x}_6) \vee v$$

holds in any complemented modular lattice.   —

Proof. Consider the following lattice property:

(P)   If $\{0, a, b, c, i\}$ is a diamond, then $(a]$ is an Arguesian lattice.

We claim that if a lattice has property (P), then it satisfies the identity $\bar{\varepsilon}$.
Let t, r, $s \in L$, and let us consider the substitution $x = t$, $y = r$, $z = s$.
There are two cases to consider:
Case 1.

$$(t \vee r) \wedge (r \vee s) \wedge (s \vee t) = (t \wedge r) \vee (r \wedge s) \vee (s \wedge t) = \alpha.$$

Then $\bar{x}_i = x_i \wedge \alpha$, and so $\bar{\varepsilon}$ becomes

$$p(x_1 \wedge \alpha, \dots, x_6 \wedge \alpha) \vee \alpha \le q(x_1 \wedge \alpha, \dots, x_6 \wedge \alpha) \vee \alpha.$$

But

$$p(x_1 \wedge \alpha, \dots, x_6 \wedge \alpha) \le p(\alpha, \dots, \alpha) = \alpha,$$

and similarly for q, so $\bar{\varepsilon}$ becomes $\alpha \le \alpha$, a triviality.
Case 2.

$$i = (t \vee r) \wedge (r \vee s) \wedge s \vee t) \ne (t \wedge r) \vee (r \wedge s) \vee (s \wedge t) = o.$$

Then it follows from the diagram of $\mathbf{F_M}(3)$, that o and i along with

$$a = (i \wedge t) \vee o,$$
$$b = (i \wedge r) \vee o,$$
$$c = (i \wedge s) \vee o$$

form a diamond. Thus property (P) applies and $(a]$ is Arguesian. Since $x_1 \wedge a$,
$\dots$, $x_6 \wedge a \in (a]$, we obtain

$$p(x_1 \wedge a, \dots, x_6 \wedge a) \le q(x_1 \wedge a, \dots, x_6 \wedge a);$$

joining both sides by o we get that $\bar{\varepsilon}$ holds in L.
    Next we claim that property (P) holds for the two types of lattices listed in
Corollary 18. Naturally, (P) holds for an $L(D, \mathfrak{m})$ because $L(D, \mathfrak{m})$ is Arguesian.
Let L be the subspace lattice of a projective plane. If $\{o, a, b, c, i\}$ is a diamond,
then a is either a point or a line. In either case, $(a]$ is Arguesian.
    By Corollary 18, every complemented modular lattice K can be embedded in
a direct product of lattices having property (P) and therefore satisfying $\bar{\varepsilon}$. Thus
K also satisfies $\bar{\varepsilon}$.   □

It is easy to construct a modular lattice in which $\bar{\varepsilon}$ fails. Let L be the subspace lattice of a projective plane in which Desargues' Theorem fails. By our discussion in Lemma 9, there are seven atoms of L, a_0, a_1, a_2, b_0, b_1, b_2, and d such that

$$p(a_0, a_1, a_2, b_0, b_1, b_2) = d,$$
$$q(a_0, a_1, a_2, b_0, b_1, b_2) = 0.$$

Let e be a complement of d; e is a dual atom of L. Let $K = L \cup \{b, c, i\}$ such that with $o = e$ and $a = 1$ (the unit of L), $\{o, a, b, c, i\}$ is a diamond. It is immediate that K is a modular lattice. Substitute $x_1 = a_0$, $x_1 = a_1$, $x_2 = a_2$, $x_3 = b_0$, $x_4 = b_1$, $x_5 = b_2$, $x = a$ ($= 1$), $y = b$, $z = c$. Then $u = i$, $v = 0$, $(u \wedge x) \vee o = a$ ($= 1$), hence $x_i = \bar{x}_i$, for $i = 1, 2, \dots, 6$. Thus $\bar{\varepsilon}$ yields $d \vee o \leq 0 \vee o$, that is, $1 \leq o$, which is not true. We conclude:

Corollary 22 (R. P. Dilworth and M. Hall [1944]). *There exists a modular lattice that cannot be embedded in a complemented modular lattice.*

The modular lattice K we constructed above is of length 4. The next result shows that this is best possible.

Theorem 23. Every modular lattice B of length at most three can be embedded in a complemented modular lattice C. The embedding can always be constructed to be a $\{0, 1\}$-embedding and C can be chosen to have the length of B.   —

We shall prove Theorem 23 using the concept of a projective plane.

Definition 24. *A projective plane $\langle A, L \rangle$ is a set A and a collection L of subsets of A satisfying:*

(i) *Every $l \in L$ has at least two elements.*

(ii) *For any two points p_0, $p_1 \in A$, $p_0 \neq p_1$, there is exactly one $l \in L$ satisfying p_0, $p_1 \in l$.*

(iii) *For any two distinct lines l_0 and l_1, there is exactly one $p \in A$ satisfying $p \in l_0$ and $p \in l_1$.*

A *partial plane* satisfies (i)–(iii) with "at most one" replacing "exactly one" in (ii) and (iii). Observe that the first two of the three properties already define a partial plane.

It is easily seen that a projective plane is a projective space.

Lemma 25. *Every partial plane can be embedded in a projective plane, that is, for every partial plane $\langle A, L \rangle$, there is a projective plane $\langle A', L' \rangle$ such that $A \subseteq A'$ and, for every $l \in L$, there is an $l' \in L'$ satisfying $l \subseteq l'$.*

Proof. Let $\langle A, L \rangle$ be a partial plane. Define

$$A^+ = A \cup \{ p(l, m) \mid l, \ m \in L \text{ and } l \cap m = \varnothing \},$$
$$L^+ = \{ l \cup \{ p(l, m) \mid m \in L, \ l \cap m = \varnothing \} \mid l \in L \}$$
$$\cup \{ \{ p, q \} \mid p, \ q \in A, \ p \neq q, \ p, \ q \in l, \text{ for no } l \in L \},$$

where $p(l, m) = p(m, l)$, and the new points defined are distinct from each other and from the points in A.

Now define

$$A_0 = A, \qquad\qquad L_0 = L,$$
$$A_{n+1} = (A_n)^+, \qquad L_{n+1} = (L_n)^+,$$

for $n < \omega$, and

$$A' = \bigcup (A_n \mid n < \omega).$$

Lines are defined as follows: let l_n be a line in $\langle A_n, L_n \rangle$; then we obtain the lines l_{k+1}, for $k \geq n$:

$$l_{k+1} = l_k \cup \{ p(l_k, m) \mid m \in L_k, \ l_k \cap m = \varnothing \}.$$

L' is the collection of all sets of the form $\bigcup (l_k \mid k \geq n)$. It is obvious that $\langle A', L' \rangle$ satisfies the requirements. $\qquad\qquad\qquad\qquad\qquad\qquad\square$

Proof of Theorem 23. If the length of B is less than 3, the statement is obvious. So let B be of length 3. We define a partial plane $\langle A, L \rangle$ as follows: the points of $\langle A, L \rangle$ are of two kinds, namely, the atoms of B and for every $a \in B$, $h(a) = 2$, which contains exactly one atom, a new point $p(a)$. For every $a \in B$, $h(a) = 2$, containing at least two atoms we define a line

$$\{ p \mid p \text{ is an atom of } B, \ p \leq a \};$$

if $a \in B$, $h(a) = 2$, and a contains exactly one atom p, then we define the line $\{ p, p(a) \}$. It is clear that $\langle A, L \rangle$ is a partial plane. Embed $\langle A, L \rangle$ into a projective plane $\langle A', L' \rangle$ by Lemma 25 and let C be the subspace lattice of $\langle A', L' \rangle$. $\qquad\square$

Exercises

1. Find a geometry $\langle A, {}^- \rangle$ in which X is a subspace iff $r \leq p_0 \vee p_1 \vee p_2$ and $p_0, p_1, p_2 \in X$ imply that $r \in X$, but $\langle A, {}^- \rangle$ is not a projective geometry.

2. Why do we need $x \neq y$ in the Pasch Axiom? Phrase the Pasch Axiom so that this assumption can be dropped.

3. For points p and q of a projective geometry, define $p \equiv q$ iff there is a third point $r \leq p \vee q$. Show that this relation is an equivalence relation.

4. Derive Corollary 7 from Exercise 3.

5. Prove by direct computation that the projective geometry associated with $L(D, \mathfrak{m})$ satisfies Desargues' Theorem.

6. Show that the Pasch Axiom is equivalent to the following implication holding for all points:

    If $a_0 \leq a_1 \vee a_2$ and $a_1 \wedge (a_0 \vee a_2) \leq a_3 \vee a_4$, then

    $$a_0 \leq (a_3 \wedge (a_4 \vee (a_1 \wedge (a_0 \vee a_2))))$$
    $$\vee ((a_0 \vee a_3) \wedge ((a_1 \vee a_0) \wedge a_2))$$
    $$\vee ((a_3 \vee (a_1 \wedge (a_0 \vee a_2))) \wedge a_4).$$

7. Prove that the implication of Exercise 6 holds for all subspaces of a projective space.

*8. Prove that the algebra D constructed in the proof of Theorem 15 is a division ring.

9. Let K and L be modular geometric lattices and let φ be a one-to-one map from the atoms of K onto the atoms of L. Prove that φ can be extended to an isomorphism iff

    $$a \leq b \vee c \quad \text{iff} \quad a\varphi \leq b\varphi \vee c\varphi$$

    holds, for any atoms a, b, c of K.

*10. In the proof of Theorem 15, verify that, for atoms r, s, t of L, $r \leq s \vee t$ iff $ad_r + bd_s + cd_t = 0$ holds, for some a, b, $c \in D$.

11. Combine Exercises 9 and 10 to prove $L \cong L(D, \mathfrak{m})$ in Theorem 15.

12. Show that the identity $\bar{\varepsilon}$ is not self-dual and that the dual of $\bar{\varepsilon}$ could also be used in Theorem 21.

13. The projective plane constructed in Lemma 25 is the *free projective plane* generated by the partial plane. What does "free" mean for projective planes?

14. Investigate the lattice theoretic properties of the embedding of the subspace lattice of $\langle A, L \rangle$ into the subspace lattice of $\langle A^+, L^+ \rangle$ in the proof of Lemma 25.

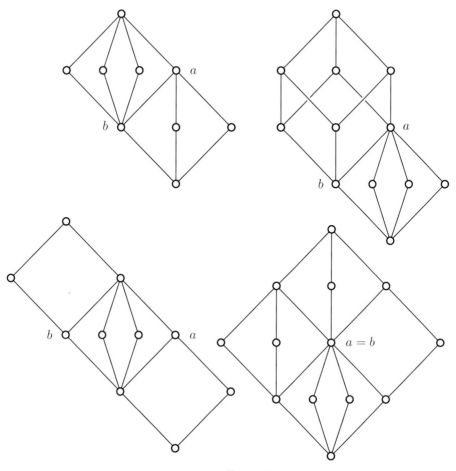

Figure 7

15. Why could Lemma 25 not be extended to provide embeddings of modular lattices of length 4?

16. Let L be a modular lattice of length 3 or 4. Let a be the join of all atoms of L and b the meet of all dual atoms of L. Consider the situations shown in Figure 7. Show that they (along with their duals) suggest a complete classification of all modular lattices of length 3 or 4 with the exception of distributive lattices and of complemented modular lattices (B. Jónsson [1959]).

17. Develop a duality theory for projective planes. The dual of $\langle A, L \rangle$ is $\langle L, A^{\mathrm{dual}} \rangle$, where $a^{\mathrm{dual}} = \{\, l \mid a \in l \,\}$, for $a \in A$.

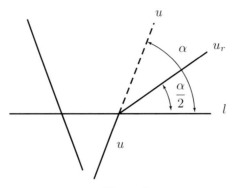

Figure 8

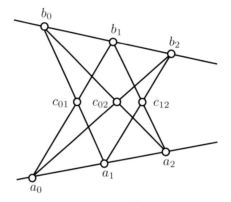

Figure 9

18. Show that $|A| = |L|$, for a finite projective plane $\langle A, L \rangle$.

19. What is the "dual" of Desargues' Theorem?

20. On the Euclidean plane, fix a line l (see Figure 8). Let u be a line and
     let α be the angle determined by u and l. A *refracted line* is defined as
     follows: if $0 \leq \alpha \leq \pi/2$, then form the line u_r with angle $\alpha/2$; below l,
     the refracted line is u, above l, it is u_r. Otherwise, u is the refracted line.
     Define a projective space as the Euclidean plane with a line at infinity; the
     lines are the refracted lines with a point at infinity and the line at infinity.
     Prove that this defines a projective plane in which Desargues' Theorem
     fails.

*21. *Pappus' Theorem* is said to hold in a projective geometry iff, whenever
     the points a_0, a_1, a_2 and b_0, b_1, b_2 are collinear and all six are in the same

plane, and

$$c_{ij} = (a_i \vee b_j) \wedge (a_j \vee b_i) \qquad (0 \le i < j < 3),$$

then c_{01}, c_{02}, and c_{12} are collinear. (See Figure 9.) Prove that if Pappus' Theorem holds in a projective space, then the division ring constructed in Theorem 15 is commutative.

*22. Prove the converse of Exercise 21.

*23. Use the Coordinatization Theorem to show that a finite projective geometry of length four or more and its dual are isomorphic.

24. Prove that in a finite complemented modular lattice, the number of k-element subsets of atoms whose join is the unit is equal to the number of k-element subsets of dual atoms whose meet is the zero.

25. In a finite modular lattice, let L_n (resp., U_n) denote the set of all $(n+1)$-element subsets $\{p, q_1, q_2, \ldots, q_n\}$ such that every q_i covers p (resp., p covers every q_i). Show that $|L_n| = |U_n|$, for every positive integer n. (Hint: use Exercise 24.)

26. Apply Exercise 25 to establish Dilworth's covering theorem for modular lattices (R. P. Dilworth [1954]): in a finite modular lattice, the number of elements covered by precisely k elements is equal to the number of elements covering precisely k elements. In particular, in a finite modular lattice, the number of join-irreducibles is equal to the number of meet-irreducibles. (This proof is due to B. Ganter and I. Rival [1973].)

Further Topics and References

For a finite modular lattice L, the posets $\mathrm{J}(L)$ and $\mathrm{M}(L)$ do not determine L as they do for a finite distributive lattice, but they have many interesting properties. The most celebrated one is a result of R. P. Dilworth [1954]: $|\mathrm{J}(L)| = |\mathrm{M}(L)|$. (See Exercises 5.23–5.26 and S. P. Avann [1958a], for a generalization.) By R. Wille [1976], $\mathrm{J}(L)$ and $\mathrm{M}(L)$ have the same width and, by B. Ganter and I. Rival [1975], the same length. In this connection, one should mention a result of R. Wille [1974a]: a finitely generated modular lattice of finite length is finite.

The Isomorphism Theorem (Theorem 1.2) is obviously equivalent to modularity. The statement: $[a, a \vee b] \cong [a \wedge b, b]$, for all a, $b \in L$, characterizes the modularity of L only in special cases: for finite lattices (M. Ward [1939]) and, more generally, for algebraic lattices (P. Crawley [1959]).

The most important result on join-representations of an element of a finite modular lattice is the Kuroš-Ore Theorem (Theorem 1.5). It holds for a finite semimodular lattice L iff L is *locally modular*, that is, for every $a \in L$, the interval $[v_a, a]$ is modular, where $v_a = \bigwedge(x \mid a \succ x)$; see R. P. Dilworth [1941a]. For generalizations to infinite lattices, see the survey article R. P. Dilworth [1961] and Chapters 5 and 6 of P. Crawley and R. P. Dilworth [1973].

In a modular lattice L, $a = x_0 \vee \cdots \vee x_{n-1}$ is a *direct join representation* iff $\{x_0, \ldots, x_{n-1}\}$ is independent. Then Theorem 1.5 holds for direct join representations in modular lattices of finite length; this is a famous theorem of O. Ore [1936], much deeper than Theorem 1.5. See also V. Havel [1955] and P. Crawley [1962a]. Such join representations are connected with direct factorizations of groups, rings, loops, and so on. A number of papers are devoted to this topic, starting with O. Ore [1937], [1938] and A. G. Kuroš [1943], [1946]; see also M. Grayev [1947], A. H. Livšic [1962], E. N. Močulskiǐ[1961], [1962], [1968].

An important new line of research in modular lattices was started in I. M. Gelfand and V. A. Ponomarev [1970]. In this paper, all 4-generated (generated by a 4-element set) subdirectly irreducible modular lattices of subspaces of a vector space are determined. This inspired R. Wille, his coworkers, and others to start a systematic search for all 4-generated subdirectly irreducible modular lattices. Partial results of this search can be found in A. Day, C. Herrmann, and R. Wille [1972], R. Freese [1973], C. Herrmann [1973], C. Herrmann, C. M. Ringel, and R. Wille [1975], C. Herrmann, M. Kindermann, and R. Wille [1975], A. Mitchke and R. Wille [1973], G. Sauer, W. Seibert, and R. Wille [1973], R. Wille [1973], [1973a].

Many of the harder computations in these papers are based on the D_2-lemma and M_3-lemma of R. Wille [1973]. Two illustrations of these results are given below.

A modular lattice completely freely generated by a finite poset P (that is, $\mathrm{F_M}(P^m)$) is finite iff P does not contain a four-element antichain nor does it contain the poset $\{x_0, y_0, y_1, z_0, z_1\}$, where $y_0 < y_1$, $z_0 < z_1$ and $x_0 \parallel y_i$, $x_0 \parallel z_j$, $y_i \parallel z_j$, for $i, j = 0, 1$ (R. Wille [1973a]). This trivially yields an earlier result of R. Wille [1973], determining when a free modular lattice generated by a set of chains is finite.

Let $P = \{o, a, b, c, d, i\}$, $o < a$, b, c, $d < i$,

$$a \wedge b = a \wedge c = a \wedge d = b \wedge c = b \wedge d = c \wedge d = o,$$
$$a \vee b = a \vee c = a \vee d = b \vee d = c \vee d = i;$$

let $\mathfrak{P}$ denote the corresponding partial lattice. Then there are exactly two modular lattices (up to isomorphism) generated by P and containing $\mathfrak{P}$ as a relative sublattice: M_4 (which is the same as M_3 but with four atoms) and an infinite (subdirectly irreducible) lattice (A. Day, C. Herrmann, and R. Wille [1972]).

Several attempts have been made to learn something about the structure of free modular lattices (see, for instance, K. Takeuchi [1951a] and [1959]). An interesting result on sublattices of $\mathrm{F_M}(4)$ can be found in R. Freese [1976].

A proper subclass of the class of finite sublattices of free modular lattices is the class of finite lattices projective in **M**. The distributive members of this class are described in A. Mitchke and R. Wille [1976].

Although it is still not known whether there is an algorithm deciding when two polynomials agree in a free modular lattice, the word problem for modular lattices has been shown to be recursively unsolvable in G. Hutchinson [1973a] and [1977]. His most recent result is that there exists a five-generated finite partial lattice Q defined by a single equation such that there is no algorithm to decide when two polynomials agree in $F_{\mathbf{M}}(Q)$. This may be best possible: if the list of known 4-generated subdirectly irreducible modular lattices is complete, then "five" cannot be reduced to "four" in Hutchinson's results.

G. M. Bergman [1969] and A. P. Huhn [1972], [1972a] defined an important class of modular lattices. A modular lattice L is called *n-distributive* iff the identity

$$x \vee \bigwedge (\, y_i \mid 0 \le i \le n \,) = \bigwedge (\, x \vee \bigwedge (\, y_i \mid 0 \le i \le n, \ i \ne j \,) \mid 0 \le j \le n \,)$$

holds in L (see Exercise V.3.22). An interesting result of A. P. Huhn states that a modular lattice L fails to be n-distributive iff L contains a sublattice $B \cong C_2^{n+1}$ with bounds o and i and an element a which is a relative complement of all dual atoms of B in $[o, i]$. This result has many applications, see, for instance, A. P. Huhn [1975], R. Freese [1976], and Section 2 of C. Herrmann [1973].

Theorem II.3.17 is also true for modular lattices, that is, every finite distributive lattice is isomorphic to the congruence lattice of a suitable modular lattice by E. T. Schmidt [1974]. The basic construction used to establish this result is further applied in E. T. Schmidt [1975a], in which a finitely generated simple modular lattice without a prime interval is exhibited.

A variety **K** of algebras is called *congruence modular* iff, for all $\mathfrak{A} \in$ **K**, the congruence lattice of $\mathfrak{A}$ is modular. Conditions for this can be found in A. Day [1969]. R. Freese [1975a] exhibits an identity ε, a weaker form of the Arguesian identity, which holds in every congruence lattice of a congruence modular variety, but ε is not implied by the modular identity.

There are two related conjectures. 1 (R. N. McKenzie). If a nontrivial lattice identity ε holds in all congruence lattices of algebras in a variety **K**, then **K** is congruence modular. 2 (J. B. Nation). If a nontrivial lattice identity ε can be expressed by a *Mal'cev type condition* (in the sense of G. Grätzer [1970a]), then ε implies the modular identity. Some results concerning the first conjecture can be found in J. B. Nation [1974] and A. Day [1976]. For some further results and an excellent review of this field, see B. Jónsson [1974a].

Necessary and sufficient conditions for the validity of the Jordan-Hölder Chain Condition in a finite poset P is given in O. Ore [1943]; see also S. MacLane [1943]. A *cell* C in P is $C = \{o, a_1, \dots, a_n, b_1, \dots, b_m, i\}$, where $\{o, a_1, \dots, a_n, i\}$

and $\{o, b_1, \ldots, b_m, i\}$ are maximal chains in the interval $[o, i]$ of P and

$$\inf\{a_k, b_j\} = o,$$
$$\sup\{a_k, b_j\} = i,$$

for $k = 1, \ldots, n, j = 1, \ldots, m$. Ore's condition is: for every cell C, $n = m$ holds (n and m depend on C). This is obviously satisfied in semimodular lattices, where, for every cell, $n = m = 1$. This result has been rediscovered (and published!) half a dozen times since 1943.

The Jordan-Hölder Chain Condition for modular elements is discussed in T. Fujiwara and K. Murata [1953].

Some 23 definitions of semimodularity are discussed in R. Croisot [1951] and [1952], especially their independence with and without chain conditions.

For further results on M-symmetric lattices, the reader should consult F. and S. Maeda [1970]; see also L. R. Wilcox [1939], [1942], [1944]; S. Maeda [1965], [1969]; and R. Padmanabhan [1974].

For a generalization of semimodularity, see O. Tamaschke [1960].

For more general discussions of independence in semimodular lattices, the reader should consult R. P. Dilworth [1944] and D. T. Finkbeiner [1951].

A sublattice of a semimodular lattice is not necessarily semimodular. In fact, R. P. Dilworth proved that every finite lattice can be embedded in a finite geometric lattice; this result was published in P. Crawley and R. P. Dilworth [1973]. For earlier related results, see R. P. Dilworth [1941a] and D. T. Finkbeiner [1960a].

The treatment of geometry from a lattice theoretic viewpoint was started in the twenties by K. Menger; his work was continued by J. von Neumann and G. Birkhoff, and later by O. Frink, L. R. Wilcox, F. Maeda, and others. The reader should consult B. Jónsson [1959b] for a general review of this field; see also the book G. Zappa [1952]. Apart from projective geometries, there are many kinds of geometries that can be classified by a lattice theoretic viewpoint: these include the classical ones such as *affine geometries* and *planar geometries* (X is a subspace iff $\overline{\{p, q, r\}} \subseteq X$ for all p, q, $r \in X$) as well as some new ones, such as the "n-stufige Geometrie" of R. Wille [1967], which has a particularly beautiful lattice theoretic characterization. For other lattice theoretic results of a geometric nature, see the books M. L. Dubreil-Jacotin, L. Lesieur, and R. Croisot [1953], F. Maeda [1958], F. and S. Maeda [1970]; the papers of F. Maeda listed in the Bibliography; W. Prenowitz [1948] and U. Sasaki [1952] and [1953].

Geometric lattices, originally called *matroids*, developed from a branch of combinatorics.

The study of matroids, in the combinatorial sense, originates with the famous Seven Bridges of Königsberg problem, solved over two hundred years ago by L. Euler. Euler demonstrated that a graph is a disjoint union of circuits iff each vertex has an even number of edges incident to it. Thus those subgraphs of a graph which are disjoint unions of circuits are closed under symmetric difference and given two circuits with nonempty intersection, the symmetric difference will always contain a circuit. This motivates the circuit definition of a matroid:

Let E be a finite set and C, the circuits, a collection of nonempty subsets of E. Then $\langle E, C \rangle$ is a *matroid* if the following two conditions are satisfied:

(C1) No circuit properly contains another.

(C2) If C and C' are distinct circuits and $e \in C \cap C'$, then there is a circuit C'' such that $C'' \subseteq (C \cup C') - \{e\}$.

There exist many matroids that do not correspond to the circuit structure of a graph. As an example, consider a *chain group*, that is, a set of maps from a finite set E to an integral domain R that is closed under pointwise addition and scalar multiplication. For any one of these maps, define the support of the map as the set of those elements of E that do not get mapped onto zero. If the minimal supports (with respect to inclusion) are defined to be the circuits on E, then a matroid results.

An equivalent approach to matroids is by the rank definition: given a finite set E and an integer valued function r on $P(E)$, $\langle E, r \rangle$ is a *matroid* iff, for $A \subseteq E$ and e, e_1, $e_2 \in E$,

(R1) $r(\varnothing) = 0$.

(R2) $r(A) \leq r(A \cup \{e\}) \leq r(A) + 1$.

(R3) $r(A) = r(A \cup \{e_1\}) = r(A \cup \{e_2\})$ implies that $r(A \cup \{e_1, e_2\}) = r(A)$.

Let $\langle E, r \rangle$ be a matroid and $E' = \{ e \in E \mid r(\{e\}) = 0 \}$. Then for any $A \subseteq E$, $r(A) = r(A - E')$ and hence $\langle E - E', r \rangle$ is a matroid with essentially the same structure as $\langle E, r \rangle$. For x, $y \in E$, let $x \equiv y$ iff $r(\{x, y\}) = 1$. Then $\equiv$ is an equivalence relation and if $[e_1], \ldots, [e_s]$ are equivalence classes, then $r([e_1] \cup \cdots \cup [e_s]) = r(\{e_1, \ldots, e_s\})$ irrespective of the choice of representatives. Thus we may identify related elements with no real loss of generality. Having done this, for $X \subseteq E$, define

$$\overline{X} = \{ e \in E \mid r(X \cup \{e\}) = r(X) \}.$$

A geometry has been formed and the rank is completely determined by its definition on the closed sets. Hence the structure of the matroid can be determined from the geometric lattice of subspaces of the geometry.

Matroids have combinatorial application in areas other than the study of the circuit structure of graphs. One such application is in transversal theory. Given sets $A_1, \ldots, A_m$, a set $X \subseteq E = A_1 \cup \cdots \cup A_m$ is called a *partial transversal* iff there is a one-to-one map $\varphi \colon X \to \{1, \ldots, m\}$ satisfying $x \in A_{x\varphi}$, for $x \in X$. The partial transversals then satisfy the independence axioms for the set E. It immediately follows, for example, that all maximal partial transversals have the same cardinality. Further results concerning the theory of transversals can be found in the definitive book of L. Mirsky [1971].

Another application of matroid theory is to the labeling of edges of graphs. Any set of labelings that are closed under addition and scalar multiplication form a chain group and hence a matroid. If the edges are labeled with integers modulo two, the set of labelings such that the sum of the labels on edges incident to any given vertex is zero forms a chain group, which yields the ordinary circuits of the graph as its matroid. More generally, a chain group can be formed by using labels from an arbitrary integral domain and including in the chain group those labelings such that, for every circuit C in the graph, the sum, $S(C)$, of the labels on the edges of C is even, that is, there is an element x in the integral domain such that $2x = S(C)$. An element in this chain group may be formed by assigning a value $g(v)$ to each vertex and then defining $f(e) = g(v_1) + g(v_2)$, where v_1 and v_2 are the two vertices incident to e. In topological terms, this is the usual *coboundary operator*. The set of all labelings that can be formed in this manner forms a chain group, and, as a matroid, it has the same rank as the original chain group. Hence every chain group element arises in this manner. If the integral domain has characteristic two, and a labeling in this chain group exists whose support is all of E, then in the associated vertex assignment no two joined vertices have the same value. If we call these assignments *colors*, then we are looking at the classical graph theoretic coloring problem. Thus these chain groups can be used to evaluate the chromatic number of graphs. Further considerations of this type are found in the book of H. H. Crapo and G.-C. Rota [1971].

The interested reader should consult the two books referred to above and W. T. Tutte [1959], [1965], [1971] and R. J. Wilson [1973].

Many papers have been published on maximum sized antichains in posets, mostly in geometric lattices. For a finite poset P and maximum sized antichains A and B, set $A \leq B$ iff for every $a \in A$ there is a $b \in B$ such that $a \leq b$ in P. Then the maximum sized antichains form a distributive lattice by R. P. Dilworth [1960]; for a simple proof and an application, see R. Freese [1974].

Many generalizations of Sperner's Lemma are known; see D. A. Drake [1971], M. J. Klass [1974], D. Kleitman, M. Edelberg, and D. Lubell [1971], D. Lubell [1966]. It is still not known whether Sperner's Lemma holds for finite partition lattices. Nevertheless, many nice results have been proved for finite geometric lattices about the Whitney numbers W_k. Let L be a finite geometric lattice of length n. Then $W_1 \leq W_k$, for $k = 2, 3, \ldots, n - 1$ (J. G. Basterfield and L. M. Kelly [1968]) and $W_1 = W_{n-1}$ iff L is modular (C. Greene [1970]); furthermore, for $1 \leq k \leq n - 2$,

$$W_1 + \cdots + W_k \leq W_{n-k} + \cdots + W_{n-1},$$

and equality holds iff L is modular (T. A. Dowling and R. M. Wilson [1975]). For finite modular geometric lattices $W_k = W_{n-k}$, for all k (R. P. Dilworth [1954]). For further results, see T. A. Dowling and R. M. Wilson [1974].

The proof of these results utilizes the Möbius function, see Exercises 3.32–37 and H. H. Crapo [1966], [1968].

Maximum sized antichains in C_n^k are determined in I. Rosenberg [1967].

The connections of geometric lattices and combinatorics are further explored in H. H. Crapo and G.-C. Rota [1970] and [1971].

Partition lattices were first studied in detail in O. Ore [1942]. Characterizations of partition lattices can be found in R. P. Dilworth [1944], U. Sasaki and S. Fujiwara [1952a], and D. Sachs [1961a].

"Which finite lattices can be embedded in a finite partition lattice?" is one of the oldest open problems of lattice theory. Corollary 4.6 tells us that we shall not obtain an answer by looking at identities. K. A. Baker observed that the same holds for all universal sentences: if a universal sentence Φ holds for all finite lattices, then Φ holds for all lattices. (The proof is easy; it uses that a lattice L can be embedded in an ultra product of its finite partial sublattices.) A. Ehrenfeucht, V. Faber, S. Fajtlowicz, and J. Mycielski [1973] consider the problem of finding, for a finite lattice L, the smallest integer $\mu(L)$ such that L can be embedded in Part X with $|X| = \mu(L)$. They obtain results for some special classes of lattices.

Lattices having type 1 representations are characterized in B. Jónsson [1959a], based on some ideas of R. C. Lyndon [1950] and [1956].

An equivalent form of the Arguesian identity that formally resembles Desargues' Theorem can be found in G. Grätzer, B. Jónsson, and H. Lakser [1973]. Applying this characterization it is proved that if $\{o, a, b, c, i\}$ is a diamond in a modular lattice, then the Arguesian identity holds in $[o, a]$; this is also used in G. Grätzer and J. Sichler [1975]. By B. Jónsson [1959], every Arguesian lattice of length ≤ 4 has a type 1 representation.

If a lattice L can be embedded in the lattice of all normal subgroups of a group, then L has a type 1 representation. The converse is false by B. Jónsson [1954]; the counterexample can be chosen to have length 5.

Representations of type 2 and 3 and congruence lattices of universal algebras are considered together in G. Grätzer and E. T. Schmidt [1962] and G. Grätzer and W. A. Lampe (Appendix 7 of G. Grätzer [1979]); more about this in the Concluding Remarks.

A very simple proof of Whitman's result, in the stronger form due to B. Jónsson (Theorem 4.4), is given in S. K. Thomason [1970] for finite lattices; in fact, the representation found is recursive. This is applied in S. K. Thomason [1970a] to prove that every finite lattice is isomorphic to a sublattice of the lattice of degrees of unsolvability. The reader should not find it difficult to generalize the reasoning of S. K. Thomason [1970] to arbitrary lattices.

An *n-partition* $(n \geq 1)$ of a set A satisfying $|A| \geq n$ is a collection of at least n element subsets of A such that every n-element subset of A is contained in exactly one member of the collection. Let Part$_n A$ denote the set of all n-partitions of A; note that Part$_1 A = $ Part A. For $P, Q \in $ Part$_n A$, let $P \leq Q$ iff every member of P is contained in a member of Q. This makes Part$_n A$ a lattice, see J. Hartmanis [1956] and [1961]. Hartmanis' main result is that every finite lattice L can be embedded in some finite $\mathrm{Part}_2\, A$. For other related results, see J. Hartmanis [1959] and B. Frontera-Marqués [1964].

The major step in the generalization of the Coordinatization Theorem was made in J. von Neumann [1936], replacing the division ring by a regular ring, $L(D, \mathfrak{m})$ by the lattice of principal right ideals of a regular ring, and directly irreducible Arguesian geometric lattices by complemented modular lattices with a homogeneous basis with 4 or more elements (a *homogeneous basis* of n elements is $a_1, \dots, a_n$ such that $\{a_1, \dots, a_n\}$ is independent, $a_1 \vee \cdots \vee a_n = 1$, and $a_i \sim a_j$, for $i \neq j$). Von Neumann's Coordinatization Theorem is considered by many the deepest lattice theoretic result.

Von Neumann's conditions have been relaxed in many ways. The reader should consult I. Amemiya [1957], I. Amemiya and I. Halperin [1959] and [1959a], K. D. Fryer and I. Halperin [1956] and [1958], B. Jónsson [1960a], B. Jónsson and G. S. Monk [1969]; the books F. Maeda [1958], L. A. Skornjakov [1961], the survey article I. Halperin [1961], and the review by I. Halperin of B. Jónsson [1960a] in the Mathematical Reviews.

M. Hall [1954] contains some useful information on projective planes; see also B. Artmann [1969] and R. Wille [1976b]. For a rather unexpected connection between projective planes and weakly associated lattices, see E. Fried and V. T. Sós [1975].

Complemented modular lattices play an important role as building stones of all modular lattices of finite length; see C. Herrmann [1973c].

Problems

1. Investigate those finite modular lattices L, for which there is a matching between $\mathrm{J}(L) \cup \{0\}$ and $\mathrm{M}(L) \cup \{1\}$. (I. Rival.)

2. Investigate those finite modular lattices L, for which there is an isomorphism φ of the posets $\mathrm{J}(L)$ and $\mathrm{M}(L)$ that is also a matching. (I. Rival.)

3. Does every finite (countable) modular lattice L have an embedding in a 4-generated modular lattice K? If L is finite, can K be chosen finite?

4. Is $\mathbf{M}$ generated by $\mathrm{F}_{\mathbf{M}}(4)$ (in the sense of Section V.1)? In other words, if an identity ε fails to hold in a modular lattice, is there a consequence ε' of ε such that ε' has only four variables and ε' also fails in some modular lattice?

5. Is $\mathbf{M}$ generated by its finite members (members of finite length)?

6. Which are the finite (simple) sublattices of $\mathrm{F}_{\mathbf{M}}(4)$ and $\mathrm{F}_{\mathbf{M}}(\aleph_0)$?

7. Is $\mathrm{F}_{\mathbf{M}}(\aleph_0)$ isomorphic to a sublattice of $\mathrm{F}_{\mathbf{M}}(4)$ or of $\mathrm{F}_{\mathbf{M}}(n)$, for some $n < \omega$? Are there n, $m < \omega$, $n < m$ such that $\mathrm{F}_{\mathbf{M}}(m)$ is isomorphic to a sublattice of $\mathrm{F}_{\mathbf{M}}(n)$?

8. Is the word problem for $F_M(4)$ decidable? Is the word problem for $F_M(\aleph_0)$ decidable?

9. Is the word problem in four variables decidable for modular lattices?

10. Describe all 4-generated subdirectly irreducible modular lattices.

11. A *modular partial lattice* is a partial lattice obtained from a modular lattice as in Definition I.5.12. Show that there is no finite set of identities characterizing the modularity of a partial lattice. (See Exercise I.5.20 for the concept of validity of an identity in a partial lattice.)

12. Is the modularity of a partial lattice undecidable?

13. Investigate modular lattices L with the following property: if a/b is a subquotient of c/d and $a/b \approx c/d$, then $a/b = c/d$. Is it possible to define in such L a "distance function" with values in a distributive lattice?

14. Characterize the congruence lattices of modular lattices. (Distributive algebraic lattices.)

15. Find a common generalization of the Kuroš-Ore Theorem (Theorem 1.5) and the Ore Theorem (on direct representations). (For modular lattices of finite length, this is done in E. T. Schmidt [1970].)

16. For algebraic lattices, is M-symmetry equivalent to semimodularity?

17. Investigate the construction of Exercise 2.12 from the point of view of M-symmetry.

18. For infinite sets A and B, are Part A and Part B elementarily equivalent? If $|A| \leq |B|$, is there an elementary embedding of Part A into Part B?

19. Characterize the congruence lattices of finite semimodular lattices. (Finite distributive lattices.)

20. Characterize the congruence lattices of weakly modular lattices. (Distributive algebraic lattices.)

21. Let the lattice L have a type 1 representation. Does L have a type 1 representation associated with a distance function?

22. Let L be a lattice having a type 1 representation. Does every homomorphic image of L have a type 1 representation?

23. Show that there is no finite set of identities defining the variety generated by lattices having type 1 representation.

24. Let L be an algebraic lattice having a type 1 representation. Does L have a type 1 complete representation?

25. Let L be an algebraic lattice having a type 1 representation. Is there a (finitary) algebra $\mathfrak{A}$ with permutable congruences such that L is isomorphic to the congruence lattice of $\mathfrak{A}$?

26. Let $\mathbf{K}$ be a variety of algebras. If there is a nontrivial lattice identity holding in all congruence lattices of algebras in $\mathbf{K}$, is then $\mathbf{K}$ congruence modular? (R. N. McKenzie.)

27. If the validity of a nontrivial lattice identity, ε, for congruence lattices of algebras in a variety is a Mal'cev type condition, does then ε imply modularity? (J. B. Nation.)

28. Find a Mal'cev type condition equivalent to the validity of the Arguesian identity for the congruence lattices of the members of a variety of algebras.

29. To what extent is a variety $\mathbf{K}$ of lattices determined by $\{\,\mathrm{Con}\,L \mid L \in \mathbf{K}\,\}$? Check $\mathbf{K} = \mathbf{D}$, $\mathbf{N}_5$, $\mathbf{M}_3$, $\mathbf{M}_{3,3}$, etc.

30. Let the lattice L have an embedding into a complemented modular lattice. Does every homomorphic image of L have such an embedding?

31. Find an effective algorithm that turns configurational conditions for geometries into equivalent identities for the subspace lattices.

32. Can every finite lattice be embedded in a finite partition lattice? In particular, does the dual of a finite partition lattice have such an embedding?

33. For a finite lattice L, let $\mu(L)$ denote the smallest integer such that L can be embedded in Part A with $|A| \le \mu(L)$. What is $\mu(L)$, where L is the dual of Part A with $|A| = 4$?

34. Let $L(K_1, \dots, K_n)$ denote the lattice with 0 and 1 consisting of n chains $c(i,1) > c(i,2) > \cdots > c(i,k_i)$, $i = 1,\,2,\,\dots,\,n$, such that if $i \ne j$, then $c(i,p)$ is a complement of $c(j,q)$. Compute $\mu(L(k_1, \dots, k_n))$. (This is done, for $n = 2$, in A. Ehrenfeucht, V. Faber, S. Fajtlowicz, and J. Mycielski [1973].)

35. Is it true in Part A, A finite, that any maximum sized antichain is of the form $E(i)$?

36. Is every finite lattice isomorphic to the congruence lattice of a finite algebra?

37. Is every finite, type 1 lattice isomorphic to the congruence lattice of a finite algebra with permutable congruences?

38. Investigate Problems 26–28 for infinitary algebras.

39. Is there a variety $\mathbf{K}$ of infinitary algebras such that

$$\mathbf{M} = \{\,\mathrm{Con}\,\mathfrak{A} \mid \mathfrak{A} \in \mathbf{K}\,\}$$

holds?

Varieties of Lattices

1. Characterizations of Varieties

In this section, we shall discuss the basic properties of varieties of lattices. Of the four characterizations and descriptions given, three apply to arbitrary varieties of universal algebras; the fourth is valid only for those varieties of universal algebras that are *congruence distributive* (that is, the congruence lattice of any algebra in the class is distributive). For the sake of simplicity, all these results are stated and proved only for lattices.

For a class $\mathbf{K}$ of lattices, let $\mathrm{Iden}(\mathbf{K})$ denote the set of all identities holding in all lattices of $\mathbf{K}$; if $\mathbf{K} = \{L\}$, we write $\mathrm{Iden}(L)$ for $\mathrm{Iden}(\mathbf{K})$. (The same convention will be used for all "operators" as well; if $\mathbf{X}$ is an operator, that is, $\mathbf{X}$ is any one of $\mathbf{H}$, $\mathbf{S}$, $\mathbf{P}$, $\mathbf{Var}$, $\mathbf{P_S}$, $\mathbf{P_U}$, $\mathbf{Si}$, and $\mathbf{I}$ and $\mathbf{K} = \{L\}$, then we write $\mathbf{X}(L)$ for $\mathbf{X}(\mathbf{K})$.) For a class Σ of identities, let $\mathbf{Mod}(\Sigma)$ denote the class of all "lattice models of Σ"—lattices in which all the identities of Σ hold. By definition, $\mathbf{K}$ is a *variety of lattices* iff $\mathbf{K} = \mathbf{Mod}(\Sigma)$, for some set Σ of identities, or equivalently, iff

$$\mathbf{K} = \mathbf{Mod}(\mathrm{Iden}(\mathbf{K})).$$

Let $\varepsilon\colon p = q$ be an identity where p and q are n-ary lattice polynomials. Let $\mathbf{K}$ be a variety and let $\mathrm{F}_{\mathbf{K}}(n)$ be the free lattice over $\mathbf{K}$ with n free generators, $u_0, \ldots, u_{n-1}$; then $\mathrm{F}_{\mathbf{K}}(n)$ exists by Corollary I.5.6. Let $\mathrm{F}_{\mathbf{K}}(\aleph_0)$ denote the free lattice over $\mathbf{K}$ with $\aleph_0$ free generators, $u_0, u_1, \ldots, u_n, \ldots$.

Let us suppose that ε holds for the free generators of $\mathrm{F}_{\mathbf{K}}(n)$, that is,

$$p(u_0, \ldots, u_{n-1}) = q(u_0, \ldots, u_{n-1}).$$

Let L be any lattice in $\mathbf{K}$ and let $a_0, \ldots, a_{n-1} \in L$. Then the map $u_i \mapsto a_i$, $i = 0, 1, \ldots, n - 1$, can be extended to a homomorphism φ of $\mathrm{F}_{\mathbf{K}}(n)$ into L, and so

$$p(a_0, \ldots, a_{n-1}) = p(u_0\varphi, \ldots, u_{n-1}\varphi) = p(u_0, \ldots, u_{n-1})\varphi$$
$$= q(u_0, \ldots, u_{n-1})\varphi = q(u_0\varphi, \ldots, u_{n-1}\varphi) = q(a_0, \ldots, a_{n-1}),$$

that is, ε holds in L. Thus

$$\varepsilon \in \mathrm{Iden}(\mathbf{K}) \quad \text{iff} \quad \varepsilon \in \mathrm{Iden}(\mathrm{F}_{\mathbf{K}}(n)).$$

If we make no restrictions on the arity of p and q, we obtain:

$$\varepsilon \in \mathrm{Iden}(\mathbf{K}) \quad \text{iff} \quad \varepsilon \in \mathrm{Iden}(\mathrm{F}_{\mathbf{K}}(\aleph_0)).$$

This means that $\mathbf{K}$ is completely determined by $\mathrm{F}_{\mathbf{K}}(\aleph_0)$.

Theorem 1.

(i) There is a one-to-one correspondence between varieties of lattices and (up to isomorphism) free lattices with $\aleph_0$ generators, wherein $\mathbf{K}$ corresponds to $\mathrm{F}_{\mathbf{K}}(\aleph_0)$.

(ii) A lattice L is a free lattice with $\aleph_0$ generators over some variety $\mathbf{K}$ of lattices iff L has a countable generating set U such that every map $U \to L$ can be extended to an endomorphism of L.   —

Remark. (i) sets up the correspondence between varieties and countably generated free lattices, while (ii) describes which lattices occur in this correspondence. It follows from the results of Chapter VI that the U in (ii) is uniquely determined as the set of doubly irreducible elements of L.

Proof. (i) has already been proved. The "only if" part of (ii) is trivial. To prove the "if" part, let L and U be given as in (ii). We wish to construct a variety $\mathbf{K}$ such that L is free over $\mathbf{K}$. Let $\mathbf{K}$ be the class of all lattices A such that any map $U \to A$ can be extended to a homomorphism of L into A. Let Σ be the set of all identities $\varepsilon: p = q$ such that $p(u_0, u_1, \ldots) = q(u_0, u_1, \ldots)$ holds in L, for all $u_0, u_1, \ldots \in U$.

We claim that $\mathbf{K} = \mathrm{Mod}(\Sigma)$. Indeed, arguing as at the beginning of this section, we obtain that $\mathbf{K} \subseteq \mathrm{Mod}(\Sigma)$. Conversely, let $A \in \mathrm{Mod}(\Sigma)$ and let $\alpha: U \to A$ be a map. If $\varepsilon: p = q$ is an identity and $p(u_0, u_1, \ldots) = q(u_0, u_1, \ldots)$, for $u_0, u_1, \ldots \in U$ (with all u_i distinct), then $\varepsilon \in \Sigma$. Since $A \in \mathrm{Mod}(\Sigma)$, we obtain that ε holds in A and therefore $p(u_0\alpha, u_1\alpha, \ldots) = q(u_0\alpha, u_1\alpha, \ldots)$. Thus (see Exercise I.5.45) α can be extended to a homomorphism and so $L \in \mathbf{K}$.

This proves that $\mathbf{K} = \mathrm{Mod}(\Sigma)$, so $\mathbf{K}$ is a variety. L is free over $\mathbf{K}$, by the definition of $\mathbf{K}$. $\square$

Let $u_0, u_1, \ldots$ be the free generators of $\mathrm{F}(\aleph_0)$ (the free lattice on $\aleph_0$ genera-tors, see Section I.5) and let $v_0, v_1, \ldots$ be the free generators of $\mathrm{F}_{\mathbf{K}}(\aleph_0)$. Then $u_i \mapsto v_i$, $i = 0, 1, \ldots$ extends to a homomorphism $\varphi = \varphi_{\mathbf{K}}$; let $\Theta = \Theta_{\mathbf{K}}$ be the kernel of φ. Since, by Theorem 1, the variety $\mathbf{K}$ is determined by $\mathrm{F}_{\mathbf{K}}(\aleph_0)$ and $\mathrm{F}_{\mathbf{K}}(\aleph_0)$ is determined by Θ, we conclude that $\mathbf{K}$ is determined by Θ. If we can ascertain which congruences arise this way, we shall have another description of varieties.

Call a congruence relation Θ of a lattice L *fully invariant* iff $a \equiv b$ (Θ) implies that $a\alpha \equiv b\alpha$ (Θ), for all $a, b \in L$ and for all endomorphisms α of L.

Theorem 2 (B. H. Neumann [1962]).   There is a one-to-one correspond-ence between varieties of lattices and fully invariant congruence relations of $\mathrm{F}(\aleph_0)$.

Proof. Let $\mathbf{K}$ be a variety and let φ, Θ ($= \Theta_{\mathbf{K}}$), u_i, v_i be as described above. We show that Θ is fully invariant. Let α be an endomorphism of $\mathrm{F}(\aleph_0)$, let a, $b \in \mathrm{F}(\aleph_0)$, and let $a \equiv b$ (Θ).

Let β be the endomorphism of $\mathrm{F}_{\mathbf{K}}(\aleph_0)$ extending $v_i \mapsto u_i \alpha \varphi$. Since $a \in \mathrm{F}(\aleph_0)$, it follows that $a = p(u_0, \ldots, u_{n-1})$, for some integer n and for some n-ary polynomial p. We compute:

$$a\varphi\beta = p(u_0, \ldots, u_{n-1})\varphi\beta = p(u_0\varphi\beta, \ldots, u_{n-1}\varphi\beta) = p(v_0\beta, \ldots, v_{n-1}\beta)$$
$$= p(u_0\alpha\varphi, \ldots, u_{n-1}\alpha\varphi) = p(u_0, \ldots, u_{n-1})\alpha\varphi = a\alpha\varphi,$$

and similarly for b. Thus,

$$a\alpha\varphi = a\varphi\beta = b\varphi\beta = b\alpha\varphi,$$

and therefore $a\alpha \equiv b\alpha$ (Θ).

Now let Θ be a fully invariant congruence relation of $\mathrm{F}(\aleph_0)$. We shall show that $\mathrm{F}(\aleph_0)/\Theta$ satisfies the condition of Theorem 1(ii). If $u_i \equiv u_j$ (Θ), for some $i \neq j$, then it is easily shown that $\Theta = \iota$ and so $\mathrm{F}(\aleph_0)/\Theta$ is the one-element lattice. So let us assume that $u_i \not\equiv u_j$ (Θ), for all $i \neq j$. We take

$$U = \{\, [u_i]\Theta \mid i = 0, 1, \ldots \,\},$$
$$\alpha \colon U \to \mathrm{F}(\aleph_0)/\Theta.$$

Let

$$[u_i]\Theta\alpha = [a_i]\Theta, \quad i = 0, 1, \ldots .$$

Then $u_i \mapsto a_i$, $i = 0, 1, \ldots$, can be extended to an endomorphism β of $\mathrm{F}(\aleph_0)$. Since $\Theta = \mathrm{Ker}\,\varphi$ is fully invariant, we have $\mathrm{Ker}\,\varphi \subseteq \mathrm{Ker}\,\beta\varphi$ and so $x\varphi \mapsto x\beta\varphi$ extends α to an endomorphism of $\mathrm{F}(\aleph_0)/\Theta$. $\square$

We have already proved, in Section I.4, that a variety is closed under the formation of homomorphic images, sublattices, and direct products. The converse, which is due to G. Birkhoff [1935], is the third description of varieties.

Theorem 3.   A class **K** of lattices is a variety iff **K** is closed under the formation of homomorphic images, sublattices, and direct products.   —

Remark. The direct product of an empty family of lattices is the one-element lattice and, therefore, if **K** is closed under the formation of direct products, then **K** is not the empty class. Observe also, that if **K** is closed under the formation of homomorphic images, then **K** is closed under the formation of isomorphic copies.

Proof. Let **K** be closed under the formation of homomorphic images, sublattices, and direct products. If **K** consists of one-element lattices only, then **K** can be defined by the identity: $x_0 = x_1$ and so **K** is equational. Now we can assume that **K** contains a lattice of more than one element. Therefore, we conclude, just as in Section I.5, that $F_{\mathbf{K}}(\mathfrak{m})$ exists for any cardinal of $\mathfrak{m}$.

Let $L \in \mathbf{Mod}(\mathrm{Iden}(\mathbf{K}))$ and take an $F_{\mathbf{K}}(\mathfrak{m}) \in \mathbf{K}$ with $|L| + \aleph_0 = \mathfrak{m}$. We denote by U the set of free generators of $F_{\mathbf{K}}(\mathfrak{m})$. Since $|L| + \aleph_0 = |U|$, there is a map α of U onto L. Let $p(u_0, u_1, \ldots, u_{n-1}) = q(u_0, u_1, \ldots, u_{n-1})$ hold in $F_{\mathbf{K}}(\mathfrak{m})$, with $u_0, u_1, \ldots, u_{n-1} \in U$. Without loss of generality, we can assume that the u_i are all distinct. Therefore, $p = q \in \mathrm{Iden}(F_{\mathbf{K}}(\mathfrak{m}))$ (as argued at the beginning of this section) and so, by the freeness of $F_{\mathbf{K}}(\mathfrak{m})$, $p = q \in \mathrm{Iden}(\mathbf{K})$. Because $L \in \mathbf{Mod}(\mathrm{Iden}(\mathbf{K}))$, it follows that $p = q \in \mathrm{Iden}(L)$; in particular, $p(u_0\alpha, u_1\alpha, \ldots, u_{n-1}\alpha) = q(u_0\alpha, u_1\alpha, \ldots, u_{n-1}\alpha)$. This shows that α satisfies the hypothesis of Exercise I.5.45 and can, therefore, be extended to a homomorphism β of $F_{\mathbf{K}}(\mathfrak{m})$ onto L. Thus L is a homomorphic image of a member of **K**, and so $L \in \mathbf{K}$.

The converse was proved in Lemma I.4.8.   □

To obtain a slightly different version of this result, we introduce some notation. For a class **K** of lattices, let $\mathbf{H}(\mathbf{K})$, $\mathbf{S}(\mathbf{K})$, and $\mathbf{P}(\mathbf{K})$ denote the class of all homomorphic images, sublattices, and direct products of members of **K**, respectively.

Corollary 4 (A. Tarski [1946]). *Let* **K** *be a class of lattices. Then* $\mathbf{HSP}(\mathbf{K})$ *is the smallest variety containing* **K**.

Proof. We start out by observing three formulas for any class **K** of lattices:

   (i) $\mathbf{SH}(\mathbf{K}) \subseteq \mathbf{HS}(\mathbf{K})$;

  (ii) $\mathbf{PH}(\mathbf{K}) \subseteq \mathbf{HP}(\mathbf{K})$;

 (iii) $\mathbf{PS}(\mathbf{K}) \subseteq \mathbf{SP}(\mathbf{K})$.

To prove (i), let $L \in \mathbf{SH(K)}$. Then there is an $A \in \mathbf{K}$ and a homomorphism α of A onto a lattice B containing L as a sublattice. Set

$$C = \{\, x \mid x \in A \text{ and } x\alpha \in L \,\}.$$

Then $C \in \mathbf{S(K)}$ and $L \in \mathbf{H}(C)$. Hence $L \in \mathbf{HS(K)}$, proving (i).

To prove (ii), let $L \in \mathbf{PH(K)}$. Then there exist $A_i \in \mathbf{K}$ and a homomorphism α_i of A_i onto B_i, for $i \in I$, such that $L = \prod(\, B_i \mid i \in I \,)$. It is clear that L is a homomorphic image of $\prod(\, A_i \mid i \in I \,) \in \mathbf{P(K)}$, proving (ii).

The proof of (iii) follows that of (ii) with A_i a sublattice of B_i.

To show that $\mathbf{HSP(K)} = \mathbf{K}_1$ is a variety using Theorem 3, we have to verify that $\mathbf{H}(\mathbf{K}_1) \subseteq \mathbf{K}_1$, $\mathbf{S}(\mathbf{K}_1) \subseteq \mathbf{K}_1$, and $\mathbf{P}(\mathbf{K}_1) \subseteq \mathbf{K}_1$. Indeed (using (i)–(iii) and that $\mathbf{XX}(\mathbf{K}_2) = \mathbf{X}(\mathbf{K}_2)$, for $\mathbf{X} \in \{\mathbf{H}, \mathbf{S}, \mathbf{P}\}$ and any class $\mathbf{K}_2$):

$$\mathbf{H}(\mathbf{K}_1) = \mathbf{HHSP(K)} = \mathbf{HSP(K)} = \mathbf{K}_1,$$
$$\mathbf{S}(\mathbf{K}_1) = \mathbf{SHSP(K)} \subseteq \mathbf{HSSP(K)} = \mathbf{HSP(K)} = \mathbf{K}_1,$$
$$\mathbf{P}(\mathbf{K}_1) = \mathbf{PHSP(K)} \subseteq \mathbf{HPSP(K)} \subseteq \mathbf{HSPP(K)} = \mathbf{HSP(K)} = \mathbf{K}_1.$$

If $\mathbf{K}_2$ is any variety containing $\mathbf{K}$, then

$$\mathbf{K}_2 = \mathbf{HSP}(\mathbf{K}_2) \supseteq \mathbf{HSP(K)} = \mathbf{K}_1. \qquad \square$$

Let $\mathbf{Var(K)}$ denote the smallest variety containing $\mathbf{K}$. Then Corollary 4 can be summarized in the formula

$$\mathbf{Var} = \mathbf{HSP}.$$

For a class $\mathbf{K}$ of lattices, let $\mathbf{P_S(K)}$ denote the class of all lattices that are isomorphic to a subdirect product of members of $\mathbf{K}$. The following variant of Corollary 4 is not especially useful, but the construction used in the proof has found some applications.

Corollary 5 (S. R. Kogalovskiĭ[1965]). *A class* $\mathbf{K}$ *of lattices is a variety iff* $\mathbf{K}$ *is closed under the formation of homomorphic images and subdirect products. Moreover, for any class* $\mathbf{K}$ *of lattices,*

$$\mathbf{Var(K)} = \mathbf{HP_S(K)}.$$

Remark. In other words,

$$\mathbf{Var} = \mathbf{HP_S}.$$

Proof. We leave it to the reader to verify that both statements follow readily from the following inequality:

(iv) $\mathbf{S(K)} \subseteq \mathbf{HP_S(K)}$.

To verify (iv), let $L \in \mathbf{S}(\mathbf{K})$, that is, let L be a sublattice of some $A \in \mathbf{K}$. Now take A^I with $|I| = \aleph_0$ and form a sublattice $B \subseteq A^I$ defined as follows: for $f \in A^I$, let $f \in B$ iff $f(i) = a$, for some $a \in L$ and for all but finitely many $i \in I$; a map $\varphi \colon B \to L$ is defined by mapping f to this a. It is trivial that $B \in \mathbf{P_S}(A)$ and $L \in \mathbf{H}(B)$, concluding that $L \in \mathbf{HP_S}(\mathbf{K})$. $\square$

We have obtained two formulas for $\mathbf{Var}$, namely,

$$\mathbf{Var} = \mathbf{HSP} = \mathbf{HP_S}.$$

Their usefulness is, however, somewhat limited when applied to describing members of varieties of lattices. A great improvement of these formulas is possible for lattices and we shall proceed to develop this.

Let L_i, $i \in I$, be lattices, let $I \neq \varnothing$, and let $\mathcal{D}$ be a dual ideal of the lattice $P(I)$ of all subsets of I. For f, $g \in \prod(L_i \mid i \in I)$, we set

$$I(f,g) = \{\, i \mid i \in I \text{ and } f(i) = g(i) \,\}.$$

(I in $I(f,g)$ stands for "identical".)

We introduce a congruence relation $\Theta = \Theta(\mathcal{D})$ on $L = \prod(L_i \mid i \in I)$. For f, $g \in L$, let

$$f \equiv g \ (\Theta) \quad \text{iff} \quad I(f,g) \in \mathcal{D}.$$

(View members of $\mathcal{D}$ as "sets of measure one". Then $f \equiv g$ (Θ) iff f and g are identical on a set of measure one.) Θ is obviously reflexive and symmetric. If f, g, $h \in L$, $f \equiv g$ (Θ), and $g \equiv h$ (Θ), then $I(f,g)$ and $I(g,h) \in \mathcal{D}$ so that $I(f,h) \supseteq I(f,g) \cap I(g,h) \in \mathcal{D}$. Thus $I(f,h) \in \mathcal{D}$ and we conclude that $f \equiv h$ (Θ). If f, g, $h \in L$ and $f \equiv g$ (Θ), then $I(f \wedge h, g \wedge h) \subseteq I(f,g) \in \mathcal{D}$ so that $I(f \wedge h, g \wedge h) \in \mathcal{D}$ and $f \wedge h \equiv g \wedge h$ (Θ). Similarly, $f \vee h \equiv g \vee h$ (Θ). Thus Θ is a congruence relation and we can form the lattice L/Θ, denoted by $\prod_{\mathcal{D}}(L_i \mid i \in I)$, called a *reduced product* of $(L_i \mid i \in I)$. When $\mathcal{D}$ is a prime dual ideal (also called, ultra filter)—briefly, $\mathcal{D}$ is *prime* over I—then $\prod_{\mathcal{D}}(L_i \mid i \in I)$ is called an *ultra product* (also called, *prime product*); see J. Łós [1955]. For a class $\mathbf{K}$ of lattices, $\mathbf{P_U}(\mathbf{K})$ will denote the class of all lattices that are isomorphic to an ultra product of members of $\mathbf{K}$.

We start out by proving some elementary properties of these constructions:

Lemma 6. *Let I be a nonempty set and let $\mathcal{D}$ be a dual ideal of $P(I)$. Let $J \in \mathcal{D}$, $J \neq \varnothing$, and define $\mathcal{E} = \{\, X \cap J \mid X \in \mathcal{D} \,\}$. Then $\mathcal{E}$ is a dual ideal of $P(J)$ and if $\mathcal{D}$ is prime, so is $\mathcal{E}$. Furthermore, for any family $(L_i \mid i \in I)$ of lattices,*

$$\prod_{\mathcal{D}}(L_i \mid i \in I) \cong \prod_{\mathcal{E}}(L_i \mid i \in J).$$

Proof. The statements concerning $\mathcal{E}$ are trivial. Let π denote the homomorphism $f \mapsto f_J$, the restriction of f to J, and let

$$\varphi: \prod(\,L_i \mid i \in I\,) \to \prod_{\mathcal{D}}(L_i \mid i \in I),$$

$$\psi: \prod(\,L_i \mid i \in J\,) \to \prod_{\mathcal{E}}(L_i \mid i \in J)$$

be the natural homomorphisms. $\Theta = \Theta(\mathcal{D})$ is the kernel of φ. Let Φ be the kernel of $\pi\psi$. Then $f \equiv g$ (Φ) iff $I(f_J, g_J) \in \mathcal{E}$, or equivalently, $J \cap I(f,g) \in \mathcal{D}$. In view of $J \in \mathcal{D}$, this is equivalent to $I(f,g) \in \mathcal{D}$. Thus $\Theta = \Phi$ and the isomorphism follows. $\qquad\square$

Corollary 7. *Let I be a nonempty set and let $\mathcal{D}$ be a dual ideal of $P(I)$. If $\mathcal{D}$ is principal, $\mathcal{D} = [J)$, then for any family $(L_i \mid i \in I)$ of lattices,*

$$\prod_{\mathcal{D}}(L_i \mid i \in I) \cong \prod(\,L_i \mid i \in J\,);$$

in particular, if $\mathcal{D}$ is principal and prime, $\mathcal{D} = [\{j\})$, then

$$\prod_{\mathcal{D}}(L_i \mid i \in I) \cong L_j.$$

Lemma 8. *Let $L_0, \ldots, L_{n-1}$ be finite lattices. Let $(L_i \mid i \in I)$, $I \neq \varnothing$, be a family of lattices, each L_i being one of $L_0, \ldots, L_{n-1}$. Let $\mathcal{D}$ be prime over I. Then there is a j, $0 \leq j < n$, such that*

$$\prod_{\mathcal{D}}(L_i \mid i \in I) \cong L_j.$$

Proof. We can assume that the lattices $L_0, \ldots, L_{n-1}$ are pairwise nonisomorphic. Thus if we define

$$I_j = \{\, i \mid i \in I \text{ and } L_i = L_j \,\},$$

for $0 \leq j < n$, then $I_0, \ldots, I_{n-1}$ are pairwise disjoint and $I_0 \cup \cdots \cup I_{n-1} = I$. Since $\mathcal{D}$ is prime, there is exactly one j, with $0 \leq j < n$, such that $I_j \in \mathcal{D}$. Applying Corollary 7 to I_j, we obtain

$$\prod_{\mathcal{D}}(L_i \mid i \in I) \cong \prod_{\mathcal{E}}(L_i \mid i \in I_j),$$

where $\mathcal{E} = \{\, X \cap I_j \mid X \in \mathcal{D} \,\}$ is prime over I_j. For $a \in L_j$, let

$$f_a \in \prod(\,L_i \mid i \in I_j\,)$$

be defined by $f_a(i) = a$, for all $i \in I$. Then, for $a, b \in L_j$, $a \neq b$, we have $I(f_a, f_b) = \varnothing \notin \mathcal{E}$, and therefore $f_a \not\equiv f_b$ $(\Theta(\mathcal{E}))$. Thus

$$\alpha: a \mapsto [f_a]\Theta(\mathcal{E})$$

embeds L_j in $\prod_{\mathcal{E}}(L_i \mid i \in I_j)$. In fact, α is an isomorphism. To see that, let $f \in \prod(L_i \mid i \in I_j)$. Then, for each $a \in L$, we define

$$I_{j,a} = \{\, i \mid i \in I_j \text{ and } f(i) = a \,\}.$$

Since the $I_{j,a}$ are pairwise disjoint and $\bigcup(I_{j,a} \mid a \in L_j) = I_j$, we conclude that there exists exactly one $a \in L_j$ such that $I_{j,a} \in \mathcal{E}$. Now $I(f, f_a) = I_{j,a} \in \mathcal{E}$, thus $f \equiv f_a \ (\Theta(\mathcal{E}))$ and α is onto. Therefore, $\prod_{\mathcal{E}}(L_i \mid i \in I_j) \cong L_j$. $\qquad \square$

Now we are ready to state the formula of B. Jónsson [1967] for **Var**:

Theorem 9.   For a class **K** of lattices,

$$\mathbf{Var}(\mathbf{K}) = \mathbf{P_S HSP_U}(\mathbf{K}).$$

———

Proof. By the Subdirect Product Representation Theorem, it is sufficient to prove that if A is a subdirectly irreducible lattice and $A \in \mathbf{Var}(\mathbf{K})$, then $A \in \mathbf{HSP_U}(\mathbf{K})$. By Corollary 4, $A \in \mathbf{HSP}(\mathbf{K})$, and therefore there exist $A_i \in \mathbf{K}$, $i \in I$, a sublattice B of $\prod(A_i \mid i \in I)$, and a congruence relation Φ on B such that $B/\Phi \cong A$.

We claim that there is a $\mathcal{D}$ prime over I such that $\Theta(\mathcal{D})$ restricted to B is contained in Φ. Indeed, if we have such a $\mathcal{D}$, then by the Second Isomorphism Theorem, B/Φ is a homomorphic image of $B/\Theta(\mathcal{D})_B$, which is, in turn, a sublattice of

$$\prod(A_i \mid i \in I)/\Theta(\mathcal{D}) = \prod_{\mathcal{D}}(A_i \mid i \in I).$$

Thus, $A \cong B/\Phi \in \mathbf{HSP_U}(\mathbf{K})$, as required.

For $J \subseteq I$, set $\Theta_J = \Theta([J))$, that is, for $f, g \in \prod(A_i \mid i \in I)$,

$$f \equiv g \ (\Theta_J) \quad \text{iff} \quad I(f, g) \supseteq J.$$

Observe that $\Theta(\mathcal{D}) = \bigcup(\Theta_J \mid J \in \mathcal{D})$; consequently, to find $\mathcal{D}$, we have to look for $\mathcal{D}$ in

$$\mathcal{E} = \{\, J \mid J \subseteq I \text{ and } (\Theta_J)_B \le \Phi \,\}.$$

$\mathcal{E}$ has the following three properties:

(i) $I \in \mathcal{E}$ and $\varnothing \notin \mathcal{E}$;

(ii) $J_0 \in \mathcal{E}$ and $J_0 \subseteq J_1 \subseteq I$ imply that $J_1 \in \mathcal{E}$;

(iii) $M, N \subseteq I$ and $M \cup N \in \mathcal{E}$ imply that $M \in \mathcal{E}$ or $N \in \mathcal{E}$.

$\Theta_I = \omega$ and $\Theta_\varnothing = \iota$, so (i) is obvious. If $J_0 \subseteq J_1$, then $\Theta_{J_0} \geq \Theta_{J_1}$, proving (ii). To prove (iii), let M and $N \subseteq I$. It is trivial that $\Theta_{M \cup N} = \Theta_M \wedge \Theta_N$ and

$$(\Theta_{M \cup N})_B = (\Theta_M)_B \wedge (\Theta_N)_B.$$

Now let $M \cup N \in \mathcal{E}$. Then $(\Theta_{M \cup N})_B \leq \Phi$, that is, $(\Theta_M)_B \wedge (\Theta_N)_B \leq \Phi$. Since B/Φ is subdirectly irreducible, Φ is meet-irreducible in Con B. So using that Con B is distributive we conclude that $(\Theta_M)_B \leq \Phi$ or $(\Theta_N)_B \leq \Phi$, that is, M or $N \in \mathcal{E}$, proving (iii).

Now let $\mathcal{D}$ be a dual ideal of P(I) maximal with respect to the property $\mathcal{D} \subseteq \mathcal{E}$. We show that $\mathcal{D}$ is prime. By (i), $\varnothing \notin \mathcal{D}$, so $\mathcal{D}$ is proper. If $\mathcal{D}$ is proper but not prime, then there exists a $J \subseteq I$ such that $J \notin \mathcal{D}$ and $I - J \notin \mathcal{D}$. If $J \cap J' \in \mathcal{E}$, for every $J' \in \mathcal{D}$, then, by (ii), $\mathcal{D}$ and J would generate a dual ideal contained in $\mathcal{E}$, contradicting $J \notin \mathcal{D}$ and the maximality of $\mathcal{D}$. Thus there exists a $J_0 \in \mathcal{D}$ such that $J \cap J_0 \notin \mathcal{E}$. Similarly, there exists a $J_1 \in \mathcal{D}$ such that $(I - J) \cap J_1 \notin \mathcal{E}$. Then

$$J_0 \cap J_1 = (J \cap (J_0 \cap J_1)) \cup ((I - J) \cap (J_0 \cap J_1)),$$

contradicting (iii). Thus $\mathcal{D}$ is prime. $\qquad\qquad\qquad\qquad\qquad\qquad\quad\square$

Let $\mathbf{Si}(\mathbf{K})$ be the class of subdirectly irreducible lattices in $\mathbf{K}$. An equivalent form of Theorem 9 is the following:

Corollary 10. *For a class $\mathbf{K}$ of lattices,*

$$\mathbf{Si\,Var(K)} \subseteq \mathbf{HSP_U(K)}.$$

We shall illustrate the power of Theorem 9 with two simple applications taken from B. Jónsson [1967].

If $\mathbf{K}$ is a finite set of finite lattices, then, by Lemma 8, $\mathbf{P_U(K)}$ is, up to isomorphic copies, the same as $\mathbf{K}$, so we conclude:

Corollary 11. *Let $\mathbf{K}$ be a finite set of finite lattices. Then*

$$\mathbf{Si\,Var(K)} \subseteq \mathbf{HS(K)}.$$

Observe that $\mathbf{HS(K)}$ is, up to isomorphic copies, a finite set of finite lattices.

Corollary 12. *Let A and B be finite nonisomorphic subdirectly irreducible lattices. If $|A| \leq |B|$, then there exists an identity ε holding in A but not holding in B.*

Proof. $B \notin \mathbf{HS}(A)$ since $|B| \geq |A|$ and B is not isomorphic to A. Hence, by Corollary 11, $B \notin \mathbf{Var}(A)$ and so some identity holding in A must fail in B. $\quad\square$

Exercises

1. Show that $\mathbf{K} \mapsto \text{Iden}(\mathbf{K})$ and $\Sigma \mapsto \mathbf{Mod}(\Sigma)$ set up a Galois connection.

2. Prove that $\mathbf{K}$ is a variety iff $\mathbf{K} = \mathbf{Mod}(\text{Iden}(\mathbf{K}))$. For a set Σ of identities, $\Sigma = \text{Iden}(\mathbf{K})$, for some class $\mathbf{K}$ of lattices, iff $\Sigma = \text{Iden}(\mathbf{Mod}(\Sigma))$.

3. Characterize sets of identities Σ that are of the form $\text{Iden}(\mathbf{K})$, for some class $\mathbf{K}$ of lattices.

4. Find properties of varieties $\mathbf{K}$ of lattices satisfying

    $$\text{Iden}(\mathbf{K}) = \text{Iden}(F_{\mathbf{K}}(n)),$$

    for some integer n.

5. Let L be a lattice and let U be a generating set of L. Show that if L is free over $\mathbf{K}$ with U as a free generating set, then the same holds over $\mathbf{Var}(\mathbf{K})$.

6. Reprove Theorem 1(ii) using Exercise 5 and Theorem 3.

7. Prove that the elements of a free generating set are doubly irreducible (B. Jónsson [1971].)

8. Let $\mathbf{K}$ be a variety of lattices. Let φ be the natural homomorphism of $F(\aleph_0)$ onto $F_{\mathbf{K}}(\aleph_0)$ and let Θ be the kernel of φ. We assign to Θ a variety $\mathbf{K}_1$ over which $F(\aleph_0)/\Theta$ is free. Prove that $\mathbf{K} = \mathbf{K}_1$.

9. Let Θ be a fully invariant congruence relation of $F(\aleph_0)$. We assign to Θ a variety $\mathbf{K}$ and to $\mathbf{K}$ a fully invariant congruence relation Θ_1 of $F(\aleph_0)$ as in the proof of Theorem 2. Prove that $\Theta = \Theta_1$.

10. Let L be a lattice. Show that the fully invariant congruence relations of L form a complete sublattice of $\text{Con}\, L$.

11. Prove that the lattice of fully invariant congruence relations of a lattice L is a distributive algebraic lattice. Characterize the compact elements.

12. A congruence relation Θ of a lattice L is called *invariant* iff $a \equiv b\ (\Theta)$ implies that $a\alpha \equiv b\alpha\ (\Theta)$, for any automorphism α of L. Do the invariant congruences form a lattice, if so, is this lattice distributive or algebraic?

13. Let $\langle A; F \rangle$ and $\langle A; G \rangle$ be algebras and $F \subseteq G$. Show that the congruence lattice of $\langle A; G \rangle$ is a complete sublattice of the congruence lattice of $\langle A; F \rangle$.

14. Derive Exercises 10–12 from Exercise 13.

15. A variety $\mathbf{K}$ of lattices is *generated by a lattice* A iff $\mathbf{K} = \mathbf{Var}(A)$. Show that every variety of lattices is generated by a lattice.

16. A variety $\mathbf{K}$ of lattices is *locally finite* iff every finitely generated member of $\mathbf{K}$ is finite. Prove that a variety generated by a finite lattice is locally finite. (Do not use Theorem 9 or any of its consequences. This result is true even in classes of algebras for which Theorem 9 fails.)

17. Is the converse of the statement of Exercise 16 true?

18. Prove that $\mathbf{P_S H(K)} \subseteq \mathbf{HP_S(K)}$, for any class $\mathbf{K}$ of lattices.

19. For a class $\mathbf{K}$ of lattices, let $\mathbf{P_R(K)}$ denote the class of reduced products (up to isomorphism) of members of $\mathbf{K}$. Prove that $\mathbf{P_R(K)} \subseteq \mathbf{P_S P_U(K)}$ (G. Grätzer and H. Lakser [1973]).

An *implication* is a sentence (all variables are universally quantified):

$$p_0(x_0, \dots, x_{n-1}) = q_0(x_0, \dots, x_{n-1})$$
$$\text{and } p_1(x_0, \dots, x_{n-1}) = q_1(x_0, \dots, x_{n-1})$$

$$\dots$$

$$\text{and } p_{m-1}(x_0, \dots, x_{n-1}) = q_{m-1}(x_0, \dots, x_{n-1})$$
$$\text{imply that}$$
$$p_m(x_0, \dots, x_{n-1}) = q_m(x_0, \dots, x_{n-1}).$$

For instance, the semidistributive law of Chapter VI is an implication: $x \wedge y = x \wedge z$ implies that $x \wedge y = x \wedge (y \vee z)$. Any identity is an implication.

An *implicational class* is the class $\mathbf{K}$ of all lattices satisfying a set of implications. For a class $\mathbf{K}$, the class of all isomorphic copies of members of $\mathbf{K}$ is denoted $\mathbf{I(K)}$.

20. Prove that $\mathbf{ISP_R(K)}$ is the smallest implicational class containing $\mathbf{K}$ (A. I. Mal'cev [1966] and G. Grätzer and H. Lakser [1973]).

21. Let $\mathcal{D}$ be prime over I. An *ultra power* L_D^I of a lattice L is an ultra product $\prod_{\mathcal{D}}(L_i \mid i \in I)$ such that $L_i = L$, for all $i \in I$. Prove that L can be embedded into L_D^I.

*22. Prove that every lattice L can be embedded into an ultra product of all finitely generated sublattices of L. (Hint: let I be the set of all nonempty finite subsets of L. For $J \in I$, let L_J be the sublattice generated by J. Choose a $\mathcal{D}$ prime over I such that, for each $J \in I$, $\{K \mid K \in I \text{ and } K \supseteq J\} \in \mathcal{D}$.)

Let us define (first-order) formulas:

(i) for the n-ary polynomials p and q, $p = q$ is a formula;

(ii) if Φ is a formula, so is $\neg \Phi$ (read: not Φ);

(iii) if Φ_0 and Φ_1 are formulas, then so is $\Phi_0 \vee \Phi_1$ (read: Φ_0 or Φ_1);

(iv) if Φ is a formula, so is $(\exists x_k)\Phi$ (read: there exists an x_k such that Φ).

Then $\Phi_0 \wedge \Phi_1$ (read: Φ_0 and Φ_1) is introduced as $\neg((\neg\Phi_0) \vee (\neg\Phi_1))$, $(x_k)\Phi$ (read: for all x_k, Φ) is introduced as $\neg((\exists x_k)\neg\Phi)$, and so on.

*23. Define inductively what it means for a formula Φ to be satisfied by certain elements of a lattice L.

24. Define precisely a *free variable* x which is not "bound" by a quantifier $\exists x$. A *sentence* is a formula without free variables. Find sentences expressing that the lattice L has at most or exactly n elements.

25. For a finite lattice K, construct a sentence Φ such that Φ holds for a lattice L iff L has a sublattice isomorphic to K.

*26. Prove that a sentence Φ holds for an ultra product $\prod_{\mathcal{D}}(L_i \mid i \in I)$ iff $\{ i \mid \Phi \text{ holds in } L_i \} \in \mathcal{D}$ (J. Łoś [1955]).

27. Prove Lemma 8 using Exercise 26.

28. A *model* of a set Σ of sentences is a lattice L in which all sentences $\Phi \in \Sigma$ are satisfied. Prove the *Compactness Theorem*: Let Σ be a set of sentences. If every finite subset of Σ has a model, then Σ has a model. (Hint: for every nonempty finite $\Omega \subseteq \Sigma$, choose a model L_Ω. Let I be the set of all finite nonempty subsets of Σ and choose a $\mathcal{D}$ prime over I containing all sets of the form $\{ X \mid X \in I \text{ and } X \supseteq J \}$, where $J \in I$. Then $L = \prod_{\mathcal{D}}(L_\Omega \mid \Omega \in I)$ is a model of Σ.)

29. Let Σ and Σ_1 be sets of sentences. Σ *implies* Σ_1 iff every model of Σ is also a model of Σ_1. Σ *is equivalent to* Σ_1 iff Σ implies Σ_1 and Σ_1 implies Σ. Prove that if Σ is equivalent to $\{\Phi\}$, then there is a finite $\Sigma_1 \subseteq \Sigma$ that is equivalent to $\{\Phi\}$. (Use the Compactness Theorem.)

30. Call a lattice L *finitely subdirectly irreducible* iff ω is meet-irreducible in $\mathrm{Con}\, L$. Prove that Corollary 10 holds also for the finitely subdirectly irreducible members of $\mathbf{Var}(\mathbf{K})$.

31. Let M_4 be the subspace lattice of a projective line with four points. Show that $\mathbf{M}_4 = \mathbf{Var}(M_4) \supset \mathbf{M}_3 = \mathbf{Var}(M_3)$ and $\mathbf{M}_4 \supset \mathbf{K} \supseteq \mathbf{M}_3$ implies that $\mathbf{K} = \mathbf{M}_3$, for every variety $\mathbf{K}$.

32. Find a variety $\mathbf{N}$ of lattices such that $\mathbf{N} \supset \mathbf{N}_5 = \mathbf{Var}(N_5)$, $\mathbf{N} \not\supseteq \mathbf{M}_3$ and $\mathbf{N} \supset \mathbf{K} \supseteq \mathbf{N}_5$ implies that $\mathbf{K} = \mathbf{N}_5$, for any variety $\mathbf{K}$.

33. Show that $\mathbf{Var}(\mathbf{K}) = \mathbf{SHPS}(\mathbf{K})$, if $\mathbf{K}$ is one of the classes of lattices listed below:

(i) $\mathbf{K} = \{L\}$, where L is a finite lattice;

(ii) $\mathbf{K} = \{M_{\aleph_0}\}$, where $M_{\aleph_0}$ is the subspace lattice of a projective line with countably many points;

(iii) for each prime number p, we choose a field F_p of characteristic p and $\mathbf{K} = \{\, L_p \mid p \text{ is a prime number}\,\}$, where L_p is the lattice of subspaces of the projective plane coordinatized by F_p.

34. Let $\mathbf{K}$ be a variety of lattices. Prove that

$$F_{\mathbf{K}}(\mathfrak{m}) \in \mathbf{ISP}(F_{\mathbf{K}}(\aleph_0)),$$

for any cardinal $\mathfrak{m}$.

35. Prove the Compactness Theorem for any type of universal algebras. Find applications of this result to lattices that go beyond the Compactness Theorem of Exercise 28. (Hint: use a type with $\wedge$ and $\vee$ and infinitely many constants.)

2.   The Lattice of Varieties of Lattices

Let $\mathbf{K}_0$ and $\mathbf{K}_1$ be varieties of lattices. Then $\mathbf{K}_0 \cap \mathbf{K}_1$ is again a variety and

$$\mathbf{K}_0 \cap \mathbf{K}_1 = \mathbf{Mod}(\mathrm{Iden}(\mathbf{K}_0) \cup \mathrm{Iden}(\mathbf{K}_1)).$$

There is also a smallest variety $\mathbf{K}_0 \vee \mathbf{K}_1$ containing both $\mathbf{K}_0$ and $\mathbf{K}_1$, namely,

$$\mathbf{K}_0 \vee \mathbf{K}_1 = \mathbf{Mod}(\mathrm{Iden}(\mathbf{K}_0) \cap \mathrm{Iden}(\mathbf{K}_1)).$$

All varieties of lattices form a lattice with the lattice operations $\mathbf{K}_0 \wedge \mathbf{K}_1 = \mathbf{K}_0 \cap \mathbf{K}_1$ and $\mathbf{K}_0 \vee \mathbf{K}_1$. (Axiomatic set theory does not permit the formation of a set whose elements are classes. Thus, formally, one cannot form the lattice of varieties. It is easy to get around this difficulty. For instance, replace a variety $\mathbf{K}$ by $\mathrm{Iden}(\mathbf{K})$, which is a subset of the countable set of all lattice identities. Then we form the lattice of all subsets of the form $\mathrm{Iden}(\mathbf{K})$ of the set of all lattice identities.)

The zero of this lattice is $\mathbf{T}$ (the trivial variety) consisting of all one-element lattices. Let $\mathbf{K}$ be a variety of lattices different from $\mathbf{T}$. Then there is a lattice L in $\mathbf{K}$ with $|L| > 1$. Therefore L has C_2 as a sublattice and so $C_2 \in \mathbf{K}$. By Corollary II.1.20, $\mathbf{Var}(C_2) = \mathbf{D}$, the class of all distributive lattices. Thus $\mathbf{K} \supseteq \mathbf{D}$. We have just verified that $\mathbf{D}$ is the only atom of the lattice of varieties of lattices and every nonzero member contains $\mathbf{D}$.

Now let $\mathbf{K}$ be a variety of lattices properly containing $\mathbf{D}$. Then there is a nondistributive lattice L in $\mathbf{K}$. By Theorem II.1.1, N_5 or M_3 is a sublattice of L, hence N_5 or $M_3 \in \mathbf{K}$. Set

$$\mathbf{N}_5 = \mathbf{Var}(N_5),$$
$$\mathbf{M}_3 = \mathbf{Var}(M_3).$$

Lemma 3. *The collection of varieties of lattices that can be generated by a single finite lattice is an ideal of the lattice of varieties of lattices. This ideal contains only elements of finite height.*

Proof. If the variety $\mathbf{K}$ is generated by a finite lattice L, then, by Corollary 1.11, up to isomorphism, $\mathbf{Si}(\mathbf{K})$ is a finite set of finite lattices. Thus if $\mathbf{K}_0 \subset \mathbf{K}$, then, up to isomorphism, $\mathbf{Si}(\mathbf{K}_0)$ must be a subset of this finite set, hence there are only finitely many such $\mathbf{K}_0$. All the statements of this corollary now follow immediately. $\qquad\square$

From the observations made above, it should be clear that the join-irreducible elements of the lattice of varieties of lattices are connected with varieties generated by a single subdirectly irreducible lattice. The following result states a number of connections of this sort (R. N. McKenzie [1972]).

Theorem 4.   Let $\mathbf{K}$ be an element of the lattice of varieties of lattices.

(i) If $[\mathbf{K})$ is a *completely prime dual ideal* (that is, $\bigvee(\mathbf{K}_i \mid i \in I) \in [\mathbf{K})$ implies that $\mathbf{K}_i \in [\mathbf{K})$, for some $i \in I$), then $\mathbf{K}$ can be generated by a finite subdirectly irreducible lattice.

(ii) If $\mathbf{K}$ can be generated by a finite subdirectly irreducible lattice, then $\mathbf{K}$ is completely join-irreducible.

(iii) If $\mathbf{K}$ is completely join-irreducible, then $\mathbf{K}$ can be generated by a subdirectly irreducible lattice.

(iv) If $\mathbf{K}$ can be generated by a subdirectly irreducible lattice, then $\mathbf{K}$ is join-irreducible. $\qquad$ —

Proof.
(i) Let $\mathbf{K}_n$ denote the variety generated by the partition lattice on an n-element set. By Corollary IV.4.6,
$$\bigvee(\mathbf{K}_n \mid n = 1, 2, 3, \dots) = \mathbf{L} \supseteq \mathbf{K};$$
since $[\mathbf{K})$ is completely prime, $\mathbf{K}_n \subseteq \mathbf{K}$, for some integer n. Thus, by Lemma 3, $\mathbf{K}$ can be generated by finitely many finite subdirectly irreducible lattices. Since $\mathbf{K}$ is join-irreducible, $\mathbf{K}$ can be generated by a single finite subdirectly irreducible lattice.
(ii) If $\mathbf{K}$ is generated by the finite subdirectly irreducible lattice L and $\mathbf{K} = \mathbf{K}_0 \vee \mathbf{K}_1$, then by Theorem 2, $L \in \mathbf{Si}(\mathbf{K}_0)$ or $L \in \mathbf{Si}(\mathbf{K}_1)$ implying that $\mathbf{K} = \mathbf{K}_0$ or $\mathbf{K} = \mathbf{K}_1$. Thus $\mathbf{K}$ is join-irreducible. By Lemma 3, $\mathbf{K}$ is of finite height, hence $\mathbf{K}$ is completely join-irreducible.
(iii) Any variety $\mathbf{K}$ is of the form
$$\bigvee(\mathbf{K}_0 \mid \mathbf{K}_0 \subseteq \mathbf{K} \text{ and } \mathbf{K}_0 \text{ is generated by a subdirectly irreducible lattice}).$$

Thus if $\mathbf{K}$ is completely join-irreducible, $\mathbf{K}$ can be generated by a subdirectly irreducible lattice.

(iv) We proceed as in (ii), by reference to Theorem 2. $\qquad\square$

The converse statements of (i)–(iv) all fail, see R. N. McKenzie [1972] and the Exercises.

Figure 1 can also be used to illustrate the very important concept of *splitting* due to R. N. McKenzie [1972]. A pair of varieties $\langle \mathbf{K}_0, \mathbf{K}_1 \rangle$ is said to be *splitting* iff, for every variety $\mathbf{K}_2$, either $\mathbf{K}_2 \subseteq \mathbf{K}_0$ or $\mathbf{K}_1 \subseteq \mathbf{K}_2$. For instance, $\langle \mathbf{M}, \mathbf{N}_5 \rangle$ is a splitting pair. ($\langle \mathbf{L}, \mathbf{T} \rangle$ is a trivial splitting pair.) Equivalently, $(\mathbf{K}_0]$ and $[\mathbf{K}_1)$ are prime, $\mathbf{K}_0 \not\supseteq \mathbf{K}_1$, and $(\mathbf{K}_0] \cup [\mathbf{K}_1) = (\mathbf{L}]$. Obviously, $\mathbf{K}_0$ determines $\mathbf{K}_1$, and conversely. Since $[\mathbf{K}_1)$ is a completely prime dual ideal, $\mathbf{K}_1$ can be generated by a finite subdirectly irreducible lattice. Finite subdirectly irreducible lattices that arise this way are called *splitting lattices* and they are characterized in R. N. McKenzie [1972].

How big is the lattice of varieties of lattices? Since there are only $\aleph_0$ identities, there are at most $2^{\aleph_0}$ varieties. Now we shall show, by construction, that there are exactly $2^{\aleph_0}$ varieties of lattices.

Let Π be the set of prime numbers and, for $p \in \Pi$, let L_p be the subspace lattice of the projective plane coordinatized by the p-element field (that is, the Galois field, $\mathrm{GF}(p)$). For a subset $S \subseteq \Pi$, set

$$\mathbf{K}(S) = \mathbf{Var}(\{\, L_i \mid i \in S \,\}).$$

We claim that S can be recovered from $\mathbf{K}(S)$, in fact,

$$L_p \in \mathbf{K}(S) \quad \text{iff} \quad p \in S.$$

Obviously, if $p \in S$, then $L_p \in \mathbf{K}(S)$. Now let $L_p \in \mathbf{K}(S)$ and $p \notin S$. Since L_p is subdirectly irreducible, we can apply Corollary 1.10:

$$L_p \in \mathbf{HSP_U}(\{\, L_i \mid i \in S \,\}),$$

that is, $L_p \in \mathbf{HS}(L)$, where $L = \prod_{\mathcal{D}}(L_i \mid i \in I)$, $\mathcal{D}$ is prime over I, and each L_i, $i \in I$, is an L_j with $j \in S$. Each L_i is a complemented modular lattice of length three, in which any two atoms are perspective. Since these properties can be expressed by (first-order) sentences, by Exercise 1.26, L has the same property and so, by the results of Section IV.5, L is the subspace lattice of a nondegenerate projective plane. Each L_i satisfies Desargues' Theorem and, by Theorem IV.5.8, this property can be expressed by a sentence. Thus L satisfies Desargues' Theorem and, by the Coordinatization Theorem, L can be coordinatized by a division ring $\mathcal{D}$. We assumed that $p \notin S$, so each L_i, $i \in I$, is coordinatized by a division ring not of characteristic p. This again can be expressed by a formula, hence $\mathcal{D}$ is not of characteristic p.

$L_p \in \mathbf{HS}(L)$, that is, L has a sublattice L' such that L_p is a homomorphic image of L'. Since both L and L_p are of length three and a proper homomorphic image of a modular lattice of length three is of length less than three, we conclude that we can assume that $L_p = L'$, that is, L_p is a sublattice of L. This is a clear contradiction: we can introduce $x + y$ for points x and y in L_p using only elements of L_p; thus for points x and y of L_p, $x + y$ in L_p is the same as $x + y$ in L. But in L_p we have $p \cdot x = 0$, while in L, $p \cdot x \neq 0$, if $x \neq 0$, which is impossible.

Thus there are at least as many varieties of lattices as there are subsets of Π, which number $2^{\aleph_0}$.

Theorem 5.   There are $2^{\aleph_0}$ varieties of (modular) lattices.   —

This result was proved by K. A. Baker [1969a] (on whose example the above discussion was based), R. N. McKenzie [1970] (without modularity), and R. Wille [1972].

The following method of constructing varieties of lattices is due to R. Wille [1972].

Let $\mathcal{P}$ be a set of finite posets. We denote by **Var**$(\mathcal{P})$ the class of all lattices that do *not* contain an isomorphic copy of a member of $\mathcal{P}$ as a primitive subset (see Definition III.1.12).

Theorem 6 (R. Wille [1972]).   For any set of finite posets $\mathcal{P}$, **Var**$(\mathcal{P})$ is a variety of lattices.   —

Proof. This is clear from Exercises III.1.22 and III.1.23.   □

The reader should verify that the construction in the proof of Theorem 5 could be verified using Theorem 6.

Exercises

1. Consider the lattices of Figures 3–11 and their duals (fifteen in all). Show that each one generates a variety covering $\mathbf{N}_5$ (R. N. McKenzie [1972]).

2. Let A be a fixed countable set. For a variety $\mathbf{K}$, let $\mathbf{Si}_A(\mathbf{K})$ denote the set of all subdirectly irreducible lattices L in $\mathbf{K}$ satisfying $L \subseteq A$. Prove that $\mathbf{K} \mapsto \mathbf{Si}_A(\mathbf{K})$ is a set representation of the lattice of all varieties of lattices. (Hint: $F_{\mathbf{K}}(\aleph_0)$ can be recovered from $\mathbf{Si}_A(\mathbf{K})$.)

3. Does the representation of Exercise 2 preserve infinite meets and joins of varieties?

4. Prove that a variety $\mathbf{K}$ of lattices can be generated by finitely many finite lattices iff $\mathbf{K}$ can be generated by a single finite lattice, which, in turn, is equivalent to the statement that all subdirectly irreducible lattices in $\mathbf{K}$ are finite and there are only finitely many nonisomorphic subdirectly irreducible lattices.

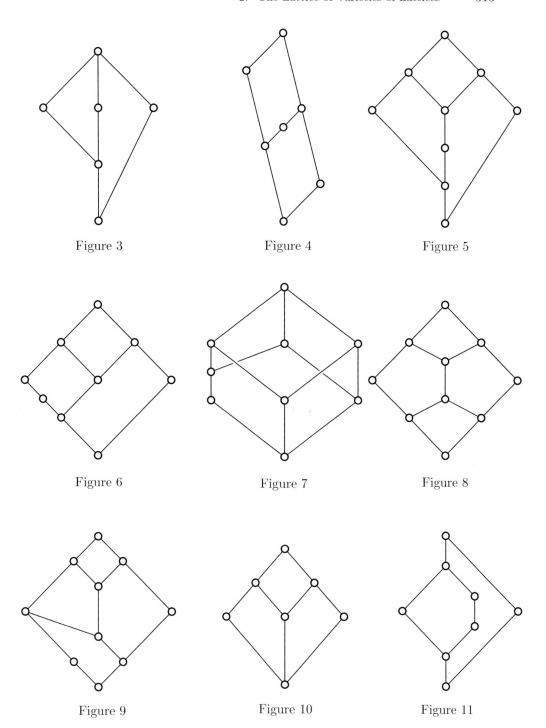

Figure 3

Figure 4

Figure 5

Figure 6

Figure 7

Figure 8

Figure 9

Figure 10

Figure 11

5. Show that **L** can be generated by a subdirectly irreducible lattice and **L** is completely join-reducible.

6. Prove that the variety generated by the lattice of Figure 12 is join-irreducible but it cannot be generated by a single subdirectly irreducible lattice (R. N. McKenzie [1972]).

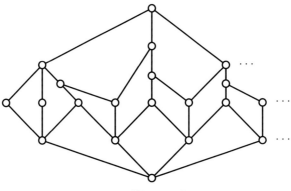

Figure 12

7. Let $\mathbf{K}_0$ and $\mathbf{K}_1$ be the varieties of lattices generated by all projective planes satisfying Desargues' Theorem and Pappus' Theorem, respectively. Prove that $\mathbf{K}_0$ and $\mathbf{K}_1$ have the same finite members (K. A. Baker [1969a]).

8. Find a variety of lattices that is not generated by its finite members (K. A. Baker [1969a] and R. Wille [1972]).

9. Find a sublattice of the lattice of varieties of lattices isomorphic to the lattice of all subsets of a countable set (K. A. Baker[1969a]).

10. Let **K** be a variety of lattices. If $\mathbf{K} \neq \mathbf{L}$, then $\mathbf{K} \prec \mathbf{K}_0$, for some variety $\mathbf{K}_0$ (B. Jónsson [1967]). (Hint: Consider $\mathbf{K} \vee \mathbf{Var}(L)$, where L is a finite lattice not in **K**.)

11. Prove that **L** is join-irreducible (B. Jónsson [1967]). (Hint: If $\mathbf{K} \subset \mathbf{L}$, then the partition lattice on an infinite set does not belong to **K**.)

12. There is no variety of lattices covered by **L** (B. Jónsson [1967]).

13. Show that every proper interval of the lattice of varieties of lattices contains a prime interval.

14. Let $\mathbf{K}_0$ and $\mathbf{K}_1$ be varieties of lattices. If $\mathbf{K}_0 \subset \mathbf{K}_1$ and $\mathbf{K}_0$ can be defined by finitely many identities, then there exists a variety **K** satisfying $\mathbf{K}_0 \prec \mathbf{K} \subseteq \mathbf{K}_1$. (Hint: use Theorems 1 and II.3.13.)

15. Prove that, in Exercise 14, we cannot require that $\mathbf{K}_0 \subseteq \mathbf{K} \prec \mathbf{K}_1$.

16. Prove that $\mathbf{Si}(\mathbf{Var}(\mathcal{P}))$ consists of those subdirectly irreducible lattices in which no $P \in \mathcal{P}$ can be embedded.

17. Let $\mathcal{P} = \{C_5\}$. Describe $\mathbf{Var}(\mathcal{P})$.

18. A lattice L is said to be of *breadth* n iff n is the smallest integer with the property that for every finite $X \subseteq L$ there exists a $Y \subseteq X$ such that $|Y| \leq n$ and $\bigvee X = \bigvee Y$. Choose a finite set of finite posets $\mathcal{P}$ such that $\mathbf{Var}(\mathcal{P})$ is generated by all subdirectly irreducible lattices of breadth at most n.

19. Let $\mathcal{P}$ consist of the n-element unordered poset. Describe $\mathbf{Var}(\mathcal{P})$ using the concept of width.

*20. Prove that $\mathbf{N}_5$ contains all lattices of width two (O. T. Nelson, Jr. [1968]).

3. Finding Equational Bases

An *equational basis* Σ of a class $\mathbf{K}$ is a set of identities such that

$$\mathbf{Var}(\mathbf{K}) = \mathbf{Mod}(\Sigma).$$

This Σ is of special interest if it is *irredundant* (that is, $\mathbf{Var}(\mathbf{K}) \neq \mathbf{Mod}(\Sigma_1)$, for any $\Sigma_1 \subset \Sigma$) or if it is *finite*. There is not much one can say about the problem of finding equational bases in general. However, if $\mathbf{K}$ has some special properties, then there is hope for some meaningful results.

A *universal disjunction of equations*, or briefly UDE, is a sentence of the form (recall, $\vee$ stands for "or ")

$$(x_0) \cdots (x_{n-1})(f_0(x_0, \ldots, x_{n-1}) = g_0(x_0, \ldots, x_{n-1})$$
$$\vee f_1(x_0, \ldots, x_{n-1}) = g_1(x_0, \ldots, x_{n-1})$$
$$\cdots$$
$$\vee f_{m-1}(x_0, \ldots, x_{n-1}) = g_{m-1}(x_0, \ldots, x_{n-1})),$$

where $f_0, \ldots, f_{m-1}$ and $g_0, \ldots, g_{m-1}$ are polynomials. In the sequel, we shall omit the quantifiers and we shall assume that $f_i \leq g_i$ holds in any lattice. Examples abound: every identity is a UDE. The following lemma yields examples of a different kind. (In the displayed formula, $\bigvee$ stands for the disjunction of the terms.)

Lemma 1. *Let $P = \{a_0, \ldots, a_{n-1}\}$ be a poset. Set $g_i = \bigvee(x_j \mid a_j \leq a_i)$ and*

$$\Phi(P): \bigvee(g_i = g_i \vee g_k \mid a_k \nleq a_i).$$

Then $\Phi(P)$ is a UDE which holds for a lattice L iff L has no subset isomorphic (as a poset) with P.

Proof. If $Q = \{b_0, \dots, b_{n-1}\} \subseteq L$ is isomorphic with P and $b_i \mapsto a_i$, $0 \le i < n$, is an isomorphism, then setting $x_i = b_i$, $0 \le i < n$, we find that $g_i(b_0, \dots, b_{n-1}) = b_i$. Thus if $a_k \not\le a_i$, then $g_k \not\le g_i$, and so $g_i \ne g_i \vee g_k$. Since all the terms of $\Phi(P)$ fail, $\Phi(P)$ itself fails in L.

Conversely, if $\Phi(P)$ fails in L, then there are elements $b_0, \dots, b_{n-1} \in L$ such that $g_k(b_0, \dots, b_{n-1}) \not\le g_i(b_0, \dots, b_{n-1})$, whenever $a_k \not\le a_i$. Since $a_k \le a_i$ obviously implies that $g_k(b_0, \dots, b_{n-1}) \le g_i(b_0, \dots, b_{n-1})$ by the definition of g_i, we conclude that $a_i \mapsto g_i(b_0, \dots, b_{n-1})$ is an isomorphism of P with

$$\{g_0(b_0, \dots, b_{n-1}), \dots, g_{n-1}(b_0, \dots, b_{n-1})\}. \qquad \square$$

Another important UDE is

$$\Phi(n): \bigvee (x_i = x_j \mid 0 \le i < j \le n)$$

which holds for L iff $|L| \le n$. (To follow the inequality convention introduced above, we should replace "$x_i = x_j$" by "$x_i \wedge x_j = x_i \vee x_j$".)

Now let us take an arbitrary UDE:

$$\Phi: \bigvee (f_i(x_0, \dots, x_{n-1}) = g_i(x_0, \dots, x_{n-1}) \mid 0 \le i < m)$$

and consider the following statement:

$S(\Phi)$: for any integer k, the polynomial

$$
\begin{aligned}
r_k = p_{2m}(x, &p_k(g_0(x_0, \dots, x_{n-1}), \quad y_0^0, \dots, \quad y_{k-1}^0), \\
&p_k(f_0(x_0, \dots, x_{n-1}), \quad y_0^0, \dots, \quad y_{k-1}^0), \\
&\cdots \\
&p_k(g_{m-1}(x_0, \dots, x_{n-1}), y_0^{m-1}, \dots, y_{k-1}^{m-1}), \\
&p_k(f_{m-1}(x_0, \dots, x_{n-1}), y_0^{m-1}, \dots, y_{k-1}^{m-1}))
\end{aligned}
$$

does not depend on x, where p_k is the polynomial introduced in Lemma III.1.1. In r_k the variables are x, x_0, $\dots$, x_{n-1} and y_j^i, $0 \le i < m$, $0 \le j < k$.

We claim that $S(\Phi)$ holds in a subdirectly irreducible lattice L iff Φ does.

Let Φ hold in L. Substituting $x_i = a_i$, for $0 \le i < n$, we obtain the elements $f_i(a_0, \dots, a_{n-1})$ and $g_i(a_0, \dots, a_{n-1})$, for $0 \le i < m$. By Φ, $f_i(a_0, \dots, a_{n-1}) = g_i(a_0, \dots, a_{n-1})$, for some i. Then two successive elements substituted in p_{2m} agree and so, by a trivial property of p_{2m} (see Exercise III.1.8), r_k does not depend on x, verifying $S(\Phi)$.

Now let Φ fail in L. Then there exist $a_0, \dots, a_{n-1} \in L$ such that

$$f_i(a_0, \dots, a_{n-1}) < g_i(a_0, \dots, a_{n-1}),$$

for $0 \le i < m$. Since L is subdirectly irreducible, by Corollary III.1.11, there are $a, b \in L$ satisfying $b < a$ and $a/b \approx_w g_i(a_0, \dots, a_{n-1})/f_i(a_0, \dots, a_{n-1})$. Thus for a suitable integer k and elements $c_0^i, \ldots, c_{k-1}^i \in L$ $(0 \le i < m)$,

$$p_k(g_i(a_0, \ldots, a_{n-1}), c_0^i, \ldots, c_{k-1}^i) = a,$$
$$p_k(f_i(a_0, \ldots, a_{n-1}), c_0^i, \ldots, c_{k-1}^i) = b.$$

Hence r_k, with $x_i = a_i$ $(0 \le i < n)$ and $y_j^i = c_j^i$ $(0 \le i < m$ and $0 \le j < k)$, takes the form

$$r_k = p_{2k}(x, a, b, \ldots, a, b).$$

But it is evident that $r_k = x$ on the interval $[b, a]$ (see Exercise III.1.9) and so r_k is dependent on x. Thus $S(\Phi)$ fails in L.

The statement that r_k does not depend on x is equivalent to the identity:

$$\varepsilon_k\colon r_k(x, \ldots) = r_k(z, \ldots)$$

where z is a variable distinct from x, x_i, and y_j^i. Set $\Sigma(\Phi) = \{\, \varepsilon_k \mid k = 1, 2, \ldots \,\}$. Now we are ready to state a solution to our problem:

Theorem 2 (K. A. Baker [1971]).   Let Φ be a UDE. Then $\Sigma(\Phi)$ is an equational basis for $\mathbf{Mod}(\Phi)$. For a set Ω of UDE-s, $\bigcup(\Sigma(\Phi) \mid \Phi \in \Omega)$ is an equational basis for $\mathbf{Mod}(\Omega)$.   —

Proof. Let $\mathbf{K}_0$ be the variety defined by $\bigcup(\Sigma(\Phi) \mid \Phi \in \Omega)$ and let $\mathbf{K}_1$ be the variety generated by $\mathbf{Mod}(\Omega)$. If $L \in \mathbf{Si}(\mathbf{K}_0)$, then, by the discussion above, $L \in \mathbf{Mod}(\Omega) \subseteq \mathbf{K}_1$. Hence $\mathbf{K}_0 \subseteq \mathbf{K}_1$. Conversely, if $L \in \mathbf{Si}(\mathbf{K}_1)$, then by Corollary 1.10, $L \in \mathbf{HSP}_{\mathbf{U}}(\mathbf{Mod}(\Omega))$. But Ω is preserved under ultra products (Exercise 1.26) and, obviously, by the formation of sublattices and homomorphic images. Hence $L \in \mathbf{Mod}(\Omega)$ and so, by the discussion above,

$$L \in \mathbf{Mod}(\bigcup(\Sigma(\Phi) \mid \Phi \in \Omega)) = \mathbf{K}_0.$$

Thus $\mathbf{K}_1 \subseteq \mathbf{K}_0$ proving that $\mathbf{K}_0 = \mathbf{K}_1$.   □

We can recast this proof using the following concept: a set or sequence of quotients of a lattice L is $(k$-$)bounded$ iff there is a proper quotient weakly projective (in k steps) into each quotient. Corollary III.1.11 and Exercise III.1.21 (slightly sharpened) then read:

Lemma 3.

(i) *Any finite set of proper quotients is bounded in a subdirectly irreducible lattice.*

(ii) *Let φ be a homomorphism of the lattice L onto L', let*

$$a_0/b_0, \quad \ldots, \quad a_{n-1}/b_{n-1}$$

be quotients of L, and let

$$a_0\varphi/b_0\varphi, \quad \ldots, \quad a_{n-1}\varphi/b_{n-1}\varphi$$

be k-bounded. Then

$$a_1/b_1, \quad \ldots, \quad a_n/b_n$$

is $(k+2)$-bounded.

Now for a UDE $\Phi\colon \bigvee f_i = g_i$, let $\alpha_k(\Phi) = \alpha_k$ denote the sentence stating that $g_0/f_0, \ldots, g_{m-1}/f_{m-1}$ is not k-bounded, for any substitution. We can write this simply by requiring that if $u/v \overset{k}{\approx}_w g_i/f_i$, for all i, then $u = v$. Then observe:

Lemma 4.

   (i) *Φ implies α_k and α_k implies α_t, for any $t \leq k$.*

   (ii) *If all α_k, $k = 1, 2, \ldots$, hold in a subdirectly irreducible lattice, then so does Φ.*

   (iii) *α_k is preserved under the formation of direct products and sublattices.*

   (iv) *For any set Ω of UDE-s,*

$$\mathbf{Var}(\mathbf{Mod}(\Omega)) = \mathbf{Mod}(\{\, \alpha_k(\Phi) \mid \Phi \in \Omega \text{ and } k = 1, 2, \ldots \}).$$

Proof.
   (i) and (iii) are trivial.
   (ii) restates Lemma 3(i).
   (iv) follows from Corollary 1.10 and (iii).   □

The reader should have no difficulty relating the sentence α_k to the identity ε_k.

We can use these ideas to give a simple proof of the following result:

Theorem 5 (R. N. McKenzie [1970]).   Any finite lattice has a finite equational basis.   —

Remark. The present proof is due to C. Herrmann [1973], which is based on K. A. Baker [1977].

Let L have n elements. If Σ is a finite equational basis for $\mathbf{Mod}(\Phi(n))$ (the class of at most n element lattices), then we can easily find a finite equational basis for L: let A be a finite set of (up to isomorphism) all finite lattices N satisfying $N \notin \mathbf{Var}(L)$ and $|N| \leq n$; for each $N \in A$, choose an identity ε_N

holding in L but not in N; then $\Sigma \cup \{\, \varepsilon_N \mid N \in A \,\}$ is a finite equational basis for L.

Any lattice, in the class $\mathbf{K} = \mathbf{Mod}(\Phi(n))$, has two properties: (i) it is defined by an at most n^2-termed UDE; (ii) every bounded set of at most n^2 quotients is n^2-bounded (because there are at most n^2 quotients in the lattice). Thus the following lemma completes the proof of Theorem 5:

Lemma 6. *Let $\mathbf{K}$ be a class of lattices and let m be an integer with the following two properties:*

(i) $\mathbf{K} = \mathbf{Mod}(\Sigma)$, *where Σ is a finite set of at most m-termed UDE-s.*

(ii) *There is an integer r such that in every subdirectly irreducible lattice $L \in \mathbf{K}$ every bounded set of m quotients is r-bounded.*

Then $\mathbf{K}$ has a finite equational basis.

Proof. Let $\varrho_{m,r}$ denote the sentence expressing that any set of m quotients that is $(r+1)$-bounded is r-bounded. A trivial induction shows that $\varrho_{m,r}$ implies that any bounded set of m quotients is r-bounded.

Let $L \in \mathbf{Var}(\mathbf{K})$. Then L is a subdirect product of subdirectly irreducible lattices L_i, $i \in I$, where each $L_i \in \mathbf{K}$ by Corollary 1.10, since UDE-s are preserved under the formation of ultra products, sublattices, and homomorphic images. Let $c \neq d$ and $c/d \approx_w a_j/b_j$, for $0 \le j < m$ in L. Then $c\varphi_i \neq d\varphi_i$ and $c\varphi_i/d\varphi_i \approx_w a_j/b_j$, for $0 \le j < m$, in L_i, for some $i \in I$, where φ_i is the projection of L onto L_i. By the second hypothesis, $\varrho_{m,r}$ holds in L_i and so, by Lemma 3(ii), $\{\, a_j/b_j \mid 0 \le j < m \,\}$ is $(r+2)$-bounded. Thus $\varrho_{m,r+2}$ holds in any $L \in \mathbf{Var}(\mathbf{K})$.

By Lemma 4(iv), $\mathbf{Var}(\mathbf{K})$ is defined by

$$\{\, \alpha_k(\Phi) \mid \Phi \in \Omega \text{ and } k = 1,\ 2,\ \dots \,\}.$$

Hence this set of sentences implies that $\varrho_{m,r+2}$ and so, by the Compactness Theorem (Exercise 1.29), finitely many $\alpha_k(\Phi)$ imply $\varrho_{m,r+2}$. Let

$$A = \{\, \alpha_k(\Phi) \mid \Phi \in \Omega \text{ and } k = 1,\ 2, \dots, t \,\}$$

imply $\varrho_{m,r+2}$ and let us further assume that $t \ge r+2$. We claim that A defines $\mathbf{Var}(\mathbf{K})$. Indeed, A holds in $\mathbf{Var}(\mathbf{K})$ by Lemma 4, hence $\mathbf{Mod}(A) \supseteq \mathbf{Var}(\mathbf{K})$. Conversely, A includes $\alpha_{r+2}(\Phi)$, for any $\Phi \in \Omega$, and A implies $\varrho_{m,r+2}$. But $\alpha_{r+2}(\Phi)$ and $\varrho_{m,r+2}$ imply $\alpha_i(\Phi)$, for any i; thus A implies $\alpha_i(\Phi)$, for any i and any $\Phi \in \Omega$, proving that $\mathbf{Mod}(A) \subseteq \mathbf{Var}(\mathbf{K})$.

We have proved that A is equivalent to $\mathrm{Iden}(\mathbf{K})$ and A is finite, hence, by the Compactness Theorem, A is equivalent to some finite $\Sigma \subseteq \mathrm{Iden}(\mathbf{K})$. Σ is a finite equational basis for $\mathbf{K}$. $\qquad\square$

The proof of Theorem 5, as exhibited above, does not give a finite equational basis; it only proves that there is one. In R. N. McKenzie [1970] and in K. A. Baker [1977], more complicated arguments are presented that actually construct a finite equational basis.

In some cases a finite equational basis for a finite lattice can be found using the following method. Let L be a finite lattice. Let us assume that we have constructed the lattices $L_0, L_1, \ldots, L_{n-1}$ such that

(i) $\mathbf{Var}(L)$ is covered by each $\mathbf{Var}(L_i)$;

(ii) if $\mathbf{K}$ is a variety of lattices and $\mathbf{Var}(L) \subset \mathbf{K}$, then $L_i \in \mathbf{K}$, for some i.

Find lattice identities $\varepsilon_0, \ldots, \varepsilon_{n-1}$ that hold in L but for which ε_i fails in L_i. Then $\{\varepsilon_0, \ldots, \varepsilon_{n-1}\}$ is a finite equational basis for L. We shall illustrate this method with an example.

Theorem 7.   In the lattice of all varieties of lattices, $\mathbf{M}_3$ is covered by $\mathbf{M}_4$ and $\mathbf{M}_{3,3}$. If $\mathbf{K}$ is any variety of modular lattices and $\mathbf{M}_3 \subset \mathbf{K}$, then $\mathbf{M}_4 \subseteq \mathbf{K}$ or $\mathbf{M}_{3,3} \subseteq \mathbf{K}$.   —

Remark. This result was proved in G. Grätzer [1966] under the additional hypothesis that $\mathbf{K}$ is generated by a finite lattice. This hypothesis was removed in B. Jónsson [1968], where the following corollary was also stated.

Corollary 8. *An equational basis for* $\mathbf{M}_3$ *is given by the modular identity and the identity*

$$x \wedge (y \vee z) \wedge (z \vee w) \wedge (w \vee y) \leq (x \wedge y) \vee (x \wedge z) \vee (x \wedge w).$$

By the discussion above, to prove this corollary it is sufficient to see that this identity holds in M_3 but fails in M_4 and $M_{3,3}$; this is left to the reader.

The proof of Theorem 7 is based on the following two lemmas.

Lemma 9. *Let L be a modular lattice and let $\{o, a, b, c, i\}$ be a diamond in L. Let $a \leq x^a < i$ and set (see Figure 1)*

$$x^b = b \vee (x^a \wedge c),$$
$$x^c = c \vee (x^a \wedge b),$$
$$o_1 = (x^a \wedge b) \vee (x^a \wedge c).$$

Then $\{o_1, x^a, x^b, x^c, i\}$ is a diamond.

Proof. Since $a \leq x^a < i$, $b \leq x^b < i$, and $c \leq x^c < i$, the relations

$$x^a \vee x^b = x^a \vee x^c = x^b \vee x^c = i$$

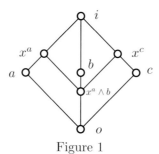

Figure 1

obviously hold. Also,

$$x^a \wedge x^b = x^a \wedge (b \vee (x^a \wedge c)) = (x^a \wedge b) \vee (x^a \wedge c) = o_1,$$

and similarly, $x^a \wedge x^c = o_1$; finally,

$$
\begin{aligned}
x^b \wedge x^c &= (b \vee (x^a \wedge c)) \wedge (c \vee (x^a \wedge b)) \\
&= ((b \vee (x^a \wedge c)) \wedge c) \vee (x^a \wedge b) \\
&= (b \wedge c) \vee (x^a \wedge c) \vee (x^a \wedge b) = o \vee o_1 = o_1.
\end{aligned}
$$
□

Lemma 10. *Let L be a subdirectly irreducible modular lattice which has no sublattice of which $M_{3,3}$ is a homomorphic image. Let $a/b \overset{n}{\approx} c/d$ in L such that a/b has no proper subquotient projective to a subquotient of c/d in fewer than n steps. Then $n \leq 3$.*

Proof. Let $x_0/y_0 \overset{4}{\approx} x_4/y_4$ be such that no proper subquotient of x_0/y_0 is projective to a proper subquotient of x_4/y_4 in fewer than four steps. By duality, we can assume that

$$x_0/y_0 \nearrow x_1/y_1 \searrow x_2/y_2 \nearrow x_3/y_3 \searrow x_4/y_4.$$

By (the proof of) Theorem IV.1.6, we can assume that this is a normal sequence. Let p_i/q_i be a proper subquotient of x_i/y_i, for $0 \leq i \leq 4$, such that

$$p_0/q_0 \nearrow p_1/q_1 \searrow p_2/q_2 \nearrow p_3/q_3 \searrow p_4/q_4.$$

We claim that $\{p_{i-1}, q_{i-1}, p_i, q_i, p_{i+1}, q_{i+1}\}$ cannot generate a distributive sublattice, for $i = 1, 2,$ or 3. Indeed, if one does, as for instance $\{p_0, q_0, p_1, q_1, p_2, q_2\}$ generates a distributive sublattice, then

$$p_0/q_0 \searrow p_0 \wedge p_2/q_0 \wedge q_2 \nearrow p_3/q_3 \searrow p_4/q_4,$$

by Theorem IV.1.7, contrary to the hypothesis. Thus the sublattice generated by $\{x_{i-1}, y_{i-1}, x_i, y_i, x_{i+1}, y_{i+1}\}$ is isomorphic to a homomorphic image of Figure IV.1.6 not collapsing M_3 or to the dual of such a lattice (see Figure 2).

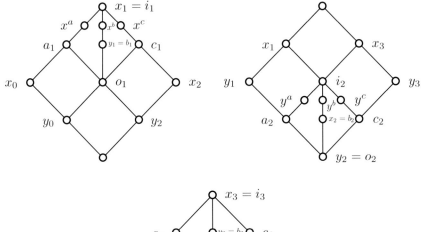

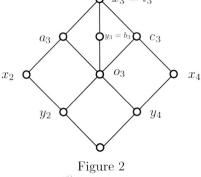

Figure 2

Now we claim that $o_1 \vee i_2 = i_1$, using the notation of Figure 2. Let $o_1 \vee i_2 \neq i_1$. Since $i_1 > o_1$ and $x_1 = i_1 > i_2$, we must have $i_1 > o_1 \vee i_2 = x^c$. Thus by Lemma 9, there are elements x^b, x^a such that

$$A_1 = \{x^b \wedge x^c, x^a, x^b, x^c, i_1\}$$

is a diamond and

$$b_1 \leq x^b < i_1,$$
$$a_1 \leq x^a < i_1.$$

Since $i_1/b_1 \searrow i_2/a_2$, there is an element y^a satisfying $i_1/x^b \searrow i_2/y^a$. Again, by Lemma 9, we find the elements y^b and y^c satisfying

$$b_2 \leq y^b < i_2,$$
$$c_2 \leq y^c < i_1,$$

and

$$A_2 = \{y^a \wedge y^b, y^a, y^b, y^c, i_2\}$$

is a diamond. Thus $i_1/x^b \searrow i_2/y^a$ and since $i_2 \leq x^c < i_1$, we also have $x^c/x^b \wedge x^c \searrow i_2/y^a$. This last relation means that

$$A_1 \cup A_2 \cong M_{3,3}$$

or if $x^c \neq i_2$ (and $x^b \wedge x^c \neq y^a$), then

$$(A_1 \cup A_2)/\Theta(x^c, i_2) \cong M_{3,3},$$

a contradiction.

Therefore, $o_1 \vee i_2 = i_1$. By duality, $o_1 \wedge i_2 = o_2$, that is

$$i_1/o_1 \searrow i_2/o_2,$$

and similarly,

$$i_2/o_3 \nearrow i_3/o_3.$$

By normality, $o_2 = o_1 \wedge o_3$, and, by definition, $i_2 = i_1 \wedge i_3$. Thus

$$\{o_1, o_2, o_3, i_1, i_2, i_3\}$$

is contained in the sublattice generated by $\{o_1, i_1, o_3, i_3\}$, which is distributive by Theorem IV.1.13. We conclude, by Theorem IV.1.7, that

$$i_1/o_1 \nearrow i_1 \vee i_3/o_1 \vee o_3 \searrow i_3/o_3,$$

which trivially implies that

$$x_0/y_0 \nearrow x_1 \vee o_3/y_1 \vee o_3 \searrow x_4/y_4,$$

contrary to the hypothesis.   □

Proof of Theorem 7. Let $\mathbf{K}$ be a variety of modular lattices such that $\mathbf{K} \supseteq M_3$ and $M_4, M_{3,3} \notin \mathbf{K}$. In order to show that $\mathbf{K} = M_3$, it is sufficient to verify that if $L \in \mathbf{Si}(\mathbf{K})$, then L is a sublattice of M_3. Assume, to the contrary, that $L \in \mathbf{Si}(\mathbf{K})$, but L is not a sublattice of M_3. Since $M_4, M_{3,3} \notin \mathbf{K}$, we must have $M_4, M_{3,3} \notin \mathbf{HS}(L)$.

Obviously, $|L| > 2$. If L is of length 2, then L must have M_4 as a sublattice, a contradiction. Thus L has a chain, $c_0 < c_1 < c_2 < c_3$, of length three. Since L is subdirectly irreducible, $\Theta(c_0, c_1) \wedge \Theta(c_1, c_2) \neq \omega$ and so, by applying Theorem III.1.2 twice, we obtain a proper quotient x/y weakly projective into c_2/c_1 and into c_1/c_0. By Theorem IV.1.6 and by the symmetry of projectivity, there is a proper subquotient a/b of c_2/c_1 and a subquotient c/d of c_1/c_0 such that $a/b \overset{n}{\approx} c/d$. Choose these a, b, c, d so that n be minimal in $a/b \approx c/d$. Then a, b, c, d, and L satisfy the conditions of Lemma 10 and therefore $n \leq 3$.

We claim that $n = 3$. Indeed, in any lattice L, we cannot have $a > b \geq c > d$ and $a/b \overset{n}{\approx} c/d$ with $n \leq 2$. (Proof. If

$$a/b \nearrow p/q \searrow c/d,$$

then

$$d = c \wedge q \geq c \wedge b = c.$$

If

$$a/b \searrow p/q \nearrow c/d,$$

then

$$a = b \vee p \leq b \vee c = b.$$

The case $n = 1$ is trivial.)

Now let

$$a/b = x_0/y_0 \sim x_1/y_1 \sim x_2/y_2 \sim x_3/y_3 = c/d.$$

If $a/b \nearrow x_1/y_1$, then $x_1 > y_1 \geq c > d$ and $x_1/y_1 \overset{2}{\approx} c/d$, which is impossible as noted in the previous paragraph. Thus

$$x_0/y_0 \searrow x_1/y_1 \nearrow x_2/y_2 \searrow x_3/y_3.$$

Applying the same arguments to $\Theta(c_3, c_2) \wedge \Theta(a, b)$, we obtain

$$z_0/u_0 \searrow z_1/u_1 \nearrow z_2/u_2 \searrow z_3/u_3,$$

where z_0/u_0 and z_3/u_3 are proper subquotients of c_3/c_2 and a/b, respectively. By making the trivial replacements

$$
\begin{array}{lll}
x_0/y_0 & \text{by} & z_3/u_3, \\
x_1/y_1 & \text{by} & x_1 \wedge z_3/x_1 \wedge u_3, \\
x_2/y_2 & \text{by} & y_2 \vee (x_1 \wedge z_3)/y_2 \vee (x_1 \wedge u_3), \\
x_3/y_3 & \text{by} & x_3 \wedge (y_2 \vee (x_1 \wedge z_3))/x_3 \wedge (y_2 \vee (x_1 \wedge u_3)),
\end{array}
$$

we can assume that x_0/y_0 equals z_3/u_3. Thus, by Theorem IV.1.7, there are diamonds

$$A_j = \{o_j, a_j, b_j, c_j, i_j\},$$

$j = 0, 1$, such that $a/b \searrow i_0/a_0$ and $a/b \nearrow a_1/o_1$. We conclude that $a_1/o_1 \searrow i_0/a_0$ and so $A_0 \cup A_1$ is a sublattice of which $M_{3,3}$ is a homomorphic image, a contradiction. $\square$

Exercises

1. A *universal sentence* Ψ is a sentence of the form $(x_0) \cdots (x_{n-1})\Phi$, where there is no quantifier in Φ and no variable other than $x_0, \ldots, x_{n-1}$. Show that every universal sentence in which no negation or implication occurs is equivalent to a finite set of UDE-s.

2. Show that a UDE is preserved under the formation of sublattices and homomorphic images.

3. Are all UDE-s preserved under direct products?

4. Is every UDE equivalent to some $\Phi(P)$?

5. Let P be the n-element unordered set. Compare $\Phi(P)$ with $\Phi(n)$.

6. Let P be a finite poset. Show that $\mathbf{Var}(\mathbf{Mod}(\Phi(P))) = \mathbf{Var}(\{P\})$, as introduced in Section 2.

7. In the proof of Theorem 5, the statement is used that in an n-element lattice every bounded set of proper quotients is n^2-bounded. Can n^2 be improved in this statement?

8. Write out the formulas α_k and $\varrho_{m,r}$ to prove formally that these are indeed (first-order) formulas.

9. An algebra $\langle A; \wedge, \vee \rangle$ is called a *weakly associative lattice* (WA lattice) iff the following are satisfied: the two binary operations satisfy the idempotent, commutative, and absorption identities, and

    $$x \leq z \text{ and } y \leq z \quad \text{imply that} \quad x \vee y \leq z$$

    and its dual hold, where $a \leq b$ means that $a = a \wedge b$ or, equivalently, $a \vee b = b$ (E. Fried [1970] and H. L. Skala [1971]). Show that all the results of this section hold for WA lattices (K. A. Baker [1977]).

*10. Take the three-element WA-lattice $T = \{0, 1, 2\}$ defined by $0 \leq 1 \leq 2 \leq 0$. Find a finite equational basis for T (E. Fried and G. Grätzer [1973]).

*11. If L is a modular lattice of length n, then any bounded set of proper quotients is k-bounded, where $k \leq \left[\frac{3n}{2}\right] + 2$ and $[x]$ stands for the largest integer $\leq x$ (C. Herrmann [1973]).

12. Let $\mathbf{M}^n$ denote the class of modular lattices of length at most n. Prove that $\mathbf{M}^n$ has a finite equational basis. (For $n = 2$, this is due to B. Jónsson [1968]; for $n = 3$, this is due to D. X. Hong [1972]. For general n, this is due to K. A. Baker; a reference to this fact and a proof of this result based on Exercise 11 is due to C. Herrmann [1973].)

13. Show that a finite equational basis for $\mathbf{M}^2$, for the notation see Exercise 12, is given by the modular identity and

$$(x \wedge (y \vee (z \wedge u))) \vee (z \wedge u) \leq y \vee (x \wedge z) \vee (x \wedge u)$$

(B. Jónsson [1968]). (Hint: Use Lemma 10 and the reasoning in the proof of Theorem 7.)

14. Let L be a subdirectly irreducible lattice of length at most 3. Show that any bounded set of proper quotients is in fact 5-bounded (C. Herrmann [1973]).

15. Let $\mathbf{L}_3$ denote the class of all lattices of length at most 3. Prove that $\mathbf{L}_3$ has a finite equational basis (C. Herrmann [1973]).

16. Show that Lemma 9 can be derived from Theorem IV.1.7 (and it is implicit in the same).

17. Using the notations of Lemma 9, project x^a into $[o, a]$, $[o, b]$, $[o, c]$, obtaining x_a, x_b, x_c, respectively. Show that $\{o, x_a, x_b, x_c, o_1\}$ is a diamond.

18. The results of this section depend heavily on Theorem 1.9. However, Theorems 5 and 7 use only Corollary 1.11. Show that the latter can be proved with no reference to ultra products, using only the primitive sets of Section III.1 (R. Wille [1972]).

*19. Let L be a subdirectly irreducible lattice of width not greater than four. Then either the lattice of Figure IV.3.4b or one of the eight lattices of Figures 3–10 is a homomorphic image of a sublattice of L (R. Freese [1972]).

20. Let $\mathbf{M}(4)$ denote the class of all modular lattices of width four. Prove that $\mathbf{M}(4)$ has a finite equational basis (R. Freese [1972]). (Hint: use Exercise 19.)

21. Show that to prove Theorem 7 for a variety generated by a finite lattice, it is sufficient to analyze projectivities of prime quotients. To what extent would this simplify the proof?

22. Let L be a modular lattice and let n be a positive integer. Call L n-distributive iff the following identity holds:

$$x \wedge \bigvee (\, y_i \mid 0 \leq i \leq n \,) = \bigvee (\, x \wedge \bigvee (\, y_i \mid 0 \leq i \leq n, \; i \neq j \,) \mid 0 \leq j \leq n \,).$$

The following form of the identity is easier to visualize:

$$x \wedge \bigvee_i y_i = \bigvee_j (x \wedge \bigvee_{i \neq j} y_i).$$

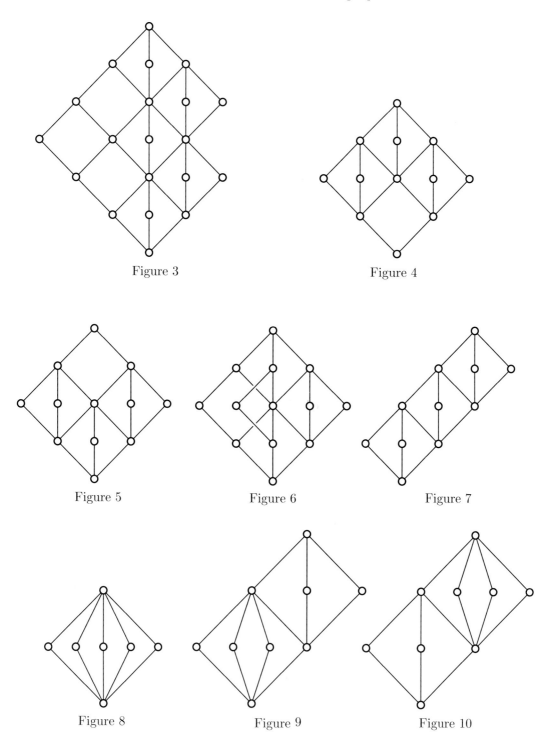

Figure 3

Figure 4

Figure 5

Figure 6

Figure 7

Figure 8

Figure 9

Figure 10

Prove that L is n-distributive iff it satisfies the following identity in the variables $x_0, x_1, \ldots, x_{n+1}$:

$$\bigwedge_j (\bigvee_i x_i) = \bigvee_k (\bigwedge_{j \neq k} (\bigvee_{i \neq j, k} x_i))$$

(G. M. Bergmann [1969] and A. P. Huhn [1972]).

23. For a positive integer n, we define a partial lattice P_n as follows: $P_n = B \cup \{w\}$, where B is a 2^{n+1}-element Boolean lattice with bounds 0 and 1; for $x, y \in B$, $x \wedge y$ and $x \vee y$ are defined as in B;

$$w \wedge 1 = 1 \wedge w = w,$$
$$w \wedge x = x \wedge w = 0,$$

for $x \in B - \{1\}$;

$$w \vee 0 = 0 \vee w = w,$$
$$w \vee d = d \vee w = 1,$$

if d is a dual atom of B or if $d = 1$; $w \vee x$ and $x \vee w$ are not defined, if $x \in B$ and $0 < h(x) < n$. Prove that a modular lattice L is n-distributive iff L does not contain P_n as a relative sublattice (A. P. Huhn [1972a]).

24. For a positive integer k and a division ring D, form the lattice $L = L(D, k)$. How is k determined by the smallest integer n such that L is n-distributive?

*25. An equational basis for $\mathbf{N}_5$ is provided by

$$x \wedge (y \vee u) \wedge (y \vee v) \leq (x \wedge (y \vee (u \wedge v))) \vee (x \wedge u) \vee (x \wedge v),$$
$$x \wedge (y \vee (u \wedge (x \vee v))) = (x \wedge (y \vee (u \wedge x))) \vee (x \wedge ((x \wedge y) \vee (u \wedge v))).$$

(See R. N. McKenzie [1972].)

*26. Consider the three-element algebra $M = \langle \{0, 1, 2\}; \cdot \rangle$ with one binary operation such that 0 is a zero ($0 \cdot x = x \cdot 0 = 0$) and

$$0 = 1 \cdot 1,$$
$$1 = 1 \cdot 2,$$
$$2 = 2 \cdot 1 = 2 \cdot 2.$$

Prove that M has no finite equational basis (V. L. Murskiĭ [1965]; see also R. C. Lyndon [1954]).

4. The Amalgamation Property

For a class $\mathbf{K}$ of lattices (or of algebras, in general), it is very important to know how members of the class can be glued together to obtain a larger member of the class. Such properties are known as amalgamation properties. We shall mention three of them.

A $\mathbb{V}$-*formation in* $\mathbf{K}$ is a pair of lattices B_0 and B_1 in $\mathbf{K}$ with a lattice $A \in \mathbf{K}$ that is a sublattice of both B_0 and B_1. More precisely, a $\mathbb{V}$-formation is a quintuplet $\langle A, B_0, B_1, \varphi_0, \varphi_1 \rangle$ such that A, B_0, $B_1 \in \mathbf{K}$ and φ_i is an embedding of A and B_i, for $i = 0$, 1. The $\mathbb{V}$-formation $\langle A, B_0, B_1, \varphi_0, \varphi_1 \rangle$ is *amalgamated by* $\langle \psi_0, \psi_1, C \rangle$ iff $C \in \mathbf{K}$, ψ_i is an embedding of B_i into C, for $i = 0$, 1, and $\varphi_0 \psi_0 = \varphi_1 \psi_1$ (see Figure 1). The $\mathbb{V}$-formation is *strongly amalgamated by* $\langle \psi_0, \psi_1, C \rangle$ iff, in addition, $B_0 \psi_0 \cap B_1 \psi_1 = A \varphi_0 \psi_0$ $(= A \varphi_1 \psi_1)$.

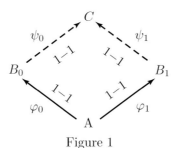

Figure 1

A class $\mathbf{K}$ is said to have the *(Strong) Amalgamation Property* iff every $\mathbb{V}$-formation can be (strongly) amalgamated.

$\mathbf{K}$ is said to have the *Weak Amalgamation Property* (also called the *Embedding Property*) iff, for any B_0, $B_1 \in \mathbf{K}$, there exists a $C \in \mathbf{K}$ into which both B_0 and B_1 can be embedded (this is the "special case" $A = \varnothing$ of the Amalgamation Property). This is of little interest for lattices: any variety of lattices has the Weak Amalgamation Property. Indeed, $C = B_0 \times B_1$ will do with the embeddings $\psi_0 \colon x \mapsto \langle x, b_1 \rangle$, $\psi_1 \colon x \mapsto \langle b_0, x \rangle$, where b_i is a fixed element of B_i, for $i = 0$, 1.

$\mathbf{L}$ has the Strong Amalgamation Property. To see this, take a $\mathbb{V}$-formation $\langle A, B_0, B_1, \varphi_0, \varphi_1 \rangle$; we can assume that $B_0 \cap B_1 = A$ and that A is a sublattice of B_0 and B_1. On the set $P = B_0 \cup B_1$ we define a partial ordering as follows:

(i) for a, $b \in B_i$, $a \leq b$ in P iff $a \leq b$ in B_i $(i = 0, 1)$;

(ii) for $a \in B_i$ and $b \in B_j$, $i \neq j$, $a \leq b$ in P iff $a \leq c$ in B_i and $c \leq b$ in B_j for some $c \in A$ $(i, j \in \{0, 1\})$.

It is easy to check that P is a poset. Then

$$a \wedge b = \inf\{a, b\},$$
$$a \vee b = \sup\{a, b\}$$

turns P into a partial lattice (see Lemma I.5.21) of which A, B_0, and B_1 are sublattices. Thus, by Theorem I.5.20, P can be embedded into a lattice, proving the Strong Amalgamation Property for **L**.

D does not have the Strong Amalgamation Property. Indeed, let $B_0 = B_1 = C_2^2$, $A = C_3$ ($A = \{0, a, 1\}$), and let $\varphi_0 = \varphi_1 = \varphi$ be given by

$$x\varphi = \begin{cases} \langle 0, 0 \rangle, & \text{for } x = 0; \\ \langle 1, 0 \rangle, & \text{for } x = a; \\ \langle 1, 1 \rangle, & \text{for } x = 1. \end{cases}$$

Let $\langle \psi_0, \psi_1, C \rangle$ strongly amalgamate $\langle A, B_0, B_1, \varphi_0, \varphi_1 \rangle$ with $C \in \mathbf{D}$. Then $B_0\psi_0 \cap B_1\psi_1 = A$ and so $\langle 0, 1 \rangle\psi_0 \neq \langle 0, 1 \rangle\psi_1$. Both of these elements are relative complements of $\langle 1, 0 \rangle\psi_0$ ($= \langle 1, 0 \rangle\psi_1$) in the interval $[\langle 0, 0 \rangle\psi_0, \langle 1, 1 \rangle\psi_0]$ of C, contradicting $C \in \mathbf{D}$ and Corollary II.1.3.

However, **D** has the Amalgamation Property. This we shall prove shortly, after some general remarks.

Let **K** be a variety of lattices (or of algebras, in general) and let

$$Q = \langle A, B_0, B_1, \varphi_0, \varphi_1 \rangle$$

be a $\mathbb{V}$-formation in **K**. We assume that $B_0 \cap B_1 = \varnothing$. We associate with Q a congruence relation $\Theta = \Theta(Q)$ as follows:

Let $F = \mathbf{F}_{\mathbf{K}}(Q)$ be the free lattice over **K** freely generated by the set $\overline{B}_0 \cup \overline{B}_1$, where

$$\overline{B}_i = \{ \overline{x} \mid x \in B_i \},$$

for $i = 0$, 1, and the sets B_0, B_1, $\overline{B}_0$, $\overline{B}_1$ are pairwise disjoint. We set

$$\Theta_A = \bigvee (\Theta(\overline{a\varphi_0}, \overline{a\varphi_1}) \mid a \in A)$$

and

$$\Theta_i = \bigvee (\Theta(\overline{a \wedge b}, \overline{a} \wedge \overline{b}) \mid a, \ b \in B_i) \vee \bigvee (\Theta(\overline{a \vee b}, \overline{a} \vee \overline{b}) \mid a, \ b \in B_i),$$

for $i = 0$, 1. Note that $\overline{a} \wedge \overline{b}$ and $\overline{a} \vee \overline{b}$ are to be performed in F. Now we define

$$\Theta = \Theta_A \vee \Theta_0 \vee \Theta_1.$$

Obviously, Θ is the smallest congruence relation of F such that

$$\alpha_i : x \mapsto [\overline{x}]\Theta, \quad i = 0, \ 1,$$

is a homomorphism of B_i into F/Θ (because $\Theta_i \leq \Theta$) and $\varphi_0\alpha_0 = \varphi_1\alpha_1$ (because $\Theta_A \leq \Theta$). Thus $\langle \alpha_0, \alpha_1, F/\Theta \rangle$ amalgamates Q iff α_0 and α_1 are one-to-one.

Therefore, if α_0 and α_1 are one-to-one, then Q can be amalgamated. Conversely, if $\langle \psi_0, \psi_1, C \rangle$ $(C \in \mathbf{K})$ amalgamates Q, then map $\overline{B}_0 \cup \overline{B}_1$ into C by

$$\overline{x} \mapsto x\psi_i, \quad \text{for } x \in B_i, \ i = 0, \ 1.$$

This map extends to a homomorphism β of F into C and obviously $\operatorname{Ker} \beta \geq \Theta$, and $\overline{x}\beta = x\psi_i$, for $x \in B_i$. Thus α_i followed by the natural homomorphism of F/Θ into C equals ψ_i, which is one-to-one by assumption. So α_i is one-to-one, for $i = 0, 1$. We have proved (Theorem 1, Corollaries 2–5, and the application to the proof of Lemma 12 are from G. Grätzer [1975]):

Theorem 1. Let $\mathbf{K}$ be a variety, let $Q = \langle A, B_0, B_1, \varphi_0, \varphi_1 \rangle$ be a $\mathbb{V}$-formation in $\mathbf{K}$, and let F and Θ be constructed as above. Then Q can be amalgamated in $\mathbf{K}$ iff, for $i = 0$ or 1 and $x, y \in B_i$,

$$\overline{x} \equiv \overline{y} \ (\Theta) \quad \text{implies that} \quad x = y.$$ —

Corollary 2. A variety $\mathbf{K}$ has the Amalgamation Property iff, for any $\mathbb{V}$-formation $\langle A, B_0, B_1, \varphi_0, \varphi_1 \rangle$ in $\mathbf{K}$ and $x, \ y \in B_0$, $x \neq y$, there exist a $C \in \mathbf{K}$ and homomorphisms $\psi_i \colon B_i \to C$ such that $\varphi_0\psi_0 = \varphi_1\psi_1$ and $x\psi_0 \neq y\psi_0$.

Proof. In order to prove the α_i (introduced in the proof of Theorem 1) one-to-one, it is sufficient to have homomorphisms ψ_i separating a pair of distinct elements. The necessity is obvious. □

Corollary 3. Let $\mathbf{K}$ be a variety and let $Q = \langle A, B_0, B_1, \varphi_0, \varphi_1 \rangle$ be a $\mathbb{V}$-formation in $\mathbf{K}$. If Q cannot be amalgamated in $\mathbf{K}$, then there are finitely generated subalgebras A' of A and B_i' of B_i, $i = 0, 1$, such that $Q' = \langle A', B_0', B_1', \varphi_0', \varphi_1' \rangle$ cannot be amalgamated in $\mathbf{K}$, where φ_i' is the restriction of φ_i to A', $i = 0, 1$.

Proof. Let F and Θ be given as in Theorem 1. If Q cannot be amalgamated in $\mathbf{K}$, then there are $i \in \{0,1\}$ and $x, \ y \in B_i$ such that $x \neq y$ and $\overline{x} \equiv \overline{y} \ (\Theta)$. By Theorem III.1.2 and Lemma III.1.3, we can select finite subsets A^* of A and B_i^* of B_i, $i = 0, 1$, such that in computing $\overline{x} \equiv \overline{y} \ (\Theta)$, we use only elements of $A^* \cup B_0^* \cup B_1^*$. Thus we can set $A' = [A^*]$, $B_i' = [B_i^*]$, $i = 0, 1$. A trivial application of Theorem 1 shows that Q' cannot be amalgamated in $\mathbf{K}$. □

Call a $\mathbb{V}$-formation $\langle A, B_0, B_1, \varphi_0, \varphi_1 \rangle$ *finitely generated* (resp, *finite*) iff A, B_0, and B_1 are finitely generated (resp., finite).

Corollary 4. A variety $\mathbf{K}$ has the Amalgamation Property iff every finitely generated $\mathbb{V}$-formation in $\mathbf{K}$ can be amalgamated in $\mathbf{K}$.

A variety $\mathbf{K}$ is called *locally finite* iff every finitely generated member of $\mathbf{K}$ is finite. It is easily seen that $\mathbf{K}$ is locally finite iff all $F_{\mathbf{K}}(n)$ are finite $(n < \omega)$. This always holds if $\mathbf{K}$ is generated by a single finite lattice L. Let $\mathbf{K}_{fin}$ denote the class of finite members of $\mathbf{K}$.

Corollary 5. *Let* **K** *be a locally finite variety. Then* **K** *has the Amalgamation Property iff all finite* $\mathbb{V}$-*formations in* **K** *can be amalgamated in* **K**, *or equivalently, iff* **K**$_{fin}$ *has the Amalgamation Property.*

The last equivalence follows from the observation that if

$$Q = \langle A, B_0, B_1, \varphi_0, \varphi_1 \rangle$$

is finite and can be amalgamated in **K**, then some quotient of $F_K(Q)$ will amalgamate Q; but $F_K(Q) = F_K(\overline{B}_0 \cup \overline{B}_1)$ is finite since K is locally finite.

Now we return to **D**. By Theorem II.2.1, $|F_D(n)| < 2^{2^n}$ and so **D** is locally finite.

Corollary 6. **D** *has the Amalgamation Property.*

Proof. Let $Q = \langle A, B_0, B_1, \varphi_0, \varphi_1 \rangle$ be a finite $\mathbb{V}$-formation in **D** and let x, $y \in B_0$, $x \neq y$. Set $C = C_2$ and let ψ_0 be a homomorphism of B_0 into C_2 such that $x\psi_0 \neq y\psi_0$ (see Corollary II.1.11). Let P be the ideal kernel of $\varphi_0\psi_0$. If $P = A$, define $\psi_1 : B_1 \to C_2$ by $x\psi_1 = 0$, for all $x \in B_1$. If $P = \varnothing$, define $\psi_1 : B_1 \to C_2$ by $x\psi_1 = 1$, for all $x \in B_1$. If $P \neq A$ and $P \neq \varnothing$, then $P = (a]$ where $a \neq 1$ is a meet-irreducible element in A. Thus there is a unique $b \in A$ such that $b \succ a$. By Corollary II.1.13, there is a meet-irreducible element p in B_1 such that $a \leq p$ and $b \not\leq p$. We then define $\psi_1 : B_1 \to C_2$ by

$$x\psi_1 = \begin{cases} 0, & \text{if } x \leq p; \\ 1, & \text{if } x \not\leq p. \end{cases}$$

It is obvious that $\langle \psi_0, \psi_1, C_2 \rangle$ satisfies the conditions of Corollary 2, thus Q can be amalgamated in **D**. $\qquad\square$

To further illustrate the usefulness of Corollary 2, for a class **K** of lattices (or algebras, in general) define $A \in$ **K** to be *injective* in **K** iff, for any B, $C \in$ **K**, B a sublattice of C, any homomorphism of B into A can be extended to a homomorphism of C into A.

Corollary 7 (R. S. Pierce [1968]). *Let* **K** *be a variety. If any member of* **K** *can be embedded in an injective member of* **K**, *then* **K** *has the Amalgamation Property.*

Proof. We apply Corollary 2. Let C be an injective member of **K** into which B_0 can be embedded; let ψ_0 be this embedding. Then $x\psi_0 \neq y\psi_0$. Set $A' = A\varphi_1$. Then $\varphi_1^{-1}\psi_0$ is a homomorphism (in fact, an embedding) of A' into C, thus this homomorphism can be extended to a homomorphism ψ_1 into B_1 into C. Obviously $\varphi_0\psi_0 = \varphi_1\psi_1$. $\qquad\square$

A further application of Corollary 3 will be given at the end of this section.

A typical application of the Amalgamation Property is the sublattice theorem of free products of lattices discussed in Section VI.1.

The next two theorems are negative. They show that certain varieties of lattices do not have the Amalgamation Property.

Theorem 8.   Let $\mathbf{K}$ be a variety generated by a finite lattice. If $\mathbf{K} \supset \mathbf{D}$, then $\mathbf{K}$ does not have the Amalgamation Property.   —

Note. This result is due to A. Day, S. D. Comer, and S. Fajtlowicz, as quoted in G. Grätzer, B. Jónsson, and H. Lakser [1973].

Proof. If $\mathbf{K}$ is generated by a finite lattice L, then by Corollary 1.11, $\mathbf{Si}(\mathbf{K}) \subseteq \mathbf{HS}(L)$. Thus no subdirectly irreducible lattice in $\mathbf{K}$ has more than $n = |L|$ elements.

Since $\mathbf{K} \supset \mathbf{D}$, there is a nondistributive lattice in $\mathbf{K}$ and thus, by Theorem II.1.1, N_5 or $M_3 \in \mathbf{K}$.

If $N_5 \in \mathbf{K}$, $N_5 = \{o, a, b, c, i\}$ as in Figure II.1.1, then consider the $\mathbb{V}$-formation $\langle C_2, N_5, N_5, \varphi_0, \varphi_1 \rangle$, where $C_2 = \{0, 1\}$ and

$$0\varphi_0 = o, \qquad 1\varphi_0 = i,$$
$$0\varphi_1 = b, \qquad 1\varphi_1 = a.$$

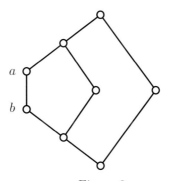

Figure 2

Let $\langle \psi_0, \psi_1, A \rangle$ amalgamate this $\mathbb{V}$-formation. Then A will have the lattice A_1 of Figure 2 as a sublattice. In fact, A_1 is the union of the two images of N_5 in A. Observe that A_1 is again subdirectly irreducible and $|A_1| = 8$. Then we take the $\mathbb{V}$-formation $\langle C_2, N_5, A_1, \varphi_0, \varphi_1 \rangle$ with

$$0\varphi_0 = o, \qquad 1\varphi_0 = i,$$
$$0\varphi_1 = b, \qquad 1\varphi_1 = a.$$

The union of the images of N_5 and A_1 will form a sublattice A_2, which is again subdirectly irreducible and $|A_2| = 11$. Proceeding by induction, we obtain the subdirectly irreducible lattice $A_k \in \mathbf{K}$ with $|A_k| = 5 + 3k$. If k is large enough so that $5 + 3k > n$, this is a contradiction.

If $M_3 \in \mathbf{K}$, $M_3 = \{o, a, b, c, i\}$, then we take $\langle C_2, M_3, M_3, \varphi_0, \varphi_1 \rangle$, where

$$0\varphi_0 = a, \qquad 1\varphi_0 = i,$$
$$0\varphi_1 = o, \qquad 1\varphi_1 = c.$$

Then the union of the images is again a sublattice of the amalgam, thus obtaining $M_{3,3} \in \mathbf{K}$. Proceeding the same way, we obtain a sequence of simple modular lattices of increasing size, leading to a contradiction as above. $\qquad\square$

Theorem 9. $\mathbf{M}$ does not have the Amalgamation Property.   —

Remark. In fact, no variety $\mathbf{K}$ of modular lattices has the Amalgamation Property unless $\mathbf{K} = \mathbf{T}$ or $\mathbf{D}$, as proved in G. Grätzer, B. Jónsson, and H. Lakser [1973].

Let us assume that $\mathbf{M}$ has the Amalgamation Property; under this assumption, we prove a few embedding theorems.

1. *Any modular lattice A can be embedded in a bounded modular lattice B that has a five-element chain.*

Proof. This is trivial. If A has no five-element chain, we add new zeros and ones until it has a five-element chain. If A is not bounded, add bounds. $\qquad\square$

2. *Any bounded modular lattice A can be embedded in a modular lattice B having the same bounds and having the property that, for every $a \in A$, $a \neq 0, 1$, B has a diamond $\{0, a, b, c, 1\}$.*

Proof. To prove this, let $A - \{0, 1\} = \{a_\gamma \mid \gamma < \alpha\}$ and we define an increasing sequence of modular lattices A_γ, $\gamma < \alpha$, with the same bounds 0 and 1. Set $\overline{A}_0 = A$. If the A_γ, $\gamma < \delta$ ($\delta < \alpha$), have already been defined, set

$$\overline{A}_\delta = \bigcup (A_\gamma \mid \gamma < \delta).$$

Let $C_3 = \{0, 1, 2\}$ and consider the $\mathbb{V}$-formation $Q_\delta = \langle C_3, M_3, \overline{A}_\delta, \varphi_0, \varphi_1 \rangle$ defined by

$$0\varphi_0 = o, \qquad 1\varphi_0 = a, \qquad 2\varphi_0 = i,$$
$$0\varphi_1 = 0, \qquad 1\varphi_1 = a_\delta, \qquad 2\varphi_1 = 1.$$

Let $\langle \psi_0, \psi_1, A \rangle$ amalgamate Q_δ. By forming the interval $[0\varphi_0\psi_0, 2\varphi_0\psi_0]$, we obtain A_δ. It is obvious that with $a = a_\delta$, A_δ has the property required of B. We define $B = \bigcup (A_\gamma \mid \gamma < \alpha)$, which obviously has the required properties. $\qquad\square$

3. *Any bounded modular lattice A can be embedded in a modular lattice B having the same bounds as A and having the following properties:*

(i) *For every $a \in B$, $a \neq 0$, 1, there is a diamond $\{0, a, b, c, 1\}$ in B.*

(ii) *B is a simple complemented modular lattice.*

Proof. Indeed, set $A_0 = A$, and if A_n is defined, then define A_{n+1} as the lattice constructed from A_n in the second embedding theorem. Then

$$B = \bigcup (\, A_n \mid n < \omega \,)$$

obviously satisfies (i).

Let Θ be a congruence relation of B, $\omega \neq \Theta$. Since B is a complemented modular lattice, B is relatively complemented; hence (see Theorem III.3.10), there is an $a \in B$ such that $a \neq 0$ and $a \equiv 0$ (Θ). If $a \neq 1$, then, by (i), there is a diamond $\{0, a, b, c, 1\}$. Since M_3 is simple, we obtain that $0 \equiv 1$ (Θ), that is, $\Theta = \iota$. If $a = 1$, again $\Theta = \iota$. Thus B is simple.   ☐

Proof of Theorem 9. Now we are ready to prove the theorem. Let $A \in \mathbf{M}$. By the embeddings 1–3, we can embed A into a simple complemented modular lattice B having a five-element chain. By Theorem IV.5.17, B can be $\{0, 1\}$-embedded into a modular geometric lattice C. The lattice C is directly indecomposable because it has a simple $\{0, 1\}$-sublattice (namely, B). C has a five-element chain because B has one. Thus Corollary IV.5.16 applies and $C \cong L(D, \mathfrak{m})$. Thus by Lemma IV.5.9 and Corollary IV.5.12, the Arguesian identity holds in C. Since A is a sublattice of C, the Arguesian identity holds in A.

A is an arbitrary member of $\mathbf{M}$ and the Arguesian identity does not hold in $\mathbf{M}$; this is a contradiction.   ☐

The Amalgamation Property fails for most varieties of lattices; so it seems reasonable to ask to what extent does it hold in general. The answer is rather surprising.

Following G. Grätzer and H. Lakser [1971], for a class $\mathbf{K}$ of lattices (or algebras, in general), let $\mathbf{Amal}(\mathbf{K})$ be the class of all those $A \in \mathbf{K}$ for which all $\mathbb{V}$-formations in $\mathbf{K}$ of the form $\langle A, \dots \rangle$ can be amalgamated in $\mathbf{K}$. Obviously, $\mathbf{K}$ has the Amalgamation Property iff $\mathbf{Amal}(\mathbf{K}) = \mathbf{K}$.

$\mathbf{Amal}(\mathbf{K}) \neq \varnothing$, for any variety $\mathbf{K}$ of lattices. Indeed, argue similarly to the proof of the Weak Amalgamation Property for $\mathbf{K}$ that $C_1 \in \mathbf{Amal}(\mathbf{K})$.

Call a subclass $\mathbf{K}_1$ of $\mathbf{K}$ *cofinal in $\mathbf{K}$* iff, for all $A \in \mathbf{K}$, there is an extension $B \in \mathbf{K}_1$.

Theorem 10 (M. Yasuhara [1974]).   For a variety $\mathbf{K}$, $\mathbf{Amal}(\mathbf{K})$ *is cofinal in $\mathbf{K}$.*
                                                                             ——

The proof of this result will be given in the next two lemmas. Some of the ideas of Lemma 11 originate in A. Robinson [1971], [1971a]. The present form of Lemma 11 is a slight variant of a lemma of M. Yasuhara and it is due to M. Makkai.

For a class $\mathbf{K}$, let $\mathbf{Ec}(K)$ stand for the class of all $A \in \mathbf{K}$ having the following property:

> For any extension $B \in \mathbf{K}$ of A and for any finite $X \subseteq A$ and $Y \subseteq B$, there is an embedding $\varphi \colon [X \cup Y] \to A$ fixing X (that is, $x\varphi = x$, for all $x \in X$).

Lemma 11. *For a variety* $\mathbf{K}$*,* $\mathbf{Ec}(\mathbf{K})$ *is cofinal in* $\mathbf{K}$*.*

Proof. To facilitate the proof, we introduce some concepts. For $A \in \mathbf{K}$, $\langle \varphi, X, C \rangle$ is a *triple over* A *in* $\mathbf{K}$ iff X is a finite subset of A, C is a finitely generated member of $\mathbf{K}$, and $\varphi \colon X \to C$ is a homomorphism of the relative sublattice X of A into C. (This last condition means that if $x \wedge y = z$ or $x \vee y = z$, for x, y, $z \in X$, then $x\varphi \wedge y\varphi = z\varphi$ or $x\varphi \vee y\varphi = z\varphi$ in C, respectively.) The triple $\langle \varphi, X, C \rangle$ *is realized over* A *by* B iff $B \in \mathbf{K}$, B is an extension of A, and there is an embedding ψ of C into B such that $x\varphi\psi = x$, for all $x \in X$.

Now let $\langle \varphi_\gamma, X_\gamma, C_\gamma \rangle$, $\gamma < \alpha$, list all the triples over A. We define two increasing sequences of lattices as follows:

$$\overline{A}_0 = A;$$

if A_γ has been defined for $\gamma < \delta \ (< \alpha)$, then set

$$\overline{A}_\delta = \bigcup (A_\gamma \mid \gamma < \delta).$$

Regard $\langle \varphi_\delta, X_\delta, C_\delta \rangle$ as a triple over $\overline{A}_\delta$. If it is realized in $\mathbf{K}$, define A_δ as any member of $\mathbf{K}$ realizing it. If it cannot be realized in $\mathbf{K}$, define $A_\delta = \overline{A}_\delta$. Then set

$$A^{(1)} = \bigcup (A_\gamma \mid \gamma < \alpha),$$
$$A^{(n+1)} = (A^{(n)})^{(1)},$$
$$A^* = \bigcup (A^{(n)} \mid n < \omega).$$

Since $\mathbf{K}$ is a variety, $A^* \in \mathbf{K}$. We claim that $A^* \in \mathbf{Ec}(K)$. Indeed, let $B \in \mathbf{K}$ be an extension of A^*, and choose finite subsets $X \subseteq A^*$ and $Y \subseteq B$. Define $C = [X \cup Y]$ and let $\varphi \colon X \to A$ be the identity map. Then $\langle \varphi, X, C \rangle$ is a triple over A^* that can be realized by B. Since $X \subseteq A^*$ and X is finite, $X \subseteq A^{(n)}$, for some $n < \omega$. Hence $\langle \varphi, X, C \rangle$ can be regarded as a triple over $A^{(n)}$. Thus $\langle \varphi, X, C \rangle$ occurs as some $\langle \varphi_\delta, X_\delta, C_\delta \rangle$ in the list of all triples over $A^{(n)}$. In the δ-th step of the construction of $A^{(n+1)} = (A^{(n)})^{(1)}$, we view $\langle \varphi, X, C \rangle$ as a triple over $(\overline{A^{(n)}})_\delta$; we observe that $\langle \varphi, X, C \rangle$ can be realized by B, and so $(A^{(n)})_\delta$ realizes $\langle \varphi, X, C \rangle$, that is, there is an embedding $\psi \colon C \to (A^{(n)})_\delta$ such that

$x\varphi\psi = x$, for $x \in X$. However, $x\varphi = x$, for all $x \in X$ and therefore, $x\varphi\psi = x\psi = x$, for all $x \in X$. Thus ψ is an embedding of $[X \cup Y]$ into A^* keeping X fixed, proving that $A^* \in \mathbf{Ec}(\mathbf{K})$. $\square$

Lemma 12. *For any variety* $\mathbf{K}$, $\mathbf{Ec}(\mathbf{K}) \subseteq \mathbf{Amal}(\mathbf{K})$.

Proof. Let $A \in \mathbf{Ec}(\mathbf{K})$ and consider a $\mathbb{V}$-formation $Q = \langle A, B_0, B_1, \varphi_0, \varphi_1 \rangle$ in $\mathbf{K}$. We can assume that $A \subseteq B_0$, $A \subseteq B_1$ and $\varphi_0 = \varphi_1$ is the identity map on A. We wish to show that Q can be amalgamated in $\mathbf{K}$. Assume, to the contrary, that Q cannot be amalgamated in $\mathbf{K}$. Then, by Corollary 3, there exist finite subsets $X \subseteq A$, $Y_0 \subseteq B_0$, and $Y_1 \subseteq B_1$ such that

$$Q' = \langle [X], [X \cup Y_0], [X \cup Y_1], \varphi_0', \varphi_1' \rangle$$

cannot be amalgamated, where φ_i' is the restriction of φ_i to $[X \cup Y_i]$, for $i = 0, 1$. Since $A \in \mathbf{Ec}(K)$, there exist embeddings $\psi_i \colon [X \cup Y_i] \to A$ keeping X fixed ($i = 0, 1$). Thus, for $x \in X$, $x\varphi_i\psi_i = x$, for $i = 0, 1$, and so $a\varphi_i\psi_i = a$, for all $a \in [X]$. This shows that $\langle \psi_0, \psi_1, A \rangle$ amalgamates Q', which is a contradiction. $\square$

Exercises

1. Show that the class of all Boolean algebras does not have the Weak Amalgamation Property.

2. Show that the class of all groups has the Strong Amalgamation Property.

3. Investigate which varieties of pseudocomplemented distributive lattices have which Amalgamation Property.

4. Show that $\mathbf{L}_{\text{fin}}$ has the Strong Amalgamation Property.

5. Define the $\mathbf{StAmal}(\mathbf{K})$ as the analogue of $\mathbf{Amal}(\mathbf{K})$ for the Strong Amalgamation Property. Determine $\mathbf{StAmal}(\mathbf{D})$.

6. Show that "$C \in \mathbf{K}$" can be changed to "$C \in \mathbf{Si}(\mathbf{K})$" in Corollary 2.

7. Did we use the Axiom of Choice in verifying Corollary 5?

8. Let L be injective in $\mathbf{K}$. If A is an extension of L in $\mathbf{K}$, then L is a *retract* of A, that is, there is a homomorphism $\varphi \colon A \to L$ such that $x\varphi = x$, for all $x \in L$.

9. Show that any retract of a complete Boolean lattice is a complete Boolean lattice.

10. A distributive lattice L is injective in $\mathbf{D}$ iff L is a complete Boolean lattice (P. R. Halmos [1963]). (Hint: use Exercises 8, 9, and Corollary II.2.7.)

11.   Prove Corollary 6 using Corollary 7.

12.   Find further examples of the phenomenon observed twice in the proof of Theorem 8, namely, that for some special $\mathbb{V}$-formation, if $\langle \psi_0, \psi_1, C \rangle$ amalgamates $\langle A, B_0, B_1, \varphi_0, \varphi_1 \rangle$ then $B_0\psi_0 \cup B_1\psi_1$ is a uniquely determined sublattice of C.

13.   Show that there are $2^{\aleph_0}$ varieties of modular lattices not having the Amalgamation Property.

14.   Let $\mathbf{K}$ be a variety. Then, for every $A \in \mathbf{K}$, we construct an extension $B \in \mathbf{Amal}(\mathbf{K})$. Give an upper bound for $|B|$.

Exercises 15–18 are from G. Grätzer, B. Jónsson, and H. Lakser [1973]. Exercise 15 is implicit in B. H. Neumann and H. Neumann [1952].

15.   Let $\mathbf{K}$ be a variety having the Amalgamation Property. Let A, $B \in \mathbf{K}$, B an extension of A, and let α be an automorphism of A. Show that there exist an extension $C \in \mathbf{K}$ of B and an automorphism β of C extending α.

16.   Let $\mathbf{K}$ be a variety of lattices. Then $C_2 \in \mathbf{Amal}(\mathbf{K})$ iff there is a $D \in \mathbf{Amal}(\mathbf{K})$ with $|D| > 1$.

17.   Let $\mathbf{K}$ be a variety of lattices. Prove that if $\operatorname{Id} L \in \mathbf{Amal}(\mathbf{K})$, then $L \in \mathbf{Amal}(\mathbf{K})$.

18.   Let $\mathbf{K} = \mathbf{Var}(L)$, where L is a finite subdirectly irreducible lattice. Then $L \in \mathbf{Amal}(\mathbf{K})$.

Further Topics and References

$\mathbf{H}$, $\mathbf{S}$, $\mathbf{P}$, $\mathbf{P_R}$, $\mathbf{P_S}$, and so on, are examples of operators: an *operator* $\mathbf{X}$ assigns to a class $\mathbf{K}$ of lattices (algebras) another class $\mathbf{X}(\mathbf{K})$ of lattices (algebras). If $\mathbf{X}$ and $\mathbf{Y}$ are operators, so is their product: $\mathbf{XY}$ defined by $\mathbf{XY}(\mathbf{K}) = \mathbf{X}(\mathbf{Y}(\mathbf{K}))$. Thus we can easily define the semigroup generated by a set of operators. The *standard semigroup* of a class $\mathbf{K}$ of algebras is the one generated by $\mathbf{H}$, $\mathbf{S}$, and $\mathbf{P}$, equality defined with respect to $\mathbf{K}$. For reference, see S. D. Comer and J. Johnson [1972], which also lists all the known results for lattices, distributive lattices, Boolean algebras, and so on; one problem they leave unresolved is our Problem 1.

Theorem 2.1 is very trivial but it has a number of useful consequences: the lattice of varieties of lattices satisfies the Meet Infinite Distributive Identity and every proper interval contains a prime interval; properties shared by all dually algebraic, distributive lattices. An additional property is that no element covers more than countably many elements, since the lattice then would have more than $2^{\aleph_0}$ elements.

For a more detailed discussion of reduced products and ultra products, see T. Frayne, A. C. Morel, and D. S. Scott [1962] and G. Grätzer [1968].

Splitting can be defined relative to a variety. Apart from the general concept, only splittings in $\mathbf{M}$ and $\mathbf{E}$ have been studied to some extent, where $\mathbf{E}$ is the variety generated by the subspace lattices of all projective planes. For instance, M_4 does not split in $\mathbf{M}$ but the subspace lattice of any projective plane splits in $\mathbf{E}$. For a discussion of these and related results see R. Wille [1976b]; see also A. Day [1973] and D. X. Hong [1972].

Wille's method of constructing varieties (Theorem 2.6) is based on his two earlier papers, R. Wille [1969] and [1969a]. For a refinement of Wille's method, see K. A. Baker [1974].

By a result of R. N. McKenzie [1970], lattices can be defined by a single identity and obviously so can $\mathbf{T}$; no other varieties of lattices can be defined by a single identity. By R. Padmanabhan [1969], any variety of lattices defined by a finite set of identities can be defined by two identities.

Combining these results with a result of A. Tarski [1968], we obtain:

$\mathbf{L}$ and $\mathbf{T}$ have an irredundant equational basis of n elements, for any $1 \leq n < \omega$. If the variety $\mathbf{K}$ of lattices has a finite equational basis and $\mathbf{K} \neq \mathbf{L}$, $\mathbf{T}$, then $\mathbf{K}$ has an irredundant equational basis of n elements iff $2 \leq n < \omega$.

Lattices are quite exceptional in that Theorem 3.5 holds. A finite algebra with no finite equational basis was found by R. C. Lyndon [1954]; this was improved in V. L. Murskiĭ [1965], in which a three-element example is found (this is optimal, see E. L. Post [1921]) and P. Perkins [1969], in which the example is a six-element semigroup. For additional references and for the main result in the field (namely, that Theorem 3.5 holds in any congruence distributive variety), see K. A. Baker [1977].

Let $\mathbf{K}_0$ and $\mathbf{K}_1$ be varieties; the identities of $\mathbf{K}_0 \vee \mathbf{K}_1$ are those that are shared by $\mathbf{K}_0$ and $\mathbf{K}_1$. Thus $\mathbf{K}_0$ and $\mathbf{K}_1$ may have finite equational bases, while $\mathbf{K}_0 \vee \mathbf{K}_1$ may not have a finite equational basis. For lattices this negative result has been verified by K. A. Baker (unpublished) and B. Jónsson [1974]. Some positive results are also given in B. Jónsson [1974]; for instance $\mathbf{M} \vee \mathbf{N}_5$ has finite equational bases.

Equational bases are known for many varieties of lattices (and other related algebras); see especially, K. A. Baker [1969a], [1971], [1974], [1974a], and [1976], S. D. Comer and D. X. Hong [1972], A. Day [1973], R. Freese [1973] and [1977], E. Fried and G. Grätzer [1973], C. Herrmann [1973] and [1973a], C. Herrmann and W. Poguntke [1974], D. X. Hong [1972], G. Hutchinson [1973b], B. Jónsson [1968], M. Makkai [1973], M. Makkai and G. McNulty [1977], R. N. McKenzie [1970], [1972], and [1973], R. Padmanabhan [1968], [1969], [1972], and [1974a], M. Schützenberger [1945], and R. Wille [1976b]; see also R. Padmanabhan [1966] and D. H. Potts [1965].

K. A. Baker's method of finding equational bases discussed in Section 3 is exploited in great detail in K. A. Baker [1977]; for further results and applications, see K. A. Baker [1974] and [1976]. C. Herrmann [1973] and M. Makkai [1973] are further variants on the same theme; see also G. Grätzer and H. Lakser [1971a].

Theorem 3.7 has been extended in B. Jónsson [1968] and D. X. Hong [1972]. Related questions are considered in C. Herrmann [1973] and in the papers of R. Freese.

The development of the Amalgamation Property is described in B. Jónsson [1965]; probably, it was B. Jónsson's work more than any other influence that convinced the algebraists of the importance of this property. The Amalgamation Property for lattices and posets is noted in B. Jónsson [1956]; it is observed for Boolean algebras in A. Daigneault [1959].

The results of Section 4 on the Amalgamation Property are mostly negative. A further negative result of G. Grätzer, B. Jónsson, and H. Lakser [1973] states that if $\mathbf{K}$ is a variety of lattices, $\mathbf{K} \subseteq \mathbf{M}$, and $\mathbf{K}$ is not Arguesian, then $C_2 \notin \mathbf{Amal}(\mathbf{K})$. This was strengthened in G. Grätzer and H. Lakser [1973a].

The existentially complete algebras of M. Yasuhara [1974] are different from members of $\mathbf{Ec}(\mathbf{K})$. Thus Lemma 4.11 is slightly stronger and Lemma 4.12 is slightly weaker than the corresponding results in M. Yasuhara [1974].

The connections of the amalgamation properties and the interpolation theorems of logic are examined in P. D. Bacsich [1975]; see also P. C. Eklof [1974].

Finite members of $\mathbf{Amal}(\mathbf{M}_n)$ are described in E. Fried, G. Grätzer, and H. Lakser [1971].

Problems

1. Is $\mathbf{Var}(\mathbf{K}) = \mathbf{SHPS}(\mathbf{K})$, for any class $\mathbf{K}$ of lattices?

2. Describe those sets of operators introduced in this chapter that generate finite semigroups for lattices.

3. Characterize splitting lattices in $\mathbf{M}$.

4. The method of R. N. McKenzie [1972] to give an equational basis for N_5 has been modified in S. D. Comer and D. X. Hong [1972] to obtain an equational basis for M_3. Explain the similarities and determine whether the method has other applications (for instance, to yield equational bases for covers of $\mathbf{N}_5$).

5. For a variety $\mathbf{K}$ of lattices, define

   $i(\mathbf{K})$ (resp., $i_g(\mathbf{K})$) the smallest integer n such that $\mathbf{F_K}(n)$ is infinite (resp., and generates $\mathbf{K}$); and if there is no such integer, then $i(\mathbf{K}) = \infty$ (resp., $i_g(\mathbf{K}) = \infty$);

   $e(\mathbf{K})$ (resp., $e_f(\mathbf{K})$) the smallest integer n such that every finite member of $\mathbf{K}$ can be embedded into an n-generated (resp., finite) member of $\mathbf{K}$; and if there is no such integer, then $e(\mathbf{K}) = \infty$ (resp., $e_f(\mathbf{K}) = \infty$);

$c(\mathbf{K})$ the smallest integer n such that every countable member of $\mathbf{K}$ can be embedded in an n-generated member of $\mathbf{K}$; and if there is no such integer, then $c(\mathbf{K}) = \infty$;

$f(\mathbf{K})$ (resp., $f_i(\mathbf{K})$) as the smallest integer n such that $\mathbf{F}_{\mathbf{K}}(n)$ contains $\mathbf{F}_{\mathbf{K}}(n+1)$ (resp, $(\mathbf{F}_{\mathbf{K}}(\aleph_0))$ as a sublattice; and $f(\mathbf{K}) = \infty$ (resp., $f_i(\mathbf{K}) = \infty$) if there is no such integer.

Are there any relationships amongst these apart from the obvious ones $(e(\mathbf{K}) \le e_f(\mathbf{K})$, $c(\mathbf{K}) \le f_i(\mathbf{K})$, and so on)?

6. Compute the functions introduced in Problem 5 for particular varieties of lattices: $\mathbf{N}_5$, $\mathbf{M}_3$, $\mathbf{M}_n$, $\mathbf{M}$, and so on. (M_n is the subspace lattice of a projective line with n points and $\mathbf{M}_n = \mathbf{Var}(M_n)$.)

7. For a variety $\mathbf{K}$ of lattices and positive integer n, define

$$\mathbf{K}_n = \mathbf{Var}(\mathbf{F}_{\mathbf{K}}(n)).$$

Characterize $I = \{\, n \mid \mathbf{K}_n \ne \mathbf{K}_{n+1} \,\}$. (Conjecture: $I = \{1, 2, 3, \dots, k\}$ or $I = \{1, 2, \dots, k, \dots\}$. For universal algebras, in general, see B. Jónsson, G. McNulty, and R. W. Quackenbush [1975].)

8. Find an $L \in \mathbf{Amal}(\mathbf{M})$, $|L| > 1$.

9. Does there exist a variety $\mathbf{K}$ satisfying $\mathbf{D} \subset \mathbf{K} \subseteq \mathbf{M}$ and $C_2 \in \mathbf{Amal}(\mathbf{K})$?

10. Does there exist a variety $\mathbf{K} \subset \mathbf{L}$ having the Strong Amalgamation Property? How many such varieties exist? (Conjecture: finite or countably infinite.)

11. Show that there are only countably many varieties of lattices having the Amalgamation Property.

12. Describe $\mathbf{Amal}(\mathbf{M}_n)$.

13. Investigate $\mathbf{Amal}(\mathbf{Var}(\text{complemented modular lattices}))$.

14. Is it true that in the lattice of varieties of lattices every proper interval contains an atom?

15. In the lattice of varieties of lattices, are the elements of finite height exactly the varieties of the form $\mathbf{Var}(L)$, where L is a finite lattice?

16. Find all covers of $\mathbf{Var}(L)$, where L is finite planar.

17. Find all the covers of $\mathbf{N}_5$.

18. Is there a variety of lattices with uncountably many covers?

19. Is there a nontrivial lattice identity holding in the lattice of varieties of commutative semigroups?

20. Find a finite equational basis for $\mathbf{M} \vee \mathbf{N}_5$.

21. Let $\mathbf{K}_0$, $\mathbf{K}_1$ be varieties of modular lattices. Is it true that if $\mathbf{K}_0$ and $\mathbf{K}_1$ have finite equational bases, then so does $\mathbf{K}_0 \vee \mathbf{K}_1$?

Free Products

1.   Free Products of Lattices

The formation of a free product of a family of lattices is one of the most fundamental constructions of lattice theory. This specializes to the construction of free lattices, which form a class of lattices that is probably the closest rival of the class of distributive lattices in the richness of its structure. Also, free products provide a very useful tool for the construction of pathological lattices.

This is made possible, in part, by the Structure Theorem of Free Products (Theorem 10 below). Since this theorem is based on a large number of long inductive definitions, we shall present first a short intuitive description of its contents. We ask the reader to follow it up with a careful reading of the theorem, bearing in mind that the final result is very simple and most efficient in use.

Let A and B be lattices and let L be a free product of A and B, in symbols, $L = A * B$; for the present discussion, this should mean that A and B are sublattices of L, $A \cup B$ generates L, and L is the "most general" lattice with these properties, as in Section I.5. Figures 1 and 2 provide two simple examples: Figure 1 is the diagram of $C_2 * C_1$ and Figure 2 that of $C_3 * C_1$.

Since L is generated by $A \cup B$, every element of L can be written in the form

$$p(a_0, \ldots, a_{n-1}, b_0, \ldots, b_{m-1}), \quad a_0, \ldots, a_{n-1} \in A, \ b_0, \ldots, b_{m-1} \in B,$$

where p is an $(n + m)$-ary lattice polynomial. For instance, if A and B are as

343

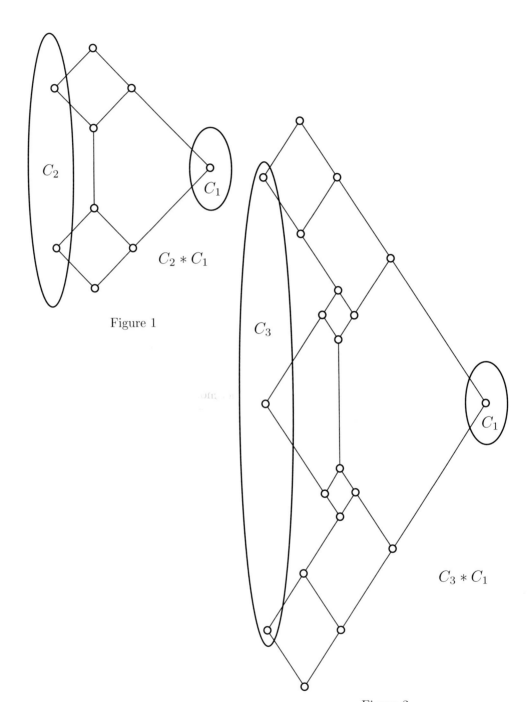

C_2

C_1

$C_2 * C_1$

Figure 1

C_3

C_1

$C_3 * C_1$

Figure 2

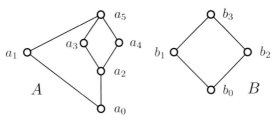

Figure 3

given in Figure 3, then

$$a_1,$$
$$b_2,$$
$$a_3 \vee b_1,$$
$$(a_3 \wedge b_2) \vee (a_0 \wedge b_0),$$
$$((a_5 \wedge b_0) \vee a_3) \wedge ((a_2 \wedge b_0) \vee a_2 \vee b_3)$$

are examples of elements of L. Observe, however, that expressions of this sort that appear to be very different may, in fact, denote the same element of L. For instance,

$$a_1 \vee ((a_3 \vee b_3) \wedge (a_4 \vee b_3)) = a_5 \vee b_3.$$

Theorem 10 will help the reader to construct more complicated examples.

The problem, then, is how to decide whether two expressions represent the same element of L. The key to this problem is the observation that if

$$p = p(a_0, \dots, a_{n-1}, b_0, \dots, b_{m-1}) \leq a,$$

for some $a \in A$, then there is a smallest element of A with this property. In fact, this smallest element (called the *upper cover* of p in A) is easy to compute knowing only p, A, and B.

So now our plan is the following: define formally the free product, expressions of the form $p(a_0, \dots, a_{n-1}, b_0, \dots, b_{m-1})$, upper and (dually) lower covers, present the algorithm deciding whether $p(a_0, \dots, a_{n-1}, b_0, \dots, b_{m-1}) \leq q(a_0, \dots, a_{n-1}, b_0, \dots, b_{m-1})$, and, finally, prove that this algorithm works.

For this whole section, let L_i, $i \in I$, $I \neq \varnothing$, be a fixed family of lattices; we assume that L_i and L_j are disjoint, for $i, j \in I$, $i \neq j$. We set $Q = \bigcup(L_i \mid i \in I)$ and we consider Q a poset under the following partial ordering:

$$a \leq b \quad \text{iff} \quad a, b \in L_i, \quad \text{for some } i \in I \text{ and } a \leq b \text{ in } L_i,$$

for $a, b \in Q$.

A free product L of the L_i, $i \in I$, is a free lattice $F(Q)$ $(= F_{\mathbf{L}}(Q))$ generated by Q in the sense of Definition I.5.2. Or, equivalently,

Definition 1. *Let* **K** *be a variety of lattices and let* L, $L_i \in$ **K**, *for* $i \in I$. *The lattice* L *is a free* **K**-*product of the lattices* L_i, $i \in I$, *iff the following conditions are satisfied:*

(i) *Each* L_i *is a sublattice of* L *and, for* i, $j \in I$, $i \neq j$, L_i *and* L_j *are disjoint.*

(ii) L *is generated by* $\bigcup(L_i \mid i \in I)$.

(iii) *For any lattice* $A \in$ **K** *and for any family of homomorphisms* $\varphi_i \colon L_i \to A$, $i \in I$, *there exists a homomorphism* $\varphi \colon L \to A$ *such that* φ *on* L_i *agrees with* φ_i, *for all* $i \in I$.

If **K** $=$ **L**, the lattice L is called a *free product*.

The next definition is a slight adaptation of Definition I.4.1.

Definition 2. *Let* X *be an arbitrary set. The set* $\mathbf{P}(X)$ *of polynomials in* X *is the smallest set satisfying* (i) *and* (ii):

(i) $X \subseteq \mathbf{P}(X)$.

(ii) *If* p, $q \in \mathbf{P}(X)$, *then* $(p \wedge q)$, $(p \vee q) \in \mathbf{P}(X)$.

The reader should keep in mind that a polynomial is a sequence of symbols and equality means formal equality. As before, parentheses will be dropped whenever there is no danger of confusion.

In what follows, we shall deal with polynomials in $Q = \bigcup(L_i \mid i \in I)$. Let a, b, $c \in L_i$, $a \vee b = c$; observe that as polynomials in Q, $a \vee b$ (which stands for $(a \vee b)$) and c are distinct.

For a lattice A, we define $A^b = A \cup \{0^b, 1^b\}$, where 0^b, $1^b \notin A$; we order A^b by the two rules:

(i) $0^b < x < 1^b$, for all $x \in A$.

(ii) $x \leq y$ in A^b iff $x \leq y$ in A, for x, $y \in A$.

Thus A^b is a bounded lattice. Note, however, that $A^b \neq A$ even if A is itself bounded. It is important to observe that 0^b is meet-irreducible and 1^b is join-irreducible. Thus if $a \wedge b = 0^b$ in A^b, then either a or b is 0^b, and dually. This will be quite important in subsequent computations.

Definition 3. *Let* $p \in \mathbf{P}(Q)$ *and* $i \in I$. *The upper* i-*cover of* p, *in notation,* $p^{(i)}$, *is an element of* L_i^b *defined as follows:*

(i) *For* $a \in Q$, *we have* $a \in L_j$, *for exactly one* j; *if* $j = i$, *then* $a^{(i)} = a$, *if* $j \neq i$, *then* $a^{(i)} = 1^b$.

(ii)

$$(p \wedge q)^{(i)} = p^{(i)} \wedge q^{(i)},$$
$$(p \vee q)^{(i)} = p^{(i)} \vee q^{(i)},$$

where $\wedge$ *and* $\vee$ *on the right side of these equations are to be taken in* L_i^b.

The definition of the *lower i-cover* of p, in notation, $p_{(i)}$, is analogous, with 0^b replacing 1^b in (i). An upper cover or a lower cover is *proper*, if it is not 0^b or 1^b. Observe that, however, no upper cover is 0^b and no lower cover is 1^b.

Corollary 4. *For any $p \in \mathbf{P}(Q)$ and $i \in I$, we have that*

$$p_{(i)} \leq p^{(i)},$$

and if $p_{(i)}$ and $p^{(j)}$ are proper, then $i = j$.

Proof. If $p \in Q$, then $p = p_{(i)} = p^{(i)}$, for $p \in L_i$, and $p_{(i)} = 0_b < 1_b = p^{(i)}$, for $p \notin L_i$, so the first statement is true. If the first statement holds for p and q, then

$$(p \wedge q)_{(i)} = p_{(i)} \wedge q_{(i)} \leq p^{(i)} \wedge q^{(i)} = (p \wedge q)^{(i)},$$

and so the first statement holds for $p \wedge q$, and similarly for $p \vee q$.

To prove the second statement, it is sufficient to verify that if $p_{(i)}$ is proper, then $p^{(j)}$ is not proper for any $j \neq i$. This is obvious for $p \in Q$, by 3(i). If $p = q \wedge r$ and $p_{(i)}$ is proper, then both $q_{(i)}$ and $r_{(i)}$ are proper, hence $q^{(j)} = r^{(j)} = 1^b$, and so $p^{(j)} = 1^b$. Finally, if $p = q \vee r$ and $p_{(i)}$ is proper, then $q_{(i)}$ or $r_{(i)}$ is proper, hence $q^{(j)} = 1^b$ or $r^{(j)} = 1^b$, ensuring that $p^{(j)} = q^{(j)} \vee r^{(j)} = 1^b$. □

Finally, we introduce a quasi-ordering of $\mathbf{P}(Q)$.

Definition 5. *For $p, q \in \mathbf{P}(Q)$, set $p \sqsubseteq q$ iff it follows from the following rules:*

(C)	$p^{(i)} \leq q_{(i)},$	*for some $i \in I$;*
($\wedge$W)	$p = p_0 \wedge p_1,$	*where* $p_0 \sqsubseteq q$ *or* $p_1 \sqsubseteq q$;
($\vee$W)	$p = p_0 \vee p_1,$	*where* $p_0 \sqsubseteq q$ *and* $p_1 \sqsubseteq q$;
(W$_\wedge$)	$q = q_0 \wedge q_1,$	*where* $p \sqsubseteq q_0$ *and* $p \sqsubseteq q_1$;
(W$_\vee$)	$q = q_0 \vee q_1,$	*where* $p \sqsubseteq q_0$ *or* $p \sqsubseteq q_1$.

Remark. In (C), C stands for Cover; in ($\wedge$W) (and in the other three conditions), W stands for P. M. Whitman. Each Whitman condition assumes a $\wedge$ or $\vee$ on the left or right of $\sqsubseteq$ in $p \sqsubseteq q$; at most two of these conditions may be applicable to a particular $p \sqsubseteq q$. Note also, that if p and $q \in Q$, then only (C) can apply.

Definition 5 gives essentially the algorithm we have been looking for. For $p, q \in \mathbf{P}(Q)$, it will be shown that p and q represent the same element of the free product iff $p \sqsubseteq q$ and $q \sqsubseteq p$. We shall show this by actually exhibiting the free product as the set of equivalence classes of $\mathbf{P}(Q)$ under this relation. To be able to do this, we have to establish a number of properties of the relation $\sqsubseteq$. All the proofs are by induction and will use the *rank* $r(p)$ of the polynomial $p \in \mathbf{P}(Q)$: for $p \in Q$, $r(p) = 1$;

$$r(p \wedge q) = r(p \vee q) = r(p) + r(q).$$

Lemma 6. *Let p, q, $r \in \mathbf{P}(Q)$ and $i \in I$.*

(i) $p \subseteq p$.

(ii) $p \subseteq q$ *implies that* $p_{(i)} \le q_{(i)}$ *and* $p^{(i)} \le q^{(i)}$.

(iii) $p \subseteq q$ *and* $q \subseteq r$ *imply that* $p \subseteq r$.

Proof.

(i) Proof by induction on $r(p)$. If $p \in Q$, then $p \in L_i$, for some $i \in I$. Hence $p = p_{(i)} = p^{(i)}$, by 3(i), and so $p \subseteq p$, by (C). Let $p = q \wedge r$. Then $q \subseteq q$ and $r \subseteq r$, by the induction hypothesis. By $(\wedge W)$, $q \wedge r \subseteq q$ and $q \wedge r \subseteq r$, hence by $(W_\wedge)$, $q \wedge r \subseteq q \wedge r$, that is, $p \subseteq p$. If $p = q \vee r$, we proceed similarly.

(ii) If $p \subseteq q$ by (C), then $p^{(j)} \le q_{(j)}$, for some $j \in I$. We conclude, on the one hand, that $p_{(j)} \le q_{(j)}$ and $p^{(j)} \le q^{(j)}$ by Corollary 4, and on the other, that $p^{(j)}$ and $q_{(j)}$ are proper; thus again by Corollary 4, $p_{(i)}$ and $q^{(i)}$ are not proper, for $i \ne j$, hence $p_{(i)} = 0^b$, $q^{(i)} = 1^b$, and so $p_{(i)} \le q_{(i)}$ and $p^{(i)} \le q^{(i)}$.

Now we induct on $r(p) + r(q)$. If p, $q \in Q$, then only (C) applies to $p \subseteq q$; in this case, and also in the induction step, if (C) is applied, we obtain the result by the last paragraph. Now if $p = p_0 \wedge p_1$ and $(\wedge W)$ applies, then $p_0 \subseteq q$ or $p_1 \subseteq q$, say $p_0 \subseteq q$. Then $(p_0)_{(i)} \le q_{(i)}$ and $(p_0)^{(i)} \le q^{(i)}$, hence

$$p_{(i)} = (p_0)_{(i)} \wedge (p_1)_{(i)} \le (p_0)_{(i)} \le q_{(i)},$$

and similarly for upper covers.

(iii) If $p \subseteq q$ by (C), then $p^{(i)} \le q_{(i)}$, for some $i \in I$. By (ii), $q_{(i)} \le r_{(i)}$, hence $p^{(i)} \le r_{(i)}$ and so by (C), $p \subseteq r$. This takes care of the base of the induction, p, q, $r \in Q$, since then only (C) applies. We induct on the sum of the ranks of p, q, and r.

If $p \subseteq q$ by (C), $p \subseteq r$ has already been proved.

If $p \subseteq q$ follows from $(\wedge W)$, then $p = p_0 \wedge p_1$ and $p_0 \subseteq q$ or $p_1 \subseteq q$. Thus, by the induction hypotheses, $p_0 \subseteq r$ or $p_1 \subseteq r$, and so by $(\wedge W)$, $p_0 \wedge p_1 = p \subseteq r$.

If $p \subseteq q$ follows from $(\vee W)$, then $p = p_0 \vee p_1$, $p_0 \subseteq q$, and $p_1 \subseteq q$, and so again $p_0 \subseteq r$ and $p_1 \subseteq r$, implying that $p_0 \vee p_1 = p \subseteq r$ by $(\vee W)$.

If $q \subseteq r$ follows from (C), $(W_\wedge)$, or $(W_\vee)$, we can proceed dually (that is, by interchanging $\wedge$ and $\vee$).

Only two cases remain; since the second is the dual of the first, we shall state only one: $q = q_0 \wedge q_1$, $(W_\wedge)$ applies to $p \subseteq q$, and $(\wedge W)$ is applied to $q \subseteq r$ (observe that $(\vee W)$ is not applicable). Then $p \subseteq q_0$ and $p \subseteq q_1$, and $q_0 \subseteq r$ or $q_1 \subseteq r$. Hence $p \subseteq q_i \subseteq r$, for $i = 0$ or 1, hence by the induction hypotheses, $p \subseteq r$. □

By Lemma 6, the relation $\subseteq$ is a quasi-ordering and so (Exercise I.1.28) we can define:

$$p \equiv q \quad \text{iff} \quad p \subseteq q \text{ and } q \subseteq p \qquad (p, q \in \mathbf{P}(Q)).$$
$$R(p) = \{\, q \mid q \in \mathbf{P}(Q) \text{ and } p \equiv q \,\} \qquad (p \in \mathbf{P}(Q)).$$

$$R(Q) = \{ R(p) \mid p \in \mathbf{P}(Q) \}.$$
$$R(p) \le R(q) \quad \text{iff} \quad p \subseteq q.$$

In other words, we split $\mathbf{P}(Q)$ into blocks under the equivalence relation $p \equiv q$; $R(Q)$ is the set of blocks which we partially order by $\le$.

Lemma 7. $R(Q)$ *is a lattice, in fact,*

$$R(p) \wedge R(q) = R(p \wedge q),$$
$$R(p) \vee R(q) = R(p \vee q).$$

Furthermore, if a, b, c, $d \in L_i$, $i \in I$, and

$$a \wedge b = c,$$
$$a \vee b = d,$$

in L_i, then

$$R(a) \wedge R(b) = R(c),$$
$$R(a) \vee R(b) = R(d);$$

and if x, $y \in Q, x \ne y$, then $R(x) \ne R(y)$.

Proof. $p \wedge q \subseteq p$ and $p \wedge q \subseteq q$ by $p \subseteq p$, $q \subseteq q$, and $(\wedge W)$. If $r \subseteq p$ and $r \subseteq q$, then $r \subseteq p \wedge q$ by $(W_\wedge)$; this argument and its dual give the first statement.

$c \subseteq a$ and $c \subseteq b$ is obvious by (C), hence $R(c) \le R(a)$ and $R(c) \le R(b)$. Now let $R(p) \le R(a)$ and $R(p) \le R(b)$, for some $p \in \mathbf{P}(Q)$. Then $p \subseteq a$ and $p \subseteq b$, and so, by Lemma 6, $p^{(i)} \le a^{(i)} = a$ and $p^{(i)} \le b^{(i)} = b$. Therefore, $p^{(i)} \le c = c_{(i)}$ and thus $p \subseteq c$ by (C). The second part follows by duality. Finally, if $R(x) = R(y)$ $(x, y \in Q)$, then $x \subseteq y$. Since only (C) applies, $x^{(i)} \le y_{(i)}$, for some $i \in I$. Thus $x^{(i)}$ and $y_{(i)}$ are proper and so $x = x^{(i)}$, $y = y_{(i)}$. We conclude that $x \le y$; similarly, $y \le x$. Thus $x = y$. □

By Lemma 7,

$$a \mapsto R(a), \quad a \in L_i,$$

is an embedding of L_i into $R(Q)$; for $i \in I$. Therefore, by identifying $a \in L_i$ with $R(a)$, we get each L_i as a sublattice of $R(Q)$ and hence $Q \subseteq R(Q)$. It is also obvious that the partial ordering induced by $R(Q)$ on Q agrees with the original partial ordering.

Theorem 8. $R(Q)$ is a free product of the L_i, $i \in I$.   —

Proof. 1(i) and 1(ii) have already been observed. Let $\varphi_i \colon L_i \to A$ be given, for all $i \in I$. We define inductively a map

$$\psi \colon \mathbf{P}(Q) \to A$$

as follows: for $p \in Q$, there is exactly one $i \in I$ with $p \in L_i$; set $p\psi = p\varphi_i$; if $p = p_0 \wedge p_1$ or $p = p_0 \vee p_1$, and if $p_0\psi$ and $p_1\psi$ have already been defined, then set $p\psi = p_0\psi \wedge p_1\psi$ or $p\psi = p_0\psi \vee p_1\psi$, respectively. Now we prove:

Lemma 9. *Let $p \in \mathbf{P}(Q)$ and $i \in I$.*

(i) *If $p_{(i)}$ is proper, then $p_{(i)}\psi \leq p\psi$.*

(ii) *If $p^{(i)}$ is proper, then $p\psi \leq p^{(i)}\psi$.*

(iii) *$p \subseteq q$ implies that $p\psi \leq q\psi$, for $p, q \in \mathbf{P}(Q)$.*

Proof.
(i) If $p \in Q$ and $p_{(i)}$ is proper, then $p \in L_i$. Hence $p = p_{(i)}$ and so $p_{(i)}\psi \leq p\psi$ is obvious. If $p = q \wedge r$, then $p_{(i)} = q_{(i)} \wedge r_{(i)}$, so $q_{(i)}$ and $r_{(i)}$ are proper. Thus $q(i)\psi \leq q\psi$ and $r_{(i)}\psi \leq r\psi$, by induction; hence

$$p_{(i)}\psi = q_{(i)}\psi \wedge r_{(i)}\psi \leq q\psi \wedge r\psi = (q \wedge r)\psi = p\psi.$$

If $p = q \vee r$, then $q_{(i)}$ or $r_{(i)}$ is proper. If both $q_{(i)}$ and $r_{(i)}$ are proper, we proceed as in the previous case. If, say, $q_{(i)}$ is proper and $r_{(i)} = 0^b$, then

$$p_{(i)}\psi = (q_{(i)} \vee r_{(i)})\psi = q_{(i)}\psi \leq q\psi \leq p\psi.$$

(ii) This follows by duality from (i).
(iii) If $p \subseteq q$ follows from (C), then, for some $i \in I$, $p^{(i)} \leq q_{(i)}$. Thus $p^{(i)}$ and $q_{(i)}$ are proper. Therefore, $p\psi \leq p^{(i)}\psi$ by (ii), $p^{(i)}\psi \leq q_{(i)}\psi$, because $p^{(i)}$ and $q_{(i)} \in Q$, and $q_{(i)}\psi \leq q\psi$ by (i), implying that $p\psi \leq q\psi$.
This takes care of $p, q \in Q$ and of the first case in the induction step.
If $p \subseteq q$ follows from ($\wedge$W), then $p = p_0 \wedge p_1$, where $p_0 \subseteq q$ or $p_1 \subseteq q$. Hence $p_0\psi \leq q\psi$ or $p_1\psi \leq q\psi$, therefore, $p\psi = p_0\psi \wedge p_1\psi \leq q\psi$.
If $p \subseteq q$ follows from ($\vee$W), ($W_\wedge$), or ($W_\vee$), the proof is analogous to the proof in the last case.   □

Now take a $p \in \mathbf{P}(Q)$ and define

$$R(p)\varphi = p\psi.$$

φ is well-defined since if $R(p) = R(q)$ $(p, q \in \mathbf{P}(Q))$, then $p \subseteq q$ and $q \subseteq p$. Hence, by Lemma 9, $p\psi \leq q\psi$ and $q\psi \leq p\psi$, and so $p\psi = q\psi$. Since

$$(R(p) \wedge R(q))\varphi = R(p \wedge q)\varphi = (p \wedge q)\varphi = p\varphi \wedge q\varphi = R(p)\varphi \wedge R(q)\varphi,$$

and similarly for $\vee$, we conclude that φ is a homomorphism. Finally, for $p \in L_i$, $i \in I$,

$$R(p)\varphi = p\psi = p\varphi_i,$$

by the definition of ψ; hence φ restricted to L_i agrees with φ_i.   □

Lemma 6(ii) implies that if $p \equiv q$ $(p, q \in \mathbf{P}(Q))$, then, for all $i \in I$, $p_{(i)} = q_{(i)}$ and $p^{(i)} = q^{(i)}$. Hence we can define

$$(R(p))_{(i)} = p_{(i)},$$
$$(R(p))^{(i)} = p^{(i)}.$$

All our results will now be summarized (G. Grätzer, H. Lakser, and C. R. Platt [1970]):

Theorem 10 (The Structure Theorem of Free Products).   Let L_i, $i \in I$, be lattices and let L be a free product of the L_i, $i \in I$. Then, for every $a \in L$ and $i \in I$, if some element of L_i is contained in a, then there is a largest one with this property, namely, $a_{(i)}$. If $a = p(a_0, \ldots, a_{n-1})$, where p is an n-ary polynomial and $a_0, \ldots, a_{n-1} \in \bigcup(L_i \mid i \in I)$, then $a_{(i)}$ can be computed by the algorithm given in Definition 3. Dually, $a^{(i)}$ can be computed. For $a, b \in L$, $a = p(a_0, \ldots, a_{n-1})$, $b = q(b_0, \ldots, b_{m-1})$, $a_0, \ldots, a_{n-1}, b_0, \ldots, b_{m-1} \in \bigcup(L_i \mid i \in I)$, we can decide whether $a \leq b$ using the algorithm of Definition 5.   —

The idea of the proof of Theorem 10 goes back to P. M. Whitman [1941] and R. P. Dilworth [1945].

We should comment on the use of the term "algorithm" in Theorem 10 and elsewhere in this section. (C) of Definition 5 brings in covers and so does the procedure described in Definition 3. This procedure is an algorithm insofar as the structure of L_i is described in an effective way. Thus if we consider the free product of finitely many finite lattices, then we really deal with an algorithm. In the general case, algorithm should be interpreted intuitively and not be given the precise meaning assigned to it in mathematical logic.

The existence of covers is not special to $\mathbf{L}$, as it was observed in B. Jónsson [1971]. To prove their existence, let L be a free $\mathbf{K}$-product of L_i, $i \in I$. Fix an $i \in I$, and, for $j \in I$, define $\psi_j \colon L_j \to L_i^b$ as follows: ψ_i is the identity map on L_i; for $j \neq i$, $a\psi_j = 0^b$, for all $a \in L_j$. By Exercise I.4.14, $L_i^b \in \mathbf{K}$ and so, by the definition of free $\mathbf{K}$-product, there is a homomorphism $\psi \colon L \to L_i^b$ extending all ψ_j, $j \in I$. For $a \in L$, $a_{(i)} = a\psi$. Now if $b \in L_i$, $b \leq a$, then $b\psi \leq a\psi$, that is, $b \leq a_{(i)}$. Thus if there is $b \in L_i$, $b \leq a$, then $a_{(i)}$ is the largest such element. Otherwise, $a\psi = 0_b$; indeed, if $a\psi \neq 0_b$, then $b = a\psi \in L_i$, a contradiction in view of $b = a\psi \leq a$. (We use the fact that $a\psi \leq a$, for all $a \in L$, since this holds for all $a \in \bigcup(L_i \mid i \in I)$.)

Free products (in fact, free **K**-products) of two lattices satisfy a special condition which is due to G. Grätzer, H. Lakser, and C. R. Platt [1970]:

Theorem 11 (The Splitting Theorem).   Let the lattice L be a free product of the lattices L_0 and L_1. For every $a \in L$, if $a_{(0)}$ is not proper, then $a^{(i)}$ is proper and conversely. Thus

$$L = (L_0] \cup [L_1),$$

where the union is a disjoint union. In other words, for every element a of L either a is contained in some element of L_0 or a contains an element of L_1.   —

Proof. $(L_0] \cup [L_1)$ is a sublattice of L containing a generating set, namely, $L_0 \cup L_1$. Hence $(L_0] \cup [L_1) = L$. Now Theorem 10 yields the statement.   □

Corollary 12. $L_0 * L_1$ can be represented as a disjoint union of four convex sublattices: the convex sublattices generated by L_0 and L_1, respectively, $(L_0] \cap (L_1]$, and $[L_0) \cap [L_1)$.

The next result is a trivial application of Theorem 10.

Corollary 13. Let K_i be a sublattice of L_i, for $i \in I$, and let L be a free product of the L_i, $i \in I$. Let K be the sublattice of L generated by $\bigcup (K_i \mid i \in I)$. Then K is a free product of the K_i, $i \in I$.

Proof. This is obvious since, for $p, q \in \mathbf{P}(\bigcup (K_i \mid i \in I))$, we have $p \subseteq q$ iff it follows from Definition 5, and in applying Definition 5, we use only elements of the K_i, $i \in I$. Thus, by the proof of Theorem 8, K is a free product of the K_i, $i \in I$.   □

This result is a special case of a result of B. Jónsson [1961] and [1965] (see Figure 4):

Theorem 14.   Let **K** be a variety satisfying the Amalgamation Property. Let $A, B, C \in \mathbf{K}$ and let C be a free **K**-product of A and B. Let A_1 be a sublattice of A, let B_1 be a sublattice of B, and let C' be the sublattice of C generated by $A_1 \cup B_1$. Then C' is a free **K**-product of A_1 and B_1.   —

Proof. Let C'' be a free **K**-product of A_1 and B_1. Thus there exists a homomorphism χ of C'' into C' such that χ is the identity on A_1 and B_1. A_1 is a sublattice of A and of C''; thus, by the Amalgamation Property, there is a lattice D in **K** containing A and C'' as sublattices, $A_1 \subseteq A \cap C''$. Similarly, B_1 is a sublattice of D and B, and thus there exists a lattice E in **K** containing B and D as sublattices, $B_1 \subseteq B \cap D$ (see Figure 5). Since C is a free product of A and B, there exists a homomorphism φ_1 of C into E such that φ_1 is the identity on

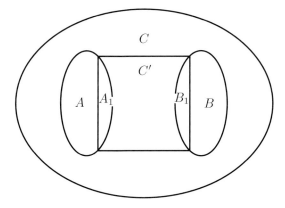

Figure 4

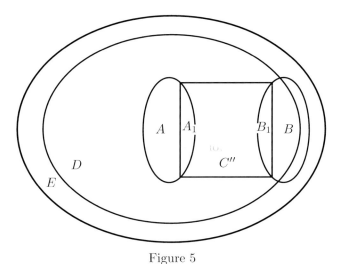

Figure 5

A and B. Let φ be the restriction of φ_1 to C'. Then φ maps C' into C'', $\varphi\chi$ is the identity on A_1 and B_1, and $\varphi\chi$ is thus the identity on C'. Similarly, $\chi\varphi$ is the identity on C'', so φ is an isomorphism between C' and C''. □

There are three properties that play an important role for free lattices. They are the following:

(W) $x \wedge y \leq u \vee v$ implies that $[x \wedge y, u \vee v] \cap \{x, y, u, v\} \neq \emptyset.$

(SD$_\wedge$) $u = x \wedge y = x \wedge z$ implies that $u = x \wedge (y \vee z).$

(SD$_\vee$) $u = x \vee y = x \vee z$ implies that $u = x \vee (y \wedge z).$

(W) is called the *Whitman condition*, (SD$_\wedge$) the *meet-semidistributive law*, and (SD$_\vee$) the *join-semidistributive law*. We can rephrase the Whitman condition: $x \wedge y \leq u \vee v$ implies that $x \leq u \vee v$ or $y \leq u \vee v$ or $x \wedge y \leq u$ or $x \wedge y \leq v$. Observe, that a free product need not satisfy (W); indeed, $x \wedge y \leq u \vee v$ may hold in one of the factors or on account of (C).

Let us say that a *subset A of a lattice L satisfies* (W) iff (W) holds in L, for $x, y, z, u \in A$. The next result (G. Grätzer and H. Lakser [1974]) shows that many subsets of a free product have (W).

Theorem 15.   Let the lattice L be a free product of the lattices $L_i, i \in I$. Let $A_i, i \in I$, be a subset of L_i^b satisfying (W). Let A be a subset of L satisfying

$$A_{(i)} = \{\, a_{(i)} \mid a \in A \,\} \subseteq A_i$$
$$A^{(i)} = \{\, a^{(i)} \mid a \in A \,\} \subseteq A_i,$$

for all $i \in I$. Then A satisfies (W) in L.   —

Proof. Let $x, y, u, v \in A$, $x = R(p)$, $y = R(q)$, $u = R(r)$, $v = R(s)$, where p, q, r, $s \in \mathbf{P}(Q)$. Let $x \wedge y \leq u \vee v$. Then $p \wedge q \subseteq r \vee s$. If (C) applies, then, for some $i \in I$, $(p \wedge q)^{(i)} \leq (r \vee s)_{(i)}$, hence $p^{(i)} \wedge q^{(i)} \leq r_{(i)} \vee s_{(i)}$ in A_i. Therefore, one of the following holds:

$$p^{(i)} \leq r_{(i)} \vee s_{(i)},$$
$$q^{(i)} \leq r_{(i)} \vee s_{(i)},$$
$$p^{(i)} \wedge q^{(i)} \leq r_{(i)},$$
$$p^{(i)} \wedge q^{(i)} \leq s_{(i)}.$$

Again, by (C), we conclude that one of the following holds:

$$p \subseteq r \vee s,$$
$$q \subseteq r \vee s,$$
$$p \wedge q \subseteq r,$$
$$p \wedge q \subseteq s.$$

($_\wedge$W) and (W$_\vee$) give exactly the same conclusions. We conclude, therefore, that (W) holds in A.   □

Observe how strong (W) really is. For instance, it follows immediately that there is no element in A which is doubly reducible.

Naturally, every $a \in L$ has infinitely many representations of the form

$$p = p(a_0, \dots, a_{n-1}) \in \mathbf{P}(Q) \quad (a_0, \, \dots, a_{n-1} \in Q).$$

If $r(p)$ is minimal, we call p a *minimal representation of a* and we call p a *minimal polynomial*. Using the notation of Theorem 10, the next result (H. Lakser [1970a]) tells us how to recognize a minimal representation:

Theorem 16.   Let $p \in \mathbf{P}(Q)$. Then p is a minimal representation iff $p \in Q$, or if $p = p_0 \vee \cdots \vee p_{n-1}$, $n > 1$, where no p_j is a join of more than one polynomial and conditions (i)–(v) below hold, or the dual of the preceding case holds.

(i) Each p_j is minimal, $0 \leq j < n$.

(ii) For each $0 \leq j < n$, $p_j \not\subseteq p_0 \vee \cdots \vee p_{j-1} \vee p_{j+1} \vee \cdots \vee p_{n-1}$.

(iii) If $0 \leq j < n$, $r(p_j) > 1$, $i \in I$, then $p_j^{(i)} \not\subseteq p_{(i)}$ in L_i^b.

(iv) If $p_j = p_j' \wedge p_j''$, where $0 \leq j < n$ and p_j', $p_j'' \in \mathbf{P}(Q)$, then $p_j' \not\subseteq p$ and $p_j'' \not\subseteq p$.

(v) If p_j, $p_k \in L_i$, where $0 \leq j \leq k < n$ and $i \in I$, then $j = k$.   —

Proof. Consider the polynomials

$$q_i = p_0 \vee \cdots \vee p_{j-1} \vee p_j' \vee p_{j+1} \vee \cdots \vee p_{n-1}, \text{ where } p_j' \equiv p_j \text{ and } r(p_j') < r(p_j),$$

$$q_{ii} = p_0 \vee \cdots \vee p_{j-1} \vee p_{j+1} \vee \cdots \vee p_{n-1},$$

$$q_{iii} = p_0 \vee \cdots \vee p_{j-1} \vee p_j^{(i)} \vee p_{j+1} \vee \cdots \vee p_{n-1},$$

$$q_{iv} = p_0 \vee \cdots \vee p_{j-1} \vee p_j' \vee p_{j+1} \vee \cdots \vee p_{n-1},$$

$$q_v = p_0 \vee \cdots \vee p_{j-1} \vee p_{j+1} \vee \cdots \vee p_{k-1} \vee p_{k+1} \vee \cdots \vee p_{n-1} \vee q,$$

where $q \in L_i$ and $q = p_j \vee p_k$ in L_i. It is obvious, that if condition (y) fails ($i \leq y \leq v$), then $p \equiv q_y$ and $r(q_y) < r(p)$, contradicting the minimality of p.

To prove the converse, we show that if $p, q \in \mathbf{P}(Q)$, $p \equiv q$, p satisfies (i)–(v), and q is minimal, then $r(p) = r(q)$. This gives us an algorithm to reduce any polynomial to a minimal one and, at the same time, verifies the converse. Indeed, if p was not minimal, there would be a q with $p \equiv q$, $r(p) > r(q)$, and q minimal, contradicting the above statement.

So let p and q be given as specified, $p = p_0 \vee \cdots \vee p_{n-1}$, $n > 1$.

Firstly, we claim that $r(q) > 1$. Indeed, if $r(q) = 1$, then $q \in L_i$, for some $i \in I$. Thus, by Lemma 6(ii), $p_{(i)} = p^{(i)} = q$. At most one component of p is in Q. Indeed, if p_j, $p_k \in Q$, $j \neq k$, $p_j \in L_{i_j}$, $p_k \in L_{i_k}$, then by (v), $i_j \neq i_k$, and so one of them is not i, say $i \neq i_j$. But then $p_{i_j} \leq p$, hence p_{i_j} is proper, which contradicts, by Corollary 4, the fact that $p^{(i)}$ is proper. Thus there is a j with $0 \leq j < n$ and $r(p_j) > 1$. Then $p_j^{(i)} \leq p^{(i)} = q$, and so $p_j^{(i)}$ is proper and $p_j^{(i)} \leq q = p_{(i)}$, contradicting the fact that p satisfies (iii).

Secondly, we claim that q is not of the form $q_0 \wedge q_1$. Indeed, let $q = q_0 \wedge q_1$. Consider $q \subseteq p$. If (C) applies, then $q^{(i)} \leq p_{(i)} = q_{(i)}$, hence $q_{(i)} = q^{(i)} \equiv q$, contradicting $r(q) > 1$ and the minimality of q. If ($\wedge$W) applies, then $q_0 \subseteq p$ or $q_1 \subseteq p$. Obviously, $p \subseteq q_i$, thus if, say $q_0 \subseteq p$, then $p \equiv q_0$, contradicting the minimality of q. Finally, if ($W_\vee$) applies, then $q \subseteq p_0 \vee \cdots \vee p_{n-2}$ or $q \subseteq p_{n-1}$.

The first possibility yields $p_{n-1} \subseteq p_0 \vee \cdots \vee p_{n-2}$, while the second gives $p_{n-1} \equiv p$ (and therefore $p_0 \subseteq p_1 \vee \cdots \vee p_{n-1}$), both contradicting the fact that p satisfies (ii). Thus q is of the form

$$q = q_0 \vee \cdots \vee q_{m-1}, \quad m > 1,$$

where no q is the join of two polynomials.

Next we show that there are functions

$$f: \{0, 1, \ldots, n-1\} \to \{0, 1, \ldots, m-1\},$$
$$g: \{0, 1, \ldots, m-1\} \to \{0, 1, \ldots, n-1\}$$

satisfying the following conditions:

(a) $g(f(j)) = j$, for $0 \le j < n$, and $f(g(j)) = j$, for $0 \le j < m$.

(b) If $0 \le j < n$ and $r(p_j) > 1$, then $q_{f(j)} \equiv p_j$ and $r(q_{f(j)}) = r(p_j)$, and similarly for any $0 \le j < m$ with $r(q_i) > 1$.

(c) If $p_j \in L_i$ ($0 \le j < n$ and $i \in I$), then $q_{f(j)} \in L_i$, and similarly for $q_j \in L_i$.

Let $0 \le j < n$ and $r(p_j) > 1$. Then $p_j \subseteq q$. If (C) is applicable, then $p_j^{(i)} \le q_{(i)}$, for some $i \in I$; since $q_{(i)} = p_{(i)}$, we obtain that $p_j^{(i)} \le p_{(i)}$, contradicting condition (iii) for p. ($\wedge$W) is not applicable either because it would contradict that p satisfies (iv). Hence only ($W_\vee$) is applicable. Thus $p_j \subseteq q_0 \vee \cdots \vee q_{m-2}$ or $p_j \subseteq q_{m-1}$. Continuing this argument, we conclude that $p_j \subseteq q_{f(j)}$, for some $0 \le f(j) < m$. If $q_{f(j)} \in L_i$, for some $i \in I$, then $p_j \subseteq q_{f(j)}$ implies that $p_j^{(i)} \le q_{f(j)}$ in L_i, thus $p_j^{(i)} \le q_{f(j)} \le q_{(i)} = p_{(i)}$, contradicting the fact that condition (iii) holds for p. Therefore, $r(q_{f(j)}) > 1$.

Since q is minimal, it satisfies (i)–(v), hence we can define the function q. Thus

$$p_j \subseteq q_{f(j)} \subseteq p_{g(f(j))},$$

and so by (ii), $g(f(j)) = j$ and $p_j \equiv q_{f(j)}$. Similarly, $f(g(j)) = j$ and $q_j \equiv p_{g(j)}$.

Now let $0 \le j < n$ and $r(p_j) = 1$. Then $p_j \in L_i$, for some $i \in I$. Since $p_j \subseteq q$, we get that $q_{(i)}$ is proper. Since $q_{(i)} = (q_0)_{(i)} \vee \cdots \vee (q_{m-1})_{(i)}$, some $(q_j)_{(i)}$ must be proper. By renumbering $q_0, \ldots, q_{m-1}$, we get that $(q_k)_{(i)}$ is proper iff $0 \le k < t$, where $t \le m$. Thus

$$q_{(i)} = (q_0)_{(i)} \vee \cdots \vee (q_{t-1})_{(i)}.$$

If $r(q_s) > 1$, for all $0 \le s < t$, then $r(p_{g(s)}) > 1$ and $(p_{g(s)})_{(i)} = (q_s)_{(i)}$. Therefore, $g(s) \ne j$, for all $0 \le s < t$, and so

$$p_j \le (q_0)_{(i)} \vee \cdots \vee (q_{t-1})_{(i)} \le (p_0 \vee \cdots \vee p_{j-1} \vee p_{j+1} \vee \cdots \vee p_{n-1})_{(i)}.$$

Thus $p_j \subseteq p_0 \vee \cdots \vee p_{j-1} \vee p_{j+1} \vee \cdots \vee p_{n-1}$, contradicting the fact that (ii) holds for p.

Consequently, we can choose $0 \leq f(j) < n$ such that $r(q_{f(j)}) = 1$ and $(q_{f(j)})_{(i)}$ is proper, that is, such that $q_{f(j)} \in L_i$. Similarly, we define $g(j)$, for $0 \leq j < m$ and $r(q_j) = 1$. It is obvious that (a), (b), and (c) are satisfied.

Now (a) implies that f and g are one-to-one, hence $n = m$. Since $r(p_j) = r(q_{f(j)})$, we conclude that $r(p) = r(q)$. $\qquad\square$

A *subset* A of a lattice L is said to *satisfy* (SD$_\vee$) iff (SD$_\vee$) holds, for all x, y, $z \in A$. The following result (G. Grätzer and H. Lakser [1974]) establishes for (SD$_\vee$) what was done for (W) in Theorem 15.

Theorem 17.   Let the lattice L be a free product of the lattices L_i, $i \in I$. For each $i \in I$, let A_i be a subset of L_i^b satisfying (SD$_\vee$). Let A be a subset of L such that $A_{(i)} \subseteq A_i$, for all $i \in I$. Then A satisfies (SD$_\vee$).   —

Proof. Let

$$x = R(p),\ y = R(q),\ z = R(s) \in A, \quad x \vee y = x \vee z,$$

and let

$$u = u_0 \vee \cdots \vee u_{n-1}, \quad n \geq 1,$$

be a representation of $p \vee q \equiv p \vee s$ satisfying (i)–(v) of Theorem 16. We show that, for each j with $0 \leq j < n$ and $r(u_j) > 1$, we have $u_j \subseteq p$ or $u_j \subseteq q$. Consider $u_j \subseteq p \vee q$. If (C) applies, then $u_j^{(i)} \leq (p \vee q)_{(i)} = u_{(i)}$, contradicting 16(iii). If ($_\wedge$W) applies, that is, if $u_j = u_j' \wedge u_j''$ and $u_j' \subseteq p \vee q$, we get a contradiction with 16(iv). Thus only (W$_\vee$) can apply, yielding $u_j \subseteq p$ or $u_j \subseteq q$. Similarly, $u_j \subseteq p$ or $u_j \subseteq s$, hence $u_j \subseteq p \vee (q \wedge s)$. Now take a j with $0 \leq j < n$ and $r(u_j) = 1$. Then $u_j = u_{(i)}$, for some $i \in I$. Therefore, $u_{(i)} = p_{(i)} \vee q_{(i)}$ and $u_{(i)} = p_{(i)} \vee s_{(i)}$ in L_i^b and $p_{(i)}, q_{(i)}, s_{(i)} \in A_i$, and so $u_j \leq p_{(i)} \vee (q_{(i)} \wedge s_{(i)})$ by (SD$_\vee$). Thus, $u_j \subseteq p \vee (q \wedge s)$ by (C). Since $u_j \subseteq p \vee (q \wedge s)$ is proved for all j with $0 \leq j < n$, we conclude that $u_0 \vee \cdots \vee u_{n-1} \subseteq p \vee (q \wedge s)$, and so $u = x \vee (y \wedge z)$, as claimed. $\qquad\square$

By dualizing Theorem 17, we can obtain the analogous result for (SD$_\wedge$).

We conclude this section by proving that the *Common Refinement Property* holds for free products (G. Grätzer and J. Sichler [1975]).

Theorem 18.   Let L be a free product of A_0 and A_1 and also of B_0 and B_1. Then L is a free product of $(A_i \cap B_j \mid i,\ j = 0,\ 1,\ A_i \cap B_j \neq \varnothing)$.   —

Proof. Let $a \in A_0$. We claim that

$$a_{B_0} \in (A_0 \cap B_0) \cup \{0^b\}.$$

Indeed, since L is generated by $B_0 \cup B_1$,

$$a = p(b_{0,0}, b_{0,1}, \ldots, b_{1,0}, b_{1,1}, \ldots),$$

where $b_{0,0}, b_{0,1}, \ldots \in B_0$ and $b_{1,0}, b_{1,1}, \ldots \in B_1$. Computing the lower A_0-covers and observing that $a_{A_0} = a$, we obtain

$$a = p((b_{0,0})_{A_0}, (b_{0,1})_{A_0}, \ldots, (b_{1,0})_{A_0}, (b_{1,1})_{A_0}, \ldots).$$

Forming lower B_0-covers in the original expression for a, we get

$$a_{B_0} = p(b_{0,0}, b_{0,1}, \ldots, 0^b, 0^b, \ldots),$$

and from the previous formula:

$$a_{B_0} = (a_{A_0})_{B_0} = p(((b_{0,0})_{A_0})_{B_0}, \ldots, ((b_{1,0})_{A_0})_{B_0}, \ldots).$$

But $b_{1,m} \geq (b_{1,m})_{A_0}$ and so $0^b = (b_{1,m})_{B_0} \geq ((b_{1,m})_{A_0})_{B_0}$, hence $((b_{1,m})_{A_0})_{B_0} = 0^b$. Thus

$$a_{B_0} = p(b_{0,0}, b_{0,1}, \ldots, 0^b, 0^b, \ldots)$$
$$= p(((b_{0,0})_{A_0})_{B_0}, ((b_{0,1})_{A_0})_{B_0}, \ldots, 0^b, 0^b, \ldots).$$

Since p is isotone and $b_{0,m} \geq (b_{0,m},)_{A_0} \geq ((b_{0,m})_{A_0})_{B_0}$, we obtain

$$a_{B_0} = p(b_{0,0}, b_{0,1}, \ldots, 0^b, 0^b, \ldots) \geq p((b_{0,0})_{A_0}, (b_{0,1})_{A_0}, \ldots, 0^b, 0^b, \ldots)$$
$$\geq p(((b_{0,0})_{A_0})_{B_0}, ((b_{0,1})_{A_0})_{B_0}, \ldots, 0^b, 0^b, \ldots) = a_{B_0},$$

and so

$$a_{B_0} = p((b_{0,0})_{A_0}, (b_{0,1})_{A_0}, \ldots, 0^b, 0^b, \ldots) \in A_0 \cup \{0^b\}.$$

By definition, $a_{B_0} \in B_0 \cup \{0^b\}$, hence $a_{B_0} \in (A_0 \cap B_0) \cup \{0^b\}$. Similarly, $a_{B_j} \in (A_i \cap B_j) \cup \{0^b\}$, for $i, j \in \{0, 1\}$. If follows immediately, that

$$a = p((b_{0,0})_{A_0}, (b_{0,1})_{A_0}, \ldots, (b_{1,0})_{A_0}, (b_{1,1})_{A_1}, \ldots)$$
$$\in (A_0 \cap B_0) \cup (A_0 \cap B_1) \cup \{0^b\}.$$

Now a simple induction on the rank of a polynomial proves that, for $a \in A_0$ and for the polynomial p of smallest rank representing a in the form

$$a = p(a_0, \ldots, a_{n-1}), \quad a_0, \quad \ldots, \quad a_{n-1} \in (A_0 \cap B_0) \cup (A_0 \cap B_1) \cup \{0^b\},$$

no a_i is 0^b. We conclude that

$$A_0 \subseteq [(A_0 \cap B_0) \cup (A_0 \cap B_1)].$$

Thus

$$L = [A_0 \cup A_1] = [\bigcup (A_i \cap B_j \mid i, \ j = 0, \ 1)].$$

Applying Corollary 13 twice, we get

$$A_0 = (A_0 \cap B_0) * (A_0 \cap B_1), \qquad A_1 = (A_1 \cap B_0) * (A_1 \cap B_1),$$
$$B_0 = (A_0 \cap B_0) * (A_1 \cap B_0), \qquad B_1 = (A_0 \cap B_1) * (A_1 \cap B_1),$$

hence

$$L = (A_0 \cap B_0) * (A_0 \cap B_1) * (A_1 \cap B_0) * (A_1 \cap B_1)$$

(to be more precise, drop all $A_i \cap B_j = \varnothing$), and this is the common refinement of $A_0 * A_1 = B_0 * B_1$. $\qquad\qquad\square$

Exercises

1. Prove that $C_2 * C_1$ is indeed the lattice of Figure 1.

2. Show that $C_3 * C_1$ is the lattice of Figure 2.

3. Find an infinite descending chain in $C_4 * C_1$. (Hint: Let $a_0 < a_1 < a_2 < a_3$ and b be the two chains. Define

$$c_1 = ((a_2 \wedge (a_1 \vee b)) \vee (a_3 \wedge b)) \wedge ((a_3 \wedge (a_0 \vee b)) \vee a_1)$$

and, for $n > 1$,

$$c_n = ((c_{n-1} \wedge a_2) \vee (a_3 \wedge b)) \wedge ((c_{n-1} \wedge (a_0 \vee b)) \vee a_1).$$

Then $c_1 > c_2 > c_3 > \cdots$.)

4. Construct an infinite descending chain in $C_2 * C_2$. (Hint: Let $a_0 < a_1$ and $b_0 < b_1$ be the two chains. Define

$$c_1 = ((a_1 \wedge (a_0 \vee b_1)) \vee b_0) \wedge ((b_1 \wedge (a_1 \vee b_0)) \vee a_0)$$

and, for $n > 1$,

$$c_n = ((c_{n-1} \wedge a_1) \vee b_0) \wedge ((c_{n-1} \wedge b_1) \vee a_0).)$$

5. $A * B$ is finite iff A or B is the one-element lattice and the other is a chain of not more than three elements (Ju. I. Sorkin [1952]).

6. Let $a_1 < a_2$ and $b_1 < b_2$ be two chains. Introduce the following notation (see Figure 6):

$$A_1 = a_2,$$
$$B_1 = b_2,$$
$$A'_1 = A_1,$$
$$B'_1 = b_1,$$

and, for $n > 1$,

$$A_n = a_2 \wedge (a_1 \vee B_{n-1}),$$
$$B_n = b_2 \wedge (b_1 \vee A_{n-1}),$$
$$C_n = a_1 \vee B_n,$$
$$D_n = b_1 \vee A_n,$$
$$P_n = A_n \vee B_n,$$
$$Q_n = C_n \wedge D_n,$$
$$M_1 = a_1 \vee b_1,$$
$$M_2 = (a_2 \wedge b_2) \vee a_1 \vee b_1,$$
$$V_1 = b_2 \wedge ((a_2 \wedge b_2) \vee a_1 \vee b_1),$$
$$V_2 = (a_2 \wedge b_2) \vee (b_2 \wedge (a_1 \vee b_1)),$$
$$V_3 = b_2 \wedge (a_1 \vee b_1),$$
$$W_1 = a_2 \wedge ((a_2 \wedge b_2) \vee a_1 \vee b_1),$$
$$W_2 = (a_2 \wedge b_2) \vee (a_2 \wedge (a_1 \vee b_1)),$$
$$W_3 = a_2 \wedge (a_1 \vee b_1).$$

Prove that these polynomials and their duals (denoted by $'$) represent distinct elements and all the elements of $C_2 * C_2$ as shown on Figure 6 (H. L. Rolf [1958]).

7. Show that Figure 7 gives a description of $C_4 * C_1$ (H. L. Rolf [1958]).

8. To what extent can Theorem 10 be simplified for free products of chains. (Hint: replace (C) in Definition 5 by "For some $i \in I$, p, $q \in L_i$ and $p \leq q$ in L_i".)

9. Define the concept of a free $\{0, 1\}$-product of bounded lattices. Develop the theory of free $\{0, 1\}$-product.

10. Let L be the free $\{0, 1\}$-product of the lattices L_i^b, $i \in I$. Show that $L - \{0, 1\}$ is a sublattice of L and it is a free product of L_i, $i \in I$.

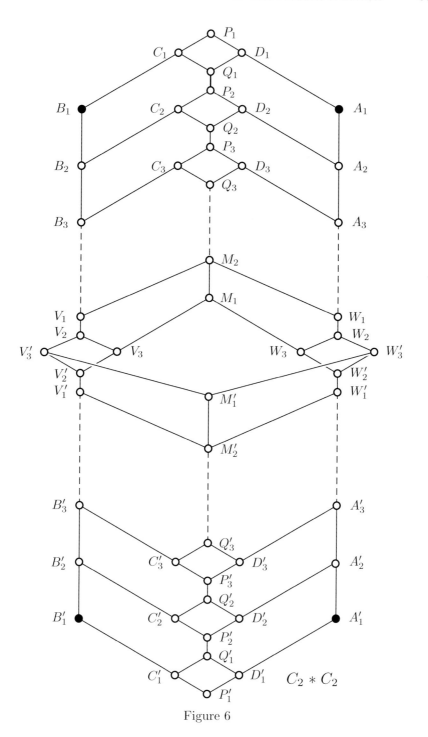

Figure 6

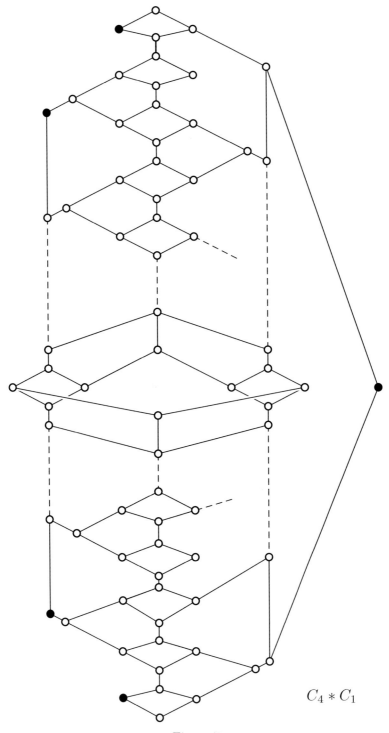

$C_4 * C_1$

Figure 7

11. Let L be a free product of the L_i, $i \in I$. Show that $a \in L_i$ is join-irreducible in L iff a is join-irreducible in L_i. (Hint: use only covers in the proof.)

12. Show that Exercise 11 holds for free **K**-products in general (B. Jónsson [1971]).

13. Let L be a free product of the L_i, $i \in I$, and let $\varphi_i \colon L_i \to A$ be an isotone map of L_i into a lattice A, for $i \in I$. Show that there is an isotone map $\varphi \colon L \to A$ extending the φ_i (Ju. I. Sorkin [1952]).

14. In a free product of chains, the minimal representation of an element is uniquely determined up to commutativity and associativity.

15. Define the *canonical representation p* of an element of a free product in the following way: $p \in Q$; or p is as in Theorem 16, all p_i are canonical and if $p_i \in L_j$, then $p_i = p_{(j)}$; and dually. Show that the canonical representation is uniquely determined up to commutativity and associativity (H. Lakser [1970a]).

16. Find a pair of lattices A, B, at least one of which is not a free product of chains and such that every minimal representation is canonical in $A * B$ (H. Lakser [1970a]).

17. Let L be a free product of L_i, $i \in I$. Under what conditions is $L - L_i$ a sublattice of L?

18. Let the set I be the disjoint union of the sets I_j, $j \in J$. Let L_i, $i \in I$, be lattices and let A_j be a free product of L_i, $i \in I_j$. Then A is a free product of L_i, $i \in I$, iff A is a free product of A_j, $j \in J$.

2. The Structure of Free Lattices

By comparing the definition of a free lattice $\mathbf{F}(\mathfrak{m})$ (see Section I.5) with the definition of a free product (see Section I.5 and Definition 1.1), we observe:

A lattice is free iff it is a free product of a family of one-element lattices.

Thus the results of the previous section can be specialized to describe the structure of free lattices.

Definition 1. *For a set X and $p, q \in \mathbf{P}(X)$, set $p \subseteq q$ iff it follows from $x \subseteq x$, for $x \in X$, and from $(\wedge W)$, $(\vee W)$, $(W_\wedge)$, and $(W_\vee)$.*
   Again we set

$$p \equiv q \quad \text{iff} \quad p \subseteq q \text{ and } q \subseteq p \qquad (p, \ q \in \mathbf{P}(X)).$$
$$R(p) = \{\, q \mid q \in \mathbf{P}(X) \text{ and } p \equiv q \,\} \qquad (p \in \mathbf{P}(X)).$$
$$R(X) = \{\, R(p) \mid p \in \mathbf{P}(X) \,\}.$$
$$R(p) \le R(q) \quad \text{iff} \quad p \subseteq q.$$

Now we can state the celebrated result of P. M. Whitman [1941]:

Theorem 2. $R(X)$ is a free lattice on $|X|$ generators. In other words, Definition 1 provides an algorithm to decide whether $a \leq b$ in the free lattice, where a and b are represented by p and q, respectively.   —

Proof. Compare Definition 1 with Definition 1.5. To make the comparison, we have to set $X = I$, $L_i = \{i\}$, for $i \in I$. It is sufficient to show that if $p \subseteq q$ by (C), then $p \subseteq q$ by Definition 1. We prove this by induction. This is obvious if $p, q \in X$. Now let $r(p) + r(q) > 2$, $p^{(i)} \leq q_{(i)}$, for some $i \in I$, that is, $p^{(i)} = q_{(i)} = i$. If $p = p_0 \wedge p_1$, then $(p_0 \wedge p_1)^{(i)} = p_0^{(i)} \wedge p_1^{(i)}$, hence $p_0^{(i)} = i$ or $p_1^{(i)} = i$. Thus $p_0^{(i)} \leq q_{(i)}$ or $p_1^{(i)} \leq q_{(i)}$, hence by the induction hypothesis, $p_0 \subseteq q$ or $p_1 \subseteq q$, and so by $(\wedge W)$, $p \subseteq q$. We proceed similarly in the other three cases.   □

From Theorem 1.16, we learned that minimal polynomials differ only in some L_i component. If all $|L_i| = 1$, they have to be identical.

Theorem 3 (P. M. Whitman [1941]).   A minimal representation of an element is unique up to commutativity and associativity. Let $p \in \mathbf{P}(X)$. Then p is minimal iff $p \in X$, or if $p = p_0 \vee \cdots \vee p_{n-1}$, $n > 1$, where no p_j is a join of more than one polynomial and conditions (i)–(iii) below hold, or dually.

(i) Each p_j is minimal, $0 \leq j < n$.

(ii) For each j with $0 \leq j < n$, $p_j \not\subseteq p_0 \vee \ldots \vee p_{j-1} \vee p_{j+1} \vee \ldots \vee p_{n-1}$.

(iii) If $p_j = p_j' \wedge p_j''$, where $0 \leq j < n$ and $p_j', p_j'' \in \mathbf{P}(X)$, then $p_j' \not\subseteq p$ and $p_j'' \not\subseteq p$.   —

Proof. In the special case considered, 1.16(v) is made superfluous by 1.16(ii). Also, 1.16(iii) does not apply since $p_j^{(i)}$, $p_{(i)} \in L_i$, hence $p_j^{(i)} = p_{(i)}$. The remaining conditions of Theorem 1.16 are identical with (i)–(iii) above.   □

A minimal representation of an element in a free lattice is called a *canonical form* of the element in the literature.

Now we shall study in some detail the structure of $F(3)$. Let x_0, x_1, and x_2 be the free generators, $X = \{x_0, x_1, x_2\}$. The zero and unit of $F(3)$ are $x_0 \wedge x_1 \wedge x_2$ and $x_0 \vee x_1 \vee x_2$, respectively. By Lemma I.5.9, $x_0 \vee x_1$, $x_0 \vee x_2$, and $x_1 \vee x_2$ generate a sublattice isomorphic to C_2^3. Since

$$F(3) = \{x_0\} * \{x_1 \wedge x_2, x_1, x_2, x_1 \vee x_2\}$$

(see Exercise 1.18), the Splitting Theorem (Theorem 1.11) gives that, for any $x \in F(3)$, either $x \geq x_0$ or $x \leq x_1 \vee x_2$. Similarly, $x \geq x_1$ or $x \leq x_0 \vee x_2$; and

$x \geq x_2$ or $x \leq x_0 \vee x_1$. From this, we can infer that

$$x_0 \vee x_1 \prec x_0 \vee x_1 \vee x_2,$$
$$(x_0 \vee x_1) \wedge (x_0 \vee x_2) \prec x_0 \vee x_1;$$

these and the symmetric cases give the nine coverings at the top of Figure 1. Let us prove the second covering; if

$$(x_0 \vee x_1) \wedge (x_0 \vee x_2) < t \leq x_0 \vee x_1,$$

then $t \not\geq x_0 \vee x_2$, hence $t \geq x_1$; similarly, $t \geq x_0$, hence $t \geq x_0 \vee x_1$, proving that $t = x_0 \vee x_1$.

If $t > x_0$, then $t \not\leq x_0$, and so $t \geq x_1 \wedge x_2$; hence

$$x_0 \prec x_0 \vee (x_1 \wedge x_2) = x_0 \vee a;$$

by symmetry and duality, this accounts for six coverings in the middle of the diagram. The other six are typified by

$$x_0 \vee a \succ (x_0 \vee a) \wedge b,$$

which is again a trivial consequence of the Splitting Theorem.

Again, easy applications of the Splitting Theorem yield that all elements of rank 4 or more lie in one of the intervals:

$$[x_0 \vee a, (x_0 \vee x_1) \wedge (x_0 \vee x_2)],$$

its two symmetric intervals, the three dual intervals, and $[a, b]$. Thus to complete the proof of Figure 1, we have to verify that

$$[x_0 \vee a, (x_0 \vee x_1) \wedge (x_0 \vee x_2)] = [x_0 \vee a, x_0 \vee b] \cup \{(x_0 \vee x_1) \wedge (x_0 \vee x_2)\}.$$

The trick (due to P. Gumm) is to verify first, by induction on $r(x)$, that if $x \in [x_0 \vee a, (x_0 \vee x_1) \wedge (x_0 \vee x_2)]$, then x is comparable with $x_0 \vee b$. Condition (W), and therefore Theorem 5, is needed in this step. Then $x_0 \vee b \prec (x_0 \vee x_1) \wedge (x_0 \vee x_2)$ follows easily. The details are left to the reader.

A word of warning about Figure 1: it gives a "generalized diagram" of F(3) but it does not contain all the elements or even all the partial orderings of the elements shown. For instance, $a < x_0 \vee a$ is not shown nor is the element $(x_0 \wedge b) \vee (x_1 \wedge b)$ (the u_0 defined below), an element of $[a, b]$. Nevertheless, Figure 1 can be very useful in visualizing F(3).

We obtain some additional coverings in F(3) from a result of R. A. Dean [1961a]:

Theorem 4.   For any

$$c \in [x_0 \vee (x_1 \wedge x_2), (x_0 \vee x_1) \wedge (x_0 \vee x_2)],$$

we have that

$$c \wedge (x_1 \vee x_2) \prec (c \wedge (x_1 \vee x_2)) \vee x_0. \qquad\qquad —$$

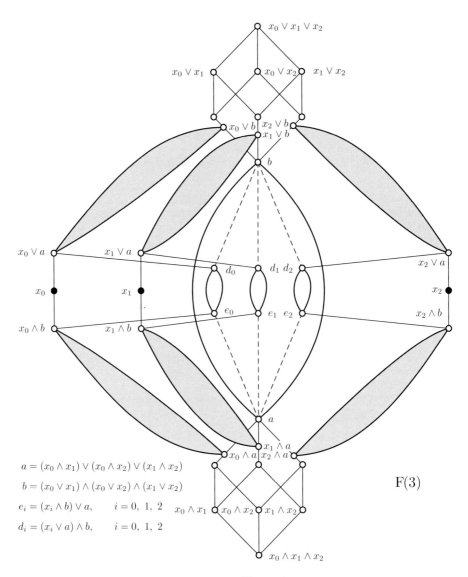

$$a = (x_0 \wedge x_1) \vee (x_0 \wedge x_2) \vee (x_1 \wedge x_2)$$
$$b = (x_0 \vee x_1) \wedge (x_0 \vee x_2) \wedge (x_1 \vee x_2)$$
$$e_i = (x_i \wedge b) \vee a, \qquad i = 0,\ 1,\ 2$$
$$d_i = (x_i \vee a) \wedge b, \qquad i = 0,\ 1,\ 2$$

F(3)

Figure 1

Proof. These two elements are obviously distinct, since they belong to disjoint intervals. If

$$c \wedge (x_1 \vee x_2) \leq t \leq (c \wedge (x_1 \vee x_2)) \vee x_0$$

and $t \geq x_0$, then obviously $t = (c \wedge (x_1 \vee x_2)) \vee x_0$. Otherwise, $t \leq x_1 \vee x_2$, hence

$$t \leq ((c \wedge (x_1 \vee x_2)) \vee x_0) \wedge (x_1 \vee x_2) \leq c \wedge (x_1 \vee x_2),$$

and so $t = c \wedge (x_1 \vee x_2)$.   □

Apart from the Splitting Theorem, the most important properties of a free lattice are given in the next result, which is due to P. M. Whitman [1941] and B. Jónsson [1961]:

Theorem 5.   Conditions (W), (SD$_\wedge$), and (SD$_\vee$) hold in any free lattice. —

Proof. If $|A_i| = 1$, then $A_i^b = C_3$, hence it satisfies all three conditions. So choose $A_i = L_i^b$, and the result follows from Theorems 1.15 and 1.17.   □

We illustrate the power of these conditions by two results.

Theorem 6 (B. Jónsson [1961]).   A sublattice A of finite length of a free lattice (of any lattice satisfying (SD$_\wedge$)) is finite.   —

Proof. Let A be of length n. We prove the statement by induction on n. If $n = 1$, the statement is trivial. Now let A be of length n and let a be an atom of A. Then

$$A = [a) \cup \{\, x \mid a \wedge x = 0 \,\}.$$

By (SD$_\wedge$), $\{\, x \mid a \wedge x = 0 \,\}$ is a sublattice. Hence A is a union of two sublattices of length less than n, and so A is a union of two finite sets. Thus A is finite. □

A sublattice of a free lattice need not be free. The next result shows how free sublattices can be found.

Theorem 7.   Let the lattice L be generated by X. If X is *irredundant*, that is,

$$x \leq \bigvee Y \qquad \text{implies that} \qquad x \in Y, \quad \text{for any } x \in X \text{ and finite } Y \subseteq X,$$

and dually, then L is a free lattice iff L satisfies (W).   —

Proof. The necessity of these conditions follows from the Splitting Theorem and from Theorem 5. Now let these conditions hold. With every $a \in L$, associate the set $S(p)$ of all $p \in \mathbf{P}(X)$ that represent a. If $a = S(p)$, $b = S(q)$, then $a \leq b$ iff $p \subseteq q$ by Definition 1. This is easily seen except if $p = p_0 \wedge p_1$ and $q = q_0 \vee q_1$, in which case it follows from (W). Thus L is a free lattice. □

Now consider in $F(3)$ the elements

$$u_0 = (x_0 \wedge (x_1 \vee x_2)) \vee (x_1 \wedge (x_0 \vee x_2)),$$
$$u_1 = (x_0 \wedge (x_1 \vee x_2)) \vee (x_2 \wedge (x_0 \vee x_1)),$$
$$u_2 = (x_0 \vee (x_1 \wedge x_2)) \wedge (x_1 \vee (x_0 \wedge x_2)),$$
$$u_3 = (x_0 \vee (x_1 \wedge x_2)) \wedge (x_2 \vee (x_0 \wedge x_1)).$$

It is an easy computation to show that $U = \{u_0, u_1, u_2, u_3\}$ is irredundant. Hence $F(3) \supset F(4)$ as a sublattice. Now define $y_0 = u_0$, $a_0 = u_1$, $b_0 = u_2$, $c_0 = u_3$. If y_n, a_n, b_n, and c_n are defined, construct y_{n+1}, a_{n+1}, b_{n+1}, and c_{n+1} from a_n, b_n, and c_n, just as the u_i were constructed from x_0, x_1, and x_2.

We claim that $Y = \{y_0, y_1, y_2, \dots\}$ is an irredundant set. To see this, we prove by induction that $y_0, \dots, y_n, a_n, b_n, c_n$ generate a free lattice on $n + 4$ generators. This is true for $n = 0$. Assume it for n. Then

$$[y_0, \dots, y_n, y_{n+1}, a_{n+1}, b_{n+1}, c_{n+1}]$$

is the sublattice generated by $[y_0, \dots, y_n]$ and $[y_{n+1}, a_{n+1}, b_{n+1}, c_{n+1}]$ in

$$[y_0, \dots, y_n, a_n, b_n, c_n] = [y_0, \dots, y_n] * [a_n, b_n, c_n].$$

Thus by Theorem 1.14,

$$[y_0, \dots, y_n, y_{n+1}, a_{n+1}, b_{n+1}, c_{n+1}] \cong [y_0, \dots, y_n] * [y_{n+1}, a_{n+1}, b_{n+1}, c_{n+1}]$$
$$\cong F(n+1) * F(4) \cong F(n+5).$$

This proves the statement. Thus Y is irredundant and so, by Theorem 7, $[Y] \cong F(\aleph_0)$. So we have proved:

Theorem 8 (P. M. Whitman [1941]). $F(3)$ contains $F(\aleph_0)$ as a sublattice.   —

A free generator of a free lattice is doubly irreducible, and, of course, no other element is such. Hence we obtain:

Theorem 9.   There is a one-to-one correspondence between automorphisms of a free lattice and permutations of its free generating set.   —

This is used in proving the following result of F. Galvin and B. Jónsson [1961].

Theorem 10.   Every chain of a free lattice is countable.   —

Proof. Let X be the free generating set of L. We can assume that X is not countable. Let X_0 be a countable subset of L, let $L_0 = [X_0]$, and let C be a chain in L.

For a, $b \in L$, let $a \equiv b$ mean that $a\varphi = b$, for some automorphism φ of L. Obviously, $\equiv$ is an equivalence relation. Now if $a \in L$, then $a \in [Y]$, for some finite $Y \subseteq X$. Take a permutation p of X with $Yp \subseteq X_0$, and let φ be the automorphism of L extending p. Then $a\varphi \in L_0$. Hence there are no more equivalence classes than $|L_0| = \aleph_0$. Therefore the proof will be complete if we show that, for any a, $b \in C$, $a \neq b$, we have $a \not\equiv b$. Indeed, let $a < b$ and $a\varphi = b$, for some automorphism φ. For some finite $Y \subseteq X$, $a \in [Y]$. There is a permutation ϱ of X such that $x\varrho = x\varphi$, for $x \in Y$ and $x\varrho = x$, for $x \in X - (Y \cup Y\varphi)$. Let ψ be the automorphism of L extending ϱ. Then $a\psi = b$ and ψ is of some finite order n (as a group element). Thus,

$$a < a\psi < a\psi^2 < \cdots < a\psi^n = a,$$

a contradiction.

□

A somewhat unexpected dividend of the study of the structure of free lattices is the following result of R. P. Dilworth [1945] and R. A. Dean [1956]:

Theorem 11.   There exists a three-generated partial lattice with infinitely many maximal elements.   —

Proof. Let x_0, x_1, x_2 be the free generators of $F(3)$ and define $y_0 = x_0$,

$$y_n = x_0 \vee (x_1 \wedge (x_2 \vee (x_0 \wedge (x_1 \vee (x_2 \wedge y_{n-1}))))),$$
$$y_{-n} = x_0 \wedge (x_1 \vee (x_2 \wedge (x_0 \vee (x_1 \wedge (x_2 \vee y_{-n+1}))))),$$
$$a_n = x_1 \vee (y_n \wedge (y_{-n} \vee x_2)).$$

It is easy to check that $[a_1, a_2, \ldots] \cong F(\aleph_0)$.

For a polynomial p, define the *component subset* of p, $\mathrm{Komp}(p)$, as follows:

$$\mathrm{Komp}(x_i) = \{x_i\};$$
$$\mathrm{Komp}(p_0 \wedge p_1) = \mathrm{Komp}(p_0) \cup \mathrm{Komp}(p_1) \cup \{p_0 \wedge p_1\},$$

and similarly for $p_0 \vee p_1$.

If $a \in F(3)$, let $a = R(p)$, where p is the canonical polynomial representing a. Define

$$\mathrm{Komp}(a) = \{\, R(q) \mid q \in \mathrm{Komp}(p) \,\}.$$

Then $\mathrm{Komp}(a) \subseteq F(3)$. Regard $\mathrm{Komp}(a)$ as a relative sublattice of $F(3)$. Then $\mathrm{Komp}(a)$ becomes a partial lattice. $\mathrm{Komp}(a)$ is generated by $\{x_0, x_1, x_2\} \cap \mathrm{Komp}(a)$.

Now define

$$A = \bigcup (\operatorname{Komp}(a_n) \mid n = 1,\ 2,\ \ldots\,).$$

All the elements of A are easily enumerated, and we observe that $a_1, a_2, \ldots$ are maximal elements of A. $\square$

Corollary 12 (Ju. I. Sorkin [1954] and R. A. Dean [1956]). *Let L be a countable lattice. Then L can be embedded in a three-generated lattice.*

Proof. Let $L = \{b_0, b_1, b_2, b_3, \ldots\}$ and, with the partial lattice A of Theorem 11, form the set $B = L \cup A$ (we assume that L and A are disjoint). We define a partial ordering $\leq$:

(i) for $u, v \in L$ or $u, v \in A$, $u \leq v$ iff $u \leq v$ in L or A, respectively;

(ii) $a_{2n+1} \leq b_n$ and $a_{2n+2} \leq b_n$;

(iii) for $u \in A$ and $v \in L$, $u \leq v$ iff there are $v_0, \ldots, v_{n-1} \in L$ such that

   (a) $v_0 \wedge \cdots \wedge v_{n-1} \leq v$ in L;
   (b) for each i, $0 \leq i < n$, there is a b_j such that $b_j \leq v_i$ in L and $u \leq a_{2j+1}$, $u \leq a_{2j+2}$ in A;

(iv) for $u \in A$ and $v \in L$, $v \leq u$ never holds.

Now it is easy to check that B is a partial lattice,

$$b_n = a_{2n+1} \vee a_{2n+2}, \quad \text{for } n = 0,\ 1,\ \ldots$$

and so B is three-generated. Every partial lattice B can be embedded in a lattice C and obviously, $[B]$ in C is three-generated and contains L as a sublattice. $\square$

Exercises

1. Let L be a lattice completely freely generated by the poset P. Show that an algorithm is provided for deciding whether $a \leq b$ in L by the rules: if $a, b \in P$, $a \subseteq b$ iff $a \leq b$ in P; and by the rules $(\wedge \mathrm{W})$, $(\vee \mathrm{W})$, $(\mathrm{W}\wedge)$, and $(\mathrm{W}\vee)$ (R. P. Dilworth [1945]).

2. Let L be a free product of L_i, $i \in I$, and let $Q = \bigcup (L_i \mid i \in I)$ (disjoint union). Show that L is completely freely generated by Q iff all L_i are chains.

3. Prove that C_2^n is a sublattice of a free lattice iff $n \leq 3$.

4. Let L be a modular lattice. Verify that if L is a sublattice of a free lattice, then L is distributive.

5. Let the lattice L be of length n. Show that if L is a sublattice of a free lattice, then $|L| \leq 2^n$.

6. Let the lattice L be generated by a finite set X and let us assume that $x \not\leq \bigvee(X - \{x\})$, for some $x \in X$. Then, for any $a \geq x$, we have that

$$a \wedge \bigvee(X - \{x\}) \prec (a \wedge \bigvee(X - \{x\})) \vee x.$$

7. Let $\mathfrak{m} \geq \aleph_0$. Show that there is no covering in $F(\mathfrak{m})$.

8. Let $\mathfrak{m} \geq \aleph_0$ and let $a, b \in F(\mathfrak{m})$, $a < b$. Show that $[a, b]$ has a sublattice isomorphic to $F(\mathfrak{m})$.

9. Let

$$f(n) = \max\{\,|L| \mid L \text{ is a sublattice of a free lattice and } L \text{ is of length } n\,\}.$$

Then $(\sqrt{2})^n \leq f(n)$ (B. Jónsson and J. E. Kiefer [1962]).

10. A *linear decomposition* A_i, $i \in I$, of a lattice A consists of a chain I, sublattices A_i of A, for $i \in I$, such that if $i, j \in I$, $i < j$, then $a < b$ in A, for all $a \in A_i$, $b \in B_j$, and $A = \bigcup(A_i \mid i \in I)$. Show that if all A_i are sublattices of a free lattice and I is countable, then A is also a sublattice of a free lattice.

11. A distributive lattice D is a sublattice of a free lattice iff D has a linear decomposition A_i, $i \in I$, such that $|I| \leq \aleph_0$ and each A_i is either C_1, or C_2^3, or of the form $C_2 \times C$ where C is a countable chain (F. Galvin and B. Jónsson [1961]).

12. Let A be an algebra with the following properties: a partial ordering $\leq$ is defined on A; A has a generating set X such that every permutation on X can be extended to an isotone automorphism of A. Prove that every chain in A is countable.

13. Enumerate in canonical form all the elements of A in the proof of Theorem 11.

14. Prove that B in the proof of Corollary 12 is a partial lattice.

15. Is there a general lemma about "gluing together" two partial lattices that contains the construction of Corollary 12?

16. Use component subsets to prove that if $a, b \in F(\aleph_0)$, $a \neq b$, then there is a homomorphism φ of $F(\aleph_0)$ onto a finite lattice such that $a\varphi \neq b\varphi$ (R. A. Dean [1956]).

17. Derive Exercise 16 from Corollary IV.4.6.

18. Prove that every finite lattice can be embedded in a finite three-generated lattice.

19. Using the notation of Figure 1, let $y_i \in F(3)$, $e_i \le y_i \le d_i$, i= 0, 1, 2. Prove that $[y_0, y_1, y_2]$ is a proper sublattice of $F(3)$ isomorphic to $F(3)$.

3.   Reduced Free Products

Let L_i, $i \in I$, be bounded lattices and let L be a free $\{0,1\}$-product of the L_i, $i \in I$ (see Definition II.5.14 and the discussion following it). As we shall see, a pair of elements x, y is *complementary in L* (that is, $x \wedge y = 0$ and $x \vee y = 1$) iff either they belong to some L_i and they are complementary in L_i or if there exist elements x_0, x_1, y_0, y_1 in some L_i such that $x_0 \le x \le y_0$, $x_1 \le y \le y_1$, and $\{x_0, x_1\}$, $\{y_0, y_1\}$ are complementary in L_i. We shall describe a construction in which there are many more complements than in the free $\{0,1\}$-product, but in which we can still keep track of the complements. We call this construction the *reduced free product*. Several applications will be given in this section and the next.

In the discussion that follows, let L_i, $i \in I$, $I \ne \varnothing$, be pairwise disjoint bounded lattices and let $Q = \bigcup(L_i \mid i \in I)$.

Definition 1. *A C-relation C on L_i, $i \in I$, is a set of two element subsets of Q such that if $\{a,b\} \in C$, then there exist i, $j \in I$, $i \ne j$, satisfying $a \in L_i - \{0_i, 1_i\}$ and $b \in L_j - \{0_j, 1_j\}$.*

Definition 2. *Let C be a C-relation on L_i, $i \in I$. A lattice L is a C-reduced free product of the L_i, $i \in I$, iff the following conditions hold:*

(i) *Each L_i, $i \in I$, is a $\{0,1\}$-sublattice of L and $L = [\bigcup(L_i \mid i \in I)]$.*

(ii) *If $\{a,b\} \in C$, then a, b is a complementary pair in L.*

(iii) *If, for $i \in I$, φ_i is a $\{0,1\}$-homomorphism of L_i into the bounded lattice A, and $\{a,b\} \in C$ ($a \in L_i$, $b \in L_j$, $i \ne j$) implies that $a\varphi_i$, $b\varphi_j$ are complementary in A, then there is a homomorphism φ of L into A extending all the φ_i, $i \in I$.*

Using the technique of Section I.5, the existence of a C-reduced free product is proved if we find a lattice satisfying (i) and (ii). Such a lattice is

$$\{0,1\} \cup \bigcup(L_i - \{0_i, 1_i\} \mid i \in I)$$

with the obvious partial ordering. The uniqueness of C-reduced free products can be established as in Section I.5. The next result shall give a description of a C-reduced free product based on the Structure Theorem of Free Products. This description is then used to describe the complementary pairs in a $\mathcal{C}$-reduced free product.

Theorems 4 and 5 appeared in their present form in G. Grätzer [1973], but earlier versions can be found in C. C. Chen and G. Grätzer [1969] and G. Grätzer [1971a]. The germ of the idea can be traced back to R. P. Dilworth [1945].

Definition 3. *We define a subset S of $\mathbf{P}(Q)$; for $p \in \mathbf{P}(Q)$, $p \in S$ is defined by induction on the rank of p:*

(i) *For $p \in Q$, $p \in S$ iff $p \in L_i - \{0_i, 1_i\}$, for some $i \in I$.*

(ii) *For $p = q \wedge r$, $p \in S$ iff q, $r \in S$ and the following two conditions hold:*

    (ii$_1$) *$p \subseteq 0_i$, for no $i \in I$;*

    (ii$_2$) *$p \subseteq x \wedge y$, for no $\{x, y\} \in C$.*

(iii) *For $p = q \vee r$, $p \in S$ iff q, $r \in S$ and the following two conditions hold:*

    (iii$_1$) *$1_i \subseteq p$, for no $i \in I$.*

    (iii$_2$) *$x \vee y \subseteq p$, for no $\{x, y\} \in C$.*

Now we set

$$L = \{0, 1\} \cup \{ R(p) \mid p \in S \},$$

and partially order L by

$$0 < R(p) < 1, \quad \text{for } p \in S,$$
$$R(p) \leq R(q) \quad \text{iff} \quad p \subseteq q.$$

If we identify $a \in L_i$ with $R(a)$, then we get the setup we need:

Theorem 4. L is a $\mathcal{C}$-reduced free product of the L_i, $i \in I$.   —

Proof. L is obviously a poset. To show that L is a lattice, we have to find the meet and the join of $R(p)$ and $R(q)$ in L (p, $q \in S$). We claim that

$$R(p) \wedge R(q) = \begin{cases} R(p \wedge q), & \text{if } p \wedge q \in S; \\ 0, & \text{otherwise.} \end{cases}$$

To verify this claim, it is sufficient to prove that, for any $u \in \mathbf{P}(Q)$, if $u \subseteq 0_i$, for some $i \in I$, or $u \subseteq x \wedge y$, for some $\{x, y\} \in C$, then $u \notin S$. We prove this by induction on the rank of u. If $u \in Q$ and $u \subseteq 0_i$, then $u = 0_i \notin S$. If $u \in Q$ and $u \subseteq x \wedge y$, for some $\{x, y\} \in C$, then $u \in L_i$, $x \in L_k$, $y \in L_n$, $k \neq n$, $u \leq x$, and $u \leq y$; these imply that $i = k$ and $i = n$, a contradiction. If $u = u_0 \wedge u_1$, and u_0 or $u_1 \notin S$, then $u \notin S$ by 3(ii); if u_0, $u_1 \in S$, then $u \notin S$ by 3(ii$_1$) or 3(ii$_2$).

Finally, if $u = u_0 \vee u_1$, then $u_0 \subseteq 0_i$ or $u_1 \subseteq 0_i$ in the first case, and $u_0 \subseteq x \wedge y$ or $u_i \subseteq x \wedge y$ in the second case, and so u_0 or $u_1 \notin S$ implying $u \notin S$ by 3(iii).

Dually,

$$R(p) \vee R(q) = \begin{cases} R(p \vee q), & \text{if } p \vee q \in S; \\ 1, & \text{otherwise.} \end{cases}$$

Now it is obvious that $a \mapsto R(a)$ is a $\{0,1\}$-embedding of L_i into L. So, after the identification, 2(i) becomes obvious. 2(ii) is clear in view of 3(ii$_1$), 3(ii$_2$), and our description of meet and join in L.

Let K be the free product of L_i, $i \in I$, as constructed in Section 1. Then $L - \{0,1\} \subseteq K$. We define a congruence Θ on K:

$$\Theta = \bigvee (\, \Theta(0_i, x) \mid i \in I, \ x \leq 0_i \,) \vee \bigvee (\, \Theta(x, 1_i) \mid i \in I, \ x \geq 1_i \,)$$

$$\vee \bigvee (\, \Theta(x, u \wedge v) \mid x \leq u \wedge v, \ \{u, v\} \in C \,)$$

$$\vee \bigvee (\, \Theta(u \vee v, x) \mid x \geq u \vee v, \ \{u, v\} \in C \,).$$

In other words, Θ is the smallest congruence relation under which all 0_i and $u \wedge v$ $(u, \, v \in C)$ are in the congruence class which is the zero of K/Θ and dually. We claim that

$$K/\Theta \cong L.$$

To see this, it is sufficient to prove that every congruence class modulo Θ, except the two extremal ones, contains one and only one element of S.

Let ε_i be the identity map as a map of L_i into L. Then there is a map φ extending all ε_i, $i \in I$, into a homomorphism of K into L. Observe, that $R(p)\varphi = R(p)$, for all $p \in S$. Indeed, $p = p(a_0, \ldots, a_{n-1})$, where $a_0, \ldots, a_{n-1} \in S \cap Q$; hence

$$R(p)\varphi = R(p\varphi) = R(p(a_0, \ldots, a_{n-1})\varphi)$$
$$= R(p(a_0\varphi, \ldots, a_{n-1}\varphi)) = R(p(a_0, \ldots, a_{n-1})) = R(p),$$

since $a_0\varphi = a_0, \ldots, a_{n-1}\varphi = a_{n-1}$.

Let Φ be the congruence kernel of φ. Since L satisfies 2(i) and 2(ii), $\Theta \leq \Phi$. Now if $p, q \in S$, and $R(p)\varphi = R(q)\varphi$, then $R(p) = R(q)$. In other words, $R(p) \equiv R(q)$ (Φ) implies that $R(p) = R(q)$. Therefore, the same holds for Θ. This proves that there is at most one $R(p)$ in the nonextremal congruence classes of Θ. To show "at least one", take a $p \in \mathbf{P}(Q)$ such that

$$R(p) \not\equiv 0_i \ (\Theta),$$
$$R(p) \not\equiv 1_i \ (\Theta)$$

(for any/all $i \in I$); we prove that there exists a $q \in S$ such that $R(p) \equiv R(q)$ (Θ).

Let $p \in L_i$, for some $i \in I$. Then, by assumption, $p \neq 0_i$ and 1_i; hence we can take $q = p$. Let

$$p = p_0 \wedge p_1,$$
$$R(p_0) \equiv R(q_0) \quad (\Theta),$$
$$R(p_1) \equiv R(q_1) \quad (\Theta),$$

where q_0, $q_1 \in S$. If $q_0 \wedge q_1 \in S$, take $q = q_0 \wedge q_1$. Otherwise, by 3(ii), $q_0 \wedge q_1 \equiv 0_i$ (Θ), hence $p \equiv 0_i$ (Θ), contrary to our assumption. The dual argument completes the proof.

Thus we have proved that $K/\Theta \cong L$.

Now we are ready to verify 2(iii). For each $i \in I$, let φ_i be a $\{0,1\}$-homomorphism of L_i into the bounded lattice A. Since K is the free product of the L_i, $i \in I$, there is a homomorphism ψ of K into A extending all the φ_i, $i \in I$. Let Ψ be the congruence kernel of ψ. It obviously follows from the definition of Θ that $\Theta \leq \Psi$. Therefore, by the Second Isomorphism Theorem, $[x]\Theta \mapsto x\psi$ is a homomorphism of K/Θ into A. Combining this with the isomorphism $L \cong K/\Theta$ as described above, we get a $\{0,1\}$-homomorphism φ of L into A extending all the φ_i, $i \in I$. □

Theorem 5.   Let a, b be a complementary pair in a $\mathcal{C}$-reduced free product L of the L_i, $i \in I$. Then there exist a_0, b_0 and a_1, b_1 satisfying

$$a_0 \leq a \leq a_1,$$
$$b_0 \leq b \leq b_1,$$

such that either $\{a_0, b_0\}$, $\{a_1, b_1\} \in C$, for some $i \in I$, or a_0, b_0 and a_1, b_1 are complementary pairs in L_i, and conversely. —

Proof. The converse is, of course, obvious. In either case, by Definition 2, a_0, b_0 and a_1, b_1 are complementary in L, hence

$$a \wedge b \leq a_1 \wedge b_1 = 0,$$
$$a \vee b \geq a_0 \vee b_0 = 1,$$

and so a and b are complementary in L.

Now to prove the main part of the theorem, take p, $q \in S$ such that $a = R(p)$ and $b = R(q)$ are complementary in L. Then $p \wedge q$ violates 3(ii$_1$) or 3(ii$_2$) and $p \vee q$ violates 3(iii$_1$) or 3(iii$_2$). The four cases will be handled separately.

Case 1. $p \wedge q$ *violates* 3(ii$_1$) *and* $p \vee q$ *violates* 3(iii$_1$). Hence, for some $i, j \in I$,

$$p \wedge q \subseteq 0_i,$$
$$1_j \subseteq p \vee q.$$

Thus in the free product K of the L_i, $i \in I$,

$$(p \wedge q)^{(i)} = 0_i,$$
$$(p \vee q)_{(j)} = 1_j.$$

Note that $q^{(i)}$ is proper, because otherwise $p^{(i)} = 0_i$, that is, $p \subseteq 0_i$, contradicting $p \in S$. Similarly, $q_{(j)}$ is proper. This is a contradiction unless $i = j$, in which case, we can put $a_0 = p_{(i)}$, $b_0 = q_{(i)}$, $a_1 = p^{(i)}$, $b_1 = q^{(i)}$ and these obviously satisfy the requirements of the theorem.

Case 2. $p \wedge q$ violates 3(ii$_1$) and $p \vee q$ violates 3(iii$_2$). Hence there exist $i \in I$ and $\{x, y\} \in C$ such that

$$p \wedge q \subseteq 0_i$$
$$x \vee y \subseteq p \vee q.$$

Let $x \in L_j$ and $y \in L_k$, j, $k \in I$, $j \neq k$. Then $i \neq j$ or $i \neq k$; let us assume that $i \neq j$. Since

$$(p \wedge q)^{(i)} = 0_i,$$
$$p^{(i)} \wedge q^{(i)} = 0_i,$$

we conclude, just as in Case 1, that $p^{(i)}$ and $q^{(i)}$ are proper. From $i \neq j$, we conclude that $p_{(j)}$ and $q_{(j)}$ are not proper, that is, $p_{(j)} = q_{(j)} = 0^b$. Thus

$$(p \vee q)_{(j)} = p_{(j)} \vee q_{(j)} = 0^b,$$

contradicting that $p \vee q \supseteq x \in L_j$. Case 2 cannot occur.

Case 3. $p \wedge q$ violates 3(ii$_2$) and $p \vee q$ violates 3(iii$_1$). This leads to a contradiction just as Case 2 does.

Case 4. $p \wedge q$ violates 3(ii$_2$) and $p \vee q$ violates 3(iii$_2$). Then there exist $\{a_0, b_0\} \in C$ and $\{a_1, b_1\} \in C$ such that $a_0 \in L_i$, $b_0 \in L_j$, $a_1 \in L_k$, $b_1 \in L_n$, i, j, k, $n \in I$, $i \neq j$, $k \neq n$,

$$a_0 \vee b_0 \subseteq p \vee q,$$
$$p \wedge q \subseteq a_1 \wedge b_1.$$

We conclude, as above, that

$$a_0 \leq p_{(i)} \vee q_{(i)},$$
$$b_0 \leq p_{(j)} \vee q_{(j)},$$
$$p^{(k)} \wedge q^{(k)} \leq a_1,$$
$$p^{(n)} \wedge q^{(n)} \leq b_1.$$

Therefore, $p_{(i)}$ or $q_{(i)}$ is proper, $p_{(j)}$ or $q_{(j)}$ is proper, $p^{(k)}$ or $q^{(k)}$ is proper, and $p^{(n)}$ or $q^{(n)}$ is proper.

We cannot have both $p_{(i)}$ and $q_{(i)}$ proper, because then neither $p^{(k)}$ nor $q^{(k)}$ could be proper unless $i = k$; and similarly, $i = n$, contradicting $k \neq n$.

So we can assume that $p_{(i)}$ is proper and $q_{(i)} = 0^b$, and therefore $p_{(i)} = (p \vee q)_{(i)} \geq a_0$. We cannot have $p_{(j)}$ proper, because then $q_{(j)} = 0^b$ and $p_{(j)} \geq b_0$; thus $p \supseteq a_0 \vee b_0$. Now $k \neq i$ and $k \neq j$ yields a contradiction (neither $p^{(k)}$ nor $q^{(k)}$ could be proper), hence $k = i$ or $k = j$. Let $k = i$ (the case $k = j$ is similar). Since $i \neq j$, $q^{(i)}$ is not proper, hence $p^{(i)}$ is proper and $p^{(i)} = (p \wedge q)^{(i)} \leq a_1$. But $i \neq n$, hence $p^{(n)}$ is not proper and so $q^{(n)}$ must be proper. Since $q_{(j)}$ is proper, we conclude that $n = j$ and $q^{(j)} = (p \wedge q)^{(j)} \leq b_1$. To sum up, we have obtained

$$a_0 \leq p_{(i)},$$
$$p^{(i)} \leq a_1,$$
$$b_0 \leq q_{(i)},$$
$$q^{(i)} \leq b_1.$$

Hence,

$$a_0 \subseteq p \subseteq a_1,$$
$$b_0 \subseteq q \subseteq b_1,$$

as required. $\square$

Let us say that a bounded lattice A *has no comparable complements* iff A contains no pentagon $\{0, a, b, c, 1\}$. Let $\mathrm{Comp}(A)$ stand for the set of complementary pairs in A.

A $\mathcal{C}$-relation C is said to have *no comparable complements* iff

$$\{a_0, b_0\}, \ \{a_1, b_1\} \in C, \ a_0 \leq a_1 \text{ and } b_0 \leq b_1 \quad \text{imply that} \quad a_0 = a_1 \text{ and } b_0 = b_1.$$

The following result is immediate from Theorem 5.

Corollary 6. *Let L_i, $i \in I$, be a family of lattices with no comparable complements and let C be a $\mathcal{C}$-relation on the L_i, $i \in I$, with no comparable complements. Let L be a $\mathcal{C}$-reduced free product of the L_i, $i \in I$. Then*

$$\mathrm{Comp}(L) = C \cup \bigcup (\mathrm{Comp}(L_i) \mid i \in I),$$

and L has no comparable complements.

Now we are ready for our first application.

Theorem 7 (C. C. Chen and G. Grätzer [1969]).   *Let L be a bounded lattice in which every element has at most one complement. Then L has a $\{0, 1\}$-embedding into a uniquely complemented lattice K (that is, into a lattice K in which every element has exactly one complement).*   —

Proof. If L is complemented, then set $K = L$. Otherwise, let $L = L_0$. We define, by induction, the lattice L_n. If L_{n-1} is defined, let I_{n-1} be the set of noncomplemented elements of L_{n-1}. For $i \in I_{n-1}$, let $L_i = \{a_i\}^b$. Define the $\mathcal{C}$-relation R_{n-1} on the family $\{L_{n-1}\} \cup (L_i \mid i \in I_{n-1})$ by the rule

$$\{a, b\} \in R_{n-1} \quad \text{iff} \quad \{a, b\} = \{i, a_i\}, \text{ for some } i \in I_{n-1}.$$

Let L_n be the $\mathcal{C}$-reduced free product with respect to the relation R_{n-1}. Since

$$L = L_0 \subseteq L_1 \subseteq L_2 \subseteq \cdots$$

and all these containments are $\{0, 1\}$-embeddings, we can form

$$K = \bigcup (L_i \mid i \in I).$$

Now observe that L_0 is a lattice with no comparable complements. By induction, if this is known of L_{n-1}, then it is true for L_n since R_{n-1} has no comparable complements and so Corollary 6 applies. Therefore, again by Corollary 6,

$$\text{Comp}(L_n) = \text{Comp}(L_{n-1}) \cup R_{n-1}.$$

Thus L_n is at most uniquely complemented and every element of L_{n-1} has a complement in L_n. It is now obvious that K is uniquely complemented.   □

Every lattice L can be embedded in a lattice in which every element has at most one complement, namely into L^b. Thus, as a special case of Theorem 7, we get the celebrated result of R. P. Dilworth [1945]:

Corollary 8. *Every lattice can be embedded into a uniquely complemented lattice.*

For a graph G (or, more precisely, $\langle G; E \rangle$), let $(L_a \mid a \in G)$ be a family of lattices, where $L_a = \{0_a, a, 1_a\}$ is a three-element lattice. Define the $\mathcal{C}$-relation C on $(L_a \mid a \in G)$ by

$$\{x, y\} \in C \quad \text{iff} \quad \{x, y\} \in E.$$

Let $L(G)$ denote the $\mathcal{C}$-reduced free product of $(L_a \mid a \in G)$. Some examples are given in Figures 1–3.

For a bounded lattice L, let $\text{End}_{0,1}(L)$ denote the *monoid* (that is, semigroup with identity) of all $\{0, 1\}$-endomorphisms of L. For a graph G, $\text{End}(G)$ denotes the monoid of endomorphisms of G; a map $\varphi \colon G \to G$ is an *endomorphism* iff $\langle a, b \rangle \in E$ implies that $\langle a\varphi, b\varphi \rangle \in E$. Observe that $G \subseteq L(G)$ and, in fact, G generates $L(G)$ as a $\{0, 1\}$-sublattice (that is, $G \cup \{0, 1\}$ generates $L(G)$). Therefore, every $\varphi \in \text{End}(G)$ has at most one extension $\overline{\varphi}$ to a $\{0, 1\}$-endomorphism of $L(G)$.

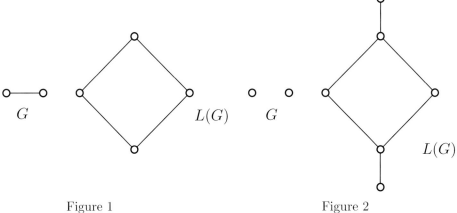

Figure 1   Figure 2

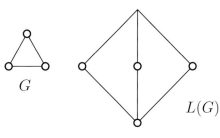

Figure 3

Corollary 9. *Every endomorphism φ of a graph G has exactly one extension $\overline{\varphi}$ to a $\{0,1\}$-endomorphism of $L(G)$. If φ is onto, so is $\overline{\varphi}$.*

Proof. This is clear from Definition 2(iii) with $A = L(G)$.   □

For an integer $n \geq 2$, an *n-cycle* of a graph G is an n-tuple of elements $\langle a_0, \ldots, a_{n-1} \rangle$ with $\{a_0, a_1\}, \ldots, \{a_{n-2}, a_{n-1}\}, \{a_{n-1}, a_0\} \in E$.

Theorem 10.   Let G be a graph satisfying the property that every element of G is contained in some cycle of odd length. Then $\mathrm{End}(G)$ is isomorphic with $\mathrm{End}_{0,1}(L(G))$.   —

Proof. It is obvious that the lattices and the $\mathcal{C}$-relation used in forming $L(G)$ satisfy the hypotheses of Corollary 6. Therefore,

$$\mathrm{Comp}(L(G)) = E \cup \{0,1\}.$$

By our assumption, every element of G is the endpoint of an edge, and G is recognized in $L(G)$ as the set of complemented elements, other than 0 and 1. Since a $\{0,1\}$-endomorphism ψ takes a complementary pair into a complementary pair, we conclude that $G\psi \subseteq G \cup \{0,1\}$. Let $g\psi \in \{0,1\}$, for $g \in G$; for instance, $g\psi = 0$. By assumption, there is a cycle of odd length $\langle g_0, \ldots, g_{2n} \rangle$ with $g = g_0$. This means that $\{g_0, g_1\}$, ..., $\{g_{2n-1}, g_{2n}\}$, and $\{g_{2n}, g_0\}$ are complementary pairs. Thus $g\psi = g_0\psi = 0$, $g_1\psi = 1$, $g_2\psi = 0$, $g_3\psi = 1$, ..., $g_{2n-1}\psi = 1$, $g_{2n}\psi = 0$, and $g_0\psi = 1$, a contradiction. This shows that every $\{0,1\}$-endomorphism ψ is the unique extension of a map φ of G into itself; this φ is obviously a graph endomorphism. Thus the map $\varphi \mapsto \overline{\varphi}$ is the required isomorphism of $\mathrm{End}(G)$ with $\mathrm{End}_{0,1}(L(G))$. $\square$

As it is shown in the Exercises, every monoid is the endomorphism semigroup of a graph in which every element lies on a cycle of odd length. Thus we conclude a result of G. Grätzer and J. Sichler [1970]:

Theorem 11.   Every monoid can be represented as the $\{0,1\}$-endomorphism semigroup of a bounded lattice.   —

Exercises

The following exercises should provide the reader with the necessary background in category theory (theory of concrete categories, graph theory) necessary for Theorem 11 and for the results of the next section. References will be given in the Further Topics and References. The present sequence of exercises is based on unpublished lecture notes of J. Sichler.

Let A be a nonempty set; in what follows, we shall deal with systems $\langle A; R \rangle$, $\langle A; R_0, \ldots, R_{n-1} \rangle$, $\langle A; R_0, \ldots, R_n, \ldots \rangle$, where R and the R_i are binary relations. Given two systems of the same type, say

$$\langle A; R_0, \ldots, R_n, \ldots \rangle,$$
$$\langle B; S_0, \ldots, S_n, \ldots \rangle,$$

a map $\varphi \colon A \to B$ is called a *homomorphism* iff

$$\langle a, b \rangle \in R_i \quad \text{implies that} \quad \langle a\varphi, b\varphi \rangle \in S_i, \text{ for all } i = 0, 1, \ldots$$

An *endomorphism* of $\langle A; R_0, \ldots \rangle$ is a homomorphism of $\langle A; R_0, \ldots \rangle$ into itself. The endomorphism semigroup $\mathrm{End}(\langle A; R_0, \ldots \rangle)$ is defined as before. $\langle A; R_0, \ldots \rangle$ is *rigid* iff the only endomorphism is the identity map.

1.   Let $\langle A; \leq \rangle$ be a well-ordered set with unit. Let A_c consist of all $a \in A$ that are cofinal with ω, that is, for which there is a sequence $a_1 < \cdots < a_n < \cdots$ such that $a = \bigvee(a_i \mid i = 1, 2, \ldots)$. For each $a \in A_c$, fix such a sequence $\langle a_1, a_2, \ldots \rangle$. Let R_0 be the relation $<$, and let $\langle x, y \rangle \in R_i$, for

$i = 1, 2, \ldots$, iff $y \in A_c$ and $x = y_i$ (that is, x is the i-th member of the sequence associated with y). Let φ be an endomorphism of $\langle A; R_0, R_1, \ldots \rangle$. Prove that $x \leq x\varphi$, for all $x \in A$.

2. Let $a < a\varphi$, for some $a \in A$. Set $a_1 = a\varphi$, $a_2 = a_1\varphi$, $\ldots$, and
$$b = \bigvee(a_n \mid n = 1, 2, \ldots).$$
Prove that $b\varphi = b$.

3. Prove that $b_n\varphi = b_n$, for all $n = 1, 2, \ldots$ Conclude that $\langle A; R_0, R_1, \ldots \rangle$ is rigid.

4. Given a system $\langle A; R_0, R_1, \ldots \rangle$, we construct a new one $\langle B; S_0, S_1, \ldots \rangle$ as follows: $B = A \times N$ ($N = \{0, 1, 2, \ldots\}$),
$$\langle \langle x, n \rangle, \langle y, m \rangle \rangle \in S_0 \quad \text{iff} \quad x = y \text{ and } m = n+1, \text{ or}$$
$$x = y \text{ and } n = 0, \ m = 2;$$
$$\langle \langle x, n \rangle, \langle y, m \rangle \rangle \in S_1 \quad \text{iff} \quad n = m \text{ and } \langle x, y \rangle \in R_n.$$
Prove that $\text{End}(\langle A; R_0, R_1, \ldots \rangle) \cong \text{End}(\langle B; S_0, S_1 \rangle)$ and if A is infinite, then $|A| = |B|$.

5. Given a system $\langle A; R_0, R_1 \rangle$, we construct a new one $\langle B; S \rangle$ as follows: $B = A \times \{0, 1, 2, 3, 4\}$,
$$\langle \langle a, i \rangle, \langle b, j \rangle \rangle \in S \quad \text{iff} \quad a = b \text{ and } j = i+1, \text{ or}$$
$$a = b \text{ and } i = 0, \ j = 4, \text{ or}$$
$$i = 0, \ j = 2, \text{ and } \langle a, b \rangle \in R_0, \text{ or}$$
$$i = 2, \ j = 4, \text{ and } \langle a, b \rangle \in R_1.$$
Prove that $\text{End}(\langle A; R_0, R_1 \rangle) \cong \text{End}(\langle B; S \rangle)$ and if A is infinite, then $|A| = |B|$.

6. Prove that for each infinite cardinal $\mathfrak{m}$, there exists a rigid $\langle A; R \rangle$ with $|A| = \mathfrak{m}$.

7. Prove the statement of Exercise 6 for finite cardinals.

8. Let $\langle A; (R_i \mid i \in I) \rangle$ be a system without any restriction on $|I|$. Let $\langle I; R \rangle$ be a connected rigid graph. We define a new system $\langle B; R_0, R_1 \rangle$ as follows: $B = A \times I$,
$$\langle \langle a, i \rangle, \langle b, j \rangle \rangle \in R_0 \quad \text{iff} \quad a = b \text{ and } \langle i, j \rangle \in R,$$
$$\langle \langle a, i \rangle, \langle b, j \rangle \rangle \in R_1 \quad \text{iff} \quad i = j \text{ and } \langle a, b \rangle \in R_i.$$
Prove that $\text{End}(\langle A; (R_i \mid i \in I) \rangle) \cong \text{End}(\langle B; R_0, R_1 \rangle)$.

9. Let M be a monoid. For every $a \in M$, define $R_a = \{\langle x, ax\rangle \mid x \in M\}$. Show that $\text{End}(\langle M; (R_a \mid a \in M)\rangle) \cong M$.

10. For any monoid M, find a system $\langle A; R\rangle$ such that $\text{End}(\langle A; R\rangle) \cong M$.

11. For a system $\langle A; R\rangle$ and $B \subseteq A$, we construct a new system $\langle A; R, S_B\rangle$ as follows: $\langle x, y\rangle \in S_B$ iff $x \neq y$ and $x, y \in B$ or $x, y \notin B$. Show that if $\langle A; R\rangle$ is rigid, then the new systems are *mutually rigid*, that is, if φ is a one-to-one homomorphism of one system to another, then the two systems are the same and φ is the identity map.

12. Prove that there are $2^{\mathfrak{m}}$ mutually rigid systems of cardinality $\mathfrak{m}$, where $\mathfrak{m}$ is any infinite cardinal.

13. Consider the graph $\langle G; E\rangle$ of Figure 4. Let $\langle A; R\rangle$ be a system satisfying $\langle a, a\rangle \in R$, for no $a \in A$. We construct a new graph $\langle H; F\rangle$ by "replacing each $\langle a, b\rangle \in R$ by a copy of $\langle G; E\rangle$ with the pair $\langle 4, 8\rangle$ signifying $\langle a, b\rangle$." Formally,

$$H = (R \times \{1, 2, 3, 5, 6, 7, 9, 10\}) \cup A,$$

and, for each $e = \langle a, b\rangle \in R$, $\langle e, 1\rangle$, $\langle e, 2\rangle$, $\langle e, 3\rangle$, a, $\langle e, 5\rangle$, $\langle e, 6\rangle$, $\langle e, 7\rangle$, b, $\langle e, 9\rangle$, $\langle e, 10\rangle$ should form a subgraph isomorphic with $\langle G; E\rangle$. Discuss the connections between $\text{End}(\langle A; R\rangle)$ and $\text{End}(\langle H; F\rangle)$.

14. A *triangle* $\{a, b, c\}$ in a graph $\langle G; E\rangle$ is a set of three elements of G such that $\{a, b\}$, $\{b, c\}$, and $\{c, a\} \in E$. A graph $\langle G; E\rangle$ is *triangle connected* iff, for any $a, b \in G$, $a \neq b$, there is a sequence $T_0, \ldots, T_{n-1}$ of triangles such that $a \in T_0$, $b \in T_{n-1}$, and $T_i \cap T_{i+1} \neq \varnothing$, for $i = 0, \ldots, n-2$. Prove that for each infinite cardinal $\mathfrak{m}$, there are $2^{\mathfrak{m}}$ mutually rigid triangle connected graphs.

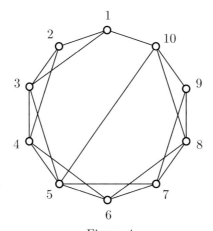

Figure 4

15. Prove that, for each infinite cardinal $\mathfrak{m}$, there are $2^{\mathfrak{m}}$ mutually rigid connected graphs in which every element lies on a cycle of length 7.

16. Let M be a monoid and let $\mathfrak{m}$ be an infinite cardinal satisfying $|M| \leq \mathfrak{m}$. Prove that there are $2^{\mathfrak{m}}$ pairwise nonisomorphic graphs $\langle G; E \rangle$ of cardinality $\mathfrak{m}$ satisfying $\operatorname{End}(\langle G; E \rangle) \cong M$.

17. Prove that the L of Theorem 4 is a $\mathcal{C}$-reduced free product of the L_i, $i \in I$, by verifying directly 2(iii). (Hint: argue as in the proof of Theorem 1.8).

Exercises 18–21 are based on C. C. Chen and G. Grätzer [1969].

18. A *bi-uniquely complemented lattice* L is a bounded lattice in which every element $x \neq 0$, 1 has exactly two complements. Define the concept of a *free bi-uniquely complemented lattice* and prove the existence and uniqueness (up to isomorphism) of a free bi-uniquely complemented lattice on $\mathfrak{m}$ generators.

19. Let F_0 and F_1 be free bi-uniquely complemented lattices and let X_0 and X_1 be the free generating sets of F_0 and F_1, respectively. Let α be a one-to-one map of X_0 onto X_1. Let $|X_0| = |X_1| = \aleph_0$. Show that there are $2^{\aleph_0}$ isomorphisms of F_0 and F_1 extending α.

20. In a bounded lattice L, the *complementation is transitive* iff whenever x, y and y, z are complementary pairs, then either $x = z$ or x, z is also a complementary pair. Prove that every lattice can be embedded into a bi-uniquely complemented lattice with transitive complementation.

21. Let L be a bounded lattice. Under what conditions is there a $\{0, 1\}$-embedding of L into a bi-uniquely complemented lattice with transitive complementation?

22. Show that a relatively complemented, uniquely complemented lattice is Boolean. (Hint: use Exercise I.6.5.)

23. Prove that a modular uniquely complemented lattice is Boolean.

24. Show that an atomic uniquely complemented lattice is Boolean (T. Ogasawara and U. Sasaki [1949]). (Hint: For an element a, let $R(a)$ denote the set of atoms contained in a; then $a \mapsto R(a)$ is a set representation.)

25. Let L be a uniquely complemented lattice and let $a \succ b$ in L. Prove that $a \wedge b'$ is an atom, where b' is the complement of b (see H.-J. Bandelt and R. Padmanabhan [1979]). (Hint: if $a \wedge b' = 0$, then b' has two complements; if $0 < x < a \wedge b'$, then $b \vee a'$ has two complements.)

26.   A lattice L is called *relatively atomic* iff, for all a, $b \in L$, if $a < b$, then there exist x and y satisfying $a \leq x \prec y \leq b$. Show that a relatively atomic uniquely complemented lattice is Boolean. (This includes the results of V. N. Saliĭ [1972] that every uniquely complemented algebraic lattice is Boolean.)

27.   A *lattice with complementation* $\langle L; \wedge, \vee, {}', 0, 1 \rangle$ is a bounded lattice in which, for each $a \in L$, a' is a complement of L. An *endomorphism* φ is a lattice endomorphism satisfying $(a\varphi)' = a'\varphi$, for all $a \in L$. Prove that, for any monoid M, there is a lattice with complementation whose endomorphisms semigroup is isomorphic to M (G. Grätzer and J. Sichler [1970]).

28.   In Theorem 11 and in Exercise 27, how many pairwise nonisomorphic lattices of a given cardinality can be constructed satisfying the requirements?

4.   Hopfian Lattices

A lattice L is *Hopfian* iff every onto endomorphism is an automorphism. Equivalently, a lattice L is Hopfian iff $L \cong L/\Theta$ implies that $\Theta = \omega$, that is, L is not isomorphic to a proper quotient of itself. The Hopfian property is similarly defined for groups, rings, and algebras, in general. The first definition of the Hopfian property also applies for graphs; one should note, however, that for a graph a one-to-one and onto endomorphism need not be an automorphism. For graphs, an automorphism has to be defined as an invertible endomorphism.

   Obviously, every finite lattice (algebra, graph) is Hopfian. In fact, the Hopfian property is a generalization of one of the most important characteristics of finiteness.

   Closest to finite lattices are *finitely presented lattices*, that is, lattices of the form $\mathrm{F}(Q)$ (see Section 1.5) where Q is a finite partial lattice. (It is easily seen that these are the lattices that can be described by finitely many generators and finitely many relations, so this concept agrees with "finitely presented" as it is used for groups and semigroups.)

Theorem 1 (T. Evans [1969]).   Every finitely presented lattice is Hopfian.   ▬

The proof of Theorem 1 is contained in the following three lemmas.

Lemma 2. *A finitely presented lattice can be represented as a subdirect product of finite lattices.*

Proof. It is sufficient to prove that if Q is a finite partial lattice, a, $b \in \mathrm{F}(Q)$, and $a \neq b$, then there exists a finite lattice K and a homomorphism $\varphi \colon \mathrm{F}(Q) \to K$

such that $a\varphi \neq b\varphi$. Just as in Section 2, we can define the component subsets of a and b and set

$$Q' = Q \cup \mathrm{Komp}(a) \cup \mathrm{Komp}(b).$$

Regarding Q' as a partial lattice, obviously $\mathrm{F}(Q) = \mathrm{F}(Q')$. By (the proof of) Theorem I.5.20, Q' can be embedded into some finite lattice K. This embedding extends to a homomorphism φ of $\mathrm{F}(Q') = \mathrm{F}(Q)$ into K. Finally, $a\varphi \neq b\varphi$, since φ is one-to-one on Q'. □

Lemma 3. *Let L be a finitely generated lattice and let Θ be a congruence relation of L. If L/Θ is finite, then there exists a fully invariant congruence relation Φ of L such that $\Phi \leq \Theta$ and L/Φ is finite.*

Proof. Set $\mathbf{K} = \mathbf{Var}(L/\Theta)$. We define

$$\Phi = \bigvee (\,\Theta(p(a_0,\dots,a_{n-1}), q(a_0,\dots,a_{n-1})) \mid p = q \in \mathrm{Iden}(\mathbf{K}),$$

$$a_0, \ \dots, \ a_{n-1} \in L\,).$$

It is easily seen that Φ is fully invariant. Since the identities used to construct Φ all hold in L/Θ, we conclude that $\Phi \leq \Theta$. Finally, L/Φ is finitely generated and $L/\Phi \in \mathbf{K}$, which is a locally finite class, hence L/Φ is finite. □

A *subdirect representation* of a lattice L is associated with a family $(\Theta_i \mid i \in I)$ of congruence relations of L such that

$$\bigwedge (\Theta_i \mid i \in I) = \omega.$$

If all Θ_i, $i \in I$, are fully invariant, we call this subdirect representation *fully invariant*.

Lemma 4. *Let the lattice L have a representation as a fully invariant subdirect product of Hopfian lattices. Then L is Hopfian.*

Proof. Let $(\Theta_i \mid i \in I)$ be a family of fully invariant congruence relations of L satisfying $\bigwedge (\Theta_i \mid i \in I) = \omega$. Let φ be an onto endomorphism of L. Since Θ_i is fully invariant, for $i \in I$, we can define

$$\varphi_i \colon [a]\Theta_i \mapsto [a\varphi]\Theta_i,$$

which is an endomorphism of L/Θ_i. Obviously, all φ_i are onto, hence they all are automorphisms. Now if a, $b \in L$ and $a\varphi = b\varphi$, then, for all $i \in I$, $([a]\Theta_i)\varphi_i = ([b]\Theta_i)\varphi_i$, hence $[a]\Theta_i = [b]\Theta_i$. Since $\bigwedge (\Theta_i \mid i \in I) = \omega$, we conclude that φ is one-to-one. □

R. Wille [1975] exhibits a finitely generated lattice that is not Hopfian. An earlier example of a finitely generated lattice that is not finitely presented (or presentable, to be more precise) is $F_M(4)$ (see T. Evans and D. X. Hong [1972]); of course, $F_M(4)$ is Hopfian.

The free product of two finitely presented lattices is again finitely presented. What about Hopfian lattices?

Theorem 5 (G. Grätzer and J. Sichler [1974]).   The free product of Hopfian lattices is not necessarily Hopfian.   —

The proof of Theorem 5 is based on the construction of the lattice $L(G)$ from the graph G introduced in Section 3 and a lattice theoretic result (Theorem 6) which is due to J. Sichler [1972] and H. Lakser [1972]. In fact, the result we shall prove will be stronger than Theorem 5 and it will be based on J. Sichler [1975].

For graphs G_i $(= \langle G_i; E_i \rangle)$, $i \in I$, we form the lattices $L(G_i)$ (with bounds 0_i, 1_i) and the free product L of the $L(G_i)$, $i \in I$.

A *triangle* of G_i is a three-element set $\{a, b, c\}$ such that $\{a, b\}$, $\{a, c\}$, $\{b, c\} \in E_i$. If $\{a, b, c\}$ is a triangle of G_i, then $\{0_i, a, b, c, 1_i\}$ is a diamond. We call this sublattice the *diamond associated with a triangle*.

Theorem 6.   Any diamond in L is associated with a triangle of some G_i, $i \in I$.   —

Proof. Let $M = \{o, a, b, c, i\}$ be a diamond in L. Let us first assume that $M \subseteq L(G_j)$, for some $j \in I$.

Case 1. $o = 0_j$ and $i = 1_j$. Then $\{a, b\}$, $\{a, c\}$, $\{b, c\} \in \mathrm{Comp}(L(G_j))$, hence by Corollary 3.6, $\{a, b\}$, $\{a, c\}$, $\{b, c\} \in E_i$, and indeed, M is associated with the triangle $\{a, b, c\}$ of G_j.

Case 2. $o = 0_j$ and $i < 1_j$. Recall that $L(G_j) - \{0_j, 1_j\} \subseteq \mathrm{F}(G_j)$ (the free lattice generated by the set G_j), hence a, b, c, $i \in \mathrm{F}(G_j)$, and there is a congruence relation Θ on $\mathrm{F}(G_j)$ such that $\mathrm{F}(G_j)/\Theta \cong L(G_j)$ and $[x]\Theta \mapsto x$ under this isomorphism, for all $x \in L(G_j) - \{0_j, 1_j\}$. Since $i = a \vee b = a \vee c$ in $\mathrm{F}(G_j)$, by $(\mathrm{SD}_\vee)$, $i = a \vee (b \wedge c)$. But $b \wedge c = 0_j$, that is, $[b \wedge c]\Theta$ is the zero of $\mathrm{F}(G_j)/\Theta$. Thus

$$[i]\Theta = [a]\Theta \vee [b \wedge c]\Theta = [a]\Theta,$$

and we conclude $a = i$, a contradiction. Hence this case is not possible.

Case 3. $o > 0_j$ and $i = 1_j$. This is impossible; argue as in Case 2.

Case 4. $o > 0_j$ and $i < 1_j$. Then $M \subseteq \mathrm{F}(G_j)$, which is impossible since there is no diamond in a free lattice.

Thus Theorem 6 is proved for sublattices of $L(G_j)$. Now let $M \subseteq L$ and consider, for each $j \in J$,

$$M_{(j)} = \{ x_{(j)} \mid x \in M \}.$$

Since $x \mapsto x_{(j)}$ is a homomorphism, $M_{(j)} \cong M$ or $M_{(j)}$ is a singleton. If, for all $j \in J$, $M_{(j)}$ is a singleton, then Theorem 1.17 yields that $(SD_\vee)$ holds in M, a contradiction. Hence there exists a $j \in I$ such that $M_{(j)}$ is a diamond and it is, therefore, associated with a triangle of G_j, in particular, $o_{(j)} = 0_j$ and $i_{(j)} = 1_j$.

By duality, there exists a $k \in I$ such that $M^{(k)}$ is a diamond; in particular, $o^{(k)} = 0_k$ and $i^{(k)} = 1_k$. This yields immediately that $j = k$ and $o = 0_j$, $i = 1_j$. Since

$$a_{(j)} \vee b_{(j)} = 1_j,$$
$$a^{(j)} \wedge b^{(j)} = 0_j,$$
$$a_{(j)} \leq a^{(j)},$$
$$b_{(j)} \leq b^{(j)},$$

we conclude that $a_{(j)}$ and $a^{(j)}$ are complements of $b_{(j)}$ in $L(G_j)$. But $L(G_j)$ is a lattice with no comparable complements, hence $a_{(j)} = a^{(j)} = a \in L(G_j)$. Similarly, b, $c \in L(G_j)$. Hence $M \subseteq L(G_j)$ and M is associated with a triangle of G_j. □

Proof of Theorem 5. Let I be a set of at least two elements. For each $i \in I$, a lattice L_i will be constructed such that the free product of all L_i, $i \in I$, is not Hopfian but the free product of L_i, $i \in I'$, is Hopfian, for all $\varnothing \neq I' \subset I$. The special case $|I| = 2$ is Theorem 5.

Let $N = \{1, 2, 3, \dots\}$, and we consider maps $\varphi \colon N \to I$. The map φ is *eventually constant* iff there exists an integer n such that $n\varphi = (n+1)\varphi = \cdots$. Let $M(I)$ denote the set of all eventually constant maps of N into I. By $M_n(I)$ we denote the set of all maps $\varphi \colon \{1, \dots, n\} \to I$ and

$$M_\omega(I) = \bigcup (M_n(I) \mid n = 1, 2, \dots).$$

For $\varphi \in M(I)$, $n \in N$, and for $\varphi \in M_m(I)$, $n \leq m$, let φ_n be the restriction of φ to $\{1, \dots, n\}$. Finally, for $\varphi \in M_n(I)$, we write $n = \mathrm{Dom}(\varphi)$ and $\mathrm{FV}(\varphi) = n\varphi$ (Dom for domain and FV for final value).

For a graph G ($= \langle G; E \rangle$), $a, b \in G$, we say that a and b are *triangle connected* (see Exercise VI.3.14) iff $a = b$ or there is a sequence $T_0, \dots, T_n$ of triangles of G such that $a \in T_0$, $b \in T_n$, and $T_i \cap T_{i+1} \neq \varnothing$, for $i = 0, \dots, n-1$. Any graph can be decomposed into a disjoint union of triangle connected components. G is *triangle connected* iff it has a single component.

Now choose a cardinal $\mathfrak{m} > |M(I)|$, a triangle connected graph G_0, and, for each $\alpha \in M(I)$, choose a triangle connected graph G_α such that $|G_0| = |G_\alpha| = \mathfrak{m}$ and the graphs G_0 and G_α, $\alpha \in M(I)$, are pairwise disjoint and mutually rigid (Exercise 3.14). We fix the elements $a_0 \in G_0$ and $a_\alpha \in G_\alpha$, for all $\alpha \in M(I)$.

For every $n \in N$ and $\varphi \in M_n(I)$, we define a graph G^φ:

$$G^\varphi = \bigcup (G_\alpha \mid \alpha \in M(I) \text{ and } \alpha_n = \varphi),$$

$$E^\varphi = \bigcup (E_\alpha \mid \alpha \in M(I) \text{ and } \alpha_n = \varphi) \cup \{ \{a_0, a_\alpha\} \mid \alpha \in M(I) \}.$$

In words, G^φ is a disjoint union of all G_α such that $\alpha_n = \varphi$ and we add the edges connecting the distinguished element of G_0 with the distinguished element of the G_α. Observe that G^φ is connected and that G_0 and the G_α are the triangle connected components of G^φ.

Let χ be a homomorphism of G^φ into G^ψ ($\varphi \in M_n(I)$ and $\psi \in M_m(I)$). Then the triangle connected components of G^φ have to be mapped into the triangle connected components of G^ψ. Thus if $\alpha \in M(I)$ with $\alpha_n = \varphi$ or $\alpha = 0$, then $G_\alpha \chi \subseteq G_\beta$, where $\beta \in M(I)$ and $\beta_m = \psi$ or $\beta = 0$. In view of the mutual rigidity of these graphs we must have $\alpha = \beta$ and χ is the identity on G_α. So we conclude that if $\alpha \in M(I)$ and $\alpha_n = \varphi$, then $\alpha_m = \psi$; this is possible iff $m \leq n$ and $\alpha_m = \beta$. Thus there is a homomorphism of G^φ into G^ψ iff $\mathrm{Dom}(\varphi) \geq \mathrm{Dom}(\psi)$ and $\varphi_{\mathrm{Dom}(\psi)} = \psi$. Moreover, in this case, there is exactly one homomorphism which is the identity map on G_0 and on all G_α ($\alpha_n = \varphi$).

Now we are ready to define L_i, for $i \in I$: let L_i be a free product of all $L(G^\varphi)$ satisfying $\mathrm{FV}(\varphi) = i$.

Let L be a free product of all L_i, $i \in I$; in other words, L is a free product of all $L(G^\varphi)$, $\varphi \in M_\omega(I)$. We verify that L is not Hopfian. We define a map β:

β restricted to G^φ is the homomorphism of G^φ into $G^{\varphi_{n-1}}$, where $n = \mathrm{Dom}(\varphi) > 1$;

β on G^φ is the identity, if $\mathrm{Dom}(\varphi) = 1$.

Thus β extends to a homomorphism γ of $L(G^\varphi)$ into $L(G^{\varphi_{n-1}})$, if $n = \mathrm{Dom}(\varphi) > 1$, and to the identity map γ on $L(G^\varphi)$, if $\mathrm{Dom}(\varphi) = 1$. Since L is a free product, γ extends to an endomorphism δ of L.

Observe that the image of β covers all G^φ. Indeed, if $a \in G^\varphi$, then $a \in G_\alpha$, for some $\alpha \in M(I)$ with $\alpha_n = \varphi$, where $n = \mathrm{Dom}(\varphi)$. Then β maps $a \in G_\alpha \subseteq G^{\alpha_{n+1}}$ onto $a \in G_\alpha \subseteq G^\varphi$. Thus δ is an onto endomorphism of L. β is not one-to-one since every element of any G^φ with $\mathrm{Dom}(\varphi) = 1$ is the image of two elements: of itself and of a suitable G^ψ with $\mathrm{Dom}(\psi) = 2$. Therefore δ is not one-to-one. We have proved that L is not Hopfian.

Now let $\varnothing \neq I' \subset I$ and let L' be the free product of the L_i, $i \in I'$. We have to show that L' is Hopfian.

Observe that L' is the free product of all $L(G^\varphi)$ satisfying $\mathrm{FV}(\varphi) \in I'$. Let δ be an onto endomorphism of L'. We shall verify that δ is the identity map, implying that L' is Hopfian.

Assume to the contrary that δ is not the identity map. Then there is an $a \in G^\varphi$ ($\mathrm{FV}(\varphi) \in I'$, $\mathrm{Dom}(\varphi) = n$) such that $a \neq a\delta$. Since a is an element of a triangle, it is an atom of the associated diamond M. If $|M\varphi| = 1$, then the bounds of $L(G^\varphi)$ are collapsed by δ, hence all of $L(G^\varphi)$ is mapped by δ onto a single element. Otherwise, $M\varphi \cong M$, hence, by Theorem 6, $a\delta \in G^\psi$, for some $\psi \in M_\omega(I)$. Since G^φ is connected, all of G^φ is mapped by δ into G^ψ. Hence $\mathrm{Dom}(\varphi) \geq \mathrm{Dom}(\psi)$ and $\psi = \varphi_m$ where $m = \mathrm{Dom}(\psi) < \mathrm{Dom}(\varphi)$, since $\mathrm{Dom}(\psi) = \mathrm{Dom}(\varphi)$ implies that $\varphi = \psi$, contradicting the rigidity of G^φ.

In either case, we see that $G^\varphi \cap G^\varphi\delta$ has at most one element. Let $i \in I - I'$ and consider the map $\beta \colon N \to I$ defined by $\beta_n = \varphi$ and $k\beta = i$, for $k > n$. Then $\beta \in M(I)$ and $G_\beta \subseteq G^\varphi$. Let $G_\beta^* = G_\beta - (G^\varphi \cap G^\varphi\delta)$. Since δ is onto, every $a \in G_\beta^*$ must be in the image of some G^ψ; but it cannot come from the homomorphism of G^ψ into G^φ, since it would imply $\mathrm{Dom}(\psi) = m > n = \mathrm{Dom}(\varphi)$ and $\beta_m = \psi$, contradicting the definition of L'. ($G\psi$ is not in L' since $\mathrm{FV}(\psi) = i \notin I'$). Thus $a \in G_\beta^*$ must come from a G^ψ collapsed by δ onto a. However, $|G_\beta^*| = \mathfrak{m}$ and $|M(I)| < \mathfrak{m}$, a contradiction. Thus δ must be the identity map. $\quad\square$

Exercises

1. Prove that, for any nontrivial variety $\mathbf{K}$ and infinite cardinal $\mathfrak{m}$, $F_\mathbf{K}(\mathfrak{m})$ is not Hopfian.

2. Prove that, for any variety $\mathbf{K}$ of lattices and natural number n, $F_\mathbf{K}(n)$ is Hopfian.

3. Find a one-to-one and onto endomorphism of a graph that is not an automorphism.

4. Can the graph in Exercise 3 be chosen to be finite?

5. Define a finitely presented lattice as the "most free" lattice generated by $x_0, \ldots, x_{n-1}$ satisfying the "relations"

$$p_i(x_0, \ldots, x_{n-1}) = q_i(x_0, \ldots, x_{n-1}), \quad i = 1, \ldots, m,$$

where p_i and q_i are n-ary polynomials. Show that this is equivalent to forming an $F(Q)$ with a suitable finite partial lattice Q.

6. Let $F(Q) \subseteq Q' \subseteq Q$, where Q and Q' are partial lattices. Show that $F(Q) \cong F(Q')$ provided that Q' is generated by Q.

7. Formulate and prove the converse of Exercise 6.

8. For a lattice L and variety $\mathbf{K}$ of lattices, we constructed, in the proof of Lemma 3, a congruence relation $\Phi = \Phi(\mathbf{K})$. Prove that $\Phi(\mathbf{K})$ is fully invariant.

9. Let Φ be a fully invariant congruence relation of a lattice L. Prove that $\Phi = \Phi(\mathbf{K})$, for some variety $\mathbf{K}$.

10. Let L be a fully invariant subdirect product of the L_i, $i \in I$. For an onto endomorphism φ, write down a formula for $\operatorname{Ker}\varphi$ from which Lemma 4 can be derived.

11. What is the analogue of Lemma 4 for graphs?

12. Show that the free product of two finitely presented lattices is finitely presented.

13. Prove that $L(G_0 \cup G_1)$ is a free $\{0,1\}$-product of $L(G_0)$ and $L(G_1)$, where $G_0 \cup G_1$ is the disjoint union of the graphs G_0 and G_1.

14. Find two bounded Hopfian lattices whose free $\{0,1\}$-product is not Hopfian (G. Grätzer and J. Sichler [1974]).

15. Let G_0 and G_1 be Hopfian graphs. Prove that a free product of $L(G_0)$ and $L(G_1)$ is Hopfian.

Further Topics and References

The idea of the covers of elements in a free product goes back to R. P. Dilworth [1945]. It becomes more explicit in C. C. Chen and G. Grätzer [1969], in which the lattice $L * \mathrm{F}(\aleph_0)$ is considered. The four W conditions, are of course, from P. M. Whitman [1941]; this should be apparent from Theorem 2.2.

The Structure Theorem has its limitations. Despite many attempts, it could not be used to verify the Common Refinement Property for free products.

The Splitting Theorem is first formulated for $\mathrm{F}(n)$ in P. M. Whitman [1941]: for $a \in \mathrm{F}(n)$, if $a \geq x_1$ does not hold, then $a \leq x_2 \vee \cdots \vee x_n$; proof by induction on $r(a)$ using Theorem 2.2. Theorem 1.11 is much more general; still, it is a very special case of a complete triviality:

The Splitting Theorem.   Let L be a lattice and let $L = [X]$. Then the prime ideals of L are in one-to-one correspondence with subsets Y of X satisfying $Y \neq \varnothing$, $X - Y \neq \varnothing$, and

$$\bigvee Y_1 \not\geq \bigwedge X_1,$$

for any finite nonempty $Y_1 \subseteq Y$ and $X_1 \subseteq X - Y$. If X is finite, this condition reduces to $\bigvee Y \not\geq \bigwedge(X - Y)$.   —

A very special case of the Common Refinement Property for free products was proved in A. Kostinsky [1971]. For any variety $\mathbf{K}$ of lattices, the Common Refinement Property holds for free $\mathbf{K}$-products by G. Grätzer and J. Sichler [1975]. B. Jónsson and E. Nelson [1974] verify the same result for *regular varieties* of algebras, that is, varieties defined by identities in which the same variables occur on both sides. In G. Grätzer and J. Sichler [1975], examples are exhibited of lattices that cannot be represented as free products of freely indecomposable lattices; the same result holds also relative to any variety $\mathbf{K}$ of lattices $\mathbf{K} \neq \mathbf{T}$.

Let

$$g(L) = \min\{\, |X| \mid X \text{ generates } L \,\}.$$

By G. Grätzer and J. Sichler [1974a], the formula

$$g(A * B) = g(A) + g(B)$$

holds; the same formula holds in any variety $\mathbf{K} \neq \mathbf{T}$ of lattices.

Many of the observations made about $F(3)$ hold for all $F_{\mathbf{K}}(3)$, where $\mathbf{K} \neq \mathbf{T}$ is any variety of lattices. An interesting example is

$$x_0 \wedge b \prec x_0 \prec x_0 \vee a.$$

This even holds for $\mathbf{K} = \mathbf{D}$, where $a = b$. Many intervals of $F(3)$ contain isomorphic copies of $F(3)$. The same problem for $F_{\mathbf{K}}(n)$ is considered in J. Berman [1972a]; this problem is closely connected with the Amalgamation Property.

Computing $F_{\mathbf{K}}(n)$ may be very difficult, even when it is finite. For instance, $F_{\mathbf{N}_5}(3)$ has 99 elements and a rather complicated structure, see A. G. Waterman [1967]. A rather effective method is worked out in R. Wille [1976a], which lends itself well to computer programming.

A curious consequence of Theorem 2.7 is that if $\mathbf{K}$ is a variety of lattices, $\mathbf{K} \neq \mathbf{T}$, and $F_{\mathbf{K}}(\aleph_0)$ satisfies (W), then $\mathbf{K} = \mathbf{L}$.

Corollary 2.12 suggests that there are very many three-generated lattices. Indeed, in P. Crawley and R. A. Dean [1959], it is proved that there are $2^{\aleph_0}$ three-generated lattices. Call a lattice L $\mathfrak{m}$-*universal* iff $|L| = \mathfrak{m}$ and every lattice of cardinality at most $\mathfrak{m}$ is isomorphic to a sublattice of L. Thus there are no $\aleph_0$-universal lattices. By B. Jónsson [1956], $\aleph_\alpha$-universal lattices exist for all $\alpha > 0$.

Corollary 2.12 has been generalized to WA lattices in E. Fried and G. Grätzer [1976]; in this case, "two-generated" suffices.

For an empty $\mathcal{C}$-relation, a $\mathcal{C}$-reduced free product becomes a free $\{0, 1\}$-product; the structure theorem for this special case is quite easy. This suggests an alternative development for this chapter. Start with establishing the structure theorem for free $\{0, 1\}$-products. Derive from this the development of Section 1 by observing that the free product of the L_i, $i \in I$, is the sublattice $[\bigcup(L_i \mid i \in I)]$ of the free $\{0, 1\}$-product of L_i^b, $i \in I$. Still one more approach to Section 1 is the one in B. Jónsson [1971]: Once we know the Structure Theorem, it is possible to give a direct proof without the use of polynomials, equivalence classes, and so on. This approach is the most economical but not necessarily the most illuminating.

The basic idea of Section 3 is in Definition 3. The credit for this should go to R. P. Dilworth [1945], which contains a significant special case of this construction. This was generalized to $L * \mathrm{F}(\aleph_0)$ in C. C. Chen and G. Grätzer [1969] under the constraints of Corollary 3.6.

An alternative proof of Corollary 3.8 is given in P. Crawley and R. P. Dilworth [1973].

The background of Corollary 3.8 is quite interesting. Results, such as the one in E. V. Huntington [1904], suggested that Boolean lattices can be characterized as uniquely complemented lattices. This was a widely accepted conjecture in the thirties, supported by a growing body of results of the type that if a lattice L is uniquely complemented and has property X, then L is Boolean. For X = modular, and X = relatively complemented, this appears to have been known already in the thirties to G. Birkhoff and J. von Neumann; for X = finite dimensional, see R. P. Dilworth [1940a]; for X = complete, atomic, and dually atomic, see G. Birkhoff and M. Ward [1939]. Then came R. P. Dilworth's result showing that without some additional property the conjecture is false. Thus the search for weaker and weaker properties X continued. For X = atomic, see T. Ogasawara and U. Sasaki [1949], see also J. E. McLaughlin [1956]; for X = the map $a \mapsto a'$ is order-inverting, see G. Birkhoff [1948]; for X = algebraic, see V. N. Saliĭ [1972]; see also Exercises 3.25 and 3.26. R. Padmanabhan has some related unpublished results: for X, we can take $L \in \mathbf{M} \vee \mathbf{N}_5$; for X, we can also take $L \in \mathbf{Var}(K)$, where K is a finite lattice satisfying $(\mathrm{SD}_\wedge)$ or an implication which is due to E. Fried and G Grätzer:

$$x \wedge y = x \wedge z \quad \text{implies that} \quad x \vee (y \wedge z) = (x \vee y) \wedge (x \vee z).$$

Many of the results of this chapter can be extended to $\mathfrak{m}$-complete lattices (lattices in which $\bigwedge X$ and $\bigvee X$ exist for all nonempty subsets of X with $|X| < \mathfrak{m}$). Some of these results can be found in P. Crawley and R. A. Dean [1959] and B. Jónsson [1962]. The Structure Theorem of Free Products and its applications have not been published for the $\mathfrak{m}$-complete case.

Some covering relations in F(3) are shown in Figure 1; some more can be found in P. M. Whitman [1941] and [1942] and in R. A. Dean [1961a]. In R. N. McKenzie [1972] an algorithm was found deciding whether $p \succ q$ holds in F(3). Some of McKenzie's results, combined with a new theorem of A. Day [1977], yields the astonishing fact that F(3) is *weakly atomic* (that is, every proper interval of F(3) contains a prime interval)!

The solution to the word problem for CF(P), the completely free lattice generated by a poset P, is very similar to Theorem 2.2; except, of course, that the relations we have to start with are $a \leq b$ where $a, b \in P$ and $a \leq b$ in P. M. E. Adams and D. Kelly [1977] prove that the free product of L_i, $i \in I$, can be embedded into CF($\bigcup (L_i \mid i \in I)$). This is the crucial step in proving that free products preserve chain conditions.

For various generalizations of the free product construction see R. Balbes and A. Horn [1967], R. A. Dean [1964], Z. Ladzianska [1974], and H. Lakser [1968].

Condition (W) is implicit in P. M. Whitman [1941]; it was first explicitly stated in B. Jónsson [1961]. A remarkable result of R. Freese [1975] states that a lattice L with no infinite chains satisfies (W) iff L is a retract of some $\mathrm{Du\,Id\,F}(n)$.

Condition (W) plays a role in the characterization of finite (R. N. McKenzie [1972]) and finitely generated (A. Kostinsky [1972]) projective lattices. (Projective lattices are characterized in R. Freese and J. B. Nation [1978].) In B. A. Davey and B. Sands [1977], it is proved that, for a lattice L with no infinite chains, (W) is equivalent to the projectivity of L in the class of all lattices with no infinite chains. B. Sands has examples showing that without any chain condition this result does not hold.

T. G. Kucera and B. Sands [1978] have considered finite lattices L such that, for all finite lattices K, the set of all homomorphisms of L into K form a lattice under pointwise ordering. Again (W) plays a role in this investigation.

An easy, but illuminating, property of (W) is pointed out in K. A. Baker and A. W. Hales [1974]: (W) holds for a lattice L iff it holds for $\mathrm{Id}\,L$.

The characterization problem of finite sublattices of a free lattice goes back, at least, to P. M. Whitman [1941]. In B. Jónsson [1961] and B. Jónsson and J. E. Kiefer [1962], it became apparent that anything that can be proved for finite sublattices of a free lattice follows already from (W), $(\mathrm{SD}_\wedge)$, and $(\mathrm{SD}_\vee)$. The converse, conjectured by B. Jónsson, is still open. For an early review article on this field, see R. A. Dean [1961]. The distributive case is settled in F. Galvin and B. Jónsson [1961]. For some recent results, see H. S. Gaskill [1978].

F(3) has many nonfinite sublattices. It was observed in P. Crawley and R. A. Dean [1959] that, for every countable poset P, $\mathrm{CF}(P)$ is a sublattice of F(3). R. N. McKenzie [1973] discusses some open problems related to free lattices.

Elementary equivalence and free products are considered in B. Jónsson and P. Olin [1976]. They prove that if $\mathbf{K} \neq \mathbf{T}$ is any variety of lattices, then free $\mathbf{K}$-products do not preserve elementary equivalence. This contrasts with Boolean algebras, see P. Olin [1976].

Automorphism groups of lattices are, up to isomorphism, arbitrary groups. This special case of Theorem 3.11 goes back to G. Birkhoff [1946]; in fact, the same result is proved there for distributive lattices. Moreover, for finite groups the lattice constructed is finite. Small lattices with given automorphism groups are considered in R. Frucht [1948] and [1950]. R. N. McKenzie and J. Sichler have some related results for lattices of finite length. Two sample results: every group is the automorphism group of a lattice of finite length; for every lattice L, there exists a lattice K such that $\mathrm{End}(L) \cong \mathrm{End}_{0,1}(K)$ and if L is finite or finite length, then so is K, where $\mathrm{End}(L)$ is the endomorphism semigroup of L. See also J. Sichler [1972].

Theorem 3.11 is one of a large body of results representing monoids as endomorphism monoids of various types of algebras. All these results are based on P. Vopenka, A. Pultr, and Z. Hedrlín [1965] proving the existence of rigid relations and on Z. Hedrlín and A. Pultr [1964] proving the representation for graphs. See also Z. Hedrlín and A. Pultr [1966] for the case of algebras with two unary operations; Z. Hedrlín and J. Lambek [1969] for the case of semigroups and for an alternate proof of the existence of rigid graphs. The result on triangle connected graphs is a special case of a result of P. Hell [1972].

Theorem 4.6 combines a result of J. Sichler [1972] with a very special case of a result of H. Lakser [1972]. The result of J. Sichler states that the diamonds in an $L(G)$ are all associated with the triangles of G. H. Lakser investigates the simple sublattices of a free product L of lattices L_i, $i \in I$; in particular, he proves that if S is a simple lattice, and S cannot be embedded in any L_i, $i \in I$, then S cannot be embedded in L. In addition, he obtains that L contains a diamond iff some L_i contains a diamond and L contains a diamond which is not in one of the L_i, $i \in I$, iff some L_i contains $M_3 \times C_2$ as a sublattice.

Problems

1. Let X be a finite smallest generating set of $A*B$. Is it true that $X \subseteq A \cup B$?

2. Does some form of the Common Refinement Property hold for free $\{0,1\}$-products? Investigate this problem also in an arbitrary variety of lattices.

3. For which variety of lattices does *Sorkin's Theorem* hold? (See Exercise 1.13.)

4. Find a condition implying the Common Refinement Property for free **K**-products, for a variety **K** of algebras, including the case of varieties of lattices and of regular varieties.

5. Which properties of homomorphism $B_1 \rightarrow B_2$ are preserved by the induced homomorphism $A * B_1 \rightarrow A * B_2$?

6. For lattices A and B, does $A * A \cong B * B$ imply that $A \cong B$? More generally, does
$$A_1 * \cdots * A_n \cong B_1 * \cdots * B_n,$$
$$A_1 \cong \cdots \cong A_n,$$
$$B_1 \cong \cdots \cong B_n$$

imply that $A_1 \cong B_1$?

7. Does
$$A_1 * \cdots * A_n * C \cong B_1 * \cdots * B_n * C,$$
$$A_1 \cong \cdots \cong A_n,$$
$$B_1 \cong \cdots \cong B_n$$

imply that $A_1 * C \cong B_1 * C$?

8. Investigate Problems 6 and 7 for free $\{0, 1\}$-products and free **K**-products.

9. Give a structure theorem of free $\mathbf{M}_3$-products. Same problem for $\mathbf{N}_5$.

10. Let **K** be a variety of lattices and let us assume that there exists an algorithm deciding identities in **K**. Is there an algorithm (relative to A and B) deciding the structure of the free **K**-product of A and B?

11. Let **K** be a variety of lattices for which the word problem for $F_\mathbf{K}(\aleph_0)$ is solved. When can we conclude that Theorem 1.14 holds for **K**?

12. Is there any variety $\mathbf{K} \supset \mathbf{D}$ of modular lattices for which the conclusion of Theorem 1.14 holds?

13. For which varieties **K** of lattices is it true that free **K**-products preserve the $\mathfrak{m}$-chain condition? ($\mathfrak{m}$ is an infinite uncountable regular cardinal; the $\mathfrak{m}$-*chain condition* means that all chains are of cardinality $< \mathfrak{m}$.)

14. Investigate varieties **K** of lattices in which the following holds: let $L_i \in \mathbf{K}$, $i \in I$, be a disjoint family of lattices and let L be a free **K**-product of L_i, $i \in I$; then L has an order embedding (that is, one-to-one and isotone map) into $F_\mathbf{K}(\bigcup(L_i \mid i \in I)^m)$, where $\bigcup(L_i \mid i \in I)$ is regarded as a poset.

15. For which varieties **K** is $\mathbf{S}(F_\mathbf{K}(\aleph_0))$ closed under the formation of free **K**-products?

16. Give necessary and sufficient conditions for an interval $[p, q]$ of $F(3)$ to contain a copy (infinitely many copies) of $F(3)$.

17. Does $F(3)$ have intervals of length n, for $n = 3,\ 4,\ \ldots$?

18. Is a finite lattice isomorphic to a sublattice of $F(3)$ iff it satisfies (W), $(\mathrm{SD}_\wedge)$, and $(\mathrm{SD}_\vee)$?

19. For $p,\ q \in F(\aleph_0)$, is it decidable whether there is an endomorphism χ of $F(\aleph_0)$ satisfying $p\chi = q$?

20. For which variety **K** of lattices does $F_\mathbf{K}(\aleph_0)$ satisfy $(\mathrm{SD}_\vee)$?

21. Does $F_\mathbf{M}(4)$ have a proper sublattice isomorphic to $F_\mathbf{M}(4)$? In general, if **K** is a variety of lattices, $n < \aleph_0$, and $F_\mathbf{K}(n)$ is infinite, does $F_\mathbf{K}(n)$ then have a proper sublattice isomorphic with $F_\mathbf{K}(n)$?

22. Find a common generalization of the McKenzie-Kostinsky Theorem and the Davey-Sands Theorem.

23. Characterize the endomorphism semigroups of lattices.

24. Describe the endomorphism semigroups and $\{0,1\}$-endomorphism semi-groups of complemented lattices.

25. Characterize the endomorphism semigroups and the $\{0,1\}$-endomorphism semigroups of complete lattices. What about the semigroups of complete endomorphisms?

26. Characterize the $\{0,1\}$-endomorphism semigroups of finite lattices, of lattices of finite length, and of lattices without infinite chains. (Consider also the categorical versions of problems 23–26.)

27. Let M be a monoid, let $\mathfrak{m}$ be an infinite cardinal, and let $|M| < \mathfrak{m} \le 2^{|M|}$. How many pairwise nonisomorphic bounded lattices exist whose $\{0,1\}$-endomorphism semigroup is isomorphic to M? (If $\mathfrak{m} \le |M|$, the answer is $2^{\mathfrak{m}}$; see G. Grätzer and J. Sichler[1970]).

28. Give a concrete example of a nondistributive uniquely complemented lattice.

29. Does there exist a complete nondistributive uniquely complemented lattice?

30. Does every uniquely complemented nondistributive lattice contain $F(3)$ as a sublattice? (An affirmative answer would also settle the next problem.)

31. Does there exist a uniquely complemented nondistributive lattice satisfying a nontrivial lattice identity?

32. For any infinite cardinal $\mathfrak{m}$, prove the existence of $\mathfrak{m}$-complete uniquely complemented nondistributive lattices. Can they be made join- or meet-continuous, or both?

33. Investigate finitely presented lattices. (See G. Grätzer and H. Lakser [1975]).

34. Is it true that the cardinality of a basis of a finitely presented lattice is uniquely determined?

35. Is the automorphism group of a finitely presented lattice finite?

36. Let L be an infinite lattice. Show that, with finitely many exceptions, if L is finitely presented, then it contains $F(3)$ as a sublattice.

37. Describe sublattices of finitely presented lattices.

38. Describe all finite partial lattices $\mathfrak{A}$ for which $\mathrm{F}(\mathfrak{A})$ is finite.

39. For a variety $\mathbf{K}$ of lattices, define "finitely $\mathbf{K}$-presented". Investigate Problems 33–38 for finitely $\mathbf{K}$-presented lattices.

40. Let $\mathbf{K}$ be a variety of lattices and let $n < \omega$. Show that if $\mathbf{K} \neq \mathbf{L}$ and $\mathrm{F}_{\mathbf{K}}(n)$ is infinite, then $\mathrm{F}_{\mathbf{K}}(n)$ is not finitely presented. (For $\mathbf{K} = \mathbf{M}$ and $n = 4$, this has already been verified; see T. Evans and D. X. Hong [1972]).

41. Under what conditions does Theorem 4.1 hold for finitely $\mathbf{K}$-presented lattices?

42. Describe Hopfian distributive lattices.

43. Improve Theorem 4.5 by requiring that one or both of L_0 and L_1 be finite, finitely presented, finitely generated, or countable. (It follows from Theorem 4.1 that both of L_0 and L_1 cannot be finite or finitely presented.)

44. Prove that the free product of two bounded Hopfian lattices is Hopfian again. (If this is true, then in Problem 43, both L_0 and L_1 cannot be finitely generated.)

45. In which variety $\mathbf{K}$ does Theorem 4.5 hold (L_0, $L_1 \in \mathbf{K}$ and we form the free $\mathbf{K}$-product).

46. Under what operators is the class of Hopfian lattices closed? For instance, is the direct product of Hopfian lattices Hopfian?

47. For what varieties $\mathbf{K}$, do there exist two cohopfian lattices in $\mathbf{K}$ whose direct product is not cohopfian? (A lattice L is *cohopfian* iff L is not isomorphic to any of its proper sublattices.)

48. Investigate first-order properties Φ that are preserved when passing from a lattice to its ideal lattice, or when passing from the ideal lattice to the lattice, or both ways.

49. Let L be a lattice satisfying (SD$_\wedge$), (SD$_\vee$), and (W). Prove that if $|L| > 1$, then L contains a prime ideal.

50. Find a variety of lattices $\mathbf{K} \neq \mathbf{L}$ such that a finite lattice is projective in $\mathbf{K}$ iff it is embeddable in $\mathrm{F}_{\mathbf{K}}(\aleph_0)$.

Concluding Remarks

What is lattice theory? Where to draw the line between lattice theory proper and its allied fields? Does a theorem characterizing a lattice associated with some object outside of lattice theory belong to lattice theory?

These questions are rather academic but they require some response and, in fact, decisions in a book of this kind.

The first question is really answered by this book just like the question, "What is an animal?" is best answered by a tour of a zoo.

The second question is more difficult. For instance, the theory of Boolean algebras develops by an interaction of lattice theory, topology, axiomatic set theory, and logic. A complete bibliography of lattice theory would be rather difficult to compile on account of jurisdictional disputes.

The third question, however, is a simple one for me: a theorem characterizing the subgroup lattices of abelian groups belongs to the theory of abelian groups and not to lattice theory.

A contentious case in point is the representation theory of algebraic lattices (resp., complete lattices) as subalgebra lattices and congruence lattices of universal algebras (resp., infinitary universal algebras). In my opinion, these results belong to universal algebra and not to lattice theory. However, in view of the close affinity between the workers in lattice theory and in universal algebra, I digress to outline this field.

Let L be an algebraic lattice. Then by Theorem II.3.13, $L \cong \operatorname{Id} C$, where C is a join-semilattice with 0. For $a \in C$, introduce on C the binary operation

$$f_a(x, y) = a \wedge (x \vee y).$$

Then the subalgebras of $\langle C; (f_a \mid a \in C) \rangle$ are exactly the ideals of C, hence the subalgebra lattice is isomorphic to L. Thus we obtain that every algebraic lattice is isomorphic to the subalgebra lattice of some algebra. The converse is even easier. Thus we have obtained a result of G. Birkhoff and O. Frink [1948]. In the same paper they conjecture that, up to isomorphism, congruence lattices of algebras can also be characterized as algebraic lattices. This is proved in

399

G. Grätzer and E. T. Schmidt [1962]. For the shortest proof, see P. Pudlák [1976]. A rather detailed history of this field, including the case of infinitary algebras and the connections between congruence lattices of algebras and type 2 and type 3 representations of lattices, can be found in the appendix of G. Grätzer and W. A. Lampe in in G. Grätzer [1979]. A detailed discussion of this material with proofs would require a monograph of about 400 pages.

The congruence lattice of the algebra $\langle A; F \rangle$ is a sublattice of Part A; if a lattice has the property that whenever it is represented as a $\{0, 1\}$-sublattice of some Part A, then that sublattice is the congruence lattice of some algebra $\langle A; F \rangle$, then the lattice is called *strongly representable*. Such lattices are described in R. W. Quackenbush and B. Wolk [1971] and P. Pudlák [1977].

Another case in point is the investigation of the lattice $\Sigma(X)$ of all topological spaces defined on a set, where for the topologies S_1 and S_2 on X, we set $S_1 \leq S_2$ iff every S_1 open set is also S_2 open. (Sample results: $\Sigma(X)$ is a complemented lattice; see A. K. Steiner [1966] and [1966a].) This point of view was introduced by G. Birkhoff [1936] and it permeates the topology book of R. Vaidyanathaswamy [1947] (reprinted in 1960).

Along with $\Sigma(X)$, about a dozen lattices are being considered in topology. For an excellent survey of this field (with 124 references), the reader should consult R. E. Larson and S. J. Andima [1975]; see also J. Rosický [1975].

The closed (and also the open) subsets of a topological space X form a lattice $\mathrm{L}(X)$, and this lattice determines the topological space (provided it is T_1). These lattices were characterized by H. Wallman [1938]. Thus certain parts of topology can be studied within lattice theory—the theory of compactifications being one example (H. Wallman [1938], O. Frink [1964]). The set of all continuous real functions on a topological space X also forms a lattice $\mathrm{C}(X)$ under pointwise ordering. The lattice $\mathrm{C}(X)$ determines X if X is compact Hausdorff (I. Kaplansky [1947]; see also L. Gilbert and M. Jerison [1960]).

A *topological lattice* L is a lattice equipped with a topology such that $\wedge$ and $\vee$ are continuous functions. The theory of topological lattices, the study of which was started by A. D. Wallace and his students, is quite extensive. The methods used are mostly topological rather than lattice theoretical in nature. The reader should consult L. W. Anderson [1961] for an early review of this field.

In contrast with topological lattices, intrinsic topologies are considered on arbitrary lattices. *Intrinsic topologies* are topologies defined on a lattice in terms of the partial ordering. For instance, the interval topology takes the intervals as a subbase for closed sets. For more details, see O. Frink [1942], G. Birkhoff [1962], and B. C. Rennie [1951].

The study of *partially ordered topological vector spaces* was pioneered by the works of F. Riesz, H. Freudenthal, S. Kakutani, L. V. Kantorovič, M. G. Krein, H. Nakano, and G. Birkhoff. The motivation came from a variety of sources— the theory of Hermitian operators in Hilbert spaces, C^* algebra theory, choquet boundary theory, and the inescapable fact that partially ordered topological vector spaces abound in analysis. The goal is to study how the presence of a partial order relation compatible with the algebraic structure of a vector space over the field of real numbers can substantially influence the topology. The earliest expositions of the subject are L. V. Kantorovič, B. Z. Vulih, and A. G. Pinkser [1950], B. Z. Vulih [1967], and H. Nakano [1950]. The modern theory of partially ordered topological vector spaces is characterized by the major role played by the concept of duality (just as in the modern theory of topological vector spaces) and there are a number of recent books stressing this aspect of the subject: A. L. Peressini [1967], G. Jameson [1970], Yau-chuen Wong and Kung-fu Ng [1973], and M. Dohoux [1974]. These books have extensive bibliographies for the interested reader.

The important work C. Carathéodory [1956] (English translation: C. Carathéodory [1963]) initiated a study of finitely additive functions and measures, and the integrals they generate on rings of elements of Boolean algebras. The concept of integral was made completely algebraic. Thus he created what, in analogy to J. von Neumann's study of Continuous Geometries, may be called "pointless" integration. Rediscovering Carathéodory's work and using the potent tools of *topological Riesz spaces* (partially ordered linear topological spaces that are also lattices with respect to the partial order), D. H. Fremlin [1974] gives a thoroughly modern account of this field.

There are many publications in quantum mechanics that border on lattice theory. A typical question is whether the lattice of propositions of quantum mechanics is modular, as postulated by G. Birkhoff and J. von Neumann [1936]. For some of the recent problems discussed in the literature, see the books: J. M. Jauch [1968] and G. W. Mackey [1963] and the papers: P. D. Finch [1969], J. M. Jauch and C. Piron [1963] and [1969], P. Mittelstaedt [1972], and C. Piron [1964]. It may be of some interest to note that of the two volume work V. S. Varadarajan [1968] and [1970] on quantum mechanics, a substantial part of the first volume deals with lattices.

Lattices and posets appear in other areas alongside with an algebraic structure and some or all the operations relate in some way to the partial ordering (for instance, they are isotone). For a survey of this field, see L. Fuchs [1963].

Many generalizations of the lattice concept appear in the literature. However, only one has been developed to some depth: weakly associative lattices, introduced independently by E. Fried [1970] and H. L. Skala [1971]. Here is a partial bibliography of this new field: P. Erdős, E. Fried, A. Hajnal, and E. C. Milner [1972]; E. Fried [1973], [1974], [1974a], [1975], and [1975]; E. Fried, G. Grätzer, and R. W. Quackenbush [1980] and [1980a]; E. Fried and V. T. Sós [1975]; K. Gladstien [1973]; G. Grätzer [1973] and [1973a]; G. Grätzer and H. Lakser [1971a]; H. L. Skala [1972]; R. W. Quackenbush [1972].

Many problems proposed in *FC* are still open. As Piet Hein [1969] writes: Problems worthy-of-attack-prove their worth-by hitting back.

The following is an account of the work done in relation to problems in *FC*; it is not claimed to be comprehensive.

3. Solved in J. Sichler [1972].

9. Solved in C. R. Platt [1976] and D. Kelly and I. Rival [1975a].

12 and 13. A negative answer is provided for the case $\mathbf{K} = \mathbf{M}$ in A. P. Huhn [1975]. The problem is still open for $\mathbf{K} = \mathbf{L}$.

14 and 15. These are discussed above.

16. $n = 2$ and 3 are done in R. Padmanabhan [1972].

18. An affirmative solution is given in R. W. Quackenbush [1974].

21 and 23. Some relevant references are given in the Further Topics and References of Chapter I.

29. Solved in R. Padmanabhan [1974a].

30. See the references given to Problem II.3.

31. According to J. D. Monk, this can be solved affirmatively under the Generalized Continuum Hypothesis.

32. An affirmative solution was found independently by B. Jónsson (unpublished) and K. D. Magill [1970].

33. References to various solutions are given in the Further Topics and References of Chapter II.

34. The question raised in the problem was answered in the negative by T. Tan. The problem remains unsolved.

35. See V. Vilhelm [1955], and J. Jakubik [1975].

39. Solved by M. E. Adams [1974] and R. Balbes [1972], independently.

47. A contribution to this problem was made in H. Lakser [1973a]. The problem is solved in M. E. Adams and D. Kelly [1977].

51. Answered in the negative in G. Grätzer, B. Jónsson, and H. Lakser [1973].

54. Solved in T. Katriňák [1972].

55. Solutions are offered in T. Katriňák [1973a] and H. A. Priestley [1974].

56. A description is given in A. Urquhart [1973].

57. See M. E. Adams [1973a], T. Katriňák [1976], and H. A. Priestley [1975].

63. This is verified in M. E. Adams [1976]. (The result of M. E. Adams was announced in Notices Amer. Math. Soc. 20 (1973): A-560. Another proof is announced by A. Wrónski [1976].)

65. Solved by T. Katriňák [1973c].

66. The case $n = 1$ is considered in T. Katriňák [1966] and [1968].

Bibliography

S. Abian and A. B. Brown

1961.   A theorem on partially ordered sets, with applications to fixed point theorems. Canad. J. Math. **13**: 78–82.

M. E. Adams

1973.   The Frattini sublattice of a distributive lattice. Algebra Universalis **3**: 216–228.

1973a. The structure of distributive lattices. Ph.D. Thesis, Bristol University.

1974.   A problem of A. Monteiro concerning relative complementation of lattices. Colloq. Math. **30**: 61–67.

1975.   The poset of prime ideals of a distributive lattice. Algebra Universalis **5**: 141–142.

1976.   Implicational classes of pseudocomplemented distributive lattices. J. London Math. Soc. (2) **13**: 381–384.

M. E. Adams and D. Kelly

1975.   Homomorphic images of free distributive lattices that are not sublattices. Algebra Universalis **5**: 143–144.

1977.   Chain conditions in free products of lattices. Algebra Universalis **7**: 235–244.

1977a. Disjointness conditions in free products of lattices. Algebra Universalis **7**: 245–258.

M. Aigner

1969.   Graphs and partial orderings. Monatsh. Math. **73**: 385–396.

J. W. Alexander

1939.   Ordered sets, complexes and the problem of compactification. Proc. Nat. Acad. Sci. U.S.A. **25**: 296–298.

M. Altwegg

1950.   Zur Axiomatik der teilweise geordneten Mengen. Comment. Math. Helv. **24**: 149–155.

I. Amemiya

1957.   On the representation of complemented modular lattices. J. Math. Soc. Japan **9**: 263–279.

I. Amemiya and I. Halperin

1959.   Coordinatization of complemented modular lattices. Nederl. Akad. Wetensch. Proc. Ser. A. **62**: 70–78.

1959a. Complemented modular lattices derived from nonassociative rings. Acta Sci. Math. (Szeged) **20**: 181–201.

F. W. Anderson and R. L. Blair

1961.   Representations of distributive lattices as lattices of functions. Math. Ann. **143**: 187–211.

L. W. Anderson

1961.   Locally compact topological lattices. In *1961 Proc. Sympos. Pure Math., Vol. II*, pp. 195–197. American Mathematical Society, Providence, R.I.

R. Antonius and I. Rival

1974.   A note on Whitman's property for free lattices. Algebra Universalis **4**: 271–272.

G. Ja. Areškin

1953.   On congruence relations in distributive lattices with zero. (Russian.) Dokl. Akad. Nauk SSSR N. S. **90**: 485–486.

1953a. Free distributive lattices and free bicompact T_0-spaces. (Russian.) Mat. Sb. N. S. **33** (**75**): 133–156.

E. Artin

1940.   Coordinates in affine geometry. Rep. Math. Colloq. (2) **2**: 15–20.

B. Artmann

1969.   Hjelmslev planes derived from modular lattices. Canad. J. Math. **21**: 76–83.

S. P. Avann

1958.   A numerical condition for modularity of a lattice. Pacific J. Math. **8**: 17–22.

1958a. Dual symmetry of projective sets in a finite modular lattice. Trans. Amer. Math. Soc. **89**: 541–558.

1960.   Upper and lower complementation in a modular lattice. Proc. Amer. Math. Soc. **11**: 17–22.

1960a. Application of the join-irreducible excess function to semi-modular lattices. Math. Ann. **142**: 345–354.

1961.   Metric ternary distributive semi-lattices. Proc. Amer. Math. Soc. **12**: 407–414.

1961a. Distributive properties in semi-modular lattices. Math. Z. **76**: 283–287.

1964.   Increases in the join-excess function in a lattice. Math. Ann. **154**: 420–426.

1964a. Dependence of finiteness conditions in distributive lattices. Math. Z. **85**: 245–256.

L. Babai

1972.   Automorphism groups of planar graphs. I. Discrete Math. **2**: 295–307.

1975.   Automorphism groups of planar graphs. II. In *Infinite and finite sets (Colloq., Keszthely, 1973; dedicated to P. Erdös on his 60th birthday), Vol. I*, pp. 29–84. Colloq. Math. Soc. János Bolyai, Vol. 10. North-Holland, Amsterdam.

P. D. Bacsich

1975.   Amalgamation properties and interpolation theorems for equational theories. Algebra Universalis **5**: 45–55.

R. Baer

1952.   *Linear Algebra and Projective Geometry.* Academic Press, New York, N.Y.

K. A. Baker

1969.   A generalization of Sperner's lemma. J. Combin. Theory **6**: 224–225.

1969a.   Equational classes of modular lattices. Pacific J. Math. **28**: 9–15.

1971.   Equational axioms for classes of lattices. Bull. Amer. Math. Soc. **77**: 97–102.

1973.   Inside free semilattices. In *Proceedings of the University of Houston Lattice Theory Conference (Houston, Tex., 1973)*, pp. 306–331. Dept. Math., Univ. Houston, Houston, Tex.

1974.   Primitive satisfaction and equational problems for lattices and other algebras. Trans. Amer. Math. Soc. **190**: 125–150.

1976.   Equational axioms for classes of Heyting algebras.

1977.   Finite equational bases for finite algebras in a congruence-distributive equational class. Algebra Universalis **6**: 105–120.

K. A. Baker, P. C. Fishburn, and F. S. Roberts

1970.   Partial orders of dimension 2, interval orders, and interval graphs. Rand Corp. P-4376.

K. A. Baker and A. W. Hales

1970.   Distributive projective lattices. Canad. J. Math. **22**: 472–475.

1974.   From a lattice to its ideal lattice. Algebra Universalis **4**: 250–258.

K. A. Baker and A. F. Pixley

1975.   Polynomial interpolation and the Chinese Remainder Theorem for algebraic systems. Math. Z. **43**: 165–174.

J. W. de Bakker

1967.   On convex sublattices of distributive lattices. Math. Centrum Amsterdam Afd. Zuivere Wisk. ZW-003.

R. Balbes

1967.   Projective and injective distributive lattices. Pacific J. Math. **21**: 405–420.

1969.   A note on distributive sublattices of a modular lattice. Fund. Math. **65**: 219–222.

1971.   On the partially ordered set of prime ideals of a distributive lattice. Canad. J. Math. **23**: 866–874.

R. Balbes

1972.   Solution to a problem concerning the intersection of maximal filters and maximal ideals in a distributive lattice. Algebra Universalis **2**: 389–392.

R. Balbes and P. Dwinger

1971.   Coproducts of Boolean algebras and chains with applications to Post algebras. Colloq. Math. **24**: 15–25.

1971a. Uniqueness of representation of a distributive lattice as a free product of a Boolean algebra and a chain. Colloq. Math. **24**: 27–35.

1974.   *Distributive Lattices.* Univ. Missouri Press, Columbia, Miss.

R. Balbes and G. Grätzer

1971.   Injective and projective Stone algebras. Duke Math. J. **38**: 339–347.

R. Balbes and A. Horn

1967.   Order sums of distributive lattices. Pacific J. Math. **21**: 421–435.

1970.   Stone lattices. Duke Math. J. **37**: 537–545.

1970a. Injective and projective Heyting algebras. Trans. Amer. Math. Soc. **148**: 549–559.

1970b. Projective distributive lattices. Pacific J. Math. **33**: 273–279.

H.-J. Bandelt and R. Padmanabhan

1979.   A note on lattices with unique comparable complements. Abh. Math. Sem. Univ. Hamburg **48**: 112–113.

J. G. Basterfield and L. M. Kelly

1968.   A characterization of sets of n points which determine n hyperplanes. Proc. Cambridge Philos. Soc. **64**: 585–588.

R. Beazer

1975.   Post-like algebras and injective Stone algebras. Algebra Universalis **5**: 16–23.

1975a. Hierarchies of distributive lattices satisfying annihilator conditions. J. London Math. Soc. (2) **11**: 216–222.

E. A. Behrens

1960.   Distributiv darstellbare Ringe. I. Math. Z. **73**: 409–432.

E. A. Behrens

1961.   Distributiv darstellbare Ringe. II. Math. Z. **76**: 367–384.

N. D. Belnap and J. H. Spencer

1966.   Intensionally complemented distributive lattices. Port. Math. **25**: 99–104.

G. M. Bergman

1969.   On 2-firs (weak Bezout rings) with distributive divisor lattices. Manuscript.

J. Berman

1972.   On the length of the congruence lattice of a lattice. Algebra Universalis **2**: 18–19.

1972a. Sublattices of intervals in relatively free lattices. Algebra Universalis **2**: 174–176.

1973.   Congruence relations of pseudocomplemented distributive lattices. Algebra Universalis **3**: 288–293.

G. Birkhoff

1933.   On the combination of subalgebras. Proc. Cambridge Philos. Soc. **29**: 441–464.

1935.   On the structure of abstract algebras. Proc. Cambridge Philos. Soc. **31**: 433–454.

1935a. Abstract linear dependence and lattices. Amer. J. Math. **57**: 800–804.

1935b. Combinatorial relations in projective geometries. Ann. of Math. **36**: 743–748.

1936.   On the combination of topologies. Fund. Math. **26**: 156–166.

1940.   *Lattice Theory.* First edition. Amer. Math. Soc., Providence, R.I.

1940a. Neutral elements in general lattices. Bull. Amer. Math. Soc. **46**: 702–705.

1944.   Subdirect unions in universal algebra. Bull. Amer. Math. Soc. **50**: 764–768.

1946.   On groups of automorphisms. (Spanish.) Rev. Un. Mat. Argentina **11**: 155–157.

1948.   *Lattice Theory.* Second edition. Amer. Math. Soc., Providence, R.I.

1962.   A new interval topology for dually directed sets. Univ. Nac. Tucumán Rev. Ser. A. **14**: 325–331.

G. Birkhoff

1967.   *Lattice Theory.* Third edition. Amer. Math. Soc., Providence, R.I.

1970.   What can lattices do for you? In *Trends in Lattice Theory (Sympos., U.S. Naval Academy, Annapolis, Md., 1966)*, pp. 1–40. Van Nostrand Reinhold, New York.

G. Birkhoff and O. Frink

1948.   Representations of lattices by sets. Trans. Amer. Math. Soc. **64**: 299–316.

G. Birkhoff and J. von Neumann

1936.   The logic of quantum mechanics. Ann. of Math. **37**: 823–843.

G. Birkhoff and M. Ward

1939.   A characterization of Boolean algebras. Ann. of Math. **40**: 609–610.

L. Blumental and K. Menger

1970.   *Studies in Geometry.* Freeman, San Francisco. Cal.

T. J. Brown

1969.   A recursion formula for finite partition lattices. Proc. Amer. Math. Soc. **22**: 124–126.

G. Bruns

1959.   Verbandstheoretische Kennzeichnung vollständinger Mengenringe. Arch. Math. (Basel) **10**: 109–112.

1961.   Distributivität und subdirekte Zerlegbarkeit vollständiger Verbände. Arch. Math. (Basel) **12**: 61–66.

1962.   Darstellungen und Erweiterungen geordneter Mengen. I, II. J. Reine Angew. Math. **209**: 167–200, **210**: 1–23.

1962a. On the representation of Boolean algebras. Canad. Math. Bull. **5**: 37–41.

1965.   Ideal representations of Stone lattices. Duke Math. J. **32**: 555–556.

G. Bruns and G. Kalmbach

1971.   Varieties of orthomodular lattices. Canad. J. Math. **23**: 802–810.

1972.   Varieties of orthomodular lattices. II. Canad. J. Math. **24**: 328–337.

G. Bruns and H. Lakser

1970.   Injective hulls of semilattices. Canad. Math. Bull. **13**: 115–118.

J. R. Büchi

1952.   Representation of complete lattices by sets. Portugal. Math. **11**: 151–167.

R. D. Byrd and R. A. Mena

1976.   Chains in generalized Boolean lattices. J. Austral. Math. Soc. (A) **21**: 231–240.

R. D. Byrd, R. A. Mena, and L. A. Troy

1975.   Generalized Boolean lattices. J. Austral. Math. Soc. **19**: 225–237.

L. Byrne

1946.   Two brief formulations of Boolean algebra. Bull. Amer. Math. Soc. **52**: 269–272.

A. D. Campbell

1943.   Set-coordinates for lattices. Bull. Amer. Math. Soc. **49**: 395–398.

C. Carathéodory

1956.   *Mass und Integral und ihre Algebraisierung.* Birkhäuser Verlag, Basel und Stuttgart.

1963.   *Algebraic Theory of Measure and Integration.* Chelsea, New York, N.Y.

L. Carroll

1865.   *Alice's Adventures in Wonderland.* McMillan and Co., London.

C. C. Chang and A. Horn

1961.   Prime ideal characterization of generalized Post algebra. In *1961 Proc. Sympos. Pure Math., Vol. II*, pp. 43–48. American Mathematical Society, Providence, R.I.

1962.   On the representation of α-complete lattices. Fund. Math. **51**: 253–258.

C. C. Chen and G. Grätzer

1969.   On the construction of complemented lattices. J. Algebra **11**: 56–63.

C. C. Chen and G. Grätzer

1969a. Stone lattices I. Construction theorems. Canad. J. Math. **21**: 884–894.

1969b. Stone lattices II. Structure theorems. Canad. J. Math. **21**: 895–903.

C. C. Chen and K. M. Koh

1972. On the lattice of convex sublattices of a finite lattice. Nanta Math. **5** No. 3: 93–95.

C. C. Chen, K. M. Koh, and S. K. Tan

1973. Frattini sublattices of distributive lattices. Algebra Universalis **3**: 294–303.

1975. On the Frattini sublattice of a finite distributive lattice. Algebra Universalis **5**: 88–97.

R. Church

1940. Numerical analysis of certain free distributive structures. Duke Math. J. **6**: 732–734.

1965. Enumeration by rank of the elements of the free distributive lattice with seven generators. Abstract. Notices Amer. Math. Soc. **12**: 724.

J. M. Cibulskis

1969. A characterization of the lattice orderings on a set which induce a given betweenness. J. London Math. Soc. (2) **1**: 480–482.

I. S. Cohen

1950. Commutative rings with restricted minimum conditions. Duke Math. J. **17**: 27–42.

P. M. Cohn

1971. *Free Rings and Their Relations*. Academic Press, New York, N.Y.

S. D. Comer and D. X. Hong

1972. Some remarks concerning the varieties generated by the diamond and the pentagon. Trans. Amer. Math. Soc. **174**: 45–54.

S. D. Comer and J. Johnson

1972. The standard semigroup of operators of a variety. Algebra Universalis **2**: 77–79.

W. H. Cornish

1972.   Normal lattices. J. Austral. Math. Soc. **14**: 200–215.

1974.   Crawley's completion of a conditionally upper continuous lattice. Pacific J. Math. **51**: 397–405.

1974a. *n*-normal lattices. Proc. Amer. Math. Soc. **45**: 48–54.

1974b. Quasicomplemented lattices. Comment. Math. Univ. Carolinae. **15**: 501–511.

M. Cotlar

1944.   A method of construction of structures and its application to topological spaces and abstract arithmetic. Univ. Nac. Tucumán Rev. Ser. A. **4**: 105–157.

H. H. Crapo

1965.   Single-element extensions of matroids. J. Res. Nat. Bur. Standards Sect. B. **69B**: 55–65.

1966.   The Möbius function of a lattice. J. Combin. Theory **1**: 126–131.

1968.   Möbius inversion in lattices. Arch. Math. (Basel) **19**: 595–607.

H. H. Crapo and G.-C. Rota

1970.   Geometric lattices. In *Trends in Lattice Theory (Sympos., U.S. Naval Academy, Annapolis, Md., 1966)*, pp. 127–172. Van Nostrand Reinhold, New York.

1971.   *Combinatorial Geometries.* M. I. T. Press, Boston, Mass.

P. Crawley

1959.   The isomorphism theorem in compactly generated lattices. Bull. Amer. Math. Soc. **65**: 377–379.

1960.   Lattices whose congruences form a Boolean algebra. Pacific J. Math. **10**: 787–795.

1961.   Decomposition theory for nonsemimodular lattices. Trans. Amer. Math. Soc. **99**: 246–254.

1962.   Regular embeddings which preserve lattice structure. Proc. Amer. Math. Soc. **13**: 748–752.

1962a. Direct decompositions with finite dimensional factors. Pacific J. Math. **12**: 457–468.

P. Crawley and R. A. Dean

1959.   Free lattices with infinite operations. Trans. Amer. Math. Soc. **92**: 35–47.

P. Crawley and R. P. Dilworth

1973.   *Algebraic Theory of Lattices.* Prentice-Hall, Englewood Cliffs, N.J.

R. Croisot

1951.   Contribution à l'étude des trellis semi-modulaires de longueur infinie. Ann. Sci. École Norm. Sup (3) **68**: 203–265.

1952.   Quelques applications et propriétés des treillis semimodulaires de longueur infinie. Ann. Fac. Sci. Univ. Toulouse (4) **16**: 11–74.

1957.   Sur l'irréductibilité dans les treillis géométriques et les géométries projectives. Convegno internazionale: Reticoli e geometrie proiettive, Palermo, 25-29 ottobre 1957; Messina, 30 ottobre 1957, pp. 3–10. Edito dalla Unione Matematica Italiana con il contributo del Consiglio Nazionale delle Ricerche Edizioni Cremonese, Rome.

H. B. Curry

1963.   *Foundations of Mathematical Logic.* McGraw-Hill, New York, N.Y.

M. Curzio

1953.   Alcune limitazioni sul minimo ordine dei reticoli modulari di lunghezza 3 contenenti sottoreticoli d'ordine dato. Ricerche Mat. **2**: 140–147.

A. Daigneault

1959.   Products of polyadic algebras and of their representations. Ph.D. Thesis, Princeton University.

B. A. Davey

1973.   A note on representable posets. Algebra Universalis **3**: 345–347.

1974.   Free products of bounded distributive lattices. Algebra Universalis **4**: 106–107.

1974a. Some annihilator conditions on distributive lattices. Algebra Universalis **4**: 316–322.

B. A. Davey, W. Poguntke, and I. Rival

1975.   A characterization of semi-distributivity. Algebra Universalis **5**: 72–
        75.

B. A. Davey and I. Rival

1976.   Finite sublattices of three-generated lattices. J. Austral. Math. Soc.
        (A) **21**: 171–178.

B. A. Davey and B. Sands

1977.   An application of Whitman's condition to lattices with no infinite
        chains. Algebra Universalis **7**: 171–178.

A. C. Davis

1955.   A characterization of complete lattices. Pacific J. Math. **5**: 311–319.

A. Day

1969.   A characterization of modularity for congruence lattices of algebras.
        Canad. Math. Bull. **12**: 167–173.

1970.   A simple solution of the word problem for lattices. Canad. Math.
        Bull. **13**: 253–254.

1970a. Injectivity in congruence distributive equational classes. Ph.D. The-
        sis, McMaster University.

1970b. Injectives in non-distributive equational classes of lattices are trivial.
        Arch. Math. (Basel) **21**: 113–115.

1973.   Splitting algebras and a weak notion of projectivity. In *Proceedings of
        the University of Houston Lattice Theory Conference (Houston, Tex.,
        1973)*, pp. 466–485. Dept. Math., Univ. Houston, Houston, Tex., and
        also Algebra Universalis **5** (1975): 153–162.

1976.   Lattice conditions implying congruence modularity. Algebra Univer-
        salis **6**: 291–303.

1977.   Splitting lattices generate all lattices. Algebra Universalis **7**: 163–170.

A. Day, C. Herrmann, and R. Wille

1972.   On modular lattices with four generators. Algebra Universalis **2**: 317–
        323.

G. W. Day

1970.   Maximal chains in atomic Boolean algebras. Fund. Math. **67**: 293–
        296.

R. A. Dean

1956.   Component subsets of the free lattice on n generators. Proc. Amer. Math. Soc. **7**: 220–226.

1956a. Completely free lattices generated by partially ordered sets. Trans. Amer. Math. Soc. **83**: 238–249.

1961.   Sublattices of free lattices. In *1961 Proc. Sympos. Pure Math., Vol. II*, pp. 31–42. American Mathematical Society, Providence, R.I.

1961a. Coverings in free lattices. Bull. Amer. Math. Soc. **67**: 548–549.

1964.   Free lattices generated by partially ordered sets and preserving bounds. Canad. J. Math. **16**: 136–148.

R. A. Dean and T. Evans

1969.   A remark on varieties of lattices and semigroups. Proc. Amer. Math. Soc. **21**: 394–396.

R. A. Dean and R. H. Oehmke

1964.   Idempotent semigroups with distributive right congruence lattices. Pacific J. Math. **14**: 1187–1209.

R. Dedekind

1900.   Über die von drei Moduln erzeugte Dualgruppe. Math. Ann. **53**: 371–403.

R. Demarr

1964.   Common fixed points for isotone mappings. Colloq. Math. **13**: 45–48.

V. Devidé

1963.   On monotonous mappings of complete lattices. Fund. Math. **53**: 147–154.

A. H. Diamond and J. C. C. McKinsey

1947.   Algebras and their subalgebras. Bull. Amer. Math. Soc. **53**: 959–962.

R. P. Dilworth

1940.   Lattices with unique irreducible decompositions. Ann. of Math. (2) **41**: 771–777.

1940a. On complemented lattices. Tôhoku Math. J. **47**: 18–23.

1941.   Ideals in Birkhoff lattices. Trans. Amer. Math. Soc. **49**: 325–353.

R. P. Dilworth

1941a. The arithmetical theory of Birkhoff lattices. Duke Math. J. **8**: 286–299.

1944. Dependence relations in a semi-modular lattice. Duke Math. J. **11**: 575–587.

1945. Lattices with unique complements. Trans. Amer. Math. Soc. **57**: 123–154.

1946. Note on the Kurosch-Ore theorem. Bull. Amer. Math. Soc. **52**: 659–663.

1950. A decomposition theorem for partially ordered sets. Ann. of Math. (2) **51**: 161–166.

1950a. The structure of relatively complemented lattices. Ann. of Math. (2) **51**: 348–359.

1954. Proof of a conjecture on finite modular lattices. Ann. of Math. (2) **60**: 359–364.

1960. Some combinatorial problems on partially ordered sets. In *1960 Proc. Sympos. Appl. Math., Vol. 10*, pp. 85–90. American Mathematical Society, Providence, R.I.

1961. Structure and decomposition theory of lattices. In *1961 Proc. Sympos. Pure Math., Vol. II*, pp. 3–16. American Mathematical Society, Providence, R.I.

R. P. Dilworth and P. Crawley

1960. Decomposition theory for lattices without chain conditions. Trans. Amer. Math. Soc. **96**: 1–22.

R. P. Dilworth and R. Freese

1966. Generators of lattice varieties. Algebra Universalis **6**: 263–267.

R. P. Dilworth and C. Greene

1971. A counterexample to the generalization of Sperner's theorem. J. Combin. Theory **10**: 18–21.

R. P. Dilworth and M. Hall

1944. The embedding problem for modular lattices. Ann. of Math. (2) **45**: 450–456.

R. P. Dilworth and J. E. McLaughlin

1952.   Distributivity in lattices. Duke Math. J. **19**: 683–693.

M. Dohoux

1974.   *Espaces vectoriels topologiques préordonnés.* Inst. de Math., Université Catholique de Louvain.

T. A. Dowling and R. M. Wilson

1974.   The slimmest geometric lattices. Trans. Amer. Math. Soc. **196**: 203–215.

1975.   Whitney number inequalities for geometric lattices. Proc. Amer. Math. Soc. **47**: 504–512.

D. A. Drake

1971.   Another note on Sperner's lemma. Canad. Math. Bull. **14**: 255–256.

H. Draškovičová

1974.   On a representation of lattices by congruence relations. Mat. Časopis Sloven. Akad. Vied. **24**: 69–75.

M. L. Dubreil-Jacotin, L. Lesieur, and R. Croisot

1953.   *Leçons sur la théorie des treillis des structures algébriques ordonnées et des treillis géométriques.* Gauthier-Villars, Paris.

J. M. Dunn and N. D. Belnap

1968.   Homomorphisms of intensionally complemented distributive lattices. Math. Ann. **176**: 28–38.

W. D. Duthie

1942.   Segments of ordered sets. Trans. Amer. Math. Soc. **51**: 1–14.

A. Ehrenfeucht, V. Faber, S. Fajtlowicz, and J. Mycielski

1973.   Representations of finite lattices as partition lattices on finite sets. In *Proceedings of the University of Houston Lattice Theory Conference (Houston, Tex., 1973)*, pp. 17–35. Dept. Math., Univ. Houston, Houston, Tex.

P. C. Eklof

1974.   Algebraic closure operators and strong amalgamation bases. Algebra Universalis **4**: 89–98.

R. Engelking

1965.   Cartesian products and dyadic spaces. Fund. Math. **57**: 287–304.

G. Epstein

1960.   The lattice theory of Post algebras. Trans. Amer. Math. Soc. **95**: 300–317.

P. Erdös, E. Fried, A. Hajnal, and E. C. Milner

1972.   Some remarks on simple tournaments. Algebra Universalis **2**: 238–245.

T. Evans

1969.   Finitely presented loops, lattices, etc. are Hopfian. J. London Math. Soc. **44**: 551–552.

1969a. Some connections between residual finiteness, finite embeddability, and the word problem. J. London Math. Soc. (2) **1**: 399–403.

T. Evans and D. X. Hong

1972.   The free modular lattice on four generators is not finitely presentable. Algebra Universalis **2**: 284–285.

N. D. Filippov

1966.   Projectivity of lattices. (Russian.) Mat. Sb. **70** (112): 36–54.

P. D. Finch

1969.   On the structure of quantum logic. J. Symbolic Logic **34**: 275–282.

D. T. Finkbeiner

1951.   A general dependence relation for lattices. Proc. Amer. Math. Soc. **2**: 756–759.

1960.   Irreducible congruence relations on lattices. Pacific J. Math. **10**: 813–821.

1960a. A semimodular imbedding of lattices. Canad. J. Math. **12**: 582–591.

I. Fleischer

    1976.   Embedding a semilattice in a distributive lattice. Algebra Universalis **6**: 85–86.

T. Frayne, A. C. Morel, and D. S. Scott

    1962.   Reduced direct products. Fund. Math. **51**: 195–228.

R. Freese

    1972.   Varieties generated by modular lattices of width four. Bull. Amer. Math. Soc. **78**: 447–450.

    1973.   Breadth two modular lattices. In *Proceedings of the University of Houston Lattice Theory Conference (Houston, Tex., 1973)*, pp. 409–451. Dept. Math., Univ. Houston, Houston, Tex.

    1974.   An application of Dilworth's lattice of maximal antichains. Discrete Math. **7**: 107–109.

    1975.   Ideal lattices of lattices. Pacific J. Math. **57**: 125–133.

    1975a.   Congruence modularity. Abstract. Notices Amer. Math. Soc. **22**: A–301.

    1976.   Planar sublattices of FM(4). Algebra Universalis **6**: 69–72.

    1977.   *The structure of modular lattices of width four with applications to varieties of lattices.* Memoirs Amer. Math. Soc. No. 181.

R. Freese and J. B. Nation

    1973.   Congruence lattices of semilattices. Pacific J. Math. **49**: 51–58.

    1978.   Projective lattices. Pacific J. Math. **75**: 93–106.

P. A. Freĭdman

    1967.   Rings with a distributive lattice of subrings. (Russian.) Mat. Sb. **73** (4): 513–534.

D. H. Fremlin

    1974.   *Topological Riesz Spaces and Measure Theory.* Cambridge University Press, Cambridge.

E. Fried

    1970.   Tournaments and nonassociative lattices. Ann. Univ. Sci. Budapest. Eötvös Sect. Math. **13**: 151–164.

E. Fried

1973. Weakly associative lattices with join and meet of several elements. Ann. Univ. Sci. Budapest. Eötvös Sect. Math. **16**: 93–98.

1974. Weakly associative lattices with congruence extension property. Algebra Universalis **4**: 151–162.

1974a. Subdirect irreducible weakly associative lattices with congruence extension property. Ann. Univ. Sci. Budapest. Eötvös Sect. Math. **17**: 59–68.

1975. Prime factorization, algebraic extension, and construction of weakly associative lattices. Acta Math. Acad. Sci. Hungar. **26**: 241–244.

1975a. Equational classes which cover the class of distributive lattices. Acta Sci. Math. (Szeged) **37**: 37–40.

E. Fried and G. Grätzer

1973. A nonassociative extension of the class of distributive lattices. Pacific J. Math. **49**: 59–78.

1973a. Some examples of weakly associative lattices. Colloq. Math. **27**: 215–221.

1976. Free and partial weakly associative lattices. Houston J. Math. **2**: 501–512.

E. Fried, G. Grätzer, and H. Lakser

1971. Amalgamation and weak injectives in the equational class of modular lattices $\mathbf{M}_n$. Abstract. Notices Amer. Math. Soc. **18**: 624.

E. Fried, G. Grätzer, and R. W. Quackenbush

1980. Uniform congruence schemes. Algebra Universalis **10**: 176–188.

1980a. The equational class generated by weakly associative lattices with the unique bound property. Ann. Univ. Sci. Budapest. Eötvös Sect. Math. **22**: 205–211.

E. Fried and E. T. Schmidt

1975. Standard sublattices. Algebra Universalis **5**: 203–211.

E. Fried and V. T. Sós

1975. Weakly associative lattices and projective planes. Algebra Universalis **5**: 114–119.

H. Friedman and D. Tamari

1967.   Problèmes d'associativité: une structure de trellis finis induite par une loi demi-associative. J. Combin. Theory **2**: 215–242.

O. Frink

1942.   Topology in lattices. Trans. Amer. Math. Soc. **51**: 569–582.

1946.   Complemented modular lattices and projective spaces of infinite dimension. Trans. Amer. Math. Soc. **60**: 452–467.

1962.   Pseudo-complements in semi-lattices. Duke Math. J. **29**: 505–514.

1964.   Compactifications and semi-normal spaces. Amer. J. Math. **86**: 602–607.

B. Frontera Marqués

1964.   Generalization of the concept of equivalence relation in a set. (Spanish.) In *1967 Proc. Fifth Annual Reunion of Spanish Mathematicians (Valencia, 1964)*, pp. 203–220. Publ. Inst. "Jorge Juan" Mat., Madrid.

R. Frucht

1948.   On the construction of partially ordered systems with a given group of automorphisms. Rev. Un. Mat. Argentina **13**: 12–18.

1950.   Lattices with a given abstract group of automorphisms. Canad. J. Math. **2**: 417–419.

K. D. Fryer and I. Halperin

1956.   The von Neumann coordinatization theorem for complemented modular lattices. Acta Sci. Math (Szeged) **17**: 450–456.

1958.   On the construction of coordinates for non-Desarguesian complemented modular lattices. I., II. Nederl. Akad. Wetensch. Proc. Ser. A. **61**: 142–161.

L. Fuchs

1949.   Über die Ideale arithmetischer Ringe. Comment. Math. Helv. **23**: 334–341.

1963.   *Partially Ordered Algebraic Systems.* Pergamon Press, New York, N.Y.

T. Fujiwara and K. Murata

1953.   On the Jordan-Hölder-Schreier theorem. Proc. Japan Acad. **29**: 151–153.

N. Funayama

1942.   On the congruence relations on lattices. Proc. Imp. Acad. Tokyo **18**: 530–531.

1944.   On the completion by cuts of distributive lattices. Proc. Imp. Acad. Tokyo **20**: 1–2.

1953.   Notes on lattice theory. IV. On partial (semi)lattices. Bull. Yamagata Univ. Natur. Sci. **2**: 171–184.

1959.   Imbedding infinitely distributive lattices completely isomorphically into Boolean algebras. Nagoya Math. J. **15**: 71–81.

N. Funayama and T. Nakayama

1942.   On the distributivity of a lattice of lattice-congruences. Proc. Imp. Acad. Tokyo **18**: 553–554.

H. Gaifman

1964.   Infinite Boolean polynomials. I. Fund. Math. **54**: 229–250.

F. Galvin and B. Jónsson

1961.   Distributive sublattices of a free lattice. Canad. J. Math. **13**: 265–272.

B. Ganter and I. Rival

1973.   Dilworth's covering theorem for modular lattices: a simple proof. Algebra Universalis **3**: 348–350.

1975.   An arithmetical theorem for modular lattices. Algebra Universalis **5**: 395–396.

H. S. Gaskill

1972.   On transferable semilattices. Algebra Universalis **2**: 303–316.

1972a. On the relation of a distributive lattice to its lattice of ideals. Bull Austral. Math. Soc. **7**: 377–385.

1978.   A note on constructing finite sublattices of free lattices. Algebra Universalis **8**: 244–255.

H. S. Gaskill, G. Grätzer, and C. R. Platt

1975.   Sharply transferable lattices. Canad. J. Math. **27**: 1247–1262.

H. S. Gaskill and C. R. Platt

1975.   Sharp transferability and finite sublattices of a free lattice. Canad. J. Math. **27**: 1036–1041.

I. M. Gelfand and V. A. Ponomarev

1970.   Problems of linear algebra and classification of quadruples of sub-spaces in a finite dimensional vector space. In *Hilbert space operators and operator algebras (Proc. Internat. Conf., Tihany, 1970)*, pp. 163–237. Colloq. Math. Soc. János Bolyai, 5. North-Holland, Amsterdam, 1972.

A. Ghouilà-Houri

1962.   Caractérisation des graphes non orientés dont on peut orienter les arêtes de manière à obtenir le graphe d'une relation d'ordre. C. R. Acad. Sci. Paris. Sér. A–B **254**: 1370–1371.

E. N. Gilbert

1954.   Lattice theoretic properties of frontal switching functions. J. Mathematical Phys. **33**: 57–67.

L. Gilbert and M. Jerison

1960.   *Rings of Continuous Functions*. University Series in Higher Mathematics. Van Nostrand, Princeton, N.J.

P. C. Gilmore and A. J. Hoffman

1964.   A characterization of comparability graphs and of interval graphs. Canad. J. Math. **16**: 539–548.

K. Gladstien

1973.   A characterization of complete trellises of finite length. Algebra Universalis **3**: 341–344.

V. Glivenko

1929.   Sur quelques points de la logique de M. Brouwer. Bull. Acad. des Sci. de Belgique **15**: 183–188.

M. M. Gluhov

1960.   On the problem of isomorphism of lattices. (Russian.) Dokl. Akad. Nauk SSSR **132**: 254–256.

G. Grätzer

1959.   Standard ideals. (Hungarian.) Magyar Tud. Akad. Mat. Fiz. Oszt. Közl. **9**: 81–97.

1962.   A characterization of neutral elements in lattices. (Notes on Lattice Theory. I.) Magyar Tud. Akad. Mat. Kutató Int. Közl. **7**: 191–192.

1962a. On Boolean functions. (Notes on Lattice Theory. II.) Rev. Roumaine Math. Pures Appl. **7**: 693–697.

1963.   A generalization of Stone's representation theorem for Boolean algebras. Duke J. Math. **30**: 469–474.

1963a. On semi-discrete lattices whose congruence relations form a Boolean algebra. Acta Math. Acad. Sci. Hungar. **14**: 441–445.

1964.   Boolean functions on distributive lattices. Acta Math. Acad. Sci. Hungar. **15**: 195–201.

1966.   Equational classes of lattices. Duke J. Math. **33**: 613–622.

1968.   *Universal Algebra.* University Series in Higher Mathematics. Van Nostrand, Princeton, N.J.

1969.   Stone algebras form an equational class. (Notes on Lattice Theory. III.) J. Austral. Math. Soc. **9**: 308–309.

1970.   Universal algebra. In *Trends in Lattice Theory (Sympos., U.S. Naval Academy, Annapolis, Md., 1966),* pp. 173–210. Van Nostrand Reinhold, New York.

1970a. Two Mal'cev type theorems in universal algebras. J. Combin. Theory **8**: 334–342.

1971.   *Lattice Theory: First Concepts and Distributive Lattices.* Freeman, San Francisco, Cal.

1971a. A reduced free product of lattices. Fund. Math. **73**: 21–27.

1973.   Free products and reduced free products of lattices. In *Proceedings of the University of Houston Lattice Theory Conference (Houston, Tex., 1973),* pp. 539–563. Dept. Math., Univ. Houston, Houston, Tex.

1974.   A property of transferable lattices. Proc. Amer. Math. Soc. **43**: 269–271.

1975.   A note on the Amalgamation Property. Abstract. Notices Amer. Math. Soc. **22**: A–453.

G. Grätzer

1979.   *Universal Algebra*, Second Edition. Springer-Verlag, New York–Heidelberg.

G. Grätzer, B. Jónsson, and H. Lakser

1973.   The Amalgamation Property in equational classes of modular lattices. Pacific J. Math. **45**: 507–524.

G. Grätzer, K. M. Koh, and M. Makkai

1972.   On the lattice of subalgebras of a Boolean algebra. Proc. Amer. Math. Soc. **36**: 87–92.

G. Grätzer and H. Lakser

1968.   Extension theorems on congruences of partial lattices. Abstract. Notices Amer. Math. Soc. **15**: 732, 785.

1969.   Chain conditions in the distributive free product of lattices. Trans. Amer. Math. Soc. **144**: 301–312.

1969a. Some applications of free distributive products. Abstract. Notices Amer. Math. Soc. **16**: 405.

1969b. Equationally compact semilattices. Colloq. Math. **20**: 27–30.

1971.   The structure of pseudocomplemented distributive lattices. II. Congruence extension and amalgamation. Trans. Amer. Math. Soc. **156**: 343–358.

1971a. Identities for equational classes generated by tournaments. Abstract. Notices Amer. Math. Soc. **18**: 794.

1972.   The structure of pseudocomplemented distributive lattices. III. Injective and absolute subretracts. Trans. Amer. Math. Soc. **169**: 475–487.

1973.   A note on the implicational class generated by a class of structures. Canad. Math. Bull. **16**: 603–605.

1973a. Three remarks on the Arguesian identity. Abstract. Notices Amer. Math. Soc. **20**: A-253, A-313.

1974.   Free lattice like sublattices of free products of lattices. Proc. Amer. Math. Soc. **44**: 43–45.

1975.   Finitely presented lattices. I. Abstract. Notices Amer. Math. Soc. **22**: A-380.

G. Grätzer, H. Lakser, and C. R. Platt

1970.   Free products of lattices. Fund. Math. **69**: 233–240.

G. Grätzer and R. N. McKenzie

1967. Equational spectra and reduction of identities. Abstract. Notices Amer. Math. Soc. **14**: 697.

G. Grätzer and E. T. Schmidt

1957. On the Jordan-Dedekind chain condition. Acta Sci. Math. (Szeged) **18**: 52–56.

1957a. On a problem of M. H. Stone. Acta Math. Acad. Sci. Hungar. **8**: 455–460.

1958. On the lattice of all join-endomorphisms of a lattice. Proc. Amer. Math. Soc. **9**: 722–726.

1958a. Characterizations of relatively complemented distributive lattices. Publ. Math. Debrecen **5**: 275–287.

1958b. Two notes on lattice-congruences. Ann. Univ. Sci. Budapest. Eötvös Sect. Math. **1**: 83–87.

1958c. On ideal theory for lattices. Acta Sci. Math. (Szeged) **19**: 82–92.

1958d. Ideals and congruence relations in lattices. Acta. Math. Acad. Sci. Hungar. **9**: 137–175.

1958e. On the generalized Boolean algebra generated by a distributive lattice. Indag. Math. **20**: 547–553.

1961. Standard ideals in lattices. Acta Math. Acad. Sci. Hungar. **12**: 17–86.

1962. On congruence lattices of lattices. Acta Math. Acad. Sci. Hungar. **13**: 179–185.

G. Grätzer and J. Sichler

1970. On the endomorphism semigroup (and category) of bounded lattices. Pacific J. Math. **34**: 639–647.

1974. Free products of Hopfian lattices. J. Austral. Math. Soc. **17**: 234–245.

1974a. On generating free products of lattices. Proc. Amer. Math. Soc. **46**: 9–14.

1975. Free decompositions of a lattice. Canad. J. Math. **27**: 276–285.

G. Grätzer and B. Wolk

1970. Finite projective distributive lattices. Canad. Math. Bull. **13**: 139–140.

G. Grätzer and M. J. Wonenburger

1962.   Some examples of complemented modular lattices. Canad. Math. Bull. **5**: 111–121.

M. Grayev

1947.   Isomorphisms of direct decompositions in Dedekind structures. (Russian.) Izv. Akad. Nauk SSSR Ser. Mat. **11**: 33–46.

C. Greene

1970.   A rank inequality for finite geometric lattices. J. Combin. Theory **9**: 357–364.

O. Hajek

1965.   Representation of finite-length modular lattices. Czechoslovak Math. J. (**90**) **15**: 503–520.

1965a. Characteristics of modular finite-length lattices. Czechoslovak Math. J. (**90**) **15**: 521–525.

A. W. Hales

1964.   On the non-existence of free complete Boolean algebras. Fund. Math. **54**: 45–66.

1970.   Partition representations of free lattices. Proc. Amer. Math. Soc. **24**: 517–520.

M. Hall

1954.   *Projective Planes and Related Topics.* California Institute of Technology.

P. Hall

1934.   A contribution to the theory of groups of prime power order. Proc. London Math. Soc. (2) **36**: 29–95.

P. R. Halmos

1961.   Injective and projective Boolean algebras. In *1961 Proc. Sympos. Pure Math., Vol. II*, pp. 114–122. American Mathematical Society, Providence, R.I.

1963.   *Lectures on Boolean algebras.* Mathematical Studies No. 1. Van Nostrand, Princeton, N.J.

I. Halperin

1961.   Complemented modular lattices. In *1961 Proc. Sympos. Pure Math.,*
        Vol. II, pp. 51–64. American Mathematical Society, Providence, R.I.

W. Hanf

1976.   Representing real numbers in denumerable Boolean algebras. Fund.
        Math. **91**: 167–170.

G. Hansoul

1967.   Problèmes de dénombrement et d'évaluation de bornes concernant les
        éléments du treillis distributif libre. Publ. Inst. Statist. Univ. Paris
        16: 219–300.

L. H. Harper

1974.   The morphology of partially ordered sets. J. Combin. Theory Ser. A.
        17: 44–58.

M. A. Harrison

1965.   *Introduction to Switching and Automata Theory.* McGraw-Hill, New
        York, N.Y.

J. Hartmanis

1956.   Two embedding theorems for finite lattices. Proc. Amer. Math. Soc.
        7: 571–577.

1959.   Lattice theory of generalized partitions. Canad. J. Math. **11**: 97–106.

1961.   Generalized partitions and lattice embedding theorems. In *1961 Proc.*
        Sympos. Pure Math., Vol. II, pp. 22–30. American Mathematical
        Society, Providence, R.I.

J. Hashimoto

1952.   Ideal theory for lattices. Math. Japon. **2**: 149–186.

1957.   Direct, subdirect decompositions and congruence relations. Osaka
        Math. J. **9**: 87–112.

1963.   Congruence relations and congruence classes in lattices. Osaka Math.
        J. **15**: 71–86.

J. Hashimoto and S. Kinugawa

1963.   On neutral elements in lattices. Proc. Japan Acad. **39**: 162–163.

G. Havas and M. Ward

1969. Lattices with sublattices of a given order. J. Combin. Theory **7**: 281–282.

V. Havel

1955. Remark on the uniqueness of direct decompositions of elements in modular lattices of finite length. (Czech.) Mat. Časopis Sloven. Akad. Vied **5**: 90–93.

T. Hecht and T. Katriňák

1972. Equational classes of relative Stone algebras. Notre Dame J. Formal Logic **13**: 248–254.

Z. Hedrlín and J. Lambek

1969. How comprehensive is the category of semigroups? J. Algebra **11**: 195–212.

Z. Hedrlín and A. Pultr

1964. Relations (graphs) with given infinite semigroups. Monatsch. Math. **68**: 421–425.

1966. On full embeddings of categories of algebras. Illinois J. Math. **10**: 392–406.

P. Hell

1972. Full embeddings into some categories of graphs. Algebra Universalis **2**: 129–141.

C. Herrmann

1973. Weak (projective) radius and finite equational bases for classes of lattices. Algebra Universalis **3**, 51–58.

1973a. On the equational theory of submodule lattices. In *Proceedings of the University of Houston Lattice Theory Conference (Houston, Tex., 1973)*, pp. 105–118. Dept. Math., Univ. Houston, Houston, Tex.

1973b. Modulare Verbände von Länge ≤ 6. In *Proceedings of the University of Houston Lattice Theory Conference (Houston, Tex., 1973)*, pp. 119–146. Dept. Math., Univ. Houston, Houston, Tex.

1973c. S-verklebte Summen von Verbänden. Math. Z. **130**: 225–274.

1975. Concerning M. M. Gluhov's paper on the word problem for free modular lattices. Algebra Universalis **5**: 445.

C. Herrmann, M. Kindermann, and R. Wille

1975. On modular lattices generated by $1 + 2 + 2$. Algebra Universalis **5**: 243–251.

C. Herrmann and W. Poguntke

1974. The class of sublattices of normal subgroup lattices is not elementary. Algebra Universalis **4**: 280–286.

C. Herrmann, C. M. Ringel, and R. Wille

1975. On modular lattices with four generators. Algebra Universalis **5**: 243–251.

D. Higgs

1971. Lattices isomorphic to their ideal lattices. Algebra Universalis **1**: 71–72.

D. X. Hong

1972. Covering relations among lattice varieties. Pacific J. Math. **40**: 575–603.

A. Horn

1962. On α-homomorphic images of α-rings of sets. Fund. Math. **51**: 259–266.

1968. A property of free Boolean algebras. Proc. Amer. Math. Soc. **19**: 142–143.

1969. Logic with truth values in a linearly ordered Heyting algebra. J. Symbolic Logic **34**: 395–408.

1969a. Free L-algebras. J. Symbolic Logic **34**: 475–480.

A. Horn and N. Kimura

1971. The category of semilattices. Algebra Universalis **1**: 26–38.

S. Huang and D. Tamari

1972. Problems of associativity: A simple proof for the lattice property of systems ordered by a semi-associative law. J. Combin. Theory Ser. A **13**: 7–13.

D. Huguet

1975.   La structure du treillis des polyèdres de paranthésages. Algebra Universalis **5**: 82–87.

A. P. Huhn

1972.   Ph.D. Thesis, University of Szeged.

1972a. Schwach distributive Verbände. I. Acta Sci. Math. (Szeged) **33**: 297–305.

1975.   On G. Grätzer's problem concerning automorphisms of a finitely presented lattice. Algebra Universalis **5**: 65–71.

E. V. Huntington

1904.   Sets of independent postulates for the algebra of logic. Trans. Amer. Math. Soc. **5**: 288–309.

G. Hutchinson

1971.   Modular lattices and abelian categories. J. Algebra **19**: 156–184.

1973.   The representation of lattices by modules. Bull. Amer. Math. Soc. **79**: 172–176.

1973a. Recursively unsolvable word problems of modular lattices and diagram-chasing. J. Algebra **26**: 385–399.

1973b. On classes of lattices representable by modules. In *Proceedings of the University of Houston Lattice Theory Conference (Houston, Tex., 1973)*, pp. 69–94. Dept. Math., Univ. Houston, Houston, Tex.

1977.   Embedding and unsolvability theorems for modular lattices. Algebra Universalis **7**: 47–84.

Iqbalunnisa

1963.   Maximal congruences on a lattice. J. Madras Univ. B. **33**: 113–128.

1964.   On neutral elements in a lattice. J. Indian Math. Soc. (N. S.) **28**: 25–31.

1965.   On some problems of G. Grätzer and E. T. Schmidt. Fund. Math. **57**: 181–185.

1965a. Neutrality in weakly modular lattices. Acta Math. Acad. Sci. Hungar. **16**: 325–326.

1966.   Normal, simple, and neutral congruences on lattices. Illinois J. Math. **10**: 227–234.

Iqbalunnisa

1971. On lattices whose lattices of congruences are Stone lattices. Fund. Math. **70**: 315–318.

S. G. Ivanov

1966. Standard subgroups. (Russian.) Ural. Gos. Univ. Mat. Zap. 5, tetrad' **3**: 49–55.

T. Iwamura

1944. A lemma on directed sets. (Japanese.) Zenkoku Shijo Sugaku Danwakai **262**: 107–111.

J. Jakubik

1954. On lattices whose graphs are isomorphic. (Russian.) Czechoslovak Math. J. **4** (**79**): 131–141.

1954a. On the graph isomorphism of semi-modular lattices. (Slovak.) Mat. Časopis Sloven. Akad. Vied. **4**: 162–177.

1955. Direct decompositions of the unity in modular lattices. (Russian.) Czechoslovak Math. J. **5** (**80**): 399–411.

1955a. Congruence relations and weak projectivity in lattices. (Slovak.) Časopis Pěst. Mat. **80**: 206–216.

1957. Remark on the Jordan-Dedekind condition in Boolean algebras. (Slovak.) Časopis Pěst. Mat. **82**: 44–46.

1958. On chains in Boolean lattices. (Slovak.) Mat. Časopis Sloven. Akad. Vied. **8**: 193–202.

1972. Conditionally α-complete sublattices of a distributive lattice. Algebra Universalis **2**: 255–261.

1975. Modular lattices of locally finite length. Acta Sci. Math. (Szeged) **37**: 79–82.

J. Jakubik and M. Kolibiar

1954. On some properties of a pair of lattices. (Russian.) Czechoslovak Math. J. **4** (**79**): 1–27.

G. Jameson

1970. *Ordered Linear Spaces.* Lecture Notes in Mathematics, Vol. 141. Springer-Verlag, Berlin-New York.

M. F. Janowitz

1964.   Projective ideals and congruence relations. I. Technical report No. 51, Univ. New Mexico.

1964a. Projective ideals and congruence relations. II. Technical report No. 63, Univ. New Mexico.

1965.   A characterisation of standard ideals. Acta Math. Acad. Sci. Hungar. **16**: 289–301.

1967.   The center of a complete relatively complemented lattice is a complete sublattice. Proc. Amer. Math. Soc. **18**: 189–190.

1968.   Perspective properties of relatively complemented lattices. J. Natur. Sci. Math. **8**: 193–210.

1968a. Section semicomplemented lattices. Math. Z. **108**: 63–76.

1975.   Annihilator preserving congruence relations of lattices. Algebra Universalis **5**: 391–394.

J. M. Jauch

1968.   *Foundations of Quantum Mechanics.* Addison-Wesley, Reading, Mass.

J. M. Jauch and C. Piron

1963.   Can hidden variables be excluded in quantum mechanics? Helv. Phys. Acta. **36**: 827–837.

1969.   On the structure of quantal proposition systems. Helv. Phys. Acta. **42**: 842–848.

R. Jegou, R. Nowakowski, and I. Rival

1987.   The diagram invariant problem for planar lattices. Acta Sci. Math. (Szeged) **51**: 103–121.

C. U. Jensen

1963.   On characterisations of Prüfer rings. Math. Scand. **13**: 90–98.

B. Jónsson

1951.   A Boolean algebra without proper automorphisms. Proc. Amer. Math. Soc. **2**: 766–770.

1953.   On the representation of lattices. Math. Scand. **1**: 193–206.

1954.   Modular lattices and Desargues' theorem. Math. Scand. **2**: 295–314.

B. Jónsson

1955. Distributive sublattices of a modular lattice. Proc. Amer. Math. Soc.
       6: 682–688.

1956. Universal relational systems. Math. Scand. **4**: 193–208.

1959. Arguesian lattices of dimension $n \leq 4$. Math. Scand. **7**: 133–145.

1959a. Representation of modular lattices and of relation algebras. Trans.
       Amer. Math. Soc. **92**: 449–464.

1959b. Lattice-theoretic approach to projective and affine geometry. In *The
       axiomatic method. With special reference to geometry and physics.
       Proceedings of an International Symposium held at the Univ. of Calif.,
       Berkeley, Dec. 26, 1957-Jan. 4, 1958 (edited by L. Henkin, P. Suppes
       and A. Tarski)* pp. 188–203. Studies in Logic and the Foundations of
       Mathematics. North-Holland Publishing Co., Amsterdam.

1960. Homogeneous universal relational systems. Math. Scand. **8**: 137–142.

1960a. Representations of complemented modular lattices. Trans. Amer.
       Math. Soc. **97**: 64–94.

1961. Sublattices of a free lattice. Canad. J. Math. **13**: 256–264.

1962. Arithmetic properties of freely α-generated lattices. Canad. J. Math.
       14: 476–481.

1965. Extensions of relational structures. In *1965 Theory of Models (Proc.
       1963 Internat. Sympos. Berkeley)*, pp. 146–157. North-Holland, Am-
       sterdam.

1967. Algebras whose congruence lattices are distributive. Math. Scand.
       21: 110–121.

1968. Equational classes of lattices. Math. Scand. **22**: 187–196.

1971. Relatively free products of lattices. Algebra Universalis **1**: 362–373.

1972. The class of Arguesian lattices is self-dual. Algebra Universalis **2**:
       396.

1974. Sums of finitely based lattice varieties. Adv. in Math. **14**: 454–468.

1974a. Varieties of algebras and their congruence varieties. In *Proceedings
       of the International Congress of Mathematicians (Vancouver, B. C.,
       1974), Vol. 1*, pp. 315–320. Canad. Math. Congress, Montreal, Que.

B. Jónsson and J. E. Kiefer

1962. Finite sublattices of a free lattice. Canad. J. Math. **14**: 487–497.

B. Jónsson, G. McNulty, and R. W. Quackenbush

1975.   The ascending and descending varietal chains of a variety. Canad.
J. Math. **27**: 25–31.

B. Jónsson and G. S. Monk

1969.   Representations of primary Arguesian lattices. Pacific J. Math. **30**:
95–139.

B. Jónsson and E. Nelson

1974.   Relatively free products in regular varieties. Algebra Universalis **4**:
14–19.

B. Jónsson and P. Olin

1976.   Elementary equivalence and relatively free products of lattices. Algebra Universalis **6**: 313–325.

J. A. Kalman

1968.   A two axiom definition for lattices. Rev. Roumaine Math. Pures Appl.
13: 669–670.

V. Kannan and M. Rajagopalan

1971.   On rigidity and groups of homeomorphisms. In *General topology and
its relations to modern analysis and algebra, III (Proc. Third Prague
Topological Sympos., 1971)*, pp. 231–234. Academia, Prague.

P. G. Kantorovič, S. G. Ivanov, and G. P. Kondrašov

1965.   Distributive pairs of elements in a lattice. (Russian.) Dokl. Akad.
Nauk. SSSR **160**: 1001–1003.

L. V. Kantorovič, B. Z. Vulih, and A. G. Pinkser

1950.   *Functional Analysis in Partially Ordered Spaces.* (Russian.) Indat.
Tehn-Teor. Lit., Moscow and Leningrad.

I. Kaplansky

1947.   Lattices of continuous functions. Bull. Amer. Math. Soc. **53**: 617–623.

D. A. Kappos

1969.   *Probability Algebras and Stochastic Spaces.* Academic Press, New
York, N.Y.

D. A. Kappos and F. Papangelou

1966.   Remarks on the extension of continuous lattices. Math. Ann. **166**: 277–283.

M. Katětov

1951.   Remarks on Boolean algebras. Colloq. Math. **2**: 229–235.

T. Katriňák

1966.   Notes on Stone lattices. I. (Russian.) Mat. Časopis Sloven. Akad. Vied. **16**: 128–142.

1967.   Notes on Stone lattices. II. (Russian.) Mat. Časopis Sloven. Akad. Vied. **17**: 20–37.

1968.   Pseudokomplementäre Halbverbände. Mat. Časopis Sloven. Akad. Vied. **18**: 121–143.

1970.   Die Kennzeichnung der distributiven pseudokomplementären Halbverbände. J. Reine Angew. Math. **241**: 160–179.

1971.   Relativ Stonesche Halbverbände sind Verbände. Bull. Soc. Roy. Sci. Liège **40**: 91–93.

1972.   Die freien Stoneschen Verbände und ihre Tripelcharakterisierung. Acta Math. Acad. Sci. Hungar. **23**: 315–326.

1972a. Über eine Konstruktion der distributiven pseudokomplementären Verbände. Math. Nachr. **53**: 85–99.

1972b. Subdirectly irreducible modular p-algebras. Algebra Universalis **2**: 166–173.

1973.   Primitive Klassen von modularen S-Algebren. J. Reine Angew. Math. **261**: 55–70.

1973a. A new proof of the construction theorem for Stone algebras. Proc. Amer. Math. Soc. **40**: 75–78.

1973b. Construction of some subdirectly irreducible modular p-algebras. Algebra Universalis **3**: 321–327.

1973c. The cardinality of the lattice of all equational classes of p-algebras. Algebra Universalis **3**: 328–329.

1974.   Injective double Stone algebras. Algebra Universalis **4**: 259–267.

1974a. Congruence extension property for distributive double p-algebras. Algebra Universalis **4**: 273–276.

1975.   A new description of the free Stone algebra. Algebra Universalis **5**: 179–189.

1976.   On a problem of G. Grätzer. Proc. Amer. Math. Soc. **56**: 19–24.

T. Katriňák

1977. Congruence lattices of distributive p-algebras. Algebra Universalis **7**: 265–272.

1978. Construction of modular double S-algebras. Algebra Universalis **8**: 15–22.

T. Katriňák and P. Mederly

1974. Construction of modular p-algebras. Algebra Universalis **4**: 301–315.

D. Kelly

1972. A note on equationally compact lattices. Algebra Universalis **2**: 80–84.

D. Kelly and I. Rival

1974. Crowns, fences, and dismantlable lattices. Canad. J. Math. **26**: 1257–1271.

1975. Certain partially ordered sets of dimension three. J. Combin. Theory Ser. A **18**: 239–242.

1975a. Planar lattices. Canad. J. Math. **27**: 636–665.

S. Kinugawa and J. Hashimoto

1966. On relative maximal ideals in lattices. Proc. Japan Acad. **42**: 1–4.

M. J. Klass

1974. Maximum antichains: A sufficient condition. Proc. Amer. Math. Soc. **45**: 28–30.

D. Kleitman

1969. On Dedekind's problem: The number of monotone Boolean functions. Proc. Amer. Math. Soc. **21**: 677–682.

D. Kleitman, M. Edelberg, and D. Lubell

1971. Maximal sized antichains in partial orders. Discrete Math. **1**: 47–53.

B. Knaster

1928. Un théorème sur les fonctions d'ensembles. Ann. Soc. Polon. Math. **6**: 133–134.

S. R. Kogalovskiĭ

1964.   On linearly complete ordered sets. (Russian.) Uspehi Mat. Nauk. **19**: 147–150.

1965.   On a theorem of Birkhoff. (Russian.) Uspehi Mat. Nauk. **20**: 206–207.

K. M. Koh

1971.   On the Frattini sublattice of a lattice. Algebra Universalis **1**: 104–116.

1971a. On the Frattini sub-semilattice of a semilattice. Nanta Math. **5**: 22–33.

1972.   On the lattice of convex sublattices of a lattice. Nanta Math. **5**: 18–37.

1973.   On sublattices of a lattice. Nanta Math. **6**: 68–79.

1973a. On the Frattini sublattice of a finite modular planar lattice. Algebra Universalis **3**: 304–317.

M. Kolibiar

1956.   On the axiomatics of modular lattices. (Russian.) Czechoslovak Math. J. **6** (**81**): 381–386.

1972.   Distributive sublattices of a lattice. Proc. Amer. Math. Soc. **34**: 359–364.

A. Komatu

1943.   On a characterization of join homomorphic transformation-lattice. Proc. Imp. Acad. Tokyo **19**: 119–124.

A. Kostinsky

1971.   Some problems for rings and lattices within the domain of general algebra. Ph.D. Thesis, Univ. Calif., Berkeley.

1972.   Projective lattices and bounded homomorphisms. Pacific J. Math. **40**: 111–119.

T. G. Kucera and B. Sands

1978.   Lattices of lattice homomorphisms. Algebra Universalis **8**, 180–190.

A. G. Kuroš

1935.   Durchschnittsdarstellungen mit irreduziblen Komponenten in Ringen und in sogenannten Dualgruppen. Mat. Sb. **42**: 613–616.

A. G. Kuroš

1943.   Isomorphisms of direct decompositions. I. (Russian.) Izv. Akad. Nauk SSSR Ser. Mat. **7**: 185–202.

1946.   Isomorphisms of direct decompositions. II. (Russian.) Izv. Akad. Nauk SSSR Ser. Mat. **10**: 47–72.

Z. Ladzianska

1974.   Poproducts of lattices and Sorkin's theorem. Mat. Časopis Sloven. Akad. Vied. **24**: 247–252.

H. Lakser

1968.   Free lattices generated by partially ordered sets. Ph.D. Thesis, University of Manitoba.

1970.   Injective hulls of Stone algebras. Proc. Amer. Math. Soc. **24**: 524–529.

1970a. Normal and canonical representations in free products of lattices. Canad. J. Math. **22**: 394–402.

1971.   The structure of pseudocomplemented distributive lattices. I. Subdirect decomposition. Trans. Amer. Math. Soc. **156**: 335–342.

1972.   Simple sublattices of free products of lattices. Abstract. Notices Amer. Math. Soc. **19**: A 509.

1973.   Principal congruences of pseudocomplemented distributive lattices. Proc. Amer. Math. Soc. **37**: 32–36.

1973a. Disjointness conditions in free products of distributive lattices: An application of Ramsey's theorem. In *Proceedings of the University of Houston Lattice Theory Conference (Houston, Tex., 1973)*, pp. 156–168. Dept. Math., Univ. Houston, Houston, Tex.

1973b. A note on the lattice of sublattices of a finite lattice. Nanta Math. **6**: 55–57.

R. E. Larson and S. J. Andima

1975.   The lattice of topologies: A survey. Rocky Mountain J. Math. **5**: 177–198.

K. B. Lee

1970.   Equational classes of distributive pseudo-complemented lattices. Canad. J. Math. **22**: 881–891.

A. H. Livšic

1962.   Direct decompositions of idempotents in semigroups. (Russian.) Trudy Moskov. Mat. Obšč. **11**: 37–98.

J. Łoś

1955.   Quelques remarques, théorèmes et problèmes sur les classes définis-sables d'algèbres. In *Mathematical Interpretation of Formal Systems*, pp. 98–113. North-Holland, Amsterdam.

H. F. J. Löwig

1943.   On the importance of the relation

$$[(A, B), (A, C)] < [(B, C), (C, A), (A, B)]$$

between three elements of a structure. Ann. of Math. (2) **44**: 573–579.

F. W. Lozier

1969.   A class of compact 0-dimensional spaces. Canad. J. Math. **21**: 817–821.

D. Lubell

1966.   A short proof of Sperner's lemma. J. Combin. Theory **1**: 299.

F. Lunnon

1971.   The IU function: The size of a free distributive lattice. In *Combinatorial Mathematics and its Applications*, pp. 173–181. Academic Press, New York, N.Y.

R. C. Lyndon

1950.   The representation of relational algebras. Ann. of Math. **51**: 707–729.

1954.   Identities in finite algebras. Proc. Amer. Math. Soc. **5**: 8–9.

1956.   The representation of relation algebras. II. Ann. of Math. **63**: 294–307.

R. N. McKenzie

1970.   Equational bases for lattice theories. Math. Scand. **27**: 24–38.

1972.   Equational bases and nonmodular lattice varieties. Trans. Amer. Math. Soc. **174**: 1–43.

R. N. McKenzie

1973. Some unsolved problems between lattice theory and equational logic. In *Proceedings of the University of Houston Lattice Theory Conference (Houston, Tex., 1973)*, pp. 564–573. Dept. Math., Univ. Houston, Houston, Tex.

R. N. McKenzie and J. D. Monk

1973. On automorphism groups of Boolean algebras. In *Coll. Math. Soc. J. Bolyai. 10. Infinite and Finite Sets*, pp. 951–988. North Holland, Amsterdam.

G. W. Mackey

1963. *Mathematical Foundations of Quantum Mechanics.* Benjamin, New York, N.Y.

S. MacLane

1938. A lattice formulation for transcendence degrees and p-bases. Duke Math. J. **4**: 455–468.

1943. A conjecture of Ore on chains in partially ordered sets. Bull. Amer. Math. Soc. **49**: 567–568.

J. E. McLaughlin

1951. Projectivities in relatively complemented lattices. Duke Math. J. **18**: 73–84.

1953. Structured theorems for relatively complemented lattices. Pacific J. Math. **3**: 197–208.

1956. Atomic lattices with unique comparable complements. Proc. Amer. Math. Soc. **7**: 864–866.

1961. The normal completion of a complemented modular point lattice. In *1961 Proc. Sympos. Pure Math., Vol. II*, pp. 78–80. American Mathematical Society, Providence, R.I.

H. M. MacNeille

1937. Partially ordered sets. Trans. Amer. Math. Soc. **42**: 416–460.

1939. Extension of a distributive lattice to a Boolean ring. Bull. Amer. Math. Soc. **45**: 452–455.

F. Maeda

1951. Lattice theoretic characterization of abstract geometries. J. Sci. Hiroshima Univ. Ser. A. **15**: 87–96.

F. Maeda

1952.   Matroid lattices of infinite length. J. Sci. Hiroshima Univ. Ser. A. **15**: 177–182.

1953.   A lattice formulation for algebraic and transcendental extensions in abstract algebras. J. Sci. Hiroshima Univ. Ser. A. **16**: 383–397.

1958.   *Kontinuierliche Geometrien.* Springer-Verlag, Berlin.

1963.   Modular centers of affine matroid lattices. J. Sci. Hiroshima Univ. Ser. A-I Math. **27**: 73–84.

1963a. Parallel mappings and comparability theorems in affine matroid lattices. J. Sci. Hiroshima Univ. Ser. A-I Math. **27**: 85–96.

1964.   Perspectivity of points in matroid lattices. J. Sci. Hiroshima Univ. Ser. A-I Math. **28**: 101–112.

F. and S. Maeda

1970.   *Theory of Symmetric Lattices.* Springer-Verlag, New York, N.Y.

S. Maeda

1965.   On the symmetry of the modular relation in atomic lattices. J. Sci. Hiroshima Univ. Ser. A-I Math. **29**: 165–170.

1966.   Infinite distributivity in complete lattices. Mem. Ehime. Univ. Sect. II Ser A. **5**, no. 3: 11–13.

1969.   Modular pairs in atomistic lattices with the covering property. Proc. Japan Acad. **45**: 149–153.

1974.   Locally modular lattices and locally distributive lattices. Proc. Amer. Math. Soc. **44**: 237–243.

K. D. Magill

1970.   The semigroup of endomorphisms of a Boolean ring. J. Austral. Math. Soc. **11**: 411–416.

M. Makkai

1973.   A proof of Baker's finite-base theorem on equational classes generated by finite elements of congruence distributive varieties. Algebra Universalis **3**: 174–181.

M. Makkai and G. McNulty

1977.   Universal Horn axioms for lattices of submodules. Algebra Universalis **7**: 25–31.

A. I. Mal'cev

1966. Several remarks on quasivarieties of algebraic systems. (Russian.) Algebra i Logika Sem. **5**, no. 3: 3–9.

M. Mandelker

1970. Relative annihilators in lattices. Duke Math. J. **37**: 377–386.

J.-M. Maranda

1964. Injective structures. Trans. Amer. Math. Soc. **110**: 98–135.

K. Menger

1936. New foundations of projective and affine geometry. Ann. of Math. (2) **37**: 456–482.

G. Michler and R. Wille

1970. Die primitiven Klassen arithmetischer Ringe. Math. Z. **113**: 369–372.

L. Mirsky

1971. *Transversal Theory.* Academic Press, New York, N.Y.

A. Mitchke and R. Wille

1973. Freie modulare Verbände $FM(_D M_3)$. In *Proceedings of the University of Houston Lattice Theory Conference (Houston, Tex., 1973)*, pp. 383–396. Dept. Math., Univ. Houston, Houston, Tex.

1976. Finite distributive lattices projective in the class of all modular lattices. Algebra Universalis **6**: 383–393.

P. Mittelstaedt

1972. On the interpretation of the lattice of subspaces of the Hilbert spaces as a propositional calculus. Z. Naturforsch. Tiel A **27**: 1358–1362.

A. F. Möbius

1832. Über eine besondere Art von Umkehrung der Reihen. J. Reine Angew. Math. **9**: 105–123.

E. N. Močulskiĭ

1961. Direct decompositions in lattices. (Russian.) Izv. Akad. Nauk SSSR Ser. Mat. **25**: 717–748.

E. N. Močulskiĭ

1962. Direct decompositions in lattices. II. (Russian.) Izv. Akad. Nauk SSSR Ser. Mat. **26**: 161–210.

1968. Isomorphisms of direct decompositions in lattices. (Russian.) Sibirsk. Mat. Ž. **9**: 1360–1385.

J. D. Monk and R. M. Solovay

1972. On the number of complete Boolean algebras. Algebra Universalis **2**: 365–368.

A. Monteiro

1947. Sur l'arithmétique des filtres premiers. C. R. Acad. Sci. Paris **225**: 846–848.

1955. Axiomes indépendants pour les algèbres de Brouwer. Rev. Un. Mat. Argentina **17**: 149–160.

M. D. Morley and R. L. Vaught

1962. Homogeneous universal models. Math Scand. **11**: 37–57.

A. Mostowski and A. Tarski

1939. Boolesche Ringe mit geordneter Basis. Fund. Math. **32**: 69–86.

V. L. Murskiĭ

1965. The existence in the three-valued logic of a closed class with a finite basis having no finite complete system of identities. (Russian.) Dokl. Akad. Nauk. SSSR **163**: 815–818.

P. V. R. Murty and V. V. R. Rao

1974. Characterization of certain classes of pseudo complemented semi-lattices. Algebra Universalis **4**: 289–300.

R. Musti and E. Buttafuoco

1956. Sui subreticoli distributivi dei reticoli modulari. Boll. Un. Mat. Ital. (3) **11**: 584–587.

L. Nachbin

1947. Une propriété caractéristique des algèbres booléiennes. Portugal. Math. **6**: 115–118.

L. Nachbin

1949.   On a characterization of the lattice of all ideals of a Boolean ring.
        Fund. Math. **36**: 137–142.

H. Nakano

1950.   *Modulared Semi-Ordered Linear Spaces.* Marazen, Tokyo.

J. B. Nation

1974.   Varieties whose congruences satisfy certain lattice identities. Algebra
        Universalis **4**:78–88.

E. Nelson

1974.   The embedding of a distributive lattice into its ideal lattice is pure.
        Algebra Universalis **4**: 135–140.

O. T. Nelson, Jr.

1968.   Subdirect decompositions of lattices of width two. Pacific J. Math.
        24: 519–523.

W. C. Nemitz

1965.   Implicative semi-lattices. Trans. Amer. Math. Soc. **117**: 128–142.

A. Nerode

1959.   Some Stone spaces and recursion theory. Duke Math. J. **26**: 397–406.

B. H. Neumann

1962.   *Universal Algebra.* Lecture Notes. Courant Inst. of Math. Sci., New
        York University.

B. H. Neumann and Hanna Neumann

1952.   Extending partial endomorphisms of groups. Proc. London Math.
        Soc. (3) **2**: 337–348.

J. von Neumann

1936.   *Lectures on Continuous Geometries.* Institute of Advanced Studies,
        Princeton, N.J.

1960.   *Continuous Geometry.* (I. Halperin, editor.) Princeton Mathematical
        Series, No. 25. Princeton Univ. Press, Princeton, N.J.

E. Noether

1926.   Abstrakter Aufbau der Idealtheorie in algebraischen Zahl- und Funk-
tionenköpern. Math. Ann. **96**: 26–61.

R. Nowakowski and I. Rival

1977.   The spectrum of a finite lattice: breadth and length techniques. Canad.
Math. Bull. **20**: 319–329.

T. Ogasawara and U. Sasaki

1949.   On a theorem in lattice theory. J. Sci. Hiroshima Univ. Ser. A **14**:
13.

P. Olin

1976.   Free products and elementary types of Boolean algebras. Math. Scand.
38: 5–23.

O. Ore

1935.   On the foundation of abstract algebra. I. Ann. of Math. **36**: 406–437.

1936.   On the foundation of abstract algebra. II. Ann. of Math. **37**: 265–292.

1937.   Structures and group theory. I. Duke Math. J. **3**: 149–174.

1938.   Structures and group theory. II. Duke Math. J. **4**: 247–269.

1940.   Remarks on structures and group relations. Vierteljschr. Naturforsch.
Ges. Zürich, 85 Beiblatt (Festschrift Rudolf Fueter), 1–4.

1942.   Theory of equivalence relations. Duke Math. J. **9**: 573–627.

1943.   Chains in partially ordered sets. Bull. Amer. Math. Soc. **49**: 558–566.

1944.   Galois connexions. Trans. Amer. Math. Soc. **55**: 493–513.

R. Padmanabhan

1966.   On axioms for semilattices. Canad. Math. Bull. **9**: 357–358.

1966a. On some ternary relations in lattices. Colloq. Math. **15**: 195–198.

1968.   A note on Kalman's paper. Rev. Roumaine Math. Pures Appl. **13**:
1149–1152.

1969.   Two identities for lattices. Proc. Amer. Math. Soc. **20**: 409–412.

1972.   On identities defining lattices. Algebra Universalis **1**: 359–361.

1974.   On M-symmetric lattices. Canad. Math. Bull. **17**: 85–86.

1983.   A self-dual equational basis for Boolean algebras. Canad. Math. Bull.
26 (1983), 9–12.

D. Papert

1964. Congruence relations in semilattices. J. London Math. Soc. **39**: 723–729.

A. Pelczar

1961. On the invariant points of a transformation. Ann. Polon. Math. **11**: 199–202.

1962. On the extremal solutions of a functional equation. Zeszyty Nauk. Uniw. Jagiello. Prace Mat. No. 7: 9–11.

A. L. Peressini

1967. *Ordered Topological Vector Spaces.* Harper and Row, New York, N.Y.

P. Perkins

1969. Bases for equational theories of semigroups. J. Algebra **11**: 293–314.

R. Permutti

1964. Sulle semicongruenze di un reticolo. Rend. Accad. Sci. Fis. Mat. Napoli (4) **31**: 160–167.

R. S. Pierce

1958. A note on complete Boolean algebras. Proc. Amer. Math. Soc. **9**: 892–896.

1961. Some questions about complete Boolean algebras. In *1961 Proc. Sympos. Pure Math., Vol. II,* pp. 129–140. American Mathematical Society, Providence, R.I.

1968. *Introduction to the Theory of Abstract Algebras.* Holt, Rinehart, and Winston, New York, N.Y.

1972. *Compact Zero-dimensional Metric Spaces of Finite Type.* Memoirs Amer. Math. Soc. No. 130.

1973. Bases of countable Boolean algebras. J. Symbolic Logic **38**: 212–214.

C. Piron

1964. Axiomatique quantique. Helv. Phys. Acta. **37**: 439–468.

E. Pitcher and M. F. Smiley

1942. Transitivities of betweenness. Trans. Amer. Math. Soc. **52**: 95–114.

C. R. Platt

1974. Iterated limits of lattices. Canad. J. Math. **26**: 1301–1320.

1976. Planar lattices and planar graphs. J. Combin. Theory, Ser. B **21**: 30–39.

W. Poguntke and I. Rival

1974. Finite sublattices generated by order-isomorphic subsets. Arch. Math. (Basel) **25**: 225–230.

E. L. Post

1921. Introduction to a general theory of elementary propositions. Amer. J. Math. **43**: 163–185.

D. H. Potts

1965. Axioms for semilattices. Canad. Math. Bull. **8**: 519.

W. Prenowitz

1948. Total lattices of convex sets and of linear spaces. Ann. of Math. (2) **49**: 659–688.

H. A. Priestley

1970. Representation of distributive lattices by means of ordered Stone spaces. Bull. London Math. Soc. **2**: 186–190.

1972. Ordered topological spaces and the representation of distributive lattices. Proc. London Math. Soc. (3) **24**: 507–530.

1974. Stone lattices: A topological approach. Fund. Math. **84**: 127–143.

1975. The construction of spaces dual to pseudocomplemented distributive lattices. Quart. J. Math. Oxford (3) **26**: 215–228.

P. Pudlák

1976. A new proof of the congruence lattice representation theorem. Algebra Universalis **6**: 269–275.

1977. Distributivity of strongly representable lattices. Algebra Universalis **7**: 85–92.

R. W. Quackenbush

1972. The triangle is functionally complete. Algebra Universalis **2**: 128.

1972a. Free products of bounded distributive lattices. Algebra Universalis **2**: 393–394.

R. W. Quackenbush

1973. Planar lattices. In *Proceedings of the University of Houston Lattice Theory Conference (Houston, Tex., 1973)*, pp. 512–518. Dept. Math., Univ. Houston, Houston, Tex.

1974. Near-Boolean algebras. I. Combinatorial aspects. Discrete Math. **10**: 301–308.

1974a. Semi-simple equational classes with distributive congruence lattices. Ann. Univ. Sci. Budapest. Eötvös Sect. Math. **17**: 15–19.

R. W. Quackenbush and H. C. Reichel

1975. Partitioning bases of Boolean lattices. Algebra Universalis **5**: 148.

R. W. Quackenbush and B. Wolk

1971. Strong representations of congruence lattices. Algebra Universalis **1**: 165–166.

G. N. Raney

1952. Completely distributive complete lattices. Proc. Amer. Math. Soc. **3**: 677–680.

1953. A subdirect-union representation for completely distributive complete lattices. Proc. Amer. Math. Soc. **4**: 518–522.

1960. Tight Galois connections and complete distributivity. Trans. Amer. Math. Soc. **97**: 418–426.

H. Rasiowa

1974. *An Algebraic Approach to Non-Classical Logics.* North-Holland, Amsterdam.

H. Rasiowa and R. Sikorski

1963. *The Mathematics of Metamathematics.* Monog. Mat., XLI. Państwowe Wydawn. Naukowe, Warsaw.

B. C. Rennie

1951. *The Theory of Lattices.* Forster and Jagg, Cambridge, England.

I. Reznikoff

1963. Chaînes de formules. C. R. Acad. Sci. Paris **256**: 5021–5023.

J. B. Rhodes

1975. Modular and distributive semilattices. Trans. Amer. Math. Soc. **201**: 31–41.

P. Ribenboim

1949. Characterization of the sup-complement in a distributive lattice with last element. Summa Brasil. Math. **2**, No. 4: 43–49.

J. Riečan

1958. To the axiomatics of modular lattices. (Slovak.) Acta Fac. Rerum Natur. Univ. Comenian. Math. **2**: 257–262.

L. Rieger

1949. A note on topological representations of distributive lattices. Časopis Pĕst. Mat. Fys. **74**: 55–61.

1951. Some remarks on automorphisms of Boolean algebras. Fund. Math. **38**: 209–216.

I. Rival

1972. Projective images of modular (distributive, complemented) lattices are modular (distributive, complemented). Algebra Universalis **2**: 395.

1973. Maximal sublattices of finite distributive lattices. Proc. Amer. Math. Soc. **37**: 417–420.

1974. Maximal sublattices of finite distributive lattices. II. Proc. Amer. Math. Soc. **44**: 263–268.

1974a. Lattices with doubly irreducible elements. Canad. Math. Bull. **17**: 91–95.

1975. Sublattices of modular lattices of finite length. Canad. Math. Bull. **18**: 95–98.

1976. A fixed point theorem for finite partially ordered sets. J. Combin. Theory, Ser. A. **21**: 309–318.

I. Rival and B. Sands

1975. Finite weakly embeddable lattices which are embeddable. Arch. Math. (Basel) **26**: 346–352.

A. Robinson

1971.   Infinite forcing in model theory. In *Proceedings of the Second Scandinavian Logic Symposium (Oslo, 1970)*, pp. 317–340. Studies in Logic and the Foundations of Mathematics, Vol. 63. North-Holland, Amsterdam.

1971a. On the notion of algebraic closedness for noncommutative groups and fields. J. Symbolic Logic **36**: 441–444.

H. L. Rolf

1958.   The free lattice generated by a set of chains. Pacific J. Math. **8**: 585–595.

1966.   The free modular lattice, $FM(2+2+2)$, is infinite. Proc. Amer. Math. Soc. **17**: 960–961.

I. Rosenberg

1967.   Maximal Gegenketten im Verband 3^n. Arch. Math. (Brno) **3**: 185–190.

P. C. Rosenbloom

1942.   Post algebras. I. Postulates and general theory. Amer. J. Math. **64**: 167–188.

J. Rosický

1975.   Sublattices of the lattice of topologies. Acta Fac. Rerum Natur. Univ. Comenian. Math., Special Number, 39–41.

G.-C. Rota

1964.   On the foundations of combinatorial theory. I. Theory of Möbius functions. Z. Wahrscheinlichkeitstheorie und Verw. Gebiete **2**: 340–368.

G. Rousseau

1970.   Post algebras and pseudo-Post algebras. Fund. Math. **67**: 133–145.

B. R.-Salinas

1969.   Fixed points in ordered sets. (Spanish.) In *In Honor of Prof. Dr. Iñíguez y Almech on the Occasion of his Seventieth Birthday and Academic Retirement,* pp. 163–172. Consejo Sup. Investigación. Ci. Fac. Ci. Zaragoza, Zaragoza.

S. Rudeanu

1963.   *Axioms for Lattices and Boolean algebras.* (Roumanian.) Mono-grafii Asupra Teoriei Algebrice A Mecanismelor Automate. Editura Academiei Republicii Populare Romine, Bucharest.

1964.   Logical dependence of certain chain conditions in lattice theory. Acta Sci. Math. (Szeged) **25**: 209–218.

1974.   *Boolean Functions and Equations.* North-Holland, Amsterdam.

D. Sachs

1961.   Identities in finite partition lattices. Proc. Amer. Math. Soc. **12**: 944–945.

1961a. Partition and modulated lattices. Pacific J. Math. **11**: 325–345.

1962.   The lattice of subalgebras of a Boolean algebra. Canad. J. Math. **14**: 451–460.

V. N. Saliĭ

1972.   A compactly generated lattice with unique complements is distributive. (Russian.) Mat. Zametki **12**: 617–620.

Y. Sampei

1953.   On lattice completions and closure operations.   Comment. Math. Univ. St. Paul **2**: 55–70; and **3**: 29–30.

U. Sasaki

1952.   Lattice theoretic characterization of an affine geometry of arbitrary dimensions. J. Sci. Hiroshima Univ. Ser. A. **16**: 223–238.

1953.   Semi-modularity in relatively atomic, upper continuous lattices. J. Sci. Hiroshima Univ. Ser. A. **16**: 409–416.

U. Sasaki and S. Fujiwara

1952.   The decomposition of matroid lattices. J. Sci. Hiroshima Univ. Ser. A. **15**: 183–188.

1952a. The characterization of partition lattices. J. Sci. Hiroshima Univ. Ser. A. **15**: 189–201.

G. Sauer, W. Seibert, and R. Wille

1973.   On free modular lattices over partial lattices with four generators. In *Proceedings of the University of Houston Lattice Theory Conference (Houston, Tex., 1973),* pp. 332–382. Dept. Math., Univ. Houston, Houston, Tex.

B. M. Schein

1970.   Relation algebras and function semigroups. Semigroup Forum **1**: 1–62.

1972.   On the definition of distributive semilattices. Algebra Unviersalis **2**: 1–2.

1972a. A representation theorem for lattices. Algebra Universalis **2**: 177–178.

E. T. Schmidt

1962.   Über die Kongruenzverbände der Verbände. Publ. Math. Debrecen **9**: 243–256.

1968.   Zur Charakterisierung der Kongruenzverbände der Verbände. Mat. Časopis Sloven. Akad. Vied. **18**: 3–20.

1969.   *Kongruenzrelationen algebraischer Strukturen.* VEB Deutscher Verlag der Wissenschaften, Berlin.

1970.   Eine Verallgemeinerung des Satzes von Schmidt-Ore. Publ. Math. Debrecen **17**: 283–287.

1974.   Every finite distributive lattice is the congruence lattice of some modular lattice. Algebra Universalis **4**: 49–57.

1975.   On the length of the congruence lattice of a lattice. Algebra Universalis **5**: 98–100.

1975a. On finitely generated simple modular lattices. Period. Math. Hungar. **6**: 213–216.

M. Schützenberger

1945.   Sur certains axiomes de la théorie des structures. C. R. Acad. Sci. Paris **221**: 218–220.

W. Schwan

1948.   Perspektivitäten in allgemeinen Verbänden. Math. Z. **51**: 126–134.

1948a. Ein allgemeiner Mengenisomorphiesatz der Theorie der Verbände. Math. Z. **51**: 346–354.

1949.   Zusammensetzung von Schwesterperspektivitäten in Verbänden. Math. Z. **52**: 150–167.

1949a. Ein Homomorphiesatz der Theorie der Verbände. Math. Z. **52**: 193–201.

L. N. Ševrin

1964.   Projectivities of semilattices. (Russian.) Dokl. Akad. Nauk SSSR **154**: 538–541.

H. Sharp

  1968.   Cardinality of finite topologies. J. Combin. Theory **5**: 82–86.

M. Sholander

  1951.   Postulates for distributive lattices. Canad. J. Math. **3**: 28–30.

J. Sichler

  1972.   Non-constant endomorphisms of lattices. Proc. Amer. Math. Soc. **34**: 67–70.

  1975.   Note on free products of Hopfian lattices. Algebra Universalis **5**: 145–146.

R. Sikorski

  1964.   *Boolean algebras.* Second edition. Ergebnisse der Mathematik und ihrer Grenzgebiete, Band 25. Academic Press, New York, N.Y.

H. L. Silcock

  1977.   Generalized wreath products and the lattice of normal subgroups of a group. Algebra Universalis **7**: 361–372.

F. M. Sioson

  1964.   Equational bases of Boolean algebras. J. Symbolic Logic **29**: 115–124.

H. L. Skala

  1971.   Trellis theory. Algebra Universalis **1**: 218–233.

  1972.   *Trellis Theory.* Memoirs Amer. Math. Soc. No. 121.

L. A. Skornjakov

  1961.   *Complemented Modular Lattices and Regular Rings.* (Russian.) Gosudarstv. Izdat. Fiz.-Mat. Lit., Moscow. (English translation: Oliver and Boyd, 1964.)

M. F. Smiley and W. R. Transcue

  1943.   Applications of transitivites of betweenness in lattice theory. Bull. Amer. Math. Soc. **49**: 280–287.

F. A. Smith

  1974.   Projectivity of prime quotients and simple lattices. J. Algebra **31**: 257–261.

Ju. I. Sorkin

1951.   Independent systems of axioms defining a lattice. (Russian.) Ukrain. Mat. Ž. **3**: 85–97.

1952.   Free unions of lattices. (Russian.) Mat. Sb. N. S. **30** (72): 677–694.

1954.   On the imbedding of latticoids in lattices. (Russian.) Doklady Akad. Nauk SSSR. N. S. **95**: 931–934.

T. P. Speed

1969.   On Stone lattices. J. Austral. Math. Soc. **9**: 293–307.

1969a. Spaces of ideals of distributive lattices. I. Prime ideals. Bull. Soc. Roy. Sci. Liège **38**: 610–628.

1972.   On the order of prime ideals. Algebra Universalis **2**: 85–87.

1972a. Profinite posets. Bull. Austral. Math. Soc. **6**: 177–183.

1974.   Spaces of ideals of distributive lattices. II. Minimal prime ideals. J. Austral. Math. Soc. **18**: 54–72.

E. Sperner

1928.   Ein Satz über Untermengen einer endlichen Menge. Math. Z. **27**: 544–548.

R. P. Stanley

1971.   Modular elements of geometric lattices. Algebra Universalis **1**: 214–217.

1972.   Supersolvable lattices. Algebra Universalis **2**: 197–217.

1973.   An extremal problem for finite topologies and distributive lattices. J. Combin. Theory Ser. A **14**: 209–214.

1974.   Finite lattices and Jordan-Hölder sets. Algebra Universalis **4**: 361–371.

A. K. Steiner

1966.   The lattice of topologies: Structure and complementation. Trans. Amer. Math. Soc. **122**: 379–398.

1966a. Complementation in the lattice of T_1-topologies. Proc. Amer. Math. Soc. **17**: 884–886.

D. Steven

1968.   Topology on finite sets. Amer. Math. Monthly **75**: 739–741.

M. H. Stone

1936.   The theory of representations for Boolean algebras. Trans. Amer. Math. Soc. **40**: 37–111.

1937.   Topological representations of distributive lattices and Brouwerian logics. Časopis Pěst. Mat. **67**: 1–25.

U. M. Swamy and V. V. R. Rao

1975.   Triple and sheaf representations of Stone lattices. Algebra Universalis **5**: 104–113.

K. Takeuchi

1951.   On maximal proper sublattices. J. Math. Soc. Japan **2**: 228–230.

1951a. On free modular lattices. Japan J. Math. **21**: 53–65.

1959.   On free modular lattices. II. Tôhoku Math. J. (2) **11**: 1–12.

D. Tamari

1951.   Monoides préordonnés et chaînes de Malcev. Thése, Université de Paris.

O. Tamaschke

1960.   Submodulare Verbände. Math. Z. **74**: 186–190.

S. Tamura

1971.   On distributive sublattices of a lattice. Proc. Japan Acad. **47**: 442–446.

1971a. A note on distributive sublattices of a lattice. Proc. Japan Acad. **47**: 603–605.

T. Tan

1975.   On transferable distributive lattices. Nanta Math. **8**, No. 2: 50–60.

T. Tanaka

1952.   Canonical subdirect factorizations of lattices. J. Sci. Hiroshima Univ. Ser. A. **16**: 239–246.

A. Tarski

1930.   Sur les classes d'ensembles closes par rapport à certaines opérations élémentaires. Fund. Math. **16**: 181–304.

A. Tarski

1946. A remark on functionally free algebras. Ann. of Math. (2) **47**: 163–165.

1955. A lattice-theoretical fixpoint theorem and its applications. Pacific J. Math. **5**: 285–309.

1968. Equational logic and equational theories of algebras. In *1968 Contributions to Math. Logic (Colloquium, Hannover, 1966)*, pp. 275–288. North-Holland, Amsterdam.

S. K. Thomason

1970. A proof of Whitman's representation theorem for finite lattices. Proc. Amer. Math. Soc. **25**: 618–619.

1970a. Sublattices and initial segments of the degrees of unsolvability. Canad. J. Math. **22**: 569–581.

R. M. Thrall and D. G. Duncan.

1953. Note on free modular lattices. Amer. J. Math. **75**: 627–632.

J. R. R. Tolkien

1954. *The Lord of the Rings.* George Allen and Unwin, London.

T. Traczyk

1963. Axioms and some properties of Post algebras. Colloq. Math. **10**: 193–209.

1967. On Post algebras with uncountable chain of constants. Algebras of homomorphisms. Bull. Acad. Polon. Sci. Sér. Sci. Math. Astronom. Phys. **15**: 673–680.

W. T. Trotter, Jr. and J. I. Moore

1975. The dimension of planar posets. Abstract. Notices Amer. Math. Soc. **22**: A-44.

W. T. Tutte

1959. Matroids and graphs. Trans. Amer. Math. Soc. **90**: 527–552.

1965. Lectures on matroids. J. Res. Nat. Bur. Standards Sect. B **69B**: 1–47.

1971. *Introduction to the Theory of Matroids.* American Elsevier, New York, N.Y.

H. Tverberg

1967. On Dilworth's decomposition theorem for partially ordered sets. J. Combin. Theory **3**: 305–306.

A. Urquhart

1973. Free distributive pseudo-complemented lattices. Algebra Universalis **3**: 13–15.

R. Vaidyanathaswamy

1947. *Treatise on Set Topology.* Part I. Indian Math. Soc., Madras.

1960. *Set Topology.* Second edition. Chelsea, New York, N.Y.

V. S. Varadarajan

1968. *Geometry of Quantum Theory.* I. The University Series in Higher Mathematics. Van Nostrand, Princeton, N.J.

1970. *Geometry of Quantum Theory.* II. The University Series in Higher Mathematics. Van Nostrand, Princeton, N.J.

J. C. Varlet

1963. Contribution à l'étude des treillis pseudo-complémentés et des treillis de Stone. Mém. Soc. Roy. Sci. Liège Coll. in-8^0, Vol. 8, No. 4.

1965. Congruences dans les demi-lattis. Bull. Soc. Roy. Sci. Liège **34**: 231–240.

1966. On the characterization of Stone lattices. Acta Sci. Math. (Szeged) **27**: 81–84.

1968. Algèbres de Lukasiewicz trivalentes. Bull. Soc. Roy. Sci. Liège **37**: 399–408.

1968a. A generalization of the notion of pseudocomplementedness. Bull. Soc. Roy. Sci. Liège **37**: 149–158.

O. Veblen and W. H. Young

1910. *Projective Geometry.* 2 volumes. Ginn and Co., Boston.

V. Vilhelm

1955. The selfdual kernel of Birkhoff's conditions in lattices with finite chains. (Russian.) Czechoslovak Math. J. **5** (**80**): 439–450.

P. Vopěnka, A. Pultr, and Z. Hedrlín

1965.   A rigid relation exists on any set. Comment. Math. Univ. Carolinae **6**: 149–155.

B. Z. Vulih

1967.   *Introduction to the Theory of Partially Ordered Spaces.* Woltors-Noordhoff, Groningen.

B. L. van der Waerden

1931.   *Modern Algebra.* First edition, 1931. First English translation, 1949. Springer-Verlag.

H. Wallman

1938.   Lattices and topological spaces. Ann. of Math. **39**: 112–126.

Shih-chiang Wang

1953.   Notes on the permutability of congruence relations. (Chinese.) Acta Math. Sinica **3**: 133–141.

M. Ward

1939.   A characterization of Dedekind structures. Bull. Amer. Math. Soc. **45**: 448–451.

A. G. Waterman

1967.   The free lattice with 3 generators over N_5. Portugal. Math. **26**: 285–288.

L. Weisner

1935.   Abstract theory of inversion of finite series. Trans. Amer. Math. Soc. **38**: 474–484.

H. Werner and R. Wille

1970.   Characterisierungen der primitiven Klassen arithmetischer Ringe. Math. Z. **115**: 197–200.

T. P. Whaley

1969.   Large sublattices of a lattice. Pacific J. Math. **28**: 477–484.

P. M. Whitman

1941.   Free lattices. Ann. of Math. (2) **42**: 325–330.

1942.   Free lattices. II. Ann. of Math. (2) **43**: 104–115.

1943.   Splittings of a lattice. Amer. J. Math. **65**: 179–196.

1946.   Lattices, equivalence relations, and subgroups. Bull. Amer. Math. Soc. **2**: 507–522.

1961.   Status of word problems for lattices. In *1961 Proc. Sympos. Pure Math., Vol. II*, pp. 17–21. American Mathematical Society, Providence, R.I.

H. Whitney

1935.   On the abstract properties of linear dependence. Amer. J. Math. **57**: 509–533.

L. R. Wilcox

1939.   Modularity in the theory of lattices. Ann. of Math. **40**: 490–505.

1942.   A note on complementation in lattices. Bull. Amer. Math. Soc. **48**: 453–458.

1944.   Modularity in Birkhoff lattices. Bull. Amer. Math. Soc. **50**: 135–138.

1969a. Variety invariants for modular lattices. Canad. J. Math. **21**: 279–283.

R. Wille

1967.   Verbandstheoretische Charakterisierung n-stufiger Geometrien. Arch. Math. (Basel) **18**: 465–468.

1969.   Primitive Länge und primitive Weite bei modularen Verbänden. Math. Z. **108**: 129–136.

1969a. Variety invariants for modular lattices. Canad. J. Math. **21**: 279–283.

1970.   *Kongruenzklassengeometrien.* Lecture Notes in Mathematics. 113. Springer-Verlag, Berlin.

1972.   Primitive subsets of lattices. Algebra Universalis **2**: 95–98.

1973.   On free modular lattices generated by finite chains. Algebra Universalis **3**: 131–138.

1973a. Über modulare Verbände, die von einer endlichen halbgeordneten Menge frei erzeugt werden. Math. Z. **131**: 241–249.

1974.   On modular lattices of order dimension two. Proc. Amer. Math. Soc. **43**: 287–292.

1974a. Jeder endlich erzeugte, modulare Verband endlicher Weite ist endlich. Mat. Časopis Sloven. Akad. Vied. **24**: 77–80.

R. Wille

1975.   An example of a finitely generated non-Hopfian lattice. Algebra Universalis **5**: 101–103.

1976.   On the width of irreducibles in finite modular lattices. Algebra Universalis **6**: 257–258.

1976a. Subdirekte Produkte vollständiger Verbände. J. Reine. Angew. Math. **283/284**: 53–70.

1976b. Finite projective planes and equational classes of modular lattices. In *Colloquio Internazionale sulle Teorie Combinatorie (Rome, 1973), Tomo II,* pp. 167–172. Atti dei Convegni Lincei, No. 17, Accad. Naz. Lincei, Rome, 1976.

R. J. Wilson

1973.   An introduction to matroid theory. Amer. Math. Monthly **80**: 500–525.

E. S. Wolk

1957.   Dedekind completeness and a fixed-point theorem. Canad. J. Math. **9**: 400–405.

Yau-chuen Wong and Kung-fu Ng

1973.   *Partially Ordered Topological Vector Spaces.* Oxford Mathematical Monographs. Clarendon Press, Oxford.

A. Wrónski

1976.   The number of quasivarieties of distributive lattices with pseudocomplementation. Pol. Acad. Sci. Bull. Sect. Logic **5**: 115–121.

M. Yaushara

1974.   The Amalgamation Property, the Universal-Homogeneous Models, and the Generic Models. Math. Scand. **34**: 5–36.

G. Zacher

1952.   Sugli emimorfismi superiori ed inferiori tra reticoli. Rend. Accad. Sci. Fis. Mat. Napoli (4) **19**: 45–56.

G. Zappa

1952.   *Reticoli e geometrie finite.* Lezioni raccolte dal dott. Giovanni Zacher. Libreria Editrice Liguori, Napoli.

Table of Notation

Symbol	Page	Symbol	Page	Symbol	Page
Posets		**Lattices**		**Special Lattices and algebras**	
$\leq$	1	$\langle L; \wedge, \vee \rangle$	6		
$\geq$	2	$\langle L; \vee \rangle$	10	$B^{[2]}$	157
$\leq_B$	2	$\langle L; \vee \rangle$	10	B_2	63
$<$	8	$\langle L; \wedge, \vee, 0, 1 \rangle$	147	$B[C]$	123
$\prec$	12	$\langle L; \wedge, \vee, {}^*, 0, 1 \rangle$	147	$B[L]$	128
$\succ$	12	A^b	346	C_n	20
$\parallel$	2	$B(L)$	118	$\mathrm{CF}(P)$	56
inf	3	$\mathrm{Con}\, L$	27	$\mathrm{F}(P)$	43
$\bigwedge$	3	$\mathrm{D}(L)$	150	$\mathrm{F}_{\mathbf{K}}(P)$	43
sup	3	$\mathrm{Du}\, L$	24	$\mathrm{F}_{\mathbf{K}}(\mathfrak{A})$	56
$\bigvee$	3	$\mathrm{Du}_0\, L$	24	$\mathrm{F}_{\mathbf{K}}(A)$	56
$\wedge$	4	$\mathrm{Id}\, L$	23	$\mathrm{F}_{\mathbf{K}}(\mathfrak{m})$	43
$\vee$	4	$\mathrm{Id}_0\, L$	23	$L(D, \mathfrak{m})$	268
$\cong$	19	$\mathrm{Part}\, L$	27, 251	M_3	13, 79
$[a, b]$	21	$\mathrm{S}(L)$	63, 149	M_4	286, 309
$(a, b]$	21	$\mathrm{Sub}\, L$	31	N_5	13, 79
$[a, b)$	21	L/Θ	27	$\mathrm{P}(X)$	2
(a, b)	22	L/I	199	S_1	156
0	3	L^I	30		
1	3	$L \times K$	29		
0^b	346	$\prod(L_i \mid \in I)$	30		
1^b	346	$\prod_{\mathcal{D}}(L_i \mid i \in I)$	300		
$h(a)$	225	$A * B$	343		
$\mathrm{H}(P)$	56	$\langle B; \wedge, \vee, {}', 0, 1 \rangle$	63		
$l(P)$	2	B/I	125		

Symbol	Page	Symbol	Page	Symbol	Page
Classes and Operators		**Subsets of Lattices**		**Congruences and Quotients**	
B	92	$(a]$	22	ω	25
D	42	$[a)$	24	ι	25
L	42	$[a,b]$	21	$\Theta(a,b)$	98
M	42	$(a,b]$	21	$\Theta(H)$	98
$\mathbf{M}_3$	307	$[a,b)$	21	$\Theta[I]$	99
$\mathbf{M}_4$	306	(a,b)	22	Θ_a	182
$\mathbf{N}_5$	307	Cen	206	Φ/Θ	154
T	57	Comp	377	$\Theta \times \Phi$	30
Amal	335	$D(L)$	150	$[a]\Theta$	25
Var	299	$[H]$	21	$\operatorname{Ker}\varphi$	29
H	298	$(H]$	22	a/b	169
I	305	$[H)$	24	$a/b \sim c/d$	169
$\mathbf{K}_{fin}$	331	$[H]_R$	119	$a/b \nearrow c/d$	169
$\mathbf{Mod}(\Sigma)$	295	I^*	121	$a/b \searrow c/d$	169
P	298	$J(L)$	81	$a/b \approx c/d$	170
$\mathbf{P_R}$	305	$M(L)$	83	$c/d \searrow_w a/b$	170
$\mathbf{P_S}$	299	$\mathcal{P}(L)$	86, 131	$c/d \nearrow_w a/b$	170
$\mathbf{P_U}$	300	$r(a)$	81, 132	$c/d \sim_w a/b$	170
S	298	$r(I)$	132	$c/d \overset{k}{\approx}_w a/b$	172
Si	303	$S(L)$	63		

Symbol	Page	Symbol	Page
Properties		**Miscellaneous**	
(C)	347	$\cong$	19
(JID)	120	$\sim$	169, 235
(L1)–(L4)	5	$\approx$	170, 236
(MID)	120	L_D^I	305
(P1)–(P4)	1	$\mathcal{S}(L)$	132
(S1), (S2)	134	$\overline{X}$	142, 240
(S3)	137	φ_Θ	28
$(SD_\wedge)$	353	$x + y$	95, 104, 265
$(SD_\vee)$	353	$X + Y$	265
(W)	353	a'	63
$(_\wedge W)$	347	a^*	63
$(_\vee W)$	347	$a * b$	67
$(W_\wedge)$	347	$a \, M \, b$	230
$(W_\vee)$	347	$\Phi(P)$	315
		$\Phi(n)$	316
		$\operatorname{Iden}(\mathbf{K})$	295
		$L(D, \mathfrak{m})$	268

Retrospective

G. Grätzer

In this appendix, I will try to update this book with the developments of the last two decades. Such writing is, by its very nature, quite subjective, so I will write it in first person singular, as a reminder. To compensate for this subjective treatment, this appendix is followed by seven others—giving, I trust, objective accounts of the main developments in several major chapters of lattice theory.

Quite a few books have been published in lattice theory in the last twenty or so years. Two introductory textbooks have appeared, one in English, B. A. Davey and H. A. Priestley [106], and one in Russian, L. A. Skornjakov [478]. In addition, a Russian translation of the first edition of the present book was published, G. Grätzer [217]. One may also consider R. N. McKenzie, G. F. McNulty, and W. F. Taylor [382] as an introductory text, albeit one to a wider field.

A number of books appeared on particular chapters of lattice theory, namely, R. Freese, J. Ježek, and J. B. Nation [172] on free lattices, P. Jipsen and H. Rose [328] on varieties of lattices, V. N. Saliĭ [462] on lattices with unique complements, M. Stern [489] on semimodular lattices, and B. Ganter and R. Wille [195] on formal concept analysis. You will find many of these same authors have contributed appendices to this book. Finally, there is the book M. Mehrtens [393] on the history of lattice theory up to 1940.

I shall discuss the following topics: major developments; for each chapter, I report on the open problems proposed at the end of the chapter, and on the developments related to the topics covered in the chapter, not covered by a separate appendix. Finally, there is a section on some problems in universal algebra of a lattice theoretic nature.

Some chapters of lattice theory have become too large and specialized to be discussed in a general book on lattice theory. *Boolean algebras* are a prime example; the reader is referred to J. D. Monk and R. Bonnet [394] for a handbook. *Orthomodular lattices* are covered in two books: L. Beran [44] and G. Kalmbach [349]. *Continuous lattices* play an important role in theoretical computer science and in some borderline areas between lattices and topology. G. Gierz, K. H. Hofmann, K. Keimel, J. D. Lawson, M. W. Mislove, D. S. Scott [200] is a useful compendium on the subject. R.-E. Hoffmann and K. H. Hofmann [313] reports on the last of a series of conferences on the subject. *Frames* (*locales*) present a new view of general topology, emphasizing its constructive aspects. Consult the book P. T. Johnstone [330] and the surveys P. T. Johnstone [331] and [332].

1.   Major Advances

A large number of longstanding major problems have been resolved. In my opinion, the following results are the most important.

1.1   Finite partition lattices

The problem whether every finite lattice can be embedded in a finite partition lattice goes back to P. M. Whitman in the early forties; the problem was often mentioned by G. Birkhoff in his books and lectures. Finally, in 1977, a positive solution was obtained by P. Pudlák and J. Tůma (published in 1980 in [450]):

Theorem 1.   Every finite lattice can be embedded in a finite partition lattice.

—

This solves Problem IV.32. Interestingly, I thought that this problem would be solved by proving the stronger result: Every finite lattice can be represented as the congruence lattice of a finite algebra. This is still not known, so Problem IV.36 is open.

1.2   Modular lattices

In two remarkable papers, [155] and [157], R. Freese succeeded in applying the techniques John von Neumann developed to coordinatize complemented modular lattices to attack some deep lying problems of modular lattice theory. His major results are Theorems 2 and 4:

Theorem 2.   The variety $\mathbf{M}$ of modular lattices is not generated by its finite members.   —

Thus the answer to Problem IV.5 is in the negative.

Let $\mathbf{M}_f$ and $\mathbf{M}_{fd}$ be the variety generated by all finite and finite dimensional modular lattices, respectively. R. Freese proved a slightly stronger theorem, namely, that $\mathbf{M}_{fd}$ is not generated by its finite members. The following is an important related result of C. Herrmann [305]:

Theorem 3.   If $\mathbf{V}$ is a variety of modular lattices containing all lattices of subspaces of vector spaces over the rationals, then either $\mathbf{V}$ is not finitely based or it is not generated by its finite dimensional members.   —

In particular, $\mathbf{M}_{fd} \subset \mathbf{M}$. Combining Theorems 2 and 3, we see that

$$\mathbf{M}_f \subset \mathbf{M}_{fd} \subset \mathbf{M}.$$

This gives a very complete answer to Problem IV.5.

Also note three important corollaries to C. Herrmann's result:

1. *The variety of Arguesian lattices is not generated by its finite dimensional members.*

2. $\mathbf{M}_{fd}$ *is not finitely based.*

3. $\mathbf{M}_f$ *is not finitely based.*

R. Freese [155] gives a direct proof of Theorem 3.

Theorem 4.   The word problem for the free modular lattice on five generators is unsolvable.   —

Thus the answer to the second question in Problem IV.8 is in the negative.

C. Herrmann [304] improved "five" to "four" in this result (answering the first question in Problem IV.8).

The modular identity was introduced by R. Dedekind as a very important property of submodules of a module. There are many stronger forms, for instance, the Arguesian identity (see Section IV.4) that holds for all lattices having type 1 representations (*linear* lattices). B. Jónsson [1953] raised the problem whether the converse holds. In 1987, M. D. Haiman succeeded in answering Jónsson's problem (see M. D. Haiman [278] and [279]):

Theorem 5.   There is an Arguesian lattice that is not linear.   —

M. D. Haiman also proved that linear lattices cannot be defined by a finite set of quasi identities. His class of examples are based on the S-glued sum construction of C. Herrmann [1973c].

Many specialists searched for a stronger form of the Arguesian identity that holds for subgroup lattices for abelian groups but not, in general, for lattices of normal subgroups of a group. This long standing problem was solved in P. P. Pálfy and Cs. Szabó [418] and [419]:

Theorem 6.   The lattice variety generated by subgroup lattices of abelian groups is a proper subvariety of the variety generated by normal subgroup lattices of groups.   —

Both modular lattices and abelian categories are abstractions of modules. G. Hutchinson calls a modular lattice L *abelian* if, for any proper quotient x/y of L, there is a diamond $M_3 = \{o, a, b, c, i\}$ in L with $x/y \nearrow a/o$. In G. Hutchinson [1971], a functor $\mathbf{A}$ is given from the category of abelian lattices with lattice homomorphisms to the category of small abelian categories with exact functors. For a ring R with 1, let R-MOD be the category of all R-modules, and let $\mathcal{L}(R)$ denote the class, in fact the quasivariety, of lattices representable by R-modules, that is,

$$\mathcal{L}(R) = \mathbf{S}\{\operatorname{Sub}(M) \mid M \in R\text{-MOD}\}.$$

Based on the fact that $\mathbf{A}$ sends embeddings to embeddings, G. Hutchinson [1973b] and [318] prove the following powerful result.

Theorem 7.   Let R and S be rings with unit. Then $\mathcal{L}(R) \subseteq \mathcal{L}(S)$ iff there is an exact embedding functor R-MOD $\to S$-MOD.   —

This theorem has many consequences. If the characteristic of R, char R, is a positive square-free integer and char $R =$ char S, then $\mathcal{L}(R) = \mathcal{L}(S)$; if R is commutative, then $\mathcal{L}(R)$ is a selfdual class of lattices. On the other hand, G. Czédli and G. Hutchinson [92] developed a technique to prove that if $0 <$ char R is divisible by p^2, for some prime p, then there are continuously many rings S of the same characteristic with pairwise distinct $\mathcal{L}(S)$. These $\mathcal{L}(S)$ generate the same (congruence) variety (see G. Czédli and G. Hutchinson [91], where a complete classification of varieties of the form $\mathbf{H}\mathcal{L}(R)$ is given).

1.3   Free lattices

Jónsson's conjecture

The conjecture that finite sublattices of a free lattice can be characterized by (W), (SD$_\wedge$), and (SD$_\vee$) (see Section VI.1) goes back to B. Jónsson [1961]. About 20 years of work, first by Jónsson and then by J. B. Nation, resolved this problem (see J. B. Nation [397]):

Theorem 8'.   Finite sublattices of a free lattice are characterized by (W), (SD$_\wedge$), and (SD$_\vee$).   —

Transferability

Let us call a lattice K *transferable* iff, for any lattice L, if K has an embedding φ into $\operatorname{Id} L$, then K has an embedding ψ into L. K is called *sharply transferable*, if, in addition, this ψ can always be chosen to satisfy

$$a\psi \in a\varphi \text{ but } a\psi \notin b\varphi \text{ for any } b < a.$$

This topic did not start out with free lattices. I was invited by J. C. Abbott and G. Birkhoff to give a lecture on my forthcoming book on universal algebra (G. Grätzer [1968]) at a conference at the United States Naval Academy in May of 1966 (G. Grätzer [1970]). In preparation for this lecture, I gave a series of talks at McMaster University in December 1965 and January 1966, where I introduced the concept of a transferable lattice (see Problem I.22; this was earlier stated in G. Grätzer [1970] and as Problem 14 in G. Grätzer [1971]).

The crucial concept in dealing with transferability was introduced by H. S. Gaskill. Let K be a finite lattice, $p \in K$, and $J \subseteq K$. We call $\langle p, J \rangle$ a *pair* iff $p \notin J$, $2 \le |J|$, $p \le \bigvee J$, and $p \not\le j$, for all $j \in J$. A pair $\langle p, J \rangle$ is *minimal* iff whenever $\langle p, J' \rangle$ is also a pair and for every $a \in J'$ there exists a $b \in J$ with $a \le b$, then $J \subseteq J'$. (Minimal pairs are the most economical way of describing the join-structure of a finite lattice.) A finite lattice K satisfies the condition $(\mathrm{T}_\vee)$ iff a rank function r can be defined on $\mathrm{J}(K)$ (with integer values) such that if $\langle p, J \rangle$ is a minimal pair of K and $q \in J$, then $r(p) < r(q)$. Condition $(\mathrm{T}_\wedge)$ is the dual of $(\mathrm{T}_\vee)$.

Theorem 8".   For a finite lattice K, the following conditions are equivalent:

(i) K is transferable.

(ii) K is sharply transferable.

(iii) K satisfies $(\mathrm{T}_\wedge)$, $(\mathrm{T}_\vee)$, and (W).

(iv) K can be embedded in $F(3)$.   —

This result was discovered in about a decade as a collaborative effort involving H. S. Gaskill, C. R. Platt, B. Sands, and myself (others making a contribution to this field include K. A. Baker, G. Bruns, A. Day, A. W. Hales, E. Nelson, T. Tan, and A. G. Waterman); see G. Grätzer [1970], H. S. Gaskill [1972], H. S. Gaskill, G. Grätzer, and C. R. Platt [1975], H. S. Gaskill and C. R. Platt [1975], G. Grätzer, C. R. Platt, B. Sands [249], and G. Grätzer and C. R. Platt [248].

The equivalence of the first four of the following five conditions, combines the previous two theorems:

Theorem 8.   For a finite lattice K, the following conditions are equivalent:

(i) K is transferable.

(ii) K is sharply transferable.

(iii) K satisfies $(SD_\wedge)$, $(SD_\vee)$, and (W).

(iv) K can be embedded in $F(3)$.

(v) K is projective.   —

The equivalence of conditions (iv) and (v) is due to B. Jónsson and R. N. McKenzie, see R. N. McKenzie [1972]; in fact, this equivalence holds for finitely generated lattices, see A. Kostinsky [1972].

Even though the arguments of the two previous theorems have a lot in common, interestingly, nobody tried to synthesize the work of the two schools.

Back in 1967, I only had one nontrivial result (see p. 208 of G. Grätzer [1970] and G. Grätzer [1974]). Let (X) denote the condition that there is no doubly-reducible element; (X) is a special case of (W). Then if the finite lattice K fails (X), then K is not (sharply) transferable.   I proved this as follows: if $a \in K$ is doubly-reducible, then I double a (and all other doubly-reducible elements); from the resulting lattice, I easily construct a lattice L such that L satisfies (X) and K has a natural embedding into $\operatorname{Id} L$. This was followed in a few years by A. Day [1970], in which he corrected any (W) failure by doubling the offending interval. This technique of Day's (see Appendices F and G) is crucial in a number of his later results.

Prime intervals

In the paper A. Day [1977], a crucial result is proved: every proper interval of a finitely generated free lattice contains a prime interval (see page 392). Based on this result, R. Freese and J. B. Nation [181] described all prime intervals, in fact, all finite intervals of finitely generated free lattices. So they answered Problem VI.17 but made not very much contribution to Problem VI.16. This was done by S. T. Tschantz [502], who proved the following remarkable result:

Theorem 9.   Every infinite interval in a free lattice contains a sublattice isomorphic to $F(3)$.   —

You can read more on this in Appendix G.

1.4   Varieties of finite height

Exercise V.2.1 (R. N. McKenzie) lists a number of lattices that generate varieties covering $\mathbf{N}_5$. B. Jónsson and I. Rival [341] proved the astonishing converse.

A lot has also been learned about modular covers of the varieties generated by M_4 and $M_{3,3}$, respectively (see Section V.3 and Appendix F). This easily lead to Problem V.15: is every variety of finite height generated by a finite lattice? Quite recently, J. B. Nation [401] provided a counterexample. He also found an example of a finite lattice with the property that the variety generated by it has countably many covers. More about this in Appendix F.

1.5   The Amalgamation Property

In Section V.2.4, we have discussed the long standing problem, which varieties of lattices have the Amalgamation Property? In G. Grätzer, B. Jónsson, and H. Lakser [1973], this was resolved for modular varieties. Finally, A. Day and J. Ježek [127] also settled the nonmodular case, so we have the following result:

Theorem 10.   The only varieties having the Amalgamation Property are **T**, **D**, and **L**.   —

1.6   Congruence lattices

This field is still dominated by Problem II.7: can every distributive algebraic lattice be represented as the congruence lattice of a lattice? Despite the fact that this question was raised by R. P. Dilworth in the early forties, and that so many of the best in the field spent so much time trying to answer it, an answer still eludes us.

My two favorite results in this field are the following.

The first is a construction of E. T. Schmidt [463]:

Theorem 11.   Let L be a distributive algebraic lattice, and let S denote the distributive join-semilattice with zero of compact elements of L. If there exists a generalized Boolean lattice B and a distributive join-homomorphism $h\colon B \to S$, then L is representable.   —

Applying this construction, large classes of distributive algebraic lattices have been represented as congruence lattices of lattices, see Appendix C for these results.

For close to thirty years, I have believed that Schmidt's construction would solve this characterization problem. This belief was shattered by a recent result of F. Wehrung [517]:

Theorem 12.   There exists a distributive semilattice S of size $\aleph_2$ that cannot be represented as a distributive join-homomorphic image of a generalized Boolean lattice.   —

To add salt to the wound, F. Wehrung also proved that the approach of P. Pudlák, M. Tischendorf, and J. Tůma, attempting to construct the lattice as a direct limit of finite atomistic lattices, also fails for his example, see F. Wehrung [517].

M. Ploščica, J. Tůma, and F. Wehrung [436] strengthen Theorem 12: the distributive join-semilattice S can be taken to be the semilattice of compact congruences of the free lattice on $\aleph_2$ generators. This says that all known "methods" fail to represent certain semilattices that are known to be representable.

It seems to me that we are back to square one.

1.7   Finite congruence lattices

So much happened in this field (see Section 1 of Appendix C) that it is hard to pick favorites.

My favorite *new concept* is the following: Let L be a finite lattice. A finite lattice K is a *congruence-preserving extension* of L, if K is an extension and every congruence of L has *exactly one* extension to K. Of course, then the congruence lattice of L is isomorphic to the congruence lattice of K; we could say that the congruence lattice of K is *naturally isomorphic* to the congruence lattice of L or that the algebraic reasons determining the congruence lattice of L are carried over to K.

And here are three results to illustrate this concept:

Theorem 13.   Every finite lattice has a congruence-preserving extension to a finite

(i) *atomistic* lattice (M. Tischendorf [500]);

(ii) *sectionally complemented* lattice (G. Grätzer and E. T. Schmidt [258]);

(iii) *semimodular* lattice (G. Grätzer and E. T. Schmidt [259]).   —

The last one solves Problem IV.19 for finite lattices, first done in G. Grätzer, H. Lakser, and E. T. Schmidt [244]. Closely connected to these is the result of G. Grätzer and F. Wehrung [267]: Every lattice has a *proper* congruence-preserving extension.

My favorite *new trend* is combinatorial: giving upper and lower bounds for the size of a smallest lattice K representing a finite distributive lattice D.

For a natural number n, define $\operatorname{cr}(n)$ as the smallest integer such that, for any distributive lattice D with n join-irreducible elements, there exists a finite lattice L satisfying $\operatorname{Con} L \cong D$ and $|L| \leq \operatorname{cr}(n)$.

Theorem 14.   For any integer $n \geq 2$,

$$\frac{1}{16}\frac{n^2}{\log_2 n} < \operatorname{cr}(n) < 3(n+1)^2.$$ —

The upper bound was proved in G. Grätzer, H. Lakser, and E. T. Schmidt [241] and the lower bound in G. Grätzer and D. Wang [265]. See Section 1 of Appendix C for further references.

1.8   Lattices of complete congruences

In 1945, G. Birkhoff [51] raised the following question: is every complete lattice isomorphic to the congruence lattice of an infinitary algebra? An affirmative answer was published by W. A. Lampe and me as Appendix 7 in G. Grätzer [215].

In the early eighties, influenced by his work on formal concept analysis, R. Wille suggested (see, for instance, K. Reuter and R. Wille [455]) that one should investigate whether the infinitary universal algebra in the previous paragraph could be replaced by a complete lattice. This was verified by S.-K. Teo [498] for finite lattices and for the general case in G. Grätzer [218]:

Theorem 15.   Every complete lattice K can be represented as the lattice of complete congruence relations of a complete lattice L.   —

A number of papers have appeared improving and applying this result, for instance, constructing a distributive complete lattice L; more about this in Section 1 of Appendix C. It was quite unexpected, however, that the techniques developed to prove this result in G. Grätzer and H. Lakser [236] turned out to be very useful in improving the results for *finite* congruence lattices. For instance, it is a direct consequence of this development that we can prove planarity and minimal size in the finite case.

1.9   Equational compactness

I got interested in the characterization problem of equationally compact lattices in 1967 (see G. Grätzer and H. Lakser [1969b]). When I realized that an equationally compact distributive lattice has to satisfy (JID) and (MID), and therefore can be embedded in a complete Boolean lattice (Theorem II.4.14), I thought I was very close to a solution.

A solution, however, eluded me, as it eluded so many others (a former student of mine spent more than a decade on this problem).

For the one variable case, D. Kelly [1972] found a solution. In the next decade, all relevant publications were incorrect. Finally, the two-variable case was solved in K. A. Nauryzbaev [404].

Again a decade of incorrect solutions, and then the surprising conclusion: F. Wehrung [514], [515] constructed a complete distributive lattice satisfying (JID) and (MID) that is not equationally compact; the system of equations showing this has three variables and it is countable.

Theorem 16.   There exists a complete distributive lattice D satisfying (JID) and (MID) that is not equationally compact. In fact, there is a countable system Σ of equations in three variables such that every finite subsystem of Σ can be solved in D but Σ has no solution in D.   —

2.   Notes on Chapter I

2.1   Notes on the problems

Problem 1

K. V. Adaricheva [22] solved this problem for finite semilattices:

Theorem 17.   Let $\langle L, \sim \rangle$ be a finite set with a reflexive and symmetric binary relation $\sim$. Then $\sim$ is the *comparability relation* of some semilattice on the set L iff, for any pair $\langle a, b \rangle$ of elements of L that are not related under $\sim$, there is an element d, denoted by $a \circ b$, satisfying $d \sim a$ and $d \sim b$, and satisfying the following conditions:

(1) There are no C-cycles in L. (The sequence, $a_0, a_1, a_2, \ldots, a_n, n \geq 2$, of elements of L is called a *C-cycle*, if $a_0 = a_n$ and $a_i = a_{i+1} \circ b_i$, for all i and for some $b_i \in L$.)

(2) $\langle L; \circ, \sim \rangle$ is a structure with univocally terminating descents. (A sequence of pairs $\langle a_1, b \rangle, \langle a_2, b \rangle, \ldots, \langle a_n, b \rangle$ is a *left descent*, if i) $a_1 \sim b$; ii) $a_{k+1} = a_k \circ b_k$, for some $b_k \sim b$; iii) for any $c \in L$, if $c \sim b$, then $a_n \sim c$. Analogously, we define when a sequence $\langle b, a_1 \rangle, \langle b, a_2 \rangle, \ldots, \langle b, a_n \rangle$ is a *right descent*. $\langle L; \circ, \sim \rangle$ has *univocally terminating descents*, if any two left (or right) descents with the same first pairs have the same terminating pairs.)

(3) The left and right slaloms in $\langle L; \circ, \sim \rangle$ have different parities. (A sequence $\langle a, b \rangle, \ldots, \langle c, d \rangle$ of pairs from L is said to be a *slalom*, if it consists of alternating left and right descents the last pairs of which are the first pairs of the subsequent descents. A slalom is said to be *even* (*odd*), if the number of alternating descents in it is even (odd), and *left* (*right*), if its first descent is left (right), and it is said to be *exact*, if $c = d$. The left and right slaloms are said to have *different parities*, if there does not exist a right exact slalom and a left exact slalom with the same first pair that are both even or both odd.)   —

Note that it is easy to obtain from this result a characterization of the comparability relation for finite lattices: Let $\langle L, \sim \rangle$ satisfy the conditions in Theorem 17; then $\sim$ is a comparability relation of some lattice iff there exists $a \in L$ satisfying $a \sim x$, for all $x \in L$, and such that the restriction of $\sim$ to $L - \{a\}$ is a comparability relation of some semilattice.

Problem 2

A. Urquhart [506] proved that the lattices T_n are bounded (in the sense of R. N. McKenzie [1972]). W. Geyer [199] proved that the lattices T_n are subdirectly irreducible; have the same number of elements (the nth Catalan number, C_n) as their congruence lattice; have $((n-1)/2)C_n$ coverings. He conjectured that finite lattices embeddable in some T_n are exactly the finite bounded lattices, which would provide an interesting solution to this problem.

Problem 4

A number of sufficient conditions can be given under which Sub L determines L up to isomorphism: L is an ordinal sum of indecomposable modular lattices (N. D. Filippov [1966] and G. Takách [492]); L is relatively complemented (N. D. Filippov [1966]); L is weakly complemented (G. Takách [492]); L is uniquely complemented (N. D. Filippov [1966]); L is sectionally complemented (C. C. Chen, K. M. Koh, and K. L. Teo [72]); L is an ordinal sum of indecomposable, strongly atomic, semimodular lattices (G. Takách [492]); L is an ordinal sum of indecomposable, semimodular lattices of finite height (C. C. Chen, K. M. Koh, and K. L. Teo [72]); L is simple (G. Takách [493]).

An interesting approach to this problem is taken in A. G. Pinus [429]. An *inner isomorphism* φ of a lattice L is an isomorphism between two sublattices of L. Let Iso L be the semigroup of inner isomorphisms of L. Then the lattice Sub L is the poset of idempotents of Iso L. Pinus proves that for the finite lattices L and M, the following conditions are equivalent:

(i) L is isomorphic to M or to the dual of M;

(ii) Iso L is isomorphic to Iso M.

So, for any finite lattice L, the following conditions are equivalent:

(i) for any finite lattice M, if Sub L and Sub M are isomorphic, then L and M or L and dual of M are isomorphic;

(ii) for any finite lattice M, if Sub L and Sub M are isomorphic, then Iso L and Iso M are isomorphic.

Problem 5

A modular lattice variety is closed under the isomorphism of the sublattice lattices if and only if it is self-dual (G. Takách [493]).

Problems 6 and 7

D. X. Hong [315] proves that for any variety $\mathbf{V} \neq \mathbf{T}$ of lattices, Sub $\mathbf{V} = \mathbf{L}$, answering both problems. This was already reported, attributed to A. P. Huhn, in G. Grätzer [216].

Problem 8

G. Takách proved in [493] that all self-dual, infinitary Horn-sentences that are stronger than modularity are preserved under any isomorphism of sublattice lattices.

Problem 10

Let Csub L be the lattice of all convex sublattices (including $\varnothing$) of the lattice L. This problem asks that Problems 3–9 be investigated with Csub L replacing Sub L.

The Csub variant of Problem 3, then, asks, for a characterization of the lattice Csub L. For finite lattices, this was done in K. M. Koh and T. C. Chua [367]; this result was generalized to arbitrary lattices in M. Kolibiar [369].

The Csub variant of Problem 4 is solved in V. I. Marmazeev [49], describing lattices L that are characterized (up to isomorphism) by Csub L.

The Csub variant of Problem 5 asks, for what varieties $\mathbf{K}$ does $L \in \mathbf{K}$ and Csub $L \cong$ Csub L_1 imply that $L_1 \in \mathbf{K}$? This holds for the variety of modular lattices, see V. I. Igoshin [319]. This problem was solved by V. Slavík [482]: The necessary and sufficient condition is that $\mathbf{K}$ be self-dual.

A contribution to the Csub variant of Problem 6 is made in V. Slavík [486], in which it is proved that there are uncountably many varieties $\mathbf{V}$ for which $\mathbf{V} =$ Csub $\mathbf{V}$.

Problem 11

For finite lattices L (and some generalizations) the automorphism group of Csub L is described in terms of the automorphisms and dual automorphisms of the directly indecomposable direct factors of L in V. I. Marmazeev [50].

Problem 12

A characterization of planar ordered sets was obtained in D. Kelly [359]. The proof uses the main result in D. Kelly [358], and also applies the result of C. R. Platt [1976].

Problem 13

Solved by D. Kelly [357]: every natural number is the order dimension of a suitable finite planar poset.

Problems 22–24 and 26

The solutions to these problems were narrated in Section 1.3. The semilattice question in Problem 26 (Is every finite transferable semilattice sharply transferable?) is still open.

Problem 25

The problem asks for a direct proof that the class of (finite) sharply transferable lattices is closed under the formation of sublattices and duals. A. Day [119] provides a simple proof *via* the characterization theorem; however, this does not qualify as a direct proof.

Problem 28

According to H. S. Gaskill [1972a] (see also E. Nelson [1974]), every finite distributive lattice is sharply **D**-transferable.

G. Grätzer and T. Tan proved [264] that if K is **D**-transferable, then for every $a \in K$, the sets $(a]$ and $\{\, b \mid b \succ a \,\}$ are finite; moreover, if K is sharply **D**-transferable, then K does not contain $C_2 \times C_\omega$ as a sublattice. It is natural to conjecture that the linear sums of type ω of finite distributive lattices are the infinite sharply **D**-transferable lattices.

Problem 31

R. Padmanabhan and W. McCune [416] provide the shortest known single identity defining the variety of all lattices. This identity is dramatically shorter than previously known ones:

Reference	variables	length
R. N. McKenzie [1970]	34	300,000
R. Padmanabhan [413]	7	243
R. Padmanabhan and W. McCune [416]	7	79

The number 300,000 is from W. Taylor's Appendix 4 in G. Grätzer [215].

Problem 35

B. M. Schein pointed out (in his review of the first edition of this book) that this problem was solved in G. I. Žitomirskiĭ [529]. There is a newer paper by the same author on this topic, see G. I. Žitomirskiĭ [530].

For finite semilattices, there is a much deeper characterization of the congruence lattices in K. V. Adaricheva [23].

Problem 36

B. M. Schein reported in 1978 some unpublished work on this problem by O. M. Mamedov.

Problem 40

M. E. Adams and J. Sichler [18] proved that in every nontrivial lattice variety **V**, every lattice $L \in \mathbf{V}$ is the Frattini sublattice of a lattice $A \in \mathbf{V}$.

See M. E. Adams, R. Freese, J. B. Nation, J. Schmid [11] for some very interesting new results, for example, that there are infinitely many varieties **V** such that whenever $L \in \mathbf{V}$ is finite, then it is the Frattini sublattice of a finite $K \in \mathbf{V}$.

2.2   Bases for lattice varieties

In R. Padmanabhan [413], there is now a much stronger version of Exercise 1.26:

Theorem 18.   Any finitely based variety of algebras definable by absorption laws and having a majority polynomial is one-based.   ⸺

Self-dual independent axiom systems were found in D. Kelly and R. Padmanabhan [362]. They prove the following remarkable result:

Theorem 19.   Let **K** be any finitely based self-dual variety of lattices. Then, given any integer $n \geq 4$, there exists an independent self-dual set of n identities defining **K**.   ⸺

In D. Kelly and R. Padmanabhan [361], the following is proved:

Theorem 20.   Any self-dual variety of lattices can be defined by a set of self-dual inequalities with respect to **L**. Any finitely based self-dual variety of lattices can be defined by a single self-dual inequality with respect to **L**.   ⸺

The analogous statements for self-dual identities fail (but both hold for modular varieties).

2.3   The lattices $\operatorname{Sub} L$ and $\operatorname{Csub} L$

In general, $\operatorname{Sub} L$ has very few nice lattice theoretic properties; indeed, by D. X. Hong [315], every finite lattice is embeddable in some $\operatorname{Sub} L$, for some finite Boolean lattice L (see Problems 6 and 7). In this connection, I should mention an interesting result of G. Czédli [86] characterizing lattices L with the property that $\operatorname{Sub} L$ is 2-distributive by the exclusion of eight sublattices and the characterization by A. D. Bol'bot and V. V. Kalinin [54] of lattices L with the property that $\operatorname{Sub} L$ satisfies $(\mathrm{SD}_\vee)$; see also K. V. Adaricheva [21].

For a variety **V**, let $\operatorname{Csub} \mathbf{V}$ denote the variety generated by the lattices $\operatorname{Csub} L$, $L \in \mathbf{V}$. V. Slavík [483], [485], and [486] deal with $\operatorname{Csub} \mathbf{V}$.

2.4   Order-polynomial and affine completeness

A lattice L is *order-polynomially complete*, if every isotone function is an algebraic function (a polynomial function with constants). The first contributions to this topic were made in D. Schweigert [472], R. Wille [520], and M. Kindermann [363]; Wille characterized the finite order-polynomially complete lattices.

H. K. Kaiser and N. Sauer [348] showed that an order-polynomially complete lattice must be bounded and that it cannot be countably infinite. M. Ploščica and M. Haviar [435] proved that the cardinality of an infinite order-polynomially complete lattice must be greater than each $\beth_n$, $n \geq 0$ (where $\beth_0 = \aleph_0$ and $\beth_n = 2^{\beth_{n-1}}$, for $n \geq 1$). Finally, M. Goldstern and S. Shelah [202] proved the following result:

Theorem 21.   The cardinality of an order-polynomially complete lattice is a strongly inaccessible cardinal, so the existence of such a lattice is not provable in ZFC or in ZFC+GCH.   —

See also M. Erné and D. Schweigert [145], K. Kaarli and K. Täht [347], and I. Rival and N. Zaguia [457].

M. Goldstern [201] showed that every lattice L has an extension K such that every isotone function on L can be represented by an algebraic function on K and $|K| \leq |L| \aleph_0$.

An algebra $\mathfrak{A}$ is called *affine complete*, if the algebraic functions of $\mathfrak{A}$ are the only *compatible* (that is, congruence-preserving) functions on A. I proved in 1962 (G. Grätzer [1962a]) that every Boolean algebra is affine complete and, somewhat later, I characterized affine complete bounded distributive lattices (G. Grätzer [1964]).

The last twenty years has seen extensive work on this topic. R. Beazer [41] characterized affine complete members in large classes of Stone algebras and double Stone algebras [42]; K. Kaarli, L. Márki and E. T. Schmidt [345] characterized affine complete semilattices.

More recently, M. Ploščica [434] characterized affine complete members in **D**, M. Haviar and M. Ploščica [300] did the same in the variety of all Stone algebras and M. Haviar, K. Kaarli, and M. Ploščica [297] described local polynomial functions on Kleene algebras and affine complete algebras in the variety of Kleene algebras. M. Haviar also investigated affine completeness for various generalizations of Stone and double Stone algebras [292]–[296]. Affine complete varieties of algebras were investigated in a number of papers, see, in particular, K. Kaarli and A. F. Pixley [346] and the survey papers, A. F. Pixley [430], [431]. K. Kaarli and R. N. McKenzie [344] prove that every affine complete variety is congruence distributive.

2.5   The Glivenko-Frink theorem

For a very brief proof of the Glivenko-Frink theorem (Theorem I.6.4), see T. Katriňák [353].

3.   Notes on Chapter II

3.1   Notes on the problems

A problem in Further Topics and References

In the third paragraph of "Further Topics and References", I raise the question of generalizing the uniqueness of an irredundant join-representation of an element of a finite distributive lattice to infinite lattices. An excellent paper on this topic is V. A. Gorbunov [208].

Problem 1

A rather short identity defining Boolean algebras by a single identity is given in R. Padmanabhan and W. McCune [415].

Problem 3

R. N. McKenzie provided the following affirmative solution.

Theorem 22.   Let A and B be infinite Boolean algebras. Then there exists a Boolean algebra D such that A is a subalgebra of D and Aut D and Aut B are isomorphic.   —

Let A and B be Boolean algebras of cardinality no greater than κ, where κ is uncountable. S. Shelah [473] proved that there is a system of 2^κ Boolean algebras of cardinality κ such that no two have isomorphic nontrivial relativized subalgebras. Obviously, there must be κ of these that have no relativized subalgebra in common with relativized subalgebras of B.

Now form an algebra C. Let Q_α, $\alpha < \kappa$, be Shelah's system, and P be the product of the Q_α, with u_α being the element of P such that $P|u_\alpha$ is (isomorphic to) Q_α. We can assume that A is a subalgebra of the algebra of all subsets of κ. Thus A is embedded into P as the set A' of all x such that, for some X in A, x is the sum of all u_α with $\alpha \in X$. Take C to be the subalgebra of P containing A' consisting of those y such that for some $x \in A'$, the symmetric difference of x and y is contained in the sum of finitely many u_α. This algebra C is rigid and includes A'. The cardinality of C is κ. The algebra C has no relativized subalgebras isomorphic to relativized subalgebras of B, and hence Aut($C \times B$) is naturally isomorphic to Aut B (or rather to Aut $C \times$ Aut B).

Now we put $D = C \times B$ and observe that Aut D and Aut B are isomorphic, while A is a subalgebra of C and hence isomorphic to a subalgebra of D.

Problem 5

I. Düntsch [138] gives an example of a poset P in which every element is contained in a maximal one, P is representable as $\mathcal{P}(L)$ by a distributive lattice L but not by a distributive lattice L with a unit. The existence of such a poset has been observed by T. Tan (see p. 127 of the first edition of this book).

Problem 7

The characterization problem of congruence lattices of lattices is discussed in Section 1.6 and in Section 2 of Appendix C.

Problem 8

A remarkable result of F. Wehrung [517] (based on a construction in F. Wehrung [516]) shows that not every distributive algebraic lattice is representable as the congruence lattice of a sectionally complemented lattice.

An even stronger result of M. Ploščica, J. Tůma, and F. Wehrung [436] shows that there exists a lattice whose congruence lattice cannot be represented as the congruence lattice of a sectionally complemented lattice.

On the positive side, every finite lattice has a congruence-preserving embedding to a finite sectionally complemented lattice (G. Grätzer and E. T. Schmidt [258]); see Section 1.7.

Problem 17

S.-K. Teo [498] proves that if D is a finite distributive lattice with n dual atoms, then there exists a finite lattice L of length $5n$ such that the congruence lattice of L is isomorphic to D; $5n$ is the lowest possible bound.

Problem 19

The *Independence Theorem for the congruence lattice and the automorphism group of a finite lattice* was proved independently by V. A. Baranskiĭ [38] and A. Urquhart [506]. Both proofs utilize the characterization theorem of congruence lattices of finite lattices (as finite distributive lattices) and the characterization theorem of automorphism groups of finite lattices (as finite groups).

In two recent papers, G. Grätzer and E. T. Schmidt [260] and [262], we proved that the Independence Theorem for finite congruence lattices and finite automorphism groups is also true for modular lattices; of course, the modular lattice showing the independence is no longer finite.

A different approach is taken in G. Grätzer and E. T. Schmidt [257]; to explain the new approach, we need to introduce two concepts.

Let L be a finite lattice. A finite lattice K is a *congruence-preserving extension* of L, if K is an extension and every congruence of L has exactly one extension to K. Of course, then the congruence lattice of L is isomorphic to the congruence lattice of K.

A finite lattice K is an *automorphism-preserving extension* of L, if K is an extension and every automorphism of L has exactly one extension to K, and in addition, every automorphism of K is the extension of an automorphism of L. Of course, then the automorphism group of L is isomorphic to the automorphism group of K.

Theorem 23.   Let L_C and L_A be finite lattices, $L_C \cap L_A = \{0\}$. Then there exists a finite atomistic lattice K that is a congruence-preserving extension of L_C and an automorphism-preserving extension of L_A. In fact, K can be chosen such that both extensions preserve the zero.   —

Problem 18 (the infinite case) is still open.

Problem 20

This problem was solved by W. A. Lampe. In R. Freese, W. A. Lampe, and W. Taylor [176] the following result is proved:

Theorem 24.   Let L be the congruence lattice of an infinite dimensional vector space over a field of cardinality $\mathfrak{m} > \aleph_0$. Then L is not isomorphic to the congruence lattice of an algebra with fewer than $\mathfrak{m}$ operations.   —

Problem 22

An affirmative solution was obtained in P. Olin [410]: there exists an infinite distributive lattice A such that for every distributive lattice B, the distributive free product of A and B is an elementary extension of A.

Problem 23

Solved negatively in F. Wehrung [515]; see Section 1.9.

Problem 27

Amal $\mathbf{B}_n$, for $n > 2$, is described in C. Bergman [45].

Problem 30

A solution is proposed in E. Capińska and A. Wroński [71].

Problem 31

The lattice of quasivarieties of distributive lattices with pseudocomplementation is investigated in V. A. Gorbunov [207] and G. Grätzer, H. Lakser, and R. W. Quackenbush [239]. In the latter it is proved that the principal ideal generated by $\mathbf{B}_3$ has $2^{\aleph_0}$ members and it is not distributive.

H. Gaitan [194] found a quasivariety that does not lie between two consecutive varieties. He proved that the quasivariety generated by the finite subdirectly irreducible algebras is the whole variety.

For more on this topic, see Section 8.3.

Problem 32

D. Thomas [499] characterized bounded distributive lattices with ideal lattices in $\mathbf{B}_n$, $n \geq 1$.

Problem 34

This problem asks to identify maximal classes $\mathbf{K} \subset \mathbf{D}$ such that, for L_1, $L_2 \in \mathbf{K}$, $\mathcal{P}(L_1) \cong \mathcal{P}(L_2)$ implies that $L_1 \cong L_2$.

R. Balbes [1971] proved that if L_1 is a free distributive lattice, then $\mathcal{P}(L_1) \cong \mathcal{P}(L_2)$ implies that $L_1 \cong L_2$. This was generalized in M. E. Adams [1975]: the same conclusion holds if L_1 is generated by its doubly-irreducible elements. M. Gehrke [197] continues this work. Some of Gehrke's results are too technical to state here, but the following is easy to formulate: if L_1 is generated by its join-irreducible elements as well as by its meet-irreducible elements, then $\mathcal{P}(L_1) \cong \mathcal{P}(L_2)$ implies that $L_1 \cong L_2$.

3.2 Variety independence

A lattice property P is said to be *variety independent*, if whenever P holds in $\mathbf{D}$, then it holds in $\mathbf{L}$. For example, R. N. McKenzie proved that the validity of an absorption law is variety independent. D. Kelly and R. Padmanabhan [360] gave some syntactic conditions for an identity to be variety independent.

4.   Notes on Chapter III

4.1 Notes on the problems

Problem 1

The concepts of distributive and neutral ideals are generalized to convex sublattices in J. Nieminen [409]. Another generalization of distributive ideals to convex sublattices is in C. Malliah and P. Bhatta [385].

Problem 5

Lattices with Stone congruence lattices were first characterized by T. Katriňák [1967] (see also T. Katriňák and S. El-Assar [355] and [356], D. Thomas [499]). M. Haviar and T. Katriňák [298] simplified this characterization and described lattices with relative Stone congruence lattices. Further, they showed that a distributive lattice L has a relative Stone congruence lattice iff L is discrete. See also Z. Heleyová [302].

Problem 6

M. Haviar [291] characterized lattices whose congruence lattices belong to $\mathbf{B}_n$, $n \geq 1$; for semi-discrete lattices this was done by M. Haviar and T. Katriňák [299].

5. Notes on Chapter IV

5.1 Notes on the problems

Problem 1

J. P. S. Kung [373] and [374] proves that in every finite modular lattice L, there is a matching between $J(L) \cup \{0\}$ and $M(L) \cup \{1\}$. Under the additional assumption that $L = (a] \cup [b)$ holds for no $a < b \in L$, K. Reuter [454] obtained a sharper form by establishing a matching $J(L) \to M(L)$.

Problem 5

A negative answer was obtained in R. Freese [155], see Section 1.2.

Problem 7

See B. Jónsson [336] for a discussion of some related problems. R. Freese [155] proved that the free distributive lattice on $\aleph_0$ generators can be embedded in the free modular lattice with five or more generators.

Problems 8 and 9

A negative answer was obtained in R. Freese [157] and C. Herrmann [304], see Section 1.2.

Problems 11 and 12

A positive answer to Problem 11 would imply one to Problem 12, which is ruled out by the negative solution to Problem 9.

Problem 15

For a common generalization of the Kurosh-Ore theorem and the Schmidt-Ore theorems for arbitrary modular lattices, see A. Walendziak [510] and G. Richter [456].

Problem 16

C. Malliah and P. Bhatta [386] provide a negative solution: they construct a semimodular and algebraic lattice that is not M-symmetric.

Problem 18

J. Ježek [323] answered the first question in the negative: there are infinite sets A and B such that the lattices Part A and Part B are not elementarily equivalent; in fact, if $|A| = \aleph_n$, for some $n < \omega$, and $|B| \geq \aleph_\omega$, then Part A and Part B are not elementarily equivalent. There are stronger results in C. Naturman and H. Rose [402] and A. G. Pinus [428].

Problem 19

In G. Grätzer, H. Lakser, and E. T. Schmidt [244], we prove that every finite distributive lattice can be represented as the congruence lattice of a finite (planar) *semimodular* lattice, solving the problem for finite lattices.

For infinite lattices the problem does not make very much sense since semimodularity is not an appropriate property for infinite lattices that do not have many prime intervals. In fact, in G. Grätzer and E. T. Schmidt [258], we prove that every distributive algebraic lattice that can be represented as the congruence lattice of a lattice can also be represented as the congruence lattice of a semimodular lattice, where semimodularity is guaranteed by the nonexistence of prime intervals.

Problem 23

M. D. Haiman [278] and [279] provides a positive solution: the class of lattices having type 1 representations cannot be defined by a finite set of identities.

Problem 25

The solution is in the negative by R. Freese, C. Herrmann, and A. P. Huhn [170].

Problem 26

A negative solution was found by S. V. Polin [438]; Polin found a variety **V** of algebras such that a nontrivial lattice identity holds in the congruence lattices of members of **V** but **V** is not congruence modular. See also Section 8.5.

Problem 28

It is proved in R. Freese and B. Jónsson [174] that modularity implies the Arguesian identity for congruence lattices of all members of a variety.

Problem 30

According to C. Herrmann [307], the answer is positive for a finite lattice L.

Problem 31

The problem should read: "Find an effective algorithm that turns configurational conditions for geometries into equivalent identities for the subspace lattices, whenever possible." R. Freese pointed out that, for instance, the Pappian law is not equivalent to a lattice identity; however, by A. Day [120], it is equivalent to a lattice identity for projective planes.

Problem 32

Solved by P. Pudlák and J. Tůma [450], see Section 1.1.

Problem 35

A negative solution was obtained in E. R. Canfield [70]: there is a finite partition lattice with an antichain that is longer than the longest antichain of elements of the same height.

Problem 36

P. P. Pálfy and P. Pudlák [417] reduced the question:

> Is every finite lattice isomorphic to the congruence lattice of a finite (universal) algebra?

to a group theoretic one:

> Is every finite lattice isomorphic to an interval in the subgroup lattice of a finite group?

The problem seems to be group theoretical, and many experts in the field expect the solution to be negative. Further important references are R. W. Baddeley and A. Lucchini [32], W. Feit [154], T. Ihringer [320], A. Lucchini [380], and J. Tůma [503].

Problem 37

By C. Herrmann [306], the solution is in the negative.

5.2　Combinatorics

There are two books on the combinatorial aspects of this chapter (Sperner's Lemma, antichains, and so on): I. Anderson [28] and K. Engel and H.-D. O. F. Gronau [143].

A large number of publications are devoted to Whitney numbers. We only mention one significant direction.

By a sophisticated counting method, L. M. Butler [66] proved that the subgroup lattice of a finite abelian group is always unimodal (see also [67] and F. Regonati [452]). F. Regonati extended this result [460]:

Theorem 25.　Each interval of a finite modular lattice L is symmetric unimodal if and only if L is a direct product of primary q-lattices (where q may vary). ───

In this result, a finite lattice L is a q-lattice, if all rank 2 intervals of L have exactly 1 or $q + 1$ atoms; a unimodal lattice L is *symmetric*, if $W_i = W_{n-i}$, for all i with $i \leq n/2$, where n denotes the length of L.

A lot has been written on the combinatorics of partition lattices. H. Strietz [490] proved that all finite partition lattices are four-generated. A number of papers followed up this result, see G. Czédli [87], [88], [89], H. Strietz [491], L. Zádori [527].

J. Dudek [137] contains a surprising characterization of modular lattices: a bisemilattice is a nondistributive modular lattice iff it has exactly 19 essentially ternary polynomials.

5.3   Arguesian lattices

M. D. Haiman [277] and D. Pickering [426] characterize Arguesian lattices with a selfdual identity.

The formulation of Desargues' Theorem (as illustrated by Figure IV.5.4) carries over to arbitrary modular lattices.

In a modular lattice, a *triangle* is a triple of elements. The triangles $\langle a_0, a_1, a_2 \rangle$ and $\langle b_0, b_1, b_2 \rangle$ are *centrally perspective*, if

$$(a_0 \vee b_0) \wedge (a_1 \vee b_1) \leq a_2 \vee b_2,$$

and *axially perspective*, if

$$(a_0 \vee a_1) \wedge (b_0 \vee b_1) \leq ((a_1 \vee a_2) \wedge (b_1 \vee b_2)) \vee ((a_2 \vee a_0) \wedge (b_2 \vee b_0)).$$

A. Day [121], [122] proved the following result:

Theorem 26.   A modular lattice is Arguesian if and only if every centrally perspective pair of triangles is also axially perspective.   —

A. Day also showed that every 2-distributive modular lattice is Arguesian.

A. Day and B. Jónsson [128]–[131] characterized Arguesian modular lattices L by excluding from the ideal lattice of L a configuration that is analogous to the classical non-Arguesian position of ten points and ten lines (see Figure IV.5.4); the configuration is obtained by gluing twenty (not necessarily distinct) projective planes. Note that the characterization of Arguesian lattices is much more difficult than that of modular lattices because there is a minimal non-Arguesian variety all of whose members of finite height are Arguesian by D. Pickering [425]. See also A. Day, C. Herrmann, B. Jónsson, J. B. Nation, and D. Pickering [126] and Problem 28 in Section 1.2.

5.4   Hermitian forms

H. Gross, C. Herrmann, and R. Moresi [276] use modular lattice theory to solve an important classification problem in classical algebra. Many problems in the theory of Hermitian forms are related to the problem of classifying indecomposable pairs $F \subseteq E$, where E is a vector space over a skew field k with an Hermitian form $\Phi(x, y)$. The lattice $L(E)$ of subspaces of E together with the unary operation $X^\perp$ forms a *polarity lattice*. There is a natural additive subgroup T of k.

If Φ is a symmetric bilinear form and k is commutative, T is just $2k$, which is k unless the characteristic of k is 2, in which case it is 0.

Define $X^* = \{\, x \in X \mid \Phi(x,x) \in T \,\}$. Now $X^* = E^* \cap X$, so $X \mapsto X^*$ is an induced operation on the polarity lattice if we include E^* as a constant. E^* plays an important role because $X \cup X^\perp \subseteq E^*$. The classification problem is closely related to the sub polarity lattice $P(F, E^*)$ of $L(E)$ generated by E^* and F. In fact, the pair $\langle E, F \rangle$ is "indecomposable" if and only if the sublattice is subdirectly irreducible. So the main part of the problem boils down to classifying subdirectly irreducible, modular, 2-generated polarity lattices subject to the relations $p(F, E^*) \cap p(F, E^*)^\perp \leq E^*$, for every polynomial p. (It turns out that the free lattice subject to these relations has 13,080 elements.) There are 14 subdirectly irreducibles; the largest one has 11 elements.

5.5   A geometric description of modular lattices

C. Herrmann, D. Pickering, and M. Roddy [309] develops a geometry closely connected to modular lattices. This has two interesting applications. It shows that the lattices that can be can be embedded into the lattice of subspaces of a vector space of arbitrary characteristic are precisely the modular, 2–distributive lattices, extending the result of B. Jónsson and J. B. Nation [340], where the result was proved for finite dimensional 2–distributive lattices. The second application is mentioned in Section 8.1.

6.   Notes on Chapter V

6.1   Notes on the problems

Problem 9

J. Berman [48] constructs $2^{\aleph_0}$ varieties $\mathbf{K}$ with $C_2 \in \mathbf{Amal(K)}$. However, none of these varieties are modular, so this problem is still open.

Problems 10 and 11

These problems were solved in A. Day and J. Ježek [127], see Section 1.5.

Problem 12

$\mathbf{Amal(M}_n)$ was characterized in C. Bergman [45]. See E. W. Kiss, L. Márki, P. Pröhle, and W. Tholen [364] on amalgamation, congruence extension, epimorphisms, residual smallness, and injectivity.

Problem 15

Solved by J. B. Nation [401], see Section 1.4.

Problem 17

B. Jónsson and I. Rival [341] proved the converse of Exercise V.2.1, see Section 1.4.

Problem 20

A very elegant finite equational basis for $\mathbf{M} \vee \mathbf{N}_5$ was exhibited by B. Jónsson [334], providing a solution to this problem.

Problem 21

K. A. Baker [34] exhibited two varieties, $\mathbf{V}_0$ and $\mathbf{V}_1$, of modular lattices such that $\mathbf{V}_0$ and $\mathbf{V}_1$ have finite equational bases but $\mathbf{V}_0 \vee \mathbf{V}_1$ does not, providing a negative solution to the problem.

6.2 Amalgamation class

The amalgamation class of the variety generated by the pentagon was described in B. Jónsson [338].

For a variety $\mathbf{V}$, is $\mathbf{Amal}(\mathbf{V})$ an elementary class? C. Bergman [46] proved that the answer is negative for the variety $\mathbf{V}$ generated by a suitable finite modular lattice.

6.3 Exercise V.4.12 and pasting

In Exercise V.4.12, the following research topic is suggested: "Find further examples of the phenomenon observed twice in the proof of Theorem 8, namely, that for some special $\mathbb{V}$-formation, if $\langle \psi_0, \psi_1, C \rangle$ amalgamates $\langle A, B_0, B_1, \varphi_0, \varphi_1 \rangle$ then $B_0\psi_0 \cup B_1\psi_1$ is a uniquely determined sublattice of C."

Of course, I had in mind to look for generalizations of gluing and of the constructions in Section V.4. This topic was taken up in V. Slavík [480], [481] and later in E. Fried and G. Grätzer [182], [183] (in which this special amalgamation was called *pasting*), [184], [186], [187], and [188], and E. Fried, G. Grätzer, and E. T. Schmidt [190]. The main result of E. Fried and G. Grätzer [182] and [183] is that $\mathbf{M}$ is closed under pasting.

See also A. Day and C. Herrmann [125].

6.4 Products of varieties

Let $\mathbf{V}$ and $\mathbf{W}$ be classes of lattices. The *product* of $\mathbf{V}$ and $\mathbf{W}$, denoted by $\mathbf{V} \circ \mathbf{W}$, consists of all those lattices L for which there exists a congruence Θ such that all Θ-classes of L are in $\mathbf{V}$ and L/Θ is in $\mathbf{W}$.

The corresponding concept for varieties of groups was extensively studied in H. Neumann [406] and generalized to universal algebras in A. I. Mal'cev [384].

The first lattice theoretic result on products is in V. B. Lender [379]: nontrivial prevarieties (classes closed under $\mathbf{I}$, $\mathbf{S}$, and $\mathbf{P}$) of lattices under product form a free groupoid. A. Day [118] showed that **L** and **T** are the only idempotent elements of this groupoid; in other words, $\mathbf{X} \circ \mathbf{X} = \mathbf{X}$ only has trivial solutions. This was generalized in G. Grätzer and D. Kelly [227]: $\mathbf{X} \circ \mathbf{Y} = \mathbf{X} \vee \mathbf{Y}$ only has trivial solutions.

Products of varieties have been extensively studied. G. Grätzer and D. Kelly [228], [231]–[234] and E. Fried and G. Grätzer [185], [186]. It was proved that $\mathbf{D} \circ \mathbf{D}$ is a variety (with continuumly many subvarieties) and every variety can be uniquely decomposed into a product of indecomposable varieties.

The deepest result of this field is in T. Harrison [290]:

Theorem 27.   If **V** is any nonmodular variety of lattices, **W** is any nontrivial variety of modular lattices, and $\mathbf{V} \circ \mathbf{W}$ is a variety, then $\mathbf{V} = \mathbf{L}$.   —

7.   Notes on Chapter VI

7.1   Notes on the problems

Problem 1

A negative solution is in M. E. Adams [1]: if $A = C_2 \times C_3$ and $B = C_1$, then the free product of A and B has a minimal generating set not included in $A \cup B$. For a related result, see G. Grätzer and A. P. Huhn [221] and [222].

Problem 2

The answer to the first question is in the affirmative: G. Grätzer and A. P. Huhn [221] establishes the Common Refinement Property for amalgamated free products over finite lattices.

Problem 3

Using the solution of the word problem for the free distributive product of a family of distributive lattices given by G. Grätzer and H. Lakser [1969], it is observed in M. E. Adams and D. Kelly [1977] that Sorkin's Theorem holds in **D**.

Problems 6 and 7

J. Ježek [324] constructs countable lattices A and B such that $A * A \cong B * B$ but $A \not\cong B$, providing a negative answer to Problem 6. He also provides a negative answer to Problem 7.

Problem 16

Solved by S. T. Tschantz [502], see Section 1.3.

Problem 17

R. Freese and J. B. Nation [181] solve this problem by proving that there are no finite intervals of length greater than 4 in the free lattice on three generators. See Section 3 of Appendix G for more detail.

Problem 18

Solved in J. B. Nation [397], see Section 1.3.

Problem 19

This problem is trivial: for any p, $q \in F(\aleph_0)$, there is an endomorphism χ of $F(\aleph_0)$ satisfying $p\chi = q$.

Problem 24

In G. Grätzer and J. Sichler [263], the following result is proved:

Theorem 28.   Every monoid $\mathcal{M}$ can be represented as $\mathrm{End}_{\{0,1\}} L$, for a suitable *complemented* lattice L. Moreover, if M is finite, then L can be chosen as a finite complemented lattice.   ––

   This solves the second part of this problem.

Problem 26

Of the three problems: characterize the $\{0, 1\}$-endomorphism semigroups of finite lattices, of lattices of finite length, and of lattices without infinite chains, M. E. Adams and J. Sichler [14] completely solve the first. They show that every finite monoid is the $\{0, 1\}$-endomorphism monoid of a finite lattice, in fact, of countably many nonisomorphic finite lattices.

   Adams and Sichler make some interesting contributions to the other two questions. For each $n \geq 1$, a monoid is exhibited that is isomorphic to the $\{0, 1\}$-endomorphism monoid of a finite lattice of height $n + 1$, but not of any finite lattice of height $\leq n$. Finally, an infinite monoid is exhibited that is not isomorphic to the $\{0, 1\}$-endomorphism monoid of any lattice in which all chains are finite.

   If $\mathbf{V}$ is a nontrivial variety of $\{0, 1\}$-lattices, then every (finite) monoid is isomorphic to the $\{0, 1\}$-endomorphism monoid of a (finite) lattice from $\mathbf{V}$ exactly when $\mathbf{V}$ contains a (finite) lattice with no prime ideal, see P. Goralčík, V. Koubek and J. Sichler [206].

Problem 29

Although the problem is still open, I would like to mention the relevant result of J. Harding [289]: a uniquely complemented lattice is constructed whose MacNeille completion is not complemented.

Problems 30 and 31

M. E. Adams and J. Sichler [19] found $2^{\aleph_0}$ varieties of lattices in which every lattice is embeddable in a uniquely complemented lattice (in fact, Theorem VI.3.7 holds).

Problem 32

In a series of papers, I showed with D. Kelly that many of the results of lattice theory can be extended to the $\mathfrak{m}$-complete case (see G. Grätzer and D. Kelly [225]–[232] and G. Grätzer, A. Hajnal, and D. Kelly [219]). In particular, there are $\mathfrak{m}$-complete, uniquely complemented lattices. The second part of the problem is still open, namely, whether similar results can be proved in the presence of join- and meet-continuity.

Problem 33

I proved the following structure theorem for finitely presented lattices with A. P. Huhn and H. Lakser [223]:

Theorem 29.   Let L be a finitely presented lattice. Then there is a congruence relation Θ on L such that L/Θ is finite and every congruence class of Θ is embeddable in a free lattice.   —

R. Freese investigated finitely presented lattices in a series of papers, see [161], [162] and R. Freese and J. B. Nation [180], dealing with covering relations, canonical form, and proving that finitely presented lattices are both upper and lower continuous.

Problem 35

An affirmative answer is provided in R. Freese and J. B. Nation [180]: the automorphism group of a finitely presented lattice is finite. See also G. Grätzer and A. P. Huhn [220].

Problem 38

The problem is to describe all finite partial lattices $\mathfrak{A}$ for which $F(\mathfrak{A})$ is finite. For large classes of finite partial lattices this has been done in J. Ježek and V. Slavík [325] and [326], V. Slavík [484].

V. Slavík [487] has shown that it is decidable whether $F(\mathfrak{A})$ is finite.

Problem 45

Solved in M. E. Adams and J. Sichler [13]: for every nontrivial variety **V** of lattices, there exist two Hopfian lattices whose **V**-free product is not Hopfian.

In M. E. Adams and J. Sichler [15], $\mathcal{C}$-reduced free products were used to establish that every finitely generated lattice L can be embedded in $2^{\aleph_0}$ pairwise nonisomorphic finitely generated Hopfian lattices and into $2^{\aleph_0}$ pairwise nonisomorphic finitely generated non-Hopfian lattices.

See R. Wille [1975] for an example of a finitely generated non-Hopfian lattice.

Problem 49

The problem raises the question whether in a lattice L satisfying $|L| > 1$, (SD$_\wedge$), (SD$_\vee$), and (W), there exists a prime ideal. This was proved for finitely generated lattices in W. Poguntke [437], while V. A. Gorbunov and V. I. Tumanov [214] provide an example that the answer is in the negative for infinite lattices.

7.2 $\mathcal{C}$-reduced free products, applications and generalizations

In M. E. Adams and J. Sichler [16], $\mathcal{C}$-reduced free products were used to show that, for any lattice L, the category of bounded lattices that have L as a $\{0,1\}$-sublattice is universal. V. Koubek and J. Sichler [372] went on to show that, for a finitely generated lattice K, the category of bounded lattices that have K as a quotient is universal if and only if K has no prime ideal, and V. Koubek removed the restriction that K be finitely generated, see V. Koubek [370].

In showing that every finite monoid is isomorphic to the $\{0,1\}$-endomorphism monoid of a finite lattice (see Problem 26) in M. E. Adams and J. Sichler [A5], certain lattices were introduced that acted as "testing" lattices for $\mathcal{C}$-reduced free products for varieties other than **L**. In M. E. Adams and J. Sichler [17], it was observed that there exists a locally finite variety in which enough of these testing lattices exist so that the variety itself is $\{0,1\}$-universal (in the categorical sense).

Finite combinations of these "testing" lattices were used in V. Koubek and J. Sichler [371] to obtain infinitely many *finitely generated* varieties **V** representing every finite monoid by *nonconstant* endomorphisms of their finite members.

7.3 $\mathcal{R}$-reduced V-free products

$\mathcal{R}$-reduced free products, introduced in M. E. Adams and J. Sichler [17], generalize $\mathcal{C}$-reduced free products. Informally, in the same way that a $\mathcal{C}$-reduced free product is a $\{0,1\}$-free product, where pairs of elements are chosen from different components and are required to be complementary, an $\mathcal{R}$-reduced free product is a $\{0,1\}$-free product, where finite sets of elements from different components are chosen and are required to join to 1 or meet to 0. Similarly to $\mathcal{C}$-reduced free

products, elements of the $\mathcal{R}$-reduced free product that join to 1 or meet to 0 are only those that are obviously forced to.

For any variety $\mathbf{V}$, Adams and Sichler introduce the concept of $\mathcal{R}$-reduced $\mathbf{V}$-free products and prove that there are $2^{\aleph_0}$ varieties (the, so called, *reduction varieties*) in which the sets of elements joining to 1 or meeting to 0 can be controlled.

$\mathcal{R}$-reduced $\mathbf{V}$-free products were used in [17] to construct, for any $\{0,1\}$-lattice L and any submonoid M of $\mathrm{End}_{\{0,1\}}L$, a $\{0,1\}$-extension K with the property that $\mathrm{End}_{\{0,1\}}K$ is formed by unique extensions of the members of M.

In G. Grätzer and A. P. Huhn [221], it is shown that $\{0,1\}$-free products have the *Common Refinement Property*. In M. E. Adams and J. Sichler [20], it is shown that in the reduction varieties, $\mathcal{R}$-reduced $\mathbf{V}$-free products exist and the Common Refinement Property holds. Since $\mathbf{L}$ is a reduction variety, this generalizes the result of Grätzer and Huhn. It is also shown that there are $2^{\aleph_0}$ varieties in which $\mathcal{R}$-reduced $\mathbf{V}$-free products exist but fail to satisfy the Common Refinement Property.

8.   Lattices and Universal Algebras

In the Concluding Remarks, I raise the questions: "Where to draw the line between lattice theory proper and its allied fields? Does a theorem characterizing a lattice associated with some object outside of lattice theory belong to lattice theory?"

And I comment: "A contentious case in point is the representation theory of algebraic lattices (complete lattices) as subalgebra lattices and congruence lattices of universal algebras (infinitary universal algebras). In my opinion, these results belong to universal algebra and not to lattice theory. However, in view of the close affinity between the workers in lattice theory and in universal algebra, I digress to outline this field."

In the meanwhile, a lot of work has been done on congruence lattices, subalgebra lattices, the lattice of varieties, the lattice of subquasivarieties of a quasivariety, and endomorphism semigroups. I shall very briefly outline the developments in these fields. It should be clear from the discussion that the borderline has become even blurrier than before. I would recommend the excellent introduction to these topics in R. Freese, K. Kearnes, and J. B. Nation [175]; this paper discusses lower bounded lattices, congruence lattices of semilattices, tame congruences, commutators, the lattice of subvarieties of a variety (many of these topics are mentioned briefly in subsequent sections). It is clear from their exposition how much these topics are intertwined.

8.1   Congruence lattices

In my joint work with W. A. Lampe (see Appendix 7 of G. Grätzer [215]), we provide an infinitary version of the congruence lattice characterization theorem.

In the finitary version, the type of the algebra gets very large with the size of the algebraic lattice. See the discussion of Problem II.20.

Congruence lattices of finite algebras are used to study their structure in D. Hobby and R. N. McKenzie [312]; this is called *tame congruence theory*, and it has developed into a field of its own.

If a variety **V** is congruence modular, then *commutators* can be introduced. See R. Freese and R. N. McKenzie [177] for an early review of this field.

The result of Section 5.5 is applied in C. Herrmann and M. Wild [310] to give a polynomial time algorithm for testing whether $\mathrm{Con}\,A$ is modular, for a finite algebra A. This is a surprisingly difficult problem even though the corresponding problem of testing distributivity is straightforward. For an interesting insight, see R. Freese's review of this article in the Mathematical Reviews.

8.2 Subalgebra lattices

For an arbitrary algebra $\mathfrak{A}$, let $\mathrm{Sub}\,\mathfrak{A}$ denote the lattice of subalgebras of $\mathfrak{A}$ (if it is necessary, for instance, for semigroups and lattices, the empty set is regarded as a subalgebra).

There are excellent and up-to-date books on groups and semigroups, see R. Schmidt [471] and L. N. Shevrin and A. J. Ovsyannikov [474].

We illustrate this field with two concepts:

A lattice L is *representable* by the lattice $\mathrm{Sub}\,\mathfrak{A}$, if L is embeddable in $\mathrm{Sub}\,\mathfrak{A}$. A class **K** of algebras is *lattice-universal*, if every lattice is represented by some $\mathrm{Sub}\,\mathfrak{A}$, $\mathfrak{A} \in \mathbf{K}$.

Here is a very recent result (V. B. Repnitskiĭ [453]):

Theorem 30.   Let $k \geq 665$ be an odd integer. Then the group variety defined by $x^k = 1$ is lattice-universal; moreover, every lattice L is representable by the subgroup lattice of the free Burnside group $B(\alpha, k)$, where $\alpha = \max\{\aleph_0, |L|\}$.———

8.3 Subquasivariety lattices

The study of the lattice of subquasivarieties of a quasivariety (Q-lattices, for short) started with V. A. Gorbunov [207] and V. A. Gorbunov and V. I. Tumanov [212]. They discovered an important class of Q-lattices: lattices of the form $\mathrm{S}_\mathrm{p}(A)$, where A is an algebraic lattice and $\mathrm{S}_\mathrm{p}(A)$ stands for the lattice of subsets of A closed under arbitrary meets and chain joins.

In V. A. Gorbunov and V. I. Tumanov [213], the construction $\mathrm{S}_\mathrm{p}(A)$ was generalized: it was shown that any Q-lattice can be represented as $\mathrm{S}_\mathrm{p}(A, e)$, where A is an algebraic lattice with a quasiorder e on A, and $\mathrm{S}_\mathrm{p}(A, e)$ consists of elements X of $\mathrm{S}_\mathrm{p}(A)$ which are *e-hereditary*, that is, $a\,e\,b$ and $a \in X$ imply that $b \in X$. In V. A. Gorbunov [209], it was proved that any Q-lattice is a complete sublattice of a suitable $\mathrm{S}_\mathrm{p}(A)$.

In K. V. Adaricheva and V. A. Gorbunov [26], the concept of an *equaclosure operator* (equational closure operator) on a complete lattice L is introduced as a closure operator satisfying the following four conditions: (1) $h(0) = 0$; (2) if $h(a) = h(b)$, then $h(a) = h(a \wedge b)$; (3) $h(a) \wedge (b \vee c) = (h(a) \wedge b) \vee (h(a) \wedge c)$; (4) every h-closed element can be represented as a meet of cocompact elements of L. This concept is very important for the characterization problem of Q-lattices; see also R. Freese, K. Kearnes, and J. B. Nation [175].

Finite atomistic lattices that can be represented as lattices of quasivarieties were characterized in K. V. Adaricheva, W. Dziobiak, and V. A. Gorbunov [24]. The following result, K. V. Adaricheva, W. Dziobiak, and V. A. Gorbunov [25], may be the deepest result of this field:

Theorem 31.   The following conditions on an algebraic atomistic lattice L are equivalent:

(i) L is a Q-lattice.

(ii) L is isomorphic to the lattice of meet subsemilattices of some algebraic lattice in which any proper ideal satisfies both chain conditions.

(iii) L is a dually algebraic lattice admitting an equaclosure operator.   —

The proof uses a number of earlier results, including K. V. Adaricheva [22]. A forthcoming book by V. A. Gorbunov [211] gives a rather complete theory of Q-lattices.

8.4   Subvariety lattices

The lattice of subvarieties of a variety of algebras satisfies an interesting property, called *Zipper condition*, discovered in W. A. Lampe [375]:

$$y \vee x_0 = \cdots = y \vee x_{n-1} = z \text{ and } x_0 \wedge \cdots \wedge x_{n-1} = 0 \text{ imply that } y = z.$$

Stronger forms were found in M. Erné [144] and W. A. Lampe [376]. See also V. Diercks, M. Erné, and J. Reinhold [134].

The following result was proved in V. A. Gorbunov [210]:

Theorem 32.   The lattice of subvarieties of a locally finite variety of algebras is pro-finite, that is, it can be represented as an inverse limit of a family of finite lattices (satisfying the Zipper condition). In particular, such a lattice is residually finite.   —

8.5   Congruence varieties

Let **V** be a variety of algebras. Let Con **V** denote the lattice variety generated by the lattices Con $\mathfrak{A}$, $\mathfrak{A} \in \mathbf{V}$; Con **V** is called a *congruence variety*.

For an early review of this field, see B. Jónsson's Appendix 3 in G. Grätzer [215] and B. Jónsson [335]. R. Freese [165] provides an up-to-date review. For more detail see Section 5 of Appendix G.

8.6 Endomorphism semigroups

There is an excellent and current survey paper (with more than a hundred references) covering this field: M. E. Adams, S. Bulman-Fleming, and M. I. Gould [10].

Distributive Lattices and Duality

B. A. Davey and H. A. Priestley

1. Introduction

Any distributive lattice is isomorphic to a lattice of sets. This famous result, due to Garrett Birkhoff, gives a flavor to the theory which is quite different from that of modular lattices or of general lattices. Distributive lattices are closer in character to Boolean algebras than to general lattices. For historical reasons, the theory of Boolean algebras got a head start, with distributive lattices regarded in the early days of lattice theory as 'Boolean algebras' weak sisters', to quote G.-C. Rota [459]. The existence of this appendix is proof that such a view can not be sustained today. As Rota foresaw in 1973, and as we affirm, with less than total objectivity, it is duality theory that has in the past 25 years driven forward the theory of distributive lattices and of distributive-lattice-ordered algebras (DLOAs). The core lattice theoretic properties of distributive lattices were already in place when the first edition of *General Lattice Theory* was published. Developments since then have principally involved DLOAs of various kinds, and these figure prominently in this appendix.

DLOAs occur very widely, especially in nonclassical logic. They provide a rich source of congruence-distributive varieties, exhibiting varied algebraic behaviors. Among these, the variety of Boolean algebras is atypically special—it is primal and arithmetical, in particular. Accordingly, the study of varieties such as de Morgan algebras and Stone algebras has proved very worthwhile. They have attracted interest in their own right (for the description of free algebras and injective lattices, for example). They have also pointed the way to results of much greater generality, thereby vindicating their investigators, some of whom might otherwise be accused of mathematical narrow-mindedness.

Against this background, our aims in this survey are

- to outline Priestley duality for distributive lattices and its application to classes of distributive lattices with additional operations;

- to indicate very briefly the role of natural duality theory;

- to illustrate a symbiotic relationship between the particular and the general, especially as regards properties and problems coming from universal algebra.

We note that a full bibliography of Priestley duality up to 1995 (but excluding unpublished theses and preprints) was compiled by M. E. Adams and W. Dzobiak and appears in the Special Issue of *Studia Logica*, Vol. 56, devoted to duality. We also draw attention to the earlier survey articles B. A. Davey and D. Duffus [100], H. A. Priestley [441] (on distributive lattices, ordered sets, and Priestley duality) and B. A. Davey [98], H. A. Priestley [445] (on natural dualities, of which a more detailed account is given in D. M. Clark and B. A. Davey [79]).

Since 1978, the circle of consumers of lattice theory has enlarged considerably. Most notably, lattice theory has played a part in the rapid growth of theoretical computer science. In particular, computational models and distributive lattices and their dual spaces come together in the theory of continuous lattices and their generalizations on which domain theory rests. These connections cannot be satisfactorily presented within the confines of this appendix, and below we mention them only in passing.

2.   Basic Duality

Lattice theory may be regarded as a branch of algebra, and has attracted many researchers who are by inclination algebraists. Notwithstanding, a highly significant advance in lattice theory was made by M. H. Stone, who approached from a functional analytic viewpoint. He showed that the definitive representation theorem for Boolean algebras involves topology. Likewise, for distributive lattices, topology encroaches on algebra: Stone's representation for distributive lattices in terms of T_0-spaces is a special case of that for distributive semilattices presented in Chapter II. This representation, though definitive from a theoretical point of view, has practical disadvantages. To reconcile it with Birkhoff's Representation Theorem for finite distributive lattices, one needs the correspondence between finite T_0-spaces and finite posets. Priestley duality retains the partial order relation of Birkhoff's representation, and adds a compact Hausdorff topology which is discrete, and so essentially disappears, in the finite case. A partially ordered topological space $(X; \leqslant, \mathcal{T})$ is a *Priestley space* if $\mathcal{T}$ is compact and, given $x \nleqslant y$ in X, there exists a clopen upset (an upset is also called an increasing set or an order filter) U such that $x \in U$ and $y \notin U$ (*total order-disconnectedness*). The latter condition implies that the family $\mathcal{V}(X)$ of clopen upsets of X determines $\leqslant$:

$$x \leqslant y \text{ if and only if } (\forall U \in \mathcal{V}(X)) \, x \in U \implies y \in U.$$

The category $\mathbf{P}$ has as its objects Priestley spaces and as its morphisms the continuous order-preserving maps.

Unless otherwise stated, we deal with distributive lattices with 0 and 1, and homomorphisms are taken to preserve these bounds. We denote the resulting category by $\mathbf{D}$. Let $L \in \mathbf{D}$. The dual space X_L of L is defined as follows. We take the underlying set to be the set of prime filters (prime dual ideals) of L, ordered by set inclusion. We put a topology $\mathcal{T}$ on X_L by taking as a subbasis the sets

$$X_a = \{ x \in X_L \mid a \in x \} \quad (a \in L),$$

and their complements. (When L is a Boolean lattice, $X - X_a = X_{a'}$.) Each set X_a is an upset for the order $\subseteq$, while each set $X - X_a$ is a downset. Further, by definition, each subbasic set is $\mathcal{T}$-clopen. The fundamental properties of the dual space of L are given in the following theorem.

Theorem 1.   Let $L \in \mathbf{D}$ and define $\langle X_L; \subseteq, \mathcal{T} \rangle$ as above.

(i) $\langle X_L; \subseteq, \mathcal{T} \rangle$ is totally order-disconnected and $\mathcal{T}$ is compact.

(ii) The clopen upsets of $\langle X_L; \subseteq, \mathcal{T} \rangle$ are precisely the sets of the form X_a $(a \in L)$, and the clopen subsets are precisely the finite unions of sets of the form $X_a \cap (X_L - X_b)$ $(a, b \in L)$.   —

For the proofs, see B. A. Davey and H. A. Priestley [106]. Some comments on the provenance of these results are in order here. The statement $x \nleqslant y$ in X_L means that there exists $a \in L$ such that $a \in x$ and $a \notin y$, so that $x \in X_a$ and $y \notin X_a$. Thus total order-disconnectedness is immediate. The compactness is, as in the case of a Boolean lattice, a version of the Compactness Theorem for a classical propositional calculus. It is harder to prove in the distributive lattice case because a subbasis rather than a basis for the topology is involved. This means that Alexander's Subbasis Lemma is needed. Part (ii) depends on (and implies)

(DPI) $a \nleqslant b$ in $L \implies (\exists x \in X_L) \, a \in x, \, b \notin x.$

Thus the Prime Ideal Theorem for $\mathbf{D}$ is required.

Theorem 2 (Representation Theorem).   Let $L \in \mathbf{D}$. Then

$$e_L : a \mapsto X_a \quad (a \in L)$$

is an isomorphism of L onto the lattice $\mathcal{V}(X_L)$ of clopen upsets of X_L.   —

The power of the duality is increased by the availability of its dual analogue, asserting that every Priestley space arises as a dual space. In the finite case, this is (equivalent to) the easy result that $x \mapsto {\downarrow}x$ is an order-isomorphism from a poset to the join-irreducible elements of its lattice of downsets. In the general case, it is harder to prove; see B. A. Davey and H. A. Priestley [106], Theorem 10.19, and Chapter II, Section 5 of this book.

Theorem 3.   Let $X \in \mathbf{P}$ and let L be its lattice of clopen upsets. Then X_L is homeomorphic and order-isomorphic to X.   —

To get full value from the Representation Theorem, we must also represent homomorphisms between members of $\mathbf{D}$, so as to set up a duality, or more precisely a categorical dual equivalence, between $\mathbf{D}$ and $\mathbf{P}$. To this end, we now let $D \colon \mathbf{D} \to \mathbf{P}$ and $E \colon \mathbf{P} \to \mathbf{D}$ be the maps for which $D(L) = X_L$ and $E(X) = \mathcal{V}(X)$, the lattice of clopen upsets of X.

Theorem 4.   Let $L, K \in \mathbf{D}$ and $X, Y \in \mathbf{P}$. Then to each map $f \in \mathbf{D}(L, K)$ is associated a map $D(f) \in \mathbf{P}(D(K), D(L))$ and to each map $\varphi \in \mathbf{P}(Y, X)$ is associated a map $E(\varphi) \in \mathbf{D}(E(X), E(Y))$. The maps $D \colon \mathbf{D}(L, K) \to \mathbf{P}(D(K), D(L))$ and $E \colon \mathbf{P}(Y, X) \to \mathbf{D}(E(X), E(Y))$ are bijections and

$$a \in (D(f))(y) \iff f(a) \in y, \quad \text{for all } a \in L, \, y \in Y.$$

Further,

(i) f is one-to-one if and only if $D(f)$ is onto;

(ii) f is onto if and only if $D(f)$ is an order-embedding.   —

Henceforth, we shall, where expedient, identify L with the lattice, $ED(L)$, of clopen upsets of its dual space X_L. When L is finite, there is an order-anti-isomorphism from $J(L)$, the join-irreducible elements of L with the induced order, onto X_L. This is given by $a \mapsto {\uparrow}a = \{\, c \in L \mid c \geqslant a \,\}$. Also, because the topology is discrete, the clopen upsets are all the upsets. We thus see that the Representation Theorem reduces in the finite case to Birkhoff's Theorem in its familiar form: every finite distributive lattice is isomorphic to the lattice of downsets (order ideals) of its join-irreducibles; see Chapter II, Section 1.

One particular good feature of the duality for $\mathbf{D}$ is that it is logarithmic, in the sense that the functor D converts products to sums. For a finite set of lattices $L_1, \ldots, L_m$ in $\mathbf{D}$ this means $D(L_1 \times \cdots \times L_n)$ is just the disjoint union, topologically and order-theoretically, of $D(L_1), \ldots, D(L_m)$. Of course, this implies that the dual of the Boolean lattice $\mathbf{2}^m$ is just the m-element antichain. In this special case, the functor D acts logarithmically as regards cardinality. At the other extreme, the dual $D(\mathbf{m})$ of the m-element chain $\mathbf{m}$ is the $(m-1)$-element chain. However, since 'most' n-element posets are of height at most 3 and width about

$n/2$, see D. J. Kleitman and B. R. Rothschild [365], the former situation is the more typical. In moving to the dual setting, we can therefore expect a worthwhile reduction in cardinality, and a resultant increase in the size of examples on which calculations can be performed by computer. The idea of a logarithm, in a structural sense, has also been exploited in the study of the arithmetic of ordered sets; see J. D. Farley [152]. Informally, the aim is to convert a problem about exponents into a more tractable one about products.

Turning briefly to infinite products, we note that the dual space of the product of an arbitrary nonempty family $\{L_i\}_{i \in I}$ in $\mathbf{D}$ can be viewed as an ordered Stone–Čech compactification of the disjoint union of $\{D(L_i)\}_{i \in I}$; see J. D. Farley [150] and G. Hansoul [288] for discussions of such compactifications. Duals of Boolean products, in the setting of Stone duality, are investigated in M. Gehrke [196]. We consider coproducts in Section 7.

The dual space $\mathcal{S}(L)$ employed in Stone duality, as described in Section II.5, consists of the set of prime ideals $\mathcal{P}(L)$, equipped with a topology, $\mathcal{T}^{\uparrow}$, having open sets $r(I) = \{\, P \in \mathcal{P}(L) \mid P \not\supseteq I \,\}$. Set-theoretic complementation gives an order-isomorphism between $\mathcal{P}(L)$ ordered by reverse inclusion and X_L. Modulo this isomorphism, the topology $\mathcal{T}^{\uparrow}$ consists just of the $\mathcal{T}$-open upsets of X_L; in the other direction, $\mathcal{T}$ is the patch topology, or co-compact topology, derived from $\mathcal{T}^{\uparrow}$, and $\leqslant$ on X_L is the specialization order coming from $\mathcal{T}^{\uparrow}$. For further details on these relationships, see H. A. Priestley [442] and [443]. The advantage of the formulation of the representation given in Theorem 2 over that given by Stone is that, up to reversal of the order, it naturally extends Birkhoff's representation. This is significant in two respects. First, thanks to the order, the representation remains pictorial. Second, the topology in a sense plays second fiddle to the order. The underlying reason for this is that each object in $\mathbf{P}$ is profinite (which is easily seen dually from the fact that each $L \in \mathbf{D}$ is the directed union of its finite sublattices). This is at the root of an operational principle for Priestley spaces: take care of the order and the topology will take care of itself—provided the necessary topological restrictions are imposed to make things make sense.

Let us illustrate by an example what we mean. Take a finite distributive lattice L, concretely represented as the lattice of upsets of a finite poset X. Then L is pseudocomplemented, with $a^* = \max\{c \in L \mid c \wedge a = 0\}$ given by $X - {\downarrow}a$ (which is the largest upset disjoint from a). Now let $L \in \mathbf{D}$ be arbitrary and take $a \in L$. The upset $X - {\downarrow}a$ need no longer be clopen (although it is open). It is not hard to prove (see B. A. Davey and H. A. Priestley [106], 10.23) that a^* exists in L if and only if ${\downarrow}a$ is open. That is, the pseudocomplementedness property holds precisely when a natural topological condition is superimposed on the purely order-theoretic behavior of the finite case. It is trivial to see that the dual space of the lattice L^d is obtained by reversing the order on X_L. Therefore, $L \in \mathbf{D}$ is a double p-algebra if and only if for each clopen upset a and clopen downset b in X_L, both ${\downarrow}a$ and ${\uparrow}b$ are $\mathcal{T}$-(cl)open. In a similar way the Heyting implication $a \to b$ exists in L if and only if ${\downarrow}(a - b)$ is open. Using Theorem 1(ii), it is an easy exercise to show that L is a Heyting algebra if and only if $\downarrow$ maps

$\mathcal{T}$ into $\mathcal{T}$. In each case—p-algebra, double p-algebra and Heyting algebra—the appropriate morphisms have neat dual descriptions. The resulting dualities have proved especially valuable in the analysis of endomorphism monoids and of universality with which the names of M. E. Adams, V. Koubek, and J. Sichler are principally associated.

The Stone duality presented in Chapter II is less restrictive than that between $\mathbf{D}$ and $\mathbf{P}$ in not demanding that the lattices have 0 and 1. For reference, we record the extension of Priestley duality to unbounded lattices. In this form it was worked out by N. G. Martinez in connection with his study of MV-algebras and ℓ-groups (see Section 3). For a self-contained account, see D. M. Clark and B. A. Davey [79], Section 1.2. The subcategory $\mathbf{P}^b$ of $\mathbf{P}$ of *bounded Priestley spaces* has objects $\langle X; \leqslant, \mathcal{T} \rangle$ such that there are points $p_0, p_1 \in X$ satisfying $p_0 \leqslant x \leqslant p_1$, for all $x \in X$; morphisms are $\mathbf{P}$-morphisms that preserve both bounds. Given $\langle L; \wedge, \vee \rangle$ belonging to the category $\mathbf{D}'$ of all distributive lattices, define X_L to be the set of prime filters of L augmented by $\varnothing$ and L, and ordered by inclusion. This becomes a bounded Priestley space when given the topology with basis

$$\{X_a\}_{a \in L} \cup \{X_L - X_a\}_{a \in L} \cup \{\varnothing, L\},$$

where $X_a = \{x \in X_L \mid a \in L\}$. Then L is isomorphic to the lattice of proper nonempty $\mathcal{T}$-clopen upsets of X_L. A dual equivalence between $\mathbf{D}'$ and $\mathbf{P}^b$ can then be established. The lattice $\langle L; \wedge, \vee \rangle$ is the $\mathbf{D}'$-reduct of a member of $\mathbf{D}$ if and only if the bounds in its dual space are isolated points for the topology $\mathcal{T}$. For more details see N. G. Martinez [390].

The topology $\mathcal{T}$ of a dual space is, of course, itself a lattice of sets. Likewise, the topologies $\mathcal{T}^{\uparrow}$ of open upsets and $\mathcal{T}^{\downarrow}$ of open downsets are lattices.

Theorem 5.   Let $L \in \mathbf{D}$ and let $\langle X; \leqslant, \mathcal{T} \rangle$ be its dual space. Then

(i) $\operatorname{Con} L \cong \mathcal{T}$;

(ii) $\operatorname{Id} L \cong \mathcal{T}^{\uparrow}$;

(iii) $\operatorname{Filt} L \cong \mathcal{T}^{\downarrow}$.   —

The correspondence in (i) derives from that between closed subsets Y of X and congruences $\Theta(Y)$ on $\mathcal{V}(X)$ by

$$\langle U, V \rangle \in \Theta(Y) \iff U \cap Y = V \cap Y,$$

which can be shown to be an order-anti-isomorphism, with the aid of Theorem 4(ii) ($\Theta(Y)$ is the kernel of the $\{0, 1\}$-homomorphism f from $\mathcal{V}(X)$ onto $\mathcal{V}(Y)$ given by $f(U) = U \cap Y$) combined with a topological separation argument. More detail can be found in B. A. Davey and H. A. Priestley [106].

The Representation Theorem captures the finitary operations of $\langle L; \vee, \wedge \rangle$ as $\cup$ and $\cap$. Arbitrary joins and meets, when these exist, can also be described: $\bigvee_{i \in I} a_i = \mathrm{cl}_{\mathcal{T}^{\downarrow}} \bigcup_{i \in I} a_i$, provided this closure is $\mathcal{T}^{\uparrow}$-open, and similarly for meets. For L to be complete, we require that the dual space is *extremally order-disconnected*, in the sense that $\mathrm{cl}_{\mathcal{T}^{\downarrow}}$ maps $\mathcal{T}^{\uparrow}$ into $\mathcal{T}^{\uparrow}$. Here we have another instance of the way that the condition dual to a restriction on the algebraic side is exactly the order-topological condition which is necessary for the formal translation of the required property to make sense in the dual space.

It is tempting to think that it may be profitable to investigate the dual spaces of distributive lattices satisfying infinite distributive identities, of distributive algebraic lattices, and so forth. This can indeed be done, though the resulting representations are, not surprisingly, not very easy to work with. A duality for frames is presented by A. Pultr and J. Sichler in [451], and a duality for ideal lattices is given by J. D. Farley in [151].

3. Distributive Lattices with Additional Operations

Much of the importance of distributive lattices comes from their occurrence as reducts of algebras arising in algebra, logic, and elsewhere. The best known of these are of course the Boolean algebras, which model classical propositional calculus. Nonclassical propositional calculi were extensively studied from an algebraic point of view by the Argentinian, Polish, and Romanian schools from the 1930s. The associated varieties of algebras—de Morgan algebras, Kleene algebras, Ockham algebras, Post algebras, Łukasiewicz algebras, and so on, have been exhaustively investigated. The book R. Balbes and P. Dwinger [1974] gives a crisp account of the algebraic fundamentals. It pre-dates the systematic use of duality methods in the study of such algebras, which has proved very fruitful. Assume we have a class $\mathcal{A}$ (usually a variety) of algebras $\mathbf{A} = \langle A; \wedge, \vee, 0, 1, F \rangle$ of the same similarity type, where F is a family of finitary operations. Assume that $\langle A; \wedge, \vee, 0, 1 \rangle \in \mathbf{D}$ and that this lattice is identified with the clopen upsets of its dual space $\langle X_A; \leqslant, \mathcal{T} \rangle$. Then we may ask the following questions.

(1) How are the operations F captured in X_A? More explicitly, can we obtain a structure $\langle X_A; \leqslant, R_F, \mathcal{T} \rangle$, where R_F is some set of relations, for example, in terms of which the operations F can be described?

(2) Can we describe $\mathcal{A}(\mathbf{A}, \mathbf{B})$ in terms of $\mathbf{P}(X_B, X_A)$ and the set of relations (or whatever) R_F?

(3) Can we give a useful characterization of the class $\mathbf{P}_{\mathcal{A}}$ of enriched Priestley spaces of type $\langle X_A; \leqslant, R_F, \mathcal{T} \rangle$?

More formally, in (1) and (2) we are seeking a dual equivalence between $\mathcal{A}$, *qua* category, and a subcategory $\mathbf{P}_{\mathcal{A}}$ of the category of enriched Priestley spaces; (3) seeks to describe the dual category $\mathbf{P}_{\mathcal{A}}$.

We illustrate, briefly. Consider the case where F consists of a single unary operation, $\neg$, so that we have algebras of type $\langle 2, 2, 1, 0, 0 \rangle$. Assume that $\neg$ satisfies de Morgan's identities and interchanges 0 and 1 (in other words, $\neg$ is a dual endomorphism of the **D**-reduct). This gives the variety **O** of Ockham algebras. Within this we focus on two very well-known subvarieties:

- the variety **M** of de Morgan algebras, characterized within **O** by the identity $\neg^2 a = a$, and

- the variety **K** of Kleene algebras, characterized within **M** by the inequality $a \wedge \neg a \leqslant b \vee \neg b$.

Note also that the variety **S** of Stone algebras is a subvariety of **O**, with * serving as $\neg$. Dualities for **M** and **K** were discovered by W. H. Cornish and P. R. Fowler ([84] and [85]) (and, in equivalent forms, by the Argentinian school).

The class $\mathbf{P_O}$ of *Ockham spaces* consists of structures $\langle X; \leqslant, g, \mathcal{T} \rangle$, where $\langle X; \leqslant, \mathcal{T} \rangle \in \mathbf{P}$ and $g \colon X \to X$ is a continuous order-reversing map. The morphisms of $\mathbf{P_O}$ are the **P**-morphisms commuting with the g-map. The negation on $\mathbf{A} \in \mathbf{O}$ and the g-map on X_A are linked by

$$(\forall a \in A)(\forall x \in X_A)\, \neg a = X_a - g^{-1}(a).$$

This duality was obtained by A. Urquhart [505]. Observe that, as with the pseudocomplement a^* discussed above, the conditions on g are exactly those which ensure that this makes sense. The dual category $\mathbf{P_M}$ is the full subcategory of $\mathbf{P_O}$ of spaces $\langle X; \leqslant, g, \mathcal{T} \rangle$ in which $g^2 = \mathrm{id}$. Within $\mathbf{P_M}$ sits $\mathbf{P_K}$: spaces in which the condition that x are $g(x)$ are comparable with respect to $\leqslant$ is satisfied. It is instructive to see how this comes from the contrapositive of total order-disconnectedness. We have, in an Ockham algebra A with dual $Y = X_A$,

$(\forall a)(\forall b)\, a \wedge \neg a \leqslant b \vee \neg b$
$\iff (\forall a)(\forall b)(\forall x \in Y)\, x \in (a \wedge \neg a) \implies x \in (b \vee \neg b)$
$\iff (\forall x \in Y)(\forall a)(\forall b)\, ((x \in a) \wedge (x \in \neg a)) \implies ((x \in b) \vee (x \in \neg b))$
$\iff (\forall x \in Y)(\forall a)(\forall b)\, ((x \in a) \wedge (g(x) \notin a)) \implies ((x \in b) \vee (g(x) \notin b))$
$\iff (\forall x \in Y)(\forall a)(\forall b)\, (((x \in a \implies g(x) \in a) \vee ((g(x) \in b) \implies (x \in b)))$
$\iff (\forall x \in Y)((x \leqslant g(x)) \vee (g(x) \leqslant x)).$

Calculations such as this point the way to a general process for dualizing algebraic identities in Ockham algebras (and in other varieties of DLOAs too).

The **O**-congruences of $\mathbf{A} \in \mathbf{O}$ are, by restriction of the correspondence in Theorem 5, just those open subsets of X_A whose complements are g-closed. This leads very easily to the characterization of an algebra $\mathbf{A} \in \mathbf{O}$ as subdirectly irreducible if and only if X_A is the closure of the g-orbit of a single point. As usual, the topology disappears if A is finite. For **K** the dual spaces of the subdirectly algebras are those with at most two points (total order-disconnectedness again in play here!). Translated back into algebraic terms, this implies that only the 2- and 3-element Kleene chains are subdirectly irreducible. The latter is $\mathbf{3} = (\{0, a, 1\}; \wedge, \vee, \neg, 0, 1)$, with $0 < a < 1$ and $\neg a = a$. We deduce that $\mathbf{K} = \mathbf{ISP}(\mathbf{3})$. We may think of $\mathbf{K}$ as the variety associated with a 3-valued logic with truth values 'true', 'false' and 'don't know'. The variety $\mathbf{B}$ of Boolean algebras sits inside $\mathbf{K}$ as $\mathbf{ISP}(\mathbf{2})$. Further, the nontrivial subvarieties of $\mathbf{M}$ are just $\mathbf{K}$ and $\mathbf{B}$. We make no attempt here to survey the very large literature of the past 25 years relating to Ockham algebras and especially to varieties low down in its subvariety lattice. An account of some of this theory appears in the book [53] of T. S. Blyth and J. C. Varlet, who, with their collaborators, have published a long series of papers in this area. Here we merely draw attention to the fact that the well-studied special varieties have led the way to results which are not confined to these varieties. In particular, the calculations above whereby the Kleene identity is converted into its dual equivalent can be carried out, algorithmically, for any finitely generated Ockham variety. W. H. Cornish realized that it was possible to develop a general theory for classes $\mathcal{A}$ of algebras $\langle A; \wedge, \vee, 0, 1, F \rangle$ where each operation in F is either an endomorphism or a dual endomorphism of $\langle A; \wedge, \vee, 0, 1 \rangle$ and the operations arise by a monoid action on A. We shall refer to such algebras as *Cornish algebras*. For example, the variety $\mathbf{O}$ is associated with the free monoid on one generator, $1, \alpha, \ldots, \alpha^k, \ldots$, with monoid action $(a, \alpha^k) \mapsto \neg^k a$. Cornish worked out his ideas in [82] (unpublished), and went on to set this theory in a general categorical framework, covering other structures besides distributive-lattice-ordered algebras. This resulted in the monograph W. H. Cornish [83]. In this, the abstract setting gives way in the later chapters to a study of the duality of Cornish algebras, and its application to such varietal questions as local finiteness, amalgamation, and so on. This analysis subsumed and unifies many earlier results for particular varieties. The duality aspects are pursued further, with fewer restrictions, in H. A. Priestley [446].

Once we leave the cosy environment of operations which are endomorphisms or dual endomorphisms we encounter more challenges. Binary operations, of course, occur widely: implication (Heyting implication or an MV-algebra implication, for example), or the group operation of a lattice-ordered group. The duality theory for Heyting algebras mentioned above has proved a useful adjunct to other techniques but has not made hard problems easy. For illustrations of work in this area, we draw attention to L. Vrancken-Mawet [508], L. Vrancken-Mawet and G. Hansoul [509] and to the thesis of J. P. G. Pretorius [439] (which, in part, builds on the study of free Heyting algebras presented by A. Urquhart [504]) and F. Bellisima [43]). Likewise, the duality for frames presented by A. Pultr and J. Sichler in [451] is only one facet of the far-reaching theory of frames and locales, to which P. T. Johnstone [329] provides a good introduction. A duality theory for lattice-ordered groups and for MV-algebras (*alias* Wajsberg algebras) has been developed in a series of papers by N. G. Martınez. In summary, Martınez has obtained a duality for a class of distributive lattices with a binary operation of implication, which were called in [389] implicative lattices; see [389], [390]. This setting allows a unified treatment of ℓ-groups and MV-algebras, as well as subsuming Ockham algebras and linear Heyting algebras. Moreover, in N. G. Martınez and H. A. Priestley [392] a mechanical procedure is presented whereby the objects of the dual category $\mathbf{P}_{\mathcal{A}}$ can be written down provided the equations of the variety $\mathcal{A}$ can be expressed in a suitable 'normal implicative' form. Martınez's duality has been applied to study the problem of uniqueness of the structure for MV-algebras and related structures in [388] and in N. G. Martınez and H. A. Priestley [391]. More recently, he has shown that for implicative structures a representation can be based on a well-behaved subspace of the Priestley dual. This subspace, which is generated from the subset of minimal prime lattice filters, is introduced in N. G. Martınez and H. A. Priestley [392] under the name of *reduced spectrum*. This is evidence for the unsurprising fact that for algebras which are distributive lattices with a rich additional structure we should generally not regard the extra operations merely as an overlay to the lattice structure. However, even within the framework of enriched Priestley spaces much work remains to be done concerning DLOAs with operations of arity > 1.

4. Distributive Lattices with $\vee$-preserving Operators, and Beyond

In a famous 1951 paper, B. Jónsson and A. Tarski [342] and [343] proved the Canonical Extension Theorem for Boolean algebras with operators and applied it to many of the specific varieties of such algebras occurring in algebraic logic. Here operators are operations that preserve disjunction in each coordinate. The theorem shows that any Boolean algebra with operators can be embedded in a power set algebra A^* in a canonical way, so that the operators become complete, and so that all positive identities (that is, ones not involving the Boolean negation) satisfied in the original algebra are also satisfied in the extension. The fact that only positive identities are preserved encouraged M. Gehrke and B. Jónsson to search for corresponding distributive lattice results. Their paper [198] gave the basic construction and a preservation theorem for identities involving operators, but did not encompass operators that switched $\wedge$ and $\vee$ in one or more coordinates. Very recently, they have obtained significant further results, applicable *inter alia* to distributive p-algebras, Ockham algebras, de Morgan algebras, and Stone algebras.

   One may regard the Jónsson–Tarski constructions as an algebraic version of the Stone duality: A^* is the algebra of *all* subsets of the dual space $D(A)$ of A, as compared with A, the algebra of *clopen* subsets of $D(A)$, with the operations built by 'power operations' (see C. Brink and I. M. Rewitzky [61]). Likewise, for the Jónsson–Gehrke extension: a lattice $L \in \mathbf{D}$ extends to a complete ring of sets, L^*, by 'dropping the topology' from its dual space: L^* is the lattice of all order-preserving maps from $D(L)$ to $\underset{\sim}{\mathbf{2}}$.

The extension theorems above 'jump over' the dual space. It is natural to ask how the operators are represented dually. This was investigated in the Boolean case by P. R. Halmos and B. F. Wright and for distributive lattices by R. Cignoli, S. Lafalce, and A. Petrovich (see [75]). Cignoli *et al.* considered Priestley relations on a Priestley space X, *viz.* binary relations R on X such that $(\forall x)\, R(x)$ a closed down-set and $U \in \mathcal{V} \implies R^{-1}(U) \in \mathcal{V}$. He established a bijective correspondence between $\vee$- and 0-preserving operators on $L \in \mathbf{D}$ and a Priestley relation R_f on X_L, as follows:

$$f \mapsto R_f = \{\, \langle y, x \rangle \in X \times X \mid x \subseteq f^{-1}(y) \,\},$$
$$R \mapsto f_R, \text{ where } f_R \colon U \mapsto R^{-1}(U).$$

This correspondence (in an order-dual form) has been exploited by C. Brink and I. M. Rewitzky in their work on program semantics, specifically in the modeling of predicate transformers (see C. Brink and I. M. Rewitzky [61] and [62]). In part motivated by the problem of modeling predicate transformers which are merely monotone, J. D. Farley has investigated how monotone maps $f \colon L \to M$ can be dualized. Now it is necessary to go up another level, from relations to multirelations, namely to subsets of $X_L \times \mathcal{V}(X_N)$. Accounts are presented by J. D. Farley in [151] and in the Ph.D. thesis of A. Petrovich [422].

In a similar way, R. Cignoli has considered quantifiers on distributive lattices. Here the dual concept for an (existential) quantifier on L is an equivalence relation on X_L, suitably compatible with the structure of X_L (see R. Cignoli [73] and [74]).) Further work in the same spirit has been carried out by A. Petrovich [421] and [422].

5.   The Natural Perspective

As a Ph.D. student of George Grätzer in the early 1970s, the first author of this appendix explored categorical dualities for prevarieties of algebras (see B. A. Davey [94]). A key example driving this investigation was the second author's duality for distributive lattices, along with the Stone duality for Boolean algebras and Pontryagin duality for abelian groups. The theory of what are now called natural dualities has since developed rapidly; B. A. Davey [97], B. A. Davey and H. Werner [113], and the surveys [98] and H. A. Priestley [445] provide stepping stones to the current state of the art as presented in D. M. Clark and B. A. Davey [79]. It would not be appropriate here to summarize natural duality theory, even in restricted generality, but we should indicate its impact on the theory of distributive lattices and DLOAs, and *vice versa*.

The variety $\mathbf{D}$ of bounded distributive lattices has the two element chain $\underline{\mathbf{2}} = \{0, 1\}$ as its only subdirectly irreducible algebra, and so by Birkhoff's Subdirect

Product Theorem, we have

$$\mathbf{D} = \mathbf{ISP}(\underline{\mathbf{2}}),$$

where the underlining is to serve as a reminder that we are dealing with an algebraic structure. Now let $\underset{\sim}{\mathbf{2}}$ be the 2-element chain, *qua* poset, with the discrete topology. Then, for any set S, the space $\underset{\sim}{\mathbf{2}}^S$, with product topology and pointwise order, belongs to $\mathbf{P}$. In fact Theorem 3 says that the objects in $\mathbf{P}$ are precisely those ordered spaces which can be identified with some closed subspace of a space $\underset{\sim}{\mathbf{2}}^S$ (with induced topology and order). We may therefore write, in shorthand,

$$\mathbf{P} = \mathbf{IS_cP}(\underset{\sim}{\mathbf{2}}),$$

the class of isomorphic copies of closed substructures of powers of $\underset{\sim}{\mathbf{2}}$.

We set up functors as follows:

(On objects:) $\qquad\qquad\qquad\qquad\qquad D: A \mapsto \mathbf{D}(A, \underline{\mathbf{2}}) \leq \underset{\sim}{\mathbf{2}}^A,$

$\qquad\qquad\qquad\qquad\qquad\qquad\quad E: X \mapsto \mathbf{P}(X, \underset{\sim}{\mathbf{2}}) \leq \underline{\mathbf{2}}^X.$

(On morphisms:) $\qquad\qquad\qquad\quad D: f \mapsto - \circ f,$

$\qquad\qquad\qquad\qquad\qquad\qquad\quad E: \varphi \mapsto - \circ \varphi.$

(Here $\leq$ means 'is a substructure of'.) These functors are well defined—because $\leqslant$ is a $\mathbf{D}$-subalgebra of $\mathbf{2}^2$! Further, D and E define an adjunction between $\mathbf{D}$ and $\mathbf{P}$ with unit and co-unit the evaluation maps e_A and ε_X ($A \in \mathbf{D}$ and $X \in \mathbf{P}$). Further,

$$(\forall A \in \mathbf{D})\ e_A: A \rightarrow ED(A) \text{ is an isomorphism,}$$
$$(\forall X \in \mathbf{P})\ \varepsilon_X: X \rightarrow DE(X) \text{ is an isomorphism,}$$

and we have a categorical equivalence between $\mathbf{D}$ and $\mathbf{P}$. The first of the above isomorphism assertions is the statement that we have a *duality*, the second that this duality is *full*. We also note that $\underset{\sim}{\mathbf{2}} = D(F_{\mathbf{D}}(1))$ and, because products in $\mathbf{P}$ are concrete, that $\underset{\sim}{\mathbf{2}}^\kappa = D(F_{\mathbf{D}}(\kappa))$.

Many varieties of DLOAs arise as $\mathbf{ISP}(\underline{M})$, where $\underline{M}$ is some 'truth value' algebra paralleling the algebra $\underline{\mathbf{2}}$ of classical propositional calculus without negation. For example, for the variety of Stone algebras we may take $\underline{M}$ to be the algebra whose $\mathbf{D}$-reduct is the 3-element chain $0 < a < 1$, with additional operation $*$ satisfying the identities $b \wedge b^* = 0$ and $b^* \vee b^{**} = 1$, and given explicitly by $0^* = 1$ and $a^* = 1^* = 0$.

Let $\underline{M}$ be a finite algebra and let R be a set of algebraic relations on M, an n-ary relation being called *algebraic*, if it is a subalgebra of some $\underline{M}^n$. Let τ denote the discrete topology. We may seek a topological relational structure $\underset{\sim}{\mathbf{M}} = \langle M; R, \tau \rangle$ on the underlying set M of $\underline{M}$ so that a dual equivalence is set up by hom-functors between $\mathcal{A}$ and the topological quasivariety $\mathcal{X}$ generated by $\underset{\sim}{\mathbf{M}}$, *viz.* the class of all those structures of the same type as $\underset{\sim}{\mathbf{M}}$ that are embeddable as closed substructures in powers of $\underset{\sim}{\mathbf{M}}$. The assumptions made above ensure that the category theory goes through smoothly. In particular, we always have well-defined hom-functors D and E given on objects by

$$D\colon A \mapsto \mathcal{A}(A, \underline{M}) \leq \underset{\sim}{\mathbf{M}}^A \in \mathcal{X} = \mathbf{IS_cP}(\underset{\sim}{\mathbf{M}}),$$
$$E\colon X \mapsto \mathcal{X}(X, \underset{\sim}{\mathbf{M}}) \leq \underline{M}^X \in \mathcal{A} = \mathbf{ISP}(\underline{M}),$$

and on morphisms by composition. Here operations (in $\underline{M}$) and relations (in $\underset{\sim}{\mathbf{M}}$) are extended pointwise to subsets of powers, and the topology on such sets is that induced by the product topology derived from τ. The aim is then to choose R in such a way that

$$(\forall A \in \mathcal{A})\ A \cong ED(A),$$

so we have a *natural duality* and say that $\underline{M}$ is *dualizable*, and, better still, also in such a way that

$$(\forall X \in \mathcal{X})\ X \cong DE(X),$$

so the duality is *full.* In general, to achieve fullness we must add operations and even partial operations to the type of $\underset{\sim}{\mathbf{M}}$.

Finite algebras need not, in general, be dualizable; see B. A. Davey and H. Werner [113], p. 151, B. A. Davey [98], p. 107, or Chapter 10 of D. M. Clark and B. A. Davey [79]. However, every finite DLOA has a lattice reduct, on which the median function provides a 3-ary near-unanimity term. This brings such an algebra $\underline{M}$ within the scope of the NU Duality Theorem of B. A. Davey and H. Werner [113], and ensures that $\underline{M}$ is dualized by $\mathbb{S}(\underline{M}^2)$. Many of the illustrative DLOA examples in [113] show that a much more economical set of relations than this may suffice. Consider, for example, de Morgan algebras that are generated by the 4-element lattice $\mathbf{2}^2 = \{0, a, b, 1\}$ with negation $\sim$ interchanging the pairs 0, 1 and a, b. In this case $|\mathbb{S}(\underline{M}^2)| = 55$ (not 45 as stated in [113]), but a duality is obtained by taking $R = \{\alpha, \preccurlyeq\}$, where α is the non-identity automorphism of $\underline{M}$ and $\preccurlyeq$ is the partial order in which $a < 0 < b$ and $a < 1 < b$. From the early 1980s, understanding of examples and theory advanced in a leapfrogging fashion, with many of the key trial examples being of DLOAs. Notable landmarks have included the following.

- The piggyback method (B. A. Davey and H. Werner [114] and [115]), which was motivated by studies of, among other varieties, $\mathbf{M}$, $\mathbf{S}$, and $\mathbf{B}_2$;

- A generalized version of duality theory, using multi-sorted dual structures B. A. Davey and H. A. Priestley [105], which was motivated by a study of Kleene algebras.

- The exploitation of restricted Priestley duality, backed up by computer programs, to find dualities: describing the members of $\mathbf{S}(\underline{M}^2)$, or of particular subsets of it, such as piggybacking relations or the graphs of endomorphisms. Here B. A. Davey and H. A. Priestley [107] led the way with a study of the varieties $\mathbf{B}_n$ ($n \geqslant 3$); see also H. A. Priestley [445] and B. A. Davey and H. A. Priestley [110].

- The notion of a test algebra, introduced in B. A. Davey and H. A. Priestley [108] in order to optimize the dualities for the varieties $\mathbf{B}_n$; the idea is that each algebraic relation r, *qua algebra*, lives in $\mathcal{A}$, and serves as just the right algebra on which to test whether r, *qua relation*, can be discarded from a duality which initially includes it.

- The concepts of a failset and of entailment, discussed respectively in B. A. Davey and H. A. Priestley [109] and B. A. Davey, M. Haviar and H. A. Priestley [103]. Here the objective was to understand just which relations it is necessary to include in a duality, and the mechanism by which these 'entail' superfluous algebraic relations, syntactically or semantically. For this project, full understanding of the Stone, Kleene, and de Morgan algebra dualities was important, and, for entailment, a computer analysis of a duality for $\mathbf{D}$, regarded as the quasivariety generated by the 3-element chain. As is shown in B. A. Davey, M. Haviar and H. A. Priestley [103], the central idea is, once again, that of a test algebra.

- Major progress in the understanding of full dualities was made in D. M. Clark and P. H. Krauss [80]. Their results were later refined and extended in D. M. Clark and B. A. Davey [77]. In these papers, the notion of a *strong* duality was introduced and it was proved that a natural duality is strong if and only if it is full and $\underline{\mathbf{M}}$ is injective in $\mathcal{X}$. Further, in the latter paper, it was shown how given dualities could be upgraded to strong dualities. For example, if $\underline{\mathbf{M}}$ dualizes a finite DLOA, $\underline{M}$, all of whose nontrivial subalgebras are subdirectly irreducible, then adding the partial endomorphisms of $\underline{M}$ to the type of $\underline{\mathbf{M}}$ gives a structure which yields a strong and therefore full duality on $\mathcal{A} = \mathbf{ISP}(\underline{M})$.

- It was clear from the early days that certain dualities were particularly nice: specifically that for $\mathbf{D}$ and those for varieties which are arithmetical or which satisfy some primality condition (as many varieties of DLOAs do—notably the n-valued Łukasiecz–Moisil algebras and any variety of MV-algebras generated by a finite chain). A systematic study of 'goodness' is undertaken in D. M. Clark and B. A. Davey [78].

A significant application of natural duality theory is to the determination of free algebras. One of the basic facts of natural duality theory is that if $\underline{\mathbf{M}}$ yields a duality on $\mathcal{A}$, then each $\mathcal{X}$-morphism from $\underline{\mathbf{M}}^S$ to $\underline{\mathbf{M}}$ is an S-ary *term function* (which, in this text, are referred to as *polynomial functions*): see, for example, B. A. Davey [98], p. 87. Conversely, every S-ary term function from $\underline{M}^S$ to $\underline{M}$ is an $\mathcal{X}$-morphism provided $\mathbf{M}$ is algebraic over $\underline{M}$. Since, of course, a free S-generated algebra $F_{\mathcal{A}}(S)$ is isomorphic to the algebra of all term functions from $\underline{M}^S$ to $\underline{M}$ (with projections as free generators), we immediately see that $F_{\mathcal{A}}(S)$ is isomorphic to the algebra of all $\mathcal{X}$-morphisms from $\mathbf{M}^S$ to $\mathbf{M}$, or, in case S is finite, just the algebra of all R-preserving functions from M^n to M. Expressing this another way, we have that $D(F_{\mathcal{A}}(S)) \cong \mathbf{M}^S$, with, in particular, $D(F_{\mathcal{A}}(1)) \cong \mathbf{M}$. If $\mathbf{M}$ yields a duality on $\mathcal{A} = \mathbf{ISP}(\underline{M})$, then $D(F_{\mathcal{A}}(S)) \cong \mathbf{M}^S$, for any nonempty set S (see, for example, B. A. Davey [98], p. 87). For $\mathbf{D}$, this tells us that the $|S|$-fold power, $\mathbf{2}^S$, of $\mathbf{2}$ is the dual space of $F_{\mathbf{D}}(S)$. Therefore, $F_{\mathbf{D}}(S)$ is the set of all continuous order-preserving maps from $\mathbf{2}^S$ to $\mathbf{2}$, with the algebraic structure coming pointwise from $\mathbf{2}$. A comparison of the Hasse diagrams of $\mathbf{2}^3$ and the 20-element lattice $F_{\mathbf{D}}(3)$ (B. A. Davey and H. A. Priestley [106], p. 173) and of $\mathbf{2}^4$ and the 168-element lattice $F_{\mathbf{D}}(4)$ (see R. Wille [523]), and the fact that $|F_{\mathbf{D}}(8)| = 56,130,437,228,687,557,907,788$ whereas its dual $\mathbf{2}^8$ has 256 elements (D. Wiedemann [519]) highlights the logarithmic nature of the duality. Nevertheless, the continuing challenge (presently for $n \geqslant 9$) of Dedekind's Problem of determining $|F_{\mathbf{D}}(n)|$ shows that casting a problem in dual form does not, by itself, necessarily makes it easier to solve.

An algebra A is said to be *endoprimal*, if every finitary function on A commuting with the endomorphisms of A is a term function and k-*endoprimal* ($k \geqslant 1$), if every function of arity not greater than k that commutes with the endomorphisms is a term function. L. Márki and R. Pöschel [387] characterized the endoprimal distributive lattices as those that are not relatively complemented. Duality theory provides a natural source of endoprimal algebras: a finite algebra $\underline{M}$ is *endodualizable*, if the (graphs of) the endomorphisms of $\underline{M}$ yield a duality on the quasivariety $\mathbf{ISP}(\underline{M})$, and any such algebra is necessarily endoprimal. In [102], B. A. Davey, M. Haviar, and H. A. Priestley investigated endodualizability for finite distributive lattices and showed that it is equivalent to endoprimality. Subsequently, the concepts of endoprimality and endodualizability have been explored in other varieties, and shown to be equivalent in some (Stone algebras, for example), but not in general (see B. A. Davey [99], B. A. Davey and J. G. Pitkethly [104]).

Assume that $\mathbf{M}$ yields a duality on $\mathcal{A} = \mathbf{ISP}(\underline{M})$. What this achieves, whether the duality is full or not, is to represent each algebra $\underline{A}$ in the quasivariety as an algebra of continuous structure-preserving maps from a suitable structure $\mathbf{X} \in \mathcal{X}$ into $\mathbf{M}$. When $|M| = 2$, as it is for Boolean algebras and distributive lattices, this gives a representation by sets. In general, a representation by functions is less appealing, though there are significant instances of a natural (strong) duality being directly used to solve problems on DLOAs. For example, D. M. Clark has studied the algebraically closed and existentially closed Stone and double Stone algebras. D. M. Clark [76] and D. M. Clark and

J. Schmid [81] have described the countable homogeneous universal model of $\mathbf{B}_2$. Frequently, we have available for a given variety $\mathcal{A}$ of DLOAs both a natural duality and a restricted Priestley duality, and a mechanism for going backwards and forwards between these; this allows us, in particular, to derive the Priestley dual of the free algebra $F_{\mathcal{A}}(S)$ from the natural dual $\underset{\sim}{\mathbf{M}}^S$. This has been thoroughly worked out in the context of Cornish algebras (see H. A. Priestley [446]), for example. Thereby we get the advantages of both approaches. This is particularly helpful when dealing with large finite algebras (as free DLAOs often are), where the Priestley duality space may be much more tractable than the corresponding algebra. For more on freeness, see Section 7.

6.   Congruence Properties

The median term on any algebra with a lattice reduct is a majority term. Hence the variety $\mathbf{D}$ and all varieties of DLOAs are congruence distributive. The latter accordingly often provide evidence for conjectures which may hold more generally in congruence-distributive varieties. Also, Jónsson's Lemma (Theorem V.1.9) is available when studying these classes and their lattices of subvarieties, which makes them particularly amenable.

In a general algebra, principal congruences can be described via Mal'cev's lemma (see G. Grätzer [215], Theorem 3, p. 54). The universal algebraic import of the shape of the resulting formula (for example, consequences concerning the Congruence Extension Property (CEP) and congruence distributivity) is studied, for example, in E. Fried, G. Grätzer, and H. Lakser [189]. As the shape of the formula describing $\Theta(a, b)$ becomes more uniform over the variety $\mathbf{K}$, further restrictions are imposed upon the variety. Descriptions of $\Theta(a, b)$ for certain varieties of DLOAs can be found in H. Lakser [1973], J. Berman [47].

The congruences of distributive lattices, and of DLOAs, in general do not possess many of the desirable properties of permutability, regularity, etc. It is therefore natural to ask which algebras in a given variety of this sort do have permutable congruences, uniform congruences, and so on. In M. E. Adams and R. Beazer [8], distributive lattices having n-permutable congruences are characterized as those with no n-element chain in their poset of prime ideals—a key theorem towards this appears in M. E. Adams and R. Beazer [7] (Theorem 3.17). A characterization of countable congruence-uniform distributive lattices is given in M. E. Adams and R. Beazer [6]. This is shown not to extend to distributive lattices of arbitrary cardinality. These results are noteworthy in that they concern the theory of distributive lattices *per se*, an area too well worked over for many worthwhile unsolved problems to remain. The names of Adams and Beazer are also associated with the study of congruence properties on DLOAs. Building on earlier work, they reveal in [4] interdependencies between various congruence properties for de Morgan algebras.

Related to this, but not fitting neatly into the interdependency pattern, is the characterization of de Morgan algebras satisfying the condition that the intersection of any two principal congruences be principal (this is true for all Kleene algebras): see M. E. Adams and R. Beazer [5] and M. E. Adams [2]. Congruence properties of distributive double p-algebras have also received considerable attention (see M. E. Adams, M. Atallah, and R. Beazer [3], M. E. Adams and R. Beazer [7], and M. E. Adams and R. Beazer [9]).

Both algebraic and duality-theoretic methods have been used to study simple and subdirectly irreducible DLOAs, and there is a rich literature. For example, opening the issue of Algebra Universalis that contains the paper E. Fried, G. Grätzer, and H. Lakser [189] on uniform congruence schemes cited above, we find the papers P. Köhler [368], T. Katriňák [352], R. Beazer [39] and [40], all of which use algebraic methods to describe subdirectly irreducible DLOAs. More recently, duality methods have been used widely; see, for example A. Urquhart [505], B. A. Davey [95], R. Cignoli [73], and D. Hobby [311]. The first of these foreshadows results for Cornish algebras; see W. H. Cornish [83] and H. A. Priestley [446].

Once the subdirectly irreducible algebras in a variety $\mathbf{K}$ of DLOAs have been obtained, then Jónsson's Lemma gives access to the lattice of subvarieties of $\mathbf{K}$. When $\mathbf{K}$ is finitely generated, this lattice is isomorphic to the downset lattice of $\mathbf{Si}(\mathbf{K})$ of subdirectly irreducible algebras (up to isomorphism), ordered by $B \leqslant C$ if and only if $B \in \mathbf{HS}(C)$; see B. A. Davey [96]. Especially when combined with duality, this makes it easy to determine subvariety lattices for particular varieties of Cornish algebras; see, for example, T. S. Blyth and J. C. Varlet [53]. Natural duality theory can then be exploited to find, in an algorithmic way, canonical identities for $\mathbf{K} = \mathbf{HSP}(A)$, for any given finite Cornish algebra A. This process is described and illustrated in M. E. Adams and H. A. Priestley [12] and H. A. Priestley and R. Santos [448]. Recently, a transparent relationship has been revealed between identities in $\mathbf{K}$ and formulae satisfied in the Priestley dual $D(A) \in \mathcal{Y}_{\mathbf{K}}$ (H. A. Priestley [447]).

7.   Freeness, Coproducts, and Injectivity

Some further comments should be made on the determination of $F_{\mathcal{A}}(S)$, where $\mathcal{A} = \mathbf{ISP}(\underline{M})$ and $\underline{M}$ is a finite DLOA. Assume that a restricted Priestley duality has been determined for $\mathcal{A}$. So long as products in $\mathbf{P}_{\mathcal{A}}$ are concrete products, we have immediate access to (the duals of) free algebras in $\mathcal{A}$. This happens, for example, for the varieties $\mathbf{N}$ and $\mathbf{S}$. It fails to happen either for $\mathbf{K}$ or for the varieties $\mathbf{B}_n$ ($n \geqslant 2$) in the ω-chain of proper subvarieties of distributive p-algebras. In these cases, the free algebras were first described by examining the adjoint to the forgetful functor from $\mathcal{A}$ to $\mathbf{D}$, exploiting the Priestley duality for $\mathcal{A}$; see W. H. Cornish and P. R. Fowler [85] and B. A. Davey and M. Goldberg [101]. In [83] W. H. Cornish went on to extend his arguments to other varieties of Cornish algebras.

In many instances, a better strategy is to work with a natural duality for $\mathcal{A}$ which we assume to be set up by functors D and E. The dual structure $D(F_{\mathcal{A}}(S))$ is then just $D(F_{\mathcal{A}}(1))^S$. This may not be a particularly workable description, and we draw attention to two situations (not mutually exclusive) in which it can be improved. First we may hope that there is a mechanical translation process taking the natural dual of $F_{\mathcal{A}}(S)$ (or of any other algebra in $\mathcal{A}$) to its restricted Priestley dual, so that the finite free algebras can be obtained as the lattices of upsets of their $\mathbf{P}_{\mathcal{A}}$ duals, with the additional operations also being obtained from the duality. This process is illustrated in M. E. Adams and H. A. Priestley [12], H. A. Priestley [446], and H. A. Priestley and R. Santos [448] (for Ockham varieties, and more generally varieties of Cornish algebras) and for n-valued Łukasiewicz algebras in H. A. Priestley [444]. Alternatively, we may have varieties which are of such a special type that their natural dualities take a particularly simple form. This occurs, for example, for the n-valued Łukasiewicz algebras and for the varieties of MV-algebras generated by finite chains, which are discriminator varieties. Any such variety has a particularly pleasing natural duality based on Boolean spaces with a semigroup action (see B. A. Davey and H. Werner [113], D. M. Clark and B. A. Davey [79]).

Consideration of a natural categorical property allows a large class of examples, including many of those mentioned above, to be viewed in a uniform way. In every category with finite products and coproducts, there is a natural morphism from $A \sqcup (B \times C)$ to $(A \sqcup B) \times (A \sqcup C)$. If this map is an isomorphism for all $A, B, C \in \mathcal{V}$, we say that coproduct distributes over products in $\mathcal{V}$. Using duality theory, this property was established for various classes of DLOAs, for example, distributive p-algebras by B. A. Davey and M. Goldberg [101] and Kleene algebras by W. H. Cornish and P. R. Fowler [85]. Based on these results a general theorem was obtained in B. A. Davey and H. Werner [112] from which it followed, for example, that coproduct distributes over products in every variety of distributive p-algebras, distributive double p-algebras, Heyting algebras, or Ockham algebras (in fact, in every variety of Cornish algebras). If free products distribute over products and the free algebra on one generator is known as a product of directly irreducible algebras, then the free algebra on n generators (the coproduct of n copies of the free algebra on one generator) is easily described: for example, see B. A. Davey and D. Duffus [100].

Every free algebra is projective. Projectives, in general, are more elusive than injectives, though it has long been known that the finite projective algebras in $\mathbf{D}$ are those whose join-irreducibles form a lattice, and quite a lot of information is available on properties of projectives (see R. Balbes [36]). For DLOAs the best that can reasonably be expected is a characterization modulo that for $\mathbf{D}$; see, for example, I. Düntsch[139].

A member L of a variety $\mathbf{K}$ of lattices is said to be $\mathbf{K}$-*catalytic* if $\mathbf{K}(L, M)$ is a lattice under the pointwise order for every $M \in \mathbf{K}$. A purely algebraic characterization of the $\mathbf{D}$-catalytic lattices was given by R. Balbes in [35] and pursued further in [36]. Balbes showed, in particular, that the finite $\mathbf{D}$-catalytic lattices are those whose posets of join-irreducible elements are lattices, or, equivalently, the finite projectives in $\mathbf{D}$. In general, L is $\mathbf{D}$-catalytic if and only if X_L is a topological lattice—another instance of the transition from the finite case to the infinite case being the most natural one possible. For more information see H. A. Priestley [440].

More generally we may consider the distributive lattice freely generated by a poset P. This has dual space $\underset{\sim}{\mathbf{2}}^P$, consisting of all order-preserving maps from P into $\mathbf{2}$ regarded as a $\mathbf{P}$-substructure of $\underset{\sim}{\mathbf{2}}^{|P|}$; see B. A. Davey and D. Duffus [100] and H. A. Priestley [440].

In the 1960s and 1970s, a spate of papers appeared that described injectives, and also weak injectives, absolute subretracts, and injective hulls, in varieties of DLOAs, using mainly algebraic methods. In every case, the injectives are finite products of bounded Boolean powers, by complete Boolean algebras, of the injective subdirectly irreducible algebras, although the results were not always stated this way explicitly. (The first use of bounded Boolean powers to describe injectives is in A. Day [116].) This flourishing industry came to an abrupt halt when all previous results of this type were brought under one umbrella in the paper of B. A. Davey and H. Werner [111].

Injectivity and congruence properties come together in various ways. In particular, if a variety $\mathbf{K}$ has enough injectives (that is, every algebra in $\mathbf{K}$ may be embedded into an injective algebra in $\mathbf{K}$), then $\mathbf{K}$ satisfies both the Congruence Extension Property and the Amalgamation Property: see P. D. Bacsich [31], B. Banaschewski [37], W. Taylor [495]. The latter properties have been studied intensely: see, for example, G. Grätzer and H. Lakser [1972] and T. Katriňák [1974a]. Based on the results for DLOAs and other congruence-distributive varieties, a general theory embracing such varieties was given in B. A. Davey [93]. Once again, we see a beneficial relationship between universal algebra and an area of lattice theory. Long may this symbiosis develop!

Congruence Lattices

G. Grätzer and E. T. Schmidt

In the early sixties, we characterized congruence lattices of universal algebras as algebraic lattices; then our interest turned to the characterization of congruence lattices of lattices. Since these lattices are distributive, we figured that this must be an easier job. After more then 35 years, we know that we were wrong.

In this Appendix, we give a brief overview of the results and methods. In Section 1, we deal with finite distributive lattices; their representation as congruence lattices raises many interesting questions and needs specialized techniques. In Section 2, we discuss the general case.

A related topic is the lattice of complete congruences of a complete lattice. In contrast to the previous problem, we can characterize complete congruence lattices, as outlined in Section 3.

1. The Finite Case

The congruence lattice, $\operatorname{Con} L$, of a finite lattice L is a finite distributive lattice (N. Funayama and T. Nakayama, see Theorem II.3.11). The converse is a result of R. P. Dilworth (see Theorem II.3.17), first published G. Grätzer and E. T. Schmidt [1962]. Note that the lattice we construct is sectionally complemented.

In Section II.1, we learned that a finite distributive lattice D is determined by the poset $\operatorname{J}(D)$ of its join-irreducible elements, and every finite poset can be represented as $\operatorname{J}(D)$, for some finite distributive lattice D. So for a finite lattice L, we can reduce the characterization problem of finite congruence lattices to the representation of finite posets as $\operatorname{J}(\operatorname{Con} L)$.

1.1 Planar lattices, small lattices

We start with the following result (G. Grätzer, H. Lakser, and E. T. Schmidt [241]):

Theorem 1.   Let D be a finite distributive lattice with n join-irreducible elements. Then there exists a planar lattice L of $O(n^2)$ elements with $\operatorname{Con} L \cong D$.

The original constructions (R. P. Dilworth's and also our own) produced lattices of size $O(2^{2n})$ and of order dimension $O(2n)$. In G. Grätzer and H. Lakser [238], this was improved to size $O(n^3)$ and order dimension 2 (planar).

We sketch the construction for Theorem 1.

Let $P = \mathrm{J}(D)$, $P = \{p_1, p_2, \ldots, p_n\}$, and we take a chain

$$C = \{c_0, c_1, \ldots, c_{2n}\}, \quad c_0 \prec c_1 \prec \cdots \prec c_{2n}.$$

To every prime interval $[c_i, c_{i+1}]$, we assign an element of P as its "color", so that each element of P is the color of two adjacent prime intervals: let the color of $[c_0, c_1]$ and $[c_1, c_2]$ be p_1; of $[c_2, c_3]$ and $[c_3, c_4]$ be p_2, and so on, of $[c_{2n-2}, c_{2n-1}]$ and $[c_{2n-1}, c_{2n}]$ be p_n. Follow this on the two examples in Figures 1 and 2; in Figure 1, $P = \{p_1, p_2\}$ and $p_1 < p_2$, while in Figure 2, $P = \{p_1, p_2, p_3\}$ and $p_1 < p_2, p_3 < p_2$. The colors are indicated on the diagrams.

In C^2, we fill in a "covering square" C_2^2 with one more element so that we obtain an M_3, if the two sides have the same color, see Figure 3. Moreover, if $p, q \in P$ and $p < q$, then we take the "double covering square" $C_3 \times C_2$, where the longer side has two prime intervals of color q and the shorter side is of color p, and we add one more element, as illustrated in Figure 3, to obtain the sublattice $N_{5,5}$.

It is an easy computation to show that $|L| \leqslant kn^2$, for some constant k, and that $D \cong \operatorname{Con} L$; this isomorphism is established by assigning to $p \in P$ the congruence of L generated by collapsing any (all) prime intervals of color p.

For a natural number n, define $\operatorname{cr}(n)$ as the smallest integer such that, for any distributive lattice D with n join-irreducible elements, there exists a finite lattice L satisfying $\operatorname{Con} L \cong D$ and $|L| \leqslant \operatorname{cr}(n)$. From the construction sketched above, it follows that

$$\operatorname{cr}(n) < 3(n+1)^2.$$

In Section A.1.7, we discussed the lower bound $\dfrac{n^2}{16 \log_2 n}$ for $\operatorname{cr}(n)$. Here is how it evolved: G. Grätzer, I. Rival, and N. Zaguia [250] proved that Theorem 1 is "best possible" in the sense that size $O(n^2)$ cannot be replaced by size $O(n^\alpha)$, for any $\alpha < 2$; that is,

$$kn^\alpha < \operatorname{cr}(n),$$

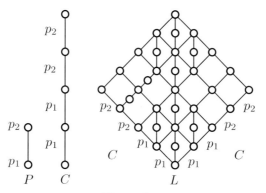

Figure 1

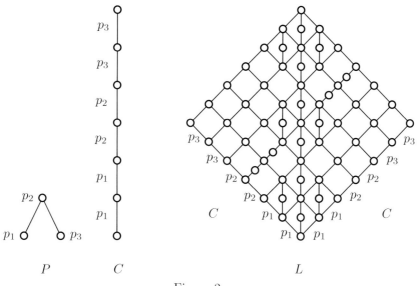

Figure 2

for any constant k, for any $\alpha < 2$, and for any sufficiently large integer n. Y. Zhang [528] noticed that the proof of this inequality can be improved to obtain the following result: for $n \geqslant 64$,

$$\frac{1}{64} \frac{n^2}{(\log_2 n)^2} < \mathrm{cr}(n).$$

The lower bound $\frac{n^2}{16 \log_2 n}$ for $\mathrm{cr}(n)$ is, of course, much stronger than the last one.

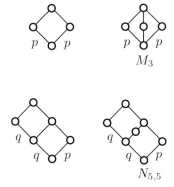

Figure 3

A different kind of lower bound is obtained in R. Freese [166]; it is shown that if $J(\operatorname{Con} L)$ has e edges ($e > 2$), then

$$\frac{e}{2\log_2 e} \leqslant |L|.$$

R. Freese also proves that $J(\operatorname{Con} L)$ can be computed in time $O(|L|^2 \log_2 |L|)$. See also Section 5.2 of Appendix G.

Consider the optimal *length* of L. E. T. Schmidt [1975] constructs a finite lattice L of length $5m$, where m is the number of dual atoms of D; S.-K. Teo [498] proves that this result is best possible. (For finite chains, this was done in J. Berman [1972].)

1.2   Modular and semimodular lattices

E. T. Schmidt [1974] proves that *Every finite distributive lattice D can be represented as the congruence lattice of a* modular *lattice M.* It follows from Theorem III.4.9 that the congruence lattice of a finite modular lattice is a Boolean lattice, therefore, we cannot expect M to be finite. For a short proof of this result, see [467].

A much deeper result was proved in E. T. Schmidt [465]:

Theorem 2.   Every finite distributive lattice can be represented as the congruence lattice of a *complemented modular lattice.*   ——

It was pointed out by F. Wehrung that the ring theoretic and ordered group-theoretic results of G. A. Elliott [142], P. A. Grillet [275], E. G. Effros, D. E. Handelman, and Chao Liang Shen [141] (along with some elementary results in F. Wehrung [517]) contain this theorem; see K. R. Goodearl and F. Wehrung [205] for more detail.

In G. Grätzer, H. Lakser, and E. T. Schmidt [244], we constructed a *finite semimodular* lattice L:

Theorem 3.   Every finite distributive lattice D can be represented as the congruence lattice of a *finite semimodular* lattice S. In fact, S can be constructed as a *planar* lattice of size $O(n^3)$, where n is the number of join-irreducible elements of D.   —

1.3   Congruence-preserving extensions

In Section A.1.7, we discussed the concept of congruence-preserving extensions and some of the major results concerning it. Many of the older results use the following construction, see E. T. Schmidt [1968]. Let D be a bounded distributive lattice; let $M_3[D]$ (the extension of M_3 by D) consist of all triples $\langle x, y, z \rangle \in D^3$ satisfying $x \wedge y = x \wedge z = y \wedge z$. Then $M_3[D]$ is a (modular) lattice, $x \mapsto \langle x, 0, 0 \rangle$, $x \in D$, is an embedding of D into $M_3[D]$ and by identifying x with $\langle x, 0, 0 \rangle$, we obtain that $M_3[D]$ is a congruence-preserving extension of D. (See a variant of this construction in Section 2.)

In G. Grätzer, H. Lakser, and R. W. Quackenbush [240], it is proved that the $M_3[D]$ construction is a special case of *tensor products*. If A and B are lattices with zero, then $A \otimes B$, the tensor product of A and B, is the join-semilattice freely generated by the poset $(A - \{0\}) \times (B - \{0\}) \cup \{0\}$ subject to the relations: $\langle a_0, b \rangle \vee \langle a_1, b \rangle = \langle a_0 \vee a_1, b \rangle$ and $\langle a, b_0 \rangle \vee \langle a, b_1 \rangle = \langle a, b_0 \vee b_1 \rangle$ (a, a_0, $a_1 \in A$, b, b_0, $b_1 \in B$).

Let A and B be finite lattices. Then $A \otimes B$ is obviously a lattice, and the following isomorphism holds (see [240]):

Theorem 4.

$$\operatorname{Con} A \otimes \operatorname{Con} B \cong \operatorname{Con}(A \otimes B).$$ —

For any finite simple lattice S, this isomorphism implies that $\operatorname{Con} A \cong \operatorname{Con}(A \otimes S)$. The case $A = D$, $S = M_3$ is the $M_3[D]$ result.

Theorem 4 has been extended to wide classes of infinite lattices (substituting $\operatorname{Con_c}$, the semilattice of compact congruences, for Con) in G. Grätzer and F. Wehrung [266] and to arbitrary lattices with zero using "box products" (a variant of tensor products) in G. Grätzer and F. Wehrung [268].

Many of the newer results utilize a new technique in G. Grätzer and E. T. Schmidt [258] (applied also in G. Grätzer and E. T. Schmidt [259] and [261]).

The *rectangular extension* $\mathbb{R}(K)$ of a finite lattice K is defined as the direct product of *all* subdirect factors of K, that is,

$$\mathbb{R}(K) = \prod (K/\Phi \mid \Phi \in \operatorname{M}(\operatorname{Con} K)),$$

where $\operatorname{M}(\operatorname{Con} K)$ is the set of all meet-irreducible congruences of K.

K has a natural embedding into $\mathbb{R}(K)$ by

$$\psi\colon a \mapsto a^{\mathbb{R}} = \langle [a]\Phi \mid \Phi \in \mathrm{M}(\mathrm{Con}\,K)\rangle.$$

Let $K\psi = K^{\mathbb{R}}$.

Theorem 5.   Let K be a finite lattice. Then $K^{\mathbb{R}}$ has the Congruence Extension Property in $\mathbb{R}(K)$ (that is, every congruence of $K^{\mathbb{R}}$ can be extended to $\mathbb{R}(K)$).
                                                                              —

1.4   p-algebras

T. Katrinák [354] characterized the congruence lattices of finite p-algebras (that is, lattices with pseudocomplementation, see Section II.6):

Theorem 6.   Every finite distributive lattice is isomorphic to the congruence lattice of a finite p-algebra.
                                                                              —

In [254], we give a more elementary proof of this theorem. In fact, we prove the following generalization:

Theorem 7.   Let D be an algebraic distributive lattice in which the unit element of D is compact and every compact element of D is a finite join of join-irreducible compact elements. Then D can be represented as the congruence lattice of a p-algebra P.
                                                                              —

1.5   Simultaneous representation

It is well-known that, given a lattice L and a convex sublattice K, the restriction map $\mathrm{Con}\,L \to \mathrm{Con}\,K$ is a $\{0,1\}$-preserving lattice homomorphism. In G. Grätzer and H. Lakser [235], see also E. T. Schmidt [466], the converse is proved: any $\{0,1\}$-preserving homomorphism of finite distributive lattices can be realized as such a restriction and, indeed, as a restriction to an ideal of a finite lattice.

If the sublattice K is not a convex sublattice, then the restriction map $\mathrm{Con}\,L \to \mathrm{Con}\,K$ need not preserve join, but it still preserves meet, 0, and 1.

Similarly, we can *extend* congruences from the sublattice K to L by minimal extension. This extension map need not preserve meet, but it does preserve join and 0. Furthermore, it *separates* 0, that is, nonzero congruences extend to nonzero congruences. We now formalize this.

Let K and L be lattices, and let φ be a homomorphism of K into L. Then φ induces a map $\mathrm{ext}\,\varphi$ of $\mathrm{Con}\,K$ into $\mathrm{Con}\,L$: for a congruence relation Θ of K, let the image Θ under $\mathrm{ext}\,\varphi$ be the congruence relation of L generated by the set $\Theta\varphi = \{\,\langle a\varphi, b\varphi\rangle \mid a \equiv b\ (\Theta)\,\}$.

The following result was proved by A. P. Huhn in [317] in the special case when ψ is an embedding and was proved for arbitrary ψ in G. Grätzer, H. Lakser, and E. T. Schmidt [243]:

Theorem 8.   Let D and E be finite distributive lattices, and let

$$\psi\colon D \to E$$

be a $\{0, \vee\}$-homomorphism. Then there are finite lattices K and L, a lattice homomorphism $\varphi\colon K \to L$, and isomorphisms

$$\alpha\colon D \to \operatorname{Con} K,$$
$$\beta\colon E \to \operatorname{Con} L$$

with

$$\psi\beta = \alpha(\operatorname{ext} \varphi).$$

Furthermore, φ is an embedding iff ψ separates 0.   —

Theorem 8 concludes that the following diagram is commutative:

$$
\begin{array}{ccc}
D & \xrightarrow{\ \psi\ } & E \\
{\scriptstyle\cong}\Big\downarrow{\scriptstyle\alpha} & & {\scriptstyle\cong}\Big\downarrow{\scriptstyle\beta} \\
\operatorname{Con} K & \xrightarrow{\ \operatorname{ext} \varphi\ } & \operatorname{Con} L
\end{array}
$$

In G. Grätzer, H. Lakser, and E. T. Schmidt [245], the following stronger version is proved (see G. Grätzer, H. Lakser, and E. T. Schmidt [242] for a short proof):

Theorem 9.   Let K be a finite lattice, let E be a finite distributive lattice, and let $\psi\colon \operatorname{Con} K \to E$ be a $\{0, \vee\}$-homomorphism. Then there is a finite lattice L, a lattice homomorphism $\varphi\colon K \to L$, and an isomorphism $\beta\colon E \to \operatorname{Con} L$ with $\operatorname{ext} \varphi = \psi\beta$. Furthermore, φ is an embedding iff ψ separates 0.   —

If L is a lattice and L_1, L_2 are sublattices of L, then there is a map

$$\operatorname{Con} L_1 \to \operatorname{Con} L_2$$

obtained by first extending each congruence relation of L_1 to L and then restricting the resulting congruence relation to L_2. All we can say about this map is that it is isotone and that it preserves 0. The main result of G. Grätzer, H. Lakser, and E. T. Schmidt [243] (see G. Grätzer, H. Lakser, and E. T. Schmidt [242] for a short proof) is that this is, in fact, a characterization of 0-preserving isotone maps between *finite* distributive lattices:

Theorem 10.   Let D_1 and D_2 be finite distributive lattices, and let

$$\psi \colon D_1 \to D_2$$

be an isotone map that preserves 0. Then there is a finite lattice L with sublattices L_1 and L_2 and there are isomorphisms

$$\alpha_1 \colon D_1 \to \operatorname{Con} L_1,$$
$$\alpha_2 \colon D_2 \to \operatorname{Con} L_2$$

such that the diagram

$$
\begin{array}{ccc}
D_1 & \xrightarrow{\ \psi\ } & D_2 \\[2pt]
{\scriptstyle\cong}\Big\downarrow{\scriptstyle\alpha_1} & & {\scriptstyle\cong}\Big\downarrow{\scriptstyle\alpha_2} \\[2pt]
\operatorname{Con} L_1 \xrightarrow{\ \text{extension}\ } & \operatorname{Con} L & \xrightarrow{\ \text{restriction}\ } \operatorname{Con} L_2
\end{array}
$$

is commutative.   —

   G. Grätzer, H. Lakser, and E. T. Schmidt [246] and [247] prove that if ψ is a 0-preserving join-homomorphism of D into E (where D and E are finite distributive lattices) and $n = \max(|\mathrm{J}(D)|, |\mathrm{J}(E)|)$, then we can construct the finite lattices K and L and the lattice homomorphism $\varphi \colon K \to L$ that represent ψ so that the size of K and L is $O(n^5)$ and the order dimensions of K and L are 3. We conjecture that this result is the best.

1.6   The D-relation

In a finite lattice L, every element a is determined by the set of join-irreducible elements contained in a. Therefore, it is quite natural that we can characterize congruences in terms of the join-irreducible elements. This idea was developed in the papers B. Jónsson and J. B. Nation [339], A. Day [119], and R. Wille [522]. On the set $\mathrm{J}(L)$, a binary relation D is defined as follows: for $p, q \in \mathrm{J}(L)$, let $p \, D \, q$ iff there exists an $x \in L$ such that $p \leqslant q \vee x$ and $p \nleqslant q_* \vee x$, where q_* denotes the lower cover of q. This relation D defines a closure operator $\mathcal{D}$ as follows: for $A \subseteq \mathrm{J}(L)$, let A be $\mathcal{D}$-closed iff $p \, D \, q$ and $q \in A$ implies that $p \in A$. The following result describes congruence lattices.

Theorem 11.   The congruence lattice of a finite lattice L is isomorphic to the lattice of $\mathcal{D}$-closed subsets of $\mathrm{J}(L)$. This isomorphism is given by

$$\Theta \mapsto \{\, p \in \mathrm{J}(L) \mid \Theta(p_*, p) \leqslant \Theta \,\}.$$ —

   This topic is treated in depth in the book R. Freese, J. Ježek, and J. B. Nation [172]; see also M. Tischendorf [500] and Appendix G.

2. The General Case

In the general case, there are two approaches to try to solve the characterization problem of congruence lattices of lattices. The first is due to the second author and the second was suggested by P. Pudlák [449].

2.1 The first approach: distributive join-homomorphisms

Let $\mathrm{Con}_c\, L$ denote the semilattice of compact congruences of the lattice L. Let us call a semilattice S *representable* iff there is a lattice L with $\mathrm{Con}_c\, L \cong S$.

Let D be a finite distributive lattice, and take the Boolean lattice B generated by D (see Section II.4). Then for every $d \in B$, there exists a smallest element $s(d) \in D$ such that $d \leqslant s(d)$. The map $s\colon B \to D$ is a closure operator (see Definition IV.3.8(i)). We construct a lattice K, whose congruence lattice is D with a new unit element adjoined. Take $B \times C_2$ and its meet-subsemilattice K, consisting of all elements of the form $\langle b, 0\rangle$, $b \in B$ and $\langle d, 1\rangle$, $d \in D$. Then K is a lattice and B can be identified with an ideal of K under the map $b \mapsto \langle b, 0\rangle$, $b \in B$. Consider a congruence $\langle x, 0\rangle \equiv \langle 0, 0\rangle$ in B. Joining both sides with $\langle 0, 1\rangle$, we get $\langle s(x), 1\rangle = \langle x, 0\rangle \vee \langle 0, 1\rangle \equiv \langle 0, 0\rangle \vee \langle 0, 1\rangle = \langle 0, 1\rangle$. Thus $\langle s(x), 0\rangle \equiv \langle 0, 0\rangle$. It is easy to see now that the congruence relation $\Theta(\langle s(x), 0\rangle, \langle 0, 0\rangle)$ of B can be extended to K and $\Theta(\langle 0, 1\rangle, \langle 0, 0\rangle)$ is the unit congruence. Now we have to identify the new unit element with the unit of $\mathrm{Con}\, K$. The following construction solves this problem (see E. T. Schmidt [1968]).

Let L consist of all triples $\langle x, y, z\rangle \in B^3$ satisfying $x \wedge y = x \wedge z = y \wedge z$ and $x \in D$. Then L is a lattice and $D \cong \mathrm{Con}\, L$.

Note that the elements $\langle 0, 0, z\rangle$, $z \in B$, and $\langle x, 1, x\rangle$, $x \in D$, form a sublattice of L isomorphic to K.

This construction also works for certain infinite distributive lattices, namely, for dually relatively pseudocomplemented lattices; indeed, for every $d \in D$, then there exists a smallest $s(d) \in B$ such that $d \leqslant s(d)$.

Let B be a generalized Boolean lattice and let $h\colon B \to S$ be an onto join-homomorphism. We will say that h is *weakly distributive* iff for all x, y, $z \in B$ with $x \vee y \leqslant z$ and $h(x \vee y) = h(z)$, there are x_1, $y_1 \in B$ such that $x_1 \vee y_1 = z$, $h(x) = h(x_1)$, and $h(y) = h(y_1)$. If B is the Boolean lattice generated by D and $s(x)$ exists for every $x \in D$, then the map $h\colon x \mapsto s(x)$ is a weakly distributive join-homomorphism.

We call h *distributive* iff $\mathrm{Ker}\, h = \bigvee(\mathrm{Ker}\, s_i \mid i \in I)$, where each s_i, $i \in I$, is a closure operator on B.

Theorem 12.   Let S be a semilattice with zero. If there is a generalized Boolean lattice B and a distributive join-homomorphism h from B onto S, then S is representable.   —

We apply the previous construction for every closure operator s_i to obtain the lattices L_i, $i \in I$. We glue these lattices to each other to construct L with $\mathrm{Con}_c\, L \cong S$.

By construction, every L_i has an ideal B_i isomorphic to B. We define a lattice M_I whose elements are vectors, $\langle \ldots, x_i, \ldots \rangle$, $x_i \in B$, $i \in I$, satisfying the condition that for three distinct indices i, j, and k, $x_i \wedge x_j = x_i \wedge x_k = x_j \wedge x_k$. For every $i \in I$, the elements $\langle x, \ldots, x, 1, x, \ldots, x \rangle$, where the 1 is i-th entry, form a dual ideal B_i' of M_I, which is isomorphic to B and, consequently, to the ideal B_i of L_i. Now we glue together M_I and L_i by identifying B_i and B_i'. This way, we get a meet semilattice. The lattice L is the lattice of all finitely generated ideals of this semilattice.

Theorem 12 is sufficient to obtain most representation theorems:

Theorem 13.   Let S be a semilattice with zero. Each one of the following conditions implies that S is representable:

 (i) $\mathrm{Id}\, S$ is completely distributive (equivalently, S is isomorphic to the semilattice of all finitely generated hereditary subsets of some partially ordered set).

 (ii) S is a lattice.

 (iii) S is locally countable (that is, for every $s \in S$, $(s]$ is countable).

 (iv) $|S| \leqslant \aleph_1$.   —

(i) was first obtained by R. P. Dilworth. Proofs can be found in G. Grätzer and E. T. Schmidt [1962], H. Dobbertin [135], and P. Crawley and R. P. Dilworth [1973].

(ii) was proved in E. T. Schmidt [464]. P. Pudlák [449] provides another proof, in a more general, categorical context. Another proof can be found in H. Dobbertin [136].

(iii) was first obtained in A. P. Huhn [317] under the condition that $|S| \leqslant \aleph_0$. The general result for locally countable semilattices was obtained in H. Dobbertin [135]. Dobbertin proved that if B is a locally countable generalized Boolean semilattice and if S is a distributive join-semilattice, then every weakly distributive homomorphism from B to S is distributive. Further, he proved that every locally countable distributive semilattice with zero is the weakly distributive image of some locally countable generalized Boolean algebra.

(iv) was obtained in A. P. Huhn [317]. One of the main tools used by Huhn is the notion of *frame* introduced in H. Dobbertin [135], which is a special sort of lattice with zero used for building transfinite direct limits of direct systems having up to $\aleph_1$ objects.

Some of these results are presented in an axiomatic form in M. Tischendorf [501].

2.2   The second approach

We have seen that the representation is relatively easy for finite distributive lattices. Let D be an arbitrary distributive semilattice with zero; P. Pudlák [449] proved that for each finite subset F of D, there is a *finite* subsemilattice of D containing F, therefore, D is a direct limit of all the finite distributive lattices contained in it as distributive join-semilattices with zero.

Let S be a distributive semilattice with zero, and let $\mathcal{S}$ be the set of finite subsets of $S - \{0\}$. P. Pudlák's approach (see [449]) is the following. Choose an order preserving function that assigns to each $F \in \mathcal{S}$ a finite distributive subsemilattice S_F of S containing F; thus S is the direct limit of the S_F, $F \in \mathcal{S}$. For each $F \in \mathcal{S}$, construct a finite atomistic lattice L_F whose congruence lattice is isomorphic to S_F and such that if $F \subseteq G$, then L_F embeds into L_G with the Congruence Extension Property. Then if $L_{\mathcal{S}}$ is the limit of the L_F, $F \in \mathcal{S}$, then the congruence lattice of $L_{\mathcal{S}}$ is isomorphic to Id S.

Applying this method, P. Pudlák [449] gave a new proof of Theorem 13(ii).

2.3   Negative results

There are a number of related negative results.

H. Dobbertin [136] constructs a distributive semilattice with zero S and a semilattice homomorphism f of a Boolean algebra B (of size $\aleph_1$) onto S such that f is weakly distributive but not distributive.

F. Wehrung [516] constructs a bounded distributive semilattice S (of size $\aleph_2$) that is not isomorphic to any weak distributive image of a generalized Boolean algebra. Note that the $\aleph_2$ size is optimal. This shows that we cannot obtain a positive solution of the congruence lattice characterization problem of lattices by the first approach.

The second approach was also ruled out in F. Wehrung [517]. The semilattice S of the previous paragraph is not isomorphic to $\mathrm{Con}_c L$, for any lattice L which is a direct limit of lattices that are either atomistic or sectionally complemented.

M. Ploščica, J. Tůma, and F. Wehrung [436] prove that there exists a distributive semilattice S that is representable as $\mathrm{Con}_c L$, for a lattice L, but S cannot be represented using the first or the second approach, that is, S is not a distributive join-homomorphic image of a generalized Boolean lattice nor is S a direct limit of lattices that are either atomistic or sectionally complemented. In fact, one can take $S = \mathrm{Con}_c F$, where F is the free lattice on $\aleph_2$ generators in any nondistributive variety of lattices.

3.   Complete Congruences

In Section A.1.8, we briefly mentioned the result (G. Grätzer [218]):

Theorem 14.   Every complete lattice K can be represented as the lattice of complete congruence relations of a complete lattice L.   —

In a series of papers, much sharper results have been obtained.

G. Grätzer and H. Lakser [236] had the first published proof of Theorem 14; in fact, it already contained more: L was constructed as a planar lattice.

Let $\mathfrak{m}$ be an infinite regular cardinal, and let K be an $\mathfrak{m}$-complete lattice. Then the lattice $\text{Con}_{\mathfrak{m}}\, K$ of all $\mathfrak{m}$-complete congruence relations of K is $\mathfrak{m}$-algebraic (this concept is the obvious modification of Definition II.3.12).

G. Grätzer and H. Lakser [237] proved a partial converse:

Let $\mathfrak{m}$ be a regular cardinal $> \aleph_0$, and let L be an $\mathfrak{m}$-algebraic lattice with an $\mathfrak{m}$-compact unit element. Then L is isomorphic to the lattice of $\mathfrak{m}$-algebraic congruences of an $\mathfrak{m}$-algebraic lattice K.

A much sharper form of the original result was proved in the paper R. Freese, G. Grätzer, and E. T. Schmidt [169]:

Every complete lattice L is isomorphic to the lattice of complete congruence relations of a complete modular lattice K.

The $\mathfrak{m}$-algebraic direction and the modular direction were combined by the present authors in [253]:

Let $\mathfrak{m}$ be a regular cardinal $> \aleph_0$. Every $\mathfrak{m}$-algebraic lattice L is isomorphic to the lattice of $\mathfrak{m}$-complete congruence relations of a suitable $\mathfrak{m}$-complete modular lattice K.

G. Grätzer, P. Johnson, and E. T. Schmidt [224] presents the same construction with a simplified proof.

The sharpest result is the following (G. Grätzer and E. T. Schmidt [255]):

Theorem 15.   Let $\mathfrak{m}$ be a regular cardinal $> \aleph_0$. Every $\mathfrak{m}$-*algebraic lattice* L can be represented as the lattice of $\mathfrak{m}$-complete congruence relations of an $\mathfrak{m}$-*complete distributive lattice* K.   —

In the construction, we use infinite complete-simple complete distributive lattices (a complete lattice is *complete-simple* if it has only the two trivial complete congruences). Such lattices were constructed in G. Grätzer and E. T. Schmidt [251] and G. Grätzer and E. T. Schmidt [252]. It can be shown, see G. Grätzer and E. T. Schmidt [256], that the representation of the three-element chain must contain such a lattice.

Continuous Geometry

F. Wehrung

Continuous Geometries represent a crucial aspect of the development of modular lattice theory. The mathematics they convey are formidable, on the one hand, because of the depth of the main theorems of the early theory, the most prominent of which is the von Neumann Coordinatization Theorem, and, on the other, because of the *tools* they provide, especially, in light of all the deep results this field inspired in modular lattice theory.

J. von Neumann [1936], [1960], [407], and [408] are the basic references for von Neumann's work on continuous geometry.

1.   The von Neumann Coordinatization Theorem

A (unital, associative) ring R is said to be (von Neumann) *regular* if, for any element $x \in R$, there exists an element $y \in R$ such that $xyx = x$. In this case, the set $\mathcal{L}(R)$ of all principal right ideals of R is a complemented, modular sublattice of the lattice of all right ideals of R. A lattice L is *coordinatizable*, if it is isomorphic to $\mathcal{L}(R)$, for some regular ring R, and *uniquely coordinatizable*, if there exists a regular ring R, unique up to isomorphism, such that L is isomorphic to $\mathcal{L}(R)$. If L is coordinatizable, then, necessarily, L is complemented and modular. Not every complemented modular lattice is coordinatizable. For example, for every integer $n \geqslant 2$, M_n is coordinatizable iff there exists a prime power q such that $n = q+1$. B. Jónsson [333] observes that this already shows that *the class of non-coordinatizable lattices is not finitely axiomatizable*, in fact, not axiomatizable, because otherwise, it would be closed under ultraproducts; however, there are infinitely many n such that M_n is not coordinatizable, and any ultraproduct of those is a modular lattice of height 2 with infinitely many atoms, which is coordinatizable. As far as I know, it is an open the problem whether the class of coordinatizable lattices is axiomatizable.

However, the representation problem of complemented modular lattices by regular rings has an "essentially positive" answer. The corresponding result, the *von Neumann Coordinatization Theorem*, is probably the deepest result in modular lattice theory. It generalizes to a "point-free" context the classical coordinatization theorem of projective geometry, see the Further Topics and References for Chapter IV.

To state this result precisely, one needs to introduce the notion of a *homogeneous basis*. If L is a complemented modular lattice and n is a positive integer, a *homogeneous basis of order n* of L is a finite independent sequence $\langle a_i \mid i < n \rangle$ of join 1 such that, for all $i < n$, a_i and a_j are perspective, in symbol, $a_i \sim a_j$ (see Section IV.3). Homogeneous bases arise naturally from regular rings, as follows: if R is a regular ring and n is a positive integer, then the ring $S = \mathrm{M}_n(R)$ of all $n \times n$ square matrices over R is regular and $\mathcal{L}(S)$ has a homogeneous basis of order n. More precisely, let ε_i, $i < n$, denote the matrix of the ith canonical projection of R^n to itself; then $\langle \varepsilon_i S \mid i < n \rangle$ is a homogeneous basis of order n of $\mathcal{L}(S)$. In particular, every coordinatizable lattice embeds as an ideal into a coordinatizable lattice with a homogeneous basis of arbitrary order.

A simplified statement of the von Neumann Coordinatization Theorem is the following:

The von Neumann Coordinatization Theorem. *Let L be a complemented modular lattice with a homogeneous basis of order at least 4. Then L is uniquely coordinatizable.*

There is more to it: if $\langle a_i \mid i < n \rangle$ is the homogeneous basis, then the regular ring R is constructed as a matrix ring $\mathrm{M}_n(S)$, where S is the so-called *auxiliary ring* (or ring of L-numbers). The underlying set of S is the set of all sectional complements of a_0 in $a_0 \oplus a_1$, endowed with lattice-theoretically defined addition and multiplication. These operations are defined in terms of the homogeneous basis $\langle a_i \mid i < n \rangle$ (see Figures IV.5.5 and 6 for the projective geometric inspiration), but also of additional parameters defining an *n-frame*. A very interesting alternative construction of the auxiliary ring can be found in I. Amemiya [1957], see also L. A. Skornjakov [1961].

Much work has been done since 1936 on the von Neumann Coordinatization Theorem, see for example K. D. Fryer and I. Halperin [192], [1956], [1961], [286], A. Day [123]. A different exposition of the results of von Neumann can be found in F. Maeda [1958].

The hypotheses on the lattice in the von Neumann Coordinatization Theorem have been relaxed to a homogeneous basis of order 3; while full coordinatization is hopeless (because of the existence of non-Arguesian projective planes), weaker versions, involving nonassociative rings, have been found, see I. Amemiya and I. Halperin [1959], [1959a], K. D. Fryer and I. Halperin [192], [1958], [193].

The von Neumann Coordinatization Theorem has been significantly improved by B. Jónsson [1960a]. If L is a complemented modular lattice and n is a positive integer, Jónsson defines a *large partial n-frame* on L. One can easily verify that L has a large partial n-frame iff there exists an element u of L satisfying the following conditions:

(i) The interval $[0, u]$ has a homogeneous basis of order n.

(ii) The greatest element 1 of L belongs to the smallest standard ideal of L generated by u.

Then Jónsson's result can be formulated as follows:

Theorem 1.   Let L be a complemented Arguesian lattice. If L has a large partial 3-frame, then L is uniquely coordinatizable.   —

Furthermore, it follows from Frink's embedding theorem (Theorem IV.5.17) and from the classical coordinatization theorem of projective geometry, that every complemented modular lattice having a large partial 4-frame is Arguesian; thus it is coordinatizable (this result is also contained in B. Jónsson [1960a]).

A closer examination of von Neumann's proof shows further that the hypothesis that the lattice is complemented has not been fully used. Let L be a modular lattice and let $a_0, a_1, \ldots, a_n \in L$, $n \geqslant 2$. After A. P. Huhn [1972a], $\mathbf{a} = \langle a_0, a_1, \ldots, a_n \rangle$ is an *n-diamond,*, if there are $0_{\mathbf{a}}, 1_{\mathbf{a}} \in L$ satisfying the following conditions:

(i) $1_{\mathbf{a}} = \bigvee(a_i \mid i \neq j)$, for all j with $0 \leqslant j \leqslant n$;

(ii) $0_{\mathbf{a}} = a_j \wedge \bigvee(a_i \mid i \notin \{j, k\})$, for all $j \neq k$ in $\{0, 1, \ldots, n\}$.

If, in addition, $0_{\mathbf{a}} = 0_L$ and $1_{\mathbf{a}} = 1_L$, then $\mathbf{a}$ is called a *spanning diamond.*

A. Day and D. Pickering [133] proved that if an Arguesian lattice L has a spanning n-diamond $\langle a_i \mid i \leqslant n \rangle$, with $n \geqslant 3$, and if L satisfies a natural additional condition, then at least the so-called hyperplane can be coordinatized by an associative ring.

These methods lead to important structural results about modular lattices, discussed in Section 1.2.

2.   Continuous Geometries and Related Topics

A lattice L is *meet-continuous*, if for every directed set $\langle I, \leqslant \rangle$ and every *increasing* family $\langle b_i \mid i \in I \rangle$ of elements of L, the supremum $\bigvee(b_i \mid i \in I)$ exists, and, for every $a \in L$,

(1) $$a \wedge \bigvee(b_i \mid i \in I) = \bigvee(a \wedge b_i \mid i \in I).$$

Join-continuous lattices are defined dually. By definition, a *continuous geometry* is a meet-continuous, join-continuous, complemented, modular lattice. Continuous geometries arise, in particular, in the following well-known result of I. Kaplansky [351]:

Theorem 2.   Every complete, modular, orthocomplemented lattice is a continuous geometry.   —

Note that the original proof of Theorem 2 uses the von Neumann Coordinatization Theorem. Kaplansky's result was generalized later to the countably complete case, and to even more general situations, in I. Amemiya and I. Halperin [27].

The most fundamental result about continuous geometries is probably the following, due to J. von Neumann [1960]:

Theorem 3.   Let L be a continuous geometry. Then the relation of perspectivity on L is transitive.   —

One cannot capture the proof of this very difficult result in a few lines. The proof is presented in two parts. In the first part, one assumes that the continuous geometry L is *directly indecomposable*. In this case, any two elements a and b of L are "comparable" with respect to the relation of perspectivity: either $b \lesssim a$, that is, there exists $x \leqslant a$ such that $x \sim b$, or $a \lesssim b$. Furthermore, von Neumann establishes a very special property, called the *finiteness* of L in I. Amemiya and I. Halperin [27]:

Finiteness Property. *In any continuous geometry, there exists no nontrivial infinite homogeneous sequence.*

With these two ingredients, the conclusion follows in the directly indecomposable case. In the general case, the full strength of the completeness assumptions is required to establish an analysis of the *center* of L (see Section III.4), which turns out to be a complete Boolean algebra (and every complete Boolean algebra can be obtained this way, see I. Halperin [284]). Every element of the center of L gives a direct decomposition of L. This allows, more or less, to reduce the problem to the directly indecomposable case.

On every directly indecomposable continuous geometry L, von Neumann constructs a *dimension function* $D\colon L \to [0,1]$ satisfying the following properties:

(i) $D(0_L) = 0$ and $D(1_L) = 1$.

(ii) If $x \wedge y = 0$ in L, then $D(x \vee y) = D(x) + D(y)$.

(iii) For all x, $y \in L$, $D(x) = D(y)$ if and only if $x \sim y$.

In the general case, von Neumann also constructs a non-numerical dimension function $D\colon L \to M$, where M is the partial commutative monoid of perspectivity classes. It turns out, for example, that M is cancellative. Furthermore, the structure of M is analogous to the structure of Dedekind-complete lattice-ordered groups (recall that every Dedekind-complete lattice-ordered group is Abelian). T. Iwamura [322] embeds this partial structure into a total structure, in fact, into a Dedekind-complete lattice-ordered group G. Although it is not explicitly stated in the paper, M is isomorphic to a closed interval of G. The dimension function D behaves as one would expect, for example, it is *continuous* (infinite independent joins are turned into infinite sums). There are extensions of this result in J. von Neumann [1960] and I. Halperin [280]–[282].

Although Theorem 3 is a formidable achievement, von Neumann's proof does not easily lend itself to further generalizations. The problem does not lie in the Finiteness Property, but in the decomposition of the lattice over its center, which requires the full strength of the completeness assumptions. However, I. Halperin [280] could relax the assumptions underlying Theorem 3. Let L be $\aleph_0$-*meet-continuous*, if every countable subset of L admits a supremum, and (1) holds for all increasing *sequences* $\langle b_n \mid n \in \omega \rangle$. We dually define $\aleph_0$-join-continuity. An $\aleph_0$-*continuous geometry* is an $\aleph_0$-meet-continuous, $\aleph_0$-join-continuous, complemented, modular lattice.

Theorem 4.   Let L be a $\aleph_0$-continuous geometry. Then the relation of perspectivity on L is transitive.   —

If L is a coordinatizable $\aleph_0$-continuous geometry, then Theorem 4 follows from known results about regular rings. Indeed, L is isomorphic to $\mathcal{L}(R)$ for some $\aleph_0$-continuous regular ring R. By Handelman's theorem (see K. R. Goodearl [203], Theorem 14.24), R must be unit-regular. It then follows easily that perspectivity in $\mathcal{L}(R)$ is the same as isomorphism of modules, so it is transitive.

This does not look, at first sight, as such an impressive generalization of Theorem 3. In particular, one could expect that a careful examination of von Neumann's proof of Theorem 3 would be sufficient to prove Theorem 4, by trying to argue that all infinite meets and joins involved in the proof should be, really, countable meets and joins. However, this is not so. The most dramatic reason for this is probably the fact that, *for $\aleph_0$-continuous geometries, there may be no nontrivial center*. In particular, the regular ring given in Example 14.35 of K. R. Goodearl [203] shows that there exists a $\aleph_0$-continuous geometry L such that L is *prime*, that is, for all non zero $x, y \in L$, there are non zero $x' \leqslant x$ and $y' \leqslant y$ in L such that $x' \sim y'$, but L *is not simple*. Another question is whether there are situations where $\aleph_0$-continuous geometries may appear or may be more natural than continuous geometries. Again, the answer to this question cannot be summarized in a few lines. Two directions can be outlined:

(i) In an astonishing paper, I. Amemiya and I. Halperin [27] extensively study *independence* in a complemented modular lattice, with some completeness assumptions but without meet- or join-continuity. I. Amemiya and I. Halperin [27] prove that *every complete complemented modular lattice admits a direct decomposition into an $\aleph_0$-meet-continuous lattice, an $\aleph_0$-join-continuous lattice, and a "finite" complete lattice*, as defined above. Interestingly, it does not seem to be known whether one can replace, for example, $\aleph_0$-meet-continuity by meet-continuity. A related open problem in ring theory has been posed by D. E. Handelman, see Problem 54 in K. R. Goodearl [203].

(ii) In ring theory, I. Kaplansky introduced a broad class of C*-algebras, the *Rickart C*-algebras*. With every "finite" Rickart C*-algebra, D. E. Handelman associates an $\aleph_0$-continuous geometry; see K. R. Goodearl [203].

The method of proof of Theorem 4 is quite different from von Neumann's proof of Theorem 3 (one could say that it is simpler because there is no structural analysis of L). Again, the proof is too technical to be summarized in a few lines, but some of its patterns are present in the outline to follow (see also F. Maeda [1958]). If a, b, and c are elements of a complemented modular lattice L such that $a \sim b$ and $b \sim c$, it is easy to prove directly that $a \sim c$, for an *independent* triple $\langle a, b, c \rangle$. On the other hand, if $a \wedge b = b \wedge c = 0$, then $a \wedge (b \vee c) \sim c \wedge (a \vee b)$. By combining these two cases and by using *both* continuity assumptions, one obtains that, in the general case, a and c are "perspective by countable decomposition", that is, there are countable decompositions

$$a = \oplus_{n \in \omega} a_n,$$
$$c = \oplus_{n \in \omega} c_n$$

($\oplus$ means independent join) such that $a_n \sim c_n$, for all $n \in \omega$. It follows easily that the same conclusion holds if a and c are projective. As a corollary, one deduces the following very particular statement, which holds in every $\aleph_0$-continuous geometry; it is called *normality* in F. Wehrung [518].

Normality Property. *If a and b are projective and $a \wedge b = 0$, then a and b are perspective.*

Accordingly, we will say that a lattice is *normal*, if it satisfies the Normality Property. The final stage is to consider any two projective elements a and b, and prove that $a \sim b$. To go further, Halperin uses arguments whose monoid-theoretical counterparts are reminiscent of A. Tarski [494] (published about a decade later). These arguments show that the caveat for a and b to be perspective lies in a certain infinite independent sequence (constructed from a and b), any two elements of which are projective. By the Normality Property, this sequence is homogeneous. By the Finiteness Property (still valid in $\aleph_0$-continuous geometries, with the same proof as for continuous geometries), this sequence is trivial, from which it follows that a and b are perspective.

A further result of I. Halperin and J. von Neumann [287] is that *every meet-continuous complemented modular lattice is normal* (see also Satz II.3.7 in F. Maeda [1958]). The proof does not generalize to $\aleph_0$-meet-continuity, as one needs to consider transfinite independent sequences. This suggests that there should exist a dimension theory of meet-continuous complemented modular lattices (this is being done by the author and K. R. Goodearl). However, a ring-theoretical analogue has been established in C. Busqué [68], under a relatively restrictive countability assumption. The lattice-theoretical analogue and the removal of the countability assumption remain to be done.

The generalization of the Halperin-von Neumann normality result to $\aleph_0$-meet-continuous complemented modular lattices is proved in F. Wehrung [518]. The proof is quite technical, and it uses Theorem 1, followed by a reduction to the 3-distributive case (in the sense of A. P. Huhn [1972a]). This gives a common generalization of the case of $\aleph_0$-continuous geometries and of the meet-continuous case.

This is the first step required in order to associate, with every $\aleph_0$-meet-continuous complemented modular lattice, a "dimension monoid", denoted by $\mathrm{Dim}\,L$, which turns out to be a *generalized cardinal algebra*, in the sense of A. Tarski [494]. Generalized cardinal algebras satisfy many interesting nontrivial properties, in particular, they embed, with their "algebraic" preordering (defined by $x \leqslant y$ iff $x + z = y$, for some z) into powers of the extended real line $\langle [0, +\infty]; +, 0, \leqslant \rangle$, see F. Wehrung [512], [513]. In particular, for every positive integer m, they satisfy the following statement ("unperforation"), see A. Tarski [494]:

$$(\forall x, y)(mx \leqslant my \Rightarrow x \leqslant y).$$

$\mathrm{Dim}\,L$ can, in fact, be defined, in a very simple fashion, for *any* lattice L. For example, for a finite lattice L, $\mathrm{Dim}\,L$ embeds, with its algebraic preordering, into a power of $\mathbb{Z}^+ \cup \{+\infty\}$, where $\mathbb{Z}^+$ is the monoid of non-negative integers.

For relatively complemented modular lattices, a "dimension-theoretic" interpretation of Theorems 3 and 4 can be given as follows: *for a relatively complemented lattice L, $\mathrm{Dim}\,L$ is cancellative if and only if L is modular and perspectivity is transitive in L*, see F. Wehrung [518]. Here, two elements a and b of a relatively complemented lattice L are said to be *perspective*, if there exists $x \in L$ such that $a \wedge x = b \wedge x$ and $a \vee x = b \vee x$. If L is a directly indecomposable continuous geometry, then $\mathrm{Dim}\,L$ is isomorphic to the positive cone of either the integers $\mathbb{Z}$ or the reals $\mathbb{R}$. If L is an arbitrary continuous geometry, then $\mathrm{Dim}\,L$ is isomorphic to the positive cone of the Dedekind-complete lattice-ordered group constructed by T. Iwamura [322]. If L is an $\aleph_0$-continuous geometry, then $\mathrm{Dim}\,L$ is isomorphic to the positive cone of a *monotone σ-complete dimension group*. We refer to K. R. Goodearl [203] for the ring-theoretic analogues of these results and to F. Wehrung [518].

As mentioned earlier, the ring-theoretic and operator-theoretic aspects of continuous geometries and related structures have become a very active focus of research, for which an excellent reference is K. R. Goodearl [203]. It contains many open problems, quite a number of which can be translated into lattice-theoretical terminology. The survey article K. R. Goodearl [204] contains more recent dimension-theoretical results and problems in the theory of regular rings.

The ring-theoretical structure most directly related to the structure of a complemented modular lattice is, of course, the structure of a regular ring. Table 1 lists the correspondence between various basic concepts of regular rings and complemented modular lattices.

Rings	Lattices
continuous rings	continuous geometries
$\aleph_0$-continuous rings	$\aleph_0$-continuous geometries
right continuous	meet-continuous
$\aleph_0$-right continuous	$\aleph_0$-meet-continuous
$V(R)$	Dim L
two-sided ideal	congruence

Table 1: Complemented modular lattices and regular rings

Specific methods in one of the fields arise from concepts *without* analogue in the other field. A good example for this is provided by the concept of right self-injectivity for rings, which does not have any analogue for lattices, see Chapter 13 in K. R. Goodearl [203], especially, Theorem 13.13. Conversely, many lattice-theoretical notions do not have any analogue in ring theory. In A. Day, C. Herrmann, and R. Wille [1972], C. Herrmann [303], then G. Bruns and M. Roddy [63], very interesting lattices are constructed. They are modular ortholattices, thus, in particular, they are complemented and modular. In particular, the lattice L of G. Bruns and M. Roddy [63] is a lattice of closed subspaces of the Hilbert space (that is, the unique infinite-dimensional separable Hilbert space), and the join is given by the sum of subspaces; in particular, L is Arguesian. However, L is not coordinatizable: it does not even satisfy the Normality Property above (this remark is due to C. Herrmann, see F. Wehrung [518]). This shows, in particular, that a complemented sublattice of a coordinatizable lattice may not be coordinatizable.

Projective Lattice Geometries

M. Greferath and S. E. Schmidt

The first lattice theoretic approach to projective geometry was developed independently by K. Menger and G. Birkhoff in the early thirties. It was based on the fact that every *projective space* is associated with a *projective geometry* defined as the partially ordered set of its linear subspaces; indeed, (classical) projective geometries can be characterized as algebraic modular lattices that are atomistic and irreducible (see F. and S. Maeda [1970]).

Modularity is one of the fundamental properties of classical projective geometry. It is responsible for the fact that projections are join preserving maps and that perspectivities are (interval) isomorphisms. Since the partially ordered set of all submodules of a module (or even more generally, any sublattice of the lattice of all normal subgroups of a group) forms a modular lattice, it is natural that order-theoretic generalizations of projective geometry are based on modular lattices.

The set of all 1-generated submodules of a module contains important information about the structure of the given module but, in general, cannot be recovered from the structure of the submodule lattice. To keep this additional information available, it is therefore natural to consider a modular lattice together with subset(s) of *points* satisfying some axioms that synthetically characterize general properties of 1-generated submodules of a module.

K. Faltings [149] was the first to axiomatically introduce *points* and *hyperplanes* to modular lattices. A comparable approach, establishing a lattice-geometric axiomatization of torsion-free modules over Ore domains, is due to

U. Brehm [57], [58]. Introducing the concept of *projective lattice geometry*, M. Greferath and S. E. Schmidt [272] unified these two approaches.

In Section 1, we sketch the historical background of projective lattice geometries originating with the famous work of J. von Neumann, B. Jónsson, R. Baer, and others. We then summarize the results of K. Faltings and U. Brehm to lay the foundation for the axiomatics presented in Section 2. Then we concentrate on representation results for projective lattice geometries; it turns out that most classical results can be reformulated and improved and algebraic representations of new classes of projective lattice geometries can be given. Section 3 concludes our presentation by showing how residuated maps between modular lattices yield *projective maps* between projective lattice geometries.

1. Background

In the 1930s, projective geometries over vector spaces were axiomatically described by a special class of complemented modular lattices. J. von Neumann [1960] (in connection with his work on rings of operators) vastly extended this result with his famous representation of arbitrary complemented modular lattices of order $\geqslant 4$ by *von Neumann regular rings* (see Chapter IV and Appendix D).

Another prominent class of lattices, the *primary lattices*, arose from finitely generated modules over Artinian chain rings (that is, rings whose one-sided ideals are two-sided and form a finite chain). A modular lattice of finite length is *primary*, if every interval is either a chain or contains at least three atoms and at least three coatoms (dual atoms). R. Baer [33] (in 1942) and E. Inaba [321] (in 1948) proved representation results for primary lattices.

Now we state a version of these representation theorems using A. P. Huhn's *spanning n-diamond* (that is, a family of $n + 1$ elements of L such that any n-element subfamily is independent and spanning in L).

Theorem 1.   For a bounded poset L and a natural number $n \geqslant 4$, the following are equivalent:

(i) L is a modular lattice containing a spanning n-diamond, and either
   (A) L is complemented or
   (B) L is primary.

(ii) L is isomorphic to the poset of all finitely generated submodules of a free R-module of rank n. Here, in case **(A)**, R is a von Neumann regular ring or, in case **(B)**, R is an Artinian chain ring.   —

There is no unified proof for cases **(A)** and **(B)**. Replacing the modular law by the Arguesian identity, B. Jónsson [1960a] and B. Jónsson and G. Monk [1969] succeeded in extending this theorem to the planar case using the *large n-diamonds* of B. Jónsson (for an overview, see B. Jónsson [337]).

The purely lattice-theoretic approach to projective geometry has been continued by several mathematicians, including the construction of a coordinate ring for modular lattices with spanning n-diamond, $n \geqslant 4$, by B. Artmann [30] and for Arguesian lattices with spanning 3-diamond by A. Day and D. Pickering [133].

Theorem 2.   In an Arguesian lattice with an upper complementable spanning n-diamond, $n \geqslant 3$, the lower interval of every hyperplane can be homomorphically mapped into the submodule lattice of a free module of rank $n-1$.   —

In this result, a spanning n-diamond $\langle x_1, \dots, x_{n+1} \rangle$ of a lattice L is *upper complementable*, if for every $t \in L$ and $C \subseteq \{x_1, \dots, x_{n+1}\}$ with $t \vee \bigvee C = 1$, there exists $t_0 \leqslant t$ with $t_0 \vee \bigvee C = 1$ and $t_0 \wedge \bigvee C = 0$; a *hyperplane* of a spanning n-diamond is the join of $n-1$ of its elements. See A. Day and D. Pickering [133].

1.1   Faltings' and Brehm's approaches to lattice geometry

K. Faltings [149] derives a representation of certain *modular lattices with point system* by modules. He defines a *point system* of a complete modular lattice L as a subset E of L such that the following axioms are satisfied:

(P1) $t \vee a = b \vee a$ and $t \wedge a = 0$ imply that $t \in E$, for all $a, b \in E$ and $t \in L$.

(P2) E is join-dense in L.

(P3) For all $x, y \in L$ and $e \in E$ with $e \leqslant x \vee y$, there exist $c, d \in E$ with $c \leqslant x$ and $d \leqslant y$ such that $c \vee d = d \vee e = e \vee c$.

(P4) If T is any nonempty subset of E such that $e \leqslant c \vee d$ with $e \in E$ and c, $d \in T$ implies that $e \in T$, then $T = \{ e \in E \mid e \leqslant \bigvee T \}$.

Faltings defines a *hyperplane* of a modular lattice L with point system E as an element h of L satisfying the following axioms:

(H1) There exists a nonzero $p \in E$ with $h \vee p = 1$ and $h \wedge p = 0$.

(H2) For $q \in E$ with $h \vee q = 1$, there exists $p \in E$ with $p \leqslant q$ such that $h \vee p = 1$ and $h \wedge p = 0$.

(H3) For all $e \in E$ with $e \leqslant h$ and $p \in E$ with $h \vee p = 1$ and $h \wedge p = 0$, there exists $q \in E$ such that $h \vee q = 1$, $h \wedge q = 0$ and $e \vee p = p \vee q = q \vee e$.

The lattice of all submodules of a module $_R M$ together with the set of all 1-generated submodules of $_R M$ is a fundamental example of a modular lattice with point system. We shall call a modular lattice L with point system E *module-induced*, if there exists a module $_R M$ and a lattice isomorphism between

L and the submodule lattice of $_RM$ that maps E onto the set of all 1-generated submodules of $_RM$. If H and P are submodules of $_RM$ such that $H \oplus P = M$ and $_RP \cong {_R}R$, then H is a hyperplane of the modular lattice with point system induced by $_RM$.

For his main result concerning the representation of modular lattices with point system, Faltings needs the existence of hyperplanes having some additional strong properties: A hyperplane of a modular lattice with point system $\langle L, E \rangle$ is called *regular*, if it satisfies the conditions (R0)–(R4) of K. Faltings [149]. Condition (R0) states for a hyperplane h of $\langle L, E \rangle$ that every 3-generated element of $\langle L, E \rangle$ (that is, every element that is the join of three elements of E) is disjoint from some complement of h in L. A sufficient condition for a hyperplane h to be regular is that every 5-generated element of $\langle L, E \rangle$ is disjoint from a suitable complement of h. Another sufficient condition is that every element of E is *uniform* in L (that is, for x, $y \in L - \{0\}$ and $e \in E$, $x \vee y \leqslant e$ implies $x \wedge y \neq 0$) and that (R0) be satisfied. A modular lattice with point system that is induced by a free module of infinite rank over an arbitrary ring always contains regular hyperplanes.

Faltings' main representation result states the following:

Theorem 3.   Every modular lattice with point system containing a regular hyperplane is induced by a module.   —

U. Brehm [57], [58] considers modular lattices together with a distinguished subset of *points* to derive a lattice-geometric axiomatization of torsion-free modules over left Ore domains. Recall that a nonzero ring R is called a *left Ore domain*, if it has no zero divisors and if $Ra \cap Rb \neq \{0\}$, for arbitrary nonzero a, $b \in R$. A module $_RM$ over a left Ore domain is *torsion-free*, if $rx = 0$ implies that $r = 0$ or $x = 0$, for all $r \in R$ and $x \in M$. It is well known that each left Ore domain R has a *skew field of left quotients* $K \supseteq R$ with $K = \{a^{-1}b \mid a,\ b \in R,\ a \neq 0\}$. If $_RM$ is a torsion-free module over a left Ore domain R and if K is the skew field of left quotients of R, then the canonical map $i \colon M \to K \otimes_R M$ is injective and thus M can be regarded as an R-submodule of a K-vector space. This fact makes the torsion-free left modules over left Ore domains an interesting class of modules.

Let L be a complete modular lattice, let P be a subset of L, and let $\varrho(L)$ denote the uniform rank of L, that is, the maximal cardinality of an independent family of L.

We consider the following axioms:

(A1) P is join-dense in L.

(A2) All $p \in P$ are compact in L.

(A3) For all p_1, $p_2 \in P$ with $p_1 \wedge p_2 = 0$, there exists $p_3 \in P$ with $p_1 \vee p_2 = p_1 \vee p_3 = p_2 \vee p_3$.

(A4) For all p, $q \in P$, $a \in L - \{0\}$ with $p \leqslant q \vee a$, there exists $q' \in P$ with $q' \leqslant a$ and $p \leqslant q \vee q'$.

(A5) Either $\varrho(L) = 3$ and L is Arguesian, or $\varrho(L) \geqslant 4$.

(A6) All $p \in P$ are uniform.

(A7) For all a, b, $c \in L$ with $a \wedge c = 0$ and $(a \vee b) \wedge c \neq 0$, there exists $c' \in L$ with $a \wedge c' = 0$ and $b \wedge c' \neq 0$.

Now Brehm's representation theorem is the following:

Theorem 4. Let L be a complete modular lattice and P a subset of L. Then $\langle L, P \rangle$ satisfies the axioms (A1)–(A7) if and only if there exists a left Ore domain R, a torsion-free R-module M of uniform rank $\geqslant 3$, and a lattice isomorphism $f \colon L \to L(_R M)$ with $f(P) = \{\, Rx \mid x \in M - \{0\} \,\}$.   —

2.   A Unified Approach to Lattice Geometry

Faltings' and Brehm's approaches to projective geometry are unified by the concept of *projective lattice geometry*, introduced in M. Greferath and S. E. Schmidt [272]. Concentrating on the representation aspects, we shall give a short survey of this field on the boundary of geometry, lattice theory, and algebra.

Definition. *A projective (lattice) geometry is a triple $G = \langle L, E, F \rangle$, where L is a complete lattice and F, E are sets of compact elements of L, satisfying $F \subseteq E$ and the following axioms:*

(E1) *$0 \in E$ and E is join-dense in L, that is, $x = \bigvee (e \in E \mid e \leqslant x)$, for all $x \in L$.*

(E2) *For x, $y \in L$ and $e \in E$ with $e \leqslant x \vee y$, there exist $c \in E$ and $d \in E$ with $c \leqslant x$ and $d \leqslant y$ such that c, d, e are balanced, that is, $c \vee d = d \vee e = e \vee c$.*

(E3) *For c, $d \in E$, there exists $e \in E$ such that c, d, e are balanced.*

(F1) *For a, $e \in E$ and $f \in F$ with $f \vee a = c \vee a$ and $f \wedge a = 0$, there exists $p \in F$ with $p \leqslant e$, $p \wedge a = 0$ and $f \vee a = p \vee a$.*

(F2) *For $e \in E$ and $f \in F$ with $e \wedge f = 0$, there exists $g \in F$ with $e \wedge g = 0$ such that e, f, g are balanced.*

(F3) *For f, $g \in F$ with $f \wedge g = 0$, there exists $p \in F$ such that f, g, p are directly balanced, that is, f, g, p are balanced and $f \wedge g = g \wedge p = p \wedge f = 0$.*

It follows easily from the axioms that L is algebraic and modular. The elements of E and F are called *points* and *free points*, respectively. Complements of free points (if they exist) are called *hyperplanes*.

n-*generated* elements of G are joins of n points; G itself is said to be n-generated, if the unit element 1 is n-generated. G is *finitely generated*, if 1 is compact in L.

A *basis* of G is an independent spanning set of free points of G; a projective geometry is said to be of *dimension* n if it has an $(n+1)$-element basis. It is clear that a projective geometry may be of various different dimensions or even of no dimension at all.

Examples

(a) For a unitary left R-module M, let $L(_RM)$ denote the submodule lattice of $_RM$, let $E(_RM)$ be the set of all 1-generated submodules of $_RM$, and let $F(_RM)$ consist of those submodules of $_RM$ that are isomorphic to $_RR$. Then $G(_RM) = \langle L(_RM), E(_RM), F(_RM) \rangle$ is a projective geometry. $G(_RM)$ is finitely generated iff $_RM$ is and it has a basis iff $_RM$ is a free module.

A projective geometry G is said to be *module-induced*, if there exists a module $_RM$ such that $G \cong G(_RM)$ (that is, there exists an order isomorphism between the underlying lattices that induces a bijection between the sets of points and also between the sets of free points). We emphasize that the underlying ring of a module-induced projective geometry G is determined up to isomorphism, whenever the geometry contains at least 3 independent free points. Furthermore, if G is of dimension at least 2, then the underlying module is determined up to isomorphism.

(b) For a modular algebraic lattice L, a very general example of a projective geometry is given by $\langle L, E, \varnothing \rangle$, where E denotes the set of all compact elements of L. For lattice theoretic constructions of various point sets, see S. E. Schmidt [469]. If, in addition, L is atomistic and irreducible and A denotes the set of all atoms of L, then the resulting geometry $\langle L, A \cup \{0\}, A \rangle$ will be called a *classical projective geometry*. Indeed a projective geometry is classical if and only if all of its nonzero points are free and form an antichain.

(c) Let $G = \langle L, E, F \rangle$ and $G' = \langle L', E', F' \rangle$ be projective geometries. Then for all $x \in L$ and every nonzero natural number n, one can construct the following projective geometries:

$G(x) = \langle L(x), E(x), F(x) \rangle$, the *subgeometry of x in G*,

$G/x = \langle L/x, E/x, F/\!/x \rangle$, the *quotient geometry of G over x*,

$G^{(n)} = \langle L, E^{(n)}, F^{[n]} \rangle$, the *$n$-th cluster geometry of G*,

$G_{\mathrm{red}} = \langle L, E, \varnothing \rangle$, the *reduced geometry of G*,

$G \times G' = \langle L \times L', E \times E', F \times F' \rangle$, the *direct product of G and G'*.

Here $T(x) = \{\, t \in T \mid t \leqslant x \,\}$ and $T/x = \{\, t \vee x \mid t \in T \,\}$; furthermore, $T/\!/x = \{\, t \vee x \mid t \in T,\ t \wedge x = 0 \,\}$, for $T \subseteq L$ and $x \in L$; the set of all (freely)

n-generated elements of G is denoted by $E^{(n)}$ (and $F^{[n]}$). If G and G' are module-induced, then so are all the geometries defined above.

U. Faigle and C. Herrmann [148] have shown that the well-known correspondence between classical projective spaces (of points and lines) and classical projective geometries (based on lattices of *flats*) is preserved in a more general situation. They designated the set of all join-irreducibles of a modular lattice of finite length as a point set of a *projective space on a partially ordered set* and showed that the lattice of all linear sets of this space is isomorphic to the original lattice. See also C. Herrmann, D. Pickering, and M. Roddy [309].

S. E. Schmidt [468] extended the considerations of U. Faigle and C. Herrmann [148] to more general point sets: each projective lattice geometry $\langle L, E, F \rangle$ is naturally associated with an intrinsically axiomatizable *projective space* $\langle E^{(2)}, E, F \rangle$. Here a projective space in the classical sense occurs when the nonzero points are free and trivially ordered. Furthermore, we point out that both in U. Faigle and C. Herrmann [148] and in S. E. Schmidt [468], the main axiom for projective spaces on partially ordered sets is a generalized version of *Pasch's axiom* (in geometry, it is usually called the *Veblen-Young axiom*, see [427]).

Projective spaces and projective lattice geometries naturally form categories.

Proposition. *The category of all projective spaces on partially ordered sets is equivalent to the category of all projective lattice geometries. Under this equivalence, projective spaces in the classical sense correspond to classical projective geometries.*

There is a very general concept of lattice geometry discussed by R. Piziak and M. K. Bennett [432].

Desargues' postulate

Next we extend Desargues' postulate to projective lattice geometries (see S. E. Schmidt [469] and M. Greferath [271]), as illustrated in Figure 1.

(D) If a^0, a^1, a^2, b^0, b^1, b^2, $e \in E$ satisfy $e \leqslant a^i \vee b^i$, for all $i \in \{0, 1, 2\}$, then there exist a_i, $b_i \in E$ and $c_i \in L$ with $a_i \leqslant a^i$, $b_i \leqslant b^i$, for all $i \in \{0, 1, 2\}$, such that $\langle e, a_i, b_i \rangle$, $\langle c_i, a_j, a_k \rangle$, $\langle c_i, b_j, b_k \rangle$, and $\langle c_0, c_1, c_2 \rangle$ are balanced triples for all $\{i, j, k\} = \{0, 1, 2\}$.

A projective geometry is *Desarguesian*, if Desargues' postulate holds. Obviously, every module-induced projective geometry is Desarguesian, and from B. Jónsson [1954] it follows that a classical projective geometry is Desarguesian exactly when the Arguesian identity holds in the underlying lattice. The next result (M. Greferath [271] and an unpublished work of C. Klee and S. E. Schmidt) establishes the geometric meaning of the Arguesian identity in full generality.

Proposition. *A projective geometry is Desarguesian if and only if its underlying lattice is Arguesian.*

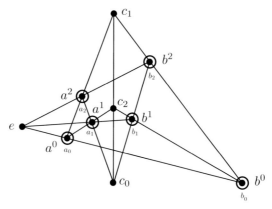

Figure 1

Property U

For many representation results, the following property has turned out to be crucial. A projective geometry is said to have the *Property* U, if each of its points is contained in a complemented free point.

The algebraic counterpart of Property U is the following property of a ring R: For all α, $\beta \in R$, there exist γ, δ, $\varepsilon \in R$ such that $\gamma\delta = \alpha$, $\gamma\varepsilon = \beta$, and $\delta R + \varepsilon R = R$; in other words, every pair $\langle \alpha, \beta \rangle$ is a left multiple of a unimodular pair $\langle \delta, \varepsilon \rangle$. We call rings with this property *proper right Bezout rings*, since for domains and local rings the properties to be right Bezout and proper right Bezout are equivalent (see P. M. Cohn [1971]). The following straightforward proposition illustrates an important connection.

Proposition. *Let R be a ring.*

(i) *R is a proper right Bezout ring iff $G(_RR^2)$ has Property* U.

(ii) *If R is a proper right Bezout ring, then the projective geometry of every free module over R has Property* U.

Representation results

The fundamental representation result of K. Faltings stated in Theorem 3 leads to a representation theorem for projective lattice geometries with interesting consequences (see M. Greferath and S. E. Schmidt [274]):

Theorem 5.   A projective geometry is module-induced, whenever it contains a complemented element the subgeometry of which satisfies $(\mathbf{C}_5^1)$, or $(\mathbf{C}_4^2)$, or $(\mathbf{C}_3^3)$.   —

Here a projective geometry G satisfies condition $(\mathbf{C_n^k})$, if there exists a hyperplane h in G such that for every family $x_1, \ldots, x_k$ of n-generated elements of G, there exists a complement of h which is disjoint from each of the $x_1, \ldots, x_k$.

Corollary 1.   Every projective geometry with an infinite basis is module-induced.   —

A general representation of finite dimensional projective geometries seems to be out of reach. An application of A. Day and D. Pickering [133], however, allows at least a representation of hyperplanes in Desarguesian projective geometries with finite basis, see M. Greferath [269].

Theorem 6.   In a Desarguesian projective geometry the subgeometry of every at least 1-dimensional hyperplane is module-induced.   —

Another application of [133] gives an algebraic representation of the affine part of a projective geometry (see S. E. Schmidt [470]); we point out that each projective geometry with a given hyperplane is naturally associated with an *affine geometry* (where the hyperplane is regarded as the *hyperplane at infinity*).

Theorem 7.   For a Desarguesian projective geometry with a hyperplane of dimension at least 1, its associated affine geometry is module-induced.   —

Comparing the last two theorems with Theorem 2, we see how an enriched geometric set-up gives rise to much stronger consequences than a purely lattice theoretic one.

Complete modular lattices with a point system—in the sense of U. Brehm [57], [58]—are closely connected with *Ore geometries*, which are defined as projective geometries where every nonzero point is free and uniform (the latter means that any two nonzero points contained in a common point have a nonzero intersection). Now, Brehm's result—as stated in Theorem 4—has a nice reformulation within projective lattice geometry.

Theorem 8.   For a projective geometry G that contains a subgeometry of dimension 3, or 2 if G is Desarguesian, the following are equivalent:

(i) G is an Ore geometry.

(ii) G is induced by a torsion-free left module over a left Ore domain.   —

Define a *point-irreducible geometry* as a projective geometry where each nonzero point is join-irreducible. Then any module over a local ring induces a point-irreducible geometry. An algebraic representation of these geometries is hard, in general, but the large class of point-irreducible geometries having Property U allows such representations. For the next two results, see M. Greferath and S. E. Schmidt [273].

Theorem 9.   For a finitely generated but not 5-generated projective geometry G, the following are equivalent:

  (i) G is a point-irreducible geometry having Property U.

  (ii) G is induced by a free left module over a right chain ring.   —

To give a purely lattice-theoretic application of this theorem, call a lattice *relatively irreducible*, if no interval of the lattice is directly decomposable. A relatively irreducible modular lattice of finite length can easily be completed to a point-irreducible geometry; together with Theorem 9, this yields the following characterization of free left modules over Artinian right chain rings. Recall that the Kurosh-Ore dimension of a lattice L is the minimal number of join-irreducibles needed to span the unit element of L.

Theorem 10.   For a poset L and a natural number $n \geqslant 6$, the following are equivalent:

  (i) L is a relatively irreducible modular lattice of finite length, the Kurosh-Ore dimension of L is n, and in L all maximal join-irreducibles are complemented and of equal rank;

  (ii) $L \cong L(_R R^n)$, for some Artinian right chain ring R.   —

We emphasize that the last two theorems give a substantial extension of the results of R. Baer [33] and E. Inaba [321] in that a larger class of rings is involved. Indeed, from ring theory one knows the big difference between (Artinian) right chain rings and Artinian chain rings! Our next aim is an extension of Theorem 9 to an even larger class of geometries.

For a nonzero natural number k, a projective geometry is called *k-stable*, if any k of its complemented free points have a common complement; a ring R is called k-stable, if the following holds:

> Whenever α_i, β_i, γ_i, $\delta_i \in R$ $(i = 1, \dots, k)$ satisfy the condition that all $\alpha_i \gamma_i + \beta_i \delta_i$ are units of R, then there exist γ, $\delta \in R$ such that all $\alpha_i \gamma + \beta_i \delta$ are units of R.

The class of k-stable geometries having Property U allows an algebraic representation in the following way.

Theorem 11.   For a projective geometry G and natural numbers $k \geqslant 2$ and $n \geqslant 3$ with $k + n \geqslant 7$, the following are equivalent:

(i) G is k-stable of dimension n and has Property U.

(ii) $G \simeq G(_R R^{n+1})$, where R is a k-stable proper right Bezout ring.   —

The proof of the above theorem is based on Theorem 5 and uses lattice-geometric considerations of independent interest (see M. Greferath and S. E. Schmidt [274]).

A *von Neumann geometry* is a projective geometry where every point has a complement and the meet of any two compact elements is again compact. It is easily verified that every (submodule of a) free left module over a von Neumann regular ring induces a von Neumann geometry.

The well known representation theorems of J. von Neumann [1960] (see also Theorem 1) and B. Jónsson [333] can be reformulated for von Neumann geometries (see M. Greferath and S. E. Schmidt [272]).

Theorem 12.   For a von Neumann geometry G of dimension n with $n \geqslant 3$, or $n = 2$ and G is Desarguesian, there exists a von Neumann regular ring R such that the reduced geometries of $G^{(n+1)}$ and $G(_R R)$ are isomorphic.   —

A large subclass of von Neumann geometries allows a more satisfactory result in light of Theorem 11. Recall that a ring R is *unit-regular*, if for every $x \in R$, there exists a unit $y \in R$ such that $xyx = x$. Unit-regular rings are 2-stable and proper (right) Bezout.

Theorem 13.   For a natural number $n \geqslant 5$ and a projective geometry G, the following are equivalent:

(i) G is a 2-stable von Neumann geometry of dimension n and has Property U.

(ii) $G \simeq G(_R R^{n+1})$, where R is a unit-regular ring.   —

We ask: Is every von Neumann geometry of dimension at least 3 module-induced?

3.   Residuated Maps

Homomorphisms between algebraic structures often fail to induce lattice homomorphisms between their subalgebra lattices; however, they induce residuated maps in the following sense (see T. S. Blyth and M. F. Janowitz [52]): for the partially ordered sets P and Q, a map $f \colon P \to Q$ is *residuated*, if there exists a map $g \colon Q \to P$ such that $f(x) \leqslant y$ is equivalent to $x \leqslant g(y)$, for all $x \in P$ and $y \in Q$. The map g (which is uniquely determined by f) is called the *residual* of f and will be denoted by f^-. A map between complete lattices is residuated if and only if it is completely join preserving; then its residual map is completely meet preserving.

Now, for a homomorphism $\varphi \colon A \to B$ between (non-indexed) algebras, let f map every subalgebra S of A to the least full subalgebra of B containing $\varphi(S)$, and let g map every subalgebra of B to its preimage under φ. Obviously $f(S) \leqslant T$ if and only if $S \leqslant g(T)$, for all subalgebras S of A and T of B. Thus, f is a residuated map between the subalgebra lattices of A and B satisfying $f^- = g$.

The problem to give an intrinsic characterization of algebraically induced residuated maps is known to be very hard even for subalgebra lattices of modules. And even for vector spaces (over possibly different division rings) a complete classification of algebraically induced residuated maps between their subspace lattices is restricted to the case, where atoms are mapped to atoms or 0 (see C.-A. Faure and A. Frölicher [153] and also H. Havlicek [301]).

By definition, it is clear that the bijective residuated maps are exactly the order isomorphisms. Thus the prominent problem of characterizing algebraically induced order isomorphisms (and automorphisms) is an important specialization. For vector spaces such a characterization is known as the *fundamental theorem of projective geometry*, see E. Artin [29] and R. Baer [1952].

A residuated map $f \colon P \to Q$ induces an order isomorphism between $f^-(Q)$ and $f(P)$. In case $f(P)$ is an order ideal or $f^-(Q)$ is an order dual ideal, f is said to be *range closed* or *dually range closed*, respectively. A *regular residuated* map is residuated and both range closed and dually range closed.

Within the context of projective lattice geometry, T. Pfeiffer and S. E. Schmidt [424] investigate regular residuated maps that preserve points. This means that the images of all points form the point set of a subgeometry, the free points of which are exactly the images of all those free points disjoint from the *kernel*. (The kernel of a residuated map is the largest element mapped to 0.) These *projective maps* play an important role for projective geometry on modular lattices. A projective map decomposes into a quotient map, an isomorphism, and a natural embedding. The idempotent projective maps are exactly the projections. In the classical situation of projective geometry, it turns out that every projective surjection is the composition of a projection and an isomorphism; here a residuated map with an at least 3-element image is already projective, if it is totally range closed (see H. Brauner [56], T. Pfeiffer and S. E. Schmidt [424]).

The algebraic background of projective maps is as follows. Every linear map between modules induces a projective map between the corresponding projective lattice geometries. An intrinsic characterization of algebraically induced projective maps, however, is a hard problem which has been only partially solved. Via the decomposition property, this problem can be reduced to projective isomorphisms. T. Pfeiffer [423] yields a vast expansion of the classical fundamental theorem of projective geometry.

Theorem 14.   Let $_RM$ and $_SN$ be modules such that on the projective line $\mathrm{PG}(_RR^2)$ any point is disjoint from some free point. Let $\varphi \colon L(_RM) \to L(_SN)$

be any map such that $\varphi(M)$ contains a free direct summand of rank 3. Then the following statements are equivalent:

(i) φ is a projective map from $\mathrm{PG}(_RM)$ to $\mathrm{PG}(_SN)$.

(ii) φ is induced by a semilinear map from $_RM$ to $_SN$ the underlying ring homomorphism of which is an isomorphism.   —

Note that the class of rings sharing this projective line property is closed under ring-direct products and contains the class of all domains and all left (or right) chain rings.

A different approach to the characterization of residuated maps between projective geometries is due to M. Greferath [271]. He defines a *Barbilian morphism* as a residuated map preserving complemented point-hyperplane pairs. *Barbilian isomorphisms* are defined in an obvious way, and it turns out that these coincide with the projective isomorphisms if a certain richness condition is satisfied. In case $_RM$ and $_SN$ are free modules and $\varphi\colon {}_RM \to {}_SN$ is a semilinear map that maps a basis of $_RM$ to one of $_SN$, then the map $\overline{\varphi}\colon L(_RM) \to L(_SN)$ with $U \mapsto {}_S\langle\varphi(U)\rangle$ is a Barbilian morphism between the induced projective geometries. It can be shown that Barbilian morphisms preserve what could be called the *arithmetics on a line* (which, in the classical case, is illustrated in Figures 5 and 6 of Section IV.5).

For every Barbilian morphism between projective geometries over free modules of rank $\geqslant 3$, this yields the construction of a ring homomorphism which extends to a semilinear map between the modules in question. We state one related result (see [270], [271]):

Theorem 15.   Let $_RM$ and $_SN$ be unital modules such that $G(_RM)$ contains a freely generated hyperplane the subgeometry of which satisfies that every point is disjoint from a free point. Then every Barbilian morphism between $G(_RM)$ and $G(_SN)$ is induced by a basis preserving semilinear map (which is uniquely determined up to a semilinear dilatation).   —

Within the context of projective lattice geometry, residuated maps and even order isomorphisms need not necessarily preserve points. Questions concerning representations of such maps lead to quite general investigations of residuated maps between submodule lattices. We will report a result of this kind due to U. Brehm [60].

Recall that it is not difficult to see that tensoring all submodules of a module $_RM$ with a bimodule $_SB_R$, which is flat as a right R-module, induces a lattice homomorphism from $L(_RM)$ to $L(_SB \otimes_R M)$ (see N. Bourbaki [55]). Thus for any S-module monomorphism $g\colon {}_SB \otimes_R M \to {}_SN$, the assignment $U \mapsto g(B \otimes_R U)$ defines the map $L(_RM) \to L(_SN)$, which is a residuated lattice homomorphism.

Under certain conditions on the module $_RM$, the converse is also true, namely, each lattice homomorphism that preserves arbitrary joins has an essentially unique representation of this kind.

In [59], [60], U. Brehm introduces two geometric conditions for modules, the *triangle property* and the *splitting property*.

The splitting property: A module $_RM$ has the *splitting property*, if there exist submodules H_i of $_RM$ and elements x_i of H_i, $(i = 0, 1, 2)$ such that the following conditions hold, where $W = (H_0 + H_1) \cup (H_1 + H_2) \cup (H_2 + H_0)$.

(i) $M = H_0 + H_1 + H_2$, $H_j \cap H_k = 0$, $\mathrm{ann}(x_i) = 0$, and $Rx_i \cap (H_j \oplus H_k) = 0$, for all $\{i, j, k\} = \{0, 1, 2\}$.

(ii) $Rx = \sum\{Rrx \mid$ there exist $y, z \in W$ with $rx - y \in W$ and $rx - z \in W$ such that $Ry \cap (Rrx + Rz) = 0\}$, for all $x \in M$.

The triangle property: A module $_RM$ has the *triangle property*, if the following holds, for arbitrary elements x_1, x_2, x_3, x_4 of M.

(i) There exists an element y of M with $\mathrm{ann}(y) = 0$ and $Ry \cap (Rx_1 + Rx_2) = 0$.

(ii) If $\mathrm{ann}(x_i) = 0$ and $Rx_i \cap Rx_{i+1} = 0$, for all $i = 1, 2, 3$, then there exists an element y of M with $\mathrm{ann}(y) = 0$ such that $Ry \cap Rx_i = 0$, for all $i = 1, 2, 3$.

(iii) If $\mathrm{ann}(x_1) = 0 = \mathrm{ann}(x_2)$ and $Rx_i \cap Rx_j = 0$, for all $i \neq j \in \{1,2,3\}$ but $Rx_3 \cap (Rx_1 + Rx_2) \neq 0$, then there exists an element y of M with $\mathrm{ann}(y) = 0$ such that $Ry \cap (Rx_i + Rx_j) = 0$, for all $i, j = 1, 2, 3$.

Theorem 16.   If $_RM$ has the triangle property or the splitting property and if $f: L(_RM) \rightarrow L(_SN)$ is a map, then the following are equivalent.

(i) f is a residuated lattice homomorphism.

(ii) There exists a bimodule $_SB_R$ and an S-module monomorphism

$$g: {_S}B \otimes_R M \rightarrow {_S}N$$

such that B_R is flat and $f(U) = g(B \otimes U)$, for all $U \in L(_RM)$.

Furthermore, if (ii) also holds with a map k and a bimodule $_SC_R$ in place of g and $_SB_R$, then there exists a unique bimodule isomorphism $\varphi: {_S}C_R \rightarrow {_S}B_R$ such that $k = g \circ (\varphi \otimes_R \mathrm{id}_M)$.   —

As an application, Brehm gives an algebraic characterization of lattice isomorphisms between submodule lattices.

The splitting property holds for every free module of rank at least 3 over a von Neumann regular ring (or a Baer ring), while the triangle property is satisfied for a module over a left Ore domain, provided that the module contains a free submodule of rank 3. Slight variations of the triangle property have been used in W. Stephenson's thesis [488] and in L. A. Skornjakov [477]. V. P. Camillo [69] considered lattice isomorphisms between free modules of finite rank $\geqslant 3$ and used a condition resembling the splitting property. His condition is satisfied whenever the underlying ring R is a serial ring or a semi-hereditary ring or a domain.

Varieties of Lattices

P. Jipsen and H. Rose

In this appendix, we discuss some of the more recent results and give a general overview of what is currently known about lattice varieties. Of course, it is impossible to give a comprehensive account. Often we only cite recent or survey papers, which themselves have many more references. We would like to apologize in advance for any errors, omissions, or miscrediting of results.

For proofs of the results mentioned here, we refer the reader to the original papers. Details of many of the results from before 1992 can also be found in our monograph, P. Jipsen and H. Rose [328].

1. The Lattice Λ

Recall from Section V.2 that the lattice Λ of all lattice varieties is a dually algebraic, distributive lattice that has the variety $\mathbf{L}$ of all lattices at the top, the variety $\mathbf{T}$ of all trivial lattices at the bottom, and the variety $\mathbf{D} = \mathbf{Var}(C_2)$ of all distributive lattices as the unique atom. To conclude that $\mathbf{L}$ is join-irreducible and has no coatoms, B. Jónsson [1967] argued as follows: Let $\mathbf{V}$, $\mathbf{W}$ be proper subvarieties of $\mathbf{L}$ and choose lattices $K \notin \mathbf{V}$, $L \notin \mathbf{W}$. Using P. M. Whitman's [1946] result that every lattice can be embedded in a partition lattice, one obtains a subdirectly irreducible lattice S that extends $K \times L$. Since $S \notin \mathbf{Si}(\mathbf{V}) \cup \mathbf{Si}(\mathbf{W}) = \mathbf{Si}(\mathbf{V} \vee \mathbf{W})$, it follows that $\mathbf{V} \vee \mathbf{W}$ is a proper subvariety as well, hence $\mathbf{L}$ is join-irreducible. By R. A. Dean [1956], $\mathbf{L}$ is generated by its finite members, so we may assume that K is finite. The distributivity of Λ and Jónsson's Lemma imply that the interval from $\mathbf{V}$ to $\mathbf{V} \vee \mathbf{Var}(K)$ is finite, so every proper subvariety has at least one cover in Λ, and $\mathbf{L}$ has no co-atoms since $\mathbf{V} < \mathbf{V} \vee \mathbf{Var}(K) < \mathbf{L}$ (by join-irreducibility).

A substantial amount of research has been done on the structure near the bottom of Λ. One of the aims was to investigate this lattice by finding all varieties of a given finite height. By Jónsson's Lemma (Theorem V.1.9), a finite lattice generates a variety of finite height. The converse assertion, called the Finite Height Conjecture, was a longstanding open problem. Finally, J. B. Nation [401] found a counterexample, which we discuss after presenting results about the known coverings near the bottom of Λ.

Specific lattices are labeled by German capital letter and the varieties they generate are referred to by the corresponding boldface letter (for example, $\mathbf{N}_5 = \mathbf{Var}(N_5)$). We say that a variety $\mathbf{V}$ is *strongly covered* by a collection $\mathcal{C}$ of varieties, if every variety that properly contains $\mathbf{V}$ also contains at least one member of $\mathcal{C}$.

The first few levels above the trivial variety are described in Sections V.2 and V.3 (see Figure V.2.1). B. Jónsson [1968] showed that for any variety $\mathbf{V}$ of modular lattices, $M_{3^2} \notin \mathbf{V}$ if and only if every subdirectly irreducible member of $\mathbf{V}$ has length $\leqslant 2$ (see also G. Grätzer [1966]). From this result, he deduced the following general form of Theorem V.3.6.

Theorem 1.   For $n \geqslant 3$, the covers of $\mathbf{M}_n$ are $\mathbf{M}_{n+1}$, $\mathbf{M}_n \vee \mathbf{M}_{3^2}$ and $\mathbf{M}_n \vee \mathbf{N}_5$. The variety $\mathbf{M}_\omega$ is strongly covered by $\mathbf{M}_\omega \vee \mathbf{M}_{3^2}$ and $\mathbf{M}_\omega \vee \mathbf{N}_5$.   —

Here $\mathbf{M}_\omega$ is the variety generated by M_ω, the countable lattice of length 2.

Let M_{3^n}, A_1, A_2, A_3 be the lattices in Figures 1, V.3.5, V.3.4 and suppose that M is a subdirectly irreducible modular lattice. The main technical result of D. X. Hong [1972] is that if M_{3^n}, A_1, A_2, $A_3 \notin \mathbf{HS}\{M\}$, then M has length at most n. This is a typical *exclusion result*, which is very useful when it comes to finding covers of varieties.

Let $\mathbf{M}_w^l$ be the variety generated by all modular lattices of length at most l and of width at most w ($1 \leq l,\ w \leq \infty$). For example, $\mathbf{M}_\infty^2 = \mathbf{M}_\omega$ and $\mathbf{M}_\infty^3$ is the variety generated by all subspace lattices of projective planes (see the proof of Theorem IV.5.23). With this notation, Hong's result implies that for any variety $\mathbf{V}$ of modular lattices, M_{3^3}, A_1, A_2, $A_3 \notin \mathbf{V}$ if and only if $\mathbf{V} \subseteq \mathbf{M}_\infty^3$. It follows immediately that $\mathbf{M}_\infty^3$ has exactly five covers in Λ, given by $\mathbf{M}_\infty^3 \vee \mathbf{V}$, where $\mathbf{V} \in \{\mathbf{M}_{3^3}, \mathbf{A}_1, \mathbf{A}_2, \mathbf{A}_3, \mathbf{N}_5\}$.

It is easy to check that the varieties $\mathbf{A}_1$, $\mathbf{A}_2$, $\mathbf{A}_3$, $\mathbf{M}_{3^3}$, $\mathbf{F}_2$ (generated by the corresponding lattices in Figures 1, V.3.5, V.3.4, V.3.7, IV.3.4b, respectively) each cover the variety $\mathbf{M}_{3^2}$. Using the above exclusion result and some added detail, D. X. Hong [314] proves that they are the only join-irreducible covers. More generally, he shows the following.

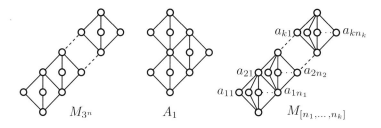

Figure 1

Theorem 2.

(i) For $n \geqslant 2$, the covers of $\mathbf{M}_{3^n}$ are $\mathbf{M}_{3^{n+1}}$ and $\mathbf{M}_{3^n} \vee \mathbf{V}$, where $\mathbf{V} \in \{\mathbf{M}_4, \mathbf{A}_1, \mathbf{A}_2, \mathbf{A}_3, \mathbf{F}_2, \mathbf{N}_5\}$.

(ii) Let $\mathbf{V}$ be a variety generated by a finite collection of finite modular lattices of length $\leqslant 3$ and let $\mathbf{W}$ be a variety generated by a finite collection of lattices of the form $M_{[n_1,\dots,n_k]}$ (see Figure 1). Then each of the following varieties is strongly covered by finitely many varieties that can be effectively found:

$$\mathbf{V} \vee \mathbf{W}, \quad \mathbf{M}_\infty^2 \vee \mathbf{V} \vee \mathbf{W}, \quad \mathbf{M}_\infty^3 \vee \mathbf{V} \vee \mathbf{W}.$$

—

This result gives a fairly good description of the bottom of Λ on the modular side.

Problem 1. Find the covers of $\mathbf{A}_1$, $\mathbf{A}_2$, $\mathbf{A}_3$.

Problem 2. Does the Finite Height Conjecture hold for modular varieties? Does it hold for the variety of modular 2-distributive lattices?

Problem 3. Does the variety of modular lattices or the variety of Arguesian lattices have any dual covers?

B. Jónsson and I. Rival [341] proved that R. N. McKenzie's [1972] list of 15 nonmodular covers of $\mathbf{N}_5$ is complete. The lattices which generate these covers are called $L_1, \dots, L_{15}$ and are shown in Figures V.2.3–10 in the order

L_5 (L_4 is dual),	L_3,	L_7 (L_8 is dual),
L_9 (L_{10} is dual),	L_{13} (L_{14} is dual),	L_{15},
L_{11} (L_{12} is dual),	L_1 (L_2 is dual),	L_6.

Theorem 3. The covers of $\mathbf{N}_5$ are $\mathbf{M}_3 \vee \mathbf{N}_5$, $\mathbf{L}_1, \dots, \mathbf{L}_{15}$.   —

Theorem 4.   The covers of $\mathbf{N}_5$ are $\mathbf{M}_3 \vee \mathbf{N}$, $\mathbf{L}_1, \ldots , \mathbf{L}_{15}$.   —

The above result makes use of the semidistributive implications $(\mathrm{SD}_\vee)$ and $(\mathrm{SD}_\wedge)$ (see Section VI.1). A variety of lattices is said to be *semidistributive*, if every member satisfies both laws. The *standard meet-sequence terms* y_n, z_n, for variables x, y, z are defined by

$$y_0 = y,$$
$$z_0 = z,$$
$$y_{n+1} = y \wedge (x \vee z_n),$$
$$z_{n+1} = z \wedge (x \vee y_n).$$

The key exclusion result by B. Jónsson and I. Rival [341] is the following.

Theorem 5.   For any variety $\mathbf{V}$, the following are equivalent.

(i) $\mathbf{V}$ is semidistributive.

(ii) $M_3, L_1, L_2, L_3, L_4, L_5 \notin \mathbf{V}$.

(iii) For some n, the equation

$(\mathrm{SD}_\vee^n)$ $x \vee (y \wedge z) = x \vee y_n$

and its dual $(\mathrm{SD}_\wedge^n)$ hold in $\mathbf{V}$.   —

It follows from this result that semidistributivity is not an equational property.

The above equations define an increasing sequence of semidistributive varieties $\mathbf{SD}_n = \mathbf{Mod}((\mathrm{SD}_\vee^n), (\mathrm{SD}_\wedge^n))$. Obviously, $\mathbf{SD}_0 = \mathbf{T}$ and $\mathbf{SD}_1 = \mathbf{D}$. Lattices and subvarieties of $\mathbf{SD}_2$ are called *near distributive*. A useful characterization is given by the next exclusion result.

Theorem 6.   A lattice variety $\mathbf{V}$ is neardistributive if and only if it is semidistributive and $L_{11}, L_{12} \notin \mathbf{V}$. (J. G. Lee [377].)   —

A lattice is said to be *almost distributive* if it is near distributive and satisfies the inequality

$(\mathrm{AD}_\vee)$ $u \wedge (w \vee (v \wedge ((x \vee y) \wedge (x \vee z)))) \leqslant v \vee (u \wedge w),$

where $w = x \vee (y \wedge (x \vee z))$, and its dual $(\mathrm{AD}_\wedge)$. The variety $\mathbf{AD}$ of all almost distributive lattices is studied by H. Rose [458] and J. G. Lee [377].

The main structural results about subdirectly irreducible almost distributive lattices require (a special case of) A. Day's doubling construction.

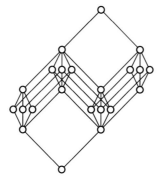

Figure 2: $(C_2 \times C_3) \star_C M_3$

The version described here is a generalization due to R. Freese, G. McNulty, and J. B. Nation [178], which will also be used later in the description of inherently nonfinitely based lattices and Nation's counterexample to the Finite Height Conjecture. Given a lattice L, a convex subset C of L and a $\{0, 1\}$-lattice K, one defines a lattice $L \star_C K$, called the *inflation of L at C by K*, as follows. The underlying set is $(L - C) \cup (C \times K)$, and for elements x, y in this set, put $x \leqslant y$ if one of the following conditions hold:

(i) x, $y \in L - C$ and $x \leqslant y$ holds in L;

(ii) x, $y \in C \times K$ and $x \leqslant y$ holds in $C \times K$;

(iii) $x \in L - C$, $y = \langle c, k \rangle \in C \times K$, and $x \leqslant c$ holds in L;

(iv) $x = \langle c, k \rangle \in C \times K$, $y \in L - C$, and $c \leqslant y$ holds in L.

Day's original doubling construction is obtained when $K = C_2$, in which case $L \star_C C_2$ is denoted by $L[C]$, and when $C = \{c\}$ this is further simplified to $L[c]$. For example, if we take $L = C_2 \times C_3$ and $C = L - \{0, 1\}$ then $L \star_C M_3$ is the lattice in Figure 2, and $(C_3 \times C_3)[d]$ gives the lattice L_{15} (Figure V.2.8). The doubling construction for single elements was actually used in the context of transferable lattices before Day's construction (see Appendix A.1.3).

For a variety $\mathbf{V}$, let $\Lambda_{\mathbf{V}}$ be the lattice of subvarieties of $\mathbf{V}$. If $\mathbf{V}$ is a lattice variety, then $\Lambda_{\mathbf{V}}$ is, of course, a principal ideal of Λ.

Theorem 7.

(i) A subdirectly irreducible lattice L is almost distributive if and only if $L \cong D[d]$, for some distributive lattice D and $d \in D$.

(ii) A lattice variety $\mathbf{V}$ is almost distributive if and only if it is semidistributive and $L_6, \ldots, L_{12} \notin \mathbf{V}$.

(iii) **AD** is locally finite (that is, every finitely generated member is finite), hence the Finite Height Conjecture holds for almost distributive varieties and **AD** is generated by its finite members.

(iv) The cardinality of $\Lambda_{\mathbf{AD}}$ is $2^{\aleph_0}$.

(v) There exists an infinite descending chain in $\Lambda_{\mathbf{AD}}$.

(vi) There exists an almost distributive variety with infinitely many covers in $\Lambda_{\mathbf{AD}}$ and one with infinitely many dual covers.

(H. Rose [458], J. G. Lee [377].)   —

Judging from the above results and additional details by Rose and Lee, one might say that the structure of $\Lambda_{\mathbf{AD}}$ is fairly well understood.

Problem 4. Is there a variety with uncountably many covers (or dual covers) in Λ or $\Lambda_{\mathbf{AD}}$?

Problem 5. Does **AD** have any dual covers?

We list below additional results about covers in Λ. In each case these results are established by long technical computations and the original papers contain further results that are of interest in their own right.

Theorem 8.   For $i = 6, 7, 8, 9, 10, 13, 14, 15$ and $n \geqslant 0$, the variety $\mathbf{L}_i^{n+1}$ is the only join-irreducible cover of $\mathbf{L}_i^n$ (where $\mathbf{L}_i^0 = \mathbf{L}_i$, see Figure 3).
(H. Rose [458].)   —

Theorem 9. $\mathbf{L}_{12}$ has exactly two join-irreducible covers $\mathbf{L}_{12}^1$ and $\mathbf{G}^1$. For $n \geqslant 1$, $\mathbf{L}_{12}^{n+1}$ is the only join-irreducible cover of $\mathbf{L}_{12}^n$, and $\mathbf{G}^{n+1}$ is the only join-irreducible cover of $\mathbf{G}^n$. Above $\mathbf{L}_{11}$, the dual results hold (see Figure 3).
(J. B. Nation [398].)   —

Theorem 10.   The join-irreducible covers of $\mathbf{L}_1$ are $\mathbf{L}_{16}, \ldots, \mathbf{L}_{25}$. The covers of $\mathbf{L}_2$ are dual (see Figure 4). (J. B. Nation [399].)   —

An approach to finding covers in Λ has been developed by J. B. Nation [400] (see also A. Day and J. B. Nation [132]).

C. Y. Wong [526] investigates weakened forms of distributivity similar to semidistributivity to find the covers of $\mathbf{L}_3$, $\mathbf{L}_4$, and $\mathbf{L}_5$. A lattice is said to be *weakly distributive* if it satisfies the following implications:

$(\mathrm{WD}_\vee)$ $x \wedge y = x \wedge z$ implies that $x \vee (y \wedge z) = (x \vee y) \wedge (x \vee z)$,

$(\mathrm{WD}_\wedge)$ $x \vee y = x \vee z$ implies that $x \wedge (y \vee z) = (x \wedge y) \vee (x \wedge z)$.

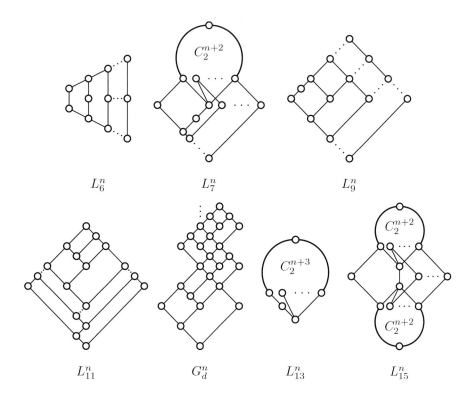

Figure 3: Sequences of lattices generating join-irreducible varieties. L_8^n, L_{10}^n, L_{12}^n, L_{14}^n are dual to L_7^n, L_9^n, L_{11}^n, L_{13}^n respectively. (Here n is a superscript label, whereas $C_2{}^{n+2}$ is a power of C_2.)

A variety of lattices is said to be *weakly distributive*, if this is true for every member. This property can also be characterized by an exclusion result.

Theorem 11.   For any variety **V**, the following are equivalent.

(i) **V** is weakly distributive.

(ii) $M_3, L_1, L_2, L_4, L_5, L_{11}, L_{12}, L_{13}, L_{14}, T_1, T_2, T_3, T_4, P_4, P_5, P_{10} \notin$ **V** (see remark before Theorem 3 and Figures 7, 5).

(iii) For some n, the equation $x \wedge (y_n \vee z_n) \leqslant (x \wedge y) \vee (x \wedge z)$ and its dual hold in **V** (y_n, z_n are the standard meet sequence terms defined on page 558).

———

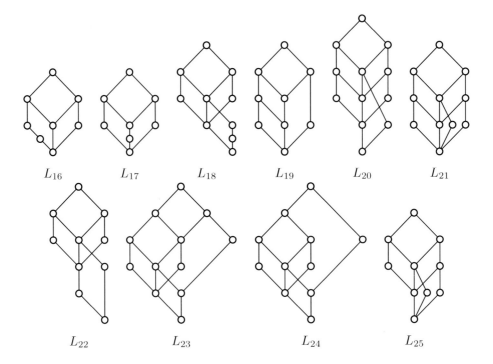

Figure 4: Lattices that generate covers of $\mathbf{L}_1$

Note that $\mathbf{L}_4$ is not weakly distributive, but does satisfy $(WD_\wedge)$. Wong shows that $(WD_\wedge)$ cannot be characterized by an exclusion result, that is, there is no finite list of finite subdirectly irreducible lattices such that a variety satisfies $(WD_\wedge)$ if and only if it contains none of these lattices. He then goes on to prove that $(WD_\wedge)$ is *weakly finitely definable with respect to* L_4, which means that there is a finite list of finite subdirectly irreducible lattices such that if $(WD_\wedge)$ fails in a variety then it contains one of these lattices. Using this result together with the approach from J. B. Nation [400] and (lots of) additional details, he succeeds in proving the following.

Theorem 12.   The join-irreducible covers of $\mathbf{L}_3$ are $\mathbf{P}_1$, ..., $\mathbf{P}_{10}$ (see Figure 5). The join-irreducible covers of $\mathbf{L}_4$ are $\mathbf{K}_1$, ..., $\mathbf{K}_6$ (see Figure 6). The covers of $\mathbf{L}_5$ are dual. (C. Y. Wong [526].)   —

For the variety $\mathbf{M}_3 \vee \mathbf{N}_5$, only the finitely generated covers are known at this point.

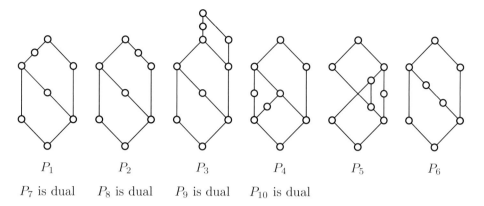

P_1 P_2 P_3 P_4 P_5 P_6

P_7 is dual P_8 is dual P_9 is dual P_{10} is dual

Figure 5: Lattices that generate covers of $\mathbf{L}_3$

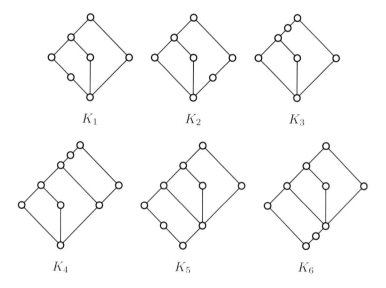

K_1 K_2 K_3

K_4 K_5 K_6

Figure 6: Lattices that generate covers of $\mathbf{L}_4$

Theorem 13.   The finitely generated join-irreducible covers of $\mathbf{M}_3 \vee \mathbf{N}_5$ are $\mathbf{V}_1, \ldots, \mathbf{V}_8$ (see Figure 8). (W. Ruckelshausen [461].)   —

Problem 6. Does $\mathbf{M}_3 \vee \mathbf{N}_5$ have any nonfinitely generated covers?

All the preceding results support the Finite Height Conjecture in that every finitely generated lattice variety of height at most 4 has only finitely many finitely generated covers (see Figure 9). Recently, however, J. B. Nation [401]

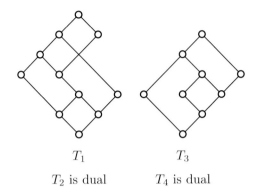

$$T_1$$

$$T_3$$

T_2 is dual T_4 is dual

Figure 7

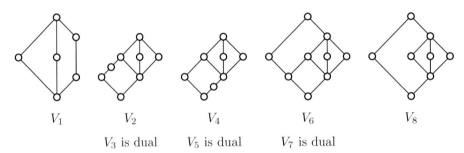

V_1 V_2 V_4 V_6 V_8

V_3 is dual V_5 is dual V_7 is dual

Figure 8: Lattices that generate finitely generated covers of $\mathbf{M}_3 \vee \mathbf{N}_5$

showed that the conjecture fails for lattices in general. Consider the lattice J in Figure 10, with the convex subset C given by the elements of height 2 and 3. Note that the elements of equal height on the left and the right of C should be identified, so C has actually 14 elements.

Let B be a chain isomorphic to the integers with top and bottom elements added. The subdirectly irreducible lattice L is obtained by replacing each element in C with a copy of B, and defining the ordering as indicated in the figure.

Theorem 14.   The infinite subdirectly irreducible lattice L in Figure 10 generates a variety of finite height.   —

This beautiful counterexample shows that even varieties of finite height are highly nontrivial, and has already inspired new results about inherently non-finitely based varieties (see Section 3). Note that the lattice has a cyclic automorphism of order 7 and is best visualized by rolling the page up into a cylinder. In essence, L is a slightly modified version of $J \star_C B$, so that the prime

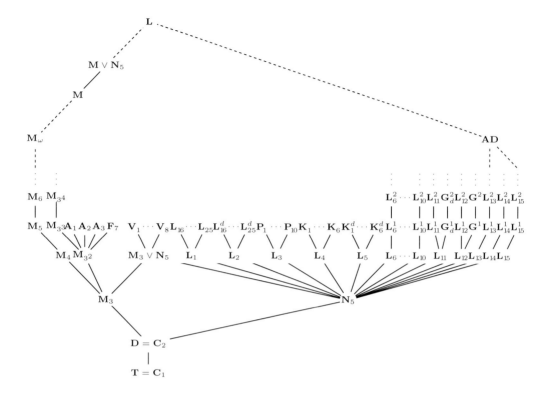

Figure 9: Known join-irreducible covers near the bottom of Λ

quotients in the 14 chains in L transpose in a spiral up and down this lattice and are all collapsed by the smallest nontrivial congruence μ on L. The lattice L/μ is isomorphic to $J \star_C C_3$, which is a subdirect product of two copies of the finite subdirectly irreducible lattice $F = J \star_C C_2$. The proof proceeds by showing that every finitely generated lattice in $\mathbf{SiVar}(L)$ is in $\mathbf{H}(L)$, and that $\mathbf{H}(L) - \mathbf{I}(L) \subseteq \mathbf{Var}(F)$, whence $\mathbf{Var}(L)$ is a cover of the variety $\mathbf{Var}(F)$.

In an unpublished note, Nation points out that lattices other than J can serve as the basis for the construction. For example, the Boolean lattice C_2^5 gives a narrower (but less easily visualizable) example with only 10 chains.

J. B. Nation [401] also shows that there is a variety of finite height that has countably infinitely many covers.

Problem 7. Is every variety of finite height finitely based?

Problem 8. Is every variety of finite height generated by a lattice of finite width?

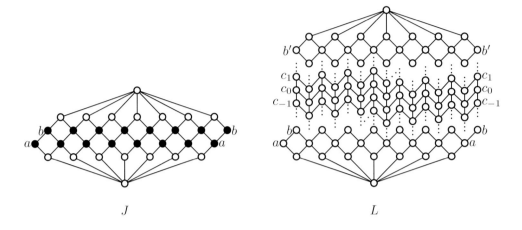

$$J \qquad\qquad\qquad\qquad\qquad\qquad L$$

Figure 10: Nation's counterexample to the Finite Height Conjecture

Problem 9. Is there an algorithm to find the covers of a finitely generated variety?

2. Generating Sets of Varieties

It is well known that the variety of all lattices is generated by its finite members (R. A. Dean [1956]). Using the doubling construction and R. N. McKenzie's [1972] characterization of splitting lattices as finite subdirectly irreducible bounded lattices, A. Day [1977] was able to prove the following sharper version of Dean's result.

Theorem 15.   The variety **L** of all lattices is generated by the class of all lattices.   —

The significance of this result is enhanced by the fact that it implies that every finitely generated free lattice is weakly atomic (R. N. McKenzie [1972] and A. Kostinsky [1972] proved this condition equivalent to Day's theorem).

More recently, R. N. McKenzie [381] showed that **L** is also generated by the collection of all finite minimal simple lattices. (A simple lattice L is *minimal* if $L \ncong C_2$ and no simple lattice other than C_2 generates a proper subvariety of **Var**(L).)

For the variety of modular lattices, the situation is quite different.

Theorem 16.

(i) The variety **M** of all modular lattices is not generated by its finite members. (R. Freese [155].)

(ii) Neither **M** nor the variety **A** of all Arguesian lattices is generated by its members of finite length. (C. Herrmann [305].)   —

Using P. Pudlák and J. Tůma's [450] result that every finite lattice can be embedded into a finite partition lattice, P. Bruyns and H. Rose [64] show that every lattice is embeddable into an ultraproduct of finite partition lattices, hence $\mathbf{L} = \mathbf{S}_U(\{\mathrm{Part}\, n \mid n \in \omega\})$. Furthermore, since any lattice variety **V** satisfies the Embedding Property (see Section V.4), there exists a lattice $L \in \mathbf{V}$ such that every member of **V** is embeddable into an ultrapower of **L**, that is, $\mathbf{V} = \mathbf{S}_U(L)$. Such lattices L are referred to as *ultra-universal* (see also C. Naturman and H. Rose [403]).

R. N. McKenzie [1972] showed that splitting lattices in Λ are finite. However, splitting lattices can be defined in any lattice of varieties.

Problem 10. Is every splitting lattice in $\Lambda_{\mathbf{M}}$ finite?

Problem 11. Is **M** generated by all the splitting lattices in $\Lambda_{\mathbf{M}}$?

If L is a splitting lattice in Λ, then the largest variety that does not contain L is called the *conjugate variety* of L.

Problem 12. Is there a nontrivial conjugate variety in Λ that is generated by its finite members?

Problem 13. Is there a conjugate variety **V** with infinite subdirectly irreducible members that are projective in **V**?

Note that if a variety **V** is generated by its finite members then every subdirectly irreducible projective member is finite. Thus a positive answer to the previous problem implies that **V** is not generated by its finite members.

3. Equational Bases

Recall that an algebra is said to be *finitely based* if the variety which it generates is determined by finitely many equations. Nonfinitely based lattices were constructed by K. A. Baker [1969a], [34], R. Freese [1977], C. Herrmann [305], R. N. McKenzie [1970] and R. Wille [1972]. One such lattice, due to McKenzie, is shown in Figure 11.

An algebra A is said to be *inherently nonfinitely based* if $\mathbf{Var}(A)$ is locally finite, and any locally finite variety to which A belongs is not finitely based. This concept was introduced independently by V. L. Murskiĭ [395] and P. Perkins [420]. Inspired by J. B. Nation's [401] counterexample to the Finite Height Conjecture, R. Freese, G. McNulty, and J. B. Nation [178] construct inherently nonfinitely based lattices. Here we only state a special case of their main result (see page 559 for the definition of $L \star_C K$).

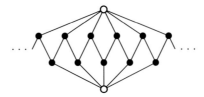

Figure 11: The lattice L_f

Theorem 17.   Let L_f be the lattice in Figure 11 and define $C = L_f - \{0,1\}$. Let K be a $\{0,1\}$-lattice which belongs to a locally finite variety, and assume that K has an automorphism with an infinite orbit. Then $L_f \star_C K$ is inherently nonfinitely based.   —

The two least complicated lattices K with the required automorphism are M_ω and B (a chain isomorphic to the integers with top and bottom elements added). The resulting lattices $L_f \star_C K$ are given in Figure 11. In the same paper, it is also shown that the lattice L_f is not inherently nonfinitely based.

Problem 14. Are there any modular lattices that are inherently nonfinitely based?

Analogously to the varieties $\mathbf{M}_w^l$, one defines $\mathbf{V}_w^l$ to be the variety generated by all lattices of height l and width w. We allow $l = \infty$ or $w = \infty$, in which case the respective parameter is not restricted. For $l, w < \infty$, the varieties $\mathbf{M}_w^l$ and $\mathbf{V}_w^l$ are finitely generated and hence finitely based.

Note that if a variety $\mathbf{V}$ is strongly covered by a finite set of varieties, then it is finitely based. Results about whether the varieties $\mathbf{M}_w^l$ and $\mathbf{V}_w^l$ are finitely based for $l = \infty$ or $w = \infty$, are as follows:

Theorem 18.

(i) $\mathbf{M}_\infty^2$ $(= \mathbf{V}_\infty^2)$ and $\mathbf{M}_\infty^3$ are finitely based (see Theorems 1 and 2), $\mathbf{M}_\infty^n$ is finitely based, for all n (K. A. Baker, see C. Herrmann [1973]).

(ii) $\mathbf{V}_\infty^3$ is finitely based (C. Herrmann [1973]), $\mathbf{V}_\infty^n$ is not finitely based, for $n \geqslant 4$ (K. A. Baker [34]).

(iii) $\mathbf{M}_1^\infty = \mathbf{M}_2^\infty = \mathbf{D}$, $\mathbf{M}_3^\infty = \mathbf{M}_3$ (see Theorem 1, since both its modular covers are generated by lattices of width 4), $\mathbf{M}_4^\infty$ is strongly covered by 10 varieties (each generated by $\mathbf{M}_4^\infty$ together with one of the lattices in Figures V.3.3, ... , V.3.10, IV.3.4b, or N_5) and hence finitely based (R. Freese [1977]), $\mathbf{M}_n^\infty$ is not finitely based for $n \geqslant 5$ (K. A. Baker [34]).

(iv) $\mathbf{V}_2^\infty = \mathbf{N}_5$ (O. T. Nelson, Jr. [1968]), hence finitely based, $\mathbf{V}_n^\infty$ is not finitely based for $n \geqslant 3$ ($n \geqslant 4$ due to K. A. Baker [34], $n = 3$ due to Y.-C. Hsueh [316]). —

B. Jónsson [1974] showed that the join of two finitely based lattice varieties need not be finitely based, and K. A. Baker [34] did the same for two finitely based modular varieties. In view of these result, it is natural to look for sufficient conditions under which the join of two finitely based varieties remains finitely based.

Theorem 19. Suppose that $\mathbf{V}$ and $\mathbf{W}$ are finitely based lattice varieties. If one of the following conditions holds, then $\mathbf{V} \vee \mathbf{W}$ is finitely based.

(i) $\mathbf{V}$ is modular and $\mathbf{W}$ is generated by a finite lattice that excludes M_3.

(ii) $\mathbf{V}$ and $\mathbf{W}$ are locally finite and the projective radius of $\mathbf{V} \cap \mathbf{W}$ is finite.

(iii) $\mathbf{V}$ and $\mathbf{W}$ are modular and $\mathbf{W}$ is generated by a lattice of finite length.

(iv) $\mathbf{V}$ is modular and $\mathbf{W}$ is generated by a finite lattice with finite projective radius.

(v) $\mathbf{V} \cap \mathbf{W} = \mathbf{D}$, the variety of all distributive lattices. —

(i) and (ii) are due to J. G. Lee [378], (iii) is due to Jónsson and the remaining statements are due to Y. Y. Kang [350].

Note that it follows from part (i) above that $\mathbf{M} \vee \mathbf{N}_5$ is finitely based. B. Jónsson [334] constructed an explicit basis for this variety of eight identities. The following problem was inspired by this result.

Problem 15. Is the unique cover of a conjugate variety in Λ always finitely based? (A. Day.)

4. Amalgamation and Absolute Retracts

G. Grätzer, B. Jónsson, and H. Lakser [1973] showed that, besides the varieties $\mathbf{T}$ and $\mathbf{D}$, no modular variety has the Amalgamation Property (see Section V.4 for a discussion). A. Day and J. Ježek [127] finally extended this result to all lattice varieties.

Theorem 20. $\mathbf{T}$, $\mathbf{D}$, and $\mathbf{L}$ are the only lattice varieties with the Amalgamation Property. —

For other varieties of algebras, the Amalgamation Property also turned out to be rarely satisfied. A comprehensive survey about amalgamation for various types of algebras can be found in E. W. Kiss, L. Márki, P. Pröhle, and

W. Tholen [364]. These results indicate that the concept of amalgamation does not mesh well with that of a variety. However the amalgamation class $\mathbf{Amal}(\mathbf{V})$ of a variety $\mathbf{V}$, introduced by G. Grätzer and H. Lakser [1971], has proved to be very fruitful. M. Yasuhara [1974] showed that for any variety $\mathbf{V}$ of algebras, each member of $\mathbf{V}$ has an extension in $\mathbf{Amal}(\mathbf{V})$, hence $\mathbf{Amal}(\mathbf{V})$ is a proper class (Theorem V.4.10). At present, the main directions of study are to characterize the amalgamation class of a given variety and to decide whether it is (strictly) elementary, that is, if it can be defined by a (finite) collection of first order sentences. Although we do not know anything about a single nontrivial member of $\mathbf{Amal}(\mathbf{M})$, significant progress has been made with residually small lattice varieties. This started with a characterization of the amalgamation class of finitely generated lattice varieties by B. Jónsson [338], and was generalized by P. Jipsen and H. Rose [327] (see also P. Ouwehand and H. Rose [411]).

Many of the results below are valid for various congruence distributive varieties (not only lattice varieties), so we will state the more general results where applicable. A *retraction* of an embedding $f \colon A \to B$ is a homomorphism $g \colon B \to A$ such that $g \circ f = \mathrm{id}_A$. An algebra A in a class $\mathbf{K}$ is said to be an *absolute retract in* $\mathbf{K}$ if for every embedding $f \colon A \hookrightarrow B \in \mathbf{K}$, there is a retraction. The class of all absolute retracts of $\mathbf{K}$ is denoted by $\mathbf{Ar}(\mathbf{K})$. The concept of absolute retract is of interest here since C. Bergman [45] observed that for any variety $\mathbf{V}$ we have $\mathbf{Ar}(\mathbf{V}) \subseteq \mathbf{Amal}(\mathbf{V})$.

A variety is said to be *residually small*, if there is an upper bound on the cardinality of its subdirectly irreducible members. W. Taylor [495] proved that a variety $\mathbf{V}$ is residually small if and only if $\mathbf{V} = \mathbf{S}(\mathbf{V})$.

Theorem 21.   Let $\mathbf{V}$ be a residually small congruence distributive variety in which every member has a one-element subalgebra. Then $A \in \mathbf{Amal}(\mathbf{V})$ if and only if, for any embedding $f \colon A \hookrightarrow B \in \mathbf{V}$ and any homomorphism $h \colon A \to M \in \mathbf{Si}(\mathbf{Ar}(\mathbf{V}))$, there exists a homomorphism $g \colon B \to M$ such that $h = fg$.   ▬

The reverse implication is due to C. Bergman [45] and the forward direction is from P. Jipsen and H. Rose [327]. A useful corollary is that for finite algebras the condition in the preceding theorem can be checked (see B. Jónsson [338], P. Jipsen and H. Rose [327]).

Corollary 1. *Let* $\mathbf{V}$ *be a finitely generated congruence distributive variety in which every member has a one-element subalgebra. For finite algebras in* $\mathbf{V}$, *membership in* $\mathbf{Amal}(\mathbf{V})$ *is decidable.*

Since the amalgamation class of a variety is, in general, a proper subclass, it is interesting to ask whether it is an elementary class. Even for a finitely generated lattice variety this is a nontrivial problem.

Theorem 22.   The amalgamation class of any finitely generated nondistributive modular lattice variety is not elementary. (C. Bergman [46].)   —

Problem 16. For which finitely generated varieties is the amalgamation class elementary?

In P. Ouwehand and H. Rose [412], some progress has been made on this problem.

Theorem 23.   Let $\mathbf{V}$ be a finitely generated variety of lattices. Suppose that there is a lattice $L \in \mathbf{Amal}(\mathbf{V})$ with either a bottom or a top element, which does not have C_2 as homomorphic image, but some ultrapower $L^I/\mathcal{U}$ does have C_2 as homomorphic image. Then $L^I/\mathcal{U} \notin \mathbf{Amal}(\mathbf{V})$, and hence neither $\mathbf{Amal}(\mathbf{V})$ nor its complement are elementary.   —

The lattice L is usually constructed by gluing countably many copies of a maximal subdirectly irreducible member on top of each other (identifying the top

of one member with the bottom of the next). Applications of this result include a simple proof of Theorem 22 as well as the result that any lattice variety generated by a finite simple lattice has a nonelementary amalgamation class. Further generalizations to nonfinitely generated varieties imply, for example, that $\mathbf{M}_\omega$ does not have an elementary amalgamation class.

Problem 17. If $\mathbf{Amal}(\mathbf{V})$ is an elementary class, does it follow that it is a Horn class?

P. Ouwehand and H. Rose [411] show that if an elementary class $\mathbf{K}$ is closed under updirected unions, then it is closed under finite direct products if and only if it is closed under reduced products (and hence definable by Horn sentences). This result applies to elementary amalgamation classes since M. Yasuhara [1974] showed that they are closed under updirected unions. Hence the above problem is equivalent to asking if every elementary amalgamation class is closed under finite products.

Problem 18. Is there a nonfinitely generated variety other than $\mathbf{L}$ whose amalgamation class is elementary? In particular, is $\mathbf{Amal}(\mathbf{M})$ an elementary class?

Absolute retracts. We now consider the problem of how the class of all absolute retracts of a variety can be constructed from its subdirectly irreducible members. Even for congruence distributive varieties, the product of two absolute retracts need not be an absolute retract (W. Taylor [496]), but fortunately lattices are well behaved.

Theorem 24.   Let $\mathbf{V}$ be a congruence distributive variety in which every member has a one-element subalgebra. Then the class of absolute retracts of

V is closed under direct products and direct factors, that is, $\prod_{i\in I} A_i \in \mathbf{Ar}(\mathbf{V})$ iff $\{\, A_i \mid i \in I \,\} \subseteq \mathbf{Ar}(\mathbf{V})$. (P. Jipsen and H. Rose [327], P. Ouwehand and H. Rose [411].)   —

In fact, P. Ouwehand and H. Rose [411] show that for congruence distributive varieties, all finite absolute retracts can be obtained as products of subdirectly irreducible absolute retracts. The general case is more complicated and requires the concept of equational compactness (see also Section 1.9 of Appendix A). Here we only need the algebraic formulation: an algebra A is *equationally compact* if for every diagonal embedding of A into an ultrapower of A, there is a retraction. Clearly every finite algebra and every absolute retract with respect to some variety is equationally compact. Ouwehand and Rose also observe that equationally compact lattices are complete (a result implicit in B. Węglorz [511]). Hence absolute retracts in a lattice variety are complete lattices.

Consider the following characterization:

($*$) An algebra A is in $\mathbf{Ar}(\mathbf{V})$ if and only if A is a product of equationally compact reduced powers of $\mathbf{Si}(\mathbf{Ar}(\mathbf{V}))$.

Theorem 25.   Let **V** be a finitely generated variety of lattices.

(i) Every equationally compact reduced power of a finite absolute retract in **V** is an absolute retract in **V** (hence the reverse implication of ($*$) holds).

(ii) If none of the subdirectly irreducible absolute retracts in **V** are homomorphic images of each other then **V** satisfies ($*$).

(iii) Assume every proper subvariety satisfies ($*$). If **V** is the join of its proper subvarieties or contains only one subdirectly irreducible absolute retract, then **V** satisfies ($*$).

(P. Ouwehand and H. Rose [411].)   —

Note that the previous theorem is a generalization of the well known result that the absolute retracts in **D** are precisely the complete Boolean lattices (since every complete Boolean lattice is a reduced power of C_2, which is the only subdirectly irreducible in **D**). All finite lattices in $\mathbf{Si}(\mathbf{M})$ are simple, hence (ii) implies that every finitely generated modular variety satisfies ($*$). It follows from Theorem 7(i) that any homomorphic image of a lattice in $\mathbf{Si}(\mathbf{AD})$ is distributive, whence ($*$) also holds for all finitely generated almost distributive varieties.

5.   Congruence Varieties

A congruence variety is a variety of lattices which is generated by the congruence lattices of some variety of algebras. An account of this area of research can be found in B. Jónsson's appendix to G. Grätzer [215] (see also B. Jónsson [335]). In this section, we mention some more recent results and some additional results not included there.

5.1   The nonmodular case: Polin's variety

Contrary to the belief of many researchers, S. V. Polin [437] constructed a variety of algebras whose congruence variety is a proper nonmodular subvariety of **L**. In the reconstruction of Polin's proof, A. Day showed that there are infinitely many distinct nonmodular congruence varieties, each of which contains no nondistributive modular lattices. Since the join of congruence varieties is again a congruence variety, there are infinitely many nonmodular congruence varieties. Moreover, we have the following results.

Theorem 26.

(i) Any nonmodular congruence variety contains the variety of all almost distributive lattices. (A. Day [1977].)

(ii) Polin's congruence variety is the unique minimal nonmodular congruence variety. (A. Day and R. Freese [124].)   —

For further information about Polin's variety, see R. Freese [165].

Theorem 27.   Each minimal modular nondistributive congruence variety is determined by one of the varieties generated by all vector spaces of characteristic p (a prime or 0). (R. Freese, C. Herrmann, and A. P. Huhn [170].)   —

Since **D** is meet-irreducible in the lattice of modular varieties, it follows from this result that the meet of two congruence varieties does not have to be a congruence variety.

Corollary 2. *The set of all congruence varieties is not a sublattice of* Λ.

Problem 19. Is there a unique largest modular congruence variety?

We now turn to the question of congruence identities. Among the most significant results are the following.

Theorem 28.

(i) There is a lattice identity strictly weaker than the modular identity such that any congruence variety which satisfies this identity is a modular variety. (J. B. Nation [1974].)

(ii) Every modular congruence variety is Arguesian. (R. Freese and B. Jónsson [174].)

(iii) No modular nondistributive congruence variety is finitely based. (R. Freese [164])

(iv) For each $n \geqslant 0$, the congruence lattice $\mathrm{Con}\, F_n$ of the free n-generated Polin algebra is a splitting lattice. Thus (by Theorem 26(ii)) a variety is congruence modular if and only if it satisfies the conjugate equation of one of these splitting lattices. (A. Day and R. Freese [124].)

(v) It is decidable whether a lattice equation implies congruence modularity (or distributivity). (G. Czédli and R. Freese [90].)   —

Problem 20. Is there a nondistributive congruence variety which is finitely based?

Free Lattices

R. Freese

In this appendix, we survey some of the major developments in the theory of free lattices and certain related topics. J. Ježek, J. B. Nation, and I have completed a monograph on the subject [172], which contains the proofs of most the results here. The first chapter of this monograph is suitable as an introduction to the theory of free lattices. It also contains more details of the interesting history of the subject, as well as a chapter on algorithms for finite and free lattices.

1. Whitman's Solutions; Basic Results

The classical papers P. M. Whitman [1941] and [1942] solved the word problem for free lattices: Whitman gave an algorithm for determining if two lattice terms (polynomials) were equal in all lattices. He showed that each element of the free lattice has a shortest term representing it, known as the *canonical form*, and gave an algorithm to put an arbitrary term into canonical form (see Section VI.2). This canonical form is closely connected with the arithmetic of the free lattice and Whitman exploited this connection to obtain important results about free lattices.

The word problem for free lattices, in fact for finitely presented lattices, was first solved by T. Skolem in 1920 [475], [476]. (This went unnoticed until it was recently discovered by S. Burris [65].) What was interesting about Skolem's solution is that it is polynomial time, unlike some of the later solutions [383], [147], [146]. The computational complexity of Whitman's algorithm is discussed in [160] and Chapter XI of [172]. Even though there is such a nice canonical form in free lattices and an easy algorithm for obtaining it, there is no term rewrite system for lattice theory, see [171]. This is also proved in Chapter XII of [172] along with some further results in this area.

The key ingredient of Whitman's solution is the following condition known as *Whitman's condition* (see Section VI.1):

(W)
$$\text{If } v = v_1 \wedge \cdots \wedge v_r \leqslant u_1 \vee \cdots \vee u_s = u,$$
$$\text{then either } v_i \leqslant u, \text{ for some } i, \text{ or } v \leqslant u_j, \text{ for some } j.$$

Let $w \in F(X)$ be join-reducible and let us suppose that $t = t_1 \vee \cdots \vee t_n$ (with $n > 1$) is the canonical form of w. Let w_i be the function associated with T_i in $F(X)$. Then $\{w_1, \ldots, w_n\}$ are called the *canonical joinands* of w. We also say $w = w_1 \vee \cdots \vee w_n$ *canonically* and that $w_1 \vee \cdots \vee w_n$ is the *canonical join-representation* (or *canonical join-expression*) of w. If w is join-irreducible, we define the canonical joinands of w to be the set $\{w\}$. Of course, the *canonical meet-representation* and *canonical meetands* of an element in a free lattice are defined dually.

The aforementioned connection between the canonical form and the arithmetic of free lattices is summarized in the following theorem which shows that the canonical form corresponds to the best way to express an element of a free lattice as a join or meet. For any lattice L and finite subsets A and B of L, we say that A *join-refines* B, and we write $A \ll B$, if for each $a \in A$, there is a $b \in B$ with $a \leqslant b$.

The dual notion is called *meet-refinement* and is denoted $A \gamma B$.

Theorem 1.   Let $w = w_1 \vee \cdots \vee w_n$ canonically in $F(X)$. If also $w = u_1 \vee \cdots \vee u_m$, then

$$\{w_1, \ldots, w_n\} \ll \{u_1, \ldots, u_m\}.$$

Thus $w = w_1 \vee \cdots \vee w_n$ is the unique minimal join-representation of w.   —

B. Jónsson observed that there is a close connection between canonical form and a weak form of distributivity known as semidistributivity. A lattice is *join-semidistributive* if it satisfies the following condition (see Definition VI.1.5):

(SD$_\vee$) $a \vee b = a \vee c$ implies $a \vee b = a \vee (b \wedge c)$.

Of course *meet-semidistributivity* is defined dually and a lattice is *semidistributive* if it satisfies both conditions. B. Jónsson and J. E. Kiefer [1962] showed that lattices with canonical forms, in particular free lattices, are semidistributive; see Theorem 1.21 and Lemma 2.22 of [172]. Thus all sublattices of free lattices are semidistributive.

2.   Classical Results

The early deep work on free lattices centered on sublattices. P. M. Whitman showed that $F(\aleph_0)$ can be embedded into $F(3)$ (see Theorem VI.2.8). B. Jónsson and J. E. Kiefer [1962] proved the following result.

Theorem 2.   Let L be a lattice satisfying (W). Suppose that the elements a_1, a_2, a_3, and $v \in L$ satisfy

(i) $a_i \nleqslant a_j \vee a_k \vee v$, whenever $\{i, j, k\} = \{1, 2, 3\}$;

(ii) $v \nleqslant a_i$ for $i = 1$, 2, 3;

(iii) v is meet-irreducible.

Then L contains a sublattice isomorphic to $F(3)$.   —

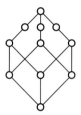

Figure 1: The bottom of $F(3)$

They used this result to prove some interesting corollaries. If $x \in X$, then $\underline{x} = \bigwedge(X - \{x\})$ is an atom of $F(X)$ and every atom has this form. Let u be the join of the atoms. When $|X| = 3$, then the interval $u/0$ has 11 elements; see Figure 1. When $|X| = 4$, this interval is infinite but the sublattice generated by the atoms is finite; see Figure 1.1 and 3.5 of [172]. But when $|X| \geqslant 5$, we have the following:

Corollary 3. *If $n \geqslant 5$, the sublattice of $F(n)$ generated by the atoms is infinite.*

Even though every nontrivial lattice equation (identity) fails in a finite lattice, finite sublattices of free lattices do satisfy an equation.

Corollary 4. *If L is a finite lattice satisfying (W), then the breadth of L is at most 4. The variety generated by finite lattices that satisfy (W) is not the variety of all lattices. In particular, finite sublattices of a free lattice have breadth at most four and satisfy a nontrivial lattice equation.*

Theorem VI.2.11 and Corollary VI.2.12 give some applications of free lattice techniques to lattice theory due to R. A. Dean, R. P. Dilworth, and Ju. I. Sorkin.

2.1 Jónsson's conjecture

For about 20 years, work on free lattices was dominated by a problem posed by Jónsson. He had observed that every sublattice of a free lattice satisfies (W) and is semidistributive and asked if every finite lattice satisfying these properties is a sublattice of a free lattice. This problem was finally solved by J. B. Nation in [397].

Theorem 3.   A finite lattice is isomorphic to a sublattice of a free lattice if and only if it is semidistributive and satisfies (W).   —

2.2 Dean's Problem

In a recent communication, R. A. Dean informed me of an old problem in free lattice theory: can a free lattice have an ascending chain of sublattices all isomorphic to $F(3)$? This problem did not appear in print and I was unaware of it. See [168] for some partial results on this interesting problem.

3. Covers in Free Lattices

In his original papers, Whitman observed that even though $F(X)$ is infinite if $|X| \geqslant 3$, it did have some elements which covered others. Additional covers were discovered in R. A. Dean [1961a]; see Theorem VI.2.4.

R. N. McKenzie's studies of lattice varieties naturally lead him to covers in free lattices. A lattice of the form $L = F(X)/\Psi$, where $u \succ v$ and Ψ is the unique largest congruence separating u and v, is called a *splitting lattice*. Such lattices are finite and subdirectly irreducible and satisfy a strong form of Jónsson's Theorem (Theorem V.1.9): if a splitting lattice lies in an arbitrary join of lattice varieties, it is in one of them. Splitting lattices naturally lead to a splitting of the lattice of all lattice varieties into a principal ideal and a principal filter. This motivated A. Day [1977] to prove the following theorem which is one of the most important results on covers in free lattices. It is so important that it is proved twice in [172].

Theorem 4.   Finitely generated free lattices are weakly atomic, that is, every nontrivial interval contains a cover.   —

A systematic theory of covers in free lattices was developed in [181] and is covered more thoroughly in Chapter III of [172]. Here we give a brief overview of the main ideas of the theory.

A join-irreducible element of a lattice is *completely join-irreducible* when it has a lower cover, which of course must be unique. When w is completely join-irreducible, its lower cover is denoted w_*. The unique upper cover of a completely meet-irreducible element q is denoted q^*.

Theorem 5.   Let w be a completely join-irreducible element of $F(X)$. Then there is a unique canonical meetand $\kappa(w)$ of w_* which is not above w. Every element of $F(X)$ which is above w_* is either above w or below $\kappa(w)$. Dually, if v is completely meet-irreducible in $F(X)$, then $\kappa^d(v)$ is the unique canonical joinand of v^* which is not below v and every element below v^* is either below v or above $\kappa^d(v)$.   —

The element $\kappa(w)$ is always completely meet-irreducible and the association $w \longleftrightarrow \kappa(w)$ is a bijection from the completely join-irreducible elements to the completely meet-irreducible elements; see Theorem 3.3 of [172].

The lower covers of arbitrary elements in $F(X)$ are determined in a straightforward manner from the lower covers of their canonical joinands; see Theorem 3.5 of [172]. Thus we can concentrate on join-irreducible elements.

We associate with each join-irreducible element $w \in F(X)$ a finite set $J(w)$ of join-irreducible elements. If $w \in X$, then $J(w) = \{w\}$. Otherwise, let

$$w = \bigwedge_i \bigvee_j w_{ij} \wedge \bigwedge_k x_k$$

be the canonical representation of w, where $x_k \in X$. Then

$$J(w) = \{w\} \cup \bigcup_{i,j} J(w_{ij}).$$

Essentially, $J(w)$ consists of the join-irreducible subterms of w. If we take the join-closure (including the empty join) of $J(w)$ in $F(X)$, we get a finite lattice $L(w)$ and w is completely join-irreducible if and only if $L(w)$ is semidistributive; see Theorem 3.26 of [172]. This gives a nice visual way of testing if w is completely join-irreducible; see Figure 3.3 and Table 3.1 of [172] for several examples. However, because we have to take the join-closure, this method can be exponential. The next theorem gives a syntactic algorithm which is polynomial time. Let

$$w_\dagger = \bigvee \{u \in J(w) \mid u < w\},$$
$$K(w) = \{v \in J(w) \mid w_\dagger \vee v \not\geq w\}.$$

Theorem 6.   Let w be a join-irreducible element of $F(X)$, for a finite set X. Then w is completely join-irreducible in $F(X)$ if and only if the following two conditions are satisfied:

(i) every $u \in J(w) - \{w\}$ is completely join-irreducible;

(ii) $w \not\leq \bigvee K(w)$.   —

What is even more interesting is that there is a simple recursive formula for $\kappa(w)$ and hence for w_*.

Theorem 7.   Let w be a completely join-irreducible element of $F(X)$. Then

$$\kappa(w) = \bigvee \{x \in X \mid w_\dagger \vee x \not\geqslant w\}$$
$$\vee \bigvee \{k^\dagger \wedge \kappa(v) \colon v \in J(w) - \{w\}, \ w \not\leqslant \kappa(v)\},$$

where

$$k^\dagger = \bigwedge \{\kappa(v) \mid v \in J(w) - \{w\}, \ \kappa(v) \geqslant \bigvee K(w)\}.$$

3.1   Chains of covers

Using these results, we can answer basic question about free lattices such as the existence of chains of covers, finite intervals, and the connected components of the covering relation. We start with a chain of length two. Since by (W) every element is either join- or meet-irreducible, we may assume that the middle element w of this chain is join-irreducible and hence completely join-irreducible. Let u be the top of this chain, so that $w_* \prec w \prec u$. Let q be the canonical meetand of w not above u. The uniqueness of q follows from the dual of Theorem 1. The same theorem also shows that $q \vee u \succ q$; that is, q is completely meet-irreducible. By Theorem 6, each of the canonical joinands of q is completely join-irreducible, and this implies (by Theorem 3.5 of [172]) that q is *lower atomic*, that is, every element properly below q is below some lower cover of q. Of course, q is upper atomic since it is completely meet-irreducible. Elements which are both lower and upper atomic are called *totally atomic*. Thus associated with a three-element chain of two covers $w_* \prec w \prec u$ is a totally atomic element.

Now it turns out that there are a limited number of totally atomic elements in free lattices and their form can be completely characterized; see Theorem 6.10 and Corollary 6.11 of [172] for the details. A more detailed analysis of totally atomic elements allows us to greatly restrict the length of a chain of covers in free lattices. For an element a in a lattice, define the *connected component* of a to be the set of those b's such that there is a sequence $a = a_0, a_1, \ldots, a_n = b$ where, for each i, either $a_i \prec a_{i+1}$ or $a_i \succ a_{i+1}$. The connected component of 0 in $F(3)$ is the interval diagrammed in Figure 1. This interval has chains of four covers. The connected component of 0 in $F(n)$, $n \geqslant 4$, has chains of covers of length 3 and no longer.

Theorem 8.   A chain of covers in a free lattice can have length at most 4. Chains of covers of length 3 and 4 in $F(n)$ occur only in the connected component of 0 or of 1. On the other hand, $F(n)$ has infinitely many chains of covers of length 2, for all $n \geqslant 3$.

However, nothing like Day's Theorem holds for chains of covers of length 2: there are infinite intervals in F(3) that do not contain any chain of covers of length 2. (The reader might try to prove this using the connection of totally atomic elements to chains of two covers described above and the fact that there are only finitely many totally atomic elements in F(n).)

3.2 Finite intervals

With a little extra work, it is possible to find all finite intervals. The bottom of F(3), diagrammed in Figure 1, is an interval and so every subinterval of it is also an interval. It turns out these and their duals are the only intervals in free lattices. So the lattices of Figure 2 and their duals are the only nontrivial finite intervals in free lattices.

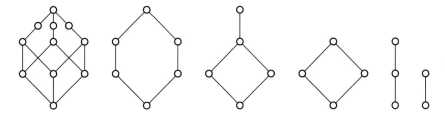

Figure 2: Finite intervals of free lattices

Actually, most of these are rare. The first two (and the dual of the first) only occur in F(3) and only at the bottom or top. The next two occur in every F(n) but only in the connected component of 0 or 1. But the three-element interval occurs infinitely often. This answers Problem VI.17.

3.3 Connected components

Using similar ideas, it is possible to characterize the connected components of the covering relation in free lattice. This was first done in [158] and is presented in Chapter VII of [172]. The connected component of 0 is easy to describe. For F(3) it is diagrammed in Figure 1 and for F(n), $n > 3$, it is described in Example 3.45 and Figure 3.5 of [172]. The fact that outside these components chains of covers can have length at most 2 severely restricts the possible connected components. Nevertheless, connected components that do occur are interesting. There are two types. The first consists of any number of copies of N_5 with a common least element and one common atom. Figure 3 shows one with three copies of N_5. The dotted lines indicate a noncover while the solid lines indicate a cover.

The only other type of connected component consists of a collection of chains each of length one or two with a common top element but otherwise disjoint. One example is given in Figure 3. It is interesting that all possibilities occur

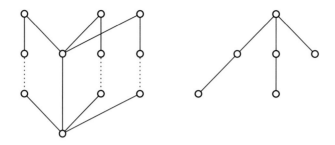

Figure 3: Two connected components

except the three-element chain and two three-element chains with a common top.

4.   Semisingular Elements and Tschantz's Theorem

Suppose w is a completely join-irreducible element. Then $w_* = w \wedge \kappa(w)$ and, by its definition, $\kappa(w)$ is a canonical meetand of w_*. If $w = \bigwedge w_i$ canonically, then

$$w_* = \kappa(w) \wedge w = \kappa(w) \wedge \bigwedge w_i.$$

Is this the canonical form of w_*? Surprisingly, it turns out that it usually is.

Theorem 9.   Let w be a completely join-irreducible element of $F(X)$ with $w = w_1 \wedge \cdots \wedge w_m$ canonically. Then

$$\{\kappa(w)\} \cup \{w_i \mid w_i \not\geqslant \kappa(w)\}$$

is the set of canonical meetands of w_*.   —

We originally proved this in [181] to simplify making a table of examples. But it turns out this result plays an important role in the proofs of several of the deeper theorems, including the characterization of chains of covers and of finite intervals given above. To explore some of these, we define a completely join-irreducible element of a free lattice to be *semisingular*, if $\kappa(w) \leqslant w_i$, for at least one of the canonical meetands w_i of w. If every $\kappa(w) \leqslant w_i$, for *every* i, we call w *singular*. In this case, $w_* = \kappa(w)$. Singular elements are characterized in [173]. They are very rare: if w is singular, then either w is the meet of two coatoms or w_* is the join of two atoms; see Theorem 8.6 of [172].

Semisingular elements play an important role in free lattices, for example, in finding maximal chains without covers. The following theorem gives a very nice characterization of semisingular elements.

Theorem 10.   A completely join-irreducible element $w \in F(X)$ is semisingular if and only if $w_* = \kappa(w)$ (that is, w is singular) or w is the middle element of a three-element interval.   —

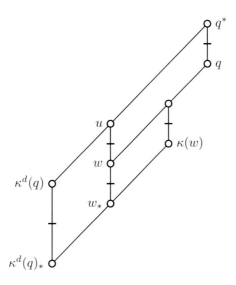

Figure 4: The form of a three-element interval u/w_*

Theorems 7.9 and 7.10 of [172] give strong characterizations of three-element intervals. Figure 4 illustrates some of the key information in these theorems. (The crosshatches indicate coverings.) It is an important fact that the canonical form of the element w of that figure is $w = u \wedge q$. This has the following useful corollary.

Corollary 5. *Let w be a completely join-irreducible element of $F(X)$ satisfying $\kappa(w) \neq w_*$. If u is an element such that $w_* = u \wedge \kappa(w)$, then either*

$$u = w_* \quad or \quad u = w \quad or \quad u \succ w.$$

Moreover, the last case only occurs if w is semisingular.

4.1   Tschantz's Theorem

Nation and I felt that it would be relatively easy to show that every infinite interval of a free lattice would contain $F(\omega)$ as a sublattice. We were surprised when we were not even able to rule out the possibility of an infinite interval which is a chain. (See the introduction and Chapter IX of [172] for more of the interesting history of this problem.) This problem was solved by S. T. Tschantz [502] with the following deep theorem.

Theorem 11.   Every infinite interval of a free lattice contains a sublattice isomorphic to $F(\omega)$.   —

The major part of the proof is showing that, in fact, there is no infinite interval which is a chain. This can be derived easily from Corollary 5: Suppose that the interval a/b is an infinite chain. By Day's Theorem and the fact that there are no long chains of covers, this interval contains a completely join-irreducible element w with $b < w_* \prec w < a$ and both a/w and w_*/b are infinite (or the dual situation holds). But this clearly cannot happen by Corollary 5.

Tschantz raised the following problem: can a free lattice have elements $a > c > b$ such that both a/c and c/b are infinite and every element of a/b is comparable with c. It is shown that no such element exists in Theorem 10.1 of [172], using rather different techniques.

4.2   Maximal chains

The characterization of semisingular elements allows us to answer some important questions about maximal chains in free lattices. Indeed, this was our original motivation for characterizing them. The problem is this: for which pairs $a > b$ in $F(X)$ is there a maximal chain in a/b without any covers? If a/b is atomic, then, since an element of a free lattice can have at most finitely many covers, there is no such maximal chain. It turns out (Theorem 9.15 of [172]) that in all other cases there is such a dense maximal chain.

Theorem 12.   If $a > b$ in $F(X)$ and a/b is neither atomic nor dually atomic, then there is a maximal chain from b to a without covers.   —

Actually, Theorem 9.15 proves more: we can find such a chain each of whose elements, except possibly a and b, is coverless, that is, has no upper and no lower cover in $F(X)$.

5.   Applications and Related Areas

Free lattices and the techniques used to study them have applications in several areas of lattice theory and even in universal algebra. This section explores these areas. We begin by introducing some basic concepts.

5.1   Bounded homomorphisms

Theorem 6 gives a very easy (and very efficient) way to test if an element w is completely join-irreducible in a free lattice and Theorem 7 shows how to find $\kappa(w)$ and therefore w_*. While these theorems are very nice, they obscure some of the details behind them and since these methods are applicable in several other areas, we describe them now.

Let $h\colon K \to L$ be a lattice epimorphism. Then we say that h is a *lower bounded homomorphism* if $h^{-1}(a)$ has a least element for each $a \in L$. An *upper bounded homomorphism* is defined dually and a homomorphism is *bounded* if it is both lower and upper bounded. When h is a lower bounded homomorphism, we let $\beta(a)$ denote the least preimage of a. $\alpha(a)$ is the dual. This concept was introduced in R. N. McKenzie [1972]. In this paper, McKenzie makes a thorough study of lattice varieties and free lattices play an important role. Lattices which are bounded homomorphic images of free lattices (*bounded lattices*, for short) are closely connected with covers in free lattices. In fact, a join-irreducible element $w \in \mathrm{F}(X)$ is completely join-irreducible if and only if the finite lattice $L(w)$ defined in the previous section is bounded and this fact is used in the proofs of Theorems 6 and 7.

B. Jónsson, in his studies of projective lattices and sublattices of free lattices, defined a dependency relation D on a lattice L:

$$a \, D \, b \quad \text{if } a \neq b, \ b \text{ is join-irreducible, and there is a}$$
$$p \in L \text{ with } a \leqslant b \vee p \text{ and } a \nleqslant c \vee p \text{ for } c < b.$$

We say that a *depends on* b. Of course, (the transitive closure of) D defines a quasiorder on L. We view D as 'less than or equal to,' that is, if $a \, D \, b$ then a is less than or equal to b in this quasiorder.

Although this dependency relation and the notion of bounded homomorphisms seem quite different, they are actually closely related as the following theorem (Corollary 2.39 of [172]) shows.

Theorem 13.   A finite lattice L is lower bounded if and only if it contains no D–cycle.   —

Having no D–cycle means that the D–relation determines a partial order. When L is finite, this means the elements can be ranked by their depth in this ordered set. The elements of rank 0 are the maximal elements of this order; these elements are the join-prime elements. Using this ranking, one can inductively calculate $\beta(a)$ (for any homomorphism h onto L). Even when there is a D–cycle one can calculate a sequence $\beta_n(a)$, $n \geqslant 0$, which is coinitial in $h^{-1}(a)$.

These concepts arose in connection with free lattices and play an important role there but they also play an important role in several other areas.

5.2   Congruence lattices

There is a strong connection between the D-relation and congruences on L, especially when L is finite. Indeed, an easy calculation shows that if $a \, D \, b$ for join-irreducibles a and b in a finite lattice, then $\Theta(a, a_*) \leqslant \Theta(b, b_*)$.

Let L be a finite lattice. Even if the D-relation contains a cycle, it still determines a quasiorder on the set of nonzero join-irreducible elements. Of course, a quasiorder induces an equivalence relation ($a \sim b$ if and only if $a \leqslant b$ and $b \leqslant a$) and if we factor by this equivalence relation, we get an ordered set. By Theorem 2.35 of [172], this ordered set is isomorphic to the ordered set of join-irreducible congruences on L. This is the basis of a very efficient method of finding the ordered set of join-irreducible congruences; see [167] and Section 5 of Chapter XI of [172].

Theorem 14.   Let L be a lattice with n elements. Then each of the following can be determined in time $O(n^2)$:

  (i) If L is simple.

 (ii) If L is subdirectly irreducible.

(iii) If L is directly indecomposable.

 (iv) If L is a bounded homomorphic image of a free lattice.

  (v) If L is semidistributive.

 (vi) If L is a splitting lattice.   —

Most of the above are based on an $O(n^2)$ algorithm that finds the ordered set of join-irreducible congruences. This algorithm finds the underlying set and a relation whose transitive closure is the order relation and this is good enough for the above facts. But for others things we need to find the full $\leqslant$ relation. In [167], I give an $O(n^2 \log n)$ algorithm for doing this. This is based on the fact that given an ordered set P having e covers, a lattice whose join-irreducible congruences is isomorphic to P cannot be too small. Specifically, we have:

Theorem 15.   If the ordered set of join-irreducible congruences on a finite lattice L has $e_{\prec}$ covers with $e_{\prec} > 2$, then

$$|L| \geqslant \frac{e_{\prec}}{2 \log_2 e_{\prec}}. \qquad —$$

Investigations into the relation of $|L|$ to the number of join-irreducible congruences were done by G. Grätzer, H. Lakser, and E. T. Schmidt [241], G. Grätzer, I. Rival, and N. Zaguia in [250], Y. Zhang [528], and G. Grätzer and D. Wang [265]; see Section C.1.

5.3   Sublattices of free lattices

In proving his theorem characterizing finite sublattices of free lattices, Theorem 3, Nation used two relations, A and B, originating with B. Jónsson, which refine the D–relation. If Nation's Theorem failed, there would be a finite semidistributive lattice containing a D–cycle. Such a lattice would contain a cycle of join-irreducible elements such that consecutive elements were related by either

A or B. Nation's proof begins with a beautiful duality result [396] (Theorem 2.63 of [172]):

(1)
$$a \, A \, b \iff \kappa(a) \, B^{\mathrm{d}} \, \kappa(b),$$
$$a \, B \, b \iff \kappa(a) \, A^{\mathrm{d}} \, \kappa(b).$$

Now $a \, A \, b$ implies that $a < b$, so clearly there can be no cycle of all A's; the duality result immediately implies there is no cycle of all B's, and this plays a key role in the proof of Nation's Theorem. It also immediately gives an important result of A. Day [119]:

Theorem 16.   A finite lattice which is a lower bounded homomorphic image of a free lattice is bounded if and only if it is semidistributive.   —

This result is used in the fast algorithm to determine if an element w of a free lattice has a lower cover given in Theorem 6.

5.4   Projective lattices

Projective lattices, that is, retracts of free lattices, and projective configurations play and important role in the study of lattice varieties. Of course, projective lattices are sublattices of free lattices. B. Jónsson, A. Kostinsky, and R. N. McKenzie (see R. N. McKenzie [1972] and A. Kostinsky [1972]) proved the converse for finitely generated lattices.

Theorem 17.   A finitely generated lattice is projective if and only if it is isomorphic to a sublattice of a free lattice. These conditions hold if and only if the lattice satisfies (W) and is a bounded homomorphic image of a free lattice.   —

Thus by Nation's Theorem a finite lattice is projective if and only if it is semidistributive and satisfies (W).

As mentioned above, a finite lattice is a lower bounded homomorphic image of a free lattice if and only if the D-relation is acyclic. In this case, the D-relation defines a partial order. For a finitely generated lattice to be lower bounded we need that all chains above each element of this order are finite. The converse also holds.

Arbitrary projective lattices were characterized in [179]. The most difficult and perhaps most surprising part of the proof consists of showing that the above condition on the D-relation holds in all projective lattices.

Theorem 18.   A lattice L is projective if and only if it satisfies the following conditions:

(i) Whitman's condition (W);

(ii) The D relation is acyclic and every D–chain above each element is finite;

(iii) The D^{d} relation is acyclic and every D^{d}–chain above each element is finite;

(iv) L has the minimal join-cover refinement property and the dual property;

(v) L is finitely separable.   —

A finite subset S of L is a *join-cover* of $a \in L$ if $a \leqslant \bigvee S$. We say L has the *minimal join-cover refinement property* if, for each $a \in L$, there is a finite set, $\mathrm{M}(a)$, of join-covers of a such that if T is a join-cover of a, then there is an $S \in \mathrm{M}(a)$ with $S \ll T$. The lattice L is *finitely separable* if, for each $a \in L$, there are two finite sets $\mathrm{A}(a)$ and $\mathrm{B}(a)$ such that $\mathrm{A}(a) \subseteq \{c \in L \mid c \geqslant a\}$, $\mathrm{B}(a) \subseteq \{c \in L \mid c \leqslant a\}$, and if $a \leqslant b$, then $\mathrm{A}(a) \cap \mathrm{B}(b)$ is nonempty.

5.5 Sharply transferable lattices

Transferable and sharply transferable lattices are discussed in the text and in Appendix A. As pointed out there, for finite lattices these concepts coincide and, in fact, are precisely the finite sublattices of free lattices. So Nation's Theorem gives a strong characterization of these lattices. Although there is no good characterization of general transferable lattices, there is a very nice characterization of sharply transferable lattices due to G. Grätzer and C. R. Platt [248]. (Note the similarity to arbitrary sublattices of free lattice for which there is no good characterization and projective lattices for which there is.)

Theorem 19.   A lattice L is sharply transferable if and only if it satisfies the following four conditions:

(i) Whitman's condition (W).

(ii) The D relation is acyclic and every D–chain *below* each element is finite.

(iii) The D^{d} relation is acyclic and every D^{d}–chain above each element is finite.

(iv) For each $x \in L$, the set $\{y \mid y \not\geqslant x\}$ is finite.   —

Notice how similar Theorems 18 and 19 are. The last condition is a finiteness condition as are the two last conditions of Theorem 18, which are, in fact, implied by the last condition of Theorem 19. The first and third conditions of the two theorems are the same but note the curious difference in the second conditions.

Nation used Theorem 19 and his duality theorem given in formula (1) to show that strongly transferable lattices are projective [396]. Although his duality theorem depended on L being finite, he showed that for a lattice satisfying the last condition of Theorem 19 that at least the forward implications of (1) hold. Now the crux of proving that sharply transferable lattices are projective is showing that they satisfy the second condition of Theorem 18. But the third condition of Theorem 19 together with the forward direction of his duality theorem show this:

Theorem 20. Every sharply transferable lattice is projective. ⎯

Nation also gives an example of a lattice which is projective and even satisfies the last condition of Theorem 19 but is not transferable.

5.6 Varieties of lattices and congruence varieties

Both of these topics are covered in Appendix F, so we will just make a few comments. Jónsson's Theorem (B. Jónsson [1967]) motivated McKenzie to make a thorough study of lattice varieties in R. N. McKenzie [1972]. The concept of a bounded homomorphism began with this paper. A finite, subdirectly irreducible lattice L which is a bounded homomorphic image of a free lattice is called a *splitting lattice*. By Jónsson's Theorem, if a subdirectly irreducible lattice is in the join of finitely many varieties, it is in one of them. For a splitting lattice this is true for an arbitrary join of lattice varieties and this leads to a 'splitting' of the lattice of all lattice varieties. So if L is a splitting lattice, there is a conjugate equation ε such that every variety either contains L or satisfies ε, but not both.

As we pointed out above, splitting lattices are closely associated with covers in free lattices. If w is a join-irreducible element of a free lattice $F(X)$ with a lower cover w_*, then $F(X)/\Psi(w, w_*)$ is a splitting lattice with conjugate equation $w \approx w_*$, where $\Psi(w, w_*)$ is the unique largest congruence separating w and w_*. Moreover, every splitting lattice has this form. Day's Theorem, Theorem 4, shows that there are many splitting lattices; in fact, they generate the variety of all lattices.

A *congruence variety* is a variety of lattices generated by all the congruence lattices of the members of some variety of algebras. Bounded homomorphisms and splitting lattices played an essential role in this area; see [335] for a survey of the results up to about 1979 and [165] for subsequent developments. One of the major questions of this area was the existence of nonmodular congruence variety other than the variety of all lattices. S. V. Polin [437] answered this question by constructing a variety, $\mathcal{P}$, whose associated congruence variety was nonmodular but did satisfy a nontrivial lattice equation. A thorough analysis of Polin's congruence variety is carried out in [124]. It is there shown that the congruence lattice of finitely generated free algebra in $\mathcal{P}$ is a splitting lattice and the associated splitting equations are given. Using these splitting equations, I proved with A. Day that Polin's congruence variety is the unique minimal nonmodular congruence variety. With G. Czédli [90], I gave an effective characterization of lattice equations which imply congruence modularity (and distributivity).

Theorem 21.   Polin's congruence variety is the unique minimal nonmodular congruence variety.   There is an effective procedure to determine if a lattice equation implies congruence modularity.   —

As a second example, consider the variety of all (meet) semilattices. The two-element semilattice, C_2, is the only subdirectly irreducible (even finitely subdirectly irreducible), hence the meet-irreducible congruences of a semilattice are all coatoms. If a is a nonzero element of a semilattice S, then let Ψ_a be the congruence with two equivalence classes, $\{x \mid x \geqslant a\}$ and its complement. This is always a coatom in $\operatorname{Con} S$ and when S is finite, these are the only coatoms. Now it is not hard to show that if $\Psi_a \, D^{\mathrm{d}} \, \Psi_b$, then $a > b$ in S; see Lemma 2.86 of [172]. Thus D^{d} is acyclic. In particular, we get the theorem of K. V. Adaricheva [22]:

Theorem 22.   If S is a finite semilattice, then $\operatorname{Con} S$ is an upper bounded homomorphic image of a free lattice.   —

D. Papert [1964] had shown that the congruence lattice of a semilattice satisfies (SD$_\wedge$) but being upper bounded is much stronger. In fact, this relation between (SD$_\wedge$) and upper boundedness holds much more generally; see [175].

Theorem 23.   If $\mathcal{V}$ is a variety of algebras such that the congruence lattices of the members of $\mathcal{V}$ satisfy (SD$_\wedge$), then the congruence lattices of the finite algebras in $\mathcal{V}$ are upper bounded homomorphic images of a free lattice. If the congruence lattices of the algebras in a variety $\mathcal{W}$ satisfy (SD$_\vee$), then the congruence lattice of the finite algebras in $\mathcal{W}$ are both upper and lower bounded homomorphic images of a free lattice.   —

Applied Lattice Theory: Formal Concept Analysis

B. Ganter and R. Wille

The "Formal Concept Analysis" project was born around 1980, when a research group in Darmstadt, Germany, began to develop systematically a framework for the application of lattice theory. It was first presented to the mathematical public in a lecture at the 1981 Banff conference on Ordered Sets by R. Wille, see [521]. Since then, several hundred articles have been published, see the list of publications at the WEB site

> http://www.mathematik.th-darmstadt.de/ags/ag1

including a textbook [195] on the mathematical foundations by the present authors. The Darmstadt group has participated in more than a hundred application projects. Former members of the team have founded a small company and now make their living from such applications.

The name, "Formal Concept Analysis" requires some explanation. The method is mainly used for the *analysis* of data, that is, for investigating and processing information. The data is structured into units that are formal abstractions of *concepts* of human thought, allowing for meaningful and comprehensible interpretation. We use the adjective *formal* to emphasize that these formal concepts are mathematical entities and should not be identified with the concepts of human thought. The same adjective indicates that the basic data

591

format, the *formal context*, formalizes only an aspect of what is usually referred to as "context".

Much of the mathematics required for the applications comes from lattice theory. The basic construction of a complete lattice from a binary relation can already be found in the first edition of G. Birkhoff's *Lattice Theory* (G. Birkhoff [1940]). The new goals made it necessary to extend this theory. Formal Concept Analysis has created results that may be of interest even outside the applications from which the motivation was derived.

For proofs, citations, and further details, we refer the reader to the book [195].

1.   Formal Contexts and Concept Lattices

A triple $\langle G, M, I \rangle$ is called a *formal context*, if G and M are sets and $I \subseteq G \times M$ is a binary relation between G and M. We call the elements of G *objects*, those of M *attributes*, and I the *incidence* of the context $\langle G, M, I \rangle$. For $A \subseteq G$, we define

$$A' = \{m \in M \mid \langle g, m \rangle \in I, \text{ for all } g \in A\}$$

and dually, for $B \subseteq M$,

$$B' = \{g \in G \mid \langle g, m \rangle \in I, \text{ for all } m \in B\}.$$

It is easy to prove that these *derivation operators* satisfy the following simple rules (for all A_1, A_2, $A \subseteq G$ and all B_1, B_2, $B \subseteq M$):

1) $A_1 \subseteq A_2 \Rightarrow A_2' \subseteq A_1'$;   1') $B_1 \subseteq B_2 \Rightarrow B_2' \subseteq B_1'$;

2) $A \subseteq A''$ and $A' = A'''$;   2') $B \subseteq B''$ and $B' = B'''$;

3) $A \subseteq B' \iff B \subseteq A'$.

The experienced reader will notice that these derivation operators establish a Galois connection between the power set lattices on G and M and, thereby, a dual isomorphism between the two closure systems. It is natural to consider the elements of this dual isomorphism, as is done in the following definition.

A pair $\langle A, B \rangle$ is a *formal concept* of $\langle G, M, I \rangle$ if and only if

$$A \subseteq G, \ B \subseteq M, \ A' = B, \text{ and } A = B'.$$

A is called the *extent* and B the *intent* of the concept $\langle A, B \rangle$. The concepts of a given context are naturally ordered by the *subconcept-superconcept relation* defined by

$$\langle A_1, B_1 \rangle \leq \langle A_2, B_2 \rangle \iff A_1 \subseteq A_2 \ (\iff B_2 \subseteq B_1).$$

The ordered set of all formal concepts of $\langle G, M, I \rangle$ is denoted by $\mathfrak{B}(G, M, I)$ and it is called the *concept lattice* of $\langle G, M, I \rangle$.

Concept lattices are indeed lattices. The following theorem shows that, more precisely, the concept lattices are, up to isomorphism, exactly the complete lattices: every concept lattice is complete, and every complete lattice is isomorphic to some concept lattice. (Using topological contexts, noncomplete lattices can also be so represented.)

Theorem 1 (The basic theorem on concept lattices).   The concept lattice $\mathfrak{B}(G, M, I)$ is a complete lattice in which infimum and supremum are given by:

$$\bigwedge_{t \in T} \langle A_t, B_t \rangle = \left(\bigcap_{t \in T} A_t, \left(\bigcup_{t \in T} B_t \right)'' \right),$$

$$\bigvee_{t \in T} \langle A_t, B_t \rangle = \left(\left(\bigcup_{t \in T} A_t \right)'', \bigcap_{t \in T} B_t \right).$$

A complete lattice L is isomorphic to $\mathfrak{B}(G, M, I)$ if and only if there are maps $\tilde{\gamma} \colon G \to L$ and $\tilde{\mu} \colon M \to L$ such that $\tilde{\gamma}(G)$ is supremum-dense in L, $\tilde{\mu}(M)$ is infimum-dense in L and $g \, I \, m$ is equivalent to $\tilde{\gamma}g \leqslant \tilde{\mu}m$, for all $g \in G$ and all $m \in M$. In particular, $L \cong \mathfrak{B}(L, L, \leqslant)$.   —

Concept lattices can be visualized with the usual lattice diagrams. It would however be too messy to label each concept by its extent and its intent. A much simpler *reduced labeling* is achieved, if each object and each attribute is marked only once in the diagram. The name of object g is attached to the lower half of the corresponding *object concept* $\gamma g = \langle \{g\}'', \{g\}' \rangle$, while the name of the attribute m is located at the upper half of the *attribute concept* $\mu m = \langle \{m\}', \{m\}'' \rangle$. It is then still possible to read off all extents and all intents from the diagram: for any concept $\langle A, B \rangle$, we have

$$g \in A \iff \gamma g \leq \langle A, B \rangle \quad \text{and} \quad m \in B \iff \langle A, B \rangle \leq \mu m.$$

In other words, extent and intent of an arbitrary concept can be found as the set of objects in the principal ideal and as the set of attributes in the principal filter generated by that concept.

Figure 1 shows an example of such a concept lattice diagram with reduced labeling. The formal context is not explicitly given but can easily be read off from the diagram. The objects here are seven finite lattices (given by their diagrams), the attributes are lattice properties. The incidence is as expected: a lattice L is incident with a property P if and only if L has property P. This can be read from the diagram according to the general rule

$$\langle g, m \rangle \in I \iff \gamma g \leq \mu m.$$

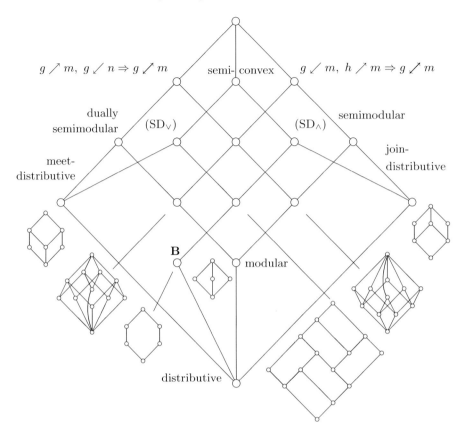

$$g \nearrow m, \; g \swarrow n \Rightarrow g \nearrow\!\!\swarrow m \qquad \text{semi-convex} \qquad g \swarrow m, \; h \nearrow m \Rightarrow g \nearrow\!\!\swarrow m$$

dually
semimodular   (SD$_\vee$)

(SD$_\wedge$)   semimodular

meet-
distributive

join-
distributive

B

modular

distributive

Figure 1: Generalizations of the distributive law. **B** abbreviates "bounded homomorphic image of a free lattice". A lattice is *semiconvex* if $x \wedge y = x \wedge z$ and $x \vee z = y \vee z$ imply that $x \leqslant z$. The arrow relations are defined in the text. The implications in this diagram hold for any finite lattice.

The *arrow relations* of a context $\langle G, M, I \rangle$ are defined as follows: for $g \in G$, $m \in M$, let

$$g \swarrow m \iff \langle g, m \rangle \notin I \text{ and if } \{g\}' \subset \{h\}', \text{ then } h \, I \, m;$$
$$g \nearrow m \iff \langle g, m \rangle \notin I \text{ and if } \{m\}' \subset \{n\}', \text{ then } g \, I \, n;$$
$$g \nearrow\!\!\swarrow m \iff g \swarrow m \text{ and } g \nearrow m.$$

For given $g \in G$, there exists an attribute $m \in M$ with $g \swarrow m$ if and only if γg is $\bigvee$-irreducible. Dually, $g \nearrow m$ holds for some $g \in G$ if and only if μm is $\bigwedge$-irreducible. We define a formal context to be *doubly founded* if

$$\langle g, m \rangle \notin I \quad \Rightarrow \quad \exists_{n \in M} \; g \nearrow n \text{ and } m' \subseteq n';$$
$$\langle g, m \rangle \notin I \quad \Rightarrow \quad \exists_{h \in G} \; h \swarrow m \text{ and } g' \subseteq h'.$$

The concept lattice of a doubly founded context contains many irreducibles. It can, in fact, be shown that if L is such a lattice, then the set $J_c(L)$ of $\bigvee$-irreducibles is $\bigvee$-dense and the set $M_c(L)$ of $\bigwedge$-irreducibles is $\bigwedge$-dense, as in the case of finite lattices. By the basic theorem, then, L is isomorphic to the concept lattice of its *standard context* $\langle J_c(L), M_c(L), \leqslant \rangle$. Every doubly founded context contains a subcontext isomorphic to the standard context; such a subcontext is called *reduced*. A sufficient condition for $\langle G, M, I \rangle$ to be doubly founded is that its concept lattice $\mathfrak{B}(G, M, I)$ is a *doubly founded lattice*, that is, that for any two elements $x < y$, there are elements s and t such that s is minimal with respect to $s \leqslant y$, $s \not\leqslant x$, and t is maximal with respect to $t \geqslant x$, $t \not\geqslant y$.

In order to construct all concepts of a finite formal context, we recall that the extents form a closure system. The corresponding closure operator $A \mapsto A''$ can easily be computed from the formal context. Assuming that $G = \{1, \ldots, n\}$, a strict linear order $<$ on the subsets of G is defined by

$$A < B \iff A <_i B, \quad \text{for some } i \in G,$$

where

$$A <_i B \iff i \in B - A \text{ and } A \cap \{1, \ldots, i-1\} = B \cap \{1, \ldots, i-1\}.$$

Moreover, let

$$A \oplus i = ((A \cap \{1, \ldots, i-1\}) \cup \{i\})''.$$

Theorem 2. For $A \subseteq G$, $A \neq G$, the smallest closed set that is larger (with respect to $<$) than A, is $A \oplus i$, where i is the largest element of G satisfying $A <_i A \oplus i$. —

The theorem explains how to find the next closed set, and thereby all closed sets: One starts with the closure $\varnothing''$ of the empty set and then repeatedly computes the next one, until the largest closed set, G, is reached. For a given formal context $\langle G, M, I \rangle$, this computes all extents, that is, subsets $A \subseteq G$ satisfying $A = A''$, and thereby all concepts $\langle A, A' \rangle$ of $\langle G, M, I \rangle$.

An *implication* $A \to B$ (between sets A, $B \subseteq M$ of attributes) *holds* in a formal context if and only if $B \subseteq A''$, that is, if every object that has all attributes in A also has all attributes in B. The implications determine the concept lattice up to isomorphism and, therefore, offer an additional interpretation of the lattice structure. For practical purposes, it is desirable to reduce the set of all valid implications (which is usually huge) to a small set from which the others can be derived by inference. There is, in fact, a nice theory of *implication bases* for finite lattices. Particularly useful is the fact that such bases can be computed using the above algorithm.

2. Applications

It is our belief that the potential of Formal Concept Analysis as a branch of Applied Mathematics is just beginning to show. Concept lattices are used to analyze data, making their conceptual structure visible and accessible, in order to find patterns, regularities, exceptions, and so on.

Most other techniques of data analysis have as their aim to drastically reduce the amount of information given and to obtain a few "significant parameters". By contrast, a concept lattice does not reduce complexity since it contains all the details of the data represented by the formal context. It is usually smaller than the power sets of G and M, since the two closure systems of extents and intents contain only those sets that are, in this specific sense, meaningful. Nevertheless, it may be exponential in size, compared to the formal context. Complexity, therefore, is a problem, even though there are efficient algorithms and advanced application programs [507]. Even a formal context of moderate size may have more concepts than one would like to see individually. The mathematical theory, therefore, must provide tools to decompose, structure, browse through, and navigate in a concept lattice. For reasons of efficiency, it is desirable to develop methods that perform such lattice manipulations by working directly on the formal context.

The basic data type of a formal context is not the one occurring most frequently in the applications of Formal Concept Analysis. Often data is recorded in the form of a tabular or relational data base, as described by the following definition.

A *many-valued context* $\langle G, M, W, I \rangle$ consists of a set G (of "objects"), a set M (of "many-valued attributes"), a set W (of "attribute values"), and a ternary relation $I \subseteq G \times M \times W$ satisfying

$$\langle g, m, v \rangle \in I, \ \langle g, m, w \rangle \in I \Rightarrow v = w.$$

A many-valued context is studied by first transforming it into a "one-valued" context, that is, into a formal context in the above sense, and then constructing the concept lattice of the latter. The transformation procedure is called *conceptual scaling*; it is not unique, but contains several degrees of freedom that allow different interpretations. The simplest version, *plain scaling*, will be introduced here.

A *conceptual scale* for a many-valued attribute m is a one-valued context which has the attribute values of m among its objects. With plain scaling, a scale is associated to each many-valued attribute m, and m is replaced by the set of its scale attributes. Each value of m is substituted by the corresponding row of the scale. Formally, let $\langle G, M, W, I \rangle$ be a many-valued context and, for each $m \in M$, let $\mathbb{S}_m = \langle G_m, M_m, I_m \rangle$ be a scale for m. The *derived context* of $\langle G, M, W, I \rangle$ with respect to plain scaling with the scales $\langle S_m \mid m \in M \rangle$ is given as $\langle G, N, J \rangle$, where

$$N = \prod_{m \in M} (\{m\} \times M_m),$$

$$\langle g, \langle m, n \rangle \rangle \in J \iff \exists_{w \in W} \langle g, m, w \rangle \in I \text{ and } \langle w, n \rangle \in I_m.$$

Nested line diagrams are useful variants of lattice diagrams. They are implemented, for example, in the commercial TOSCANA program, which uses concept lattices for browsing in data bases. As a small example, we show how such a diagram is obtained for the concept lattice of the formal context $\langle G, M, I \rangle$ given in Figure 2.

At first, the attribute set M is split into the sets $M_1 = \{a, b, c, d\}$, $M_2 = \{e, f, g, h, i\}$. Then the concept lattices $\mathfrak{B}(G, M_1, I \cap (G \times M_1))$ (which is an eight-element Boolean lattice) and $\mathfrak{B}(G, M_2, I \cap (G \times M_2))$ (which is a cardinal sum of a three element chain with two singletons, with 0 and 1 adjoined) are computed. The diagram shown in Figure 2 is that of a direct product of these two lattices, in which the elements of $\mathfrak{B}(G, M, I)$ (which is isomorphic to a $\bigvee$-subsemilattice of this product) are black-filled.

Many important applications use *conceptual data systems* that consist of a many-valued context stored in a (relational) data base, together with a collection of conceptual scales with line diagrams of their concept lattices. The conceptual knowledge encoded in a conceptual data system is expressed by its concept lattice; however, this lattice is, in most cases, far too large to be represented graphically. Therefore, it must be decomposed and displayed in small "aspects", the simplest of which are the concept lattices of the given conceptual scales. Suitable collections of these can be combined and displayed as nested line diagrams. This is performed automatically by the TOSCANA application program mentioned above: the user can choose a combination of conceptual scales, and a nested line diagram is generated and displayed. It is possible to "roam" through the data system by selecting various scales.

The many tasks that can be performed with this approach, like *exploring, searching, recognizing, identifying*, and so on, have been discussed in detail elsewhere [525]. Selected projects in different disciplines, most of which have been worked out in cooperation with the Darmstadt group, may illustrate the multitude of possible applications: *medicine*: diabetes of children, *psychosomatics*: repertory grids of anorexic patients, *psychology*: children's concept development, *musicology*: audio-visual perception of music, *politics*: international cooperations, *linguistics*: semantic structures of lexical data bases, *information science*: retrieval system for a library, *computer science*: software reengineering, *electronics*: improving chip production, *civil engineering*: retrieval system concerning laws and regulations, *ecology*: water pollution, *biology*: color perception. See [524] for references.

	a	b	c	d	e	f	g	h	i
1	×	×					×		
2	×	×					×	×	
3	×	×	×				×	×	
4	×		×				×	×	×
5	×	×		×		×			
6	×	×	×	×		×			
7	×		×	×	×				
8	×		×	×		×			

Figure 2: To obtain a nested line diagram, the attribute set is divided.

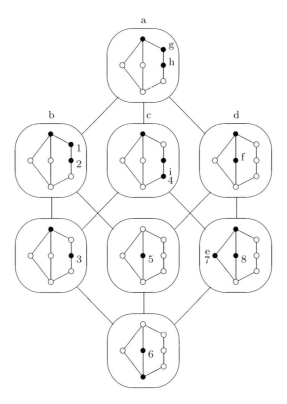

Figure 3: The filled circles represent the concept lattice of the formal context in Figure 2.

3.   Sublattices and Quotient Lattices

There are two natural ways to define substructures of a formal context $\langle G, M, I \rangle$. One is to omit some objects and attributes and to restrict the incidence to the remaining ones, that is, to form a *subcontext* $\langle H, N, I \cap (H \times N) \rangle$, for some $H \subseteq G$, $N \subseteq M$. The other is to leave G and M unaltered and to consider the context $\langle G, M, J \rangle$, for some *subrelation* $J \subseteq I$. In both cases, additional conditions are necessary to obtain an interesting construction.

A subrelation $J \subseteq I$ is *closed*, if every concept of $\langle G, M, J \rangle$ is also a concept of $\langle G, M, I \rangle$. It is easy to check whether a given subrelation $J \subseteq I$ is closed. A necessary and sufficient condition is that $\{g\}^J$ ($= \{ m \in M \mid \langle g, m \rangle \in J \}$) is an intent and $\{m\}^J$ is an extent of $\langle G, M, I \rangle$, for each object $g \in G$ and each attribute $m \in M$. Closed subrelations correspond to complete sublattices.

Theorem 3. S is a complete sublattice of $\mathfrak{B}(G, M, I)$ if and only if $S = \mathfrak{B}(G, M, J)$, for some closed subrelation $J \subseteq I$.   —

A subcontext $\langle H, N, I \cap (H \times N) \rangle$ is called *compatible*, if $\langle A \cap H, B \cap N \rangle$ is a concept of $\langle H, N, I \cap (H \times N) \rangle$, whenever $\langle A, B \rangle$ is a concept of $\langle G, M, I \rangle$. For doubly founded contexts, there is a simple way to recognize compatible subcontexts; they are precisely the *arrow-closed* subcontexts, that is, those with

$$h \in H, \ h \nearrow m \Rightarrow m \in N \quad \text{and} \quad n \in N, \ g \swarrow n \Rightarrow g \in H.$$

After reducing the formal context (as can always be done in the doubly founded case), there is an even simpler description using a transitive closure of the arrows. Define a binary relation $\varnothing \subseteq G \times M$ as the smallest relation satisfying the conditions

(i) if $g \swarrow m$, then $g \varnothing m$;

(ii) if $g \varnothing m$, $h \nearrow m$ and $h \swarrow n$, then $g \varnothing n$;

and let $\varnothing$ denote the complement of this relation.

Theorem 4.   A subcontext $\langle H, N, I \cap (H \times N) \rangle$ of a standard context $\langle G, M, I \rangle$ is compatible if and only if $\langle G - H, N \rangle$ is a concept of $\langle G, M, \varnothing \rangle$.   —

As the next theorem shows, compatible subcontexts correspond to quotient lattices and thereby to complete congruences of $\mathfrak{B}(G, M, I)$. In general, this correspondence needs to be neither one-to-one nor onto. The situation is simpler for doubly founded contexts and, in particular, for reduced ones.

Theorem 5.   A subcontext $\langle H, N, I \cap (H \times N) \rangle$ of $\langle G, M, I \rangle$ is compatible if and only if

$$\langle A, B \rangle \overset{\Pi_{H,N}}{\longrightarrow} \langle A \cap H, B \cap N \rangle$$

defines a complete homomorphism from $\mathfrak{B}(G, M, I)$ onto $\mathfrak{B}\langle H, N, I \cap (H \times N) \rangle$.

Theorem 6. If $\langle G, M, I \rangle$ is the standard context of a doubly founded complete lattice L, then for every complete congruence Θ of $\mathfrak{B}(G, M, I)$, there is a unique subcontext $\langle H, N, I \cap (H \times N) \rangle$ such that

$$\Theta = \mathrm{Ker}\, \Pi_{H,N}.$$

A combination of these results yields the following statement:

Theorem 7. Let $\langle G, M, I \rangle$ be doubly founded and reduced. The lattice of complete congruences of $\mathfrak{B}(G, M, I)$ is isomorphic to $\mathfrak{B}(G, M, \nnearrow)$.

This makes it very easy to compute the congruence lattice of any finite lattice L.

The concepts of $\mathfrak{B}(J_c(L), M_c(L), \nnearrow)$, which can be generated with the above algorithm, are in bijective correspondence with the (complete) congruences.

The quotient lattice corresponding to a concept $\langle A, B \rangle$ of

$$\mathfrak{B}(J_c(L), M_c(L), \nnearrow)$$

is isomorphic to

$$\mathfrak{B}(G - A, B, I \cap ((G - A) \times B)).$$

An alternative characterization of congruences is obtained using the fact that certain superrelations of the incidence relation I, the so-called *block relations*, are in one-to-one correspondence with the complete tolerance relations of the concept lattice.

4.   Subdirect Products and Tensor Products

Closed subrelations and compatible subcontexts can be combined neatly to describe subdirect products. Note that we only consider *complete* subdirect products, defined as complete sublattices of the direct product for which the projections are surjective. The first result shows that such decompositions can be read off from the formal context:

Theorem 8. The subdirect decompositions of a doubly founded complete lattice L correspond (bijectively) to the families of arrow-closed subcontexts

$$\langle G_t, M_t, I \cap (G_t \times M_t) \rangle$$

of its standard context $\langle G, M, I \rangle$ satisfying $\bigcup G_t = G$ and $\bigcup M_t = M$.

As a corollary, we obtain the following statement:

Theorem 9.   A doubly founded complete lattice is subdirectly irreducible if and only if its standard context contains an attribute m with $g \nearrow\hspace{-0.9em}\nearrow m$, for all objects g.   —

These characterizations allow us to transfer a well known result to complete lattices:

Theorem 10.   Every doubly founded complete lattice has a subdirect decomposition into subdirectly irreducible factors.   —

It is also worthwhile to consider the complete subdirect product as a *construction* method. First, we mention that a direct product of complete lattices translates to a sum of formal contexts, defined as follows. Let T be some index set and let $\langle G_t, M_t, I_t \rangle$, $t \in T$, be a family of formal contexts. To simplify, let us assume that $G_s \cap G_t = \varnothing = M_s \cap M_t$, for all $s \neq t$. We define the *sum* of these contexts to be

$$\sum_{t \in T} \langle G_t, M_t, I_t \rangle = \left(\bigcup_{t \in T} G_t, \bigcup_{t \in T} M_t, \bigcup_{t \in T} I_t \cup \bigcup_{s \neq t} (G_s \times M_t) \right).$$

Theorem 11.   The following isomorphism holds:

$$\mathfrak{B} \left(\sum_{t \in T} (G_t, M_t, T_t) \right) \cong \prod_{t \in T} \mathfrak{B}(G_t, M_t, I_t).$$ —

The isomorphism is canonical. Therefore, to every complete sublattice of the direct product, there corresponds a closed subrelation of the sum context. These can be characterized as follows.

Theorem 12.   The complete subdirect products of the concept lattices

$$\mathfrak{B}(G_t, M_t, I_t), \quad t \in T,$$

correspond bijectively to those closed subrelations J of the sum context

$$\sum_{t \in T} \langle G_t, M_t, I_t \rangle$$

that satisfy $J \cap (G_t \times M_t) = I_t$, for all $t \in T$.   —

There is an even more operational characterization, for which we refer to [195]. It uses the notion of bonds between formal contexts. A *bond* from a context $\langle G_1, M_1, I_1 \rangle$ to $\langle G_2, M_2, I_2 \rangle$ is a relation $J \subseteq G_1 \times M_2$ such that

(i) for each $g \in G_1$, the set $\{g\}^J$ is an intent of $\langle G_2, M_2, I_2 \rangle$;

(ii) for each $m \in M_2$, $\{m\}^J$ is an extent of $\langle G_1, M_1, I_1 \rangle$.

In order to turn the subdirect product construction into an operation with a unique result, we introduce the notion of a P-lattice. Let P be some set and define a complete P-*lattice* as a pair $\langle L, \alpha \rangle$, where L is a complete lattice and α is a function that maps P onto a generating set in L. For a family $\langle L_t, \alpha_t \rangle$, $t \in T$, of P-lattices the P-*product* is defined as the complete sublattice of the direct product that is generated by $\{ \langle \alpha_t(p) \mid t \in T \rangle \mid p \in P \}$. If each L_t is represented as a concept lattice $\mathfrak{B}(G_t, M_t, I_t)$, then each $\alpha_t(p)$ is some concept, say, $\alpha_t(p) = \langle A_t, B_t \rangle$.

Theorem 13.   The closed subrelation J of the sum context

$$\sum_{t \in T} \mathfrak{B}(G_t, M_t, I_t)$$

that corresponds to a P-product of P-concept lattices $\mathfrak{B}(G_t, M_t, I_t)$ is characterized by the following properties:

(i) $J \cap (G_t \times M_t) = I_t$, for all $t \in T$.

(ii) For $s, t \in T$, $s \neq t$, $J \cap (G_s \times M_t)$ is the smallest bond from $\langle G_s, M_s, I_s \rangle$ to $\langle G_t, M_t, I_t \rangle$ containing the sets $A_s^p \times B_t^p$, for all $p \in P$.   —

There is also a natural definition of a direct product of formal contexts. The corresponding lattice construction is the tensor product of complete lattices. The definitions are easy to give.

The *direct product* of the formal contexts $\langle G_t, M_t, I_t \rangle$, $t \in T$, is given by

$$\prod_{t \in T} \langle G_t, M_t, I_t \rangle = \left(\prod_{t \in T} G_t, \prod_{t \in T} M_t, \nabla \right),$$

where, for $g = \langle g_t \mid t \in T \rangle$ and $m = \langle m_t \mid t \in T \rangle$, we have

$$\langle g, m \rangle \in \nabla \iff \langle g_t, m_t \rangle \in I_t, \quad \text{for some } t \in T.$$

The *tensor product* of complete lattices L_t, $t \in T$, is defined as

$$\bigotimes_{t \in T} L_t = \mathfrak{B} \left(\prod_{t \in T} \langle L_t, L_t, \leqslant \rangle \right).$$

For a complete lattice L, there are usually many formal contexts with $L \cong \mathfrak{B}(G, M, I)$. It was mentioned in the basic theorem that $\langle L, L, \leqslant \rangle$ is one of them. But it is not crucial that this particular context was used in the definition of the tensor product, as the next theorem shows.

Theorem 14.   The following isomorphism holds:

$$\bigotimes_{t\in T}\mathfrak{B}(G_t,M_t,I_t)\cong\mathfrak{B}\Big(\prod_{t\in T}\langle G_t,M_t,I_t\rangle\Big).\qquad\text{—}$$

The arrow relations of a product context are given by those of the quotients. This yields the following isomorphism:

Theorem 15.   The congruence lattice of a tensor product of finitely many doubly founded complete lattices is isomorphic to the tensor product of the congruence lattices of the quotients:

$$\operatorname{Con}\bigotimes_{i=1}^{n}L_i\cong\bigotimes_{i=1}^{n}\operatorname{Con}L_i.\qquad\text{—}$$

For example, the congruence lattice of the free bounded distributive lattice $\mathrm{F}_{\mathbf{D}_{\{0,1\}}}(n)$ with n generators is the Boolean lattice with 2^n atoms, since $\mathrm{F}_{\mathbf{D}_{\{0,1\}}}(n)$ is isomorphic with the n-th tensor power of the three element chain $\mathrm{F}_{\mathbf{D}_{\{0,1\}}}(1)$. The standard contexts of $C_2\times C_2$ (which is the congruence lattice of $\mathrm{F}_{\mathbf{D}_{\{0,1\}}}(1)$) is

	×
×	

Its n-th power is isomorphic to $\langle S,S,\neq\rangle$, where $|S|=2^n$, and $\mathfrak{B}(S,S,\neq)\cong\langle\mathrm{P}(S),\subseteq\rangle$.

In the distributive case, our definition of a tensor product coincides with those of other authors.

Theorem 16.   The tensor product of completely distributive complete lattices is completely distributive.   —

Many further properties of the tensor product can be found in [195]. The fact that the tensor product is distributive over the P-product can be utilized for constructions, for instance, of subdirect representations of free distributive lattices. It is possible to define a subdirect product of formal contexts and thereby to obtain a lattice construction of the *subtensorial product*.

5. Lattice Properties

A concept lattice is uniquely determined by its formal context. Thus, in principle, every structural property can be read off from the incidence relation. This is of importance, for instance, for algorithms: a lattice feature that can easily be detected in a formal context, is "algorithmically harmless". Fortunately, this is the case for many of the commonly used properties, at least for doubly founded lattices. Most of the results which we formulate below for the finite case generalize to doubly founded lattices, see [195].

The arrow relations $\nearrow$ and $\swarrow$ and their "transitive closure" $\nearrow\!\!\!\!\swarrow$ are easy to compute. Once this is done, the following theorem can be applied without much effort:

Theorem 17.   A finite lattice with standard context $\langle G, M, I \rangle$ is

distributive   if and only if

$$g \swarrow m,\ g \nearrow n \Rightarrow m = n;$$

join distributive[1]   if and only if

$$g \nearrow m,\ h \nearrow n \Rightarrow g = h;$$

join semidistributive[2]   if and only if

$$g \swarrow m,\ h \swarrow m \Rightarrow g = h;$$

semimodular   if and only if

$$g \swarrow m,\ g \swarrow n,\ \langle h, m \rangle \in I,\ \langle h, n \rangle \notin I$$

implies that

$$\exists_{p \in M} \langle h, p \rangle \notin I,\ \langle g, p \rangle \in I,\ m' \cap n' \subseteq p';$$

simple   if and only if

$$g \nearrow\!\!\!\!\swarrow m,\ \text{for all } g \in G \text{ and all } m \in M;$$

subdirectly irreducible   if and only if

$$g \nearrow\!\!\!\!\swarrow m,\ \text{for all } g \in G \text{ and some } m \in M.$$

More precise information on the complexity of these and similar properties are compiled in M. Skorsky's thesis [479].

Determining the order dimension of an ordered set is known to be a NP-complete problem. Since the dimension of $\langle P, \leqslant \rangle$ is equal to that of its MacNeille-completion $\mathfrak{B}(P, P, \leqslant)$, it is not an easy task to compute the dimension of a lattice from its context. But there is at least a nice reformulation that

---

[1]that is, locally distributive
[2]that is, satisfies (SD$_\vee$)

sometimes is quite illuminating. It uses the notion of a *Ferrers relation*, that is, a relation $F \subseteq G \times M$ satisfying

$$\langle g, m \rangle \in F, \ \langle h, n \rangle \in F, \ \langle h, m \rangle \notin F \ \Rightarrow \ \langle g, n \rangle \in F.$$

The *Ferrers dimension* of a formal context $\langle G, M, I \rangle$ is defined as the smallest number of Ferrers relations whose intersection is I. Since the complement of a Ferrers relation is again a Ferrers relation, the Ferrers dimension is also the smallest number of Ferrers relations the union of which is $(G \times M) - I$. The latter is more convenient for practical work, because the Ferrers relations can be entered into the "empty cells" of the formal context.

Theorem 18.   The order dimension of a concept lattice $\mathfrak{B}(G, M, I)$ equals the Ferrers dimension of $\langle G, M, I \rangle$.   —

A relation $F \subseteq G \times M$ is Ferrers if and only if $\mathfrak{B}(G, M, F)$ is a chain. Using the length of these chains as parameters, one can obtain an analogous theorem for the *k-dimension* of finite lattices.

New Bibliography

[1] M. E. Adams, *Generators of free products of lattices*, Algebra Universalis **7** (1977), 409–410.

[2] ———, *Principal congruences in de Morgan algebras*, Proc. Edinburgh Math. Soc. **30** (1987), 415–421.

[3] M. E. Adams, M. Atallah, and R. Beazer, *Congruence coherent distributive double p-algebras*, Proc. Edinburgh Math. Soc. **39** (1996), 71–80.

[4] M. E. Adams and R. Beazer, *Congruence relations on de Morgan algebras*, Algebra Universalis **26** (1989), 103–125.

[5] ———, *The intersection of principal congruences on de Morgan algebras*, Houston J. Math. **16** (1990), 59–70.

[6] ———, *Congruence uniform distributive lattices*, Acta Math. Hungar. **57** (1991), 41–52.

[7] ———, *Congruence properties of distributive double p-algebras*, Czechoslovak Math. J. **41** (1991), 216–231.

[8] ———, *Distributive lattices having n-permutable congruences*, Proc. Amer. Math. Soc. **113** (1991), 41–45.

[9] ———, *Double p-algebras with Stone congruence lattices*, Czechoslovak Math. J. **41** (1991), 395–404.

[10] M. E. Adams, S. Bulman-Fleming, and M. I. Gould, *Endomorphism properties of algebraic structures*, Proc. Tennessee Topology Conference (1996), World Scientific Pub. Co., NJ.

[11] M. E. Adams, R. Freese, J. B. Nation, and J. Schmid, *Maximal sublattices and Frattini sublattices of bounded lattices*, J. Austral. Math. Soc. (Ser. A.) **63** (1997), 110-127.

[12] M. E. Adams and H. A. Priestley, *Equational bases for varieties of Ockham algebras*, Algebra Universalis **32** (1994), 368–397.

[13] M. E. Adams and J. Sichler, **V**-*free products of Hopfian lattices*, Math. Z. **151** (1976), 259–262.

[14] _____, *Bounded endomorphisms of lattices of finite height*, Canad. J. Math. **29** (1977), 1254–1263.

[15] _____, *A note on finitely generated Hopfian lattices*, Algebra Universalis **8** (1978), 381-383.

[16] _____, *Homomorphisms of bounded lattices with a given sublattice*, Arch. Math. (Basel) **30** (1978), 122-128.

[17] _____, *Cover set lattices*, Canad. J. Math. **32** (1980), 1177–1205.

[18] _____, *Frattini sublattices in varieties of lattices*, Colloq. Math. **44** (1981), 181–184.

[19] _____, *Lattices with unique complementation*, Pacific J. Math. **92** (1981), 1–13.

[20] _____, *Refinement property of reduced free products in varieties of lattices*, Contributions to lattice theory (Szeged, 1980), pp. 19–53, Colloq. Math. Soc. János Bolyai, **33**, North-Holland, Amsterdam-New York, 1983.

[21] K. V. Adaricheva, *Semidistributive and co-algebraic lattices of subsemilattices* (Russian), Algebra i Logika **27** (1988), 625–640, 736; translation in Algebra and Logic **27** (1988), 385–395.

[22] _____, *The structure of finite lattices of subsemilattices* (Russian), Algebra i Logika **30** (1991), 385–404, 507; translation in Algebra and Logic **30** (1991), 249–264.

[23] _____, *The structure of congruence lattices of finite semilattices*, Algebra and Logic **35** (1996), 3–30; translation in Algebra and Logic **35** (1996), 1–15.

[24] K. V. Adaricheva, W. Dziobiak, and V. A. Gorbunov, *Finite atomistic lattices that can be represented as lattices of quasivarieties*, Fund. Math. **142** (1993), 19–43.

[25] _____, *Algebraic atomistic lattices of quasivarieties*, Algebra i Logika, to appear.

[26] K. V. Adaricheva and V. A. Gorbunov, *Equaclosure operator and forbidden semidistributive lattices* (Russian), Sibirsk. Mat. Zh. **30** (1989), 7–25; translation in Siberian Math. J. **30** (1989), 831–849.

[27] I. Amemiya and I. Halperin, *Complemented modular lattices*, Canad. J. Math. **11** (1959), 481–520.

[28] I. Anderson, *Combinatorics of finite sets*, Oxford Science Publications. The Clarendon Press, Oxford University Press, New York, 1987. xvi+250 pp.

[29] E. Artin, *Geometric algebra*, Interscience Publishers, New York, London, 1957.

[30] B. Artmann, *On coordinates in modular lattices with a homogeneous basis*, Illinois J. Math. **12** (1968), 626–648.

[31] P. D. Bacsich, *Injectivity in model theory*, Colloq. Math. **25** (1972), 165–176.

[32] R. W. Baddeley and A. Lucchini, *On representing finite lattices as intervals in subgroup lattices of finite groups*, J. Algebra **196** (1997), 1–100.

[33] R. Baer, *A unified theory of projective spaces and finite abelian groups*, Trans. Amer. Math. Soc. **52** (1942), 283–343.

[34] K. A. Baker, *Some non-finitely-based varieties of lattices*, Universal algebra (Esztergom, 1977), pp. 53–59, Colloq. Math. Soc. János Bolyai, **29**, North-Holland, Amsterdam-New York, 1982.

[35] R. Balbes, *Catalytic distributive lattices*, Algebra Universalis **11** (1980), 334–340.

[36] ———, *Generating sets for catalytic and projective distributive lattices*, Algebra Universalis **30** (1993), 262–268.

[37] B. Banaschewski, *Injectivity and essential extensions in equational classes of algebras*, in 'Proc. of the Conference on Universal Algebra' (1969), Queen's papers in pure and applied mathematics, Kingston, Ontario, **25** (1970), pp. 131–147.

[38] V. A. Baranskiĭ, *On the independence of the automorphism group and the congruence lattice for lattices*, Abstracts of lectures of the 15th All-Soviet Algebraic Conference, Krasnojarsk, July 1979, vol. 1, p. 11.

[39] R. Beazer, *Subdirectly irreducible double Heyting algebras*, Algebra Universalis **10** (1980), 220–224.

[40] ———, *Subdirectly irreducibles for various pseudocomplemented algebras*, Algebra Universalis **10** (1980), 225–231.

[41] ———, *Affine complete Stone algebras*, Acta. Math. Acad. Sci. Hungar. **39** (1982), 169–174.

[42] R. Beazer, *Affine complete double Stone algebras with bounded core*, Algebra Universalis **16** (1983), 237–244.

[43] F. Bellisima, *Finitely generated free Heyting algebras*, J. Symbolic Logic **51** (1986), 152–165.

[44] L. Beran, *Orthomodular lattices. Algebraic approach*, Mathematics and its Applications (East European Series), D. Reidel Publishing Co., Dordrecht-Boston, Mass., 1985. xix+394 pp.

[45] C. Bergman, *Amalgamation classes of some distributive varieties*, Algebra Universalis **20** (1985), 143–165.

[46] _____, *Non-axiomatizability of the amalgamation class of modular lattice varieties*, Order **6** (1989), 49–58.

[47] J. Berman, *Distributive lattices with an additional unary operation*, Aequationes Math. **16** (1977), 165–171.

[48] _____, *Interval lattices and the amalgamation property*, Algebra Universalis **12** (1981), 360–375.

[49] V. I. Bhatta, *The lattice of convex sublattices of a lattice* (Russian), Ordered sets and lattices, No. 9 (Russian), 50–58, 110–111, Saratov. Gos. Univ., Saratov, 1986.

[50] _____, *The group of automorphisms of the lattice of convex sublattices* (Russian), Vestsī Akad. Navuk BSSR Ser. Fīz.-Mat. Navuk 1988, no. 6, 110–112, 128.

[51] G. Birkhoff, *Universal algebra*, Proc. First Canadian Math. Congress, Montreal, 1945, pp. 310–326. University of Toronto Press, Toronto, 1946.

[52] T. S. Blyth and M. F. Janowitz, *Residuation theory*, Pergamon Press, Oxford, 1972.

[53] T. S. Blyth and J. C. Varlet, *Ockham algebras*, Oxford University Press, Oxford (1994).

[54] A. D. Bol'bot and V. V. Kalinin, *Lattices with constraints for the sublattices* (Russian), Algebra i Logika **17** (1978), 127–133, 241.

[55] N. Bourbaki, *Commutative algebra. Elements of mathematics*, Addison-Wesley Publ. Comp., Reading, MA, 1972.

[56] H. Brauner, *Zur Theorie linearer Abbildungen*, Abh. Math. Sem. Univ. Hamburg **53** (1983), 154–169.

[57] U. Brehm, *Untermodulverbände torsionsfreier Moduln*, Dissertation, Universität Freiburg, 1983.

[58] ———, *Coordinatization of lattices*, in: R. Kaya et al. (ed.), Rings and Geometry. NATO ASI-C160, Istanbul 1984, Reidel Dordrecht, 511–550.

[59] ———, *Representation of sum preserving mappings between submodule lattices by R-balanced mappings*, manuscript, 1984.

[60] ———, *Representation of homomorphisms between submodule lattices*, Resultate Math. **12** (1987), 62–70.

[61] C. Brink and I. M. Rewitzky, *Power structures and program semantics*, submitted.

[62] ———, *Unification of four versions of program semantics*, submitted.

[63] G. Bruns and M. Roddy, *A finitely generated modular ortholattice*, Canadian Math. Bull. **35** (1) (1992), 29–33.

[64] P. Bruyns and H. Rose, *Varieties with cofinal sets: examples and amalgamation*, Proc. Amer. Math. Soc. **111** (1991), 833–840.

[65] S. Burris, *Polynomial time uniform word problems*, Math. Logic Quart. **41** (1995), 173–182.

[66] L. M. Butler, *A unimodality result in the enumeration of subgroups of a finite abelian group*, Proc. Amer. Math. Soc. **101** (1987), 771–775.

[67] ———, *Subgroup lattices and symmetric functions*, Mem. Amer. Math. Soc. **112** (1994).

[68] C. Busqué, *Two-sided ideals in right self-injective regular rings*, J. Pure Appl. Algebra **67** (1990), 209-245.

[69] V. P. Camillo, *Inducing lattice maps by semilinear isomorphisms*, Rocky Mount. J. Math. **14** (1984), 475–486.

[70] E. R. Canfield, *On a problem of Rota*, Adv. in Math. **29** (1978), 1–10.

[71] E. Capińska and A. Wroński, *On classes of distributive lattices with pseudocomplementation definable by conditional identities*, Rep. Math. Logic **20** (1986), 93–97.

[72] C. C. Chen, K. M. Koh, and K. L. Teo, *On the sublattice-lattice of a lattice*, Algebra Universalis **19** (1984), 61–73.

[73] R. Cignoli, *Quantifiers on distributive lattices*, Discrete Math. **9** (1991), 183–197.

[74] R. Cignoli, *Free Q-distributive lattices*, Studia Logica **56** (1996), 23–29.

[75] R. Cignoli, S. Lafalce, and A. Petrovich, *Remarks on Priestley duality for distributive lattices*, Order **8** (1991), 299–315.

[76] D. M. Clark, *The structure of algebraically and existentially closed Stone and double Stone algebras*, J. Symbolic Logic **54** (1989), 363–375.

[77] D. M. Clark and B. A. Davey, *The quest for strong dualities*, J. Austral. Math. Soc. Ser. A **58** (1995), 248–280.

[78] ———, *When is a natural duality 'good'?*, Algebra Universalis **35** (1996), 265–295.

[79] ———, *Natural dualities for the working algebraist*, Cambridge University Press (1998).

[80] D. M. Clark and P. H. Krauss, *Topological quasi varieties*, Acta Sci. Math. (Szeged) **47** (1984), 3–39.

[81] D. M. Clark and J. Schmid, *The countable homogeneous universal model of B_2*, Studia Logica **56** (1996), 31–66.

[82] W. H. Cornish, *Monoids acting on distributive lattices*, manuscript (invited talk at the annual meeting of the Austral. Math. Soc., May 1977).

[83] ———, *Antimorphic action. Categories of algebraic structures with involutions or anti-endomorphisms*, Research and Exposition in Mathematics, Heldermann Verlag **12** (1986).

[84] W. H. Cornish and P. R. Fowler, *Coproducts of de Morgan algebras*, Bull. Austral. Math. Soc. **16** (1977), 1–13.

[85] ———, *Coproducts of Kleene algebras*, J. Austral. Math. Soc. Ser. A **27** (1979), 209–220.

[86] G. Czédli, *On the 2-distributivity of sublattice lattices*, Acta Math. Acad. Sci. Hungar. **36** (1980), 49–55.

[87] ———, *Lattice generation of small equivalences of a countable set*, Order **13** (1996), 11–16.

[88] ———, *Four-generated large equivalence lattices*, Acta Sci. Math. (Szeged) **62** (1996), 47–69.

[89] ———, *$(1 + 1 + 2)$-generated equivalence lattices*, J. Algebra, submitted.

[90] G. Czédli and R. Freese, *On congruence distributivity and modularity*, Algebra Universalis **17** (1983), 216–219.

[91] G. Czédli and G. Hutchinson, *A test for identities satisfied in lattices of submodules*, Algebra Universalis **8** (1978), 269–309.

[92] ———, *Submodule lattice quasivarieties and exact embedding functors for rings with prime power characteristic*, Algebra Universalis **35** (1996), 425–445.

[93] B. A. Davey, *Weak injectivity and congruence extension in congruence-distributive equational classes*, Canad. J. Math. **29** (1977), 449–459.

[94] ———, *Topological duality for pre-varieties of universal algebras*, in 'Studies in Foundations and Combinatorics' (G.-C. Rota, ed) Advances in Mathematics Supplementary studies, **1** (1978), pp. 61–99.

[95] ———, *Subdirectly irreducible distributive double p-algebras*, Algebra Universalis **8** (1978), 73–88.

[96] ———, *On the lattice of subvarieties*, Houston J. Math. **5** (1979), 183–192.

[97] ———, *Dualities for Stone algebras, double Stone algebras, and relative Stone algebras*, Colloq. Math. **46** (1982), 1–14.

[98] ———, *Duality theory on ten dollars a day*, in 'Algebras and Orders' (I. G. Rosenberg and G. Sabidussi, eds), NATO Advanced Study Institute Series, Series C, Kluwer, **389** (1993), pp. 71–111.

[99] ———, *Dualisability in general and endodualizability in particular*, in 'Logic and Algebra' (A. Ursini and P. Aglianó, eds), Lecture Notes in Pure and Applied Mathematics, Marcel Dekker, New York, **180** (1996), pp. 437–454.

[100] B. A. Davey and D. Duffus, *Exponentiation and duality*, in 'Ordered Sets' (I. Rival, ed) NATO Advanced Study Institute Series, Series C, D. Reidel, Dordrecht, **83** (1982), pp. 43–94.

[101] B. A. Davey and M. Goldberg, *The free p-algebra generated by a distributive lattice*, Algebra Universalis **11** (1980), 90–100.

[102] B. A. Davey, M. Haviar, and H. A. Priestley, *Endoprimal distributive lattices are endodualizable*, Algebra Universalis **34** (1995), 444–453.

[103] ———, *The syntax and semantics of entailment in duality theory*, J. Symbolic Logic **60** (1995), 1087–1114.

[104] B. A. Davey and J. G. Pitkethly, *Endoprimal algebras*, Algebra Universalis, to appear.

[105] B. A. Davey and H. A. Priestley, *Generalized piggyback dualities and applications to Ockham algebras*, Houston J. Math. **13** (1987), 151–197.

[106] B. A. Davey and H. A. Priestley, *Introduction to lattices and order*, Cambridge Mathematical Textbooks, Cambridge University Press, Cambridge, 1990. viii+248 pp.

[107] ———, *Partition-induced natural dualities for varieties of pseudocomplemented distributive lattices*, Discrete Math. **113** (1993), 41–58.

[108] ———, *Optimal natural dualities*, Trans. Amer. Math. Soc. **338** (1993), 655–677.

[109] ———, *Optimal natural dualities II: General theory*, Trans. Amer. Math. Soc. **348** (1996), 3673–3711.

[110] ———, *Optimal natural dualities for varieties of Heyting algebras*, Studia Logica **56** (1996), 67–96.

[111] B. A. Davey and H. Werner, *Injectivity and Boolean powers*, Math. Zeit. **166** (1979), 205–223.

[112] ———, *Distributivity of coproducts over products*, Algebra Universalis **12** (1981), 387–394.

[113] ———, *Dualities and equivalences for varieties of algebras*, in 'Contributions to Lattice Theory' (Szeged, 1980) (A. P. Huhn and E. T. Schmidt, eds) Colloq. Math. Soc. János Bolyai North-Holland, Amsterdam, **33** (1983), pp. 101–275.

[114] ———, *Piggyback-dualitäten*, Bull. Austral. Math. Soc. **32** (1985), 1–32.

[115] ———, *Piggyback dualities*, in 'Lectures in Universal Algebra' (Szeged, 1983) (L. Szabó and A. Szendrei, eds) Coll. Math. Soc. János Bolyai, North-Holland, Amsterdam, **43** (1986), pp. 61–83.

[116] A. Day, *Injectivity in equational classes of algebras*, Canad. J. Math. **24** (1972), 209–220.

[117] ———, *Splitting lattices and congruence-modularity*, Contributions to universal algebra (Colloq., József Attila Univ., Szeged, 1975), pp. 57–71. Colloq. Math. Soc. János Bolyai, Vol. 17, North-Holland, Amsterdam, 1977.

[118] ———, *Idempotents in the groupoid of all* **SP** *classes of lattices*, Canad. Math. Bull. **21** (1978), 499–501.

[119] ———, *Characterizations of finite lattices that are bounded-homomorphic images of sublattices of free lattices*, Canad. J. Math. **31** (1979), 69–78.

[120] ———, *In search of a Pappian lattice identity*, Canad. Math. Bull. **24** (1981), 187–198.

[121] A. Day, *A note on Arguesian lattices*, Arch. Math (Brno) **19** (1983), 117–123.

[122] _____, *On some geometric properties defining classes of rings and varieties of modular lattices*, Algebra Universalis **17** (1983), 21–33.

[123] _____, *Applications of coordinatization in modular lattice theory: the legacy of J. von Neumann*, Order **1** (1985), 295–300.

[124] A. Day and R. Freese, *A characterization of identities implying congruence modularity*, I, Canad. J. Math. **32** (1980), 1140–1167.

[125] A. Day and C. Herrmann, *Gluings of modular lattices*, Order **5** (1988), 85–101.

[126] A. Day, C. Herrmann, B. Jónsson, J. B. Nation, and D. Pickering *Small non-Arguesian lattices*, Algebra Universalis **31** (1994), 66–94.

[127] A. Day and J. Ježek, *The amalgamation property for varieties of lattices*, Trans. Amer. Math. Soc. **286** (1984), 251–256.

[128] A. Day and B. Jónsson, *The structure of non-Arguesian lattices*, Bull. Amer. Math. Soc. (N.S.) **13** (1985), 157–159.

[129] _____, *A structural characterization of non-Arguesian lattices*, Order **2** (1986), 335–350.

[130] _____, *Non-Arguesian configurations in a modular lattice*, Acta Sci. Math. (Szeged) **51** (1987), 309–318.

[131] _____, *Non-Arguesian configurations and gluings of modular lattices*, Algebra Universalis **26** (1989), 208–215.

[132] A. Day and J. B. Nation, *Congruence normal covers of finitely generated lattice varieties*, Canad. Math. Bull. **35** (1992), 311–320.

[133] A. Day and D. Pickering, *The coordinatization of Arguesian lattices*, Trans. Amer. Math. Soc. **278** (1983), 507–522.

[134] V. Diercks, M. Erné, and J. Reinhold, *Complements in lattices of varieties and equational theories*, Algebra Universalis **31** (1994), 506–515.

[135] H. Dobbertin, *Vaught measures and their applications in lattice theory*, J. Pure Appl. Algebra **43** (1986), 27–51.

[136] _____, *Boolean Representations of Refinement Monoids and their Applications*, manuscript.

[137] J. Dudek, *A polynomial characterization of nondistributive modular lattices*, Colloq. Math. **55** (1988), 195–212.

[138] I. Düntsch, *On a problem of Chen and Grätzer*, Algebra Universalis **14** (1982), 401.

[139] ———, *A description of the projective Stone algebras*, Glasgow. Math. J. (1983), 75–82.

[140] W. Dziobiak, *On atoms in the lattice of quasivarieties*, Algebra Universalis **24** (1987), 32–35.

[141] E. G. Effros, D. E. Handelman, and Chao Liang Shen, *Dimension groups and their affine representations*, Amer. J. Math. **102** (1980), 385–407.

[142] G. A. Elliott, *On the classification of inductive limits of sequences of semisimple finite-dimensional algebras*, J. Algebra **38** (1976), 29–44.

[143] K. Engel and H.-D. O. F. Gronau, *Sperner theory in partially ordered sets*, Teubner-Texte zur Mathematik, **78**. BSB B. G. Teubner Verlagsgesellschaft, Leipzig, 1985. 232 pp.

[144] M. Erné, *Weak distributive laws and their role in lattices of congruences and equational theories*, Algebra Universalis **25** (1988), 290–321.

[145] M. Erné and D. Schweigert, *Pre-fixed points of polynomial functions of lattices*, Algebra Universalis **31** (1994), 298–300.

[146] T. Evans, *On multiplicative systems defined by generators and relations, I*, Proc. Cambridge Philos. Soc. **47** (1951), 637–649.

[147] ———, *The word problem for abstract algebras*, J. London Math. Soc. **26** (1951), 64–71.

[148] U. Faigle and C. Herrmann, *Projective geometry on partially ordered sets*, Trans. Amer. Math. Soc. **266** (1981), 319–332.

[149] K. Faltings, *Modulare Verbände mit Punktsystem*, Geom. Dedicata **4** (1975), 105–137.

[150] J. D. Farley, *Priestley duality for order-preserving maps into distributive lattices*, Order **13** (1996), 65–98.

[151] ———, *Priestley powers of lattices and their congruences: a problem of E. T. Schmidt*, Acta Sci. Math. (Szeged) **62** (1996), 3–45.

[152] ———, *The automorphism group of a function lattice: a problem of Jónsson and McKenzie*, Algebra Universalis **36** (1996), 8–45.

[153] C.-A. Faure and A. Frölicher, *Morphisms of projective geometries and of corresponding lattices*, Geom. Dedicata **47** (1993), 25–40.

[154] W. Feit, *An interval in the subgroup lattice of a finite group which is isomorphic to M_7*, Algebra Universalis **17** (1983), 220–221.

[155] R. Freese, *The variety of modular lattices is not generated by its finite members*, Trans. Amer. Math. Soc. **255** (1979), 277–300.

[156] _____, *Projective geometries as projective modular lattices*, Trans. Amer. Math. Soc. **251** (1979), 329–342.

[157] _____, *Free modular lattices*, Trans. Amer. Math. Soc. **261** (1980), 81–91.

[158] _____, *Connected components of the covering relation in free lattices*, Universal algebra and lattice theory (Charleston, S.C., 1984), 82–93, Lecture Notes in Math., 1149, Springer, Berlin-New York, 1985.

[159] _____, *A decomposition theorem for modular lattices containing an n-frame*, Acta Sci. Math. (Szeged) **51** (1987), 57–71.

[160] _____, *Free lattice algorithms*, Order **3** (1987), 331–344.

[161] _____, *Finitely presented lattices: canonical forms and the covering relation*, Trans. Amer. Math. Soc. **312** (1989), 841–860.

[162] _____, *Finitely presented lattices: continuity and semidistributivity*, Lattices, semigroups, and universal algebra (Lisbon, 1988), 67–70, Plenum, New York, 1990.

[163] _____, *Free and finitely presented lattices*, Proceedings of the International Conference on Algebra, Part 3 (Novosibirsk, 1989), 85–97, Contemp. Math. **131**, Part 3, Amer. Math. Soc., Providence, RI, 1992.

[164] _____, *Finitely based modular congruence varieties are distributive*, Algebra Universalis **32** (1994), 104–114.

[165] _____, *Alan Day's early work: congruence identities*, Algebra Universalis **34** (1995), 4–23.

[166] _____, *Computing congruence lattices of finite lattices*, Proc. Amer. Math. Soc., (1997) **125**, 3457–3463.

[167] _____, *Computing congruences of finite lattices*, Proc. Amer. Math. Soc., to appear.

[168] _____, *Notes on Dean's problem*, Online manuscript available at: `http://math.hawaii.edu/~ralph/Notes`.

[169] R. Freese, G. Grätzer, and E. T. Schmidt, *On complete congruence lattices of complete modular lattices*, Internat. J. Algebra Comput. **1** (1991), 147–160.

[170] R. Freese, C. Herrmann, and A. P. Huhn, *On some identities valid in modular congruence varieties*, Algebra Universalis **12** (1981), 322–334.

[171] R. Freese, J. Ježek, and J. B. Nation, *Term rewrite systems for lattice theory*, J. Symbolic Computation **16** (1993), 279–288.

[172] _____, *Free lattices*, Mathematical Surveys and Monographs, Vol. 42, American Mathematical Society, Providence, RI, 1995. viii+293 pp.

[173] R. Freese, J. Ježek, J. B. Nation, and V. Slavík, *Singular covers in free lattices*, Order **3** (1986), 39–46.

[174] R. Freese and B. Jónsson, *Congruence modularity implies the Arguesian identity*, Algebra Universalis **6** (1976), 225–228.

[175] R. Freese, K. Kearnes, and J. B. Nation, *Congruence lattices of congruence semidistributive algebras*, Lattice theory and its applications (Darmstadt, 1991), 63–78, Res. Exp. Math., **23**, Heldermann, Lemgo, 1995.

[176] R. Freese, W. A. Lampe, and W. Taylor, *Congruence lattices of algebras of fixed similarity type. I*, Pacific J. Math. **82** (1979), 59–68.

[177] R. Freese and R. N. McKenzie, *Commutator theory for congruence modular varieties*, London Mathematical Society Lecture Note Series, **125**. Cambridge University Press, Cambridge-New York, 1987. iv+227 pp.

[178] R. Freese, G. McNulty, and J. B. Nation, *Inherently nonfinitely based lattices*, manuscript.

[179] R. Freese and J. B. Nation, *Projective lattices*, Pacific J. Math. **75** (1978), 93–106.

[180] _____, *Finitely presented lattices*, Proc. Amer. Math. Soc. **77** (1979), 174–178.

[181] _____, *Covers in free lattices*, Trans. Amer. Math. Soc. **288** (1985), 1–42.

[182] E. Fried and G. Grätzer, *Pasting infinite lattices*, J. Austral. Math. Soc. Ser. A **47** (1989), 1–21.

[183] _____, *Pasting and modular lattices*, Proc. Amer. Math. Soc. **106** (1989), 885–890.

[184] _____, *The unique amalgamation property for lattices*, Ann. Univ. Sci. Budapest. Eötvös Sect. Math. **33** (1990), 167–176. (Correction: **35** (1992), 271.)

[185] _____, *Generalized congruences and products of lattice varieties*, Acta Sci. Math. (Szeged) **54** (1990), 21–36.

[186] E. Fried and G. Grätzer, *Strong amalgamation of distributive lattices*, J. Algebra **128** (1990), 446–455.

[187] _____, *Notes on tolerance relations of lattices: On a conjecture of R. N. McKenzie*, J. Pure Appl. Algebra **68** (1990), 127–134.

[188] _____, *Unique Envelope Property*, Studia Sci. Math. Hungar. **27** (1992), 183–187.

[189] E. Fried, G. Grätzer, and R. W. Quackenbush, *Uniform congruence scheme*, Algebra Universalis **10** (1980), 176–188.

[190] E. Fried, G. Grätzer, and E. T. Schmidt, *Multipasting of lattices*, Algebra Universalis **30** (1993), 241–261.

[191] K. D. Fryer and I. Halperin, *Coordinates in Geometry*, Trans. Roy. Soc. Canada, Third Series, Section III, **48** (1954), 11–26.

[192] _____, *On the coordinatization theorem of J. von Neumann*, Canad. J. Math. **7** (1955), 432–444.

[193] _____, *Coordinatization of complemented modular lattices*, Nederl. Akad. Wetensch. Proc. Ser. A **62** (1958), 70–78.

[194] H. Gaitan, *Quasivarieties of distributive p-algebras*, Algebra Universalis **29** (1992), 484–494.

[195] B. Ganter and R. Wille, *Formale Begriffsanalyse—Mathematische Grundlagen*, Springer-Verlag, Berlin-New York, 1996. 296 pp.

[196] M. Gehrke, *The order structure of Stone spaces and the T_D-separation axiom*, Z. Logik Grund. Math. **37** (1991) , 5–15.

[197] _____, *Uniquely representable posets*, Papers on General Topology and Applications, Eighth Summer Conference at Queen's College (G. Itzkowitz *et al.*, eds), Annals of the New York Academy of Sciences, Vol. 728 (1994), pp. 32–41.

[198] M. Gehrke and B. Jónsson, *Bounded distributive lattices with operator*, Math. Japon. **40** (1994), 207–215.

[199] W. Geyer, *On Tamari lattices*, Discrete Math. **133** (1994), 99–122.

[200] G. Gierz, K. H. Hofmann, K. Keimel, J. D. Lawson, M. W. Mislove, and D. S. Scott, *A compendium on continuous lattices*, Springer-Verlag, Berlin-New York, 1980. xx+371 pp.

[201] M. Goldstern, *Interpolation of monotone functions in lattices*, Algebra Universalis **36** (1996), 108–121.

[202] M. Goldstern and S. Shelah, *Order-polynomially complete lattices must be large*, Algebra Universalis, to appear.

[203] K. R. Goodearl, *Von Neumann regular rings*, Second edition. Robert E. Krieger Publishing Co., Inc., Malabar, FL, 1991. xviii+412

[204] ———, *Von Neumann regular rings and direct sum decomposition problems*, Abelian Groups and Modules, Padova 1994 (A. Facchini and C. Menini, eds.), Dordrecht (1995) Kluwer, 249–255.

[205] K. R. Goodearl and F. Wehrung, *Representations of distributive semilattices by dimension groups, regular rings, C*-algebras, and complemented modular lattices*, manuscript, 1997.

[206] P. Goralčík, V. Koubek, and J. Sichler, *Universal varieties of {0, 1}-lattices*, Canad. J. Math. **42** (1990), 470–490.

[207] V. A. Gorbunov, *Lattices of quasivarieties* (Russian), Algebra i Logika **15** (1976), 436–457, 487–488.

[208] ———, *Canonical decompositions in complete lattices* (Russian), Algebra i Logika **17** (1978), 495–511, 622.

[209] ———, *The structure of the lattices of quasivarieties*, Algebra Universalis **32** (1994), 493–530.

[210] ———, *The structure of lattices of varieties and lattices of quasivarieties: their similarity and difference, III* (Russian), Algebra i Logika **34** (1995), 646–666, 728–729; translation in Algebra and Logic **34** (1995), 359–370 (1996).

[211] ———, *Algebraic theory of quasivarieties*, Plenum Press, New York, 1997.

[212] V. A. Gorbunov and V. I. Tumanov, *A class of lattices of quasivarieties*, Algebra i Logika **19** (1980), 59–80, 132–133.

[213] ———, *Construction of lattices of quasivarieties* (Russian), Mathematical logic and the theory of algorithms, 12–44, Trudy Inst. Mat., 2, "Nauka" Sibirsk. Otdel., Novosibirsk, 1982.

[214] ———, *On the existence of prime ideals in semidistributive lattices*, Algebra Universalis **16** (1983), 250–252.

[215] G. Grätzer, *Universal Algebra*, Second Edition, Springer-Verlag, Berlin-New York, 1979. xviii+581 pp.

[216] ———, *General Lattice Theory: 1979 Problem Update*, Algebra Universalis **11** (1980), 396–402.

[217] G. Grätzer, *Obshchaya teoriya reshetok* (Russian), Russian translation of *General Lattice Theory*, translated from the English by A. D. Bol'bot, V. A. Gorbunov, and V. I. Tumanov. Translation edited and with a preface by D. M. Smirnov. "Mir", Moscow, 1982. 454 pp.

[218] ———, *The complete congruence lattice of a complete lattice*. Lattices, semigroups, and universal algebra (Lisbon, 1988), 81–87, Plenum, New York, 1990.

[219] G. Grätzer, A. Hajnal, and D. Kelly, *Chain conditions in free products of lattices with infinitary operations*, Pacific J. Math. **83** (1979), 107–115.

[220] G. Grätzer and A. P. Huhn, *A note on finitely presented lattices*, C. R. Math. Rep. Acad. Sci. Canada **2** (1980), 291–296.

[221] ———, *Amalgamated free product of lattices. I. The common refinement property*, Acta Sci. Math. (Szeged) **44** (1982), 53–66.

[222] ———, *Amalgamated free product of lattices. II. Generating sets*, Studia Sci. Math. Hungar. **16** (1981), 141–148.

[223] G. Grätzer, A. P. Huhn, and H. Lakser, *On the structure of finitely presented lattices*, Canad. J. Math. **33** (1981), 404–411.

[224] G. Grätzer, P. Johnson, and E. T. Schmidt, *A representation of $\mathfrak{m}$-algebraic lattices*, Algebra Universalis **32** (1994), 1–12.

[225] G. Grätzer and D. Kelly, *When is the free product of lattices complete?* Proc. Amer. Math. Soc. **66** (1977), 6–8.

[226] ———, *A normal form theorem for lattices completely generated by a subset*, Proc. Amer. Math. Soc. **67** (1977), 215–218.

[227] ———, *On a special type of subdirectly irreducible lattice with an application to products of varieties*, C. R. Math. Rep. Acad. Sci. Canada **2** (1980/81), 43–48.

[228] ———, *A survey of products of lattice varieties*, Contributions to lattice theory (Szeged, 1980), pp. 457–472, Colloq. Math. Soc. János Bolyai, **33**, North-Holland, Amsterdam-New York, 1983.

[229] ———, *The construction of some free $\mathfrak{m}$-lattices on posets*. Orders: description and roles (L'Arbresle, 1982), pp. 103–117, North-Holland Math. Stud., **99**, North-Holland, Amsterdam-New York, 1984.

[230] ———, *Free $\mathfrak{m}$-products of lattices. I*, Colloq. Math. **48** (1984), 181–192.

[231] ———, *Products of lattice varieties*, Algebra Universalis **21** (1985), 33–45.

[232] G. Grätzer and D. Kelly, *Free m-products of lattices. II*, Colloq. Math. **50** (1986), 155–166.

[233] _____, *The lattice variety* **D** ∘ **D**, Acta Sci. Math. (Szeged) **51** (1987), 73–80. *Addendum*, **52** (1988), 465.

[234] _____, *Subdirectly irreducible members of products of lattice varieties*, Proc. Amer. Math. Soc. **102** (1988), 483–489.

[235] G. Grätzer and H. Lakser, *Homomorphisms of distributive lattices as restrictions of congruences*, Can. J. Math. **38** (1986), 1122–1134.

[236] _____, *On complete congruence lattices of complete lattices*, Trans. Amer. Math. Soc. **327** (1991), 385–405.

[237] _____, *On congruence lattices of m-complete lattices*, J. Austral. Math. Soc. Ser. A **52** (1992), 57–87.

[238] _____, *Congruence lattices of planar lattices*, Acta Math. Hungar. **60** (1992), 251–268.

[239] G. Grätzer, H. Lakser, and R. W. Quackenbush, *On the lattice of quasivarieties of distributive lattices with pseudocomplementation*, Acta Sci. Math. (Szeged) **42** (1980), 257–263.

[240] _____, *The structure of tensor products of semilattices with zero*, Trans. Amer. Math. Soc. **267** (1981), 503–515.

[241] G. Grätzer, H. Lakser, and E. T. Schmidt, *Congruence lattices of small planar lattices*, Proc. Amer. Math. Soc. **123** (1995), 2619–2623.

[242] _____, *Congruence representations of join homomorphisms of distributive lattices: A short proof*, Math. Slovaca **46** (1996), 363–369.

[243] _____, *Isotone maps as maps of congruences. I. Abstract maps*, Acta Math. Hungar. **75** (1997), 81-111.

[244] _____, *Congruence lattices of finite semimodular lattices*, Canad. Math. Bull., to appear. (AMS Abstract 97T-06-56.)

[245] _____, *Isotone maps as maps of congruences. II. Concrete maps*, manuscript.

[246] _____, *Congruence representations of join-homomorphisms of distributive lattices. I. Size and breadth*, manuscript.

[247] _____, *Congruence representations of join-homomorphisms of distributive lattices. II. Order dimension*, manuscript.

[248] G. Grätzer and C. R. Platt, *A characterization of sharply transferable lattices*, Canad. J. Math. **32** (1980), 145–154.

[249] G. Grätzer, C. R. Platt, and B. Sands, *Embedding lattices into lattices of ideals*, Pacific J. Math. **85** (1979), 65–75.

[250] G. Grätzer, I. Rival, and N. Zaguia, *Small representations of finite distributive lattices as congruence lattices*, Proc. Amer. Math. Soc. **123** (1995), 1959–1961.

[251] G. Grätzer and E. T. Schmidt, *"Complete-simple" distributive lattices*, Proc. Amer. Math. Soc. **119** (1993), 63–69.

[252] _____, *Another construction of complete-simple distributive lattices*, Acta Sci. Math. (Szeged) **58** (1993), 115-126.

[253] _____, *Algebraic lattices as congruence lattices: The m-complete case*, Lattice theory and its applications (Darmstadt, 1991), pp. 91–101, Res. Exp. Math., **23**, Heldermann, Lemgo, 1995.

[254] _____, *Congruence lattices of p-algebras*, Algebra Universalis **33** (1995), 470–477.

[255] _____, *Complete congruence lattices of complete distributive lattices*, J. Algebra **170** (1995), 204–229.

[256] _____, *Do we need complete-simple distributive lattices?* Algebra Universalis **33** (1995), 140–141.

[257] _____, *The Strong Independence Theorem for automorphism groups and congruence lattices of finite lattices*, Beiträge Algebra Geom. **36** (1995), 97–108.

[258] _____, *Congruence-preserving extensions of finite lattices into sectionally complemented lattices*, Proc. Amer. Math. Soc., to appear.

[259] _____, *Congruence-preserving extensions of finite lattices to semimodular lattices*, manuscript. Abstract 97T-06-69.

[260] _____, *On finite automorphism groups of simple Arguesian lattices*, Studia Sci. Math. Hungar., to appear.

[261] _____, *Sublattices and standard congruences*, manuscript.

[262] _____, *On the Independence Theorem of related structures for modular (Arguesian) lattices*, manuscript.

[263] G. Grätzer and J. Sichler, *On the endomorphism monoids of complemented lattices*, manuscript. (AMS Abstract 97T-06-98.)

[264] G. Grätzer and T. Tan, *On **D**-transferability*, Notices Amer. Math. Soc. **23** (1976), A-268.

[265] G. Grätzer and D. Wang, *A lower bound for congruence representations*, Order **14** (1997), 67–74.

[266] G. Grätzer and F. Wehrung, *Tensor products of lattices with zero, revisited*, manuscript.

[267] ———, *Proper congruence-preserving extensions of lattices*, AMS Abstract 97T-06-189.

[268] ———, *A new lattice construction: the box product*, manuscript.

[269] M. Greferath, *Zur Hyperebenenalgebraisierung in Desarguesschen Projektiven Verbandsgeometrien*, J. Geometry **42** (1991), 100–108.

[270] ———, *Global-affine morphisms of projective lattice geometries*, Resultate Math. **24** (1993), 76–83.

[271] ———, *Zur Strukturtheorie der Projektiven Verbandsgeometrie*, Mitt. Math. Sem. Gießen **217** (1994).

[272] M. Greferath and S. E. Schmidt, *A unified approach to projective lattice geometries*, Geom. Dedicata **43** (1992), 243–264.

[273] ———, *On point-irreducible projective geometries*, J. Geometry **50** (1994), 73–83.

[274] ———, *On stable geometries*, Geom. Dedicata **51** (1994), 181–199.

[275] P. A. Grillet, *Directed colimits of free commutative semigroups*, J. Pure Appl. Algebra **9** (1976/77), 73–87.

[276] H. Gross, C. Herrmann, R. Moresi, *The classification of subspaces in Hermitian vector spaces*, J. Algebra, **105** (1987), 516–541.

[277] M. D. Haiman, *The theory of linear lattices*, Ph.D. thesis, MIT, 1984.

[278] ———, *Arguesian lattices which are not linear*, Bull. Amer. Math. Soc. (N.S.) **16** (1987), 121–123.

[279] ———, *Arguesian lattices which are not type-1*, Algebra Universalis **28** (1991), 128–137.

[280] I. Halperin, *On the transitivity of perspectivity in continuous geometries*, Trans. Amer. Math. Soc. **44** (1938), 537–562.

[281] ———, *Dimensionality in reducible geometries*, Ann. of Math. **40** (1939), 581–599.

[282] I. Halperin, *Additivity and continuity of perspectivity*, Duke Math. J. **5** (1939), 503–511.

[283] _____, *Introduction to von Neumann algebras and continuous geometries*, Canad. Math. Bull. **3** (1960), 273–288.

[284] _____, *Reducible von Neumann geometries*, Trans. Amer. Math. Soc. **107** (1963), 347–359.

[285] _____, *Continuous geometries and continuous rings*, Trans. Royal Soc. Canada **5** (1967), 221–226.

[286] _____, *Von Neumann's coordinatization theorem*, Acta Sci. Math. (Szeged) **45** (1983), 213–218.

[287] I. Halperin and J. von Neumann, *On the transitivity of perspective mappings*, Ann. of Math. **41** (1940), 87–93.

[288] G. Hansoul, *The Stone–Čech compactification of a pospace*, in 'Lectures in Universal Algebra' (Szeged, 1983) (L. Szabó and A. Szendrei, eds) Coll. Math. Soc. János Bolyai, Vol. 43, North-Holland, Amsterdam, (1986), pp. 161–176.

[289] J. Harding, *The MacNeille completion of a uniquely complemented lattice*, Canad. Math. Bull. **37** (1994), 222–227.

[290] T. Harrison, *A problem concerning the lattice varietal product*, Algebra Universalis **25** (1988), 40–84.

[291] M. Haviar, *Lattices whose congruence lattices satisfy Lee's identities*, Universal algebra, quasigroups and related systems (Jadwisin, 1989). Demonstratio Math. **24** (1991), 247–261.

[292] _____, *On affine completeness of distributive p-algebras*, Glasgow Math. J. **34** (1992), 365–368.

[293] _____, *Affine complete algebras abstracting Kleene and Stone algebras*, Acta Math. Univ. Comenianae **2** (1993), 179–190.

[294] _____, *Affine complete Stone and Post algebras of order n*, Acta Univ. Mathaei Belii Nat. Sci. Ser. Ser. Math. **2** (1994), 17–28.

[295] _____, *Construction and affine completeness of principal p-algebras*, Tatra Mt. Math. Publ. **5** (1995), 217–228.

[296] _____, *The study of affine completeness for quasi-modular double p-algebras*, Acta Math. Univ. Comenianae **2** (1995), 179–190.

[391] N. G. Martınez and H. A. Priestley, *Uniqueness of the implication in MV-algebras*, MathWare and Soft Computing **2** (1995), 229–245.

[392] ———, *On the Priestley space of lattice-ordered algebraic structures*, to appear.

[393] M. Mehrtens, *Die Entstehung der Verbandstheorie* (German), Arbor Scientiarum: Beiträge Wissenschaftsgeschichte, Reihe A: Abhandlungen, Vol. VI, Gerstenberg Verlag, Hildesheim, 1990.

[394] J. D. Monk and R. Bonnet (editors), *Handbook of Boolean algebras*, North-Holland Publishing Co., Amsterdam-New York, 1989. Vol. 1 xx+312 pp., Vol. 2 xx+313–716 pp., and Vol. 3 xx 717–1367 pp.

[395] V. L. Murskiĭ, *The number of k-element algebras with a binary operation which do not have a finite basis of identities* (Russian), Problemy Kibernet. No. 35 (1979), 5–27, 208.

[396] J. B. Nation, *Bounded finite lattices*, Universal algebra (Esztergom, 1977), pp. 531–533, Colloq. Math. Soc. János Bolyai, **29**, North-Holland, Amsterdam-New York, 1982.

[397] ———, *Finite sublattices of a free lattice*, Trans. Amer. Math. Soc. **269** (1982), 311–337.

[398] ———, *Some varieties of semidistributive lattices*, Universal algebra and lattice theory (Charleston, S.C., 1984), 198–223, Lecture Notes in Math., 1149, Springer, Berlin-New York, 1985.

[399] ———, *Lattice varieties covering* $\mathbf{V}(L_1)$, Algebra Universalis **23** (1986), 132–166.

[400] ———, *An approach to lattice varieties of finite height*, Algebra Universalis **27** (1990), 521–543.

[401] ———, *A counterexample to the finite height conjecture*, Order **13** (1996), 1–9.

[402] C. Naturman and H. Rose, *Elementary equivalent pairs of algebras associated with sets*, Algebra Universalis **28** (1991), 324–338.

[403] ———, *Ultra-universal models*, Quaestiones Math. **15** (1992), 189–195.

[404] K. A. Nauryzbaev, *On equationally compact distributive lattices* (Russian), Algebra i Logika **25** (1986), 584–599, 614.

[405] O. T. Nelson,Jr., *Subdirect decompositions of lattices of width two*, Pacific J. Math. **24** (1968), 519–523.

[406] H. Neumann, *Varieties of groups*, Springer-Verlag, Berlin-New York, 1967. x+192 pp.

[407] J. von Neumann, *On regular rings*, Proc. Nat. Acad. Sci. USA **22** (1936), 707–713.

[408] ———, *Algebraic theory of continuous geometries*, Proc. Nat. Acad. Sci. USA **23** (1937), 16–22.

[409] J. Nieminen, *Distributive, standard and neutral convex sublattices of a lattice*, Comment. Math. Univ. St. Paul. **33** (1984), 87–93.

[410] P. Olin, *Elementary properties of distributive lattice free products*, Houston J. Math. **3** (1977), 247–259.

[411] P. Ouwehand and H. Rose, *Small congruence distributive varieties: retracts, injectives, equational compactness and amalgamation*, Period. Math. Hungar. **33** (**3**) (1996), 207–228.

[412] ———, *Lattice varieties with non-elementary amalgamation classes*, Algebra Universalis, to appear.

[413] R. Padmanabhan, *Equational theory of algebras with a majority polynomial*, Algebra Universalis **7** (1977), 273–275.

[414] ———, *A self-dual equational basis for Boolean algebras*, Canad. Math. Bull. **26** (1983), 9–12.

[415] R. Padmanabhan and W. McCune, *Single identities for ternary Boolean algebras*, Comput. Math. Appl. **29** (1995), 13–16.

[416] ———, *Single identities for lattice theory and for weakly associative lattices*, Algebra Universalis **36** (1996), 436–449.

[417] P. P. Pálfy and P. Pudlák, *Congruence lattices of finite algebras and intervals in subgroup lattices of finite groups*, Algebra Universalis **11** (1980), 22–27.

[418] P. P. Pálfy and Cs. Szabó, *congruence varieties of groups and Abelian groups*, Lattice Theory and Its Applications, a volume in honor of Garrett Birkhoff's 80th birthday, Darmstadt, 1991 (K. A. Baker, R. Wille editors), pp. 163–183. Heldermann Verlag (Lemgo, 1995).

[419] ———, *An identity for subgroup lattices of Abelian groups*, Algebra Universalis **33** (1995), 191–195.

[420] P. Perkins, *Basis questions for general algebras*, Algebra Universalis **19** (1984), 16–23.

[421] A. Petrovich *Distributive lattices with an operator*, Studia Logica **56** (1996), 205–224.

[422] _____ *Reticulados distributivos con un operador y algebras de De Morgan monadicas*, Ph.D. thesis, University of Buenos Aires, 1997.

[423] T. Pfeiffer, *Zur Abbildungstheorie von affinen und projektiven Verbandsgeometrien*, Dissertation, Universität Mainz, 1997.

[424] T. Pfeiffer and S. E. Schmidt, *Projective mappings between projective lattice geometries*, J. Geometry **54** (1995), 105–114.

[425] D. Pickering, *Minimal non-Arguesian lattices*, Ph.D. thesis, University of Hawaii, 1984.

[426] _____ , *A self-dual Arguesian inequality*, Algebra Universalis **22** (1986) 99.

[427] G. Pickert, *Pasch- oder Veblen-Axiom?*, Geom. Dedicata **50** (1994), 81–86.

[428] A. G. Pinus, *Elementary equivalence of lattices of partitions* (Russian), Sibirsk. Mat. Zh. **29** (1988), no. 3, 211–212, 223; translation in Siberian Math. J. **29** (1988), 507–508.

[429] _____ , *Definition of finite algebras by derivate algebraic structures* (Russian), Algebra i Logika, to appear.

[430] A. F. Pixley, *A survey of interpolation in universal algebra*, Universal algebra (Esztergom, 1977), pp. 583–607, Colloq. Math. Soc. János Bolyai, **29**, North-Holland, Amsterdam-New York, 1982.

[431] _____ , *Functional and affine completeness and arithmetical varieties*, Algebras and orders (Montreal, PQ, 1991), 317–357, NATO Adv. Sci. Inst. Ser. C Math. Phys. Sci., **389**, Kluwer Acad. Publ., Dordrecht, 1993.

[432] R. Piziak and M. K. Bennett, *Lattice geometries*, Algebra Universalis **30** (1993), 451–462.

[433] C. R. Platt, *Finite transferable lattices are sharply transferable*, Proc. Amer. Math. Soc. **81** (1981), 355–358.

[434] M. Ploščica, *Affine complete distributive lattices*, Order **11** (1994), 385–390.

[435] M. Ploščica and M. Haviar, *Order polynomial completeness of lattices*, manuscript.

[436] M. Ploščica, J. Tůma, and F. Wehrung, *Congruence lattices of free lattices in non-distributive varieties*, Colloq. Math, to appear.

[437] W. Poguntke, *A note on semi-distributive lattices*, Algebra Universalis **9** (1979), 398–399.

[438] S. V. Polin, *Identities in congruence lattices of universal algebras* (Russian), Mat. Zametki **22** (1977), 443–451.

[439] J. P. G. Pretorius, *On the structure of* (*free*) *Heyting algebras*, Ph.D. thesis, University of Cambridge, 1993.

[440] H. A. Priestley, *Catalytic distributive lattices and compact zero-dimensional topological lattices*, Algebra Universalis **19** (1984), 322–329.

[441] ———, *Ordered sets and duality for distributive lattices*, in 'Orders, Descriptions and Roles' (M. Pouzet and D. Richard, eds) Annals of Discrete Mathematics, North-Holland, Amsterdam, New York, **23** (1984), pp. 39–60.

[442] ———, *Spectral sets*, J. Pure Appl. Algebra **94** (1994), 101–114.

[443] ———, *Intrinsic spectral topologies*, in 'Papers on General Topology and Applications', Eighth Summer Conference at Queen's College (G. Itzkowitz *et al.*, eds), Annals of the New York Academy of Sciences, **728** (1994), 78–95.

[444] ———, *Natural dualities for varieties of n-valued Łukasiewicz algebras*, Studia Logica **54** (1995), 333–370.

[445] ———, *Natural dualities*, in 'Lattice Theory and its Applications–a Volume in Honor of Garrett Birkhoff's 80th Birthday' (K. A. Baker and R. Wille, eds) Helderman, Berlin, (1995), pp. 185–209.

[446] ———, *Distributive lattices with additional operations I*, J. Austral. Math. Soc. Ser. A (1997), to appear.

[447] ———, *Identities in varieties of distributive lattices with additional unary operations*, manuscript.

[448] H. A. Priestley and R. Santos, *Distributive lattices with additional operations II*, Portugal. Math. , to appear.

[449] P. Pudlák, *On congruence lattices of lattices*, Algebra Universalis **20** (1985), 96–114.

[450] P. Pudlák and J. Tůma, *Every finite lattice can be embedded in a finite partition lattice*, Algebra Universalis **10** (1980), 74–95.

[451] A. Pultr and J. Sichler, *Frames in Priestley's duality*, Cahiers Topologie Géom. Différentielle Catég. **24** (1988), 193–201.

[452] F. Regonati, *On the number of subgroups of given order of finite abelian p-groups*, Istit. Lombardo Accad. Sci. Lett. Rend. A **122** (1988), 369–380.

[453] V. B. Repnitskiĭ, *The lattice universality of free Burnside's groups* (Russian), Algebra i Logika **35** (1996), 587–611.

[454] K. Reuter, *Matchings for linearly indecomposable modular lattices*, Discrete Math. **63** (1987), 245–247.

[455] K. Reuter and R. Wille, *Complete congruence relations of concept lattices*, Acta Sci. Math. (Szeged) **51** (1987), 319–327.

[456] G. Richter, *Verallgemeinerungen der Satze von Kurosh-Ore und Schmidt-Ore*, Beiträge Algebra Geom. No. 26 (1988), 89–100.

[457] I. Rival and N. Zaguia, *Images of simple lattice polynomials*, Algebra Universalis **33** (1995), 10–14.

[458] H. Rose, *Nonmodular lattice varieties*, Mem. Amer. Math. Soc. **292** (1984).

[459] G.-C. Rota, *The valuation ring of a distributive lattice*, in 'Proceedings of the University of Houston Lattice Theory Conference', (1973), pp. 574–628.

[460] _____ , *Whitney numbers of the second kind of finite modular lattices*, J. Combin. Theory Ser. A **6-** (1992), 34–39.

[461] W. Ruckelshausen, *Zur hierarchie kleiner verbandsvarietäten*, Studentenwerk Darmstadt, Darmstadt, Dissertation, 1983.

[462] V. N. Saliĭ, *Lattices with unique complements*, translated from the Russian by G. A. Kandall, Translations of Mathematical Monographs, American Mathematical Society, Providence, RI, 1988.

[463] E. T. Schmidt, *Zur Charakterisierung der Kongruenzverbände der Verbände*, Math. Časopis **18** (1968), 3–20.

[464] _____ , *The ideal lattice of a distributive lattice with 0 is the congruence lattice of a lattice*, Acta Sci. Math. (Szeged) **43** (1981), 153–168.

[465] _____ , *Congruence lattices of complemented modular lattices*, Algebra Universalis **18** (1984), 386–395.

[466] _____ , *Homomorphisms of distributive lattices as restrictions of congruences*, Acta Sci. Math. (Szeged) **51** (1987), 209–215.

[467] _____ , *Congruence lattices of modular lattices*, Publ. Math. Debrecen **42** (1993), 129–134.

[468] S. E. Schmidt, *Projektive Räume mit geordneter Punktmenge*, Mitt. Math. Sem. Gießen **182** (1987).

[469] S. E. Schmidt, *Projective spaces on partially ordered sets and Desargues' postulate*, Geom. Dedicata **37** (1991), 233–243.

[470] _____, *On the algebraic representation of projectively embeddable affine geometries*, J. Geometry **54** (1995), 155–160.

[471] R. Schmidt, *Subgroup lattices of groups*, Expositions in Mathematics, **14**. Walter de Gruyter & Co., Berlin, 1994. xvi+572 pp.

[472] D. Schweigert, *Über endliche, ordnungspolynomvollständige Verbände*, Monatsh. f. Math. **78** (1974), 68–76.

[473] S. Shelah, *Non structure theory*, Oxford University Press, to appear.

[474] L. N. Shevrin and A. J. Ovsyannikov, *Semigroups and their subsemigroup lattices*, translated and revised from the 1990/1991 Russian originals by the authors. Mathematics and its Applications, **379**. Kluwer Academic Publishers Group, Dordrecht, 1996. xii+378 pp.

[475] T. Skolem, *Logisch-kombinatorische Untersuchungen über die Erfüllbarkeit und Beweisbarkeit mathematischen Sätze nebst einem Theoreme über dichte Mengen*, Videnskapsselskapets skrifter I, Matematisk-naturvidenskabelig klasse, Videnskabsakademiet i Kristiania **4** (1920), 1–36.

[476] _____, *Select works in logic*, Edited by Jens Erik Fenstad Universitetsforlaget, Oslo 1970. 732 pp.

[477] L. A. Skornjakov, *Projective mappings of modules*, Izv. Akad. Nauk USSR Ser. Math **24** (1969), 511–520.

[478] _____, *Elements of lattice theory* (Russian), Second edition, Nauka, Moscow, 1982. 160 pp.

[479] M. Skorsky, *Endliche Verbände—Diagramme und Eigenschaften*. Ph.D. thesis, TH Darmstadt, 1992. Shaker Publications.

[480] V. Slavík, *The amalgamation property of varieties determined by primitive lattices*, Comment. Math. Univ. Carolin. **21** (1980), 473–478.

[481] _____, *A note on the amalgamation property in lattice varieties*, Contributions to lattice theory (Szeged, 1980), pp. 723–736, Colloq. Math. Soc. János Bolyai, **33**, North-Holland, Amsterdam-New York, 1983.

[482] _____, *A note on the lattice of intervals of a lattice*, Algebra Universalis **23** (1986), 22–23.

[483] _____, *On the variety* C sub **D**, Comment. Math. Univ. Carolin. **32** (1991), 431–434.

[484] V. Slavík, *Finiteness of finitely presented lattices*, Lattice theory and its applications (Darmstadt, 1991), 219–227, Res. Exp. Math., **23**, Heldermann, Lemgo, 1995.

[485] _____, *A note on subvarieties of varieties* Csub **K**, Riv. Mat. Pura Appl. No. 16 (1995), 9–11.

[486] _____, *A note on convex sublattices of lattices*, Comment. Math. Univ. Carolin. **36** (1995), 7–9.

[487] _____, *Lattices with finite W-covers*, Algebra Universalis, to appear.

[488] W. Stephenson, *Characterizations of rings and modules by lattices*, Ph.D. thesis, University of London, 1969.

[489] M. Stern, *Semimodular lattices*, Teubner-Texte zur Mathematik, **125**, B. G. Teubner Verlagsgesellschaft mbH, Stuttgart, 1991. 235 pp.

[490] H. Strietz, *Finite partition lattices are four-generated*, Proceedings of the Lattice Theory Conference (Ulm, 1975), pp. 257–259. Univ. Ulm, Ulm, 1975.

[491] _____, *Über Erzeugendenmengen endlicher Partitionenverbände*, Studia Sci. Math. Hungar. **12** (1977), 1–17 (1980).

[492] G. Takách, *Lattices characterized by their sublattice-lattices*, Algebra Universalis **37** (1997), 422–425.

[493] _____, *On the sublattice-lattices of lattices*, manuscript.

[494] A. Tarski, *Cardinal Algebras*, New York: Oxford, 1949.

[495] W. Taylor, *Residually small varieties*, Algebra Universalis **2** (1972), 33–53.

[496] _____, *Products of absolute retracts*, Algebra Universalis **3** (1973), 400–401.

[497] S.-K. Teo, *Representing finite lattices as complete congruence lattices of complete lattices*, Ann. Univ. Sci. Budapest. Eötvös Sect. Math. **33** (1990), 177–182.

[498] _____, *On the length of the congruence lattice of a lattice*, Period. Math. Hungar. **21** (1990), 179–186.

[499] D. Thomas, *Problems in functional analysis*, Ph.D. thesis, Oxford, 1976.

[500] M. Tischendorf, *The representation problem for algebraic distributive lattices*, Ph. D. thesis, TH Darmstadt, 1992.

[501] M. Tischendorf, *On the representation of distributive semilattices*, Algebra Universalis **31** (1994), 446–455.

[502] S. T. Tschantz, *Infinite intervals in free lattices*, Order **6** (1990), 367–388.

[503] J. Tůma, *Intervals in subgroup lattices of infinite groups*, J. Algebra **125** (1989), 367–399.

[504] A. Urquhart, *Free Heyting algebras*, Algebra Universalis **3** (1973), 94–97.

[505] _____, *Distributive lattices with a dual homomorphic operation*, Studia Logica **18** (1978), 201–209.

[506] _____, *A topological representation theory for lattices*, Algebra Universalis **8** (1978), 45–58.

[507] F. Vogt, *Formale Begriffsanalyse mit C^{++}: Datenstrukturen und Algorithmen*, Springer-Verlag, Berlin-New York, 1996. 323 pp.

[508] L. Vrancken-Mawet, *The 0-distributivity in the class of subalgebra lattices of Heyting algebras and closure algebras*, Comment. Math. Univ. Carolin. **28** (1987), 387–396.

[509] L. Vrancken-Mawet and G. Hansoul, *The subalgebra lattice of a Heyting algebra*, Czechoslovak Math. J. **37** (1987), 34–41.

[510] A. Walendziak, *Solution of Grätzer's problem*, Comment. Math. Prace Mat. **26** (1986), 349–359.

[511] B. Węglorz, *Equationally compact algebras* (I), Fund. Math. **59** (1966), 289–298.

[512] F. Wehrung, *Injective positively ordered monoids I*, J. Pure Appl. Algebra **83** (1992), 43–82.

[513] _____, *Injective positively ordered monoids II*, J. Pure Appl. Algebra **83** (1992), 83–100.

[514] _____, *Equational compactness of bi-frames and projection algebras*, Algebra Universalis **33** (1995), 478–515.

[515] _____, *Treillis bi-locaux équationnellement compacts*, C. R. Acad. Sci. Paris Sér. I Math. **318** (1994), 5–9.

[516] _____, *Non-measurability properties of interpolation vector spaces*, Israel J. Math., to appear.

[517] _____, *A uniform refinement property of certain congruence lattices*, Proc. Amer. Math. Soc., to appear.

[518] F. Wehrung, *The dimension monoid of a lattice*, Algebra Universalis, to appear.

[519] D. Wiedemann, *A computation of the eighth Dedekind number*, Order **8** (1991), 5–6.

[520] R. Wille, *Eine Charakterisierung endlicher ordnungspolynomvollständiger Verbände*, Arch. d. Math. **28** (1977), 557–560.

[521] _____, *Restructuring lattice theory: an approach based on hierarchies of concepts*, Ordered sets. Edited by Ivan Rival. Proceedings of a NATO Advanced Study Institute held in Banff, Alta., August 28–September 12, 1981. NATO Advanced Study Institute Series C: Mathematical and Physical Sciences, 83. D. Reidel Publishing Co., Dordrecht-Boston, Mass., 1982. xviii+966 pp.

[522] _____, *Subdirect decomposition of concept lattices*, Algebra Universalis **17** (1983), 275 – 287.

[523] _____, *Finite distributive lattices as concept lattices*, Atti Inc. Logica Matematica (Siena) **2** (1985), 635–648.

[524] _____, *Introduction to Formal Concept Analysis*, manuscript.

[525] _____, *Conceptual landscapes of knowledge: a pragmatic paradigm for knowledge processing*, manuscript.

[526] C. Y. Wong, *Lattice varieties with weak distributivity*, Ph.D. thesis, University of Hawaii, 1989.

[527] L. Zádori, *Generation of finite partition lattices*, Lectures in universal algebra (Szeged, 1983), pp. 573–586, Colloq. Math. Soc. János Bolyai, **43**, North-Holland, Amsterdam-New York, 1986.

[528] Y. Zhang, *A note on "Small representations of finite distributive lattices as congruence lattices"*, Order **13** (1996), 365–367.

[529] G. I. Žitomirskiĭ, *The lattice of all congruence relations on a semilattice* (Russian), Ordered sets and lattices, No. 1 (Russian), pp. 11–21. Izdat. Saratov. Univ., Saratov, 1971.

[530] _____, *The characterization of lattices of congruence relations on semilattices* (Russian), VINITI, Dep. No. 8142-84 (1984).

Index